W9-CSD-279

# SELECTED TABLES IN MATHEMATICAL STATISTICS

## Volume V

This volume was prepared with the aid of:

C. Bingham, University of Minnesota
R. F. Gunst, Southern Methodist University
K. Hinkelmann, Virginia Polytechnic Institute and State University
W. Kennedy, Iowa State University
S. Pearson, Southern Methodist University
N. S. Urquhart, New Mexico State University
R. H. Wampler, National Bureau of Standards
E. J. Wegman, University of North Carolina

# SELECTED TABLES IN MATHEMATICAL STATISTICS

**Volume V**

**Edited by the Institute of Mathematical Statistics**

*Coeditors*

**D. B. Owen**
*Southern Methodist University*

*and*

**R. E. Odeh**
*University of Victoria*

*Managing Editor*

**J. M. Davenport**
*Texas Tech University*

**AMERICAN MATHEMATICAL SOCIETY**

**PROVIDENCE, RHODE ISLAND**

*AMS (MOS) subject classifications* (1970)
Primary 62Q05; Secondary 62G30, 62G35, 62E99, 62F25

International Standard Book Number 0-8218-1905-4
Library of Congress Card Number 74-6283

# PREFACE

This volume of mathematical tables has been prepared under the aegis of the Institute of Mathematical Statistics. The Institute of Mathematical Statistics is a professional society for mathematically oriented statisticians. The purpose of the Institute is to encourage the development, dissemination, and application of mathematical statistics. The Committee on Mathematical Tables of the Institute of Mathematical Statistics is responsible for preparing and editing this series of tables. The Institute of Mathematical Statistics has entered into an agreement with the American Mathematical Society to jointly publish this series of volumes. At the time of this writing, submissions for future volumes are being solicited. No set number of volumes has been established for this series. The editors will consider publishing as many volumes as are necessary to disseminate meritorious material.

Potential authors should consider the following rules when submitting material.

1. The manuscript must be prepared by the author in a form acceptable for photo-offset. This includes both the tables and introductory material. The author should assume that nothing will be set in type although the editors reserve the right to make editorial changes.

2. While there are no fixed upper and lower limits on the length of tables, authors should be aware that the purpose of this series is to provide an outlet for tables of high quality and utility which are too long to be accepted by a technical journal but too short for separate publication in book form.

3. The author must, wherever applicable, include in his introduction the following:

(a) He should give the formula used in the calculation, and the computational procedure (or algorithm) used to generate his tables. Generally speaking, FORTRAN or ALGOL programs will not be included but the description of the algorithm used should be complete enough that such programs can be easily prepared.

(b) A recommendation for interpolation in the tables should be given. The author should give the number of figures of accuracy which can be obtained with linear (and higher degree) interpolation.

(c) Adequate references must be given.

(d) The author should give the accuracy of the table and his method of rounding.

(e) In considering possible formats for his tables, the author should attempt to give as much information as possible in as little space as possible. Generally speaking, critical values of a distribution convey more information than the distribution itself, but each case must be judged on its own merits. The text portion of the tables (including column headings, titles, etc.) must be proportional to the size 5–1/4″ by 8–1/4″. Tables may be printed proportional to the size 8–1/4″ by 5–1/4″ (i.e., turned sideways on the page) when absolutely necessary; but this should be avoided and every attempt made to orient the tables in a vertical manner.

(f) The table should adequately cover the entire function. Asymptotic results should be given and tabulated if informative.

(g) An example or examples of the use of the tables should be included.

4. The author should submit as accurate a tabulation as he can. The table will be checked before publication, and any excess of errors will be considered grounds for rejection. The manuscript introduction will be subjected to refereeing and an inadequate introduction may also lead to rejection.

5. Authors having tables they wish to submit should send two copies to:

Dr. Robert E. Odeh, Coeditor
Department of Mathematics
University of Victoria
Victoria, B. C., Canada V8W 2Y2

At the same time, a third copy should be sent to:

William J. Kennedy
117 Snedecor Hall
Statistical Laboratory
Iowa State University
Ames, Iowa 50011

Additional copies may be required, as needed for the editorial process. After the editorial process is complete, a camera-ready copy must be prepared for the publisher.

Authors should check several current issues of *The Institute of Mathematical Statistics Bulletin* and *The AMSTAT News* for any up-to-date announcements about submissions to this series.

## ACKNOWLEDGMENTS

The tables included in the present volume were checked at the University of Victoria. Dr. R. E. Odeh arranged for, and directed this checking with the assistance of Mr. Bruce Wilson. The editors and the Institute of Mathematical Statistics wish to express their great appreciation for this invaluable assistance. So many other people have contributed to the instigation and preparation of this volume that it would be impossible to record their names here. To all these people, who will remain anonymous, the editors and the Institute also wish to express their thanks.

## TABLE OF CONTENTS

## Contents of VOLUMES I, II, III and IV of this Series

Selected Tables in Mathematical Statistics
Volume V, 1977

# VARIANCES AND COVARIANCES OF THE NORMAL ORDER STATISTICS FOR SAMPLE SIZES 2 TO 50.

G. L. Tietjen Los Alamos Scientific Laboratory
D. K. Kahaner Los Alamos Scientific Laboratory
R. J. Beckman Los Alamos Scientific Laboratory

## ABSTRACT

Tables of the variances and covariances of the normal order statistics for samples of size $N \leq 20$ were given by Sarhan and Greenberg (1956), based on tables of expected values given by Teichroew (1956). Table I extends these results to $N \leq 50$. Table II is a revised table of means of the normal order statistics.

## INTRODUCTION

Suppose that a sample of N observations from a N(0,1) population is arranged so that $x_1 \geq x_2 \geq \cdots \geq x_N$. The variances and covariances of these normal order statistics may be obtained from the following relationships:

$$E(x_j;N) = \frac{N!}{(j-1)!(N-j)!} \int_{-\infty}^{\infty} xf(x)[F(x)]^{j-1} [1-F(x)]^{N-j} dx \tag{1}$$

$$E(x_j^2;N) = \frac{N!}{(j-1)!(N-j)!} \int_{-\infty}^{\infty} x^2 f(x)[F(x)]^{j-1} [1-F(x)]^{N-j} dx \tag{2}$$

$$E(x_i x_j;N) = c\int_{-\infty}^{\infty}\int_{-\infty}^{y} xyf(x)f(y)[F(x)]^{i-1} [1-F(y)]^{N-j} [F(y)-F(x)]^{j-i-1} dxdy \tag{3}$$

where $f(x) = \frac{e^{-x^2/2}}{\sqrt{2\pi}}$, $F(x) = \int_{-\infty}^{x} f(t)dt$

and $c = \frac{N!}{(i-1)!(N-j)!(j-i-1)!}$ .

---

Received by the editors October 1975 and in revised form July 1976, November 1976 and March 1977.
AMS(MOS) Subject Classifications (1970): Primary 62Q05; Secondary 62G30, 62E99.
Work performed under the auspices of U.S. Energy Research & Devel. Admin.

Teichroew (1956) gave tables of all three quantities for N = 1-20.

In 1972 the Japanese Standards Association included in their _Statistical Tables and Formulas with Computer Applications_, edited by Ziro Yamauti, an 8-place table of $E(x_i x_j;N)$ for N = 1-30. Tables of $E(x_j;N)$ have been given in many places (e.g., Pearson and Hartley (1972)).

## APPLICATIONS OF THE TABLES

Unbiased linear estimates of the mean $\hat{\mu}$ and the standard deviation $\hat{\sigma}$, based on the order statistics, are found from $\hat{\theta} = (\hat{\mu},\hat{\sigma}) = (B'V^{-1}B)^{-1}B'V^{-1}y$ where y is the vector of ordered observations and B is the (n x 2) matrix in which the first column are 1's and the second column is made up of the expected values $E(x_i;N)$ obtained from Table II. V is the variance-covariance matrix obtained from Table I. The (2 x n) array $(B'V^{-1}B)^{-1}\ B'V^{-1}$ yields the coefficients $a_i$ and $b_i$ needed to obtain $\hat{\mu} = \Sigma a_i y_i$ and $\hat{\sigma} = \Sigma b_i y_i$. In this situation, of course, $a_i = 1/N$. In the case of censored observations (in which the k largest observations of the set are of unknown magnitude), one obtains linear estimates of the mean and standard deviation by omitting the last k rows and columns of V and the last k rows of B and y and applying the above formulas. The coefficients $a_i^*$ and $b_i^*$ for calculating $\hat{\mu} = \Sigma a_i^* y_i$ and $\hat{\sigma} = \Sigma b_i^* y_i$ are given by Sarhan and Greenberg (1962, pp. 218-251) for N = 2-20. Those for N = 21-50 were not previously available.

The W-test given by Shapiro and Wilk (1965) has been recognized as one of the most powerful of the tests for normality. The coefficients used in calculating the numerator of the W-test statistic are the (normalized) coefficients used for calculating the linear estimate of the standard deviation. Using our previous notation, these coefficients are $b_i/\sqrt{\Sigma b_i^2}$, hence they may be obtained from Table I. They have been calculated for N = 3-20, by Shapiro and Wilk, but for N = 21-50 "various approximations" have been used by the latter authors.

Harter (1969) has explored various applications of quasi-ranges from normal populations. The rth quasi-range $w_r$ is the range of (n-2r) sample values, the r largest and r smallest sample values having been omitted. The range has long been used as an estimate of the standard deviation, but its efficiency decreases rapidly with the sample size. Quasi-ranges retain the simplicity of computation of the range but may yield much higher efficiencies than the range. In order to calculate the efficiency of one quasi-range (with respect to another) it is necessary to know the variances and the covariances of the two quasi-ranges. The covariance of two quasi-ranges is

$$\operatorname{cov}(w_r, w_{r'}) = \operatorname{cov}(x_{r+1}, x_{r'+1}) - \operatorname{cov}(x_{r+1}, x_{n-r'}) + E(x_{r+1})E(x_{r'+1})$$
$$- E(x_{r+1})E(x_{n-r'}) - E(w_r)E(w_{r'}).$$

The first two terms of this expression may be obtained from Table I, the next two terms from Table II, and the last two terms from Harter (1969, Table A1). Harter has calculated some of the necessary covariances, but found the limitations on computer time too great to investigate all those of interest. Table I gives the variances and covariances of the order statistics for N = 2(1)50 and Table II gives the means for the same values of N. Missing values may be obtained from the following relationships:

(1) $\operatorname{Cov}(x_i x_j; N) = \operatorname{Cov}(x_{N+1-i} x_{N+1-j}; N)$

(2) $E(x_i; N) = -E(x_{N+1-i}; N)$

## COMPUTATION AND CHECKS

Values of $E(x_i x_j; N)$ for N = 20-50 were first calculated so that checks could be made.

The linear transformation $y = \lambda + \mu$, $x = \lambda - \mu$ was used to transform (3) to an integral of the form

$$K \int_{-\infty}^{\infty} e^{-\lambda^2} \int_{0}^{\infty} e^{-\mu^2} g(\lambda, \mu) d\mu d\lambda.$$

The value of the integral (calculated independently for each N, i, and j) was obtained by Gaussian quadrature. A 128-point Gaussian-Hermite formula was used on the outer integral. The inner integral was approximated by 40-point Gaussian quadrature with weight function $e^{-x^2}$ on $[0,\infty]$. The nodes and weights for this quadrature were generated using a variant of the method given by Gautschi (1968). Values of F(x) were obtained from a rational approximation accurate to 24 decimal places. The calculations were carried out in double precision on the CDC 7600 (98 binary digits). The value in such an approach lies in the fact that most of the computations for the integrand can be made once and for all and used for every sample size. Otherwise, computational time is beyond reason.

The values of $E(x_i x_j;N)$ and $E(x_j^2;N)$ for N = 20 were calculated and checked against the table given by Teichroew. Our values check with his tables to at least ten decimal places.

If we denote the integral in Eqn.(3) (without its coefficient) by $G(i-1, N-j, j-i-1) = G(m,n,p)$, we have the recursive relation: $G(m,n,p+1) = G(m,n,p) - G(m+1,n,p) - G(m,n+1,p)$. For N = 20 and N = 50, we checked this relationship and found it to be satisfied (for all $i \neq j$ in the tables) to at least 12 decimal places. For each N, the columns of the matrix of values of $E(x_i x_j;N)$ (including the variances) should sum to 1, and the values of $E(x_j^2;N)$ for any given N should sum to N. These relationships were satisfied for N = 20, 30, 40, and 50 to at least 7 decimal places. We are, therefore, confident that the computational method involved will yield, at the very worst, seven decimal place accuracy, and, in most places, much more than that. Seven place tables of $E(x_i x_j;N)$ for N = 21, 30 were checked against the Japanese tables and found to agree exactly.

We then calculated the necessary values of $E(x_i;N)$ and used Eqn. (3) to obtain the covariances directly from the formula $E[(x_i - E(x_i))(x_j)]$ rather than from the formula $E(x_i x_j) - E(x_i)E(x_j)$ used by Sarhan and Greenberg. Due to the loss of accuracy from subtraction of small numbers, our formula should yield more accurate results. Variances were obtained from Eqn. (2) in the same way.

The values of $E(x_i;N)$ in Table II were calculated by the Gaussian - Hermite quadrature described above. Exact values for $(x_1;2)$, $(x_1;3)$ and $(x_2;3)$ are known to be $1/\sqrt{\pi}$, $1.5/\sqrt{\pi}$ and 0. The calculated values, when compared to the exact values, are accurate to at least 20 decimal places. Also, the center order statistic for N odd was calculated and found to agree with the exact value (zero) to 27 decimal places for N = 2-20. By comparison, a double-precision seven-point adaptive Newton-Cotes formula gives only 12 decimal place accuracy for these known values. A similar method was probably used by Teichroew, since it checks with his values closely. While most of the values of $E(x_i;N)$ given by Teichroew agree with our values to within 2 units in the tenth digit, values of N near 20 and i near N/2 yield values of $E(x_i;N)$ which differ in the eighth digit. Finally, in checking our table of variances and covariances against those given by Sarhan and Greenberg, we found only one case (N=20, i=7, j=7) in which we differed with their values in the eighth digit and 34 cases in which the ninth digit differed. These differences seem to be due mostly to differences in the means. We believe our values are more accurate.

## REFERENCES

[1] Gautschi, W. (1968). Construction of Gauss-Christoffel quadrature formulas. *Math. of Comput.* 22 251-70.

[2] Harter, H. L. (1969). *Order Statistics and Their Use in Testing and Estimation*, v. 2, Aerospace Research Laboratories, United States Air Force.

[3] Pearson, E. S. and Hartley, H. O. (1972). *Biometrika Tables for Statisticians*. Cambridge Univ. Press.

[4] Sarhan, A. E. and Greenberg, B. G. (1956). Estimation of location and scale parameters by order statistics from singly and doubly censored samples. *Annals of Math. Stat.* 27 427-451.

[5] Sarhan, A. E. and Greenberg, B. G. (1962). *Contributions to Order Statistics*. John Wiley and Sons, New York.

[6] Shapiro, S. S. and Wilk, M. B. (1965). An analysis of variance test for normality. *Biometrika* 52 591-611.

[7] Teichroew, D. (1956). Tables of expected values of order statistics and products of order statistics for samples of size twenty and less from the normal distribution. *Annals of Math. Stat.* 27 410-426.

TABLE I-COVARIANCES OF NORMAL ORDER STATISTICS

| N | I | J | COVARIANCE | N | I | J | COVARIANCE | N | I | J | COVARIANCE |
|---|---|---|---|---|---|---|---|---|---|---|---|
| 2 | 1 | 1 | .6816901138 | 8 | 2 | 5 | .0975647193 | 10 | 4 | 6 | .1057858169 |
| 2 | 1 | 2 | .3183098862 | 8 | 2 | 6 | .0787224682 | 10 | 4 | 7 | .0889462026 |
| 3 | 1 | 1 | .5594672038 | 8 | 2 | 7 | .0632466119 | 10 | 5 | 5 | .1510539039 |
| 3 | 1 | 2 | .2756644477 | 8 | 3 | 3 | .2007687900 | 10 | 5 | 6 | .1255989678 |
| 3 | 1 | 3 | .1648683485 | 8 | 3 | 4 | .1523584312 | 11 | 1 | 1 | .3332474427 |
| 3 | 2 | 2 | .4486711046 | 8 | 3 | 5 | .1209637555 | 11 | 1 | 2 | .1653647712 |
| 4 | 1 | 1 | .4917152369 | 8 | 3 | 6 | .0978171355 | 11 | 1 | 3 | .1123584351 |
| 4 | 1 | 2 | .2455926930 | 8 | 4 | 4 | .1871862195 | 11 | 1 | 4 | .0855170596 |
| 4 | 1 | 3 | .1580080701 | 8 | 4 | 5 | .1491754908 | 11 | 1 | 5 | .0688483065 |
| 4 | 1 | 4 | .1046840000 | 9 | 1 | 1 | .3573533264 | 11 | 1 | 6 | .0572007586 |
| 4 | 2 | 2 | .3604553434 | 9 | 1 | 2 | .1781434239 | 11 | 1 | 7 | .0483754063 |
| 4 | 2 | 3 | .2359438935 | 9 | 1 | 3 | .1207454442 | 11 | 1 | 8 | .0412423472 |
| 5 | 1 | 1 | .4475340690 | 9 | 1 | 4 | .0913071400 | 11 | 1 | 9 | .0351103357 |
| 5 | 1 | 2 | .2243309596 | 9 | 1 | 5 | .0727422354 | 11 | 1 | 10 | .0294198503 |
| 5 | 1 | 3 | .1481477252 | 9 | 1 | 6 | .0594831125 | 11 | 1 | 11 | .0233152868 |
| 5 | 1 | 4 | .1057719776 | 9 | 1 | 7 | .0490764061 | 11 | 2 | 2 | .2051975798 |
| 5 | 1 | 5 | .0742152685 | 9 | 1 | 8 | .0400936928 | 11 | 2 | 3 | .1403096511 |
| 5 | 2 | 2 | .3115189521 | 9 | 1 | 9 | .0310552188 | 11 | 2 | 4 | .1071492595 |
| 5 | 2 | 3 | .2084354440 | 9 | 2 | 2 | .2256968778 | 11 | 2 | 5 | .0864430257 |
| 5 | 2 | 4 | .1499426667 | 9 | 2 | 3 | .1541163526 | 11 | 2 | 6 | .0719205024 |
| 5 | 3 | 3 | .2868336616 | 9 | 2 | 4 | .1170056917 | 11 | 2 | 7 | .0608869662 |
| 6 | 1 | 1 | .4159271090 | 9 | 2 | 5 | .0934477393 | 11 | 2 | 8 | .0519504506 |
| 6 | 1 | 2 | .2085030023 | 9 | 2 | 6 | .0765461432 | 11 | 2 | 9 | .0442549455 |
| 6 | 1 | 3 | .1394352565 | 9 | 2 | 7 | .0632354695 | 11 | 2 | 10 | .0371029977 |
| 6 | 1 | 4 | .1024293940 | 9 | 2 | 8 | .0517146092 | 11 | 3 | 3 | .1657242880 |
| 6 | 1 | 5 | .0773637839 | 9 | 3 | 3 | .1863826133 | 11 | 3 | 4 | .1269672925 |
| 6 | 1 | 6 | .0563414544 | 9 | 3 | 4 | .1420779776 | 11 | 3 | 5 | .1026407291 |
| 6 | 2 | 2 | .2795777392 | 9 | 3 | 5 | .1137680176 | 11 | 3 | 6 | .0855178832 |
| 6 | 2 | 3 | .1889859560 | 9 | 3 | 6 | .0933625385 | 11 | 3 | 7 | .0724741050 |
| 6 | 2 | 4 | .1396640604 | 9 | 3 | 7 | .0772351805 | 11 | 3 | 8 | .0618873278 |
| 6 | 2 | 5 | .1059054582 | 9 | 4 | 4 | .1705588454 | 11 | 3 | 9 | .0527550070 |
| 6 | 3 | 3 | .2462125354 | 9 | 4 | 5 | .1369913669 | 11 | 4 | 4 | .1479546565 |
| 6 | 3 | 4 | .1832727978 | 9 | 4 | 6 | .1126671842 | 11 | 4 | 5 | .1198752861 |
| 7 | 1 | 1 | .3919177761 | 9 | 5 | 5 | .1661012814 | 11 | 4 | 6 | .1000346585 |
| 7 | 1 | 2 | .1961990246 | 10 | 1 | 1 | .3443438233 | 11 | 4 | 7 | .0848765182 |
| 7 | 1 | 3 | .1321155811 | 10 | 1 | 2 | .1712629030 | 11 | 4 | 8 | .0725451434 |
| 7 | 1 | 4 | .0984868607 | 10 | 1 | 3 | .1162590989 | 11 | 5 | 5 | .1396410803 |
| 7 | 1 | 5 | .0765598346 | 10 | 1 | 4 | .0882494247 | 11 | 5 | 6 | .1167449805 |
| 7 | 1 | 6 | .0599187124 | 10 | 1 | 5 | .0707413677 | 11 | 5 | 7 | .0991935961 |
| 7 | 1 | 7 | .0448022105 | 10 | 1 | 6 | .0583987134 | 11 | 6 | 6 | .1371624335 |
| 7 | 2 | 2 | .2567328862 | 10 | 1 | 7 | .0489206279 | 12 | 1 | 1 | .3236363870 |
| 7 | 2 | 3 | .1744833274 | 10 | 1 | 8 | .0410844589 | 12 | 1 | 2 | .1602373762 |
| 7 | 2 | 4 | .1307298656 | 10 | 1 | 9 | .0340406470 | 12 | 1 | 3 | .1089309642 |
| 7 | 2 | 5 | .1019550089 | 10 | 1 | 10 | .0266989352 | 12 | 1 | 4 | .0830686766 |
| 7 | 2 | 6 | .0799811748 | 10 | 2 | 2 | .2145241430 | 12 | 1 | 5 | .0670884464 |
| 7 | 3 | 3 | .2197215626 | 10 | 2 | 3 | .1466226180 | 12 | 1 | 6 | .0559933694 |
| 7 | 3 | 4 | .1655598429 | 10 | 2 | 4 | .1117015962 | 12 | 1 | 7 | .0476620974 |
| 7 | 3 | 5 | .1296048425 | 10 | 2 | 5 | .0897428245 | 12 | 1 | 8 | .0410208554 |
| 7 | 4 | 4 | .2104468615 | 10 | 2 | 6 | .0741995414 | 12 | 1 | 9 | .0354439060 |
| 8 | 1 | 1 | .3728971433 | 10 | 2 | 7 | .0622278486 | 12 | 1 | 10 | .0305012591 |
| 8 | 1 | 2 | .1863073995 | 10 | 2 | 8 | .0523067221 | 12 | 1 | 11 | .0257945391 |
| 8 | 1 | 3 | .1259660298 | 10 | 2 | 9 | .0433711561 | 12 | 1 | 12 | .0206221232 |
| 8 | 1 | 4 | .0947230277 | 10 | 3 | 3 | .1750032834 | 12 | 2 | 2 | .1972646039 |
| 8 | 1 | 5 | .0747650242 | 10 | 3 | 4 | .1338022448 | 12 | 2 | 3 | .1349020328 |
| 8 | 1 | 6 | .0602075170 | 10 | 3 | 5 | .1077445336 | 12 | 2 | 4 | .1031959206 |
| 8 | 1 | 7 | .0482985509 | 10 | 3 | 6 | .0892254012 | 12 | 2 | 5 | .0835045822 |
| 8 | 1 | 8 | .0368353075 | 10 | 3 | 7 | .0749183943 | 12 | 2 | 6 | .0697859657 |
| 8 | 2 | 2 | .2394010457 | 10 | 3 | 8 | .0630332449 | 12 | 2 | 7 | .0594590652 |
| 8 | 2 | 3 | .1631958726 | 10 | 4 | 4 | .1579389144 | 12 | 2 | 8 | .0512113198 |
| 8 | 2 | 4 | .1232633317 | 10 | 4 | 5 | .1275089295 | 12 | 2 | 9 | .0442747124 |

TABLE I-COVARIANCES OF NORMAL ORDER STATISTICS

| N | I | J | COVARIANCE | N | I | J | COVARIANCE | N | I | J | COVARIANCE |
|---|---|---|---|---|---|---|---|---|---|---|---|
| 12 | 2 | 10 | .0381191478 | 13 | 4 | 9 | .0591628730 | 14 | 5 | 10 | .0576401464 |
| 12 | 2 | 11 | .0322507340 | 13 | 4 | 10 | .0517328051 | 14 | 6 | 6 | .1115324579 |
| 12 | 3 | 3 | .1579786877 | 13 | 5 | 5 | .1232503256 | 14 | 6 | 7 | .0961405595 |
| 12 | 3 | 4 | .1212063211 | 13 | 5 | 6 | .1037367701 | 14 | 6 | 8 | .0839617110 |
| 12 | 3 | 5 | .0982605602 | 13 | 5 | 7 | .0890434754 | 14 | 6 | 9 | .0739069221 |
| 12 | 3 | 6 | .0822228461 | 13 | 5 | 8 | .0773552864 | 14 | 7 | 7 | .1090269480 |
| 12 | 3 | 7 | .0701213964 | 13 | 5 | 9 | .0676230994 | 14 | 7 | 8 | .0953087256 |
| 12 | 3 | 8 | .0604384621 | 13 | 6 | 6 | .1183175326 | 15 | 1 | 1 | .3010415703 |
| 12 | 3 | 9 | .0522825611 | 13 | 6 | 7 | .1016824204 | 15 | 1 | 2 | .1481297708 |
| 12 | 3 | 10 | .0450357614 | 13 | 6 | 8 | .0884194610 | 15 | 1 | 3 | .1007223449 |
| 12 | 4 | 4 | .1398109405 | 13 | 7 | 7 | .1167989950 | 15 | 1 | 4 | .0770594060 |
| 12 | 4 | 5 | .1135687821 | 14 | 1 | 1 | .3077301025 | 15 | 1 | 5 | .0625845850 |
| 12 | 4 | 6 | .0951645279 | 14 | 1 | 2 | .1517203663 | 15 | 1 | 6 | .0526530128 |
| 12 | 4 | 7 | .0812419809 | 14 | 1 | 3 | .1031719531 | 15 | 1 | 7 | .0453078886 |
| 12 | 4 | 8 | .0700795832 | 14 | 1 | 4 | .0788715915 | 15 | 1 | 8 | .0395736673 |
| 12 | 4 | 9 | .0606620874 | 14 | 1 | 5 | .0639657428 | 15 | 1 | 9 | .0349035905 |
| 12 | 5 | 5 | .1306137358 | 14 | 1 | 6 | .0537064714 | 15 | 1 | 10 | .0309614122 |
| 12 | 5 | 6 | .1096212247 | 14 | 1 | 7 | .0460899190 | 15 | 1 | 11 | .0275211039 |
| 12 | 5 | 7 | .0936951520 | 14 | 1 | 8 | .0401141687 | 15 | 1 | 12 | .0244126313 |
| 12 | 5 | 8 | .0808972960 | 14 | 1 | 9 | .0352141760 | 15 | 1 | 13 | .0214819828 |
| 12 | 6 | 6 | .1266377911 | 14 | 1 | 10 | .0310371163 | 15 | 1 | 14 | .0185333263 |
| 12 | 6 | 7 | .1083945831 | 14 | 1 | 11 | .0273362865 | 15 | 1 | 15 | .0151137071 |
| 13 | 1 | 1 | .3152053842 | 14 | 1 | 12 | .0239061001 | 15 | 2 | 2 | .1791215291 |
| 13 | 1 | 2 | .1557272905 | 14 | 1 | 13 | .0205080256 | 15 | 2 | 3 | .1224176952 |
| 13 | 1 | 3 | .1058908842 | 14 | 1 | 14 | .0166279802 | 15 | 2 | 4 | .0939067143 |
| 13 | 1 | 4 | .0808649736 | 14 | 2 | 2 | .1844200252 | 15 | 2 | 5 | .0763912337 |
| 13 | 1 | 5 | .0654634498 | 14 | 2 | 3 | .1260791990 | 15 | 2 | 6 | .0643390895 |
| 13 | 1 | 6 | .0548221796 | 14 | 2 | 4 | .0966524633 | 15 | 2 | 7 | .0554074400 |
| 13 | 1 | 7 | .0468833088 | 14 | 2 | 5 | .0785202980 | 15 | 2 | 8 | .0484238833 |
| 13 | 1 | 8 | .0406132549 | 14 | 2 | 6 | .0660028340 | 15 | 2 | 9 | .0427294113 |
| 13 | 1 | 9 | .0354226462 | 14 | 2 | 7 | .0566896715 | 15 | 2 | 10 | .0379177516 |
| 13 | 1 | 10 | .0309322744 | 14 | 2 | 8 | .0493708147 | 15 | 2 | 11 | .0337151721 |
| 13 | 1 | 11 | .0268537250 | 14 | 2 | 9 | .0433617156 | 15 | 2 | 12 | .0299152347 |
| 13 | 1 | 12 | .0228858068 | 14 | 2 | 10 | .0382337404 | 15 | 2 | 13 | .0263303885 |
| 13 | 1 | 13 | .0184348220 | 14 | 2 | 11 | .0336863221 | 15 | 2 | 14 | .0227213594 |
| 13 | 2 | 2 | .1904130721 | 14 | 2 | 12 | .0294681314 | 15 | 3 | 3 | .1407322503 |
| 13 | 2 | 3 | .1302055829 | 14 | 2 | 13 | .0252863928 | 15 | 3 | 4 | .1082138452 |
| 13 | 2 | 4 | .0997262695 | 14 | 3 | 3 | .1457045666 | 15 | 3 | 5 | .0881605755 |
| 13 | 2 | 5 | .0808785939 | 14 | 3 | 4 | .1119816876 | 15 | 3 | 6 | .0743268436 |
| 13 | 2 | 6 | .0678145832 | 14 | 3 | 5 | .0911181271 | 15 | 3 | 7 | .0640558182 |
| 13 | 2 | 7 | .0580457285 | 14 | 3 | 6 | .0766754957 | 15 | 3 | 8 | .0560136122 |
| 13 | 2 | 8 | .0503167946 | 14 | 3 | 7 | .0659084825 | 15 | 3 | 9 | .0494485110 |
| 13 | 2 | 9 | .0439095087 | 14 | 3 | 8 | .0574341188 | 15 | 3 | 10 | .0438960669 |
| 13 | 2 | 10 | .0383601798 | 14 | 3 | 9 | .0504677802 | 15 | 3 | 11 | .0390426915 |
| 13 | 2 | 11 | .0333147765 | 14 | 3 | 10 | .0445169192 | 15 | 3 | 12 | .0346513381 |
| 13 | 2 | 12 | .0284018130 | 14 | 3 | 11 | .0392352317 | 15 | 3 | 13 | .0305060359 |
| 13 | 3 | 3 | .1513917013 | 14 | 3 | 12 | .0343322070 | 15 | 4 | 4 | .1222328270 |
| 13 | 3 | 4 | .1162698131 | 14 | 4 | 4 | .1272273070 | 15 | 4 | 5 | .0997323941 |
| 13 | 3 | 5 | .0944566603 | 14 | 4 | 5 | .1036931108 | 15 | 4 | 6 | .0841705696 |
| 13 | 3 | 6 | .0792922993 | 14 | 4 | 6 | .0873562483 | 15 | 4 | 7 | .0725946868 |
| 13 | 3 | 7 | .0679282354 | 14 | 4 | 7 | .0751519909 | 15 | 4 | 8 | .0635175907 |
| 13 | 3 | 8 | .0589221432 | 14 | 4 | 8 | .0655310935 | 15 | 4 | 9 | .0560990512 |
| 13 | 3 | 9 | .0514460446 | 14 | 4 | 9 | .0576120957 | 15 | 4 | 10 | .0498187836 |
| 13 | 3 | 10 | .0449637542 | 14 | 4 | 10 | .0508402241 | 15 | 4 | 11 | .0443247452 |
| 13 | 3 | 11 | .0390643799 | 14 | 4 | 11 | .0448243469 | 15 | 4 | 12 | .0393501819 |
| 13 | 4 | 4 | .1330111819 | 14 | 5 | 5 | .1171012461 | 15 | 5 | 5 | .1118698986 |
| 13 | 4 | 5 | .1082512666 | 14 | 5 | 6 | .0987747550 | 15 | 5 | 6 | .0945206004 |
| 13 | 4 | 6 | .0909855605 | 14 | 5 | 7 | .0850536546 | 15 | 5 | 7 | .0815891122 |
| 13 | 4 | 7 | .0780173339 | 14 | 5 | 8 | .0742181416 | 15 | 5 | 8 | .0714331681 |
| 13 | 4 | 8 | .0677217142 | 14 | 5 | 9 | .0652867776 | 15 | 5 | 9 | .0631224389 |

TABLE I-COVARIANCES OF NORMAL ORDER STATISTICS

| N | I | J | COVARIANCE | N | I | J | COVARIANCE | N | I | J | COVARIANCE |
|---|---|---|---|---|---|---|---|---|---|---|---|
| 15 | 5 | 10 | .0560795064 | 16 | 4 | 11 | .0436607328 | 17 | 3 | 8 | .0534208202 |
| 15 | 5 | 11 | .0499127743 | 16 | 4 | 12 | .0391112668 | 17 | 3 | 9 | .0474555487 |
| 15 | 6 | 6 | .1058666366 | 16 | 4 | 13 | .0349253748 | 17 | 3 | 10 | .0424726884 |
| 15 | 6 | 7 | .0914683203 | 16 | 5 | 5 | .1073517089 | 17 | 3 | 11 | .0381925587 |
| 15 | 6 | 8 | .0801407559 | 16 | 5 | 6 | .0908232621 | 17 | 3 | 12 | .0344194567 |
| 15 | 6 | 9 | .0708582100 | 16 | 5 | 7 | .0785480533 | 17 | 3 | 13 | .0310047771 |
| 15 | 6 | 10 | .0629824402 | 16 | 5 | 8 | .0689488802 | 17 | 3 | 14 | .0278210708 |
| 15 | 7 | 7 | .1026916922 | 16 | 5 | 9 | .0611364182 | 17 | 3 | 15 | .0247342095 |
| 15 | 7 | 8 | .0900499964 | 16 | 5 | 10 | .0545638941 | 17 | 4 | 4 | .1140068197 |
| 15 | 7 | 9 | .0796738323 | 16 | 5 | 11 | .0488684328 | 17 | 4 | 5 | .0931620339 |
| 15 | 8 | 8 | .1016946521 | 16 | 5 | 12 | .0437882959 | 17 | 4 | 6 | .0788266621 |
| 16 | 1 | 1 | .2950098090 | 16 | 6 | 6 | .1010461906 | 17 | 4 | 7 | .0682298908 |
| 16 | 1 | 2 | .1448881688 | 16 | 6 | 7 | .0874627155 | 17 | 4 | 8 | .0599826092 |
| 16 | 1 | 3 | .0985009765 | 16 | 6 | 8 | .0768239668 | 17 | 4 | 9 | .0533057575 |
| 16 | 1 | 4 | .0754040024 | 16 | 6 | 9 | .0681545540 | 17 | 4 | 10 | .0477239973 |
| 16 | 1 | 5 | .0613086724 | 16 | 6 | 10 | .0608534805 | 17 | 4 | 11 | .0429261817 |
| 16 | 1 | 6 | .0516624963 | 16 | 6 | 11 | .0545210724 | 17 | 4 | 12 | .0386942630 |
| 16 | 1 | 7 | .0445503705 | 16 | 7 | 7 | .0974026614 | 17 | 4 | 13 | .0348624030 |
| 16 | 1 | 8 | .0390194715 | 16 | 7 | 8 | .0856181916 | 17 | 4 | 14 | .0312881042 |
| 16 | 1 | 9 | .0345378158 | 16 | 7 | 9 | .0760015577 | 17 | 5 | 5 | .1034004378 |
| 16 | 1 | 10 | .0307810093 | 16 | 7 | 10 | .0678931921 | 17 | 5 | 6 | .0875729931 |
| 16 | 1 | 11 | .0275353611 | 16 | 8 | 8 | .0957213007 | 17 | 5 | 7 | .0758534534 |
| 16 | 1 | 12 | .0246479006 | 16 | 8 | 9 | .0850291218 | 17 | 5 | 8 | .0667204245 |
| 16 | 1 | 13 | .0219956754 | 17 | 1 | 1 | .2895330037 | 17 | 5 | 9 | .0593187706 |
| 16 | 1 | 14 | .0194585037 | 17 | 1 | 2 | .1419424629 | 17 | 5 | 10 | .0531257771 |
| 16 | 1 | 15 | .0168710289 | 17 | 1 | 3 | .0964748737 | 17 | 5 | 11 | .0477987292 |
| 16 | 1 | 16 | .0138287377 | 17 | 1 | 4 | .0738849615 | 17 | 5 | 12 | .0430970793 |
| 16 | 2 | 2 | .1743940788 | 17 | 1 | 5 | .0601272302 | 17 | 5 | 13 | .0388375657 |
| 16 | 2 | 3 | .1191409286 | 17 | 1 | 6 | .0507326947 | 17 | 6 | 6 | .0968824667 |
| 16 | 2 | 4 | .0914359918 | 17 | 1 | 7 | .0438236491 | 17 | 6 | 7 | .0839811736 |
| 16 | 2 | 5 | .0744591145 | 17 | 1 | 8 | .0384672833 | 17 | 6 | 8 | .0739130258 |
| 16 | 2 | 6 | .0628093909 | 17 | 1 | 9 | .0341441055 | 17 | 6 | 9 | .0657442736 |
| 16 | 2 | 7 | .0542033940 | 17 | 1 | 10 | .0305389549 | 17 | 6 | 10 | .0589030404 |
| 16 | 2 | 8 | .0475009770 | 17 | 1 | 11 | .0274465527 | 17 | 6 | 11 | .0530137276 |
| 16 | 2 | 9 | .0420638230 | 17 | 1 | 12 | .0247237145 | 17 | 6 | 12 | .0478122601 |
| 16 | 2 | 10 | .0375018251 | 17 | 1 | 13 | .0222620771 | 17 | 7 | 7 | .0929031782 |
| 16 | 2 | 11 | .0335574913 | 17 | 1 | 14 | .0199690650 | 17 | 7 | 8 | .0818194608 |
| 16 | 2 | 12 | .0300461298 | 17 | 1 | 15 | .0177476891 | 17 | 7 | 9 | .0728154074 |
| 16 | 2 | 13 | .0268189579 | 17 | 1 | 16 | .0154552070 | 17 | 7 | 10 | .0652667273 |
| 16 | 2 | 14 | .0237301562 | 17 | 1 | 17 | .0127264751 | 17 | 7 | 11 | .0587626217 |
| 16 | 2 | 15 | .0205785433 | 17 | 2 | 2 | .1701426763 | 17 | 8 | 8 | .0907361649 |
| 16 | 3 | 3 | .1363385613 | 17 | 2 | 3 | .1161866734 | 17 | 8 | 9 | .0808000267 |
| 16 | 3 | 4 | .1048706757 | 17 | 2 | 4 | .0891982556 | 17 | 8 | 10 | .0724599964 |
| 16 | 3 | 5 | .0855189036 | 17 | 2 | 5 | .0726970385 | 17 | 9 | 9 | .0900465814 |
| 16 | 3 | 6 | .0722075087 | 17 | 2 | 6 | .0613998459 | 18 | 1 | 1 | .2845301297 |
| 16 | 3 | 7 | .0623568515 | 17 | 2 | 7 | .0530761573 | 18 | 1 | 2 | .1392501620 |
| 16 | 3 | 8 | .0546749107 | 17 | 2 | 8 | .0466140918 | 18 | 1 | 3 | .0946172636 |
| 16 | 3 | 9 | .0484366096 | 17 | 2 | 9 | .0413928192 | 18 | 1 | 4 | .0724851730 |
| 16 | 3 | 10 | .0431979378 | 17 | 2 | 10 | .0370349110 | 18 | 1 | 5 | .0590304274 |
| 16 | 3 | 11 | .0386652994 | 17 | 2 | 11 | .0332940892 | 18 | 1 | 6 | .0498600635 |
| 16 | 3 | 12 | .0346277256 | 17 | 2 | 12 | .0299982826 | 18 | 1 | 7 | .0431302310 |
| 16 | 3 | 13 | .0309149134 | 17 | 2 | 13 | .0270170379 | 18 | 1 | 8 | .0379260195 |
| 16 | 3 | 14 | .0273595377 | 17 | 2 | 14 | .0242386813 | 18 | 1 | 9 | .0337388141 |
| 16 | 4 | 4 | .1178657554 | 17 | 2 | 15 | .0215459396 | 18 | 1 | 10 | .0302610667 |
| 16 | 4 | 5 | .0962513414 | 17 | 2 | 16 | .0187658306 | 18 | 1 | 11 | .0272938041 |
| 16 | 4 | 6 | .0813480448 | 17 | 3 | 3 | .1324207975 | 18 | 1 | 12 | .0247002471 |
| 16 | 4 | 7 | .0703000911 | 17 | 3 | 4 | .1018792434 | 18 | 1 | 13 | .0223801572 |
| 16 | 4 | 8 | .0616728990 | 17 | 3 | 5 | .0831421716 | 18 | 1 | 14 | .0202537421 |
| 16 | 4 | 9 | .0546595026 | 17 | 3 | 6 | .0702850403 | 18 | 1 | 15 | .0182488619 |
| 16 | 4 | 10 | .0487647746 | 17 | 3 | 7 | .0607964413 | 18 | 1 | 16 | .0162850442 |

TABLE I-COVARIANCES OF NORMAL ORDER STATISTICS

| N | I | J | COVARIANCE | N | I | J | COVARIANCE | N | I | J | COVARIANCE |
|---|---|---|---|---|---|---|---|---|---|---|---|
| 18 | 1 | 17 | .0142368875 | 18 | 6 | 12 | .0467370899 | 19 | 3 | 13 | .0307215918 |
| 18 | 1 | 18 | .0117719054 | 18 | 6 | 13 | .0423879850 | 19 | 3 | 14 | .0279835020 |
| 18 | 2 | 2 | .1662929294 | 18 | 7 | 7 | .0890167026 | 19 | 3 | 15 | .0254424108 |
| 18 | 2 | 3 | .1135058132 | 18 | 7 | 8 | .0785179678 | 19 | 3 | 16 | .0230195062 |
| 18 | 2 | 4 | .0871597604 | 18 | 7 | 9 | .0700199026 | 19 | 3 | 17 | .0206214645 |
| 18 | 2 | 5 | .0710825990 | 18 | 7 | 10 | .0629269074 | 19 | 4 | 4 | .1074740835 |
| 18 | 2 | 6 | .0600975753 | 18 | 7 | 11 | .0568501034 | 19 | 4 | 5 | .0879051964 |
| 18 | 2 | 7 | .0520217422 | 18 | 7 | 12 | .0515199090 | 19 | 4 | 6 | .0745033876 |
| 18 | 2 | 8 | .0457683625 | 18 | 8 | 8 | .0864960641 | 19 | 4 | 7 | .0646406186 |
| 18 | 2 | 9 | .0407317967 | 18 | 8 | 9 | .0771762287 | 19 | 4 | 8 | .0570032284 |
| 18 | 2 | 10 | .0365451034 | 18 | 8 | 10 | .0693891331 | 19 | 4 | 9 | .0508572608 |
| 18 | 2 | 11 | .0329704894 | 18 | 8 | 11 | .0627116904 | 19 | 4 | 10 | .0457576598 |
| 18 | 2 | 12 | .0298442464 | 18 | 9 | 9 | .0853127878 | 19 | 4 | 11 | .0414165091 |
| 18 | 2 | 13 | .0270462261 | 18 | 9 | 10 | .0767442322 | 19 | 4 | 12 | .0376368753 |
| 18 | 2 | 14 | .0244806359 | 19 | 1 | 1 | .2799358049 | 19 | 4 | 13 | .0342765542 |
| 18 | 2 | 15 | .0220607111 | 19 | 1 | 2 | .1367768168 | 19 | 4 | 14 | .0312262552 |
| 18 | 2 | 16 | .0196894667 | 19 | 1 | 3 | .0929061763 | 19 | 4 | 15 | .0283944529 |
| 18 | 2 | 17 | .0172154924 | 19 | 1 | 4 | .0711902425 | 19 | 4 | 16 | .0256935152 |
| 18 | 3 | 3 | .1288998943 | 19 | 1 | 5 | .0580094835 | 19 | 5 | 5 | .0967944749 |
| 18 | 3 | 4 | .0991828539 | 19 | 1 | 6 | .0490405678 | 19 | 5 | 6 | .0821055698 |
| 18 | 3 | 5 | .0809899791 | 19 | 1 | 7 | .0424705246 | 19 | 5 | 7 | .0712796745 |
| 18 | 3 | 6 | .0685324700 | 19 | 1 | 8 | .0374006329 | 19 | 5 | 8 | .0628870097 |
| 18 | 3 | 7 | .0593598602 | 19 | 1 | 9 | .0333319395 | 19 | 5 | 9 | .0561272026 |
| 18 | 3 | 8 | .0522488413 | 19 | 1 | 10 | .0299634144 | 19 | 5 | 10 | .0505141639 |
| 18 | 3 | 9 | .0465162120 | 19 | 1 | 11 | .0271011338 | 19 | 5 | 11 | .0457330143 |
| 18 | 3 | 10 | .0417473296 | 19 | 1 | 12 | .0246129452 | 19 | 5 | 12 | .0415681232 |
| 18 | 3 | 11 | .0376730987 | 19 | 1 | 13 | .0224037540 | 19 | 5 | 13 | .0378636085 |
| 18 | 3 | 12 | .0341080171 | 19 | 1 | 14 | .0204007371 | 19 | 5 | 14 | .0344995258 |
| 18 | 3 | 13 | .0309157650 | 19 | 1 | 15 | .0185431531 | 19 | 5 | 15 | .0313752924 |
| 18 | 3 | 14 | .0279875014 | 19 | 1 | 16 | .0167731147 | 19 | 6 | 6 | .0900218692 |
| 18 | 3 | 15 | .0252244786 | 19 | 1 | 17 | .0150223068 | 19 | 6 | 7 | .0782029063 |
| 18 | 3 | 16 | .0225161109 | 19 | 1 | 18 | .0131789994 | 19 | 6 | 8 | .0690294360 |
| 18 | 4 | 4 | .1105660330 | 19 | 1 | 19 | .0109382528 | 19 | 6 | 9 | .0616336896 |
| 18 | 4 | 5 | .0903973785 | 19 | 2 | 2 | .1627856651 | 19 | 6 | 10 | .0554877905 |
| 18 | 4 | 6 | .0765579277 | 19 | 2 | 3 | .1110590145 | 19 | 6 | 11 | .0502493168 |
| 18 | 4 | 7 | .0663522085 | 19 | 2 | 4 | .0852931052 | 19 | 6 | 12 | .0456834841 |
| 18 | 4 | 8 | .0584310521 | 19 | 2 | 5 | .0695970758 | 19 | 6 | 13 | .0416203596 |
| 18 | 4 | 9 | .0520394282 | 19 | 2 | 6 | .0588910196 | 19 | 6 | 14 | .0379290225 |
| 18 | 4 | 10 | .0467183403 | 19 | 2 | 7 | .0510351092 | 19 | 7 | 7 | .0856172970 |
| 18 | 4 | 11 | .0421694862 | 19 | 2 | 8 | .0449652248 | 19 | 7 | 8 | .0756153406 |
| 18 | 4 | 12 | .0381869633 | 19 | 2 | 9 | .0400891754 | 19 | 7 | 9 | .0675433158 |
| 18 | 4 | 13 | .0346192645 | 19 | 2 | 10 | .0360490040 | 19 | 7 | 10 | .0608297030 |
| 18 | 4 | 14 | .0313452501 | 19 | 2 | 11 | .0326137544 | 19 | 7 | 11 | .0551032227 |
| 18 | 4 | 15 | .0282548287 | 19 | 2 | 12 | .0296258235 | 19 | 7 | 12 | .0501089632 |
| 18 | 5 | 5 | .0999084324 | 19 | 2 | 13 | .0269716592 | 19 | 7 | 13 | .0456621845 |
| 18 | 5 | 6 | .0846879171 | 19 | 2 | 14 | .0245641908 | 19 | 8 | 8 | .0828339976 |
| 18 | 5 | 7 | .0734460812 | 19 | 2 | 15 | .0223306886 | 19 | 8 | 9 | .0740273553 |
| 18 | 5 | 8 | .0647101858 | 19 | 2 | 16 | .0202017248 | 19 | 8 | 10 | .0666958229 |
| 18 | 5 | 9 | .0576543521 | 19 | 2 | 17 | .0180952194 | 19 | 8 | 11 | .0604372716 |
| 18 | 5 | 10 | .0517756674 | 19 | 2 | 18 | .0158767294 | 19 | 8 | 12 | .0549752068 |
| 18 | 5 | 11 | .0467468132 | 19 | 3 | 3 | .1257138905 | 19 | 9 | 9 | .0812876323 |
| 18 | 5 | 12 | .0423415561 | 19 | 3 | 4 | .0967367097 | 19 | 9 | 10 | .0732703911 |
| 18 | 5 | 13 | .0383932043 | 19 | 3 | 5 | .0790298793 | 19 | 9 | 11 | .0664202905 |
| 18 | 5 | 14 | .0347682766 | 19 | 3 | 6 | .0669273697 | 19 | 10 | 10 | .0807909751 |
| 18 | 6 | 6 | .0932407327 | 19 | 3 | 7 | .0580336125 | 20 | 1 | 1 | .2756966156 |
| 18 | 6 | 7 | .0809202642 | 19 | 3 | 8 | .0511541418 | 20 | 1 | 2 | .1344941714 |
| 18 | 6 | 8 | .0713338043 | 19 | 3 | 9 | .0456228816 | 20 | 1 | 3 | .0913234064 |
| 18 | 6 | 9 | .0635829688 | 19 | 3 | 10 | .0410365629 | 20 | 1 | 4 | .0699879991 |
| 18 | 6 | 10 | .0571197288 | 19 | 3 | 11 | .0371346427 | 20 | 1 | 5 | .0570566385 |
| 18 | 6 | 11 | .0515868554 | 19 | 3 | 12 | .0337391171 | 20 | 1 | 6 | .0482701093 |

TABLE I-COVARIANCES OF NORMAL ORDER STATISTICS

| N | I | J | COVARIANCE | N | I | J | COVARIANCE | N | I | J | COVARIANCE |
|---|---|---|---|---|---|---|---|---|---|---|---|
| 20 | 1 | 7 | .0418437826 | 20 | 4 | 16 | .0258897457 | 21 | 1 | 17 | .0157929442 |
| 20 | 1 | 8 | .0368937057 | 20 | 4 | 17 | .0235070347 | 21 | 1 | 18 | .0143793492 |
| 20 | 1 | 9 | .0329296302 | 20 | 5 | 5 | .0939960005 | 21 | 1 | 19 | .0129577833 |
| 20 | 1 | 10 | .0296562522 | 20 | 5 | 6 | .0797773754 | 21 | 1 | 20 | .0114366399 |
| 20 | 1 | 11 | .0268838808 | 20 | 5 | 7 | .0693175755 | 21 | 1 | 21 | .0095549300 |
| 20 | 1 | 12 | .0244839566 | 20 | 5 | 8 | .0612251428 | 21 | 2 | 2 | .1566164697 |
| 20 | 1 | 13 | .0223649803 | 20 | 5 | 9 | .0547222526 | 21 | 2 | 3 | .1067457022 |
| 20 | 1 | 14 | .0204584277 | 20 | 5 | 10 | .0493374275 | 21 | 2 | 4 | .0819892255 |
| 20 | 1 | 15 | .0187096782 | 20 | 5 | 11 | .0447662310 | 21 | 2 | 5 | .0669523756 |
| 20 | 1 | 16 | .0170711407 | 20 | 5 | 12 | .0408014074 | 21 | 2 | 6 | .0567254558 |
| 20 | 1 | 17 | .0154951854 | 20 | 5 | 13 | .0372948401 | 21 | 2 | 7 | .0492440895 |
| 20 | 1 | 18 | .0139227071 | 20 | 5 | 14 | .0341351571 | 21 | 2 | 8 | .0434834365 |
| 20 | 1 | 19 | .0122530117 | 20 | 5 | 15 | .0312332041 | 21 | 2 | 9 | .0388743175 |
| 20 | 1 | 20 | .0102047204 | 20 | 5 | 16 | .0285109202 | 21 | 2 | 10 | .0350736979 |
| 20 | 2 | 2 | .1595731635 | 20 | 6 | 6 | .0871511273 | 21 | 2 | 11 | .0318613555 |
| 20 | 2 | 3 | .1088143706 | 20 | 6 | 7 | .0757703374 | 21 | 2 | 12 | .0290884030 |
| 20 | 2 | 4 | .0835758043 | 20 | 6 | 8 | .0669555799 | 21 | 2 | 13 | .0266494931 |
| 20 | 2 | 5 | .0682247553 | 20 | 6 | 9 | .0598659775 | 21 | 2 | 14 | .0244667120 |
| 20 | 2 | 6 | .0577699655 | 20 | 6 | 10 | .0539910641 | 21 | 2 | 15 | .0224795250 |
| 20 | 2 | 7 | .0501109522 | 20 | 6 | 11 | .0490008078 | 21 | 2 | 16 | .0206378031 |
| 20 | 2 | 8 | .0442041191 | 20 | 6 | 12 | .0446702765 | 21 | 2 | 17 | .0188959841 |
| 20 | 2 | 9 | .0394693443 | 20 | 6 | 13 | .0408385539 | 21 | 2 | 18 | .0172062933 |
| 20 | 2 | 10 | .0355565554 | 20 | 6 | 14 | .0373845180 | 21 | 2 | 19 | .0155066659 |
| 20 | 2 | 11 | .0322405467 | 20 | 6 | 15 | .0342111005 | 21 | 2 | 20 | .0136875474 |
| 20 | 2 | 12 | .0293684960 | 20 | 7 | 7 | .0826123919 | 21 | 3 | 3 | .1201588366 |
| 20 | 2 | 13 | .0268315105 | 20 | 7 | 8 | .0730383651 | 21 | 3 | 4 | .0924582740 |
| 20 | 2 | 14 | .0245479493 | 20 | 7 | 9 | .0653307650 | 21 | 3 | 5 | .0755852250 |
| 20 | 2 | 15 | .0224526610 | 20 | 7 | 10 | .0589387423 | 21 | 3 | 6 | .0640877082 |
| 20 | 2 | 16 | .0204888032 | 20 | 7 | 11 | .0535056771 | 21 | 3 | 7 | .0556654428 |
| 20 | 2 | 17 | .0185994024 | 20 | 7 | 12 | .0487882271 | 21 | 3 | 8 | .0491736697 |
| 20 | 2 | 18 | .0167136502 | 20 | 7 | 13 | .0446121115 | 21 | 3 | 9 | .0439754258 |
| 20 | 2 | 19 | .0147107672 | 20 | 7 | 14 | .0408460024 | 21 | 3 | 10 | .0396862595 |
| 20 | 3 | 3 | .1228134690 | 20 | 8 | 8 | .0796309779 | 21 | 3 | 11 | .0360590616 |
| 20 | 3 | 4 | .0945049011 | 20 | 8 | 9 | .0712591619 | 21 | 3 | 12 | .0329265958 |
| 20 | 3 | 5 | .0772355100 | 20 | 8 | 10 | .0643103379 | 21 | 3 | 13 | .0301704234 |
| 20 | 3 | 6 | .0654510179 | 20 | 8 | 11 | .0583997306 | 21 | 3 | 14 | .0277028731 |
| 20 | 3 | 7 | .0568056677 | 20 | 8 | 12 | .0532644482 | 21 | 3 | 15 | .0254557702 |
| 20 | 3 | 8 | .0501310270 | 20 | 8 | 13 | .0487159812 | 21 | 3 | 16 | .0233726059 |
| 20 | 3 | 9 | .0447763202 | 20 | 9 | 9 | .0778118324 | 21 | 3 | 17 | .0214019641 |
| 20 | 3 | 10 | .0403482354 | 20 | 9 | 10 | .0702526466 | 21 | 3 | 18 | .0194898677 |
| 20 | 3 | 11 | .0365934286 | 20 | 9 | 11 | .0638176732 | 21 | 3 | 19 | .0175661123 |
| 20 | 3 | 12 | .0333397949 | 20 | 9 | 12 | .0582229126 | 21 | 4 | 4 | .1021303742 |
| 20 | 3 | 13 | .0304645791 | 20 | 10 | 10 | .0769474345 | 21 | 4 | 5 | .0835814082 |
| 20 | 3 | 14 | .0278756579 | 20 | 10 | 11 | .0699266209 | 21 | 4 | 6 | .0709191324 |
| 20 | 3 | 15 | .0254994380 | 21 | 1 | 1 | .2717684437 | 21 | 4 | 7 | .0616315839 |
| 20 | 3 | 16 | .0232716370 | 21 | 1 | 2 | .1323788072 | 21 | 4 | 8 | .0544658004 |
| 20 | 3 | 17 | .0211277371 | 21 | 1 | 3 | .0898537327 | 21 | 4 | 9 | .0487233941 |
| 20 | 3 | 18 | .0189874446 | 21 | 1 | 4 | .0688680970 | 21 | 4 | 10 | .0439822525 |
| 20 | 4 | 4 | .1046766239 | 21 | 1 | 5 | .0561650671 | 21 | 4 | 11 | .0399707477 |
| 20 | 4 | 5 | .0856442352 | 21 | 1 | 6 | .0475447416 | 21 | 4 | 12 | .0365048686 |
| 20 | 4 | 6 | .0726321557 | 21 | 1 | 7 | .0412486423 | 21 | 4 | 13 | .0334541815 |
| 20 | 4 | 7 | .0630731773 | 21 | 1 | 8 | .0364063819 | 21 | 4 | 14 | .0307220559 |
| 20 | 4 | 8 | .0556855080 | 21 | 1 | 9 | .0325356366 | 21 | 4 | 15 | .0282332864 |
| 20 | 4 | 9 | .0497539272 | 21 | 1 | 10 | .0293462246 | 21 | 4 | 16 | .0259254790 |
| 20 | 4 | 10 | .0448455403 | 21 | 1 | 11 | .0266521183 | 21 | 4 | 17 | .0237418028 |
| 20 | 4 | 11 | .0406811669 | 21 | 1 | 12 | .0243276996 | 21 | 4 | 18 | .0216225255 |
| 20 | 4 | 12 | .0370709494 | 21 | 1 | 13 | .0222841795 | 21 | 5 | 5 | .0914639429 |
| 20 | 4 | 13 | .0338793393 | 21 | 1 | 14 | .0204559555 | 21 | 5 | 6 | .0776649264 |
| 20 | 4 | 14 | .0310045148 | 21 | 1 | 15 | .0187921085 | 21 | 5 | 7 | .0675303529 |
| 20 | 4 | 15 | .0283650520 | 21 | 1 | 16 | .0172505172 | 21 | 5 | 8 | .0597032383 |

TABLE I-COVARIANCES OF NORMAL ORDER STATISTICS

| N | I | J | COVARIANCE | N | I | J | COVARIANCE | N | I | J | COVARIANCE |
|---|---|---|---|---|---|---|---|---|---|---|---|
| 21 | 5 | 9 | .0534259094 | 22 | 1 | 16 | .0173508487 | 22 | 4 | 19 | .0199843802 |
| 21 | 5 | 10 | .0482397873 | 22 | 1 | 17 | .0159803725 | 22 | 5 | 5 | .0891592181 |
| 21 | 5 | 11 | .0438494382 | 22 | 1 | 18 | .0146741239 | 22 | 5 | 6 | .0757374130 |
| 21 | 5 | 12 | .0400545258 | 22 | 1 | 19 | .0133978177 | 22 | 5 | 7 | .0658940940 |
| 21 | 5 | 13 | .0367129179 | 22 | 1 | 20 | .0121051902 | 22 | 5 | 8 | .0583035144 |
| 21 | 5 | 14 | .0337192277 | 22 | 1 | 21 | .0107122475 | 22 | 5 | 9 | .0522261702 |
| 21 | 5 | 15 | .0309913643 | 22 | 1 | 22 | .0089758218 | 22 | 5 | 10 | .0472150082 |
| 21 | 5 | 16 | .0284611560 | 22 | 2 | 2 | .1538834348 | 22 | 5 | 11 | .0429825069 |
| 21 | 5 | 17 | .0260664414 | 22 | 2 | 3 | .1048313072 | 22 | 5 | 12 | .0393342304 |
| 21 | 6 | 6 | .0845709380 | 22 | 2 | 4 | .0805177919 | 22 | 5 | 13 | .0361328985 |
| 21 | 6 | 7 | .0735768812 | 22 | 2 | 5 | .0657686798 | 22 | 5 | 14 | .0332776235 |
| 21 | 6 | 8 | .0650771409 | 22 | 2 | 6 | .0557496572 | 22 | 5 | 15 | .0306911232 |
| 21 | 6 | 9 | .0582547328 | 22 | 2 | 7 | .0484296475 | 22 | 5 | 16 | .0283112216 |
| 21 | 6 | 10 | .0526144799 | 22 | 2 | 8 | .0428010533 | 22 | 5 | 17 | .0260844764 |
| 21 | 6 | 11 | .0478369933 | 22 | 2 | 9 | .0383046960 | 22 | 5 | 18 | .0239602586 |
| 21 | 6 | 12 | .0437054701 | 22 | 2 | 10 | .0346039022 | 22 | 6 | 6 | .0822361546 |
| 21 | 6 | 13 | .0400659531 | 22 | 2 | 11 | .0314828599 | 22 | 6 | 7 | .0715864306 |
| 21 | 6 | 14 | .0368041864 | 22 | 2 | 12 | .0287960386 | 22 | 6 | 8 | .0633659623 |
| 21 | 6 | 15 | .0338310889 | 22 | 2 | 13 | .0264409419 | 22 | 6 | 9 | .0567791925 |
| 21 | 6 | 16 | .0310726097 | 22 | 2 | 14 | .0243424165 | 22 | 6 | 10 | .0513445403 |
| 21 | 7 | 7 | .0799322254 | 22 | 2 | 15 | .0224430216 | 22 | 6 | 11 | .0467519339 |
| 21 | 7 | 8 | .0707312834 | 22 | 2 | 16 | .0206966508 | 22 | 6 | 12 | .0427914887 |
| 21 | 7 | 9 | .0633395388 | 22 | 2 | 17 | .0190637763 | 22 | 6 | 13 | .0393148917 |
| 21 | 7 | 10 | .0572241851 | 22 | 2 | 18 | .0175070608 | 22 | 6 | 14 | .0362130614 |
| 21 | 7 | 11 | .0520411359 | 22 | 2 | 19 | .0159856941 | 22 | 6 | 15 | .0334023748 |
| 21 | 7 | 12 | .0475565713 | 22 | 2 | 20 | .0144445464 | 22 | 6 | 16 | .0308154993 |
| 21 | 7 | 13 | .0436042866 | 22 | 2 | 21 | .0127834441 | 22 | 6 | 17 | .0283945078 |
| 21 | 7 | 14 | .0400608197 | 22 | 3 | 3 | .1177174938 | 22 | 7 | 7 | .0775230987 |
| 21 | 7 | 15 | .0368298179 | 22 | 3 | 4 | .0905728888 | 22 | 7 | 8 | .0686507899 |
| 21 | 8 | 8 | .0767996256 | 22 | 3 | 5 | .0740610778 | 22 | 7 | 9 | .0615358817 |
| 21 | 8 | 9 | .0688016030 | 22 | 3 | 6 | .0628241881 | 22 | 7 | 10 | .0556615230 |
| 21 | 8 | 10 | .0621794637 | 22 | 3 | 7 | .0546036829 | 22 | 7 | 11 | .0506945487 |
| 21 | 8 | 11 | .0565631822 | 22 | 3 | 8 | .0482764657 | 22 | 7 | 12 | .0464092181 |
| 21 | 8 | 12 | .0517010213 | 22 | 3 | 9 | .0432181436 | 22 | 7 | 13 | .0426458719 |
| 21 | 8 | 13 | .0474138510 | 22 | 3 | 10 | .0390522471 | 22 | 7 | 14 | .0392869862 |
| 21 | 8 | 14 | .0435684716 | 22 | 3 | 11 | .0355371766 | 22 | 7 | 15 | .0362423918 |
| 21 | 9 | 9 | .0747720428 | 22 | 3 | 12 | .0325098531 | 22 | 7 | 16 | .0334394223 |
| 21 | 9 | 10 | .0676001826 | 22 | 3 | 13 | .0298553234 | 22 | 8 | 8 | .0742741389 |
| 21 | 9 | 11 | .0615132362 | 22 | 3 | 14 | .0274892349 | 22 | 8 | 9 | .0666013909 |
| 21 | 9 | 12 | .0562403023 | 22 | 3 | 15 | .0253470669 | 22 | 8 | 10 | .0602618654 |
| 21 | 9 | 13 | .0515883920 | 22 | 3 | 16 | .0233769898 | 22 | 8 | 11 | .0548983179 |
| 21 | 10 | 10 | .0736292629 | 22 | 3 | 17 | .0215345307 | 22 | 8 | 12 | .0502684365 |
| 21 | 10 | 11 | .0670228535 | 22 | 3 | 18 | .0197776401 | 22 | 8 | 13 | .0462006795 |
| 21 | 10 | 12 | .0612958925 | 22 | 3 | 19 | .0180603083 | 22 | 8 | 14 | .0425686641 |
| 21 | 11 | 11 | .0732597550 | 22 | 3 | 20 | .0163203204 | 22 | 8 | 15 | .0392753382 |
| 22 | 1 | 1 | .2681144875 | 22 | 4 | 4 | .0998003945 | 22 | 9 | 9 | .0720851097 |
| 22 | 1 | 2 | .1304111315 | 22 | 4 | 5 | .0816898374 | 22 | 9 | 10 | .0652455979 |
| 22 | 1 | 3 | .0884843241 | 22 | 4 | 6 | .0693437016 | 22 | 9 | 11 | .0594551817 |
| 22 | 1 | 4 | .0678216925 | 22 | 4 | 7 | .0603004923 | 22 | 9 | 12 | .0544539693 |
| 22 | 1 | 5 | .0553287760 | 22 | 4 | 8 | .0533335026 | 22 | 9 | 13 | .0500577849 |
| 22 | 1 | 6 | .0468607712 | 22 | 4 | 9 | .0477595955 | 22 | 9 | 14 | .0461308068 |
| 22 | 1 | 7 | .0406834410 | 22 | 4 | 10 | .0431663352 | 22 | 10 | 10 | .0707277909 |
| 22 | 1 | 8 | .0359389291 | 22 | 4 | 11 | .0392887514 | 22 | 10 | 11 | .0644711601 |
| 22 | 1 | 9 | .0321521843 | 22 | 4 | 12 | .0359478248 | 22 | 10 | 12 | .0590638487 |
| 22 | 1 | 10 | .0290376670 | 22 | 4 | 13 | .0330172628 | 22 | 10 | 13 | .0543080654 |
| 22 | 1 | 11 | .0264125927 | 22 | 4 | 14 | .0304043228 | 22 | 11 | 11 | .0700770602 |
| 22 | 1 | 12 | .0241538434 | 22 | 4 | 15 | .0280380148 | 22 | 11 | 12 | .0642191580 |
| 22 | 1 | 13 | .0221747937 | 22 | 4 | 16 | .0258612690 | 23 | 1 | 1 | .2647037742 |
| 22 | 1 | 14 | .0204119854 | 22 | 4 | 17 | .0238250746 | 23 | 1 | 2 | .1285746124 |
| 22 | 1 | 15 | .0188169580 | 22 | 4 | 18 | .0218830469 | 23 | 1 | 3 | .0872042823 |

TABLE I-COVARIANCES OF NORMAL ORDER STATISTICS

| N | I | J | COVARIANCE | N | I | J | COVARIANCE | N | I | J | COVARIANCE |
|---|---|---|---|---|---|---|---|---|---|---|---|
| 23 | 1 | 4 | .0668411850 | 23 | 4 | 4 | .0976581348 | 23 | 8 | 12 | .0489516566 |
| 23 | 1 | 5 | .0545425000 | 23 | 4 | 5 | .0799474393 | 23 | 8 | 13 | .0450712784 |
| 23 | 1 | 6 | .0462147967 | 23 | 4 | 6 | .0678887639 | 23 | 8 | 14 | .0416190521 |
| 23 | 1 | 7 | .0401463995 | 23 | 4 | 7 | .0590669572 | 23 | 8 | 15 | .0385031101 |
| 23 | 1 | 8 | .0354910834 | 23 | 4 | 8 | .0522793614 | 23 | 8 | 16 | .0356508223 |
| 23 | 1 | 9 | .0317805148 | 23 | 4 | 9 | .0468567531 | 23 | 9 | 9 | .0696884508 |
| 23 | 1 | 10 | .0287334028 | 23 | 4 | 10 | .0423954631 | 23 | 9 | 10 | .0631376667 |
| 23 | 1 | 11 | .0261698810 | 23 | 4 | 11 | .0386364763 | 23 | 9 | 11 | .0576033482 |
| 23 | 1 | 12 | .0239690152 | 23 | 4 | 12 | .0354051362 | 23 | 9 | 12 | .0528350643 |
| 23 | 1 | 13 | .0220459835 | 23 | 4 | 13 | .0325786322 | 23 | 9 | 13 | .0486559924 |
| 23 | 1 | 14 | .0203389979 | 23 | 4 | 14 | .0300673026 | 23 | 9 | 14 | .0449365466 |
| 23 | 1 | 15 | .0188013602 | 23 | 4 | 15 | .0278032393 | 23 | 9 | 15 | .0415782244 |
| 23 | 1 | 16 | .0173963331 | 23 | 4 | 16 | .0257329068 | 23 | 10 | 10 | .0681633883 |
| 23 | 1 | 17 | .0160935489 | 23 | 4 | 17 | .0238119527 | 23 | 10 | 11 | .0622063854 |
| 23 | 1 | 18 | .0148661309 | 23 | 4 | 18 | .0220010258 | 23 | 10 | 12 | .0570709626 |
| 23 | 1 | 19 | .0136877822 | 23 | 4 | 19 | .0202615159 | 23 | 10 | 13 | .0525678611 |
| 23 | 1 | 20 | .0125286785 | 23 | 4 | 20 | .0185495057 | 23 | 10 | 14 | .0485582528 |
| 23 | 1 | 21 | .0113471622 | 23 | 5 | 5 | .0870502669 | 23 | 11 | 11 | .0672946093 |
| 23 | 1 | 22 | .0100657102 | 23 | 5 | 6 | .0739698020 | 23 | 11 | 12 | .0617560836 |
| 23 | 1 | 23 | .0084568651 | 23 | 5 | 7 | .0643891594 | 23 | 11 | 13 | .0568967972 |
| 23 | 2 | 2 | .1513472666 | 23 | 5 | 8 | .0570110727 | 23 | 12 | 12 | .0670122963 |
| 23 | 2 | 3 | .1030530274 | 23 | 5 | 9 | .0511125400 | 24 | 1 | 1 | .2615100245 |
| 23 | 2 | 4 | .0791483745 | 23 | 5 | 10 | .0462569073 | 24 | 1 | 2 | .1268551922 |
| 23 | 2 | 5 | .0646640578 | 23 | 5 | 11 | .0421637065 | 24 | 1 | 3 | .0860042863 |
| 23 | 2 | 6 | .0548357214 | 23 | 5 | 12 | .0386436499 | 24 | 1 | 4 | .0659200075 |
| 23 | 2 | 7 | .0476631362 | 23 | 5 | 13 | .0355635378 | 24 | 1 | 5 | .0538016071 |
| 23 | 2 | 8 | .0421546360 | 23 | 5 | 14 | .0328260566 | 24 | 1 | 6 | .0456037148 |
| 23 | 2 | 9 | .0377602426 | 23 | 5 | 15 | .0303574544 | 24 | 1 | 7 | .0396357291 |
| 23 | 2 | 10 | .0341490860 | 23 | 5 | 16 | .0280995506 | 24 | 1 | 8 | .0350622659 |
| 23 | 2 | 11 | .0311093185 | 23 | 5 | 17 | .0260041106 | 24 | 1 | 9 | .0314212296 |
| 23 | 2 | 12 | .0284983465 | 23 | 5 | 18 | .0240283046 | 24 | 1 | 10 | .0284352471 |
| 23 | 2 | 13 | .0262160614 | 23 | 5 | 19 | .0221300722 | 24 | 1 | 11 | .0259271100 |
| 23 | 2 | 14 | .0241894785 | 23 | 6 | 6 | .0801108125 | 24 | 1 | 12 | .0237778306 |
| 23 | 2 | 15 | .0223633938 | 23 | 6 | 7 | .0697700220 | 24 | 1 | 13 | .0219041379 |
| 23 | 2 | 16 | .0206943446 | 23 | 6 | 8 | .0617991761 | 24 | 1 | 14 | .0202456072 |
| 23 | 2 | 17 | .0191463743 | 23 | 6 | 9 | .0554220845 | 24 | 1 | 15 | .0187568828 |
| 23 | 2 | 18 | .0176876306 | 23 | 6 | 10 | .0501693770 | 24 | 1 | 16 | .0174027208 |
| 23 | 2 | 19 | .0162869164 | 23 | 6 | 11 | .0457392558 | 24 | 1 | 17 | .0161546294 |
| 23 | 2 | 20 | .0149088122 | 23 | 6 | 12 | .0419278569 | 24 | 1 | 18 | .0149883775 |
| 23 | 2 | 21 | .0135037976 | 23 | 6 | 13 | .0385916111 | 24 | 1 | 19 | .0138818427 |
| 23 | 2 | 22 | .0119796544 | 23 | 6 | 14 | .0356255543 | 24 | 1 | 20 | .0128126293 |
| 23 | 3 | 3 | .1154626096 | 23 | 6 | 15 | .0329500831 | 24 | 1 | 21 | .0117544395 |
| 23 | 3 | 4 | .0888288867 | 23 | 6 | 16 | .0305023574 | 24 | 1 | 22 | .0106694123 |
| 23 | 3 | 5 | .0726480567 | 23 | 6 | 17 | .0282302416 | 24 | 1 | 23 | .0094855943 |
| 23 | 3 | 6 | .0616492103 | 23 | 6 | 18 | .0260874033 | 24 | 1 | 24 | .0079894815 |
| 23 | 3 | 7 | .0536123113 | 23 | 7 | 7 | .0753429148 | 24 | 2 | 2 | .1489854382 |
| 23 | 3 | 8 | .0474341928 | 23 | 7 | 8 | .0667626712 | 24 | 2 | 3 | .1013955419 |
| 23 | 3 | 9 | .0425019784 | 23 | 7 | 9 | .0598927410 | 24 | 2 | 4 | .0778698243 |
| 23 | 3 | 10 | .0384464573 | 23 | 7 | 10 | .0542305301 | 24 | 2 | 5 | .0636302595 |
| 23 | 3 | 11 | .0350309710 | 23 | 7 | 11 | .0494525283 | 24 | 2 | 6 | .0539776542 |
| 23 | 3 | 12 | .0320960781 | 23 | 7 | 12 | .0453400020 | 24 | 2 | 7 | .0469404665 |
| 23 | 3 | 13 | .0295297523 | 23 | 7 | 13 | .0417387852 | 24 | 2 | 8 | .0415418155 |
| 23 | 3 | 14 | .0272502613 | 23 | 7 | 14 | .0385360820 | 24 | 2 | 9 | .0372402315 |
| 23 | 3 | 15 | .0251957430 | 23 | 7 | 15 | .0356462884 | 24 | 2 | 10 | .0337101800 |
| 23 | 3 | 16 | .0233174598 | 23 | 7 | 16 | .0330017787 | 24 | 2 | 11 | .0307434169 |
| 23 | 3 | 17 | .0215750622 | 23 | 7 | 17 | .0305464023 | 24 | 2 | 12 | .0281999669 |
| 23 | 3 | 18 | .0199327781 | 23 | 8 | 8 | .0720039053 | 24 | 2 | 13 | .0259817826 |
| 23 | 3 | 19 | .0183555401 | 23 | 8 | 9 | .0646171799 | 24 | 2 | 14 | .0240176602 |
| 23 | 3 | 20 | .0168034967 | 23 | 8 | 10 | .0585249274 | 24 | 2 | 15 | .0222541141 |
| 23 | 3 | 21 | .0152208846 | 23 | 8 | 11 | .0533811432 | 24 | 2 | 16 | .0206495539 |

TABLE I-COVARIANCES OF NORMAL ORDER STATISTICS

| N | I | J | COVARIANCE | N | I | J | COVARIANCE | N | I | J | COVARIANCE |
|---|---|---|---|---|---|---|---|---|---|---|---|
| 24 | 2 | 17 | .0191703335 | 24 | 5 | 20 | .0205275879 | 25 | 1 | 4 | .0650524574 |
| 24 | 2 | 18 | .0177878163 | 24 | 6 | 6 | .0781658681 | 25 | 1 | 5 | .0531020135 |
| 24 | 2 | 19 | .0164758353 | 24 | 6 | 7 | .0681041220 | 25 | 1 | 6 | .0450247092 |
| 24 | 2 | 20 | .0152078761 | 24 | 6 | 8 | .0603580063 | 25 | 1 | 7 | .0391496922 |
| 24 | 2 | 21 | .0139527743 | 24 | 6 | 9 | .0541689508 | 25 | 1 | 8 | .0346517190 |
| 24 | 2 | 22 | .0126656262 | 24 | 6 | 10 | .0490786419 | 25 | 1 | 9 | .0310745120 |
| 24 | 2 | 23 | .0112610456 | 24 | 6 | 11 | .0447927115 | 25 | 1 | 10 | .0281443334 |
| 24 | 3 | 3 | .1133718169 | 24 | 6 | 12 | .0411126219 | 25 | 1 | 11 | .0256864182 |
| 24 | 3 | 4 | .0872096423 | 24 | 6 | 13 | .0378988969 | 25 | 1 | 12 | .0235835394 |
| 24 | 3 | 5 | .0713335032 | 24 | 6 | 14 | .0350499826 | 25 | 1 | 13 | .0217537916 |
| 24 | 3 | 6 | .0605531736 | 24 | 6 | 15 | .0324894109 | 25 | 1 | 14 | .0201379035 |
| 24 | 3 | 7 | .0526842778 | 24 | 6 | 16 | .0301575807 | 25 | 1 | 15 | .0186915930 |
| 24 | 3 | 8 | .0466420696 | 24 | 6 | 17 | .0280061596 | 25 | 1 | 16 | .0173807311 |
| 24 | 3 | 9 | .0418242839 | 24 | 6 | 18 | .0259939042 | 25 | 1 | 17 | .0161781223 |
| 24 | 3 | 10 | .0378683595 | 24 | 6 | 19 | .0240830172 | 25 | 1 | 18 | .0150612184 |
| 24 | 3 | 11 | .0345421192 | 24 | 7 | 7 | .0733581380 | 25 | 1 | 19 | .0140103204 |
| 24 | 3 | 12 | .0316893618 | 24 | 7 | 8 | .0650394867 | 25 | 1 | 20 | .0130069046 |
| 24 | 3 | 13 | .0292005904 | 24 | 7 | 9 | .0583881581 | 25 | 1 | 21 | .0120316120 |
| 24 | 3 | 14 | .0269962332 | 24 | 7 | 10 | .0529144065 | 25 | 1 | 22 | .0110609823 |
| 24 | 3 | 15 | .0250164815 | 24 | 7 | 11 | .0483033538 | 25 | 1 | 23 | .0100603317 |
| 24 | 3 | 16 | .0232148004 | 24 | 7 | 12 | .0443424451 | 25 | 1 | 24 | .0089625537 |
| 24 | 3 | 17 | .0215535205 | 24 | 7 | 13 | .0408822435 | 25 | 1 | 25 | .0075666181 |
| 24 | 3 | 18 | .0200005601 | 24 | 7 | 14 | .0378138685 | 25 | 2 | 2 | .1467788564 |
| 24 | 3 | 19 | .0185265826 | 24 | 7 | 15 | .0350552828 | 25 | 2 | 3 | .0998458242 |
| 24 | 3 | 20 | .0171018373 | 24 | 7 | 16 | .0325425050 | 25 | 2 | 4 | .0766726079 |
| 24 | 3 | 21 | .0156913277 | 24 | 7 | 17 | .0302236165 | 25 | 2 | 5 | .0626601640 |
| 24 | 3 | 22 | .0142445918 | 24 | 7 | 18 | .0280542823 | 25 | 2 | 6 | .0531701977 |
| 24 | 4 | 4 | .0956800128 | 24 | 8 | 8 | .0699492829 | 25 | 2 | 7 | .0462579396 |
| 24 | 4 | 5 | .0783358839 | 24 | 8 | 9 | .0628163410 | 25 | 2 | 8 | .0409602833 |
| 24 | 4 | 6 | .0665400103 | 24 | 8 | 10 | .0569425517 | 25 | 2 | 9 | .0367436649 |
| 24 | 4 | 7 | .0579199925 | 24 | 8 | 11 | .0519919109 | 25 | 2 | 10 | .0332874839 |
| 24 | 4 | 8 | .0512953118 | 24 | 8 | 12 | .0477373965 | 25 | 2 | 11 | .0303868016 |
| 24 | 4 | 9 | .0460095138 | 24 | 8 | 13 | .0440192732 | 25 | 2 | 12 | .0279040099 |
| 24 | 4 | 10 | .0416669321 | 24 | 8 | 14 | .0407210730 | 25 | 2 | 13 | .0257428796 |
| 24 | 4 | 11 | .0380139366 | 24 | 8 | 15 | .0377549827 | 25 | 2 | 14 | .0238337209 |
| 24 | 4 | 12 | .0348797505 | 24 | 8 | 16 | .0350524706 | 25 | 2 | 15 | .0221244321 |
| 24 | 4 | 13 | .0321445733 | 24 | 8 | 17 | .0325578873 | 25 | 2 | 16 | .0205748319 |
| 24 | 4 | 14 | .0297212895 | 24 | 9 | 9 | .0675339494 | 25 | 2 | 17 | .0191528822 |
| 24 | 4 | 15 | .0275443817 | 24 | 9 | 10 | .0612366675 | 25 | 2 | 18 | .0178320018 |
| 24 | 4 | 16 | .0255628439 | 24 | 9 | 11 | .0559260829 | 25 | 2 | 19 | .0165889521 |
| 24 | 4 | 17 | .0237353633 | 24 | 9 | 12 | .0513600320 | 25 | 2 | 20 | .0154018640 |
| 24 | 4 | 18 | .0220267330 | 24 | 9 | 13 | .0473679813 | 25 | 2 | 21 | .0142478622 |
| 24 | 4 | 19 | .0204047351 | 24 | 9 | 14 | .0438254941 | 25 | 2 | 22 | .0130992021 |
| 24 | 4 | 20 | .0188366718 | 24 | 9 | 15 | .0406386807 | 25 | 2 | 23 | .0119148374 |
| 24 | 4 | 21 | .0172840483 | 24 | 9 | 16 | .0377342147 | 25 | 2 | 24 | .0106153141 |
| 24 | 5 | 5 | .0851112954 | 24 | 10 | 10 | .0658761719 | 25 | 3 | 3 | .1114263056 |
| 24 | 5 | 6 | .0723415215 | 24 | 10 | 11 | .0601790701 | 25 | 3 | 4 | .0857011207 |
| 24 | 5 | 7 | .0629992602 | 24 | 10 | 12 | .0552780979 | 25 | 3 | 5 | .0701066633 |
| 24 | 5 | 8 | .0558133528 | 24 | 10 | 13 | .0509912609 | 25 | 3 | 6 | .0595278356 |
| 24 | 5 | 9 | .0500758789 | 24 | 10 | 14 | .0471856488 | 25 | 3 | 7 | .0518134179 |
| 24 | 5 | 10 | .0453596407 | 24 | 10 | 15 | .0437609001 | 25 | 3 | 8 | .0458957614 |
| 24 | 5 | 11 | .0413905274 | 24 | 11 | 11 | .0648357457 | 25 | 3 | 9 | .0411824242 |
| 24 | 5 | 12 | .0379838190 | 24 | 11 | 12 | .0595703878 | 25 | 3 | 10 | .0373169749 |
| 24 | 5 | 13 | .0350098412 | 24 | 11 | 13 | .0549624926 | 25 | 3 | 11 | .0340713356 |
| 24 | 5 | 14 | .0323742384 | 24 | 11 | 14 | .0508700391 | 25 | 3 | 12 | .0312922299 |
| 24 | 5 | 15 | .0300060061 | 24 | 12 | 12 | .0643335821 | 25 | 3 | 13 | .0288723939 |
| 24 | 5 | 16 | .0278498346 | 24 | 12 | 13 | .0593716340 | 25 | 3 | 14 | .0267340962 |
| 24 | 5 | 17 | .0258608988 | 25 | 1 | 1 | .2585107775 | 25 | 3 | 15 | .0248191897 |
| 24 | 5 | 18 | .0240009738 | 25 | 1 | 2 | .1252408324 | 25 | 3 | 16 | .0230828065 |
| 24 | 5 | 19 | .0222350552 | 25 | 1 | 3 | .0848763128 | 25 | 3 | 17 | .0214891522 |

TABLE I-COVARIANCES OF NORMAL ORDER STATISTICS

| N | I | J | COVARIANCE | N | I | J | COVARIANCE | N | I | J | COVARIANCE |
|---|---|---|---|---|---|---|---|---|---|---|---|
| 25 | 3 | 18 | .0200085126 | 25 | 7 | 10 | .0516990983 | 26 | 1 | 15 | .0186112572 |
| 25 | 3 | 19 | .0186148944 | 25 | 7 | 11 | .0472368304 | 26 | 1 | 16 | .0173379031 |
| 25 | 3 | 20 | .0172838191 | 25 | 7 | 12 | .0434102618 | 26 | 1 | 17 | .0161739672 |
| 25 | 3 | 21 | .0159896632 | 25 | 7 | 13 | .0400741052 | 26 | 1 | 18 | .0150980485 |
| 25 | 3 | 22 | .0147013265 | 25 | 7 | 14 | .0371228179 | 26 | 1 | 19 | .0140919761 |
| 25 | 3 | 23 | .0133727703 | 25 | 7 | 15 | .0344772718 | 26 | 1 | 20 | .0131394645 |
| 25 | 4 | 4 | .0938463604 | 25 | 7 | 16 | .0320762801 | 26 | 1 | 21 | .0122247649 |
| 25 | 4 | 5 | .0768398166 | 25 | 7 | 17 | .0298709263 | 26 | 1 | 22 | .0113309225 |
| 25 | 4 | 6 | .0652853686 | 25 | 7 | 18 | .0278205137 | 26 | 1 | 23 | .0104367967 |
| 25 | 4 | 7 | .0568502189 | 25 | 7 | 19 | .0258893536 | 26 | 1 | 24 | .0095103990 |
| 25 | 4 | 8 | .0503743200 | 25 | 8 | 8 | .0680787177 | 26 | 1 | 25 | .0084888807 |
| 25 | 4 | 9 | .0452129954 | 25 | 8 | 9 | .0611727315 | 26 | 1 | 26 | .0071824274 |
| 25 | 4 | 10 | .0409779301 | 25 | 8 | 10 | .0554935856 | 26 | 2 | 2 | .1447112195 |
| 25 | 4 | 11 | .0374204058 | 25 | 8 | 11 | .0507142579 | 26 | 2 | 3 | .0983927207 |
| 25 | 4 | 12 | .0343731413 | 25 | 8 | 12 | .0466140846 | 26 | 2 | 4 | .0755485214 |
| 25 | 4 | 13 | .0317189911 | 25 | 8 | 13 | .0430381013 | 26 | 2 | 5 | .0617475932 |
| 25 | 4 | 14 | .0293730121 | 25 | 8 | 14 | .0398736601 | 26 | 2 | 6 | .0524087276 |
| 25 | 4 | 15 | .0272716276 | 25 | 8 | 15 | .0370362552 | 26 | 2 | 7 | .0456122208 |
| 25 | 4 | 16 | .0253657543 | 25 | 8 | 16 | .0344605024 | 26 | 2 | 8 | .0404078436 |
| 25 | 4 | 17 | .0236162151 | 25 | 8 | 17 | .0320940997 | 26 | 2 | 9 | .0362694069 |
| 25 | 4 | 18 | .0219904703 | 25 | 8 | 18 | .0298935054 | 26 | 2 | 10 | .0328808960 |
| 25 | 4 | 19 | .0204600378 | 25 | 9 | 9 | .0655839488 | 26 | 2 | 11 | .0300404253 |
| 25 | 4 | 20 | .0189980781 | 25 | 9 | 10 | .0595112818 | 26 | 2 | 12 | .0276125492 |
| 25 | 4 | 21 | .0175764762 | 25 | 9 | 11 | .0543980886 | 26 | 2 | 13 | .0255026703 |
| 25 | 4 | 22 | .0161610835 | 25 | 9 | 12 | .0500095254 | 26 | 2 | 14 | .0236424181 |
| 25 | 5 | 5 | .0833209921 | 25 | 9 | 13 | .0461805401 | 26 | 2 | 15 | .0219808482 |
| 25 | 5 | 6 | .0708354913 | 25 | 9 | 14 | .0427910677 | 26 | 2 | 16 | .0204788931 |
| 25 | 5 | 7 | .0617107638 | 25 | 9 | 15 | .0397509794 | 26 | 2 | 17 | .0191057032 |
| 25 | 5 | 8 | .0546996986 | 25 | 9 | 16 | .0369904970 | 26 | 2 | 18 | .0178361074 |
| 25 | 5 | 9 | .0491082322 | 25 | 9 | 17 | .0344537682 | 26 | 2 | 19 | .0166487211 |
| 25 | 5 | 10 | .0445178217 | 25 | 10 | 10 | .0638200956 | 26 | 2 | 20 | .0155243652 |
| 25 | 5 | 11 | .0406601360 | 25 | 10 | 11 | .0583508743 | 26 | 2 | 21 | .0144444797 |
| 25 | 5 | 12 | .0373545603 | 25 | 10 | 12 | .0536544426 | 26 | 2 | 22 | .0133890674 |
| 25 | 5 | 13 | .0344745265 | 25 | 10 | 13 | .0495551088 | 26 | 2 | 23 | .0123331758 |
| 25 | 5 | 14 | .0319282036 | 25 | 10 | 14 | .0459249720 | 26 | 2 | 24 | .0112390259 |
| 25 | 5 | 15 | .0296468225 | 25 | 10 | 15 | .0426679538 | 26 | 2 | 25 | .0100323634 |
| 25 | 5 | 16 | .0275772637 | 25 | 10 | 16 | .0397096192 | 26 | 3 | 3 | .1096101302 |
| 25 | 5 | 17 | .0256771065 | 25 | 11 | 11 | .0626429079 | 26 | 3 | 4 | .0842913838 |
| 25 | 5 | 18 | .0239110984 | 25 | 11 | 12 | .0576141501 | 26 | 3 | 5 | .0689583381 |
| 25 | 5 | 19 | .0222483642 | 25 | 11 | 13 | .0532227052 | 26 | 3 | 6 | .0585660821 |
| 25 | 5 | 20 | .0206597902 | 25 | 11 | 14 | .0493323061 | 26 | 3 | 7 | .0509943295 |
| 25 | 5 | 21 | .0191148571 | 25 | 11 | 15 | .0458405123 | 26 | 3 | 8 | .0451913596 |
| 25 | 6 | 6 | .0763775864 | 25 | 12 | 12 | .0619667682 | 26 | 3 | 9 | .0405738623 |
| 25 | 6 | 7 | .0665693947 | 25 | 12 | 13 | .0572561797 | 26 | 3 | 10 | .0367910855 |
| 25 | 6 | 8 | .0590268856 | 25 | 12 | 14 | .0530811706 | 26 | 3 | 11 | .0336187275 |
| 25 | 6 | 9 | .0530076130 | 25 | 13 | 13 | .0617462570 | 26 | 3 | 12 | .0309061796 |
| 25 | 6 | 10 | .0480633428 | 26 | 1 | 1 | .2556867054 | 26 | 3 | 13 | .0285481800 |
| 25 | 6 | 11 | .0439064369 | 26 | 1 | 2 | .1237211561 | 26 | 3 | 14 | .0264686042 |
| 25 | 6 | 12 | .0403431309 | 26 | 1 | 3 | .0838134144 | 26 | 3 | 15 | .0246106990 |
| 25 | 6 | 13 | .0372375484 | 26 | 1 | 4 | .0642335583 | 26 | 3 | 16 | .0229309206 |
| 25 | 6 | 14 | .0344910450 | 26 | 1 | 5 | .0524401094 | 26 | 3 | 17 | .0213948680 |
| 25 | 6 | 15 | .0320297079 | 26 | 1 | 6 | .0444752321 | 26 | 3 | 18 | .0199744586 |
| 25 | 6 | 16 | .0297964143 | 26 | 1 | 7 | .0386866355 | 26 | 3 | 19 | .0186458217 |
| 25 | 6 | 17 | .0277455225 | 26 | 1 | 8 | .0342585930 | 26 | 3 | 20 | .0173875371 |
| 25 | 6 | 18 | .0258390818 | 26 | 1 | 9 | .0307402738 | 26 | 3 | 21 | .0161788622 |
| 25 | 6 | 19 | .0240438342 | 26 | 1 | 10 | .0278613302 | 26 | 3 | 22 | .0149974334 |
| 25 | 6 | 20 | .0223283970 | 26 | 1 | 11 | .0254492595 | 26 | 3 | 23 | .0138153282 |
| 25 | 7 | 7 | .0715416743 | 26 | 1 | 12 | .0233884450 | 26 | 3 | 24 | .0125902484 |
| 25 | 7 | 8 | .0634589337 | 26 | 1 | 13 | .0215982062 | 26 | 4 | 4 | .0921406302 |
| 25 | 7 | 9 | .0570041028 | 26 | 1 | 14 | .0200202735 | 26 | 4 | 5 | .0754462687 |

TABLE I-COVARIANCES OF NORMAL ORDER STATISTICS

| N | I | J | COVARIANCE | N | I | J | COVARIANCE | N | I | J | COVARIANCE |
|---|---|---|---|---|---|---|---|---|---|---|---|
| 26 | 4 | 6 | .0641145785 | 26 | 7 | 15 | .0339166632 | 27 | 1 | 13 | .0214397494 |
| 26 | 4 | 7 | .0558495832 | 26 | 7 | 16 | .0316120962 | 27 | 1 | 14 | .0198959223 |
| 26 | 4 | 8 | .0495102398 | 26 | 7 | 17 | .0295031594 | 27 | 1 | 15 | .0185200773 |
| 26 | 4 | 9 | .0444627832 | 26 | 7 | 18 | .0275516944 | 27 | 1 | 16 | .0172796749 |
| 26 | 4 | 10 | .0403256874 | 26 | 7 | 19 | .0257252007 | 27 | 1 | 17 | .0161492049 |
| 26 | 4 | 11 | .0368547456 | 26 | 7 | 20 | .0239944472 | 27 | 1 | 18 | .0151080957 |
| 26 | 4 | 12 | .0338858575 | 26 | 8 | 8 | .0663667175 | 27 | 1 | 19 | .0141392074 |
| 26 | 4 | 13 | .0313042564 | 26 | 8 | 9 | .0596651090 | 27 | 1 | 20 | .0132276562 |
| 26 | 4 | 14 | .0290268964 | 26 | 8 | 10 | .0541606724 | 27 | 1 | 21 | .0123597734 |
| 26 | 4 | 15 | .0269918323 | 26 | 8 | 11 | .0495345167 | 27 | 1 | 22 | .0115219852 |
| 26 | 4 | 16 | .0251515140 | 26 | 8 | 12 | .0455716912 | 27 | 1 | 23 | .0106992706 |
| 26 | 4 | 17 | .0234683582 | 26 | 8 | 13 | .0421214653 | 27 | 1 | 24 | .0098724130 |
| 26 | 4 | 18 | .0219116702 | 26 | 8 | 14 | .0390745133 | 27 | 1 | 25 | .0090117373 |
| 26 | 4 | 19 | .0204553465 | 26 | 8 | 15 | .0363491141 | 27 | 1 | 26 | .0080581664 |
| 26 | 4 | 20 | .0190759496 | 26 | 8 | 16 | .0338824127 | 27 | 1 | 27 | .0068320228 |
| 26 | 4 | 21 | .0177507704 | 26 | 8 | 17 | .0316246332 | 27 | 2 | 2 | .1427685155 |
| 26 | 4 | 22 | .0164553094 | 26 | 8 | 18 | .0295350431 | 27 | 2 | 3 | .0970266199 |
| 26 | 4 | 23 | .0151589581 | 26 | 8 | 19 | .0275789287 | 27 | 2 | 4 | .0744904635 |
| 26 | 5 | 5 | .0816615745 | 26 | 9 | 9 | .0638085317 | 27 | 2 | 5 | .0608871608 |
| 26 | 5 | 6 | .0694373976 | 26 | 9 | 10 | .0579364658 | 27 | 2 | 6 | .0516891654 |
| 26 | 5 | 7 | .0605121657 | 26 | 9 | 11 | .0529989059 | 27 | 2 | 7 | .0450003090 |
| 26 | 5 | 8 | .0536610128 | 26 | 9 | 12 | .0487675572 | 27 | 2 | 8 | .0398824382 |
| 26 | 5 | 9 | .0482026676 | 26 | 9 | 13 | .0450822144 | 27 | 2 | 9 | .0358162672 |
| 26 | 5 | 10 | .0437265521 | 26 | 9 | 14 | .0418266039 | 27 | 2 | 10 | .0324900621 |
| 26 | 5 | 11 | .0399696274 | 26 | 9 | 15 | .0389137598 | 27 | 2 | 11 | .0297047729 |
| 26 | 5 | 12 | .0367550048 | 26 | 9 | 16 | .0362767558 | 27 | 2 | 12 | .0273269466 |
| 26 | 5 | 13 | .0339588940 | 26 | 9 | 17 | .0338625630 | 27 | 2 | 13 | .0252634681 |
| 26 | 5 | 14 | .0314916693 | 26 | 9 | 18 | .0316277623 | 27 | 2 | 14 | .0234471399 |
| 26 | 5 | 15 | .0292864430 | 26 | 10 | 10 | .0619590981 | 27 | 2 | 15 | .0218280129 |
| 26 | 5 | 16 | .0272918470 | 26 | 10 | 11 | .0566915522 | 27 | 2 | 16 | .0203679354 |
| 26 | 5 | 17 | .0254672618 | 26 | 10 | 12 | .0521753668 | 27 | 2 | 17 | .0190369811 |
| 26 | 5 | 18 | .0237794976 | 26 | 10 | 13 | .0482403998 | 27 | 2 | 18 | .0178110052 |
| 26 | 5 | 19 | .0222003144 | 26 | 10 | 14 | .0447630802 | 27 | 2 | 19 | .0166698801 |
| 26 | 5 | 20 | .0207043430 | 26 | 10 | 15 | .0416509253 | 27 | 2 | 20 | .0155961173 |
| 26 | 5 | 21 | .0192669879 | 26 | 10 | 16 | .0388327205 | 27 | 2 | 21 | .0145736461 |
| 26 | 5 | 22 | .0178616971 | 26 | 10 | 17 | .0362520054 | 27 | 2 | 22 | .0135864968 |
| 26 | 6 | 6 | .0747263639 | 26 | 11 | 11 | .0606718682 | 27 | 2 | 23 | .0126169844 |
| 26 | 6 | 7 | .0651497927 | 26 | 11 | 12 | .0558502736 | 27 | 2 | 24 | .0116424694 |
| 26 | 6 | 8 | .0577927940 | 26 | 11 | 13 | .0516473993 | 27 | 2 | 25 | .0106279732 |
| 26 | 6 | 9 | .0519277502 | 26 | 11 | 14 | .0479319317 | 27 | 2 | 26 | .0095038373 |
| 26 | 6 | 10 | .0471156693 | 26 | 11 | 15 | .0446055255 | 27 | 3 | 3 | .1079096756 |
| 26 | 6 | 11 | .0430750543 | 26 | 11 | 16 | .0415924029 | 27 | 3 | 4 | .0829702063 |
| 26 | 6 | 12 | .0396164606 | 26 | 12 | 12 | .0598562677 | 27 | 3 | 5 | .0678806091 |
| 26 | 6 | 13 | .0366072245 | 26 | 12 | 13 | .0553631010 | 27 | 3 | 6 | .0576617405 |
| 26 | 6 | 14 | .0339512406 | 26 | 12 | 14 | .0513893786 | 27 | 3 | 7 | .0502222651 |
| 26 | 6 | 15 | .0315767487 | 26 | 12 | 15 | .0478304682 | 27 | 3 | 8 | .0445253495 |
| 26 | 6 | 16 | .0294286122 | 26 | 13 | 13 | .0594606340 | 27 | 3 | 9 | .0399962048 |
| 26 | 6 | 17 | .0274632109 | 26 | 13 | 14 | .0552036333 | 27 | 3 | 10 | .0362893636 |
| 26 | 6 | 18 | .0256448862 | 27 | 1 | 1 | .2530210738 | 27 | 3 | 11 | .0331840223 |
| 26 | 6 | 19 | .0239432821 | 27 | 1 | 2 | .1222871652 | 27 | 3 | 12 | .0305320250 |
| 26 | 6 | 20 | .0223311142 | 27 | 1 | 3 | .0828095420 | 27 | 3 | 13 | .0282299273 |
| 26 | 6 | 21 | .0207819122 | 27 | 1 | 4 | .0634589479 | 27 | 3 | 14 | .0262030341 |
| 26 | 7 | 7 | .0698713688 | 27 | 1 | 5 | .0518126949 | 27 | 3 | 15 | .0243957939 |
| 26 | 7 | 8 | .0620026771 | 27 | 1 | 6 | .0439529805 | 27 | 3 | 16 | .0227657559 |
| 26 | 7 | 9 | .0557256208 | 27 | 1 | 7 | .0382450059 | 27 | 3 | 17 | .0212796071 |
| 26 | 7 | 10 | .0505727680 | 27 | 1 | 8 | .0338820014 | 27 | 3 | 18 | .0199104582 |
| 26 | 7 | 11 | .0462441087 | 27 | 1 | 9 | .0304182529 | 27 | 3 | 19 | .0186358832 |
| 26 | 7 | 12 | .0425375824 | 27 | 1 | 10 | .0275865867 | 27 | 3 | 20 | .0174363882 |
| 26 | 7 | 13 | .0393115962 | 27 | 1 | 11 | .0252166074 | 27 | 3 | 21 | .0162940496 |
| 26 | 7 | 14 | .0364635153 | 27 | 1 | 12 | .0231941843 | 27 | 3 | 22 | .0151910460 |

TABLE I-COVARIANCES OF NORMAL ORDER STATISTICS

| N | I | J | COVARIANCE | N | I | J | COVARIANCE | N | I | J | COVARIANCE |
|---|---|---|---|---|---|---|---|---|---|---|---|
| 27 | 3 | 23 | .0141076301 | 27 | 7 | 7 | .0683289220 | 27 | 13 | 13 | .0574227467 |
| 27 | 3 | 24 | .0130185080 | 27 | 7 | 8 | .0606554996 | 27 | 13 | 14 | .0533671550 |
| 27 | 3 | 25 | .0118845838 | 27 | 7 | 9 | .0545402051 | 27 | 13 | 15 | .0497413666 |
| 27 | 4 | 4 | .0905487907 | 27 | 7 | 10 | .0495253862 | 27 | 14 | 14 | .0572472948 |
| 27 | 4 | 5 | .0741442044 | 27 | 7 | 11 | .0453175100 | 28 | 1 | 1 | .2504993106 |
| 27 | 4 | 6 | .0630188606 | 27 | 7 | 12 | .0417190413 | 28 | 1 | 2 | .1209310142 |
| 27 | 4 | 7 | .0549111338 | 27 | 7 | 13 | .0385917134 | 28 | 1 | 3 | .0818594022 |
| 27 | 4 | 8 | .0486976873 | 27 | 7 | 14 | .0358354678 | 28 | 1 | 4 | .0627247857 |
| 27 | 4 | 9 | .0437549044 | 27 | 7 | 15 | .0333757452 | 28 | 1 | 5 | .0512169254 |
| 27 | 4 | 10 | .0397075570 | 27 | 7 | 16 | .0311554675 | 28 | 1 | 6 | .0434558739 |
| 27 | 4 | 11 | .0363156062 | 27 | 7 | 17 | .0291297665 | 28 | 1 | 7 | .0378233567 |
| 27 | 4 | 12 | .0334178692 | 27 | 7 | 18 | .0272623647 | 28 | 1 | 8 | .0335210559 |
| 27 | 4 | 13 | .0309017355 | 27 | 7 | 19 | .0255229548 | 28 | 1 | 9 | .0301080781 |
| 27 | 4 | 14 | .0286858465 | 27 | 7 | 20 | .0238851425 | 28 | 1 | 10 | .0273202327 |
| 27 | 4 | 15 | .0267096657 | 27 | 7 | 21 | .0223246081 | 28 | 1 | 11 | .0249890959 |
| 27 | 4 | 16 | .0249269124 | 27 | 8 | 8 | .0647924058 | 28 | 1 | 12 | .0230019176 |
| 27 | 4 | 17 | .0233012523 | 27 | 8 | 9 | .0582759872 | 28 | 1 | 13 | .0212801515 |
| 27 | 4 | 18 | .0218033455 | 27 | 8 | 10 | .0529294031 | 28 | 1 | 14 | .0197672183 |
| 27 | 4 | 19 | .0204087126 | 27 | 8 | 11 | .0484411720 | 28 | 1 | 15 | .0184211624 |
| 27 | 4 | 20 | .0190960639 | 27 | 8 | 12 | .0446015198 | 28 | 1 | 16 | .0172100474 |
| 27 | 4 | 21 | .0178458151 | 27 | 8 | 13 | .0412635164 | 28 | 1 | 17 | .0161089525 |
| 27 | 4 | 22 | .0166384827 | 27 | 8 | 14 | .0383207721 | 28 | 1 | 18 | .0150979363 |
| 27 | 4 | 23 | .0154524654 | 27 | 8 | 15 | .0356939686 | 28 | 1 | 19 | .0141605952 |
| 27 | 4 | 24 | .0142600804 | 27 | 8 | 16 | .0333223551 | 28 | 1 | 20 | .0132829830 |
| 27 | 5 | 5 | .0801180719 | 27 | 8 | 17 | .0311581539 | 28 | 1 | 21 | .0124527270 |
| 27 | 5 | 6 | .0681351417 | 27 | 8 | 18 | .0291627186 | 28 | 1 | 22 | .0116581969 |
| 27 | 5 | 7 | .0593936815 | 27 | 8 | 19 | .0273037466 | 28 | 1 | 23 | .0108875499 |
| 27 | 5 | 8 | .0526894816 | 27 | 8 | 20 | .0255530898 | 28 | 1 | 24 | .0101273426 |
| 27 | 5 | 9 | .0473531290 | 27 | 9 | 9 | .0621835923 | 28 | 1 | 25 | .0093599748 |
| 27 | 5 | 10 | .0429814110 | 27 | 9 | 10 | .0564919277 | 28 | 1 | 26 | .0085577788 |
| 27 | 5 | 11 | .0393161626 | 27 | 9 | 11 | .0517118005 | 28 | 1 | 27 | .0076650421 |
| 27 | 5 | 12 | .0361839041 | 27 | 9 | 12 | .0476208255 | 28 | 1 | 28 | .0065112919 |
| 27 | 5 | 13 | .0334633590 | 27 | 9 | 13 | .0440631322 | 28 | 2 | 2 | .1409386260 |
| 27 | 5 | 14 | .0310668604 | 27 | 9 | 14 | .0409257890 | 28 | 2 | 3 | .0957391901 |
| 27 | 5 | 15 | .0289291467 | 27 | 9 | 15 | .0381245539 | 28 | 2 | 4 | .0734922530 |
| 27 | 5 | 16 | .0270003027 | 27 | 9 | 16 | .0355948722 | 28 | 2 | 5 | .0600741487 |
| 27 | 5 | 17 | .0252411247 | 27 | 9 | 17 | .0332859508 | 28 | 2 | 6 | .0510079026 |
| 27 | 5 | 18 | .0236199431 | 27 | 9 | 18 | .0311566837 | 28 | 2 | 7 | .0444195038 |
| 27 | 5 | 19 | .0221103229 | 27 | 9 | 19 | .0291726946 | 28 | 2 | 8 | .0393821552 |
| 27 | 5 | 20 | .0206892628 | 27 | 10 | 10 | .0602645386 | 28 | 2 | 9 | .0353830532 |
| 27 | 5 | 21 | .0193355933 | 27 | 10 | 11 | .0551769211 | 28 | 2 | 10 | .0321144742 |
| 27 | 5 | 22 | .0180282436 | 27 | 10 | 12 | .0508209504 | 28 | 2 | 11 | .0293800147 |
| 27 | 5 | 23 | .0167438376 | 27 | 10 | 13 | .0470314281 | 28 | 2 | 12 | .0270480693 |
| 27 | 6 | 6 | .0731958519 | 27 | 10 | 14 | .0436885874 | 28 | 2 | 13 | .0250268833 |
| 27 | 6 | 7 | .0638318760 | 27 | 10 | 15 | .0407030354 | 28 | 2 | 14 | .0232503174 |
| 27 | 6 | 8 | .0566447521 | 27 | 10 | 16 | .0380062337 | 28 | 2 | 15 | .0216692979 |
| 27 | 6 | 9 | .0509205613 | 27 | 10 | 17 | .0355442263 | 28 | 2 | 16 | .0202464497 |
| 27 | 6 | 10 | .0462288283 | 27 | 10 | 18 | .0332733246 | 28 | 2 | 17 | .0189525936 |
| 27 | 6 | 11 | .0422936937 | 27 | 11 | 11 | .0588879359 | 28 | 2 | 18 | .0177643697 |
| 27 | 6 | 12 | .0389296534 | 27 | 11 | 12 | .0542494679 | 28 | 2 | 19 | .0166625532 |
| 27 | 6 | 13 | .0360069443 | 27 | 11 | 13 | .0502125715 | 28 | 2 | 20 | .0156307913 |
| 27 | 6 | 14 | .0334317139 | 27 | 11 | 14 | .0466502672 | 28 | 2 | 21 | .0146545686 |
| 27 | 6 | 15 | .0311340619 | 27 | 11 | 15 | .0434677190 | 28 | 2 | 22 | .0137202322 |
| 27 | 6 | 16 | .0290605019 | 27 | 11 | 16 | .0405921730 | 28 | 2 | 23 | .0128138719 |
| 27 | 6 | 17 | .0271690074 | 27 | 11 | 17 | .0379663215 | 28 | 2 | 24 | .0119196866 |
| 27 | 6 | 18 | .0254256140 | 27 | 12 | 12 | .0579594877 | 28 | 2 | 25 | .0110169779 |
| 27 | 6 | 19 | .0238019581 | 27 | 12 | 13 | .0536564180 | 28 | 2 | 26 | .0100731934 |
| 27 | 6 | 20 | .0222733502 | 27 | 12 | 14 | .0498577969 | 28 | 2 | 27 | .0090227662 |
| 27 | 6 | 21 | .0208170562 | 27 | 12 | 15 | .0464629877 | 28 | 3 | 3 | .1063132378 |
| 27 | 6 | 22 | .0194104322 | 27 | 12 | 16 | .0433947374 | 28 | 3 | 4 | .0817287738 |

TABLE I-COVARIANCES OF NORMAL ORDER STATISTICS

| N | I | J | COVARIANCE | N | I | J | COVARIANCE | N | I | J | COVARIANCE |
|---|---|---|---|---|---|---|---|---|---|---|---|
| 28 | 3 | 5 | .0668666198 | 28 | 5 | 21 | .0193466039 | 28 | 9 | 17 | .0327278224 |
| 28 | 3 | 6 | .0568094293 | 28 | 5 | 22 | .0181153507 | 28 | 9 | 18 | .0306893397 |
| 28 | 3 | 7 | .0494930395 | 28 | 5 | 23 | .0169206043 | 28 | 9 | 19 | .0287972062 |
| 28 | 3 | 8 | .0438945742 | 28 | 5 | 24 | .0157415674 | 28 | 9 | 20 | .0270237523 |
| 28 | 3 | 9 | .0394472214 | 28 | 6 | 6 | .0717722898 | 28 | 10 | 10 | .0587134226 |
| 28 | 3 | 10 | .0358104502 | 28 | 6 | 7 | .0626042879 | 28 | 10 | 11 | .0537874508 |
| 28 | 3 | 11 | .0327667134 | 28 | 6 | 8 | .0555734272 | 28 | 10 | 12 | .0495749471 |
| 28 | 3 | 12 | .0301701243 | 28 | 6 | 9 | .0499784935 | 28 | 10 | 13 | .0459151588 |
| 28 | 3 | 13 | .0279189044 | 28 | 6 | 10 | .0453968944 | 28 | 10 | 14 | .0426917039 |
| 28 | 3 | 14 | .0259396478 | 28 | 6 | 11 | .0415579728 | 28 | 10 | 15 | .0398178933 |
| 28 | 3 | 15 | .0241778616 | 28 | 6 | 12 | .0382798430 | 28 | 10 | 16 | .0372274702 |
| 28 | 3 | 16 | .0225920250 | 28 | 6 | 13 | .0354353839 | 28 | 10 | 17 | .0348685360 |
| 28 | 3 | 17 | .0211497088 | 28 | 6 | 14 | .0329327688 | 28 | 10 | 18 | .0326994092 |
| 28 | 3 | 18 | .0198249410 | 28 | 6 | 15 | .0307037358 | 28 | 10 | 19 | .0306856751 |
| 28 | 3 | 19 | .0185963381 | 28 | 6 | 16 | .0286962081 | 28 | 11 | 11 | .0572634566 |
| 28 | 3 | 20 | .0174457036 | 28 | 6 | 17 | .0268694646 | 28 | 11 | 12 | .0527882328 |
| 28 | 3 | 21 | .0163568787 | 28 | 6 | 18 | .0251908559 | 28 | 11 | 13 | .0488987481 |
| 28 | 3 | 22 | .0153146573 | 28 | 6 | 19 | .0236334706 | 28 | 11 | 14 | .0454718652 |
| 28 | 3 | 23 | .0143035377 | 28 | 6 | 20 | .0221743796 | 28 | 11 | 15 | .0424158103 |
| 28 | 3 | 24 | .0133059019 | 28 | 6 | 21 | .0207931961 | 28 | 11 | 16 | .0396604032 |
| 28 | 3 | 25 | .0122986601 | 28 | 6 | 22 | .0194707101 | 28 | 11 | 17 | .0371506440 |
| 28 | 3 | 26 | .0112454858 | 28 | 6 | 23 | .0181873047 | 28 | 11 | 18 | .0348423377 |
| 28 | 4 | 4 | .0890588585 | 28 | 7 | 7 | .0668990906 | 28 | 12 | 12 | .0562430649 |
| 28 | 4 | 5 | .0729241687 | 28 | 7 | 8 | .0594046711 | 28 | 12 | 13 | .0521077927 |
| 28 | 4 | 6 | .0619906556 | 28 | 7 | 9 | .0534373225 | 28 | 12 | 14 | .0484630643 |
| 28 | 4 | 7 | .0540288397 | 28 | 7 | 10 | .0485484138 | 28 | 12 | 15 | .0452117155 |
| 28 | 4 | 8 | .0479319336 | 28 | 7 | 11 | .0444503691 | 28 | 12 | 16 | .0422794018 |
| 28 | 4 | 9 | .0430857927 | 28 | 7 | 12 | .0409497963 | 28 | 12 | 17 | .0396078275 |
| 28 | 4 | 10 | .0391210546 | 28 | 7 | 13 | .0379114507 | 28 | 13 | 13 | .0555913650 |
| 28 | 4 | 11 | .0358015492 | 28 | 7 | 14 | .0352375857 | 28 | 13 | 14 | .0517116572 |
| 28 | 4 | 12 | .0329687874 | 28 | 7 | 15 | .0328555048 | 28 | 13 | 15 | .0482495406 |
| 28 | 4 | 13 | .0305121312 | 28 | 7 | 16 | .0307097247 | 28 | 13 | 16 | .0451262167 |
| 28 | 4 | 14 | .0283517430 | 28 | 7 | 17 | .0287568434 | 28 | 14 | 14 | .0552741536 |
| 28 | 4 | 15 | .0264283279 | 28 | 7 | 18 | .0269620468 | 28 | 14 | 15 | .0515818355 |
| 28 | 4 | 16 | .0246966874 | 28 | 7 | 19 | .0252966283 | 29 | 1 | 1 | .2481086595 |
| 28 | 4 | 17 | .0231215061 | 28 | 7 | 20 | .0237361222 | 29 | 1 | 2 | .1196458271 |
| 28 | 4 | 18 | .0216744888 | 28 | 7 | 21 | .0222587644 | 29 | 1 | 3 | .0809583401 |
| 28 | 4 | 19 | .0203323316 | 28 | 7 | 22 | .0208440343 | 29 | 1 | 4 | .0620276761 |
| 28 | 4 | 20 | .0190751965 | 28 | 8 | 8 | .0633384886 | 29 | 1 | 5 | .0506502648 |
| 28 | 4 | 21 | .0178854586 | 28 | 8 | 9 | .0569908111 | 29 | 1 | 6 | .0429820315 |
| 28 | 4 | 22 | .0167465242 | 28 | 8 | 10 | .0517876929 | 29 | 1 | 7 | .0374203469 |
| 28 | 4 | 23 | .0156414681 | 28 | 8 | 11 | .0474244434 | 29 | 1 | 8 | .0331748877 |
| 28 | 4 | 24 | .0145510460 | 28 | 8 | 12 | .0436960116 | 29 | 1 | 9 | .0298093141 |
| 28 | 4 | 25 | .0134500251 | 28 | 8 | 13 | .0404589096 | 29 | 1 | 10 | .0270622476 |
| 28 | 5 | 5 | .0786777794 | 28 | 8 | 14 | .0376093767 | 29 | 1 | 11 | .0247671166 |
| 28 | 5 | 6 | .0669184193 | 28 | 8 | 15 | .0350702060 | 29 | 1 | 12 | .0228124609 |
| 28 | 5 | 7 | .0583469317 | 28 | 8 | 16 | .0327824452 | 29 | 1 | 13 | .0211206801 |
| 28 | 5 | 8 | .0517783551 | 28 | 8 | 17 | .0306999580 | 29 | 1 | 14 | .0196359260 |
| 28 | 5 | 9 | .0465543100 | 28 | 8 | 18 | .0287857230 | 29 | 1 | 15 | .0183168411 |
| 28 | 5 | 10 | .0422784229 | 28 | 8 | 19 | .0270092017 | 29 | 1 | 16 | .0171320109 |
| 28 | 5 | 11 | .0386970401 | 28 | 8 | 20 | .0253443552 | 29 | 1 | 17 | .0160570049 |
| 28 | 5 | 12 | .0356398210 | 28 | 8 | 21 | .0237680070 | 29 | 1 | 18 | .0150723806 |
| 28 | 5 | 13 | .0329877841 | 28 | 9 | 9 | .0606894200 | 29 | 1 | 19 | .0141622863 |
| 28 | 5 | 14 | .0306550266 | 28 | 9 | 10 | .0551609965 | 29 | 1 | 20 | .0133134385 |
| 28 | 5 | 15 | .0285777202 | 28 | 9 | 11 | .0505229155 | 29 | 1 | 21 | .0125143259 |
| 28 | 5 | 16 | .0267071901 | 28 | 9 | 12 | .0465581633 | 29 | 1 | 22 | .0117545242 |
| 28 | 5 | 17 | .0250053901 | 28 | 9 | 13 | .0431147822 | 29 | 1 | 23 | .0110240143 |
| 28 | 5 | 18 | .0234418265 | 28 | 9 | 14 | .0400828239 | 29 | 1 | 24 | .0103123501 |
| 28 | 5 | 19 | .0219913755 | 28 | 9 | 15 | .0373804355 | 29 | 1 | 25 | .0096073922 |
| 28 | 5 | 20 | .0206326409 | 28 | 9 | 16 | .0349450873 | 29 | 1 | 26 | .0088929145 |

TABLE I-COVARIANCES OF NORMAL ORDER STATISTICS

| N | I | J | COVARIANCE | N | I | J | COVARIANCE | N | I | J | COVARIANCE |
|---|---|---|---|---|---|---|---|---|---|---|---|
| 29 | 1 | 27 | .0081430072 | 29 | 4 | 9 | .0424522493 | 29 | 7 | 9 | .0524080471 |
| 29 | 1 | 28 | .0073049793 | 29 | 4 | 10 | .0385638743 | 29 | 7 | 10 | .0476345468 |
| 29 | 1 | 29 | .0062167509 | 29 | 4 | 11 | .0353111234 | 29 | 7 | 11 | .0436368949 |
| 29 | 2 | 2 | .1392110104 | 29 | 4 | 12 | .0325380073 | 29 | 7 | 12 | .0402255134 |
| 29 | 2 | 3 | .0945231690 | 29 | 4 | 13 | .0301357062 | 29 | 7 | 13 | .0372679233 |
| 29 | 2 | 4 | .0725484829 | 29 | 4 | 14 | .0280257687 | 29 | 7 | 14 | .0346684740 |
| 29 | 2 | 5 | .0593044063 | 29 | 4 | 15 | .0261500112 | 29 | 7 | 15 | .0323561328 |
| 29 | 2 | 6 | .0503617367 | 29 | 4 | 16 | .0244641793 | 29 | 7 | 16 | .0302768081 |
| 29 | 2 | 7 | .0438673744 | 29 | 4 | 17 | .0229338150 | 29 | 7 | 17 | .0283883364 |
| 29 | 2 | 8 | .0389052290 | 29 | 4 | 18 | .0215314631 | 29 | 7 | 18 | .0266570892 |
| 29 | 2 | 9 | .0349686008 | 29 | 4 | 19 | .0202347134 | 29 | 7 | 19 | .0250555879 |
| 29 | 2 | 10 | .0317535353 | 29 | 4 | 20 | .0190247654 | 29 | 7 | 20 | .0235607550 |
| 29 | 2 | 11 | .0290661087 | 29 | 4 | 21 | .0178853071 | 29 | 7 | 21 | .0221525468 |
| 29 | 2 | 12 | .0267764382 | 29 | 4 | 22 | .0168015486 | 29 | 7 | 22 | .0208127686 |
| 29 | 2 | 13 | .0247940270 | 29 | 4 | 23 | .0157592517 | 29 | 7 | 23 | .0195238827 |
| 29 | 2 | 14 | .0230537019 | 29 | 4 | 24 | .0147435482 | 29 | 8 | 8 | .0619905185 |
| 29 | 2 | 15 | .0215071721 | 29 | 4 | 25 | .0137371375 | 29 | 8 | 9 | .0557973676 |
| 29 | 2 | 16 | .0201177377 | 29 | 4 | 26 | .0127168617 | 29 | 8 | 10 | .0507253322 |
| 29 | 2 | 17 | .0188568443 | 29 | 5 | 5 | .0773298377 | 29 | 8 | 11 | .0464759764 |
| 29 | 2 | 18 | .0177017573 | 29 | 5 | 6 | .0657783940 | 29 | 8 | 12 | .0428485841 |
| 29 | 2 | 19 | .0166339330 | 29 | 5 | 7 | .0573646953 | 29 | 8 | 13 | .0397028093 |
| 29 | 2 | 20 | .0156378255 | 29 | 5 | 8 | .0509217727 | 29 | 8 | 14 | .0369372685 |
| 29 | 2 | 21 | .0146999564 | 29 | 5 | 9 | .0458015476 | 29 | 8 | 15 | .0344766376 |
| 29 | 2 | 22 | .0138081137 | 29 | 5 | 10 | .0416140201 | 29 | 8 | 16 | .0322635333 |
| 29 | 2 | 23 | .0129505542 | 29 | 5 | 11 | .0381097308 | 29 | 8 | 17 | .0302532080 |
| 29 | 2 | 24 | .0121150270 | 29 | 5 | 12 | .0351212450 | 29 | 8 | 18 | .0284099576 |
| 29 | 2 | 25 | .0112872868 | 29 | 5 | 13 | .0325316922 | 29 | 8 | 19 | .0267045981 |
| 29 | 2 | 26 | .0104482836 | 29 | 5 | 14 | .0302567748 | 29 | 8 | 20 | .0251126126 |
| 29 | 2 | 27 | .0095675863 | 29 | 5 | 15 | .0282339470 | 29 | 8 | 21 | .0236126993 |
| 29 | 2 | 28 | .0085832950 | 29 | 5 | 16 | .0264156177 | 29 | 8 | 22 | .0221855063 |
| 29 | 3 | 3 | .1048106942 | 29 | 5 | 17 | .0247647164 | 29 | 9 | 9 | .0593096508 |
| 29 | 3 | 4 | .0805594428 | 29 | 5 | 18 | .0232516967 | 29 | 9 | 10 | .0539297723 |
| 29 | 3 | 5 | .0659104005 | 29 | 5 | 19 | .0218524339 | 29 | 9 | 11 | .0494206185 |
| 29 | 3 | 6 | .0560044340 | 29 | 5 | 20 | .0205466841 | 29 | 9 | 12 | .0455700880 |
| 29 | 3 | 7 | .0488029513 | 29 | 5 | 21 | .0193168747 | 29 | 9 | 13 | .0422297823 |
| 29 | 3 | 8 | .0432961972 | 29 | 5 | 22 | .0181470618 | 29 | 9 | 14 | .0392924434 |
| 29 | 3 | 9 | .0389248495 | 29 | 5 | 23 | .0170218951 | 29 | 9 | 15 | .0366783437 |
| 29 | 3 | 10 | .0353530039 | 29 | 5 | 24 | .0159253429 | 29 | 9 | 16 | .0343267187 |
| 29 | 3 | 11 | .0323661564 | 29 | 5 | 25 | .0148387293 | 29 | 9 | 17 | .0321901639 |
| 29 | 3 | 12 | .0298205315 | 29 | 6 | 6 | .0704439922 | 29 | 9 | 18 | .0302308414 |
| 29 | 3 | 13 | .0276158888 | 29 | 6 | 7 | .0614573504 | 29 | 9 | 19 | .0284178127 |
| 29 | 3 | 14 | .0256799985 | 29 | 6 | 8 | .0545708232 | 29 | 9 | 20 | .0267250767 |
| 29 | 3 | 15 | .0239593154 | 29 | 6 | 9 | .0490950216 | 29 | 9 | 21 | .0251300251 |
| 29 | 3 | 16 | .0224131289 | 29 | 6 | 10 | .0446146780 | 29 | 10 | 10 | .0572870868 |
| 29 | 3 | 17 | .0210097518 | 29 | 6 | 11 | .0408639575 | 29 | 10 | 11 | .0525072130 |
| 29 | 3 | 18 | .0197239458 | 29 | 6 | 12 | .0376643152 | 29 | 10 | 12 | .0484240015 |
| 29 | 3 | 19 | .0185351178 | 29 | 6 | 13 | .0348910543 | 29 | 10 | 13 | .0448807288 |
| 29 | 3 | 20 | .0174259970 | 29 | 6 | 14 | .0324541882 | 29 | 10 | 14 | .0417640556 |
| 29 | 3 | 21 | .0163816034 | 29 | 6 | 15 | .0302869205 | 29 | 10 | 15 | .0389896920 |
| 29 | 3 | 22 | .0153883600 | 29 | 6 | 16 | .0283384052 | 29 | 10 | 16 | .0364933661 |
| 29 | 3 | 23 | .0144332042 | 29 | 6 | 17 | .0265690227 | 29 | 10 | 17 | .0342249170 |
| 29 | 3 | 24 | .0135025021 | 29 | 6 | 18 | .0249471851 | 29 | 10 | 18 | .0321442914 |
| 29 | 3 | 25 | .0125803916 | 29 | 6 | 19 | .0234470930 | 29 | 10 | 19 | .0302187258 |
| 29 | 3 | 26 | .0116456528 | 29 | 6 | 20 | .0220470850 | 29 | 10 | 20 | .0284206709 |
| 29 | 3 | 27 | .0106643777 | 29 | 6 | 21 | .0207283505 | 29 | 11 | 11 | .0557761755 |
| 29 | 4 | 4 | .0876605326 | 29 | 6 | 22 | .0194738238 | 29 | 11 | 12 | .0514475325 |
| 29 | 4 | 5 | .0717780102 | 29 | 6 | 23 | .0182670641 | 29 | 11 | 13 | .0476899866 |
| 29 | 4 | 6 | .0610234178 | 29 | 6 | 24 | .0170908845 | 29 | 11 | 14 | .0443838430 |
| 29 | 4 | 7 | .0531974429 | 29 | 7 | 7 | .0655690804 | 29 | 11 | 15 | .0414400374 |
| 29 | 4 | 8 | .0472088122 | 29 | 7 | 8 | .0582394669 | 29 | 11 | 16 | .0387906245 |

TABLE I-COVARIANCES OF NORMAL ORDER STATISTICS

| N | I | J | COVARIANCE | N | I | J | COVARIANCE | N | I | J | COVARIANCE |
|---|---|---|---|---|---|---|---|---|---|---|---|
| 29 | 11 | 17 | .0363825529 | 30 | 2 | 13 | .0245656528 | 30 | 4 | 21 | .0178560500 |
| 29 | 11 | 18 | .0341734418 | 30 | 2 | 14 | .0228585547 | 30 | 4 | 22 | .0168180952 |
| 29 | 11 | 19 | .0321286074 | 30 | 2 | 15 | .0213434556 | 30 | 4 | 23 | .0158263721 |
| 29 | 12 | 12 | .0546802955 | 30 | 2 | 16 | .0199842541 | 30 | 4 | 24 | .0148685226 |
| 29 | 12 | 13 | .0506944314 | 30 | 2 | 17 | .0187529324 | 30 | 4 | 25 | .0139313544 |
| 29 | 12 | 14 | .0471862216 | 30 | 2 | 18 | .0176272714 | 30 | 4 | 26 | .0129991897 |
| 29 | 12 | 15 | .0440615640 | 30 | 2 | 19 | .0165892627 | 30 | 4 | 27 | .0120506462 |
| 29 | 12 | 16 | .0412486324 | 30 | 2 | 20 | .0156239619 | 30 | 5 | 5 | .0760649078 |
| 29 | 12 | 17 | .0386913167 | 30 | 2 | 21 | .0147186206 | 30 | 5 | 6 | .0647074439 |
| 29 | 12 | 18 | .0363447703 | 30 | 2 | 22 | .0138619822 | 30 | 5 | 7 | .0564407161 |
| 29 | 13 | 13 | .0539343894 | 30 | 2 | 23 | .0130436462 | 30 | 5 | 8 | .0501146232 |
| 29 | 13 | 14 | .0502096792 | 30 | 2 | 24 | .0122533970 | 30 | 5 | 9 | .0450907322 |
| 29 | 13 | 15 | .0468911663 | 30 | 2 | 25 | .0114803385 | 30 | 5 | 10 | .0409850038 |
| 29 | 13 | 16 | .0439028974 | 30 | 2 | 26 | .0107115256 | 30 | 5 | 11 | .0375518916 |
| 29 | 13 | 17 | .0411855070 | 30 | 2 | 27 | .0099293229 | 30 | 5 | 12 | .0346266643 |
| 29 | 14 | 14 | .0535009912 | 30 | 2 | 28 | .0091051697 | 30 | 5 | 13 | .0320944042 |
| 29 | 14 | 15 | .0499726035 | 30 | 2 | 29 | .0081804717 | 30 | 5 | 14 | .0298722868 |
| 29 | 14 | 16 | .0467944015 | 30 | 3 | 3 | .1033932396 | 30 | 5 | 15 | .0278989280 |
| 29 | 15 | 15 | .0533592307 | 30 | 3 | 4 | .0794555493 | 30 | 5 | 16 | .0261276991 |
| 30 | 1 | 1 | .2458378992 | 30 | 3 | 5 | .0650067273 | 30 | 5 | 17 | .0245223736 |
| 30 | 1 | 2 | .1184255494 | 30 | 3 | 6 | .0552426065 | 30 | 5 | 18 | .0230541936 |
| 30 | 1 | 3 | .0801022447 | 30 | 3 | 7 | .0481487166 | 30 | 5 | 19 | .0216998280 |
| 30 | 1 | 4 | .0613646056 | 30 | 3 | 8 | .0427276685 | 30 | 5 | 20 | .0204398984 |
| 30 | 1 | 5 | .0501104452 | 30 | 3 | 9 | .0384271908 | 30 | 5 | 21 | .0192578608 |
| 30 | 1 | 6 | .0425297513 | 30 | 3 | 10 | .0349157287 | 30 | 5 | 22 | .0181390888 |
| 30 | 1 | 7 | .0370347379 | 30 | 3 | 11 | .0319816323 | 30 | 5 | 23 | .0170700473 |
| 30 | 1 | 8 | .0328426605 | 30 | 3 | 12 | .0294830990 | 30 | 5 | 24 | .0160374331 |
| 30 | 1 | 9 | .0295214907 | 30 | 3 | 13 | .0273213168 | 30 | 5 | 25 | .0150270416 |
| 30 | 1 | 10 | .0268125083 | 30 | 3 | 14 | .0254251385 | 30 | 5 | 26 | .0140219641 |
| 30 | 1 | 11 | .0245508877 | 30 | 3 | 15 | .0237418783 | 30 | 6 | 6 | .0692009521 |
| 30 | 1 | 12 | .0226263793 | 30 | 3 | 16 | .0222315448 | 30 | 6 | 7 | .0603827481 |
| 30 | 1 | 13 | .0209622641 | 30 | 3 | 17 | .0208630895 | 30 | 6 | 8 | .0536300361 |
| 30 | 1 | 14 | .0195033674 | 30 | 3 | 18 | .0196118819 | 30 | 6 | 9 | .0482644696 |
| 30 | 1 | 15 | .0182088734 | 30 | 3 | 19 | .0184579525 | 30 | 6 | 10 | .0438776116 |
| 30 | 1 | 16 | .0170478289 | 30 | 3 | 20 | .0173847238 | 30 | 6 | 11 | .0402081142 |
| 30 | 1 | 17 | .0159962223 | 30 | 3 | 21 | .0163780486 | 30 | 6 | 12 | .0370805319 |
| 30 | 1 | 18 | .0150350200 | 30 | 3 | 22 | .0154254313 | 30 | 6 | 13 | .0343724056 |
| 30 | 1 | 19 | .0141488013 | 30 | 3 | 23 | .0145153225 | 30 | 6 | 14 | .0319954343 |
| 30 | 1 | 20 | .0133247743 | 30 | 3 | 24 | .0136363733 | 30 | 6 | 15 | .0298841472 |
| 30 | 1 | 21 | .0125520311 | 30 | 3 | 25 | .0127764731 | 30 | 6 | 16 | .0279887923 |
| 30 | 1 | 22 | .0118209447 | 30 | 3 | 26 | .0119212254 | 30 | 6 | 17 | .0262707025 |
| 30 | 1 | 23 | .0111226245 | 30 | 3 | 27 | .0110510137 | 30 | 6 | 18 | .0246991735 |
| 30 | 1 | 24 | .0104483421 | 30 | 3 | 28 | .0101340597 | 30 | 6 | 19 | .0232492911 |
| 30 | 1 | 25 | .0097887930 | 30 | 4 | 4 | .0863449040 | 30 | 6 | 20 | .0219003540 |
| 30 | 1 | 26 | .0091329282 | 30 | 4 | 5 | .0706986608 | 30 | 6 | 21 | .0206346754 |
| 30 | 1 | 27 | .0084657026 | 30 | 4 | 6 | .0601114489 | 30 | 6 | 22 | .0194366160 |
| 30 | 1 | 28 | .0077627584 | 30 | 4 | 7 | .0524123390 | 30 | 6 | 23 | .0182917094 |
| 30 | 1 | 29 | .0069741335 | 30 | 4 | 8 | .0465246409 | 30 | 6 | 24 | .0171857215 |
| 30 | 1 | 30 | .0059454301 | 30 | 4 | 9 | .0418514047 | 30 | 6 | 25 | .0161034441 |
| 30 | 2 | 2 | .1375764527 | 30 | 4 | 10 | .0380338908 | 30 | 7 | 7 | .0643280670 |
| 30 | 2 | 3 | .0933721948 | 30 | 4 | 11 | .0348429090 | 30 | 7 | 8 | .0571507831 |
| 30 | 2 | 4 | .0716544004 | 30 | 4 | 12 | .0321248009 | 30 | 7 | 9 | .0514447641 |
| 30 | 2 | 5 | .0585742669 | 30 | 4 | 13 | .0297724325 | 30 | 7 | 10 | .0467775053 |
| 30 | 2 | 6 | .0497478171 | 30 | 4 | 14 | .0277086262 | 30 | 7 | 11 | .0428720432 |
| 30 | 2 | 7 | .0433417305 | 30 | 4 | 15 | .0258762060 | 30 | 7 | 12 | .0395423284 |
| 30 | 2 | 8 | .0384500350 | 30 | 4 | 16 | .0242317560 | 30 | 7 | 13 | .0366584286 |
| 30 | 2 | 9 | .0345717933 | 30 | 4 | 17 | .0227415576 | 30 | 7 | 14 | .0341266088 |
| 30 | 2 | 10 | .0314066030 | 30 | 4 | 18 | .0213788534 | 30 | 7 | 15 | .0318773395 |
| 30 | 2 | 11 | .0287628721 | 30 | 4 | 19 | .0201219422 | 30 | 7 | 16 | .0298577643 |
| 30 | 2 | 12 | .0265123308 | 30 | 4 | 20 | .0189528021 | 30 | 7 | 17 | .0280267902 |

TABLE I-COVARIANCES OF NORMAL ORDER STATISTICS

| N | I | J | COVARIANCE | N | I | J | COVARIANCE | N | I | J | COVARIANCE |
|---|---|---|---|---|---|---|---|---|---|---|---|
| 30 | 7 | 18 | .0263517783 | 30 | 12 | 13 | .0493976948 | 31 | 2 | 12 | .0262558530 |
| 30 | 7 | 19 | .0248062326 | 30 | 12 | 14 | .0460116153 | 31 | 2 | 13 | .0243422556 |
| 30 | 7 | 20 | .0233681276 | 30 | 12 | 15 | .0429998609 | 31 | 2 | 14 | .0226657798 |
| 30 | 7 | 21 | .0220186429 | 30 | 12 | 16 | .0402928208 | 31 | 2 | 15 | .0211794955 |
| 30 | 7 | 22 | .0207411342 | 30 | 12 | 17 | .0378362678 | 31 | 2 | 16 | .0198478398 |
| 30 | 7 | 23 | .0195202016 | 30 | 12 | 18 | .0355870467 | 31 | 2 | 17 | .0186432651 |
| 30 | 7 | 24 | .0183406805 | 30 | 12 | 19 | .0335100591 | 31 | 2 | 18 | .0175439904 |
| 30 | 8 | 8 | .0607363649 | 30 | 13 | 13 | .0524262226 | 31 | 2 | 19 | .0165324426 |
| 30 | 8 | 9 | .0546853609 | 30 | 13 | 14 | .0488392642 | 31 | 2 | 20 | .0155941440 |
| 30 | 8 | 10 | .0497336612 | 30 | 13 | 15 | .0456478929 | 31 | 2 | 21 | .0147168850 |
| 30 | 8 | 11 | .0455886164 | 30 | 13 | 16 | .0427786489 | 31 | 2 | 22 | .0138900767 |
| 30 | 8 | 12 | .0420535093 | 30 | 13 | 17 | .0401742783 | 31 | 2 | 23 | .0131042046 |
| 30 | 8 | 13 | .0389908792 | 30 | 13 | 18 | .0377891956 | 31 | 2 | 24 | .0123503109 |
| 30 | 8 | 14 | .0363015106 | 30 | 14 | 14 | .0518966767 | 31 | 2 | 25 | .0116194176 |
| 30 | 8 | 15 | .0339117772 | 30 | 14 | 15 | .0485125899 | 31 | 2 | 26 | .0109017477 |
| 30 | 8 | 16 | .0317656867 | 30 | 14 | 16 | .0454692741 | 31 | 2 | 27 | .0101854548 |
| 30 | 8 | 17 | .0298196930 | 30 | 14 | 17 | .0427062490 | 31 | 2 | 28 | .0094541289 |
| 30 | 8 | 18 | .0280391950 | 30 | 15 | 15 | .0516385597 | 31 | 2 | 29 | .0086808713 |
| 30 | 8 | 19 | .0263960924 | 30 | 15 | 16 | .0484056176 | 31 | 2 | 30 | .0078100826 |
| 30 | 8 | 20 | .0248670204 | 31 | 1 | 1 | .2436771137 | 31 | 3 | 3 | .1020531747 |
| 30 | 8 | 21 | .0234320129 | 31 | 1 | 2 | .1172648258 | 31 | 3 | 4 | .0784112533 |
| 30 | 8 | 22 | .0220734082 | 31 | 1 | 3 | .0792874695 | 31 | 3 | 5 | .0641510070 |
| 30 | 8 | 23 | .0207748390 | 31 | 1 | 4 | .0607328896 | 31 | 3 | 6 | .0545202817 |
| 30 | 9 | 9 | .0580305098 | 31 | 1 | 5 | .0495954331 | 31 | 3 | 7 | .0475274125 |
| 30 | 9 | 10 | .0527865062 | 31 | 1 | 6 | .0420974912 | 31 | 3 | 8 | .0421866919 |
| 30 | 9 | 11 | .0483950146 | 31 | 1 | 7 | .0366653883 | 31 | 3 | 9 | .0379525029 |
| 30 | 9 | 12 | .0446484506 | 31 | 1 | 8 | .0325235780 | 31 | 3 | 10 | .0344973901 |
| 30 | 9 | 13 | .0414016739 | 31 | 1 | 9 | .0292441237 | 31 | 3 | 11 | .0316123888 |
| 30 | 9 | 14 | .0385498729 | 31 | 1 | 10 | .0265708236 | 31 | 3 | 12 | .0291575468 |
| 30 | 9 | 15 | .0360152327 | 31 | 1 | 11 | .0243405034 | 31 | 3 | 13 | .0270353861 |
| 30 | 9 | 16 | .0337385516 | 31 | 1 | 12 | .0224440537 | 31 | 3 | 14 | .0251757621 |
| 30 | 9 | 17 | .0316737688 | 31 | 1 | 13 | .0208055812 | 31 | 3 | 15 | .0235267769 |
| 30 | 9 | 18 | .0297842702 | 31 | 1 | 14 | .0193705365 | 31 | 3 | 16 | .0220490867 |
| 30 | 9 | 19 | .0280403125 | 31 | 1 | 15 | .0180986000 | 31 | 3 | 17 | .0207122047 |
| 30 | 9 | 20 | .0264171606 | 31 | 1 | 16 | .0169592327 | 31 | 3 | 18 | .0194920170 |
| 30 | 9 | 21 | .0248936656 | 31 | 1 | 17 | .0159287890 | 31 | 3 | 19 | .0183690638 |
| 30 | 9 | 22 | .0234511012 | 31 | 1 | 18 | .0149885804 | 31 | 3 | 20 | .0173273082 |
| 30 | 10 | 10 | .0559701571 | 31 | 1 | 19 | .0141235351 | 31 | 3 | 21 | .0163532205 |
| 30 | 10 | 11 | .0513230380 | 31 | 1 | 20 | .0133212403 | 31 | 3 | 22 | .0154350629 |
| 30 | 10 | 12 | .0473570050 | 31 | 1 | 21 | .0125712312 | 31 | 3 | 23 | .0145622847 |
| 30 | 10 | 13 | .0439190098 | 31 | 1 | 22 | .0118644360 | 31 | 3 | 24 | .0137249502 |
| 30 | 10 | 14 | .0408984713 | 31 | 1 | 23 | .0111927067 | 31 | 3 | 25 | .0129130986 |
| 30 | 10 | 15 | .0382132523 | 31 | 1 | 24 | .0105483758 | 31 | 3 | 26 | .0121158759 |
| 30 | 10 | 16 | .0358008353 | 31 | 1 | 25 | .0099237615 | 31 | 3 | 27 | .0113201218 |
| 30 | 10 | 17 | .0336125601 | 31 | 1 | 26 | .0093105030 | 31 | 3 | 28 | .0105076089 |
| 30 | 10 | 18 | .0316097338 | 31 | 1 | 27 | .0086984739 | 31 | 3 | 29 | .0096484482 |
| 30 | 10 | 19 | .0297609156 | 31 | 1 | 28 | .0080736531 | 31 | 4 | 4 | .0851042240 |
| 30 | 10 | 20 | .0280399463 | 31 | 1 | 29 | .0074130642 | 31 | 4 | 5 | .0696799578 |
| 30 | 10 | 21 | .0264244474 | 31 | 1 | 30 | .0066692217 | 31 | 4 | 6 | .0592497623 |
| 30 | 11 | 11 | .0544081165 | 31 | 1 | 31 | .0056947839 | 31 | 4 | 7 | .0516694774 |
| 30 | 11 | 12 | .0502118968 | 31 | 2 | 2 | .1360268549 | 31 | 4 | 8 | .0458761542 |
| 30 | 11 | 13 | .0465732062 | 31 | 2 | 3 | .0922806679 | 31 | 4 | 9 | .0412806815 |
| 30 | 11 | 14 | .0433754605 | 31 | 2 | 4 | .0708058100 | 31 | 4 | 10 | .0375291527 |
| 30 | 11 | 15 | .0405320203 | 31 | 2 | 5 | .0578804786 | 31 | 4 | 11 | .0343955438 |
| 30 | 11 | 16 | .0379769114 | 31 | 2 | 6 | .0491635983 | 31 | 4 | 12 | .0317283778 |
| 30 | 11 | 17 | .0356587627 | 31 | 2 | 7 | .0428405957 | 31 | 4 | 13 | .0294220915 |
| 30 | 11 | 18 | .0335367105 | 31 | 2 | 8 | .0380150817 | 31 | 4 | 14 | .0274006861 |
| 30 | 11 | 19 | .0315775364 | 31 | 2 | 9 | .0341915711 | 31 | 4 | 15 | .0256079085 |
| 30 | 11 | 20 | .0297535890 | 31 | 2 | 10 | .0310730183 | 31 | 4 | 16 | .0240010977 |
| 30 | 12 | 12 | .0532494703 | 31 | 2 | 11 | .0284700301 | 31 | 4 | 17 | .0225471856 |

TABLE I-COVARIANCES OF NORMAL ORDER STATISTICS

| N | I | J | COVARIANCE | N | I | J | COVARIANCE | N | I | J | COVARIANCE |
|---|---|---|---|---|---|---|---|---|---|---|---|
| 31 | 4 | 18 | .0212200102 | 31 | 7 | 12 | .0388967946 | 31 | 11 | 12 | .0490687218 |
| 31 | 4 | 19 | .0199984517 | 31 | 7 | 13 | .0360804646 | 31 | 11 | 13 | .0455376707 |
| 31 | 4 | 20 | .0188650989 | 31 | 7 | 14 | .0336104360 | 31 | 11 | 14 | .0424377959 |
| 31 | 4 | 21 | .0178052575 | 31 | 7 | 15 | .0314185501 | 31 | 11 | 15 | .0396846482 |
| 31 | 4 | 22 | .0168061769 | 31 | 7 | 16 | .0294530614 | 31 | 11 | 16 | .0372140132 |
| 31 | 4 | 23 | .0158563977 | 31 | 7 | 17 | .0276738196 | 31 | 11 | 17 | .0349759941 |
| 31 | 4 | 24 | .0149451226 | 31 | 7 | 18 | .0260490316 | 31 | 11 | 18 | .0329310248 |
| 31 | 4 | 25 | .0140615108 | 31 | 7 | 19 | .0245530122 | 31 | 11 | 19 | .0310470986 |
| 31 | 4 | 26 | .0131937542 | 31 | 7 | 20 | .0231645688 | 31 | 11 | 20 | .0292977763 |
| 31 | 4 | 27 | .0123275437 | 31 | 7 | 21 | .0218657987 | 31 | 11 | 21 | .0276606950 |
| 31 | 4 | 28 | .0114430309 | 31 | 7 | 22 | .0206411473 | 31 | 12 | 12 | .0519328726 |
| 31 | 5 | 5 | .0748749147 | 31 | 7 | 23 | .0194766241 | 31 | 12 | 13 | .0482022623 |
| 31 | 5 | 6 | .0636989611 | 31 | 7 | 24 | .0183590499 | 31 | 12 | 14 | .0449262359 |
| 31 | 5 | 7 | .0555695488 | 31 | 7 | 25 | .0172751658 | 31 | 12 | 15 | .0420158831 |
| 31 | 5 | 8 | .0493524314 | 31 | 8 | 8 | .0595658143 | 31 | 12 | 16 | .0394035667 |
| 31 | 5 | 9 | .0444182341 | 31 | 8 | 9 | .0536460917 | 31 | 12 | 17 | .0370367090 |
| 31 | 5 | 10 | .0403885070 | 31 | 8 | 10 | .0488053243 | 31 | 12 | 18 | .0348736022 |
| 31 | 5 | 11 | .0370213678 | 31 | 8 | 11 | .0447562366 | 31 | 12 | 19 | .0328804926 |
| 31 | 5 | 12 | .0341546094 | 31 | 8 | 12 | .0413058165 | 31 | 12 | 20 | .0310294836 |
| 31 | 5 | 13 | .0316751300 | 31 | 8 | 13 | .0383192669 | 31 | 13 | 13 | .0510459005 |
| 31 | 5 | 14 | .0295014663 | 31 | 8 | 14 | .0356993632 | 31 | 13 | 14 | .0475823836 |
| 31 | 5 | 15 | .0275732962 | 31 | 8 | 15 | .0333740241 | 31 | 13 | 15 | .0445045760 |
| 31 | 5 | 16 | .0258448575 | 31 | 8 | 16 | .0312885064 | 31 | 13 | 16 | .0417412309 |
| 31 | 5 | 17 | .0242806673 | 31 | 8 | 17 | .0294003216 | 31 | 13 | 17 | .0392369398 |
| 31 | 5 | 18 | .0228526429 | 31 | 8 | 18 | .0276758140 | 31 | 13 | 18 | .0369477296 |
| 31 | 5 | 19 | .0215381059 | 31 | 8 | 19 | .0260877824 | 31 | 13 | 19 | .0348379963 |
| 31 | 5 | 20 | .0203183551 | 31 | 8 | 20 | .0246137729 | 31 | 14 | 14 | .0504367125 |
| 31 | 5 | 21 | .0191776093 | 31 | 8 | 21 | .0232348140 | 31 | 14 | 15 | .0471805964 |
| 31 | 5 | 22 | .0181021738 | 31 | 8 | 22 | .0219344354 | 31 | 14 | 16 | .0442564370 |
| 31 | 5 | 23 | .0170797223 | 31 | 8 | 23 | .0206978111 | 31 | 14 | 17 | .0416058245 |
| 31 | 5 | 24 | .0160986335 | 31 | 8 | 24 | .0195109348 | 31 | 14 | 18 | .0391823829 |
| 31 | 5 | 25 | .0151472600 | 31 | 9 | 9 | .0568402964 | 31 | 15 | 15 | .0500815419 |
| 31 | 5 | 26 | .0142128993 | 31 | 9 | 10 | .0517211761 | 31 | 15 | 16 | .0469839068 |
| 31 | 5 | 27 | .0132801348 | 31 | 9 | 11 | .0474376107 | 31 | 15 | 17 | .0441753773 |
| 31 | 6 | 6 | .0680345360 | 31 | 9 | 12 | .0437861884 | 31 | 16 | 16 | .0499655787 |
| 31 | 6 | 7 | .0593732841 | 31 | 9 | 13 | .0406247688 | 32 | 1 | 1 | .2416175029 |
| 31 | 6 | 8 | .0527450636 | 31 | 9 | 14 | .0378507804 | 32 | 1 | 2 | .1161589000 |
| 31 | 6 | 9 | .0474818705 | 31 | 9 | 15 | .0353881452 | 32 | 1 | 3 | .0785107683 |
| 31 | 6 | 10 | .0431816567 | 31 | 9 | 16 | .0331790632 | 32 | 1 | 4 | .0601301286 |
| 31 | 6 | 11 | .0395872651 | 31 | 9 | 17 | .0311786566 | 32 | 1 | 5 | .0491033997 |
| 31 | 6 | 12 | .0365261370 | 31 | 9 | 18 | .0293513623 | 32 | 1 | 6 | .0416838519 |
| 31 | 6 | 13 | .0338778895 | 31 | 9 | 19 | .0276684261 | 32 | 1 | 7 | .0363112466 |
| 31 | 6 | 14 | .0315557756 | 31 | 9 | 20 | .0261061105 | 32 | 1 | 8 | .0322168874 |
| 31 | 6 | 15 | .0294955354 | 31 | 9 | 21 | .0246443606 | 32 | 1 | 9 | .0289767279 |
| 31 | 6 | 16 | .0276484019 | 31 | 9 | 22 | .0232657464 | 32 | 1 | 10 | .0263369583 |
| 31 | 6 | 17 | .0259765481 | 31 | 9 | 23 | .0219545649 | 32 | 1 | 11 | .0241359691 |
| 31 | 6 | 18 | .0244500258 | 31 | 10 | 10 | .0547497123 | 32 | 1 | 12 | .0222657289 |
| 31 | 6 | 19 | .0230446471 | 31 | 10 | 11 | .0502238253 | 32 | 1 | 13 | .0206511213 |
| 31 | 6 | 20 | .0217404654 | 31 | 10 | 12 | .0463645499 | 32 | 1 | 14 | .0192381811 |
| 31 | 6 | 21 | .0205206378 | 31 | 10 | 13 | .0430222100 | 32 | 1 | 15 | .0179870472 |
| 31 | 6 | 22 | .0193705331 | 31 | 10 | 14 | .0400887415 | 32 | 1 | 16 | .0168675572 |
| 31 | 6 | 23 | .0182769892 | 31 | 10 | 15 | .0374839569 | 32 | 1 | 17 | .0158563915 |
| 31 | 6 | 24 | .0172275984 | 31 | 10 | 16 | .0351469087 | 32 | 1 | 18 | .0149351594 |
| 31 | 6 | 25 | .0162099117 | 31 | 10 | 17 | .0330302564 | 32 | 1 | 19 | .0140890800 |
| 31 | 6 | 26 | .0152103516 | 31 | 10 | 18 | .0310964723 | 32 | 1 | 20 | .0133060436 |
| 31 | 7 | 7 | .0631668086 | 31 | 10 | 19 | .0293152072 | 32 | 1 | 21 | .0125759241 |
| 31 | 7 | 8 | .0561308240 | 31 | 10 | 20 | .0276614014 | 32 | 1 | 22 | .0118900539 |
| 31 | 7 | 9 | .0505409238 | 31 | 10 | 21 | .0261138721 | 32 | 1 | 23 | .0112407992 |
| 31 | 7 | 10 | .0459718507 | 31 | 10 | 22 | .0246541953 | 32 | 1 | 24 | .0106211882 |
| 31 | 7 | 11 | .0421513982 | 31 | 11 | 11 | .0531447031 | 32 | 1 | 25 | .0100245399 |

TABLE I-COVARIANCES OF NORMAL ORDER STATISTICS

| N | I | J | COVARIANCE | N | I | J | COVARIANCE | N | I | J | COVARIANCE |
|---|---|---|---|---|---|---|---|---|---|---|---|
| 32 | 1 | 26 | .0094440313 | 32 | 3 | 26 | .0122541165 | 32 | 6 | 11 | .0389985486 |
| 32 | 1 | 27 | .0088720904 | 32 | 3 | 27 | .0115126583 | 32 | 6 | 12 | .0359989538 |
| 32 | 1 | 28 | .0082993786 | 32 | 3 | 28 | .0107700882 | 32 | 6 | 13 | .0334059967 |
| 32 | 1 | 29 | .0077127713 | 32 | 3 | 29 | .0100093933 | 32 | 6 | 14 | .0311343711 |
| 32 | 1 | 30 | .0070905288 | 32 | 3 | 30 | .0092023709 | 32 | 6 | 15 | .0291209311 |
| 32 | 1 | 31 | .0063874246 | 32 | 4 | 4 | .0839317157 | 32 | 6 | 16 | .0273178059 |
| 32 | 1 | 32 | .0054626184 | 32 | 4 | 5 | .0687164994 | 32 | 6 | 17 | .0256879210 |
| 32 | 2 | 2 | .1345550701 | 32 | 4 | 6 | .0584339730 | 32 | 6 | 18 | .0242019904 |
| 32 | 2 | 3 | .0912436381 | 32 | 4 | 7 | .0509652789 | 32 | 6 | 19 | .0228364393 |
| 32 | 2 | 4 | .0699989919 | 32 | 4 | 8 | .0452604465 | 32 | 6 | 20 | .0215719315 |
| 32 | 2 | 5 | .0572201470 | 32 | 4 | 9 | .0407377614 | 32 | 6 | 21 | .0203922804 |
| 32 | 2 | 6 | .0486068007 | 32 | 4 | 10 | .0370478725 | 32 | 6 | 22 | .0192836069 |
| 32 | 2 | 7 | .0423621836 | 32 | 4 | 11 | .0339677362 | 32 | 6 | 23 | .0182336653 |
| 32 | 2 | 8 | .0375990006 | 32 | 4 | 12 | .0313479238 | 32 | 6 | 24 | .0172312484 |
| 32 | 2 | 9 | .0338269362 | 32 | 4 | 13 | .0290843418 | 32 | 6 | 25 | .0162655990 |
| 32 | 2 | 10 | .0307521240 | 32 | 4 | 14 | .0271020887 | 32 | 6 | 26 | .0153257369 |
| 32 | 2 | 11 | .0281872499 | 32 | 4 | 15 | .0253457628 | 32 | 6 | 27 | .0143994379 |
| 32 | 2 | 12 | .0260069906 | 32 | 4 | 16 | .0237733914 | 32 | 7 | 7 | .0620773315 |
| 32 | 2 | 13 | .0241241427 | 32 | 4 | 17 | .0223524877 | 32 | 7 | 8 | .0551728444 |
| 32 | 2 | 14 | .0224760186 | 32 | 4 | 18 | .0210574077 | 32 | 7 | 9 | .0496908355 |
| 32 | 2 | 15 | .0210162895 | 32 | 4 | 19 | .0198675221 | 32 | 7 | 10 | .0452128293 |
| 32 | 2 | 16 | .0197098830 | 32 | 4 | 20 | .0187659155 | 32 | 7 | 11 | .0414710633 |
| 32 | 2 | 17 | .0185296714 | 32 | 4 | 21 | .0177384285 | 32 | 7 | 12 | .0382858228 |
| 32 | 2 | 18 | .0174542532 | 32 | 4 | 22 | .0167729267 | 32 | 7 | 13 | .0355317230 |
| 32 | 2 | 19 | .0164664208 | 32 | 4 | 23 | .0158587172 | 32 | 7 | 14 | .0331184225 |
| 32 | 2 | 20 | .0155520717 | 32 | 4 | 24 | .0149860327 | 32 | 7 | 15 | .0309790224 |
| 32 | 2 | 21 | .0146994096 | 32 | 4 | 25 | .0141454957 | 32 | 7 | 16 | .0290627872 |
| 32 | 2 | 22 | .0138983341 | 32 | 4 | 26 | .0133275052 | 32 | 7 | 17 | .0273304100 |
| 32 | 2 | 23 | .0131399465 | 32 | 4 | 27 | .0125214105 | 32 | 7 | 18 | .0257508349 |
| 32 | 2 | 24 | .0124161155 | 32 | 4 | 28 | .0117140671 | 32 | 7 | 19 | .0242990614 |
| 32 | 2 | 25 | .0117190464 | 32 | 4 | 29 | .0108869634 | 32 | 7 | 20 | .0229545769 |
| 32 | 2 | 26 | .0110407749 | 32 | 5 | 5 | .0737528428 | 32 | 7 | 21 | .0217001998 |
| 32 | 2 | 27 | .0103724593 | 32 | 5 | 6 | .0627471907 | 32 | 7 | 22 | .0205211884 |
| 32 | 2 | 28 | .0097031904 | 32 | 5 | 7 | .0547464344 | 32 | 7 | 23 | .0194045266 |
| 32 | 2 | 29 | .0090176294 | 32 | 5 | 8 | .0486312667 | 32 | 7 | 24 | .0183383300 |
| 32 | 2 | 30 | .0082903642 | 32 | 5 | 9 | .0437808399 | 32 | 7 | 25 | .0173112025 |
| 32 | 2 | 31 | .0074685213 | 32 | 5 | 10 | .0398219610 | 32 | 7 | 26 | .0163114289 |
| 32 | 3 | 3 | .1007837342 | 32 | 5 | 11 | .0365161881 | 32 | 8 | 8 | .0584702424 |
| 32 | 3 | 4 | .0774214130 | 32 | 5 | 12 | .0337036792 | 32 | 8 | 9 | .0526721940 |
| 32 | 3 | 5 | .0633391831 | 32 | 5 | 13 | .0312730288 | 32 | 8 | 10 | .0479340662 |
| 32 | 3 | 6 | .0538342084 | 32 | 5 | 14 | .0291440378 | 32 | 8 | 11 | .0439735919 |
| 32 | 3 | 7 | .0469364293 | 32 | 5 | 15 | .0272573632 | 32 | 8 | 12 | .0406012047 |
| 32 | 3 | 8 | .0416711971 | 32 | 5 | 16 | .0255680313 | 32 | 8 | 13 | .0376845779 |
| 32 | 3 | 9 | .0374991890 | 32 | 5 | 17 | .0240412237 | 32 | 8 | 14 | .0351283249 |
| 32 | 3 | 10 | .0340968230 | 32 | 5 | 18 | .0226494478 | 32 | 8 | 15 | .0328617818 |
| 32 | 3 | 11 | .0312576675 | 32 | 5 | 19 | .0213705789 | 32 | 8 | 16 | .0308313386 |
| 32 | 3 | 12 | .0288435118 | 32 | 5 | 20 | .0201864661 | 32 | 8 | 17 | .0289954504 |
| 32 | 3 | 13 | .0267581282 | 32 | 5 | 21 | .0190819154 | 32 | 8 | 18 | .0273212866 |
| 32 | 3 | 14 | .0249323070 | 32 | 5 | 22 | .0180439156 | 32 | 8 | 19 | .0257824100 |
| 32 | 3 | 15 | .0233148775 | 32 | 5 | 23 | .0170609940 | 32 | 8 | 20 | .0243571169 |
| 32 | 3 | 16 | .0218670865 | 32 | 5 | 24 | .0161226437 | 32 | 8 | 21 | .0230272131 |
| 32 | 3 | 17 | .0205589490 | 32 | 5 | 25 | .0152187855 | 32 | 8 | 22 | .0217770988 |
| 32 | 3 | 18 | .0193668010 | 32 | 5 | 26 | .0143391229 | 32 | 8 | 23 | .0205930263 |
| 32 | 3 | 19 | .0182716106 | 32 | 5 | 27 | .0134722027 | 32 | 8 | 24 | .0194624043 |
| 32 | 3 | 20 | .0172577770 | 32 | 5 | 28 | .0126038848 | 32 | 8 | 25 | .0183731256 |
| 32 | 3 | 21 | .0163122471 | 32 | 6 | 6 | .0669372502 | 32 | 9 | 9 | .0557290534 |
| 32 | 3 | 22 | .0154238401 | 32 | 6 | 7 | .0584226933 | 32 | 9 | 10 | .0507252172 |
| 32 | 3 | 23 | .0145827013 | 32 | 6 | 8 | .0519106588 | 32 | 9 | 11 | .0465411040 |
| 32 | 3 | 24 | .0137798227 | 32 | 6 | 9 | .0467428554 | 32 | 9 | 12 | .0429771728 |
| 32 | 3 | 25 | .0130065688 | 32 | 6 | 10 | .0425232292 | 32 | 9 | 13 | .0398940583 |

TABLE I-COVARIANCES OF NORMAL ORDER STATISTICS

| N | I | J | COVARIANCE | N | I | J | COVARIANCE | N | I | J | COVARIANCE |
|---|---|---|---|---|---|---|---|---|---|---|---|
| 32 | 9 | 14 | .0371912528 | 32 | 14 | 19 | .0361194683 | 33 | 2 | 22 | .0138911485 |
| 32 | 9 | 15 | .0347942680 | 32 | 15 | 15 | .0486643703 | 33 | 2 | 23 | .0131564518 |
| 32 | 9 | 16 | .0326465728 | 32 | 15 | 16 | .0456866095 | 33 | 2 | 24 | .0124580514 |
| 32 | 9 | 17 | .0307043455 | 32 | 15 | 17 | .0429905779 | 33 | 2 | 25 | .0117889278 |
| 32 | 9 | 18 | .0289329437 | 32 | 15 | 18 | .0405290796 | 33 | 2 | 26 | .0111422335 |
| 32 | 9 | 19 | .0273044561 | 32 | 16 | 16 | .0484517701 | 33 | 2 | 27 | .0105108440 |
| 32 | 9 | 20 | .0257959608 | 32 | 16 | 17 | .0455976240 | 33 | 2 | 28 | .0098867066 |
| 32 | 9 | 21 | .0243882518 | 33 | 1 | 1 | .2396512267 | 33 | 2 | 29 | .0092597200 |
| 32 | 9 | 22 | .0230648590 | 33 | 1 | 2 | .1151035317 | 33 | 2 | 30 | .0086154879 |
| 32 | 9 | 23 | .0218112491 | 33 | 1 | 3 | .0777692399 | 33 | 2 | 31 | .0079299369 |
| 32 | 9 | 24 | .0206141033 | 33 | 1 | 4 | .0595541703 | 33 | 2 | 32 | .0071526848 |
| 32 | 10 | 10 | .0536146532 | 33 | 1 | 5 | .0486326962 | 33 | 3 | 3 | .0995789468 |
| 32 | 10 | 11 | .0492000112 | 33 | 1 | 6 | .0412875616 | 33 | 3 | 4 | .0764814818 |
| 32 | 10 | 12 | .0454384948 | 33 | 1 | 7 | .0359713452 | 33 | 3 | 5 | .0625676589 |
| 32 | 10 | 13 | .0421835373 | 33 | 1 | 8 | .0319218805 | 33 | 3 | 6 | .0531814920 |
| 32 | 10 | 14 | .0393293860 | 33 | 1 | 9 | .0287188265 | 33 | 3 | 7 | .0463734300 |
| 32 | 10 | 15 | .0367976322 | 33 | 1 | 10 | .0261106496 | 33 | 3 | 8 | .0411793139 |
| 32 | 10 | 16 | .0345287474 | 33 | 1 | 11 | .0239372270 | 33 | 3 | 9 | .0370657869 |
| 32 | 10 | 17 | .0324765685 | 33 | 1 | 12 | .0220915492 | 33 | 3 | 10 | .0337129348 |
| 32 | 10 | 18 | .0306045868 | 33 | 1 | 13 | .0204992334 | 33 | 3 | 11 | .0309167215 |
| 32 | 10 | 19 | .0288833795 | 33 | 1 | 14 | .0191068628 | 33 | 3 | 12 | .0285405800 |
| 32 | 10 | 20 | .0272887820 | 33 | 1 | 15 | .0178750032 | 33 | 3 | 13 | .0264894593 |
| 32 | 10 | 21 | .0258005448 | 33 | 1 | 16 | .0167738386 | 33 | 3 | 14 | .0246950249 |
| 32 | 10 | 22 | .0244012944 | 33 | 1 | 17 | .0157803431 | 33 | 3 | 15 | .0231067822 |
| 32 | 10 | 23 | .0230756703 | 33 | 1 | 18 | .0148763890 | 33 | 3 | 16 | .0216865210 |
| 32 | 11 | 11 | .0519739653 | 33 | 1 | 19 | .0140474432 | 33 | 3 | 17 | .0204047108 |
| 32 | 11 | 12 | .0480076297 | 33 | 1 | 20 | .0132816464 | 33 | 3 | 18 | .0192380867 |
| 32 | 11 | 13 | .0445745014 | 33 | 1 | 21 | .0125691401 | 33 | 3 | 19 | .0181679854 |
| 32 | 11 | 14 | .0415634167 | 33 | 1 | 22 | .0119015628 | 33 | 3 | 20 | .0171791676 |
| 32 | 11 | 15 | .0388919188 | 33 | 1 | 23 | .0112716529 | 33 | 3 | 21 | .0162589601 |
| 32 | 11 | 16 | .0364973805 | 33 | 1 | 24 | .0106729180 | 33 | 3 | 22 | .0153966083 |
| 32 | 11 | 17 | .0343312168 | 33 | 1 | 25 | .0100993317 | 33 | 3 | 23 | .0145827660 |
| 32 | 11 | 18 | .0323549923 | 33 | 1 | 26 | .0095450183 | 33 | 3 | 24 | .0138090685 |
| 32 | 11 | 19 | .0305377257 | 33 | 1 | 27 | .0090038665 | 33 | 3 | 25 | .0130677455 |
| 32 | 11 | 20 | .0288539692 | 33 | 1 | 28 | .0084689701 | 33 | 3 | 26 | .0123512225 |
| 32 | 11 | 21 | .0272823849 | 33 | 1 | 29 | .0079316699 | 33 | 3 | 27 | .0116516183 |
| 32 | 11 | 22 | .0258046473 | 33 | 1 | 30 | .0073796319 | 33 | 3 | 28 | .0109600045 |
| 32 | 12 | 12 | .0507161600 | 33 | 1 | 31 | .0067922310 | 33 | 3 | 29 | .0102651840 |
| 32 | 12 | 13 | .0470956354 | 33 | 1 | 32 | .0061263090 | 33 | 3 | 30 | .0095512137 |
| 32 | 12 | 14 | .0439193514 | 33 | 1 | 33 | .0052470337 | 33 | 3 | 31 | .0087914050 |
| 32 | 12 | 15 | .0411006310 | 33 | 2 | 2 | .1331547632 | 33 | 4 | 4 | .0828214219 |
| 32 | 12 | 16 | .0385736134 | 33 | 2 | 3 | .0902567111 | 33 | 4 | 5 | .0678035270 |
| 32 | 12 | 17 | .0362871859 | 33 | 2 | 4 | .0692306364 | 33 | 4 | 6 | .0576602061 |
| 32 | 12 | 18 | .0342009012 | 33 | 2 | 5 | .0565906865 | 33 | 4 | 7 | .0502965672 |
| 32 | 12 | 19 | .0322821466 | 33 | 2 | 6 | .0480753771 | 33 | 4 | 8 | .0446749227 |
| 32 | 12 | 20 | .0305041211 | 33 | 2 | 7 | .0419048770 | 33 | 4 | 9 | .0402205538 |
| 32 | 12 | 21 | .0288443399 | 33 | 2 | 8 | .0372005373 | 33 | 4 | 10 | .0365884137 |
| 32 | 13 | 13 | .0497759419 | 33 | 2 | 9 | .0334769538 | 33 | 4 | 11 | .0335582708 |
| 32 | 13 | 14 | .0464239262 | 33 | 2 | 10 | .0304432776 | 33 | 4 | 12 | .0309826259 |
| 32 | 13 | 15 | .0434484151 | 33 | 2 | 11 | .0279141651 | 33 | 4 | 13 | .0287587657 |
| 32 | 13 | 16 | .0407801480 | 33 | 2 | 12 | .0257656454 | 33 | 4 | 14 | .0268128143 |
| 32 | 13 | 17 | .0383653514 | 33 | 2 | 13 | .0239114846 | 33 | 4 | 15 | .0250901611 |
| 32 | 13 | 18 | .0361614557 | 33 | 2 | 14 | .0222897174 | 33 | 4 | 16 | .0235494662 |
| 32 | 13 | 19 | .0341341215 | 33 | 2 | 15 | .0208545748 | 33 | 4 | 17 | .0221587698 |
| 32 | 13 | 20 | .0322551143 | 33 | 2 | 16 | .0195714353 | 33 | 4 | 18 | .0208928864 |
| 32 | 14 | 14 | .0491012304 | 33 | 2 | 17 | .0184135508 | 33 | 4 | 19 | .0197316091 |
| 32 | 14 | 15 | .0459594603 | 33 | 2 | 18 | .0173598556 | 33 | 4 | 20 | .0186584357 |
| 32 | 14 | 16 | .0431414086 | 33 | 2 | 19 | .0163934559 | 33 | 4 | 21 | .0176596380 |
| 32 | 14 | 17 | .0405904933 | 33 | 2 | 20 | .0155005602 | 33 | 4 | 22 | .0167235565 |
| 32 | 14 | 18 | .0382619013 | 33 | 2 | 21 | .0146697008 | 33 | 4 | 23 | .0158400535 |

TABLE I-COVARIANCES OF NORMAL ORDER STATISTICS

| N | I | J | COVARIANCE | N | I | J | COVARIANCE | N | I | J | COVARIANCE |
|---|---|---|---|---|---|---|---|---|---|---|---|
| 33 | 4 | 24 | .0150000673 | 33 | 7 | 12 | .0377066227 | 33 | 10 | 18 | .0301336835 |
| 33 | 4 | 25 | .0141951887 | 33 | 7 | 13 | .0350100679 | 33 | 10 | 19 | .0284663641 |
| 33 | 4 | 26 | .0134171912 | 33 | 7 | 14 | .0326490770 | 33 | 10 | 20 | .0269245225 |
| 33 | 4 | 27 | .0126575001 | 33 | 7 | 15 | .0305579151 | 33 | 10 | 21 | .0254886553 |
| 33 | 4 | 28 | .0119064425 | 33 | 7 | 16 | .0286867733 | 33 | 10 | 22 | .0241421993 |
| 33 | 4 | 29 | .0111518708 | 33 | 7 | 17 | .0269971089 | 33 | 10 | 23 | .0228707359 |
| 33 | 4 | 30 | .0103764514 | 33 | 7 | 18 | .0254585214 | 33 | 10 | 24 | .0216613073 |
| 33 | 5 | 5 | .0726925689 | 33 | 7 | 19 | .0240465998 | 33 | 11 | 11 | .0508858013 |
| 33 | 5 | 6 | .0618470996 | 33 | 7 | 20 | .0227413972 | 33 | 11 | 12 | .0470198558 |
| 33 | 5 | 7 | .0539671978 | 33 | 7 | 21 | .0215263155 | 33 | 11 | 13 | .0436761868 |
| 33 | 5 | 8 | .0479476646 | 33 | 7 | 22 | .0203872526 | 33 | 11 | 14 | .0407460133 |
| 33 | 5 | 9 | .0431756988 | 33 | 7 | 23 | .0193119160 | 33 | 11 | 15 | .0381487024 |
| 33 | 5 | 10 | .0392830629 | 33 | 7 | 24 | .0182892933 | 33 | 11 | 16 | .0358230694 |
| 33 | 5 | 11 | .0360345547 | 33 | 7 | 25 | .0173091980 | 33 | 11 | 17 | .0337217099 |
| 33 | 5 | 12 | .0332725549 | 33 | 7 | 26 | .0163616698 | 33 | 11 | 18 | .0318071908 |
| 33 | 5 | 13 | .0308872478 | 33 | 7 | 27 | .0154362646 | 33 | 11 | 19 | .0300494169 |
| 33 | 5 | 14 | .0287996149 | 33 | 8 | 8 | .0574423233 | 33 | 11 | 20 | .0284237680 |
| 33 | 5 | 15 | .0269512212 | 33 | 8 | 9 | .0517573970 | 33 | 11 | 21 | .0269097353 |
| 33 | 5 | 16 | .0252978170 | 33 | 8 | 10 | .0471145438 | 33 | 11 | 22 | .0254898780 |
| 33 | 5 | 17 | .0238051847 | 33 | 8 | 11 | .0432361783 | 33 | 11 | 23 | .0241489986 |
| 33 | 5 | 18 | .0224463550 | 33 | 8 | 12 | .0399359501 | 33 | 12 | 12 | .0495878485 |
| 33 | 5 | 19 | .0211996835 | 33 | 8 | 13 | .0370838320 | 33 | 12 | 13 | .0460677450 |
| 33 | 5 | 20 | .0200474830 | 33 | 8 | 14 | .0345861425 | 33 | 12 | 14 | .0429822380 |
| 33 | 5 | 21 | .0189750301 | 33 | 8 | 15 | .0323735257 | 33 | 12 | 15 | .0402466903 |
| 33 | 5 | 22 | .0179698299 | 33 | 8 | 16 | .0303934095 | 33 | 12 | 16 | .0377968529 |
| 33 | 5 | 23 | .0170210325 | 33 | 8 | 17 | .0286051010 | 33 | 12 | 17 | .0355829289 |
| 33 | 5 | 24 | .0161189179 | 33 | 8 | 18 | .0269764995 | 33 | 12 | 18 | .0335655846 |
| 33 | 5 | 25 | .0152544396 | 33 | 8 | 19 | .0254818274 | 33 | 12 | 19 | .0317131924 |
| 33 | 5 | 26 | .0144187746 | 33 | 8 | 20 | .0241000149 | 33 | 12 | 20 | .0299998722 |
| 33 | 5 | 27 | .0136027337 | 33 | 8 | 21 | .0228135063 | 33 | 12 | 21 | .0284040526 |
| 33 | 5 | 28 | .0127959203 | 33 | 8 | 22 | .0216073712 | 33 | 12 | 22 | .0269073763 |
| 33 | 5 | 29 | .0119852970 | 33 | 8 | 23 | .0204686357 | 33 | 13 | 13 | .0486018051 |
| 33 | 6 | 6 | .0659025627 | 33 | 8 | 24 | .0193856925 | 33 | 13 | 14 | .0453512108 |
| 33 | 6 | 7 | .0575255000 | 33 | 8 | 25 | .0183477394 | 33 | 13 | 15 | .0424685309 |
| 33 | 6 | 8 | .0511222182 | 33 | 8 | 26 | .0173442217 | 33 | 13 | 16 | .0398863028 |
| 33 | 6 | 9 | .0460435709 | 33 | 9 | 9 | .0546883518 | 33 | 13 | 17 | .0375522190 |
| 33 | 6 | 10 | .0418991423 | 33 | 9 | 10 | .0497913510 | 33 | 13 | 18 | .0354249489 |
| 33 | 6 | 11 | .0384393884 | 33 | 9 | 11 | .0456992533 | 33 | 13 | 19 | .0334712484 |
| 33 | 6 | 12 | .0354969810 | 33 | 9 | 12 | .0422161218 | 33 | 13 | 20 | .0316639003 |
| 33 | 6 | 13 | .0329552819 | 33 | 9 | 13 | .0392051698 | 33 | 13 | 21 | .0299802034 |
| 33 | 6 | 14 | .0307303272 | 33 | 9 | 14 | .0365678004 | 33 | 14 | 14 | .0478734865 |
| 33 | 6 | 15 | .0287600022 | 33 | 9 | 15 | .0342309893 | 33 | 14 | 15 | .0448347372 |
| 33 | 6 | 16 | .0269972576 | 33 | 9 | 16 | .0321393648 | 33 | 14 | 16 | .0421120082 |
| 33 | 6 | 17 | .0254057009 | 33 | 9 | 17 | .0302500569 | 33 | 14 | 17 | .0396503488 |
| 33 | 6 | 18 | .0239566362 | 33 | 9 | 18 | .0285292399 | 33 | 14 | 18 | .0374063230 |
| 33 | 6 | 19 | .0226270218 | 33 | 9 | 19 | .0269497378 | 33 | 14 | 19 | .0353449773 |
| 33 | 6 | 20 | .0213980287 | 33 | 9 | 20 | .0254893200 | 33 | 14 | 20 | .0334376811 |
| 33 | 6 | 21 | .0202539931 | 33 | 9 | 21 | .0241294655 | 33 | 15 | 15 | .0473682511 |
| 33 | 6 | 22 | .0191816158 | 33 | 9 | 22 | .0228544359 | 33 | 15 | 16 | .0444972795 |
| 33 | 6 | 23 | .0181693310 | 33 | 9 | 23 | .0216505490 | 33 | 15 | 17 | .0419011100 |
| 33 | 6 | 24 | .0172067770 | 33 | 9 | 24 | .0205055658 | 33 | 15 | 18 | .0395340826 |
| 33 | 6 | 25 | .0162842998 | 33 | 9 | 25 | .0194080569 | 33 | 15 | 19 | .0373594398 |
| 33 | 6 | 26 | .0153924879 | 33 | 10 | 10 | .0525552898 | 33 | 16 | 16 | .0470734253 |
| 33 | 6 | 27 | .0145215889 | 33 | 10 | 11 | .0482432153 | 33 | 16 | 17 | .0443324164 |
| 33 | 6 | 28 | .0136605013 | 33 | 10 | 12 | .0445716680 | 33 | 16 | 18 | .0418328682 |
| 33 | 7 | 7 | .0610526790 | 33 | 10 | 13 | .0413969613 | 33 | 17 | 17 | .0469776919 |
| 33 | 7 | 8 | .0542709423 | 33 | 10 | 14 | .0386154762 | 34 | 1 | 1 | .2377712736 |
| 33 | 7 | 9 | .0488895006 | 33 | 10 | 15 | .0361504380 | 34 | 1 | 2 | .1140949263 |
| 33 | 7 | 10 | .0444962412 | 33 | 10 | 16 | .0339436111 | 34 | 1 | 3 | .0770602829 |
| 33 | 7 | 11 | .0408275670 | 33 | 10 | 17 | .0319498937 | 34 | 1 | 4 | .0590030784 |

TABLE I-COVARIANCES OF NORMAL ORDER STATISTICS

| N | I | J | COVARIANCE | N | I | J | COVARIANCE | N | I | J | COVARIANCE |
|---|---|---|---|---|---|---|---|---|---|---|---|
| 34 | 1 | 5 | .0481818317 | 34 | 2 | 32 | .0075963893 | 34 | 5 | 5 | .0716887237 |
| 34 | 1 | 6 | .0409074622 | 34 | 2 | 33 | .0068598897 | 34 | 5 | 6 | .0609942658 |
| 34 | 1 | 7 | .0356447933 | 34 | 3 | 3 | .0984335205 | 34 | 5 | 7 | .0532281606 |
| 34 | 1 | 8 | .0316378932 | 34 | 3 | 4 | .0755874234 | 34 | 5 | 8 | .0472985607 |
| 34 | 1 | 9 | .0284699567 | 34 | 3 | 5 | .0618332329 | 34 | 5 | 9 | .0426002740 |
| 34 | 1 | 10 | .0258916198 | 34 | 3 | 6 | .0525595468 | 34 | 5 | 10 | .0387697457 |
| 34 | 1 | 11 | .0237441753 | 34 | 3 | 7 | .0458363161 | 34 | 5 | 11 | .0355748309 |
| 34 | 1 | 12 | .0219215845 | 34 | 3 | 8 | .0407093501 | 34 | 5 | 12 | .0328600051 |
| 34 | 1 | 13 | .0203501596 | 34 | 3 | 9 | .0366509578 | 34 | 5 | 13 | .0305169475 |
| 34 | 1 | 14 | .0189770012 | 34 | 3 | 10 | .0333447053 | 34 | 5 | 14 | .0284677460 |
| 34 | 1 | 15 | .0177630745 | 34 | 3 | 11 | .0305888269 | 34 | 5 | 15 | .0266548127 |
| 34 | 1 | 16 | .0166788846 | 34 | 3 | 12 | .0282483109 | 34 | 5 | 16 | .0250345680 |
| 34 | 1 | 17 | .0157016742 | 34 | 3 | 13 | .0262292169 | 34 | 5 | 17 | .0235733432 |
| 34 | 1 | 18 | .0148135504 | 34 | 3 | 14 | .0244640335 | 34 | 5 | 18 | .0222446376 |
| 34 | 1 | 19 | .0140001970 | 34 | 3 | 15 | .0229028980 | 34 | 5 | 19 | .0210272295 |
| 34 | 1 | 20 | .0132499654 | 34 | 3 | 16 | .0215081028 | 34 | 5 | 20 | .0199038325 |
| 34 | 1 | 21 | .0125532160 | 34 | 3 | 17 | .0202505339 | 34 | 5 | 21 | .0188601174 |
| 34 | 1 | 22 | .0119018269 | 34 | 3 | 18 | .0191072853 | 34 | 5 | 22 | .0178839959 |
| 34 | 1 | 23 | .0112888163 | 34 | 3 | 19 | .0180600176 | 34 | 5 | 23 | .0169650801 |
| 34 | 1 | 24 | .0107080367 | 34 | 3 | 20 | .0170937996 | 34 | 5 | 24 | .0160942232 |
| 34 | 1 | 25 | .0101539113 | 34 | 3 | 21 | .0161962694 | 34 | 5 | 25 | .0152631061 |
| 34 | 1 | 26 | .0096211827 | 34 | 3 | 22 | .0153570082 | 34 | 5 | 26 | .0144638737 |
| 34 | 1 | 27 | .0091046385 | 34 | 3 | 23 | .0145670561 | 34 | 5 | 27 | .0136887282 |
| 34 | 1 | 28 | .0085987636 | 34 | 3 | 24 | .0138185174 | 34 | 5 | 28 | .0129293941 |
| 34 | 1 | 29 | .0080972190 | 34 | 3 | 25 | .0131042187 | 34 | 5 | 29 | .0121763884 |
| 34 | 1 | 30 | .0075919359 | 34 | 3 | 26 | .0124173944 | 34 | 5 | 30 | .0114176296 |
| 34 | 1 | 31 | .0070712831 | 34 | 3 | 27 | .0117513467 | 34 | 6 | 6 | .0649247664 |
| 34 | 1 | 32 | .0065156449 | 34 | 3 | 28 | .0110989794 | 34 | 6 | 7 | .0566769084 |
| 34 | 1 | 33 | .0058837641 | 34 | 3 | 29 | .0104521097 | 34 | 6 | 8 | .0503756966 |
| 34 | 1 | 34 | .0050463765 | 34 | 3 | 30 | .0098003324 | 34 | 6 | 9 | .0453806145 |
| 34 | 2 | 2 | .1318202968 | 34 | 3 | 31 | .0091286524 | 34 | 6 | 10 | .0413065623 |
| 34 | 2 | 3 | .0893159714 | 34 | 3 | 32 | .0084117482 | 34 | 6 | 11 | .0379074697 |
| 34 | 2 | 4 | .0684977870 | 34 | 4 | 4 | .0817680800 | 34 | 6 | 12 | .0350183892 |
| 34 | 2 | 5 | .0559897799 | 34 | 4 | 5 | .0669368286 | 34 | 6 | 13 | .0325243814 |
| 34 | 2 | 6 | .0475674838 | 34 | 4 | 6 | .0569250220 | 34 | 6 | 14 | .0303427387 |
| 34 | 2 | 7 | .0414672092 | 34 | 4 | 7 | .0496605112 | 34 | 6 | 15 | .0284123061 |
| 34 | 2 | 8 | .0368185415 | 34 | 4 | 8 | .0441172569 | 34 | 6 | 16 | .0266867923 |
| 34 | 2 | 9 | .0331407510 | 34 | 4 | 9 | .0397271694 | 34 | 6 | 17 | .0251304270 |
| 34 | 2 | 10 | .0301458583 | 34 | 4 | 10 | .0361492774 | 34 | 6 | 18 | .0237150473 |
| 34 | 2 | 11 | .0276503914 | 34 | 4 | 11 | .0331660094 | 34 | 6 | 19 | .0224180851 |
| 34 | 2 | 12 | .0255316619 | 34 | 4 | 12 | .0306316880 | 34 | 6 | 20 | .0212211482 |
| 34 | 2 | 13 | .0237043522 | 34 | 4 | 13 | .0284449002 | 34 | 6 | 21 | .0201090000 |
| 34 | 2 | 14 | .0221071776 | 34 | 4 | 14 | .0265327328 | 34 | 6 | 22 | .0190687954 |
| 34 | 2 | 15 | .0206948926 | 34 | 4 | 15 | .0248413136 | 34 | 6 | 23 | .0180894868 |
| 34 | 2 | 16 | .0194332937 | 34 | 4 | 16 | .0233298892 | 34 | 6 | 24 | .0171613426 |
| 34 | 2 | 17 | .0182959802 | 34 | 4 | 17 | .0219669824 | 34 | 6 | 25 | .0162754952 |
| 34 | 2 | 18 | .0172621875 | 34 | 4 | 18 | .0207278206 | 34 | 6 | 26 | .0154235339 |
| 34 | 2 | 19 | .0163152966 | 34 | 4 | 19 | .0195925665 | 34 | 6 | 27 | .0145971737 |
| 34 | 2 | 20 | .0154417806 | 34 | 4 | 20 | .0185450696 | 34 | 6 | 28 | .0137876919 |
| 34 | 2 | 21 | .0146304414 | 34 | 4 | 21 | .0175719576 | 34 | 6 | 29 | .0129849166 |
| 34 | 2 | 22 | .0138718416 | 34 | 4 | 22 | .0166619530 | 34 | 7 | 7 | .0600867210 |
| 34 | 2 | 23 | .0131578667 | 34 | 4 | 23 | .0158053426 | 34 | 7 | 8 | .0534199014 |
| 34 | 2 | 24 | .0124813704 | 34 | 4 | 24 | .0149935673 | 34 | 7 | 9 | .0481324831 |
| 34 | 2 | 25 | .0118358674 | 34 | 4 | 25 | .0142188761 | 34 | 7 | 10 | .0438183390 |
| 34 | 2 | 26 | .0112152411 | 34 | 4 | 26 | .0134739579 | 34 | 7 | 11 | .0402177863 |
| 34 | 2 | 27 | .0106134233 | 34 | 4 | 27 | .0127515287 | 34 | 7 | 12 | .0371566530 |
| 34 | 2 | 28 | .0100239920 | 34 | 4 | 28 | .0120438765 | 34 | 7 | 13 | .0345135132 |
| 34 | 2 | 29 | .0094395674 | 34 | 4 | 29 | .0113421582 | 34 | 7 | 14 | .0322009561 |
| 34 | 2 | 30 | .0088507490 | 34 | 4 | 30 | .0106350863 | 34 | 7 | 15 | .0301543277 |
| 34 | 2 | 31 | .0082439770 | 34 | 4 | 31 | .0099063761 | 34 | 7 | 16 | .0283246738 |

TABLE I-COVARIANCES OF NORMAL ORDER STATISTICS

| N | I | J | COVARIANCE | N | I | J | COVARIANCE | N | I | J | COVARIANCE |
|---|---|---|---|---|---|---|---|---|---|---|---|
| 34 | 7 | 17 | .0266741515 | 34 | 10 | 20 | .0265700650 | 34 | 17 | 17 | .0456354534 |
| 34 | 7 | 18 | .0251729527 | 34 | 10 | 21 | .0251809776 | 34 | 17 | 18 | .0430974183 |
| 34 | 7 | 19 | .0237971861 | 34 | 10 | 22 | .0238812520 | 35 | 1 | 1 | .2359713518 |
| 34 | 7 | 20 | .0225273842 | 34 | 10 | 23 | .0226571914 | 35 | 1 | 2 | .1131296770 |
| 34 | 7 | 21 | .0213474252 | 34 | 10 | 24 | .0214966912 | 35 | 1 | 3 | .0763815569 |
| 34 | 7 | 22 | .0202437227 | 34 | 10 | 25 | .0203887171 | 35 | 1 | 4 | .0584751057 |
| 34 | 7 | 23 | .0192045625 | 34 | 11 | 11 | .0498713522 | 35 | 1 | 5 | .0477494553 |
| 34 | 7 | 24 | .0182195702 | 34 | 11 | 12 | .0460977101 | 35 | 1 | 6 | .0405424973 |
| 34 | 7 | 25 | .0172793484 | 34 | 11 | 13 | .0428361299 | 35 | 1 | 7 | .0353307705 |
| 34 | 7 | 26 | .0163750850 | 34 | 11 | 14 | .0399800326 | 35 | 1 | 8 | .0313643047 |
| 34 | 7 | 27 | .0154979800 | 34 | 11 | 15 | .0374504640 | 35 | 1 | 9 | .0282296734 |
| 34 | 7 | 28 | .0146386677 | 34 | 11 | 16 | .0351875587 | 35 | 1 | 10 | .0256795846 |
| 34 | 8 | 8 | .0564757612 | 34 | 11 | 17 | .0331449845 | 35 | 1 | 11 | .0235566814 |
| 34 | 8 | 9 | .0508963019 | 34 | 11 | 18 | .0312862087 | 35 | 1 | 12 | .0217558495 |
| 34 | 8 | 10 | .0463421443 | 34 | 11 | 19 | .0295819189 | 35 | 1 | 13 | .0202040612 |
| 34 | 8 | 11 | .0425400916 | 34 | 11 | 20 | .0280082019 | 35 | 1 | 14 | .0188489067 |
| 34 | 8 | 12 | .0393068035 | 34 | 11 | 21 | .0265452299 | 35 | 1 | 15 | .0176517274 |
| 34 | 8 | 13 | .0365144009 | 34 | 11 | 22 | .0251762694 | 35 | 1 | 16 | .0165833276 |
| 34 | 8 | 14 | .0340707901 | 34 | 11 | 23 | .0238869036 | 35 | 1 | 17 | .0156211975 |
| 34 | 8 | 15 | .0319078286 | 34 | 11 | 24 | .0226643875 | 35 | 1 | 18 | .0147476578 |
| 34 | 8 | 16 | .0299739010 | 34 | 12 | 12 | .0485387269 | 35 | 1 | 19 | .0139485842 |
| 34 | 8 | 17 | .0282290933 | 34 | 12 | 13 | .0451104943 | 35 | 1 | 20 | .0132125104 |
| 34 | 8 | 18 | .0266419595 | 34 | 12 | 14 | .0421078518 | 35 | 1 | 21 | .0125299796 |
| 34 | 8 | 19 | .0251872979 | 34 | 12 | 15 | .0394480351 | 35 | 1 | 22 | .0118930637 |
| 34 | 8 | 20 | .0238445797 | 34 | 12 | 16 | .0370682668 | 35 | 1 | 23 | .0112949981 |
| 34 | 8 | 21 | .0225967898 | 34 | 12 | 17 | .0349199425 | 35 | 1 | 24 | .0107298931 |
| 34 | 8 | 22 | .0214295467 | 34 | 12 | 18 | .0329647253 | 35 | 1 | 25 | .0101924939 |
| 34 | 8 | 23 | .0203304614 | 34 | 12 | 19 | .0311718515 | 35 | 1 | 26 | .0096779662 |
| 34 | 8 | 24 | .0192886225 | 34 | 12 | 20 | .0295162263 | 35 | 1 | 27 | .0091816839 |
| 34 | 8 | 25 | .0182941191 | 34 | 12 | 21 | .0279770373 | 35 | 1 | 28 | .0086989870 |
| 34 | 8 | 26 | .0173376225 | 34 | 12 | 22 | .0265367074 | 35 | 1 | 29 | .0082248635 |
| 34 | 8 | 27 | .0164098469 | 34 | 12 | 23 | .0251800661 | 35 | 1 | 30 | .0077534618 |
| 34 | 9 | 9 | .0537111245 | 34 | 13 | 13 | .0475116933 | 35 | 1 | 31 | .0072772311 |
| 34 | 9 | 10 | .0489134557 | 34 | 13 | 14 | .0443538360 | 35 | 1 | 32 | .0067851681 |
| 34 | 9 | 11 | .0449067849 | 34 | 13 | 15 | .0415558628 | 35 | 1 | 33 | .0062585762 |
| 34 | 9 | 12 | .0414985421 | 34 | 13 | 16 | .0390519509 | 35 | 1 | 34 | .0056579515 |
| 34 | 9 | 13 | .0385543442 | 34 | 13 | 17 | .0367911195 | 35 | 1 | 35 | .0048592023 |
| 34 | 9 | 14 | .0359773714 | 34 | 13 | 18 | .0347331455 | 35 | 2 | 2 | .1305466357 |
| 34 | 9 | 15 | .0336959556 | 34 | 13 | 19 | .0328457457 | 35 | 2 | 3 | .0884179174 |
| 34 | 9 | 16 | .0316557918 | 34 | 13 | 20 | .0311025777 | 35 | 2 | 4 | .0677977941 |
| 34 | 9 | 17 | .0298148762 | 34 | 13 | 21 | .0294817794 | 35 | 2 | 5 | .0554153441 |
| 34 | 9 | 18 | .0281401140 | 34 | 13 | 22 | .0279648679 | 35 | 2 | 6 | .0470814559 |
| 34 | 9 | 19 | .0266049774 | 34 | 14 | 14 | .0467390106 | 35 | 2 | 7 | .0410478468 |
| 34 | 9 | 20 | .0251878373 | 34 | 14 | 15 | .0437938926 | 35 | 2 | 8 | .0364519575 |
| 34 | 9 | 21 | .0238707515 | 34 | 14 | 16 | .0411575940 | 35 | 2 | 9 | .0328175144 |
| 34 | 9 | 22 | .0226385655 | 34 | 14 | 17 | .0387766357 | 35 | 2 | 10 | .0298592720 |
| 34 | 9 | 23 | .0214782263 | 34 | 14 | 18 | .0366088096 | 35 | 2 | 11 | .0273955391 |
| 34 | 9 | 24 | .0203782645 | 34 | 14 | 19 | .0346202260 | 35 | 2 | 12 | .0253048461 |
| 34 | 9 | 25 | .0193282741 | 34 | 14 | 20 | .0327832172 | 35 | 2 | 13 | .0235027442 |
| 34 | 9 | 26 | .0183183236 | 34 | 14 | 21 | .0310748059 | 35 | 2 | 14 | .0219285917 |
| 34 | 10 | 10 | .0515631315 | 34 | 15 | 15 | .0461776994 | 35 | 2 | 15 | .0205376355 |
| 34 | 10 | 11 | .0473460610 | 34 | 15 | 16 | .0434024885 | 35 | 2 | 16 | .0192960621 |
| 34 | 10 | 12 | .0437577168 | 34 | 15 | 17 | .0408955850 | 35 | 2 | 17 | .0181777911 |
| 34 | 10 | 13 | .0406570912 | 34 | 15 | 18 | .0386126928 | 35 | 2 | 18 | .0171623317 |
| 34 | 10 | 14 | .0379425452 | 34 | 15 | 19 | .0365182313 | 35 | 2 | 19 | .0162333093 |
| 34 | 10 | 15 | .0355388131 | 34 | 15 | 20 | .0345831398 | 35 | 2 | 20 | .0153774265 |
| 34 | 10 | 16 | .0333888459 | 34 | 16 | 16 | .0458122142 | 35 | 2 | 21 | .0145837124 |
| 34 | 10 | 17 | .0314485061 | 34 | 16 | 17 | .0431719217 | 35 | 2 | 22 | .0138429669 |
| 34 | 10 | 18 | .0296830063 | 34 | 16 | 18 | .0407672278 | 35 | 2 | 23 | .0131473379 |
| 34 | 10 | 19 | .0280644461 | 34 | 16 | 19 | .0385607619 | 35 | 2 | 24 | .0124899876 |

TABLE I-COVARIANCES OF NORMAL ORDER STATISTICS

| N | I | J | COVARIANCE | N | I | J | COVARIANCE | N | I | J | COVARIANCE |
|---|---|---|---|---|---|---|---|---|---|---|---|
| 35 | 2 | 25 | .0118648148 | 35 | 4 | 23 | .0157582837 | 35 | 6 | 29 | .0131157407 |
| 35 | 2 | 26 | .0112662051 | 35 | 4 | 24 | .0149712070 | 35 | 6 | 30 | .0123653273 |
| 35 | 2 | 27 | .0106887815 | 35 | 4 | 25 | .0142225437 | 35 | 7 | 7 | .0591740181 |
| 35 | 2 | 28 | .0101271231 | 35 | 4 | 26 | .0135056052 | 35 | 7 | 8 | .0526150785 |
| 35 | 2 | 29 | .0095754018 | 35 | 4 | 27 | .0128139734 | 35 | 7 | 9 | .0474158183 |
| 35 | 2 | 30 | .0090268166 | 35 | 4 | 28 | .0121411482 | 35 | 7 | 10 | .0431757550 |
| 35 | 2 | 31 | .0084725788 | 35 | 4 | 29 | .0114801518 | 35 | 7 | 11 | .0396388931 |
| 35 | 2 | 32 | .0078998762 | 35 | 4 | 30 | .0108228441 | 35 | 7 | 12 | .0366335858 |
| 35 | 2 | 33 | .0072869488 | 35 | 4 | 31 | .0101586883 | 35 | 7 | 13 | .0340402059 |
| 35 | 2 | 34 | .0065878045 | 35 | 4 | 32 | .0094723440 | 35 | 7 | 14 | .0317726680 |
| 35 | 3 | 3 | .0973427472 | 35 | 5 | 5 | .0707365754 | 35 | 7 | 15 | .0297673262 |
| 35 | 3 | 4 | .0747356410 | 35 | 5 | 6 | .0601847854 | 35 | 7 | 16 | .0279760183 |
| 35 | 3 | 5 | .0611330457 | 35 | 5 | 7 | .0525260673 | 35 | 7 | 17 | .0263615462 |
| 35 | 3 | 6 | .0519660557 | 35 | 5 | 8 | .0466812330 | 35 | 7 | 18 | .0248946437 |
| 35 | 3 | 7 | .0453231982 | 35 | 5 | 9 | .0420523008 | 35 | 7 | 19 | .0235518887 |
| 35 | 3 | 8 | .0402597720 | 35 | 5 | 10 | .0382801492 | 35 | 7 | 20 | .0223142320 |
| 35 | 3 | 9 | .0362534760 | 35 | 5 | 11 | .0351355259 | 35 | 7 | 21 | .0211659450 |
| 35 | 3 | 10 | .0329911848 | 35 | 5 | 12 | .0324648851 | 35 | 7 | 22 | .0200938529 |
| 35 | 3 | 11 | .0302732893 | 35 | 5 | 13 | .0301613154 | 35 | 7 | 23 | .0190867148 |
| 35 | 3 | 12 | .0279662527 | 35 | 5 | 14 | .0281479451 | 35 | 7 | 24 | .0181346776 |
| 35 | 3 | 13 | .0259771857 | 35 | 5 | 15 | .0263679800 | 35 | 7 | 25 | .0172289016 |
| 35 | 3 | 14 | .0242393537 | 35 | 5 | 16 | .0247784650 | 35 | 7 | 26 | .0163613098 |
| 35 | 3 | 15 | .0227034871 | 35 | 5 | 17 | .0233462375 | 35 | 7 | 27 | .0155241887 |
| 35 | 3 | 16 | .0213323471 | 35 | 5 | 18 | .0220452217 | 35 | 7 | 28 | .0147097004 |
| 35 | 3 | 17 | .0200972032 | 35 | 5 | 19 | .0208545702 | 35 | 7 | 29 | .0139094248 |
| 35 | 3 | 18 | .0189754764 | 35 | 5 | 20 | .0197573475 | 35 | 8 | 8 | .0555650479 |
| 35 | 3 | 19 | .0179491171 | 35 | 5 | 21 | .0187395656 | 35 | 8 | 9 | .0500841772 |
| 35 | 3 | 20 | .0170034629 | 35 | 5 | 22 | .0177894742 | 35 | 8 | 10 | .0456128148 |
| 35 | 3 | 21 | .0161264134 | 35 | 5 | 23 | .0168970414 | 35 | 8 | 11 | .0418818883 |
| 35 | 3 | 22 | .0153078215 | 35 | 5 | 24 | .0160535404 | 35 | 8 | 12 | .0387108830 |
| 35 | 3 | 23 | .0145390287 | 35 | 5 | 25 | .0152511791 | 35 | 8 | 13 | .0359739325 |
| 35 | 3 | 24 | .0138124944 | 35 | 5 | 26 | .0144827801 | 35 | 8 | 14 | .0335804268 |
| 35 | 3 | 25 | .0131214808 | 35 | 5 | 27 | .0137414701 | 35 | 8 | 15 | .0314633526 |
| 35 | 3 | 26 | .0124597763 | 35 | 5 | 28 | .0130202811 | 35 | 8 | 16 | .0295719821 |
| 35 | 3 | 27 | .0118214423 | 35 | 5 | 29 | .0123117057 | 35 | 8 | 17 | .0278671195 |
| 35 | 3 | 28 | .0112005116 | 35 | 5 | 30 | .0116070689 | 35 | 8 | 18 | .0263179154 |
| 35 | 3 | 29 | .0105905460 | 35 | 5 | 31 | .0108951106 | 35 | 8 | 19 | .0248996791 |
| 35 | 3 | 30 | .0099840021 | 35 | 6 | 6 | .0639988701 | 35 | 8 | 20 | .0235923455 |
| 35 | 3 | 31 | .0093711713 | 35 | 6 | 7 | .0558727162 | 35 | 8 | 21 | .0223793570 |
| 35 | 3 | 32 | .0087379019 | 35 | 6 | 8 | .0496675391 | 35 | 8 | 22 | .0212468010 |
| 35 | 3 | 33 | .0080601140 | 35 | 6 | 9 | .0447509837 | 35 | 8 | 23 | .0201827669 |
| 35 | 4 | 4 | .0807670197 | 35 | 6 | 10 | .0407429741 | 35 | 8 | 24 | .0191768670 |
| 35 | 4 | 5 | .0661126589 | 35 | 6 | 11 | .0374007193 | 35 | 8 | 25 | .0182198061 |
| 35 | 4 | 6 | .0562253542 | 35 | 6 | 12 | .0345615167 | 35 | 8 | 26 | .0173030552 |
| 35 | 4 | 7 | .0490545778 | 35 | 6 | 13 | .0321120260 | 35 | 8 | 27 | .0164185308 |
| 35 | 4 | 8 | .0435853569 | 35 | 6 | 14 | .0299707193 | 35 | 8 | 28 | .0155579515 |
| 35 | 4 | 9 | .0392558957 | 35 | 6 | 15 | .0280773396 | 35 | 9 | 9 | .0527914954 |
| 35 | 4 | 10 | .0357290898 | 35 | 6 | 16 | .0263862996 | 35 | 9 | 10 | .0480864320 |
| 35 | 4 | 11 | .0327898888 | 35 | 6 | 17 | .0248624000 | 35 | 9 | 11 | .0441592923 |
| 35 | 4 | 12 | .0302943400 | 35 | 6 | 18 | .0234779617 | 35 | 9 | 12 | .0408206604 |
| 35 | 4 | 13 | .0281422589 | 35 | 6 | 19 | .0222108437 | 35 | 9 | 13 | .0379384012 |
| 35 | 4 | 14 | .0262616389 | 35 | 6 | 20 | .0210430414 | 35 | 9 | 14 | .0354173529 |
| 35 | 4 | 15 | .0245992997 | 35 | 6 | 21 | .0199596823 | 35 | 9 | 15 | .0331871091 |
| 35 | 4 | 16 | .0231150344 | 35 | 6 | 22 | .0189482859 | 35 | 9 | 16 | .0311943534 |
| 35 | 4 | 17 | .0217778121 | 35 | 6 | 23 | .0179981993 | 35 | 9 | 17 | .0293978771 |
| 35 | 4 | 18 | .0205632368 | 35 | 6 | 24 | .0171001647 | 35 | 9 | 18 | .0277652456 |
| 35 | 4 | 19 | .0194518011 | 35 | 6 | 25 | .0162459175 | 35 | 9 | 19 | .0262705054 |
| 35 | 4 | 20 | .0184276581 | 35 | 6 | 26 | .0154277704 | 35 | 9 | 20 | .0248925549 |
| 35 | 4 | 21 | .0174777343 | 35 | 6 | 27 | .0146383302 | 35 | 9 | 21 | .0236139563 |
| 35 | 4 | 22 | .0165910664 | 35 | 6 | 28 | .0138702781 | 35 | 9 | 22 | .0224200522 |

TABLE I-COVARIANCES OF NORMAL ORDER STATISTICS

| N | I | J | COVARIANCE | N | I | J | COVARIANCE | N | I | J | COVARIANCE |
|---|---|---|---|---|---|---|---|---|---|---|---|
| 35 | 9 | 23 | .0212982835 | 35 | 13 | 23 | .0261694204 | 36 | 1 | 35 | .0054472623 |
| 35 | 9 | 24 | .0202377169 | 35 | 14 | 14 | .0456853962 | 36 | 1 | 36 | .0046842434 |
| 35 | 9 | 25 | .0192286733 | 35 | 14 | 15 | .0428260592 | 36 | 2 | 2 | .1293292667 |
| 35 | 9 | 26 | .0182621958 | 35 | 14 | 16 | .0402687944 | 36 | 2 | 3 | .0875594068 |
| 35 | 9 | 27 | .0173295686 | 35 | 14 | 17 | .0379614622 | 36 | 2 | 4 | .0671282756 |
| 35 | 10 | 10 | .0506308294 | 35 | 14 | 18 | .0358629607 | 36 | 2 | 5 | .0548655006 |
| 35 | 10 | 11 | .0465021304 | 35 | 14 | 19 | .0339403461 | 36 | 2 | 6 | .0466157860 |
| 35 | 10 | 12 | .0429910758 | 35 | 14 | 20 | .0321667936 | 36 | 2 | 7 | .0406455761 |
| 35 | 10 | 13 | .0399591587 | 35 | 14 | 21 | .0305201114 | 36 | 2 | 8 | .0360998161 |
| 35 | 10 | 14 | .0373065846 | 35 | 14 | 22 | .0289816299 | 36 | 2 | 9 | .0325064876 |
| 35 | 10 | 15 | .0349594896 | 35 | 15 | 15 | .0450791703 | 36 | 2 | 10 | .0295829532 |
| 35 | 10 | 16 | .0328619207 | 35 | 15 | 16 | .0423905815 | 36 | 2 | 11 | .0271492202 |
| 35 | 10 | 17 | .0309706228 | 35 | 15 | 17 | .0399642119 | 36 | 2 | 12 | .0250849790 |
| 35 | 10 | 18 | .0292515455 | 35 | 15 | 18 | .0377569888 | 36 | 2 | 13 | .0233066068 |
| 35 | 10 | 19 | .0276774306 | 35 | 15 | 19 | .0357343757 | 36 | 2 | 14 | .0217540710 |
| 35 | 10 | 20 | .0262260988 | 35 | 15 | 20 | .0338682364 | 36 | 2 | 15 | .0203830832 |
| 35 | 10 | 21 | .0248792013 | 35 | 15 | 21 | .0321352810 | 36 | 2 | 16 | .0191601960 |
| 35 | 10 | 22 | .0236213162 | 35 | 16 | 16 | .0446537883 | 36 | 2 | 17 | .0180596268 |
| 35 | 10 | 23 | .0224392935 | 35 | 16 | 17 | .0421034276 | 36 | 2 | 18 | .0170611359 |
| 35 | 10 | 24 | .0213216800 | 35 | 16 | 18 | .0397831452 | 36 | 2 | 19 | .0161485689 |
| 35 | 10 | 25 | .0202582195 | 35 | 16 | 19 | .0376567295 | 36 | 2 | 20 | .0153088312 |
| 35 | 10 | 26 | .0192394496 | 35 | 16 | 20 | .0356946694 | 36 | 2 | 21 | .0145311473 |
| 35 | 11 | 11 | .0489225685 | 35 | 17 | 17 | .0444063808 | 36 | 2 | 22 | .0138065153 |
| 35 | 11 | 12 | .0452341917 | 35 | 17 | 18 | .0419641318 | 36 | 2 | 23 | .0131272943 |
| 35 | 11 | 13 | .0420483082 | 35 | 17 | 19 | .0397256305 | 36 | 2 | 24 | .0124868852 |
| 35 | 11 | 14 | .0392603836 | 35 | 18 | 18 | .0443269162 | 36 | 2 | 25 | .0118794712 |
| 35 | 11 | 15 | .0367930163 | 36 | 1 | 1 | .2342457965 | 36 | 2 | 26 | .0112997971 |
| 35 | 11 | 16 | .0345875552 | 36 | 1 | 2 | .1122047148 | 36 | 2 | 27 | .0107429613 |
| 35 | 11 | 17 | .0325986471 | 36 | 1 | 3 | .0757309502 | 36 | 2 | 28 | .0102041979 |
| 35 | 11 | 18 | .0307905765 | 36 | 1 | 4 | .0579686713 | 36 | 2 | 29 | .0096786283 |
| 35 | 11 | 19 | .0291347351 | 36 | 1 | 5 | .0473343396 | 36 | 2 | 30 | .0091609319 |
| 35 | 11 | 20 | .0276078323 | 36 | 1 | 6 | .0401917018 | 36 | 2 | 31 | .0086448088 |
| 35 | 11 | 21 | .0261906132 | 36 | 1 | 7 | .0350285207 | 36 | 2 | 32 | .0081220082 |
| 35 | 11 | 22 | .0248669128 | 36 | 1 | 8 | .0311005347 | 36 | 2 | 33 | .0075803906 |
| 35 | 11 | 23 | .0236229282 | 36 | 1 | 9 | .0279975511 | 36 | 2 | 34 | .0069992026 |
| 35 | 11 | 24 | .0224466301 | 36 | 1 | 10 | .0254742593 | 36 | 2 | 35 | .0063343950 |
| 35 | 11 | 25 | .0213271960 | 36 | 1 | 11 | .0233745923 | 36 | 3 | 3 | .0963024233 |
| 35 | 12 | 12 | .0475611671 | 36 | 1 | 12 | .0215943182 | 36 | 3 | 4 | .0739229190 |
| 35 | 12 | 13 | .0442171913 | 36 | 1 | 13 | .0200610379 | 36 | 3 | 5 | .0604645353 |
| 35 | 12 | 14 | .0412903827 | 36 | 1 | 14 | .0187228059 | 36 | 3 | 6 | .0513989364 |
| 35 | 12 | 15 | .0386997035 | 36 | 1 | 15 | .0175413192 | 36 | 3 | 7 | .0448323714 |
| 35 | 12 | 16 | .0363837267 | 36 | 1 | 16 | .0164876630 | 36 | 3 | 8 | .0398291869 |
| 35 | 12 | 17 | .0342949359 | 36 | 1 | 17 | .0155395574 | 36 | 3 | 9 | .0358722183 |
| 35 | 12 | 18 | .0323958970 | 36 | 1 | 18 | .0146795186 | 36 | 3 | 10 | .0326514913 |
| 35 | 12 | 19 | .0306566167 | 36 | 1 | 19 | .0138935957 | 36 | 3 | 11 | .0299694481 |
| 35 | 12 | 20 | .0290526807 | 36 | 1 | 20 | .0131704822 | 36 | 3 | 12 | .0276939539 |
| 35 | 12 | 21 | .0275639127 | 36 | 1 | 21 | .0125008776 | 36 | 3 | 13 | .0257331168 |
| 35 | 12 | 22 | .0261733772 | 36 | 1 | 22 | .0118770148 | 36 | 3 | 14 | .0240209371 |
| 35 | 12 | 23 | .0248665863 | 36 | 1 | 23 | .0112923028 | 36 | 3 | 15 | .0225087045 |
| 35 | 12 | 24 | .0236308570 | 36 | 1 | 24 | .0107410510 | 36 | 3 | 16 | .0211596193 |
| 35 | 13 | 13 | .0464963931 | 36 | 1 | 25 | .0102182439 | 36 | 3 | 17 | .0199453072 |
| 35 | 13 | 14 | .0434235929 | 36 | 1 | 26 | .0097193490 | 36 | 3 | 18 | .0188434889 |
| 35 | 13 | 15 | .0407031585 | 36 | 1 | 27 | .0092401402 | 36 | 3 | 19 | .0178363777 |
| 35 | 13 | 16 | .0382707717 | 36 | 1 | 28 | .0087765147 | 36 | 3 | 20 | .0169095497 |
| 35 | 13 | 17 | .0360766613 | 36 | 1 | 29 | .0083242762 | 36 | 3 | 21 | .0160511313 |
| 35 | 13 | 18 | .0340816046 | 36 | 1 | 30 | .0078788413 | 36 | 3 | 22 | .0152512007 |
| 35 | 13 | 19 | .0322541715 | 36 | 1 | 31 | .0074347805 | 36 | 3 | 23 | .0145013393 |
| 35 | 13 | 20 | .0305687794 | 36 | 1 | 32 | .0069850004 | 36 | 3 | 24 | .0137942811 |
| 35 | 13 | 21 | .0290042817 | 36 | 1 | 33 | .0065190624 | 36 | 3 | 25 | .0131236195 |
| 35 | 13 | 22 | .0275429026 | 36 | 1 | 34 | .0060191103 | 36 | 3 | 26 | .0124835482 |

TABLE I-COVARIANCES OF NORMAL ORDER STATISTICS

| N | I | J | COVARIANCE | N | I | J | COVARIANCE | N | I | J | COVARIANCE |
|---|---|---|---|---|---|---|---|---|---|---|---|
| 36 | 3 | 27 | .0118686415 | 36 | 5 | 27 | .0137688360 | 36 | 8 | 12 | .0381455724 |
| 36 | 3 | 28 | .0112736502 | 36 | 5 | 28 | .0130791894 | 36 | 8 | 13 | .0354602728 |
| 36 | 3 | 29 | .0106932196 | 36 | 5 | 29 | .0124063045 | 36 | 8 | 14 | .0331133421 |
| 36 | 3 | 30 | .0101214656 | 36 | 5 | 30 | .0117433450 | 36 | 8 | 15 | .0310388220 |
| 36 | 3 | 31 | .0095514009 | 36 | 5 | 31 | .0110823579 | 36 | 8 | 16 | .0291868102 |
| 36 | 3 | 32 | .0089739337 | 36 | 5 | 32 | .0104127765 | 36 | 8 | 17 | .0275187774 |
| 36 | 3 | 33 | .0083756646 | 36 | 6 | 6 | .0631205067 | 36 | 8 | 18 | .0260044256 |
| 36 | 3 | 34 | .0077336485 | 36 | 6 | 7 | .0551092409 | 36 | 8 | 19 | .0246195280 |
| 36 | 4 | 4 | .0798140795 | 36 | 6 | 8 | .0489946235 | 36 | 8 | 20 | .0233444227 |
| 36 | 4 | 5 | .0653276745 | 36 | 6 | 9 | .0441520330 | 36 | 8 | 21 | .0221629353 |
| 36 | 4 | 6 | .0555584583 | 36 | 6 | 10 | .0402061505 | 36 | 8 | 22 | .0210615495 |
| 36 | 4 | 7 | .0484764921 | 36 | 6 | 11 | .0369172880 | 36 | 8 | 23 | .0200287658 |
| 36 | 4 | 8 | .0430773349 | 36 | 6 | 12 | .0341248642 | 36 | 8 | 24 | .0190546327 |
| 36 | 4 | 9 | .0388051774 | 36 | 6 | 13 | .0317170482 | 36 | 8 | 25 | .0181303252 |
| 36 | 4 | 10 | .0353265905 | 36 | 6 | 14 | .0296134233 | 36 | 8 | 26 | .0172478232 |
| 36 | 4 | 11 | .0324289181 | 36 | 6 | 15 | .0277545761 | 36 | 8 | 27 | .0163997860 |
| 36 | 4 | 12 | .0299698439 | 36 | 6 | 16 | .0260955772 | 36 | 8 | 28 | .0155791673 |
| 36 | 4 | 13 | .0278503466 | 36 | 6 | 17 | .0246017552 | 36 | 8 | 29 | .0147784454 |
| 36 | 4 | 14 | .0259992763 | 36 | 6 | 18 | .0232458716 | 36 | 9 | 9 | .0519245769 |
| 36 | 4 | 15 | .0243641035 | 36 | 6 | 19 | .0220061716 | 36 | 9 | 10 | .0473060365 |
| 36 | 4 | 16 | .0229051327 | 36 | 6 | 20 | .0208650006 | 36 | 9 | 11 | .0434531040 |
| 36 | 4 | 17 | .0215917478 | 36 | 6 | 21 | .0198078102 | 36 | 9 | 12 | .0401793256 |
| 36 | 4 | 18 | .0203999002 | 36 | 6 | 22 | .0188224280 | 36 | 9 | 13 | .0373546738 |
| 36 | 4 | 19 | .0193103824 | 36 | 6 | 23 | .0178985087 | 36 | 9 | 14 | .0348855384 |
| 36 | 4 | 20 | .0183076152 | 36 | 6 | 24 | .0170271390 | 36 | 9 | 15 | .0327026907 |
| 36 | 4 | 21 | .0173787785 | 36 | 6 | 25 | .0162005091 | 36 | 9 | 16 | .0307537346 |
| 36 | 4 | 22 | .0165131696 | 36 | 6 | 26 | .0154115192 | 36 | 9 | 17 | .0289981986 |
| 36 | 4 | 23 | .0157017011 | 36 | 6 | 27 | .0146534156 | 36 | 9 | 18 | .0274042517 |
| 36 | 4 | 24 | .0149365019 | 36 | 6 | 28 | .0139196754 | 36 | 9 | 19 | .0259464524 |
| 36 | 4 | 25 | .0142106190 | 36 | 6 | 29 | .0132038064 | 36 | 9 | 20 | .0246041596 |
| 36 | 4 | 26 | .0135177887 | 36 | 6 | 30 | .0124986176 | 36 | 9 | 21 | .0233603756 |
| 36 | 4 | 27 | .0128521860 | 36 | 6 | 31 | .0117954051 | 36 | 9 | 22 | .0222008829 |
| 36 | 4 | 28 | .0122081258 | 36 | 7 | 7 | .0583097315 | 36 | 9 | 23 | .0211135540 |
| 36 | 4 | 29 | .0115797745 | 36 | 7 | 8 | .0518523304 | 36 | 9 | 24 | .0200878547 |
| 36 | 4 | 30 | .0109607804 | 36 | 7 | 9 | .0467359547 | 36 | 9 | 25 | .0191145766 |
| 36 | 4 | 31 | .0103435959 | 36 | 7 | 10 | .0425654564 | 36 | 9 | 26 | .0181854752 |
| 36 | 4 | 32 | .0097183530 | 36 | 7 | 11 | .0390883221 | 36 | 9 | 27 | .0172927334 |
| 36 | 4 | 33 | .0090705663 | 36 | 7 | 12 | .0361352873 | 36 | 9 | 28 | .0164286909 |
| 36 | 5 | 5 | .0698319302 | 36 | 7 | 13 | .0335884194 | 36 | 10 | 10 | .0497521889 |
| 36 | 5 | 6 | .0594151926 | 36 | 7 | 14 | .0313628793 | 36 | 10 | 11 | .0457059881 |
| 36 | 5 | 7 | .0518580203 | 36 | 7 | 15 | .0293959653 | 36 | 10 | 12 | .0422670001 |
| 36 | 5 | 8 | .0460932509 | 36 | 7 | 16 | .0276402523 | 36 | 10 | 13 | .0392990604 |
| 36 | 5 | 9 | .0415297470 | 36 | 7 | 17 | .0260591375 | 36 | 10 | 14 | .0367040989 |
| 36 | 5 | 10 | .0378125932 | 36 | 7 | 18 | .0246238541 | 36 | 10 | 15 | .0344095600 |
| 36 | 5 | 11 | .0347152788 | 36 | 7 | 19 | .0233114103 | 36 | 10 | 16 | .0323605092 |
| 36 | 5 | 12 | .0320861317 | 36 | 7 | 20 | .0221031354 | 36 | 10 | 17 | .0305145051 |
| 36 | 5 | 13 | .0298195728 | 36 | 7 | 21 | .0209836374 | 36 | 10 | 18 | .0288381674 |
| 36 | 5 | 14 | .0278397111 | 36 | 7 | 22 | .0199400665 | 36 | 10 | 19 | .0273048146 |
| 36 | 5 | 15 | .0260904985 | 36 | 7 | 23 | .0189615548 | 36 | 10 | 20 | .0258927886 |
| 36 | 5 | 16 | .0245295650 | 36 | 7 | 24 | .0180386921 | 36 | 10 | 21 | .0245842253 |
| 36 | 5 | 17 | .0231242192 | 36 | 7 | 25 | .0171630982 | 36 | 10 | 22 | .0233641449 |
| 36 | 5 | 18 | .0218487749 | 36 | 7 | 26 | .0163271966 | 36 | 10 | 23 | .0222198232 |
| 36 | 5 | 19 | .0206827199 | 36 | 7 | 27 | .0155239747 | 36 | 10 | 24 | .0211402884 |
| 36 | 5 | 20 | .0196094257 | 36 | 7 | 28 | .0147465874 | 36 | 10 | 25 | .0201158612 |
| 36 | 5 | 21 | .0186152031 | 36 | 7 | 29 | .0139880506 | 36 | 10 | 26 | .0191378255 |
| 36 | 5 | 22 | .0176885959 | 36 | 7 | 30 | .0132407713 | 36 | 10 | 27 | .0181979781 |
| 36 | 5 | 23 | .0168198651 | 36 | 8 | 8 | .0547052591 | 36 | 11 | 11 | .0480320145 |
| 36 | 5 | 24 | .0160006013 | 36 | 8 | 9 | .0493167827 | 36 | 11 | 12 | .0444228050 |
| 36 | 5 | 25 | .0152233941 | 36 | 8 | 10 | .0449229121 | 36 | 11 | 13 | .0413071024 |
| 36 | 5 | 26 | .0144815475 | 36 | 8 | 11 | .0412584605 | 36 | 11 | 14 | .0385822781 |

TABLE I-COVARIANCES OF NORMAL ORDER STATISTICS

| N | I | J | COVARIANCE | N | I | J | COVARIANCE | N | I | J | COVARIANCE |
|---|---|---|---|---|---|---|---|---|---|---|---|
| 36 | 11 | 15 | .0361723744 | 36 | 16 | 20 | .0349228584 | 37 | 2 | 17 | .0179419846 |
| 36 | 11 | 16 | .0340198635 | 36 | 16 | 21 | .0331664133 | 37 | 2 | 18 | .0169592655 |
| 36 | 11 | 17 | .0320802917 | 36 | 17 | 17 | .0432763689 | 37 | 2 | 19 | .0160619260 |
| 36 | 11 | 18 | .0303186885 | 36 | 17 | 18 | .0409196483 | 37 | 2 | 20 | .0152370525 |
| 36 | 11 | 19 | .0287070853 | 36 | 17 | 19 | .0387619577 | 37 | 2 | 21 | .0144740405 |
| 36 | 11 | 20 | .0272227548 | 36 | 17 | 20 | .0367733205 | 37 | 2 | 22 | .0137640575 |
| 36 | 11 | 21 | .0258469454 | 36 | 18 | 18 | .0431284992 | 37 | 2 | 23 | .0130996380 |
| 36 | 11 | 22 | .0245639616 | 36 | 18 | 19 | .0408571014 | 37 | 2 | 24 | .0124743731 |
| 36 | 11 | 23 | .0233604736 | 37 | 1 | 1 | .2325894919 | 37 | 2 | 25 | .0118826641 |
| 36 | 11 | 24 | .0222250055 | 37 | 1 | 2 | .1113172673 | 37 | 2 | 26 | .0113195200 |
| 36 | 11 | 25 | .0211474203 | 37 | 1 | 3 | .0751065517 | 37 | 2 | 27 | .0107803778 |
| 36 | 11 | 26 | .0201183990 | 37 | 1 | 4 | .0574823405 | 37 | 2 | 28 | .0102609237 |
| 36 | 12 | 12 | .0466484258 | 37 | 1 | 5 | .0469353666 | 37 | 2 | 29 | .0097569001 |
| 36 | 12 | 13 | .0433819465 | 37 | 1 | 6 | .0398541922 | 37 | 2 | 30 | .0092638876 |
| 36 | 12 | 14 | .0405247451 | 37 | 1 | 7 | .0347373467 | 37 | 2 | 31 | .0087770038 |
| 36 | 12 | 15 | .0379973817 | 37 | 1 | 8 | .0308460423 | 37 | 2 | 32 | .0082903842 |
| 36 | 12 | 16 | .0357396732 | 37 | 1 | 9 | .0277731853 | 37 | 2 | 33 | .0077962542 |
| 36 | 12 | 17 | .0337051010 | 37 | 1 | 10 | .0252753627 | 37 | 2 | 34 | .0072830836 |
| 36 | 12 | 18 | .0318570548 | 37 | 1 | 11 | .0231977423 | 37 | 2 | 35 | .0067310417 |
| 36 | 12 | 19 | .0301662396 | 37 | 1 | 12 | .0214369353 | 37 | 2 | 36 | .0060978788 |
| 36 | 12 | 20 | .0286088464 | 37 | 1 | 13 | .0199211423 | 37 | 3 | 3 | .0953087837 |
| 36 | 12 | 21 | .0271652401 | 37 | 1 | 14 | .0185988601 | 37 | 3 | 4 | .0731463732 |
| 36 | 12 | 22 | .0258190080 | 37 | 1 | 15 | .0174321222 | 37 | 3 | 5 | .0598253994 |
| 36 | 12 | 23 | .0245562217 | 37 | 1 | 16 | .0163922793 | 37 | 3 | 6 | .0508563122 |
| 36 | 12 | 24 | .0233648041 | 37 | 1 | 17 | .0154572661 | 37 | 3 | 7 | .0443622929 |
| 36 | 12 | 25 | .0222340556 | 37 | 1 | 18 | .0146097794 | 37 | 3 | 8 | .0394163282 |
| 36 | 13 | 13 | .0455489139 | 37 | 1 | 19 | .0138360265 | 37 | 3 | 9 | .0355061557 |
| 36 | 13 | 14 | .0425542165 | 37 | 1 | 20 | .0131248440 | 37 | 3 | 10 | .0323248063 |
| 36 | 13 | 15 | .0399048384 | 37 | 1 | 21 | .0124670669 | 37 | 3 | 11 | .0296766781 |
| 36 | 13 | 16 | .0375378469 | 37 | 1 | 22 | .0118550647 | 37 | 3 | 12 | .0274309709 |
| 36 | 13 | 17 | .0354045805 | 37 | 1 | 23 | .0112823897 | 37 | 3 | 13 | .0254967410 |
| 36 | 13 | 18 | .0334667283 | 37 | 1 | 24 | .0107435086 | 37 | 3 | 14 | .0238086861 |
| 36 | 13 | 19 | .0316936277 | 37 | 1 | 25 | .0102335899 | 37 | 3 | 15 | .0223186252 |
| 36 | 13 | 20 | .0300603631 | 37 | 1 | 26 | .0097483248 | 37 | 3 | 16 | .0209901718 |
| 36 | 13 | 21 | .0285463988 | 37 | 1 | 27 | .0092837711 | 37 | 3 | 17 | .0197952841 |
| 36 | 13 | 22 | .0271345571 | 37 | 1 | 28 | .0088362056 | 37 | 3 | 18 | .0187119598 |
| 36 | 13 | 23 | .0258102276 | 37 | 1 | 29 | .0084019627 | 37 | 3 | 19 | .0177226517 |
| 36 | 13 | 24 | .0245607369 | 37 | 1 | 30 | .0079772388 | 37 | 3 | 20 | .0168131518 |
| 36 | 14 | 14 | .0447024846 | 37 | 1 | 31 | .0075578183 | 37 | 3 | 21 | .0159717868 |
| 36 | 14 | 15 | .0419221925 | 37 | 1 | 32 | .0071386400 | 37 | 3 | 22 | .0151888282 |
| 36 | 14 | 16 | .0394376392 | 37 | 1 | 33 | .0067130172 | 37 | 3 | 23 | .0144560535 |
| 36 | 14 | 17 | .0371979128 | 37 | 1 | 34 | .0062710221 | 37 | 3 | 24 | .0137664112 |
| 36 | 14 | 18 | .0351629183 | 37 | 1 | 35 | .0057955696 | 37 | 3 | 25 | .0131137498 |
| 36 | 14 | 19 | .0333005608 | 37 | 1 | 36 | .0052502833 | 37 | 3 | 26 | .0124925748 |
| 36 | 14 | 20 | .0315847594 | 37 | 1 | 37 | .0045203833 | 37 | 3 | 27 | .0118978359 |
| 36 | 14 | 21 | .0299940066 | 37 | 2 | 2 | .1281641313 | 37 | 3 | 28 | .0113247593 |
| 36 | 14 | 22 | .0285102911 | 37 | 2 | 3 | .0867376106 | 37 | 3 | 29 | .0107686758 |
| 36 | 14 | 23 | .0271182678 | 37 | 2 | 4 | .0664870826 | 37 | 3 | 30 | .0102247440 |
| 36 | 15 | 15 | .0440603469 | 37 | 2 | 5 | .0543385506 | 37 | 3 | 31 | .0096875512 |
| 36 | 15 | 16 | .0414509331 | 37 | 2 | 6 | .0461691060 | 37 | 3 | 32 | .0091506018 |
| 36 | 15 | 17 | .0390980252 | 37 | 2 | 7 | .0402592896 | 37 | 3 | 33 | .0086053498 |
| 36 | 15 | 18 | .0369596587 | 37 | 2 | 8 | .0357612264 | 37 | 3 | 34 | .0080390701 |
| 36 | 15 | 19 | .0350022322 | 37 | 2 | 9 | .0322069669 | 37 | 3 | 35 | .0074298617 |
| 36 | 15 | 20 | .0331984293 | 37 | 2 | 10 | .0293163667 | 37 | 4 | 4 | .0789055378 |
| 36 | 15 | 21 | .0315257059 | 37 | 2 | 11 | .0269110543 | 37 | 4 | 5 | .0645788804 |
| 36 | 15 | 22 | .0299651649 | 37 | 2 | 12 | .0248718271 | 37 | 4 | 6 | .0549218700 |
| 36 | 16 | 16 | .0435861640 | 37 | 2 | 13 | .0231158488 | 37 | 4 | 7 | .0479242049 |
| 36 | 16 | 17 | .0411165405 | 37 | 2 | 14 | .0215836655 | 37 | 4 | 8 | .0425914833 |
| 36 | 16 | 18 | .0388718054 | 37 | 2 | 15 | .0202314290 | 37 | 4 | 9 | .0383735990 |
| 36 | 16 | 19 | .0368167816 | 37 | 2 | 16 | .0190260366 | 37 | 4 | 10 | .0349406223 |

TABLE I-COVARIANCES OF NORMAL ORDER STATISTICS

| N | I | J | COVARIANCE | N | I | J | COVARIANCE | N | I | J | COVARIANCE |
|---|---|---|---|---|---|---|---|---|---|---|---|
| 37 | 4 | 11 | .0320821752 | 37 | 6 | 13 | .0313383849 | 37 | 8 | 23 | .0198709462 |
| 37 | 4 | 12 | .0296574967 | 37 | 6 | 14 | .0292700590 | 37 | 8 | 24 | .0189251269 |
| 37 | 4 | 13 | .0275686700 | 37 | 6 | 15 | .0274434908 | 37 | 8 | 25 | .0180297883 |
| 37 | 4 | 14 | .0257453565 | 37 | 6 | 16 | .0258143695 | 37 | 8 | 26 | .0171772974 |
| 37 | 4 | 15 | .0241356413 | 37 | 6 | 17 | .0243485157 | 37 | 8 | 27 | .0163607017 |
| 37 | 4 | 16 | .0227003086 | 37 | 6 | 18 | .0230190944 | 37 | 8 | 28 | .0155737179 |
| 37 | 4 | 17 | .0214091296 | 37 | 6 | 19 | .0218046990 | 37 | 8 | 29 | .0148100971 |
| 37 | 4 | 18 | .0202383787 | 37 | 6 | 20 | .0206879910 | 37 | 8 | 30 | .0140629056 |
| 37 | 4 | 19 | .0191691246 | 37 | 6 | 21 | .0196547166 | 37 | 9 | 9 | .0511062346 |
| 37 | 4 | 20 | .0181860301 | 37 | 6 | 22 | .0186929811 | 37 | 9 | 10 | .0465686825 |
| 37 | 4 | 21 | .0172764946 | 37 | 6 | 23 | .0177926981 | 37 | 9 | 11 | .0427851175 |
| 37 | 4 | 22 | .0164300313 | 37 | 6 | 24 | .0169452039 | 37 | 9 | 12 | .0395718736 |
| 37 | 4 | 23 | .0156377868 | 37 | 6 | 25 | .0161430010 | 37 | 9 | 13 | .0368009049 |
| 37 | 4 | 24 | .0148921490 | 37 | 6 | 26 | .0153794531 | 37 | 9 | 14 | .0343800562 |
| 37 | 4 | 25 | .0141864403 | 37 | 6 | 27 | .0146483666 | 37 | 9 | 15 | .0322412007 |
| 37 | 4 | 26 | .0135147031 | 37 | 6 | 28 | .0139437550 | 37 | 9 | 16 | .0303328017 |
| 37 | 4 | 27 | .0128715140 | 37 | 6 | 29 | .0132598675 | 37 | 9 | 17 | .0286150755 |
| 37 | 4 | 28 | .0122517499 | 37 | 6 | 30 | .0125908890 | 37 | 9 | 18 | .0270567511 |
| 37 | 4 | 29 | .0116503397 | 37 | 6 | 31 | .0119301735 | 37 | 9 | 19 | .0256328510 |
| 37 | 4 | 30 | .0110620295 | 37 | 6 | 32 | .0112696670 | 37 | 9 | 20 | .0243231369 |
| 37 | 4 | 31 | .0104809931 | 37 | 7 | 7 | .0574895679 | 37 | 9 | 21 | .0231109860 |
| 37 | 4 | 32 | .0099001825 | 37 | 7 | 8 | .0511279709 | 37 | 9 | 22 | .0219825644 |
| 37 | 4 | 33 | .0093103424 | 37 | 7 | 9 | .0460897208 | 37 | 9 | 23 | .0209261463 |
| 37 | 4 | 34 | .0086977604 | 37 | 7 | 10 | .0419847221 | 37 | 9 | 24 | .0199315550 |
| 37 | 5 | 5 | .0689710487 | 37 | 7 | 11 | .0385637567 | 37 | 9 | 25 | .0189898939 |
| 37 | 5 | 6 | .0586823914 | 37 | 7 | 12 | .0356598125 | 37 | 9 | 26 | .0180933677 |
| 37 | 5 | 7 | .0512214256 | 37 | 7 | 13 | .0331565575 | 37 | 9 | 27 | .0172348342 |
| 37 | 5 | 8 | .0455324310 | 37 | 7 | 14 | .0309703284 | 37 | 9 | 28 | .0164074388 |
| 37 | 5 | 9 | .0410307791 | 37 | 7 | 15 | .0290393120 | 37 | 9 | 29 | .0156044824 |
| 37 | 5 | 10 | .0373655519 | 37 | 7 | 16 | .0273167720 | 37 | 10 | 10 | .0489221765 |
| 37 | 5 | 11 | .0343128424 | 37 | 7 | 17 | .0257666571 | 37 | 10 | 11 | .0449532007 |
| 37 | 5 | 12 | .0317227563 | 37 | 7 | 18 | .0243606598 | 37 | 10 | 12 | .0415815989 |
| 37 | 5 | 13 | .0294909764 | 37 | 7 | 19 | .0230761858 | 37 | 10 | 13 | .0386734059 |
| 37 | 5 | 14 | .0275425389 | 37 | 7 | 20 | .0218949159 | 37 | 10 | 14 | .0361321685 |
| 37 | 5 | 15 | .0258220988 | 37 | 7 | 21 | .0208017636 | 37 | 10 | 15 | .0338865558 |
| 37 | 5 | 16 | .0242878361 | 37 | 7 | 22 | .0197841452 | 37 | 10 | 16 | .0318825854 |
| 37 | 5 | 17 | .0229075011 | 37 | 7 | 23 | .0188314717 | 37 | 10 | 17 | .0300785722 |
| 37 | 5 | 18 | .0216557700 | 37 | 7 | 24 | .0179346868 | 37 | 10 | 18 | .0284417508 |
| 37 | 5 | 19 | .0205124348 | 37 | 7 | 25 | .0170858154 | 37 | 10 | 19 | .0269459574 |
| 37 | 5 | 20 | .0194611368 | 37 | 7 | 26 | .0162776883 | 37 | 10 | 20 | .0255699972 |
| 37 | 5 | 21 | .0184884461 | 37 | 7 | 27 | .0155037754 | 37 | 10 | 21 | .0242964454 |
| 37 | 5 | 22 | .0175831658 | 37 | 7 | 28 | .0147578778 | 37 | 10 | 22 | .0231107303 |
| 37 | 5 | 23 | .0167358131 | 37 | 7 | 29 | .0140338697 | 37 | 10 | 23 | .0220004945 |
| 37 | 5 | 24 | .0159382380 | 37 | 7 | 30 | .0133255825 | 37 | 10 | 24 | .0209551322 |
| 37 | 5 | 25 | .0151833155 | 37 | 7 | 31 | .0126260974 | 37 | 10 | 25 | .0199653637 |
| 37 | 5 | 26 | .0144646867 | 37 | 8 | 8 | .0538919001 | 37 | 10 | 26 | .0190229869 |
| 37 | 5 | 27 | .0137765697 | 37 | 8 | 9 | .0485902358 | 37 | 10 | 27 | .0181205922 |
| 37 | 5 | 28 | .0131135431 | 37 | 8 | 10 | .0442690878 | 37 | 10 | 28 | .0172509028 |
| 37 | 5 | 29 | .0124701727 | 37 | 8 | 11 | .0406669381 | 37 | 11 | 11 | .0471928937 |
| 37 | 5 | 30 | .0118406936 | 37 | 8 | 12 | .0376084397 | 37 | 11 | 12 | .0436575761 |
| 37 | 5 | 31 | .0112188871 | 37 | 8 | 13 | .0349713998 | 37 | 11 | 13 | .0406073020 |
| 37 | 5 | 32 | .0105974343 | 37 | 8 | 14 | .0326679067 | 37 | 11 | 14 | .0379412268 |
| 37 | 5 | 33 | .0099663450 | 37 | 8 | 15 | .0306329904 | 37 | 11 | 15 | .0355847421 |
| 37 | 6 | 6 | .0622858512 | 37 | 8 | 16 | .0288175182 | 37 | 11 | 16 | .0334813662 |
| 37 | 6 | 7 | .0543832540 | 37 | 8 | 17 | .0271835806 | 37 | 11 | 17 | .0315874765 |
| 37 | 6 | 8 | .0483542078 | 37 | 8 | 18 | .0257013933 | 37 | 11 | 18 | .0298687833 |
| 37 | 6 | 9 | .0435814326 | 37 | 8 | 19 | .0243471611 | 37 | 11 | 19 | .0282979002 |
| 37 | 6 | 10 | .0396941225 | 37 | 8 | 20 | .0231015842 | 37 | 11 | 20 | .0268526154 |
| 37 | 6 | 11 | .0364555311 | 37 | 8 | 21 | .0219488113 | 37 | 11 | 21 | .0255146384 |
| 37 | 6 | 12 | .0337070865 | 37 | 8 | 22 | .0208756504 | 37 | 11 | 22 | .0242686877 |

TABLE I-COVARIANCES OF NORMAL ORDER STATISTICS

| N | I | J | COVARIANCE | N | I | J | COVARIANCE | N | I | J | COVARIANCE |
|---|---|---|---|---|---|---|---|---|---|---|---|
| 37 | 11 | 23 | .0231018091 | 37 | 17 | 17 | .0422346771 | 38 | 2 | 15 | .0200827996 |
| 37 | 11 | 24 | .0220029214 | 37 | 17 | 18 | .0399542987 | 38 | 2 | 16 | .0188938363 |
| 37 | 11 | 25 | .0209624268 | 37 | 17 | 19 | .0378684965 | 38 | 2 | 17 | .0178252481 |
| 37 | 11 | 26 | .0199716542 | 37 | 17 | 20 | .0359481823 | 38 | 2 | 18 | .0168572442 |
| 37 | 11 | 27 | .0190225269 | 37 | 17 | 21 | .0341693887 | 38 | 2 | 19 | .0159740563 |
| 37 | 12 | 12 | .0457940781 | 37 | 18 | 18 | .0420252738 | 38 | 2 | 20 | .0151629346 |
| 37 | 12 | 13 | .0425991627 | 37 | 18 | 19 | .0398359942 | 38 | 2 | 21 | .0144134263 |
| 37 | 12 | 14 | .0398061215 | 37 | 18 | 20 | .0378202199 | 38 | 2 | 22 | .0137168457 |
| 37 | 12 | 15 | .0373369995 | 37 | 19 | 19 | .0419592559 | 38 | 2 | 23 | .0130658766 |
| 37 | 12 | 16 | .0351327652 | 38 | 1 | 1 | .2309978044 | 38 | 2 | 24 | .0124542670 |
| 37 | 12 | 17 | .0331478177 | 38 | 1 | 2 | .1104648223 | 38 | 2 | 25 | .0118765908 |
| 37 | 12 | 18 | .0313463017 | 38 | 1 | 3 | .0745066271 | 38 | 2 | 26 | .0113280574 |
| 37 | 12 | 19 | .0296995619 | 38 | 1 | 4 | .0570148087 | 38 | 2 | 27 | .0108043506 |
| 37 | 12 | 20 | .0281843410 | 38 | 1 | 5 | .0465515163 | 38 | 2 | 28 | .0103014788 |
| 37 | 12 | 21 | .0267814839 | 38 | 1 | 6 | .0395291582 | 38 | 2 | 29 | .0098156158 |
| 37 | 12 | 22 | .0254750084 | 38 | 1 | 7 | .0344566046 | 38 | 2 | 30 | .0093429300 |
| 37 | 12 | 23 | .0242514277 | 38 | 1 | 8 | .0306003227 | 38 | 2 | 31 | .0088793955 |
| 37 | 12 | 24 | .0230991602 | 38 | 1 | 9 | .0275561923 | 38 | 2 | 32 | .0084205107 |
| 37 | 12 | 25 | .0220080562 | 38 | 1 | 10 | .0250826205 | 38 | 2 | 33 | .0079607910 |
| 37 | 12 | 26 | .0209690201 | 38 | 1 | 11 | .0230259582 | 38 | 2 | 34 | .0074928864 |
| 37 | 13 | 13 | .0446638627 | 38 | 1 | 12 | .0212836239 | 38 | 2 | 35 | .0070058222 |
| 37 | 13 | 14 | .0417408897 | 38 | 1 | 13 | .0197843916 | 38 | 2 | 36 | .0064806157 |
| 37 | 13 | 15 | .0391566271 | 38 | 1 | 14 | .0184771798 | 38 | 2 | 37 | .0058766884 |
| 37 | 13 | 16 | .0368494218 | 38 | 1 | 15 | .0173243423 | 38 | 3 | 3 | .0943584447 |
| 37 | 13 | 17 | .0347716331 | 38 | 1 | 16 | .0162974805 | 38 | 3 | 4 | .0724034088 |
| 37 | 13 | 18 | .0328857866 | 38 | 1 | 17 | .0153747314 | 38 | 3 | 5 | .0592135628 |
| 37 | 13 | 19 | .0311619200 | 38 | 1 | 18 | .0145389597 | 38 | 3 | 6 | .0503364876 |
| 37 | 13 | 20 | .0295757141 | 38 | 1 | 19 | .0137765181 | 38 | 3 | 7 | .0439115644 |
| 37 | 13 | 21 | .0281071591 | 38 | 1 | 20 | .0130763734 | 38 | 3 | 8 | .0390200420 |
| 37 | 13 | 22 | .0267395753 | 38 | 1 | 21 | .0124294802 | 38 | 3 | 9 | .0351543436 |
| 37 | 13 | 23 | .0254588630 | 38 | 1 | 22 | .0118283254 | 38 | 3 | 10 | .0320103713 |
| 37 | 13 | 24 | .0242529009 | 38 | 1 | 23 | .0112665826 | 38 | 3 | 11 | .0293943902 |
| 37 | 13 | 25 | .0231110013 | 38 | 1 | 24 | .0107388465 | 38 | 3 | 12 | .0271768729 |
| 37 | 14 | 14 | .0437825726 | 38 | 1 | 25 | .0102404303 | 38 | 3 | 13 | .0252677790 |
| 37 | 14 | 15 | .0410753065 | 38 | 1 | 26 | .0097671968 | 38 | 3 | 14 | .0236024695 |
| 37 | 14 | 16 | .0386578245 | 38 | 1 | 27 | .0093154131 | 38 | 3 | 15 | .0221332654 |
| 37 | 14 | 17 | .0364803468 | 38 | 1 | 28 | .0088816242 | 38 | 3 | 16 | .0208241710 |
| 37 | 14 | 18 | .0345037027 | 38 | 1 | 29 | .0084625259 | 38 | 3 | 17 | .0196474567 |
| 37 | 14 | 19 | .0326965675 | 38 | 1 | 30 | .0080548216 | 38 | 3 | 18 | .0185813775 |
| 37 | 14 | 20 | .0310335216 | 38 | 1 | 31 | .0076550424 | 38 | 3 | 19 | .0176086062 |
| 37 | 14 | 21 | .0294936530 | 38 | 1 | 32 | .0072592903 | 38 | 3 | 20 | .0167151306 |
| 37 | 14 | 22 | .0280595209 | 38 | 1 | 33 | .0068628284 | 38 | 3 | 21 | .0158894581 |
| 37 | 14 | 23 | .0267163470 | 38 | 1 | 34 | .0064593336 | 38 | 3 | 22 | .0151220314 |
| 37 | 14 | 24 | .0254513817 | 38 | 1 | 35 | .0060393418 | 38 | 3 | 23 | .0144047936 |
| 37 | 15 | 15 | .0431101633 | 38 | 1 | 36 | .0055864788 | 38 | 3 | 24 | .0137308647 |
| 37 | 15 | 16 | .0405738604 | 38 | 1 | 37 | .0050657686 | 38 | 3 | 25 | .0130942888 |
| 37 | 15 | 17 | .0382886918 | 38 | 1 | 38 | .0043666349 | 38 | 3 | 26 | .0124898133 |
| 37 | 15 | 18 | .0362137054 | 38 | 2 | 2 | .1270475681 | 38 | 3 | 27 | .0119126818 |
| 37 | 15 | 19 | .0343161463 | 38 | 2 | 3 | .0859499744 | 38 | 3 | 28 | .0113584648 |
| 37 | 15 | 20 | .0325694294 | 38 | 2 | 4 | .0658722716 | 38 | 3 | 29 | .0108229370 |
| 37 | 15 | 21 | .0309516719 | 38 | 2 | 5 | .0538329539 | 38 | 3 | 30 | .0103019194 |
| 37 | 15 | 22 | .0294445956 | 38 | 2 | 6 | .0457401710 | 38 | 3 | 31 | .0097909991 |
| 37 | 15 | 23 | .0280326952 | 38 | 2 | 7 | .0398879749 | 38 | 3 | 32 | .0092851727 |
| 37 | 16 | 16 | .0425983273 | 38 | 2 | 8 | .0354353686 | 38 | 3 | 33 | .0087783846 |
| 37 | 16 | 17 | .0402018782 | 38 | 2 | 9 | .0319182988 | 38 | 3 | 34 | .0082625690 |
| 37 | 16 | 18 | .0380254393 | 38 | 2 | 10 | .0290590067 | 38 | 3 | 35 | .0077256093 |
| 37 | 16 | 19 | .0360347650 | 38 | 2 | 11 | .0266806718 | 38 | 3 | 36 | .0071465715 |
| 37 | 16 | 20 | .0342020482 | 38 | 2 | 12 | .0246651490 | 38 | 4 | 4 | .0780380561 |
| 37 | 16 | 21 | .0325043833 | 38 | 2 | 13 | .0229303525 | 38 | 4 | 5 | .0638635855 |
| 37 | 16 | 22 | .0309226256 | 38 | 2 | 14 | .0214173796 | 38 | 4 | 6 | .0543133704 |

TABLE I-COVARIANCES OF NORMAL ORDER STATISTICS

| N | I | J | COVARIANCE | N | I | J | COVARIANCE | N | I | J | COVARIANCE |
|---|---|---|---|---|---|---|---|---|---|---|---|
| 38 | 4 | 7 | .0473958659 | 38 | 6 | 7 | .0536919183 | 38 | 8 | 15 | .0302446160 |
| 38 | 4 | 8 | .0421262539 | 38 | 6 | 8 | .0477438792 | 38 | 8 | 16 | .0284632014 |
| 38 | 4 | 9 | .0379598702 | 38 | 6 | 9 | .0430371281 | 38 | 8 | 17 | .0268609604 |
| 38 | 4 | 10 | .0345701221 | 38 | 6 | 10 | .0392051483 | 38 | 8 | 18 | .0254085878 |
| 38 | 4 | 11 | .0317488030 | 38 | 6 | 11 | .0360139872 | 38 | 8 | 19 | .0240826947 |
| 38 | 4 | 12 | .0293566325 | 38 | 6 | 12 | .0333069793 | 38 | 8 | 20 | .0228643209 |
| 38 | 4 | 13 | .0272967448 | 38 | 6 | 13 | .0309750734 | 38 | 8 | 21 | .0217378999 |
| 38 | 4 | 14 | .0254995718 | 38 | 6 | 14 | .0289398949 | 38 | 8 | 22 | .0206905028 |
| 38 | 4 | 15 | .0239137806 | 38 | 6 | 15 | .0271435777 | 38 | 8 | 23 | .0197112439 |
| 38 | 4 | 16 | .0225006072 | 38 | 6 | 16 | .0255423952 | 38 | 8 | 24 | .0187908605 |
| 38 | 4 | 17 | .0212301854 | 38 | 6 | 17 | .0241026305 | 38 | 8 | 25 | .0179213395 |
| 38 | 4 | 18 | .0200790901 | 38 | 6 | 18 | .0227978222 | 38 | 8 | 26 | .0170954465 |
| 38 | 4 | 19 | .0190286475 | 38 | 6 | 19 | .0216068785 | 38 | 8 | 27 | .0163064800 |
| 38 | 4 | 20 | .0180637435 | 38 | 6 | 20 | .0205127423 | 38 | 8 | 28 | .0155484676 |
| 38 | 4 | 21 | .0171719742 | 38 | 6 | 21 | .0195014256 | 38 | 8 | 29 | .0148160187 |
| 38 | 4 | 22 | .0163430344 | 38 | 6 | 22 | .0185612921 | 38 | 8 | 30 | .0141034329 |
| 38 | 4 | 23 | .0155682604 | 38 | 6 | 23 | .0176824966 | 38 | 8 | 31 | .0134043269 |
| 38 | 4 | 24 | .0148402553 | 38 | 6 | 24 | .0168565787 | 38 | 9 | 9 | .0503328408 |
| 38 | 4 | 25 | .0141525780 | 38 | 6 | 25 | .0160762377 | 38 | 9 | 10 | .0458712247 |
| 38 | 4 | 26 | .0134995170 | 38 | 6 | 26 | .0153351485 | 38 | 9 | 11 | .0421526134 |
| 38 | 4 | 27 | .0128759316 | 38 | 6 | 27 | .0146275998 | 38 | 9 | 12 | .0389959706 |
| 38 | 4 | 28 | .0122770804 | 38 | 6 | 28 | .0139481053 | 38 | 9 | 13 | .0362751173 |
| 38 | 4 | 29 | .0116984143 | 38 | 6 | 29 | .0132913581 | 38 | 9 | 14 | .0338992661 |
| 38 | 4 | 30 | .0111353968 | 38 | 6 | 30 | .0126522851 | 38 | 9 | 15 | .0318013231 |
| 38 | 4 | 31 | .0105832809 | 38 | 6 | 31 | .0120255780 | 38 | 9 | 16 | .0299305553 |
| 38 | 4 | 32 | .0100366664 | 38 | 6 | 32 | .0114050639 | 38 | 9 | 17 | .0282478227 |
| 38 | 4 | 33 | .0094889354 | 38 | 6 | 33 | .0107833359 | 38 | 9 | 18 | .0267223796 |
| 38 | 4 | 34 | .0089314101 | 38 | 7 | 7 | .0567097459 | 38 | 9 | 19 | .0253296795 |
| 38 | 4 | 35 | .0083510686 | 38 | 7 | 8 | .0504387463 | 38 | 9 | 20 | .0240498432 |
| 38 | 5 | 5 | .0681505757 | 38 | 7 | 9 | .0454743093 | 38 | 9 | 21 | .0228665636 |
| 38 | 5 | 6 | .0579835988 | 38 | 7 | 10 | .0414311336 | 38 | 9 | 22 | .0217663220 |
| 38 | 5 | 7 | .0506139453 | 38 | 7 | 11 | .0380631263 | 38 | 9 | 23 | .0207377542 |
| 38 | 5 | 8 | .0449967996 | 38 | 7 | 12 | .0352054091 | 38 | 9 | 24 | .0197710456 |
| 38 | 5 | 9 | .0405537327 | 38 | 7 | 13 | .0327431654 | 38 | 9 | 25 | .0188575725 |
| 38 | 5 | 10 | .0369376320 | 38 | 7 | 14 | .0305938440 | 38 | 9 | 26 | .0179898098 |
| 38 | 5 | 11 | .0339270678 | 38 | 7 | 15 | .0286964709 | 38 | 9 | 27 | .0171610480 |
| 38 | 5 | 12 | .0313738357 | 38 | 7 | 16 | .0270049576 | 38 | 9 | 28 | .0163649997 |
| 38 | 5 | 13 | .0291748168 | 38 | 7 | 17 | .0254837661 | 38 | 9 | 29 | .0155957483 |
| 38 | 5 | 14 | .0272559257 | 38 | 7 | 18 | .0241050104 | 38 | 9 | 30 | .0148474067 |
| 38 | 5 | 15 | .0255624813 | 38 | 7 | 19 | .0228464608 | 38 | 10 | 10 | .0481368563 |
| 38 | 5 | 16 | .0240531816 | 38 | 7 | 20 | .0216901261 | 38 | 10 | 11 | .0442402976 |
| 38 | 5 | 17 | .0226961929 | 38 | 7 | 21 | .0206212115 | 38 | 10 | 12 | .0409318177 |
| 38 | 5 | 18 | .0214665321 | 38 | 7 | 22 | .0196273729 | 38 | 10 | 13 | .0380795215 |
| 38 | 5 | 19 | .0203442727 | 38 | 7 | 23 | .0186982317 | 38 | 10 | 14 | .0355884751 |
| 38 | 5 | 20 | .0193132965 | 38 | 7 | 24 | .0178249871 | 38 | 10 | 15 | .0333884964 |
| 38 | 5 | 21 | .0183603967 | 38 | 7 | 25 | .0169999956 | 38 | 10 | 16 | .0314264962 |
| 38 | 5 | 22 | .0174746006 | 38 | 7 | 26 | .0162164489 | 38 | 10 | 17 | .0296614980 |
| 38 | 5 | 23 | .0166466531 | 38 | 7 | 27 | .0154682142 | 38 | 10 | 18 | .0280613073 |
| 38 | 5 | 24 | .0158686341 | 38 | 7 | 28 | .0147495783 | 38 | 10 | 19 | .0266002293 |
| 38 | 5 | 25 | .0151336545 | 38 | 7 | 29 | .0140549342 | 38 | 10 | 20 | .0252574759 |
| 38 | 5 | 26 | .0144356027 | 38 | 7 | 30 | .0133788393 | 38 | 10 | 21 | .0240160093 |
| 38 | 5 | 27 | .0137689818 | 38 | 7 | 31 | .0127159103 | 38 | 10 | 22 | .0228616373 |
| 38 | 5 | 28 | .0131288008 | 38 | 7 | 32 | .0120596962 | 38 | 10 | 23 | .0217823492 |
| 38 | 5 | 29 | .0125103037 | 38 | 8 | 8 | .0531208103 | 38 | 10 | 24 | .0207678617 |
| 38 | 5 | 30 | .0119085378 | 38 | 8 | 9 | .0479009284 | 38 | 10 | 25 | .0198091895 |
| 38 | 5 | 31 | .0113182019 | 38 | 8 | 10 | .0436482156 | 38 | 10 | 26 | .0188984170 |
| 38 | 5 | 32 | .0107336648 | 38 | 8 | 11 | .0401046256 | 38 | 10 | 27 | .0180286631 |
| 38 | 5 | 33 | .0101481373 | 38 | 8 | 12 | .0370971835 | 38 | 10 | 28 | .0171935347 |
| 38 | 5 | 34 | .0095521129 | 38 | 8 | 13 | .0345053802 | 38 | 10 | 29 | .0163864012 |
| 38 | 6 | 6 | .0614915442 | 38 | 8 | 14 | .0322425380 | 38 | 11 | 11 | .0463992241 |

TABLE I-COVARIANCES OF NORMAL ORDER STATISTICS

| N | I | J | COVARIANCE | N | I | J | COVARIANCE | N | I | J | COVARIANCE |
|---|---|---|---|---|---|---|---|---|---|---|---|
| 38 | 11 | 12 | .0429332079 | 38 | 15 | 16 | .0397510440 | 39 | 1 | 32 | .0073552204 |
| 38 | 11 | 13 | .0399442432 | 38 | 15 | 17 | .0375288365 | 39 | 1 | 33 | .0069810698 |
| 38 | 11 | 14 | .0373331609 | 38 | 15 | 18 | .0355126825 | 39 | 1 | 34 | .0066054075 |
| 38 | 11 | 15 | .0350266187 | 38 | 15 | 19 | .0336706009 | 39 | 1 | 35 | .0062222378 |
| 38 | 11 | 16 | .0329691150 | 38 | 15 | 20 | .0319766642 | 39 | 1 | 36 | .0058225193 |
| 38 | 11 | 17 | .0311178011 | 38 | 15 | 21 | .0304095741 | 39 | 1 | 37 | .0053905352 |
| 38 | 11 | 18 | .0294390134 | 38 | 15 | 22 | .0289516012 | 39 | 1 | 38 | .0048926155 |
| 38 | 11 | 19 | .0279058929 | 38 | 15 | 23 | .0275877703 | 39 | 1 | 39 | .0042221226 |
| 38 | 11 | 20 | .0264966958 | 38 | 15 | 24 | .0263052238 | 39 | 2 | 2 | .1259762647 |
| 38 | 11 | 21 | .0251935596 | 38 | 16 | 16 | .0416795347 | 39 | 2 | 3 | .0851941849 |
| 38 | 11 | 22 | .0239815938 | 38 | 16 | 17 | .0393502809 | 39 | 2 | 4 | .0652820795 |
| 38 | 11 | 23 | .0228481800 | 38 | 16 | 18 | .0372364476 | 39 | 2 | 5 | .0533473098 |
| 38 | 11 | 24 | .0217825165 | 38 | 16 | 19 | .0353046347 | 39 | 2 | 6 | .0453278458 |
| 38 | 11 | 25 | .0207753653 | 38 | 16 | 20 | .0335277570 | 39 | 2 | 7 | .0395307047 |
| 38 | 11 | 26 | .0198185915 | 38 | 16 | 21 | .0318835508 | 39 | 2 | 8 | .0351214872 |
| 38 | 11 | 27 | .0189046926 | 38 | 16 | 22 | .0303534596 | 39 | 2 | 9 | .0316398755 |
| 38 | 11 | 28 | .0180267244 | 38 | 16 | 23 | .0289217896 | 39 | 2 | 10 | .0288103970 |
| 38 | 12 | 12 | .0449917056 | 38 | 17 | 17 | .0412722558 | 39 | 2 | 11 | .0264577166 |
| 38 | 12 | 13 | .0418632342 | 38 | 17 | 18 | .0390602722 | 39 | 2 | 12 | .0244647012 |
| 38 | 12 | 14 | .0391296724 | 38 | 17 | 19 | .0370386665 | 39 | 2 | 13 | .0227499823 |
| 38 | 12 | 15 | .0367144513 | 38 | 17 | 20 | .0351791479 | 39 | 2 | 14 | .0212551832 |
| 38 | 12 | 16 | .0345596108 | 38 | 17 | 21 | .0334584484 | 39 | 2 | 15 | .0199372710 |
| 38 | 12 | 17 | .0326203983 | 38 | 17 | 22 | .0318571649 | 39 | 2 | 16 | .0187637786 |
| 38 | 12 | 18 | .0308616512 | 38 | 18 | 18 | .0410064978 | 39 | 2 | 17 | .0177097120 |
| 38 | 12 | 19 | .0292553008 | 38 | 18 | 19 | .0388906175 | 39 | 2 | 18 | .0167554840 |
| 38 | 12 | 20 | .0277786007 | 38 | 18 | 20 | .0369444044 | 39 | 2 | 19 | .0158854981 |
| 38 | 12 | 21 | .0264128407 | 38 | 18 | 21 | .0351435319 | 39 | 2 | 20 | .0150871544 |
| 38 | 12 | 22 | .0251424164 | 38 | 19 | 19 | .0408825322 | 39 | 2 | 21 | .0143501362 |
| 38 | 12 | 23 | .0239541873 | 38 | 19 | 20 | .0388382240 | 39 | 2 | 22 | .0136658866 |
| 38 | 12 | 24 | .0228369486 | 39 | 1 | 1 | .2294665256 | 39 | 2 | 23 | .0130272180 |
| 38 | 12 | 25 | .0217809852 | 39 | 1 | 2 | .1096450976 | 39 | 2 | 24 | .0124280117 |
| 38 | 12 | 26 | .0207777592 | 39 | 1 | 3 | .0739295987 | 39 | 2 | 25 | .0118629835 |
| 38 | 12 | 27 | .0198193928 | 39 | 1 | 4 | .0565648859 | 39 | 2 | 26 | .0113275000 |
| 38 | 13 | 13 | .0438366504 | 39 | 1 | 5 | .0461818555 | 39 | 2 | 27 | .0108174295 |
| 38 | 13 | 14 | .0409795683 | 39 | 1 | 6 | .0392158555 | 39 | 2 | 28 | .0103290132 |
| 38 | 13 | 15 | .0384549940 | 39 | 1 | 7 | .0341856995 | 39 | 2 | 29 | .0098587343 |
| 38 | 13 | 16 | .0362024595 | 39 | 1 | 8 | .0303629052 | 39 | 2 | 30 | .0094031727 |
| 38 | 13 | 17 | .0341752650 | 39 | 1 | 9 | .0273462092 | 39 | 2 | 31 | .0089588576 |
| 38 | 13 | 18 | .0323367054 | 39 | 1 | 10 | .0248957665 | 39 | 2 | 32 | .0085221046 |
| 38 | 13 | 19 | .0306574630 | 39 | 1 | 11 | .0228590634 | 39 | 2 | 33 | .0080887437 |
| 38 | 13 | 20 | .0291137650 | 39 | 1 | 12 | .0211342921 | 39 | 2 | 34 | .0076536229 |
| 38 | 13 | 21 | .0276860675 | 39 | 1 | 13 | .0196507759 | 39 | 2 | 35 | .0072097743 |
| 38 | 13 | 22 | .0263581064 | 39 | 1 | 14 | .0183578364 | 39 | 2 | 36 | .0067467323 |
| 38 | 13 | 23 | .0251161886 | 39 | 1 | 15 | .0172181326 | 39 | 2 | 37 | .0062462949 |
| 38 | 13 | 24 | .0239486515 | 39 | 1 | 16 | .0162035038 | 39 | 2 | 38 | .0056694402 |
| 38 | 13 | 25 | .0228453256 | 39 | 1 | 17 | .0152922787 | 39 | 3 | 3 | .0934483568 |
| 38 | 13 | 26 | .0217970808 | 39 | 1 | 18 | .0144674788 | 39 | 3 | 4 | .0716916844 |
| 38 | 14 | 14 | .0429203360 | 39 | 1 | 19 | .0137155911 | 39 | 3 | 5 | .0586271500 |
| 38 | 14 | 15 | .0402804890 | 39 | 1 | 20 | .0130257022 | 39 | 3 | 6 | .0498379266 |
| 38 | 14 | 16 | .0379248223 | 39 | 1 | 21 | .0123888750 | 39 | 3 | 7 | .0434789149 |
| 38 | 14 | 17 | .0358046038 | 39 | 1 | 22 | .0117976978 | 39 | 3 | 8 | .0386392751 |
| 38 | 14 | 18 | .0338815158 | 39 | 1 | 23 | .0112459485 | 39 | 3 | 9 | .0348159151 |
| 38 | 14 | 19 | .0321249412 | 39 | 1 | 24 | .0107283302 | 39 | 3 | 10 | .0317074836 |
| 38 | 14 | 20 | .0305100562 | 39 | 1 | 25 | .0102402714 | 39 | 3 | 11 | .0291220306 |
| 38 | 14 | 21 | .0290164680 | 39 | 1 | 26 | .0097777692 | 39 | 3 | 12 | .0269312450 |
| 38 | 14 | 22 | .0276272197 | 39 | 1 | 27 | .0093372522 | 39 | 3 | 13 | .0250459481 |
| 38 | 14 | 23 | .0263280238 | 39 | 1 | 28 | .0089154613 | 39 | 3 | 14 | .0234021328 |
| 38 | 14 | 24 | .0251066284 | 39 | 1 | 29 | .0085093469 | 39 | 3 | 15 | .0219525977 |
| 38 | 14 | 25 | .0239523480 | 39 | 1 | 30 | .0081159573 | 39 | 3 | 16 | .0206617178 |
| 38 | 15 | 15 | .0422193238 | 39 | 1 | 31 | .0077323090 | 39 | 3 | 17 | .0195020584 |

TABLE I-COVARIANCES OF NORMAL ORDER STATISTICS

| N | I | J | COVARIANCE | N | I | J | COVARIANCE | N | I | J | COVARIANCE |
|---|---|---|---|---|---|---|---|---|---|---|---|
| 39 | 3 | 18 | .0184521149 | 39 | 5 | 12 | .0310385050 | 39 | 7 | 14 | .0302323640 |
| 39 | 3 | 19 | .0174947631 | 39 | 5 | 13 | .0288704165 | 39 | 7 | 15 | .0283666100 |
| 39 | 3 | 20 | .0166161693 | 39 | 5 | 14 | .0269793743 | 39 | 7 | 16 | .0267042054 |
| 39 | 3 | 21 | .0158050050 | 39 | 5 | 15 | .0253113255 | 39 | 7 | 17 | .0252100930 |
| 39 | 3 | 22 | .0150518671 | 39 | 5 | 16 | .0238254568 | 39 | 7 | 18 | .0238567753 |
| 39 | 3 | 23 | .0143488457 | 39 | 5 | 17 | .0224903264 | 39 | 7 | 19 | .0226223524 |
| 39 | 3 | 24 | .0136892030 | 39 | 5 | 18 | .0212812733 | 39 | 7 | 20 | .0214891347 |
| 39 | 3 | 25 | .0130671342 | 39 | 5 | 19 | .0201786371 | 39 | 7 | 21 | .0204426085 |
| 39 | 3 | 26 | .0124775700 | 39 | 5 | 20 | .0191665198 | 39 | 7 | 22 | .0194706665 |
| 39 | 3 | 27 | .0119159865 | 39 | 5 | 21 | .0182319081 | 39 | 7 | 23 | .0185631116 |
| 39 | 3 | 28 | .0113782287 | 39 | 5 | 22 | .0173640178 | 39 | 7 | 24 | .0177113205 |
| 39 | 3 | 29 | .0108603857 | 39 | 5 | 23 | .0165537900 | 39 | 7 | 25 | .0169078853 |
| 39 | 3 | 30 | .0103586991 | 39 | 5 | 24 | .0157935109 | 39 | 7 | 26 | .0161462873 |
| 39 | 3 | 31 | .0098694002 | 39 | 5 | 25 | .0150765078 | 39 | 7 | 27 | .0154207362 |
| 39 | 3 | 32 | .0093884363 | 39 | 5 | 26 | .0143968824 | 39 | 7 | 28 | .0147259674 |
| 39 | 3 | 33 | .0089111691 | 39 | 5 | 27 | .0137493301 | 39 | 7 | 29 | .0140568226 |
| 39 | 3 | 34 | .0084319381 | 39 | 5 | 28 | .0131290969 | 39 | 7 | 30 | .0134081612 |
| 39 | 3 | 35 | .0079430948 | 39 | 5 | 29 | .0125318853 | 39 | 7 | 31 | .0127752807 |
| 39 | 3 | 36 | .0074330799 | 39 | 5 | 30 | .0119534575 | 39 | 7 | 32 | .0121534828 |
| 39 | 3 | 37 | .0068818583 | 39 | 5 | 31 | .0113892176 | 39 | 7 | 33 | .0115365584 |
| 39 | 4 | 4 | .0772086328 | 39 | 5 | 32 | .0108343092 | 39 | 8 | 8 | .0523881231 |
| 39 | 4 | 5 | .0631793669 | 39 | 5 | 33 | .0102836518 | 39 | 8 | 9 | .0472454924 |
| 39 | 4 | 6 | .0537309576 | 39 | 5 | 34 | .0097308645 | 39 | 8 | 10 | .0430573625 |
| 39 | 4 | 7 | .0468898010 | 39 | 5 | 35 | .0091668729 | 39 | 8 | 11 | .0395689789 |
| 39 | 4 | 8 | .0416802412 | 39 | 6 | 6 | .0607346186 | 39 | 8 | 12 | .0366096143 |
| 39 | 4 | 9 | .0375628139 | 39 | 6 | 7 | .0530327264 | 39 | 8 | 13 | .0340603543 |
| 39 | 4 | 10 | .0342141137 | 39 | 6 | 8 | .0471615032 | 39 | 8 | 14 | .0318356890 |
| 39 | 4 | 11 | .0314280063 | 39 | 6 | 9 | .0425172991 | 39 | 8 | 15 | .0298724552 |
| 39 | 4 | 12 | .0290666239 | 39 | 6 | 10 | .0387376804 | 39 | 8 | 16 | .0281229154 |
| 39 | 4 | 13 | .0270341020 | 39 | 6 | 11 | .0355913525 | 39 | 8 | 17 | .0265502714 |
| 39 | 4 | 14 | .0252616060 | 39 | 6 | 12 | .0329234621 | 39 | 8 | 18 | .0251256616 |
| 39 | 4 | 15 | .0236983560 | 39 | 6 | 13 | .0306262420 | 39 | 8 | 19 | .0238260792 |
| 39 | 4 | 16 | .0223060152 | 39 | 6 | 14 | .0286222585 | 39 | 8 | 20 | .0226328908 |
| 39 | 4 | 17 | .0210550574 | 39 | 6 | 15 | .0268543568 | 39 | 8 | 21 | .0215307987 |
| 39 | 4 | 18 | .0199223379 | 39 | 6 | 16 | .0252793637 | 39 | 8 | 22 | .0205071011 |
| 39 | 4 | 19 | .0188894227 | 39 | 6 | 17 | .0238640010 | 39 | 8 | 23 | .0195511167 |
| 39 | 4 | 20 | .0179414070 | 39 | 6 | 18 | .0225821579 | 39 | 8 | 24 | .0186538001 |
| 39 | 4 | 21 | .0170660685 | 39 | 6 | 19 | .0214130298 | 39 | 8 | 25 | .0178074576 |
| 39 | 4 | 22 | .0162532615 | 39 | 6 | 20 | .0203398087 | 39 | 8 | 26 | .0170052843 |
| 39 | 4 | 23 | .0154944759 | 39 | 6 | 21 | .0193487396 | 39 | 8 | 27 | .0162408965 |
| 39 | 4 | 24 | .0147824868 | 39 | 6 | 22 | .0184284213 | 39 | 8 | 28 | .0155084208 |
| 39 | 4 | 25 | .0141110530 | 39 | 6 | 23 | .0175692453 | 39 | 8 | 29 | .0148028015 |
| 39 | 4 | 26 | .0134746774 | 39 | 6 | 24 | .0167629510 | 39 | 8 | 30 | .0141192850 |
| 39 | 4 | 27 | .0128684442 | 39 | 6 | 25 | .0160023780 | 39 | 8 | 31 | .0134525615 |
| 39 | 4 | 28 | .0122878806 | 39 | 6 | 26 | .0152813675 | 39 | 8 | 32 | .0127968911 |
| 39 | 4 | 29 | .0117287921 | 39 | 6 | 27 | .0145945468 | 39 | 9 | 9 | .0496010446 |
| 39 | 4 | 30 | .0111871137 | 39 | 6 | 28 | .0139369089 | 39 | 9 | 10 | .0452107553 |
| 39 | 4 | 31 | .0106587941 | 39 | 6 | 29 | .0133035612 | 39 | 9 | 11 | .0415530756 |
| 39 | 4 | 32 | .0101395182 | 39 | 6 | 30 | .0126898369 | 39 | 9 | 12 | .0384494549 |
| 39 | 4 | 33 | .0096242151 | 39 | 6 | 31 | .0120912013 | 39 | 9 | 13 | .0357754766 |
| 39 | 4 | 34 | .0091066589 | 39 | 6 | 32 | .0115026941 | 39 | 9 | 14 | .0334416431 |
| 39 | 4 | 35 | .0085787231 | 39 | 6 | 33 | .0109186294 | 39 | 9 | 15 | .0313818309 |
| 39 | 4 | 36 | .0080279898 | 39 | 6 | 34 | .0103322368 | 39 | 9 | 16 | .0295460584 |
| 39 | 5 | 5 | .0673674830 | 39 | 7 | 7 | .0559669703 | 39 | 9 | 17 | .0278957878 |
| 39 | 5 | 6 | .0573162979 | 39 | 7 | 8 | .0497818176 | 39 | 9 | 18 | .0264007682 |
| 39 | 5 | 7 | .0500334598 | 39 | 7 | 9 | .0448872656 | 39 | 9 | 19 | .0250368617 |
| 39 | 5 | 8 | .0444845611 | 39 | 7 | 10 | .0409025702 | 39 | 9 | 20 | .0237845236 |
| 39 | 5 | 9 | .0400970871 | 39 | 7 | 11 | .0375846074 | 39 | 9 | 21 | .0226277192 |
| 39 | 5 | 10 | .0365275535 | 39 | 7 | 12 | .0347705249 | 39 | 9 | 22 | .0215531747 |
| 39 | 5 | 11 | .0335568911 | 39 | 7 | 13 | .0323469428 | 39 | 9 | 23 | .0205498183 |

TABLE I-COVARIANCES OF NORMAL ORDER STATISICS

| N | I | J | COVARIANCE | N | I | J | COVARIANCE | N | I | J | COVARIANCE |
|---|---|---|---|---|---|---|---|---|---|---|---|
| 39 | 9 | 24 | .0196081867 | 39 | 12 | 24 | .0225803988 | 39 | 18 | 18 | .0400644779 |
| 39 | 9 | 25 | .0187199569 | 39 | 12 | 25 | .0215560891 | 39 | 18 | 19 | .0380138754 |
| 39 | 9 | 26 | .0178778270 | 39 | 12 | 26 | .0205847583 | 39 | 18 | 20 | .0361293077 |
| 39 | 9 | 27 | .0170753602 | 39 | 12 | 27 | .0196591645 | 39 | 18 | 21 | .0343871393 |
| 39 | 9 | 28 | .0163066035 | 39 | 12 | 28 | .0187722660 | 39 | 18 | 22 | .0327677014 |
| 39 | 9 | 29 | .0155660692 | 39 | 13 | 13 | .0430627221 | 39 | 19 | 19 | .0398862307 |
| 39 | 9 | 30 | .0148488038 | 39 | 13 | 14 | .0402662906 | 39 | 19 | 20 | .0379132503 |
| 39 | 9 | 31 | .0141494338 | 39 | 13 | 15 | .0377965409 | 39 | 19 | 21 | .0360892398 |
| 39 | 10 | 10 | .0473932314 | 39 | 13 | 16 | .0355941053 | 39 | 20 | 20 | .0398316602 |
| 39 | 10 | 11 | .0435646425 | 39 | 13 | 17 | .0336131530 | 40 | 1 | 1 | .2279918237 |
| 39 | 10 | 12 | .0403153403 | 39 | 13 | 18 | .0318176861 | 40 | 1 | 2 | .1088560144 |
| 39 | 10 | 13 | .0375153813 | 39 | 13 | 19 | .0301789814 | 40 | 1 | 3 | .0733740273 |
| 39 | 10 | 14 | .0350712615 | 39 | 13 | 20 | .0286737749 | 40 | 1 | 4 | .0561314849 |
| 39 | 10 | 15 | .0329138783 | 39 | 13 | 21 | .0272829541 | 40 | 1 | 5 | .0458255290 |
| 39 | 10 | 16 | .0309909806 | 39 | 13 | 22 | .0259906187 | 40 | 1 | 6 | .0389135994 |
| 39 | 10 | 17 | .0292622597 | 39 | 13 | 23 | .0247833941 | 40 | 1 | 7 | .0339240809 |
| 39 | 10 | 18 | .0276960578 | 39 | 13 | 24 | .0236499580 | 40 | 1 | 8 | .0301333509 |
| 39 | 10 | 19 | .0262671140 | 39 | 13 | 25 | .0225805869 | 40 | 1 | 9 | .0271428937 |
| 39 | 10 | 20 | .0249549986 | 39 | 13 | 26 | .0215666163 | 40 | 1 | 10 | .0247145447 |
| 39 | 10 | 21 | .0237430026 | 39 | 13 | 27 | .0206002478 | 40 | 1 | 11 | .0226968809 |
| 39 | 10 | 22 | .0226172681 | 39 | 14 | 14 | .0421123336 | 40 | 1 | 12 | .0209888377 |
| 39 | 10 | 23 | .0215661138 | 39 | 14 | 15 | .0395345431 | 40 | 1 | 13 | .0195202648 |
| 39 | 10 | 24 | .0205795789 | 39 | 14 | 16 | .0372356592 | 40 | 1 | 14 | .0182408704 |
| 39 | 10 | 25 | .0196489655 | 39 | 14 | 17 | .0351679199 | 40 | 1 | 15 | .0171136046 |
| 39 | 10 | 26 | .0187665202 | 39 | 14 | 18 | .0332937965 | 40 | 1 | 16 | .0161105332 |
| 39 | 10 | 27 | .0179255665 | 39 | 14 | 19 | .0315833267 | 40 | 1 | 17 | .0152101667 |
| 39 | 10 | 28 | .0171203609 | 39 | 14 | 20 | .0300122352 | 40 | 1 | 18 | .0143956750 |
| 39 | 10 | 29 | .0163451697 | 39 | 14 | 21 | .0285605909 | 40 | 1 | 19 | .0136536692 |
| 39 | 10 | 30 | .0155939337 | 39 | 14 | 22 | .0272118432 | 40 | 1 | 20 | .0129733465 |
| 39 | 11 | 11 | .0456460583 | 39 | 14 | 23 | .0259521069 | 40 | 1 | 21 | .0123458697 |
| 39 | 11 | 12 | .0422452872 | 39 | 14 | 24 | .0247695598 | 40 | 1 | 22 | .0117639172 |
| 39 | 11 | 13 | .0393140028 | 39 | 14 | 25 | .0236539723 | 40 | 1 | 23 | .0112213559 |
| 39 | 11 | 14 | .0367546139 | 39 | 14 | 26 | .0225962973 | 40 | 1 | 24 | .0107129829 |
| 39 | 11 | 15 | .0344949701 | 39 | 15 | 15 | .0413808994 | 40 | 1 | 25 | .0102343227 |
| 39 | 11 | 16 | .0324804922 | 39 | 15 | 16 | .0389760989 | 40 | 1 | 26 | .0097814781 |
| 39 | 11 | 17 | .0306690586 | 39 | 15 | 17 | .0368125903 | 40 | 1 | 27 | .0093510046 |
| 39 | 11 | 18 | .0290275872 | 39 | 15 | 18 | .0348512220 | 40 | 1 | 28 | .0089397943 |
| 39 | 11 | 19 | .0275296974 | 39 | 15 | 19 | .0330607279 | 40 | 1 | 29 | .0085449817 |
| 39 | 11 | 20 | .0261540633 | 39 | 15 | 20 | .0314157789 | 40 | 1 | 30 | .0081638577 |
| 39 | 11 | 21 | .0248832107 | 39 | 15 | 21 | .0298955931 | 40 | 1 | 31 | .0077937687 |
| 39 | 11 | 22 | .0237026232 | 39 | 15 | 22 | .0284829174 | 40 | 1 | 32 | .0074319991 |
| 39 | 11 | 23 | .0226000223 | 39 | 15 | 23 | .0271632460 | 40 | 1 | 33 | .0070756151 |
| 39 | 11 | 24 | .0215648482 | 39 | 15 | 24 | .0259241957 | 40 | 1 | 34 | .0067212375 |
| 39 | 11 | 25 | .0205880770 | 39 | 15 | 25 | .0247549808 | 40 | 1 | 35 | .0063646728 |
| 39 | 11 | 26 | .0196619326 | 39 | 16 | 16 | .0408194738 | 40 | 1 | 36 | .0060002207 |
| 39 | 11 | 27 | .0187793836 | 39 | 16 | 17 | .0385527930 | 40 | 1 | 37 | .0056192270 |
| 39 | 11 | 28 | .0179340282 | 39 | 16 | 18 | .0364972057 | 40 | 1 | 38 | .0052065849 |
| 39 | 11 | 29 | .0171199078 | 39 | 16 | 19 | .0346200863 | 40 | 1 | 39 | .0047298451 |
| 39 | 12 | 12 | .0442348817 | 39 | 16 | 20 | .0328950027 | 40 | 1 | 40 | .0040860671 |
| 39 | 12 | 13 | .0411685104 | 39 | 16 | 21 | .0313002652 | 40 | 2 | 2 | .1249472155 |
| 39 | 12 | 14 | .0384904774 | 39 | 16 | 22 | .0298178467 | 40 | 2 | 3 | .0844681412 |
| 39 | 12 | 15 | .0361255141 | 39 | 16 | 23 | .0284325535 | 40 | 2 | 4 | .0647149029 |
| 39 | 12 | 16 | .0340166636 | 39 | 16 | 24 | .0271313840 | 40 | 2 | 5 | .0528803422 |
| 39 | 12 | 17 | .0321199625 | 39 | 17 | 17 | .0403802106 | 40 | 2 | 6 | .0449310926 |
| 39 | 12 | 18 | .0304008854 | 39 | 17 | 18 | .0382301469 | 40 | 2 | 7 | .0391866282 |
| 39 | 12 | 19 | .0288318988 | 39 | 17 | 19 | .0362665048 | 40 | 2 | 8 | .0348188853 |
| 39 | 12 | 20 | .0273907277 | 39 | 17 | 20 | .0344617139 | 40 | 2 | 9 | .0313711319 |
| 39 | 12 | 21 | .0260590841 | 39 | 17 | 21 | .0327931346 | 40 | 2 | 10 | .0285700894 |
| 39 | 12 | 22 | .0248217235 | 39 | 17 | 22 | .0312419292 | 40 | 2 | 11 | .0262418468 |
| 39 | 12 | 23 | .0236658030 | 39 | 17 | 23 | .0297922075 | 40 | 2 | 12 | .0242702411 |

TABLE I-COVARIANCES OF NORMAL ORDER STATISTICS

| N | I | J | COVARIANCE | N | I | J | COVARIANCE | N | I | J | COVARIANCE |
|---|---|---|---|---|---|---|---|---|---|---|---|
| 40 | 2 | 13 | .0225745906 | 40 | 3 | 36 | .0076447323 | 40 | 5 | 28 | .0131175596 |
| 40 | 2 | 14 | .0210970214 | 40 | 3 | 37 | .0071595412 | 40 | 5 | 29 | .0125388952 |
| 40 | 2 | 15 | .0197948797 | 40 | 3 | 38 | .0066340271 | 40 | 5 | 30 | .0119803732 |
| 40 | 2 | 16 | .0186359927 | 40 | 4 | 4 | .0764145632 | 40 | 5 | 31 | .0114381210 |
| 40 | 2 | 17 | .0175956008 | 40 | 4 | 5 | .0625240388 | 40 | 5 | 32 | .0109078326 |
| 40 | 2 | 18 | .0166543089 | 40 | 4 | 6 | .0531728229 | 40 | 5 | 33 | .0103851144 |
| 40 | 2 | 19 | .0157966813 | 40 | 4 | 7 | .0464044943 | 40 | 5 | 34 | .0098653636 |
| 40 | 2 | 20 | .0150102567 | 40 | 4 | 8 | .0412521683 | 40 | 5 | 35 | .0093424872 |
| 40 | 2 | 21 | .0142848419 | 40 | 4 | 9 | .0371813561 | 40 | 5 | 36 | .0088078436 |
| 40 | 2 | 22 | .0136119945 | 40 | 4 | 10 | .0338717009 | 40 | 6 | 6 | .0600124301 |
| 40 | 2 | 23 | .0129846387 | 40 | 4 | 11 | .0311190490 | 40 | 6 | 7 | .0524034422 |
| 40 | 2 | 24 | .0123967703 | 40 | 4 | 12 | .0287868821 | 40 | 6 | 8 | .0466051742 |
| 40 | 2 | 25 | .0118432256 | 40 | 4 | 13 | .0267802912 | 40 | 6 | 9 | .0420203178 |
| 40 | 2 | 26 | .0113194998 | 40 | 4 | 14 | .0250311417 | 40 | 6 | 10 | .0382903318 |
| 40 | 2 | 27 | .0108216047 | 40 | 4 | 15 | .0234891803 | 40 | 6 | 11 | .0351864541 |
| 40 | 2 | 28 | .0103459523 | 40 | 4 | 16 | .0221164760 | 40 | 6 | 12 | .0325555575 |
| 40 | 2 | 29 | .0098892481 | 40 | 4 | 17 | .0208838227 | 40 | 6 | 13 | .0302910959 |
| 40 | 2 | 30 | .0094483712 | 40 | 4 | 18 | .0197683373 | 40 | 6 | 14 | .0283165286 |
| 40 | 2 | 31 | .0090202414 | 40 | 4 | 19 | .0187518083 | 40 | 6 | 15 | .0265753742 |
| 40 | 2 | 32 | .0086016976 | 40 | 4 | 20 | .0178195276 | 40 | 6 | 16 | .0250249844 |
| 40 | 2 | 33 | .0081893545 | 40 | 4 | 21 | .0169594452 | 40 | 6 | 17 | .0236324982 |
| 40 | 2 | 34 | .0077793311 | 40 | 4 | 22 | .0161615624 | 40 | 6 | 18 | .0223721389 |
| 40 | 2 | 35 | .0073667669 | 40 | 4 | 23 | .0154174991 | 40 | 6 | 19 | .0212233703 |
| 40 | 2 | 36 | .0069450432 | 40 | 4 | 24 | .0147201641 | 40 | 6 | 20 | .0201696064 |
| 40 | 2 | 37 | .0065041611 | 40 | 4 | 25 | .0140634754 | 40 | 6 | 21 | .0191972903 |
| 40 | 2 | 38 | .0060266388 | 40 | 4 | 26 | .0134421205 | 40 | 6 | 22 | .0182952238 |
| 40 | 2 | 39 | .0054749092 | 40 | 4 | 27 | .0128513800 | 40 | 6 | 23 | .0174540225 |
| 40 | 3 | 3 | .0925757635 | 40 | 4 | 28 | .0122869971 | 40 | 6 | 24 | .0166656412 |
| 40 | 3 | 4 | .0710090814 | 40 | 4 | 29 | .0117450335 | 40 | 6 | 25 | .0159230640 |
| 40 | 3 | 5 | .0580644612 | 40 | 4 | 30 | .0112217331 | 40 | 6 | 26 | .0152202166 |
| 40 | 3 | 6 | .0493592341 | 40 | 4 | 31 | .0107134652 | 40 | 6 | 27 | .0145518971 |
| 40 | 3 | 7 | .0430631867 | 40 | 4 | 32 | .0102166278 | 40 | 6 | 28 | .0139134651 |
| 40 | 3 | 8 | .0382730644 | 40 | 4 | 33 | .0097272493 | 40 | 6 | 29 | .0133004602 |
| 40 | 3 | 9 | .0344900727 | 40 | 4 | 34 | .0092405056 | 40 | 6 | 30 | .0127085609 |
| 40 | 3 | 10 | .0314154916 | 40 | 4 | 35 | .0087505564 | 40 | 6 | 31 | .0121336839 |
| 40 | 3 | 11 | .0288590799 | 40 | 4 | 36 | .0082497968 | 40 | 6 | 32 | .0115716366 |
| 40 | 3 | 12 | .0266936901 | 40 | 4 | 37 | .0077263234 | 40 | 6 | 33 | .0110177141 |
| 40 | 3 | 13 | .0248309672 | 40 | 5 | 5 | .0666190232 | 40 | 6 | 34 | .0104668088 |
| 40 | 3 | 14 | .0232075070 | 40 | 5 | 6 | .0566782000 | 40 | 6 | 35 | .0099127173 |
| 40 | 3 | 15 | .0217765632 | 40 | 5 | 7 | .0494780367 | 40 | 7 | 7 | .0552584038 |
| 40 | 3 | 16 | .0205028636 | 40 | 5 | 8 | .0439940736 | 40 | 7 | 8 | .0491547405 |
| 40 | 3 | 17 | .0193592532 | 40 | 5 | 9 | .0396594461 | 40 | 7 | 9 | .0443264745 |
| 40 | 3 | 18 | .0183244539 | 40 | 5 | 10 | .0361341346 | 40 | 7 | 10 | .0403972014 |
| 40 | 3 | 19 | .0173815301 | 40 | 5 | 11 | .0332013227 | 40 | 7 | 11 | .0371266222 |
| 40 | 3 | 20 | .0165168112 | 40 | 5 | 12 | .0307159508 | 40 | 7 | 12 | .0343538111 |
| 40 | 3 | 21 | .0157191170 | 40 | 5 | 13 | .0285771272 | 40 | 7 | 13 | .0319667529 |
| 40 | 3 | 22 | .0149791852 | 40 | 5 | 14 | .0267123946 | 40 | 7 | 14 | .0298849512 |
| 40 | 3 | 23 | .0142892396 | 40 | 5 | 15 | .0250682965 | 40 | 7 | 15 | .0280489827 |
| 40 | 3 | 24 | .0136426679 | 40 | 5 | 16 | .0236044803 | 40 | 7 | 16 | .0264139552 |
| 40 | 3 | 25 | .0130337871 | 40 | 5 | 17 | .0222898735 | 40 | 7 | 17 | .0249452671 |
| 40 | 3 | 26 | .0124576640 | 40 | 5 | 18 | .0211001200 | 40 | 7 | 18 | .0236157816 |
| 40 | 3 | 27 | .0119099511 | 40 | 5 | 19 | .0200158133 | 40 | 7 | 19 | .0224038964 |
| 40 | 3 | 28 | .0113867170 | 40 | 5 | 20 | .0190212642 | 40 | 7 | 20 | .0212921902 |
| 40 | 3 | 29 | .0108843034 | 40 | 5 | 21 | .0181036328 | 40 | 7 | 21 | .0202664125 |
| 40 | 3 | 30 | .0103992412 | 40 | 5 | 22 | .0172522927 | 40 | 7 | 22 | .0193146972 |
| 40 | 3 | 31 | .0099281688 | 40 | 5 | 23 | .0164583452 | 40 | 7 | 23 | .0184270323 |
| 40 | 3 | 32 | .0094676559 | 40 | 5 | 24 | .0157142522 | 40 | 7 | 24 | .0175949377 |
| 40 | 3 | 33 | .0090139592 | 40 | 5 | 25 | .0150135411 | 40 | 7 | 25 | .0168111611 |
| 40 | 3 | 34 | .0085627768 | 40 | 5 | 26 | .0143505258 | 40 | 7 | 26 | .0160693743 |
| 40 | 3 | 35 | .0081087975 | 40 | 5 | 27 | .0137200763 | 40 | 7 | 27 | .0153640321 |

TABLE I-COVARIANCES OF NORMAL ORDER STATISTICS

| N | I | J | COVARIANCE | N | I | J | COVARIANCE | N | I | J | COVARIANCE |
|---|---|---|---|---|---|---|---|---|---|---|---|
| 40 | 7 | 28 | .0146902667 | 40 | 10 | 13 | .0369794605 | 40 | 13 | 16 | .0350212189 |
| 40 | 7 | 29 | .0140434475 | 40 | 10 | 14 | .0345792169 | 40 | 13 | 17 | .0330827634 |
| 40 | 7 | 30 | .0134187341 | 40 | 10 | 15 | .0324615916 | 40 | 13 | 18 | .0313267942 |
| 40 | 7 | 31 | .0128114739 | 40 | 10 | 16 | .0305751203 | 40 | 13 | 19 | .0297251333 |
| 40 | 7 | 32 | .0122177702 | 40 | 10 | 17 | .0288801224 | 40 | 13 | 20 | .0282549902 |
| 40 | 7 | 33 | .0116333949 | 40 | 10 | 18 | .0273454504 | 40 | 13 | 21 | .0268976641 |
| 40 | 7 | 34 | .0110522168 | 40 | 10 | 19 | .0259462546 | 40 | 13 | 22 | .0256376007 |
| 40 | 8 | 8 | .0516902736 | 40 | 10 | 20 | .0246624309 | 40 | 13 | 23 | .0244616969 |
| 40 | 8 | 9 | .0466208092 | 40 | 10 | 21 | .0234775463 | 40 | 13 | 24 | .0233588618 |
| 40 | 8 | 10 | .0424937987 | 40 | 10 | 22 | .0223780179 | 40 | 13 | 25 | .0223196921 |
| 40 | 8 | 11 | .0390576151 | 40 | 10 | 23 | .0213524496 | 40 | 13 | 26 | .0213359380 |
| 40 | 8 | 12 | .0361436652 | 40 | 10 | 24 | .0203911965 | 40 | 13 | 27 | .0204002218 |
| 40 | 8 | 13 | .0336345481 | 40 | 10 | 25 | .0194858967 | 40 | 13 | 28 | .0195059152 |
| 40 | 8 | 14 | .0314458620 | 40 | 10 | 26 | .0186289872 | 40 | 14 | 14 | .0413561610 |
| 40 | 8 | 15 | .0295152763 | 40 | 10 | 27 | .0178137906 | 40 | 14 | 15 | .0388352866 |
| 40 | 8 | 16 | .0277956909 | 40 | 10 | 28 | .0170348078 | 40 | 14 | 16 | .0365883581 |
| 40 | 8 | 17 | .0262508073 | 40 | 10 | 29 | .0162870841 | 40 | 14 | 17 | .0345685122 |
| 40 | 8 | 18 | .0248521720 | 40 | 10 | 30 | .0155653081 | 40 | 14 | 18 | .0327389480 |
| 40 | 8 | 19 | .0235771327 | 40 | 10 | 31 | .0148641595 | 40 | 14 | 19 | .0310703105 |
| 40 | 8 | 20 | .0224073692 | 40 | 11 | 11 | .0449296481 | 40 | 14 | 20 | .0295388368 |
| 40 | 8 | 21 | .0213278500 | 40 | 11 | 12 | .0415904489 | 40 | 14 | 21 | .0281250231 |
| 40 | 8 | 22 | .0203260928 | 40 | 11 | 13 | .0387135624 | 40 | 14 | 22 | .0268126709 |
| 40 | 8 | 23 | .0193915869 | 40 | 11 | 14 | .0362028872 | 40 | 14 | 23 | .0255882125 |
| 40 | 8 | 24 | .0185154197 | 40 | 11 | 15 | .0339873961 | 40 | 14 | 24 | .0244401606 |
| 40 | 8 | 25 | .0176900967 | 40 | 11 | 16 | .0320133808 | 40 | 14 | 25 | .0233586706 |
| 40 | 8 | 26 | .0169091960 | 40 | 11 | 17 | .0302394077 | 40 | 14 | 26 | .0223351707 |
| 40 | 8 | 27 | .0161667698 | 40 | 11 | 18 | .0286329420 | 40 | 14 | 27 | .0213618059 |
| 40 | 8 | 28 | .0154571172 | 40 | 11 | 19 | .0271680431 | 40 | 15 | 15 | .0405906022 |
| 40 | 8 | 29 | .0147752004 | 40 | 11 | 20 | .0258237550 | 40 | 15 | 16 | .0382449153 |
| 40 | 8 | 30 | .0141167094 | 40 | 11 | 21 | .0245829406 | 40 | 15 | 17 | .0361360003 |
| 40 | 8 | 31 | .0134772384 | 40 | 11 | 22 | .0234314244 | 40 | 15 | 18 | .0342255137 |
| 40 | 8 | 32 | .0128519058 | 40 | 11 | 23 | .0223572782 | 40 | 15 | 19 | .0324828569 |
| 40 | 8 | 33 | .0122357385 | 40 | 11 | 24 | .0213502127 | 40 | 15 | 20 | .0308832544 |
| 40 | 9 | 9 | .0489075921 | 40 | 11 | 25 | .0204013641 | 40 | 15 | 21 | .0294063877 |
| 40 | 9 | 10 | .0445844409 | 40 | 11 | 26 | .0195031610 | 40 | 15 | 22 | .0280354102 |
| 40 | 9 | 11 | .0409840439 | 40 | 11 | 27 | .0186488668 | 40 | 15 | 23 | .0267562093 |
| 40 | 9 | 12 | .0379302040 | 40 | 11 | 28 | .0178324423 | 40 | 15 | 24 | .0255568243 |
| 40 | 9 | 13 | .0353001727 | 40 | 11 | 29 | .0170486231 | 40 | 15 | 25 | .0244269056 |
| 40 | 9 | 14 | .0330056720 | 40 | 11 | 30 | .0162920553 | 40 | 15 | 26 | .0233573194 |
| 40 | 9 | 15 | .0309814925 | 40 | 12 | 12 | .0435174194 | 40 | 16 | 16 | .0400090677 |
| 40 | 9 | 16 | .0291783568 | 40 | 12 | 13 | .0405095009 | 40 | 16 | 17 | .0378012986 |
| 40 | 9 | 17 | .0275582881 | 40 | 12 | 14 | .0378836998 | 40 | 16 | 18 | .0358005333 |
| 40 | 9 | 18 | .0260914999 | 40 | 12 | 15 | .0355659726 | 40 | 16 | 19 | .0339748623 |
| 40 | 9 | 19 | .0247542448 | 40 | 12 | 16 | .0335003075 | 40 | 16 | 20 | .0322984518 |
| 40 | 9 | 20 | .0235273023 | 40 | 12 | 17 | .0316434822 | 40 | 16 | 21 | .0307501252 |
| 40 | 9 | 21 | .0223948915 | 40 | 12 | 18 | .0299615619 | 40 | 16 | 22 | .0293123233 |
| 40 | 9 | 22 | .0213439310 | 40 | 12 | 19 | .0284275004 | 40 | 16 | 23 | .0279703063 |
| 40 | 9 | 23 | .0203635562 | 40 | 12 | 20 | .0270194485 | 40 | 16 | 24 | .0267115130 |
| 40 | 9 | 24 | .0194445971 | 40 | 12 | 21 | .0257195142 | 40 | 16 | 25 | .0255250734 |
| 40 | 9 | 25 | .0185790463 | 40 | 12 | 22 | .0245128101 | 40 | 17 | 17 | .0395490127 |
| 40 | 9 | 26 | .0177598565 | 40 | 12 | 23 | .0233867869 | 40 | 17 | 18 | .0374559650 |
| 40 | 9 | 27 | .0169808402 | 40 | 12 | 24 | .0223307650 | 40 | 17 | 19 | .0355456091 |
| 40 | 9 | 28 | .0162362785 | 40 | 12 | 25 | .0213355063 | 40 | 17 | 20 | .0337910277 |
| 40 | 9 | 29 | .0155207933 | 40 | 12 | 26 | .0203931136 | 40 | 17 | 21 | .0321701455 |
| 40 | 9 | 30 | .0148298281 | 40 | 12 | 27 | .0194968649 | 40 | 17 | 22 | .0306646374 |
| 40 | 9 | 31 | .0141592814 | 40 | 12 | 28 | .0186403828 | 40 | 17 | 23 | .0292590873 |
| 40 | 9 | 32 | .0135038973 | 40 | 12 | 29 | .0178173270 | 40 | 17 | 24 | .0279403407 |
| 40 | 10 | 10 | .0466889966 | 40 | 13 | 13 | .0423370057 | 40 | 18 | 18 | .0391927460 |
| 40 | 10 | 11 | .0429242222 | 40 | 13 | 14 | .0395966566 | 40 | 18 | 19 | .0372002572 |
| 40 | 10 | 12 | .0397304102 | 40 | 13 | 15 | .0371775082 | 40 | 18 | 20 | .0353703686 |

TABLE I-COVARIANCES OF NORMAL ORDER STATISTICS

TABLE I-COVARIANCES OF NORMAL ORDER STATISISCS

| N | I | J | COVARIANCE | N | I | J | COVARIANCE | N | I | J | COVARIANCE |
|---|---|---|---|---|---|---|---|---|---|---|---|
| 40 | 18 | 21 | .0336800702 | 41 | 2 | 12 | .0240815305 | 41 | 3 | 34 | .0086648556 |
| 40 | 18 | 22 | .0321102461 | 41 | 2 | 13 | .0224040227 | 41 | 3 | 35 | .0082375709 |
| 40 | 18 | 23 | .0306448053 | 41 | 2 | 14 | .0209428204 | 41 | 3 | 36 | .0078067938 |
| 40 | 19 | 19 | .0389623691 | 41 | 2 | 15 | .0196556322 | 41 | 3 | 37 | .0073655443 |
| 40 | 19 | 20 | .0370530685 | 41 | 2 | 16 | .0185105656 | 41 | 3 | 38 | .0069032781 |
| 40 | 19 | 21 | .0352895879 | 41 | 2 | 17 | .0174830836 | 41 | 3 | 39 | .0064015763 |
| 40 | 19 | 22 | .0336519755 | 41 | 2 | 18 | .0165539729 | 41 | 4 | 4 | .0756534068 |
| 40 | 20 | 20 | .0388587592 | 41 | 2 | 19 | .0157079494 | 41 | 4 | 5 | .0618956268 |
| 40 | 20 | 21 | .0370095663 | 41 | 2 | 20 | .0149326811 | 41 | 4 | 6 | .0526373297 |
| 41 | 1 | 1 | .2265702007 | 41 | 2 | 21 | .0142180882 | 41 | 4 | 7 | .0459385716 |
| 41 | 1 | 2 | .1080956749 | 41 | 2 | 22 | .0135558317 | 41 | 4 | 8 | .0408408745 |
| 41 | 1 | 3 | .0728385975 | 41 | 2 | 23 | .0129389332 | 41 | 4 | 9 | .0368145170 |
| 41 | 1 | 4 | .0557136097 | 41 | 2 | 24 | .0123614875 | 41 | 4 | 10 | .0335420616 |
| 41 | 1 | 5 | .0454817514 | 41 | 2 | 25 | .0118184350 | 41 | 4 | 11 | .0308212508 |
| 41 | 1 | 6 | .0386217586 | 41 | 2 | 26 | .0113053803 | 41 | 4 | 12 | .0285168568 |
| 41 | 1 | 7 | .0336712388 | 41 | 2 | 27 | .0108184521 | 41 | 4 | 13 | .0265348841 |
| 41 | 1 | 8 | .0299112502 | 41 | 2 | 28 | .0103541928 | 41 | 4 | 14 | .0248078667 |
| 41 | 1 | 9 | .0269459228 | 41 | 2 | 29 | .0099094677 | 41 | 4 | 15 | .0232860543 |
| 41 | 1 | 10 | .0245387090 | 41 | 2 | 30 | .0094813711 | 41 | 4 | 16 | .0219319028 |
| 41 | 1 | 11 | .0225392355 | 41 | 2 | 31 | .0090671126 | 41 | 4 | 17 | .0207165093 |
| 41 | 1 | 12 | .0208471517 | 41 | 2 | 32 | .0086638996 | 41 | 4 | 18 | .0196172349 |
| 41 | 1 | 13 | .0193928132 | 41 | 2 | 33 | .0082688406 | 41 | 4 | 19 | .0186160744 |
| 41 | 1 | 14 | .0181262984 | 41 | 2 | 34 | .0078788137 | 41 | 4 | 20 | .0176985021 |
| 41 | 1 | 15 | .0170108369 | 41 | 2 | 35 | .0074901875 | 41 | 4 | 21 | .0168526327 |
| 41 | 1 | 16 | .0160187102 | 41 | 2 | 36 | .0070983611 | 41 | 4 | 22 | .0160686093 |
| 41 | 1 | 17 | .0151286013 | 41 | 2 | 37 | .0066970378 | 41 | 4 | 23 | .0153381699 |
| 41 | 1 | 18 | .0143238218 | 41 | 2 | 38 | .0062766460 | 41 | 4 | 24 | .0146543309 |
| 41 | 1 | 19 | .0135910985 | 41 | 2 | 39 | .0058203701 | 41 | 4 | 25 | .0140111300 |
| 41 | 1 | 20 | .0129197298 | 41 | 2 | 40 | .0052920067 | 41 | 4 | 26 | .0134034008 |
| 41 | 1 | 21 | .0123009723 | 41 | 3 | 3 | .0917381654 | 41 | 4 | 27 | .0128265938 |
| 41 | 1 | 22 | .0117275872 | 41 | 3 | 4 | .0703536772 | 41 | 4 | 28 | .0122766427 |
| 41 | 1 | 23 | .0111935169 | 41 | 3 | 5 | .0575239518 | 41 | 4 | 29 | .0117498271 |
| 41 | 1 | 24 | .0106936397 | 41 | 3 | 6 | .0488991397 | 41 | 4 | 30 | .0112426219 |
| 41 | 1 | 25 | .0102235651 | 41 | 3 | 7 | .0426633228 | 41 | 4 | 31 | .0107516375 |
| 41 | 1 | 26 | .0097794803 | 41 | 3 | 8 | .0379205272 | 41 | 4 | 32 | .0102736616 |
| 41 | 1 | 27 | .0093580397 | 41 | 3 | 9 | .0341760821 | 41 | 4 | 33 | .0098054979 |
| 41 | 1 | 28 | .0089562567 | 41 | 3 | 10 | .0311337910 | 41 | 4 | 34 | .0093434156 |
| 41 | 1 | 29 | .0085714065 | 41 | 3 | 11 | .0286050511 | 41 | 4 | 35 | .0088827848 |
| 41 | 1 | 30 | .0082009536 | 41 | 3 | 12 | .0264638296 | 41 | 4 | 36 | .0084181701 |
| 41 | 1 | 31 | .0078424769 | 41 | 3 | 13 | .0246225599 | 41 | 4 | 37 | .0079424404 |
| 41 | 1 | 32 | .0074935768 | 41 | 3 | 14 | .0230184142 | 41 | 4 | 38 | .0074441226 |
| 41 | 1 | 33 | .0071517679 | 41 | 3 | 15 | .0216050803 | 41 | 5 | 5 | .0659026945 |
| 41 | 1 | 34 | .0068143320 | 41 | 3 | 16 | .0203476223 | 41 | 5 | 6 | .0560672165 |
| 41 | 1 | 35 | .0064781035 | 41 | 3 | 17 | .0192191518 | 41 | 5 | 7 | .0489459079 |
| 41 | 1 | 36 | .0061391193 | 41 | 3 | 18 | .0181986055 | 41 | 5 | 8 | .0435238297 |
| 41 | 1 | 37 | .0057919467 | 41 | 3 | 19 | .0172692243 | 41 | 5 | 9 | .0392395228 |
| 41 | 1 | 38 | .0054282872 | 41 | 3 | 20 | .0164174877 | 41 | 5 | 10 | .0357562803 |
| 41 | 1 | 39 | .0050336022 | 41 | 3 | 21 | .0156323494 | 41 | 5 | 11 | .0328594393 |
| 41 | 1 | 40 | .0045765854 | 41 | 3 | 22 | .0149046743 | 41 | 5 | 12 | .0304054066 |
| 41 | 1 | 41 | .0039577729 | 41 | 3 | 23 | .0142268091 | 41 | 5 | 13 | .0282943286 |
| 41 | 2 | 2 | .1239576867 | 41 | 3 | 24 | .0135922580 | 41 | 5 | 14 | .0264545059 |
| 41 | 2 | 3 | .0837699308 | 41 | 3 | 25 | .0129954439 | 41 | 5 | 15 | .0248330495 |
| 41 | 2 | 4 | .0641692800 | 41 | 3 | 26 | .0124315381 | 41 | 5 | 16 | .0233900428 |
| 41 | 2 | 5 | .0524308847 | 41 | 3 | 27 | .0118963201 | 41 | 5 | 17 | .0220947601 |
| 41 | 2 | 6 | .0445489609 | 41 | 3 | 28 | .0113860308 | 41 | 5 | 18 | .0209231324 |
| 41 | 2 | 7 | .0388549630 | 41 | 3 | 29 | .0108972240 | 41 | 5 | 19 | .0198559964 |
| 41 | 2 | 8 | .0345269188 | 41 | 3 | 30 | .0104266610 | 41 | 5 | 20 | .0188778650 |
| 41 | 2 | 9 | .0311115425 | 41 | 3 | 31 | .0099712538 | 41 | 5 | 21 | .0179760626 |
| 41 | 2 | 10 | .0283376626 | 41 | 3 | 32 | .0095279717 | 41 | 5 | 22 | .0171400999 |
| 41 | 2 | 11 | .0260327360 | 41 | 3 | 33 | .0090936609 | 41 | 5 | 23 | .0163612051 |

TABLE I-COVARIANCES OF NORMAL ORDER STATISTICS

| N | I | J | COVARIANCE | N | I | J | COVARIANCE | N | I | J | COVARIANCE |
|---|---|---|---|---|---|---|---|---|---|---|---|
| 41 | 5 | 24 | .0156319764 | 41 | 7 | 22 | .0191599874 | 41 | 9 | 28 | .0161570580 |
| 41 | 5 | 25 | .0149461108 | 41 | 7 | 23 | .0182906879 | 41 | 9 | 29 | .0154639691 |
| 41 | 5 | 26 | .0142981322 | 41 | 7 | 24 | .0174767470 | 41 | 9 | 30 | .0147959745 |
| 41 | 5 | 27 | .0136831129 | 41 | 7 | 25 | .0167110360 | 41 | 9 | 31 | .0141495850 |
| 41 | 5 | 28 | .0130965290 | 41 | 7 | 26 | .0159873378 | 41 | 9 | 32 | .0135211903 |
| 41 | 5 | 29 | .0125343510 | 41 | 7 | 27 | .0153002219 | 41 | 9 | 33 | .0129053592 |
| 41 | 5 | 30 | .0119931148 | 41 | 7 | 28 | .0146450721 | 41 | 10 | 10 | .0460222359 |
| 41 | 5 | 31 | .0114695626 | 41 | 7 | 29 | .0140177956 | 41 | 10 | 11 | .0423173791 |
| 41 | 5 | 32 | .0109600584 | 41 | 7 | 30 | .0134141024 | 41 | 10 | 12 | .0391755970 |
| 41 | 5 | 33 | .0104605898 | 41 | 7 | 31 | .0128293570 | 41 | 10 | 13 | .0364705332 |
| 41 | 5 | 34 | .0099672439 | 41 | 7 | 32 | .0122595403 | 41 | 10 | 14 | .0341113048 |
| 41 | 5 | 35 | .0094757711 | 41 | 7 | 33 | .0117015199 | 41 | 10 | 15 | .0320307785 |
| 41 | 5 | 36 | .0089802796 | 41 | 7 | 34 | .0111512422 | 41 | 10 | 16 | .0301782280 |
| 41 | 5 | 37 | .0084726085 | 41 | 7 | 35 | .0106026414 | 41 | 10 | 17 | .0285145615 |
| 41 | 6 | 6 | .0593225930 | 41 | 8 | 8 | .0510240358 | 41 | 10 | 18 | .0270091156 |
| 41 | 6 | 7 | .0518020462 | 41 | 8 | 9 | .0460240463 | 41 | 10 | 19 | .0256374383 |
| 41 | 6 | 8 | .0460731683 | 41 | 8 | 10 | .0419550337 | 41 | 10 | 20 | .0243797369 |
| 41 | 6 | 9 | .0415447086 | 41 | 8 | 11 | .0385683482 | 41 | 10 | 21 | .0232198196 |
| 41 | 6 | 10 | .0378618416 | 41 | 8 | 12 | .0356974269 | 41 | 10 | 22 | .0221443198 |
| 41 | 6 | 11 | .0347982221 | 41 | 8 | 13 | .0332263073 | 41 | 10 | 23 | .0211420587 |
| 41 | 6 | 12 | .0322023690 | 41 | 8 | 14 | .0310716418 | 41 | 10 | 24 | .0202036549 |
| 41 | 6 | 13 | .0299689001 | 41 | 8 | 15 | .0291718921 | 41 | 10 | 25 | .0193211119 |
| 41 | 6 | 14 | .0280221236 | 41 | 8 | 16 | .0274805644 | 41 | 10 | 26 | .0184871798 |
| 41 | 6 | 15 | .0263061967 | 41 | 8 | 17 | .0259618301 | 41 | 10 | 27 | .0176951943 |
| 41 | 6 | 16 | .0247789674 | 41 | 8 | 18 | .0245876108 | 41 | 10 | 28 | .0169395801 |
| 41 | 6 | 17 | .0234079721 | 41 | 8 | 19 | .0233355759 | 41 | 10 | 29 | .0162157671 |
| 41 | 6 | 18 | .0221677554 | 41 | 8 | 20 | .0221876983 | 41 | 10 | 30 | .0155191853 |
| 41 | 6 | 19 | .0210380381 | 41 | 8 | 21 | .0211292096 | 41 | 10 | 31 | .0148450434 |
| 41 | 6 | 20 | .0200024387 | 41 | 8 | 22 | .0201478568 | 41 | 10 | 32 | .0141889647 |
| 41 | 6 | 21 | .0190475687 | 41 | 8 | 23 | .0192333046 | 41 | 11 | 11 | .0442474798 |
| 41 | 6 | 22 | .0181623927 | 41 | 8 | 24 | .0183767311 | 41 | 11 | 12 | .0409664266 |
| 41 | 6 | 23 | .0173377208 | 41 | 8 | 25 | .0175707161 | 41 | 11 | 13 | .0381408752 |
| 41 | 6 | 24 | .0165657299 | 41 | 8 | 26 | .0168090821 | 41 | 11 | 14 | .0356761321 |
| 41 | 6 | 25 | .0158395846 | 41 | 8 | 27 | .0160863354 | 41 | 11 | 15 | .0335022300 |
| 41 | 6 | 26 | .0151532927 | 41 | 8 | 28 | .0153971352 | 41 | 11 | 16 | .0315662825 |
| 41 | 6 | 27 | .0145017106 | 41 | 8 | 29 | .0147364803 | 41 | 11 | 17 | .0298275087 |
| 41 | 6 | 28 | .0138803956 | 41 | 8 | 30 | .0141001338 | 41 | 11 | 18 | .0282538960 |
| 41 | 6 | 29 | .0132852426 | 41 | 8 | 31 | .0134842049 | 41 | 11 | 19 | .0268199180 |
| 41 | 6 | 30 | .0127122841 | 41 | 8 | 32 | .0128844695 | 41 | 11 | 20 | .0255049510 |
| 41 | 6 | 31 | .0121577934 | 41 | 8 | 33 | .0122967175 | 41 | 11 | 21 | .0242921416 |
| 41 | 6 | 32 | .0116181352 | 41 | 8 | 34 | .0117167347 | 41 | 11 | 22 | .0231675980 |
| 41 | 6 | 33 | .0110892429 | 41 | 9 | 9 | .0482492156 | 41 | 11 | 23 | .0221197250 |
| 41 | 6 | 34 | .0105666924 | 41 | 9 | 10 | .0439894204 | 41 | 11 | 24 | .0211385473 |
| 41 | 6 | 35 | .0100460728 | 41 | 9 | 11 | .0404430189 | 41 | 11 | 25 | .0202153946 |
| 41 | 6 | 36 | .0095215852 | 41 | 9 | 12 | .0374360469 | 41 | 11 | 26 | .0193428473 |
| 41 | 7 | 7 | .0545816295 | 41 | 9 | 13 | .0348473384 | 41 | 11 | 27 | .0185143303 |
| 41 | 7 | 8 | .0485554353 | 41 | 9 | 14 | .0325897715 | 41 | 11 | 28 | .0177239107 |
| 41 | 7 | 9 | .0437901379 | 41 | 9 | 15 | .0305990019 | 41 | 11 | 29 | .0169667046 |
| 41 | 7 | 10 | .0399134706 | 41 | 9 | 16 | .0288264140 | 41 | 11 | 30 | .0162383979 |
| 41 | 7 | 11 | .0366878275 | 41 | 9 | 17 | .0272345530 | 41 | 11 | 31 | .0155337179 |
| 41 | 7 | 12 | .0339541175 | 41 | 9 | 18 | .0257940635 | 41 | 12 | 12 | .0428337756 |
| 41 | 7 | 13 | .0316016237 | 41 | 9 | 19 | .0244815682 | 41 | 12 | 13 | .0398812356 |
| 41 | 7 | 14 | .0295508000 | 41 | 9 | 20 | .0232781664 | 41 | 12 | 14 | .0373049040 |
| 41 | 7 | 15 | .0277429413 | 41 | 9 | 21 | .0221683339 | 41 | 12 | 15 | .0350318957 |
| 41 | 7 | 16 | .0261337115 | 41 | 9 | 22 | .0211391612 | 41 | 12 | 16 | .0330070940 |
| 41 | 7 | 17 | .0246889444 | 41 | 9 | 23 | .0201799200 | 41 | 12 | 17 | .0311879787 |
| 41 | 7 | 18 | .0233818446 | 41 | 9 | 24 | .0192816430 | 41 | 12 | 18 | .0295411701 |
| 41 | 7 | 19 | .0221910801 | 41 | 9 | 25 | .0184365936 | 41 | 12 | 19 | .0280400681 |
| 41 | 7 | 20 | .0210994628 | 41 | 9 | 26 | .0176379946 | 41 | 12 | 20 | .0266632024 |
| 41 | 7 | 21 | .0200929754 | 41 | 9 | 27 | .0168799540 | 41 | 12 | 21 | .0253930361 |

TABLE I-COVARIANCES OF NORMAL ORDER STATISTICS

| N | I | J | COVARIANCE | N | I | J | COVARIANCE | N | I | J | COVARIANCE |
|---|---|---|---|---|---|---|---|---|---|---|---|
| 41 | 12 | 22 | .0242150269 | 41 | 16 | 22 | .0288328022 | 42 | 1 | 31 | .0078807483 |
| 41 | 12 | 23 | .0231169337 | 41 | 16 | 23 | .0275314325 | 42 | 1 | 32 | .0075428642 |
| 41 | 12 | 24 | .0220883428 | 41 | 16 | 24 | .0263123029 | 42 | 1 | 33 | .0072132930 |
| 41 | 12 | 25 | .0211201776 | 41 | 16 | 25 | .0251649535 | 42 | 1 | 34 | .0068897486 |
| 41 | 12 | 26 | .0202045878 | 41 | 16 | 26 | .0240801911 | 42 | 1 | 35 | .0065697008 |
| 41 | 12 | 27 | .0193351083 | 41 | 17 | 17 | .0387687068 | 42 | 1 | 36 | .0062501743 |
| 41 | 12 | 28 | .0185059979 | 41 | 17 | 18 | .0367292007 | 42 | 1 | 37 | .0059274137 |
| 41 | 12 | 29 | .0177114534 | 41 | 17 | 19 | .0348688647 | 42 | 1 | 38 | .0055962297 |
| 41 | 12 | 30 | .0169460407 | 41 | 17 | 20 | .0331613780 | 42 | 1 | 39 | .0052486522 |
| 41 | 13 | 13 | .0416537061 | 41 | 17 | 21 | .0315851726 | 42 | 1 | 40 | .0048706727 |
| 41 | 13 | 14 | .0389656050 | 41 | 17 | 22 | .0301223810 | 42 | 1 | 41 | .0044320575 |
| 41 | 13 | 15 | .0365935336 | 41 | 17 | 23 | .0287580229 | 42 | 1 | 42 | .0038366169 |
| 41 | 13 | 16 | .0344801124 | 41 | 17 | 24 | .0274793604 | 42 | 2 | 2 | .1230051850 |
| 41 | 13 | 17 | .0325810697 | 41 | 17 | 25 | .0262753859 | 42 | 2 | 3 | .0830978078 |
| 41 | 13 | 18 | .0308616576 | 41 | 18 | 18 | .0383843015 | 42 | 2 | 4 | .0636438745 |
| 41 | 13 | 19 | .0292941946 | 41 | 18 | 19 | .0364441364 | 42 | 2 | 5 | .0519978699 |
| 41 | 13 | 20 | .0278563243 | 41 | 18 | 20 | .0346633248 | 42 | 2 | 6 | .0441805782 |
| 41 | 13 | 21 | .0265297376 | 41 | 18 | 21 | .0330194246 | 42 | 2 | 7 | .0385349888 |
| 41 | 13 | 22 | .0252992273 | 41 | 18 | 22 | .0314938204 | 42 | 2 | 8 | .0342449921 |
| 41 | 13 | 23 | .0241519631 | 41 | 18 | 23 | .0300708764 | 42 | 2 | 9 | .0308606177 |
| 41 | 13 | 24 | .0230770290 | 41 | 18 | 24 | .0287372828 | 42 | 2 | 10 | .0281127209 |
| 41 | 13 | 25 | .0220652124 | 41 | 19 | 19 | .0381059237 | 42 | 2 | 11 | .0258300732 |
| 41 | 13 | 26 | .0211085540 | 41 | 19 | 20 | .0362528329 | 42 | 2 | 12 | .0238983366 |
| 41 | 13 | 27 | .0202000151 | 41 | 19 | 21 | .0345425732 | 42 | 2 | 13 | .0222381200 |
| 41 | 13 | 28 | .0193334945 | 41 | 19 | 22 | .0329557375 | 42 | 2 | 14 | .0207924932 |
| 41 | 13 | 29 | .0185031064 | 41 | 19 | 23 | .0314760299 | 42 | 2 | 15 | .0195195117 |
| 41 | 14 | 14 | .0406494672 | 41 | 20 | 20 | .0379539046 | 42 | 2 | 16 | .0183875508 |
| 41 | 14 | 15 | .0381806835 | 41 | 20 | 21 | .0361675432 | 42 | 2 | 17 | .0173722853 |
| 41 | 14 | 16 | .0359811833 | 41 | 20 | 22 | .0345100367 | 42 | 2 | 18 | .0164546741 |
| 41 | 14 | 17 | .0340049357 | 41 | 21 | 21 | .0379093843 | 42 | 2 | 19 | .0156195769 |
| 41 | 14 | 18 | .0322158086 | 42 | 1 | 1 | .2251984562 | 42 | 2 | 20 | .0148547830 |
| 41 | 14 | 19 | .0305850064 | 42 | 1 | 2 | .1073623422 | 42 | 2 | 21 | .0141503185 |
| 41 | 14 | 20 | .0290892464 | 42 | 1 | 3 | .0723221041 | 42 | 2 | 22 | .0134979384 |
| 41 | 14 | 21 | .0277094334 | 42 | 1 | 4 | .0553103463 | 42 | 2 | 23 | .0128907511 |
| 41 | 14 | 22 | .0264296983 | 42 | 1 | 5 | .0451498001 | 42 | 2 | 24 | .0123229364 |
| 41 | 14 | 23 | .0252367355 | 42 | 1 | 6 | .0383397506 | 42 | 2 | 25 | .0117895242 |
| 41 | 14 | 24 | .0241192983 | 42 | 1 | 7 | .0334267004 | 42 | 2 | 26 | .0112862150 |
| 41 | 14 | 25 | .0230678057 | 42 | 1 | 8 | .0296962209 | 42 | 2 | 27 | .0108092368 |
| 41 | 14 | 26 | .0220740974 | 42 | 1 | 9 | .0267549922 | 42 | 2 | 28 | .0103552379 |
| 41 | 14 | 27 | .0211308191 | 42 | 1 | 10 | .0243680245 | 42 | 2 | 29 | .0099212023 |
| 41 | 14 | 28 | .0202310242 | 42 | 1 | 11 | .0223859553 | 42 | 2 | 30 | .0095043760 |
| 41 | 15 | 15 | .0398465165 | 42 | 1 | 12 | .0207091213 | 42 | 2 | 31 | .0091021797 |
| 41 | 15 | 16 | .0375555282 | 42 | 1 | 13 | .0192683642 | 42 | 2 | 32 | .0087120996 |
| 41 | 15 | 17 | .0354970377 | 42 | 1 | 14 | .0180141185 | 42 | 2 | 33 | .0083315896 |
| 41 | 15 | 18 | .0336334550 | 42 | 1 | 15 | .0169098820 | 42 | 2 | 34 | .0079579972 |
| 41 | 15 | 19 | .0319348051 | 42 | 1 | 16 | .0159281421 | 42 | 2 | 35 | .0075884303 |
| 41 | 15 | 20 | .0303768332 | 42 | 1 | 17 | .0150477458 | 42 | 2 | 36 | .0072194730 |
| 41 | 15 | 21 | .0289396498 | 42 | 1 | 18 | .0142521399 | 42 | 2 | 37 | .0068467592 |
| 41 | 15 | 22 | .0276067625 | 42 | 1 | 19 | .0135281618 | 42 | 2 | 38 | .0064642915 |
| 41 | 15 | 23 | .0263643726 | 42 | 1 | 20 | .0128652017 | 42 | 2 | 39 | .0060628881 |
| 41 | 15 | 24 | .0252008551 | 42 | 1 | 21 | .0122546026 | 42 | 2 | 40 | .0056263528 |
| 41 | 15 | 25 | .0241062607 | 42 | 1 | 22 | .0116892052 | 42 | 2 | 41 | .0051197627 |
| 41 | 15 | 26 | .0230718456 | 42 | 1 | 23 | .0111630189 | 42 | 3 | 3 | .0909332905 |
| 41 | 15 | 27 | .0220898893 | 42 | 1 | 24 | .0106709890 | 42 | 3 | 4 | .0697237222 |
| 41 | 16 | 16 | .0392414590 | 42 | 1 | 25 | .0102088010 | 42 | 3 | 5 | .0570042140 |
| 41 | 16 | 17 | .0370894240 | 42 | 1 | 26 | .0097727163 | 42 | 3 | 6 | .0484564837 |
| 41 | 16 | 18 | .0351405151 | 42 | 1 | 27 | .0093594643 | 42 | 3 | 7 | .0422783558 |
| 41 | 16 | 19 | .0333634932 | 42 | 1 | 28 | .0089661546 | 42 | 3 | 8 | .0375808529 |
| 41 | 16 | 20 | .0317330821 | 42 | 1 | 29 | .0085901776 | 42 | 3 | 9 | .0338732658 |
| 41 | 16 | 21 | .0302285694 | 42 | 1 | 30 | .0082291273 | 42 | 3 | 10 | .0308618209 |

TABLE I-COVARIANCES OF NORMAL ORDER STATISTICS

| N | I | J | COVARIANCE | N | I | J | COVARIANCE | N | I | J | COVARIANCE |
|---|---|---|---|---|---|---|---|---|---|---|---|
| 42 | 3 | 11 | .0283594878 | 42 | 4 | 34 | .0094224264 | 42 | 6 | 26 | .0150818466 |
| 42 | 3 | 12 | .0262413035 | 42 | 4 | 35 | .0089853220 | 42 | 6 | 27 | .0144455396 |
| 42 | 3 | 13 | .0244204568 | 42 | 4 | 36 | .0085486241 | 42 | 6 | 28 | .0138397126 |
| 42 | 3 | 14 | .0228346718 | 42 | 4 | 37 | .0081073309 | 42 | 6 | 29 | .0132604666 |
| 42 | 3 | 15 | .0214380510 | 42 | 4 | 38 | .0076547116 | 42 | 6 | 30 | .0127041170 |
| 42 | 3 | 16 | .0201959796 | 42 | 4 | 39 | .0071796555 | 42 | 6 | 31 | .0121672410 |
| 42 | 3 | 17 | .0190818235 | 42 | 5 | 5 | .0652162138 | 42 | 6 | 32 | .0116466931 |
| 42 | 3 | 18 | .0180747244 | 42 | 5 | 6 | .0554814373 | 42 | 6 | 33 | .0111390512 |
| 42 | 3 | 19 | .0171580906 | 42 | 5 | 7 | .0484354520 | 42 | 6 | 34 | .0106402177 |
| 42 | 3 | 20 | .0163185404 | 42 | 5 | 8 | .0430724430 | 42 | 6 | 35 | .0101462063 |
| 42 | 3 | 21 | .0155451491 | 42 | 5 | 9 | .0388361290 | 42 | 6 | 36 | .0096533680 |
| 42 | 3 | 22 | .0148288947 | 42 | 5 | 10 | .0353929751 | 42 | 6 | 37 | .0091560578 |
| 42 | 3 | 23 | .0141622363 | 42 | 5 | 11 | .0325303789 | 42 | 7 | 7 | .0539345996 |
| 42 | 3 | 24 | .0135387870 | 42 | 5 | 12 | .0301061504 | 42 | 7 | 8 | .0479821449 |
| 42 | 3 | 25 | .0129530695 | 42 | 5 | 13 | .0280214287 | 42 | 7 | 9 | .0432767393 |
| 42 | 3 | 26 | .0124003437 | 42 | 5 | 14 | .0262052393 | 42 | 7 | 10 | .0394500675 |
| 42 | 3 | 27 | .0118764807 | 42 | 5 | 15 | .0246052354 | 42 | 7 | 11 | .0362670939 |
| 42 | 3 | 28 | .0113778436 | 42 | 5 | 16 | .0231819141 | 42 | 7 | 12 | .0335704774 |
| 42 | 3 | 29 | .0109011540 | 42 | 5 | 17 | .0219048755 | 42 | 7 | 13 | .0312507382 |
| 42 | 3 | 30 | .0104433702 | 42 | 5 | 18 | .0207503193 | 42 | 7 | 14 | .0292292337 |
| 42 | 3 | 31 | .0100016175 | 42 | 5 | 19 | .0196993126 | 42 | 7 | 15 | .0274479408 |
| 42 | 3 | 32 | .0095731348 | 42 | 5 | 20 | .0187365655 | 42 | 7 | 16 | .0258630539 |
| 42 | 3 | 33 | .0091551574 | 42 | 5 | 21 | .0178495648 | 42 | 7 | 17 | .0244408260 |
| 42 | 3 | 34 | .0087447597 | 42 | 5 | 22 | .0170279505 | 42 | 7 | 18 | .0231547903 |
| 42 | 3 | 35 | .0083387423 | 42 | 5 | 23 | .0162630554 | 42 | 7 | 19 | .0219838675 |
| 42 | 3 | 36 | .0079334170 | 42 | 5 | 24 | .0155475748 | 42 | 7 | 20 | .0209110712 |
| 42 | 3 | 37 | .0075240107 | 42 | 5 | 25 | .0148753283 | 42 | 7 | 21 | .0199225793 |
| 42 | 3 | 38 | .0071038157 | 42 | 5 | 26 | .0142410235 | 42 | 7 | 22 | .0190069733 |
| 42 | 3 | 39 | .0066627701 | 42 | 5 | 27 | .0136399600 | 42 | 7 | 23 | .0181546492 |
| 42 | 3 | 40 | .0061831720 | 42 | 5 | 28 | .0130677822 | 42 | 7 | 24 | .0173574518 |
| 42 | 4 | 4 | .0749229582 | 42 | 5 | 29 | .0125204819 | 42 | 7 | 25 | .0166083934 |
| 42 | 4 | 5 | .0612923455 | 42 | 5 | 30 | .0119946060 | 42 | 7 | 26 | .0159013488 |
| 42 | 4 | 6 | .0521229964 | 42 | 5 | 31 | .0114872220 | 42 | 7 | 27 | .0152309049 |
| 42 | 4 | 7 | .0454907863 | 42 | 5 | 32 | .0109953682 | 42 | 7 | 28 | .0145924945 |
| 42 | 4 | 8 | .0404453045 | 42 | 5 | 33 | .0105155796 | 42 | 7 | 29 | .0139823869 |
| 42 | 4 | 9 | .0364614028 | 42 | 5 | 34 | .0100442005 | 42 | 7 | 30 | .0133970027 |
| 42 | 4 | 10 | .0332244428 | 42 | 5 | 35 | .0095777693 | 42 | 7 | 31 | .0128321365 |
| 42 | 4 | 11 | .0305339844 | 42 | 5 | 36 | .0091122252 | 42 | 7 | 32 | .0122834361 |
| 42 | 4 | 12 | .0282560361 | 42 | 5 | 37 | .0086418585 | 42 | 7 | 33 | .0117476707 |
| 42 | 4 | 13 | .0262974764 | 42 | 5 | 38 | .0081590632 | 42 | 7 | 34 | .0112222484 |
| 42 | 4 | 14 | .0245914786 | 42 | 6 | 6 | .0586629243 | 42 | 7 | 35 | .0107030255 |
| 42 | 4 | 15 | .0230887739 | 42 | 6 | 7 | .0512266864 | 42 | 7 | 36 | .0101841803 |
| 42 | 4 | 16 | .0217521884 | 42 | 6 | 8 | .0455639017 | 42 | 8 | 8 | .0503865737 |
| 42 | 4 | 17 | .0205531082 | 42 | 6 | 9 | .0410891122 | 42 | 8 | 9 | .0454527070 |
| 42 | 4 | 18 | .0194691241 | 42 | 6 | 10 | .0374510446 | 42 | 8 | 10 | .0414388645 |
| 42 | 4 | 19 | .0184824210 | 42 | 6 | 11 | .0344256631 | 42 | 8 | 11 | .0380992358 |
| 42 | 4 | 20 | .0175786412 | 42 | 6 | 12 | .0318630593 | 42 | 8 | 12 | .0352691925 |
| 42 | 4 | 21 | .0167460539 | 42 | 6 | 13 | .0296589621 | 42 | 8 | 13 | .0328341419 |
| 42 | 4 | 22 | .0159749409 | 42 | 6 | 14 | .0277384887 | 42 | 8 | 14 | .0307117400 |
| 42 | 4 | 23 | .0152571561 | 42 | 6 | 15 | .0260464023 | 42 | 8 | 15 | .0288412047 |
| 42 | 4 | 24 | .0145858114 | 42 | 6 | 16 | .0245410197 | 42 | 8 | 16 | .0271766225 |
| 42 | 4 | 25 | .0139550351 | 42 | 6 | 17 | .0231902545 | 42 | 8 | 17 | .0256826114 |
| 42 | 4 | 26 | .0133597667 | 42 | 6 | 18 | .0219689601 | 42 | 8 | 18 | .0243314414 |
| 42 | 4 | 27 | .0127955881 | 42 | 6 | 19 | .0208571084 | 42 | 8 | 19 | .0231010704 |
| 42 | 4 | 28 | .0122585959 | 42 | 6 | 20 | .0198385156 | 42 | 8 | 20 | .0219737342 |
| 42 | 4 | 29 | .0117452710 | 42 | 6 | 21 | .0188999422 | 42 | 8 | 21 | .0209349133 |
| 42 | 4 | 30 | .0112523089 | 42 | 6 | 22 | .0180304780 | 42 | 8 | 22 | .0199725937 |
| 42 | 4 | 31 | .0107764983 | 42 | 6 | 23 | .0172210824 | 42 | 8 | 23 | .0190766462 |
| 42 | 4 | 32 | .0103147876 | 42 | 6 | 24 | .0164641344 | 42 | 8 | 24 | .0182383528 |
| 42 | 4 | 33 | .0098643835 | 42 | 6 | 25 | .0157530096 | 42 | 8 | 25 | .0174502938 |

TABLE I-COVARIANCES OF NORMAL ORDER STATISTICS

| N | I | J | COVARIANCE | N | I | J | COVARIANCE | N | I | J | COVARIANCE |
|---|---|---|---|---|---|---|---|---|---|---|---|
| 42 | 8 | 26 | .0167063648 | 42 | 11 | 11 | .0435981477 | 42 | 14 | 14 | .0399890881 |
| 42 | 8 | 27 | .0160014035 | 42 | 11 | 12 | .0403719571 | 42 | 14 | 15 | .0375680430 |
| 42 | 8 | 28 | .0153305261 | 42 | 11 | 13 | .0375947994 | 42 | 14 | 16 | .0354119006 |
| 42 | 8 | 29 | .0146889621 | 42 | 11 | 14 | .0351733130 | 42 | 14 | 17 | .0334754019 |
| 42 | 8 | 30 | .0140724996 | 42 | 11 | 15 | .0330385307 | 42 | 14 | 18 | .0317230293 |
| 42 | 8 | 31 | .0134774706 | 42 | 11 | 16 | .0311383400 | 42 | 14 | 19 | .0301264973 |
| 42 | 8 | 32 | .0129000300 | 42 | 11 | 17 | .0294325779 | 42 | 14 | 20 | .0286629665 |
| 42 | 8 | 33 | .0123362422 | 42 | 11 | 18 | .0278897329 | 42 | 14 | 21 | .0273137352 |
| 42 | 8 | 34 | .0117828866 | 42 | 11 | 19 | .0264846786 | 42 | 14 | 22 | .0260632636 |
| 42 | 8 | 35 | .0112362404 | 42 | 11 | 20 | .0251971029 | 42 | 14 | 23 | .0248984966 |
| 42 | 9 | 9 | .0476226019 | 42 | 11 | 21 | .0240103939 | 42 | 14 | 24 | .0238083740 |
| 42 | 9 | 10 | .0434227738 | 42 | 11 | 22 | .0229108747 | 42 | 14 | 25 | .0227834520 |
| 42 | 9 | 11 | .0399274314 | 42 | 11 | 23 | .0218872255 | 42 | 14 | 26 | .0218158227 |
| 42 | 9 | 12 | .0369647334 | 42 | 11 | 24 | .0209297936 | 42 | 14 | 27 | .0208986247 |
| 42 | 9 | 13 | .0344150179 | 42 | 11 | 25 | .0200301608 | 42 | 14 | 28 | .0200253182 |
| 42 | 9 | 14 | .0321922624 | 42 | 11 | 26 | .0191811053 | 42 | 14 | 29 | .0191899977 |
| 42 | 9 | 15 | .0302329451 | 42 | 11 | 27 | .0183762144 | 42 | 15 | 15 | .0391482956 |
| 42 | 9 | 16 | .0284890759 | 42 | 11 | 28 | .0176095090 | 42 | 15 | 16 | .0369074831 |
| 42 | 9 | 17 | .0269236870 | 42 | 11 | 29 | .0168760236 | 42 | 15 | 17 | .0348951324 |
| 42 | 9 | 18 | .0255078193 | 42 | 11 | 30 | .0161721213 | 42 | 15 | 18 | .0330743537 |
| 42 | 9 | 19 | .0242184386 | 42 | 11 | 31 | .0154938804 | 42 | 15 | 19 | .0314157512 |
| 42 | 9 | 20 | .0230369543 | 42 | 11 | 32 | .0148358354 | 42 | 15 | 20 | .0298955613 |
| 42 | 9 | 21 | .0219481123 | 42 | 12 | 12 | .0421794820 | 42 | 15 | 21 | .0284943048 |
| 42 | 9 | 22 | .0209391871 | 42 | 12 | 13 | .0392796630 | 42 | 15 | 22 | .0271958179 |
| 42 | 9 | 23 | .0199995539 | 42 | 12 | 14 | .0367504231 | 42 | 15 | 23 | .0259865584 |
| 42 | 9 | 24 | .0191203639 | 42 | 12 | 15 | .0345199756 | 42 | 15 | 24 | .0248551383 |
| 42 | 9 | 25 | .0182940609 | 42 | 12 | 16 | .0325340536 | 42 | 15 | 25 | .0237919159 |
| 42 | 9 | 26 | .0175140810 | 42 | 12 | 17 | .0307508067 | 42 | 15 | 26 | .0227885041 |
| 42 | 9 | 27 | .0167748298 | 42 | 12 | 18 | .0291373851 | 42 | 15 | 27 | .0218375917 |
| 42 | 9 | 28 | .0160713274 | 42 | 12 | 19 | .0276676061 | 42 | 15 | 28 | .0209325156 |
| 42 | 9 | 29 | .0153984913 | 42 | 12 | 20 | .0263203356 | 42 | 16 | 16 | .0385126934 |
| 42 | 9 | 30 | .0147512850 | 42 | 12 | 21 | .0250783411 | 42 | 16 | 17 | .0364132506 |
| 42 | 9 | 31 | .0141259483 | 42 | 12 | 22 | .0239273898 | 42 | 16 | 18 | .0345132474 |
| 42 | 9 | 32 | .0135199447 | 42 | 12 | 23 | .0228555467 | 42 | 16 | 19 | .0327820776 |
| 42 | 9 | 33 | .0129296446 | 42 | 12 | 24 | .0218527063 | 42 | 16 | 20 | .0311949932 |
| 42 | 9 | 34 | .0123494754 | 42 | 12 | 25 | .0209100213 | 42 | 16 | 21 | .0297317143 |
| 42 | 10 | 10 | .0453911135 | 42 | 12 | 26 | .0200196410 | 42 | 16 | 22 | .0283754444 |
| 42 | 10 | 11 | .0417425321 | 42 | 12 | 27 | .0191750954 | 42 | 16 | 23 | .0271121572 |
| 42 | 10 | 12 | .0386495452 | 42 | 12 | 28 | .0183710288 | 42 | 16 | 24 | .0259300295 |
| 42 | 10 | 13 | .0359874481 | 42 | 12 | 29 | .0176022154 | 42 | 16 | 25 | .0248189803 |
| 42 | 10 | 14 | .0336665630 | 42 | 12 | 30 | .0168636118 | 42 | 16 | 26 | .0237702341 |
| 42 | 10 | 15 | .0316206578 | 42 | 12 | 31 | .0161508600 | 42 | 16 | 27 | .0227758565 |
| 42 | 10 | 16 | .0297996968 | 42 | 13 | 13 | .0410064737 | 42 | 17 | 17 | .0380299336 |
| 42 | 10 | 17 | .0281651376 | 42 | 13 | 14 | .0383675280 | 42 | 17 | 18 | .0360415460 |
| 42 | 10 | 18 | .0266867720 | 42 | 13 | 15 | .0360397143 | 42 | 17 | 19 | .0342289955 |
| 42 | 10 | 19 | .0253405333 | 42 | 13 | 16 | .0339665617 | 42 | 17 | 20 | .0325665117 |
| 42 | 10 | 20 | .0241069390 | 42 | 13 | 17 | .0321045092 | 42 | 17 | 21 | .0310329885 |
| 42 | 10 | 21 | .0229700280 | 42 | 13 | 18 | .0304193692 | 42 | 17 | 22 | .0296109609 |
| 42 | 10 | 22 | .0219166133 | 42 | 13 | 19 | .0288839126 | 42 | 17 | 23 | .0282858321 |
| 42 | 10 | 23 | .0209356632 | 42 | 13 | 20 | .0274761720 | 42 | 17 | 24 | .0270452562 |
| 42 | 10 | 24 | .0200179647 | 42 | 13 | 21 | .0261781980 | 42 | 17 | 25 | .0258786072 |
| 42 | 10 | 25 | .0191558410 | 42 | 13 | 22 | .0249751276 | 42 | 17 | 26 | .0247765951 |
| 42 | 10 | 26 | .0183424598 | 42 | 13 | 23 | .0238544335 | 42 | 18 | 18 | .0376306488 |
| 42 | 10 | 27 | .0175713445 | 42 | 13 | 24 | .0228053849 | 42 | 18 | 19 | .0357386384 |
| 42 | 10 | 28 | .0168367828 | 42 | 13 | 25 | .0218188906 | 42 | 18 | 20 | .0340029138 |
| 42 | 10 | 29 | .0161342131 | 42 | 13 | 26 | .0208871516 | 42 | 18 | 21 | .0324015461 |
| 42 | 10 | 30 | .0154595068 | 42 | 13 | 27 | .0200032911 | 42 | 18 | 22 | .0309163683 |
| 42 | 10 | 31 | .0148083381 | 42 | 13 | 28 | .0191615887 | 42 | 18 | 23 | .0295321648 |
| 42 | 10 | 32 | .0141768447 | 42 | 13 | 29 | .0183569557 | 42 | 18 | 24 | .0282360171 |
| 42 | 10 | 33 | .0135614989 | 42 | 13 | 30 | .0175836679 | 42 | 18 | 25 | .0270168007 |

TABLE I-COVARIANCES OF NORMAL ORDER STATISTICS

| N | I | J | COVARIANCE | N | I | J | COVARIANCE | N | I | J | COVARIANCE |
|---|---|---|---|---|---|---|---|---|---|---|---|
| 42 | 19 | 19 | .0373127969 | 43 | 2 | 7 | .0382260411 | 43 | 3 | 27 | .0118515400 |
| 42 | 19 | 20 | .0355091626 | 43 | 2 | 8 | .0339725535 | 43 | 3 | 28 | .0113634895 |
| 42 | 19 | 21 | .0338455206 | 43 | 2 | 9 | .0306179019 | 43 | 3 | 29 | .0108976996 |
| 42 | 19 | 22 | .0323029650 | 43 | 2 | 10 | .0278948927 | 43 | 3 | 30 | .0104512873 |
| 42 | 19 | 23 | .0308656456 | 43 | 2 | 11 | .0256335627 | 43 | 3 | 31 | .0100215598 |
| 42 | 19 | 24 | .0295200979 | 43 | 2 | 12 | .0237204339 | 43 | 3 | 32 | .0096059637 |
| 42 | 20 | 20 | .0371114438 | 43 | 2 | 13 | .0220767231 | 43 | 3 | 33 | .0092020084 |
| 42 | 20 | 21 | .0353809803 | 43 | 2 | 14 | .0206459435 | 43 | 3 | 34 | .0088071332 |
| 42 | 20 | 22 | .0337767722 | 43 | 2 | 15 | .0193864839 | 43 | 3 | 35 | .0084186134 |
| 42 | 20 | 23 | .0322822921 | 43 | 2 | 16 | .0182669757 | 43 | 3 | 36 | .0080335193 |
| 42 | 21 | 21 | .0370257092 | 43 | 2 | 17 | .0172632955 | 43 | 3 | 37 | .0076484296 |
| 42 | 21 | 22 | .0353455163 | 43 | 2 | 18 | .0163565660 | 43 | 3 | 38 | .0072587494 |
| 43 | 1 | 1 | .2238736555 | 43 | 2 | 19 | .0155317829 | 43 | 3 | 39 | .0068580237 |
| 43 | 1 | 2 | .1066544234 | 43 | 2 | 20 | .0147768509 | 43 | 3 | 40 | .0064366656 |
| 43 | 1 | 3 | .0718234406 | 43 | 2 | 21 | .0140818947 | 43 | 3 | 41 | .0059776259 |
| 43 | 1 | 4 | .0549208540 | 43 | 2 | 22 | .0134387571 | 43 | 4 | 4 | .0742212212 |
| 43 | 1 | 5 | .0448290089 | 43 | 2 | 23 | .0128406251 | 43 | 4 | 5 | .0607125780 |
| 43 | 1 | 6 | .0380670364 | 43 | 2 | 24 | .0122817520 | 43 | 4 | 6 | .0516284802 |
| 43 | 1 | 7 | .0331900265 | 43 | 2 | 25 | .0117572437 | 43 | 4 | 7 | .0450600071 |
| 43 | 1 | 8 | .0294879061 | 43 | 2 | 26 | .0112628839 | 43 | 4 | 8 | .0400644987 |
| 43 | 1 | 9 | .0265698154 | 43 | 2 | 27 | .0107949887 | 43 | 4 | 9 | .0361211981 |
| 43 | 1 | 10 | .0242022670 | 43 | 2 | 28 | .0103502967 | 43 | 4 | 10 | .0329181547 |
| 43 | 1 | 11 | .0222368726 | 43 | 2 | 29 | .0099258870 | 43 | 4 | 11 | .0302566722 |
| 43 | 1 | 12 | .0205746316 | 43 | 2 | 30 | .0095191145 | 43 | 4 | 12 | .0280039446 |
| 43 | 1 | 13 | .0191468535 | 43 | 2 | 31 | .0091275466 | 43 | 4 | 13 | .0260676879 |
| 43 | 1 | 14 | .0179043147 | 43 | 2 | 32 | .0087488778 | 43 | 4 | 14 | .0243816875 |
| 43 | 1 | 15 | .0168107712 | 43 | 2 | 33 | .0083808272 | 43 | 4 | 15 | .0228971354 |
| 43 | 1 | 16 | .0158389085 | 43 | 2 | 34 | .0080210648 | 43 | 4 | 16 | .0215772130 |
| 43 | 1 | 17 | .0149677288 | 43 | 2 | 35 | .0076671512 | 43 | 4 | 17 | .0203935838 |
| 43 | 1 | 18 | .0141808079 | 43 | 2 | 36 | .0073163923 | 43 | 4 | 18 | .0193240567 |
| 43 | 1 | 19 | .0134650898 | 43 | 2 | 37 | .0069655579 | 43 | 4 | 19 | .0183509927 |
| 43 | 1 | 20 | .0128100503 | 43 | 2 | 38 | .0066105002 | 43 | 4 | 20 | .0174601865 |
| 43 | 1 | 21 | .0122071092 | 43 | 2 | 39 | .0062455007 | 43 | 4 | 21 | .0166400492 |
| 43 | 1 | 22 | .0116491839 | 43 | 2 | 40 | .0058617309 | 43 | 4 | 22 | .0158809960 |
| 43 | 1 | 23 | .0111303481 | 43 | 2 | 41 | .0054435730 | 43 | 4 | 23 | .0151749981 |
| 43 | 1 | 24 | .0106456014 | 43 | 2 | 42 | .0049573103 | 43 | 4 | 24 | .0145152625 |
| 43 | 1 | 25 | .0101906935 | 43 | 3 | 3 | .0901590668 | 43 | 4 | 25 | .0138959949 |
| 43 | 1 | 26 | .0097619564 | 43 | 3 | 4 | .0691176196 | 43 | 4 | 26 | .0133122096 |
| 43 | 1 | 27 | .0093561811 | 43 | 3 | 5 | .0565039613 | 43 | 4 | 27 | .0127595773 |
| 43 | 1 | 28 | .0089705472 | 43 | 3 | 6 | .0480302039 | 43 | 4 | 28 | .0122343135 |
| 43 | 1 | 29 | .0086025408 | 43 | 3 | 7 | .0419073976 | 43 | 4 | 29 | .0117330714 |
| 43 | 1 | 30 | .0082498633 | 43 | 3 | 8 | .0372532955 | 43 | 4 | 30 | .0112527736 |
| 43 | 1 | 31 | .0079103778 | 43 | 3 | 9 | .0335809975 | 43 | 4 | 31 | .0107904153 |
| 43 | 1 | 32 | .0075820682 | 43 | 3 | 10 | .0305990594 | 43 | 4 | 32 | .0103430363 |
| 43 | 1 | 33 | .0072629671 | 43 | 3 | 11 | .0281219621 | 43 | 4 | 33 | .0099079344 |
| 43 | 1 | 34 | .0069510799 | 43 | 3 | 12 | .0260257697 | 43 | 4 | 34 | .0094827260 |
| 43 | 1 | 35 | .0066442937 | 43 | 3 | 13 | .0242243965 | 43 | 4 | 35 | .0090647886 |
| 43 | 1 | 36 | .0063402438 | 43 | 3 | 14 | .0226560954 | 43 | 4 | 36 | .0086505801 |
| 43 | 1 | 37 | .0060361255 | 43 | 3 | 15 | .0212753658 | 43 | 4 | 37 | .0082358737 |
| 43 | 1 | 38 | .0057283713 | 43 | 3 | 16 | .0200478996 | 43 | 4 | 38 | .0078161162 |
| 43 | 1 | 39 | .0054120134 | 43 | 3 | 17 | .0189473055 | 43 | 4 | 39 | .0073848791 |
| 43 | 1 | 40 | .0050793866 | 43 | 3 | 18 | .0179529217 | 43 | 4 | 40 | .0069313739 |
| 43 | 1 | 41 | .0047169787 | 43 | 3 | 19 | .0170483162 | 43 | 5 | 5 | .0645574970 |
| 43 | 1 | 42 | .0042955634 | 43 | 3 | 20 | .0162202387 | 43 | 5 | 6 | .0549191155 |
| 43 | 1 | 43 | .0037220400 | 43 | 3 | 21 | .0154578748 | 43 | 5 | 7 | .0479451831 |
| 43 | 2 | 2 | .1220874314 | 43 | 3 | 22 | .0147523029 | 43 | 5 | 8 | .0426386393 |
| 43 | 2 | 3 | .0824501756 | 43 | 3 | 23 | .0140960818 | 43 | 5 | 9 | .0384481687 |
| 43 | 2 | 4 | .0631374629 | 43 | 3 | 24 | .0134829257 | 43 | 5 | 10 | .0350432777 |
| 43 | 2 | 5 | .0515803179 | 43 | 3 | 25 | .0129074538 | 43 | 5 | 11 | .0322133383 |
| 43 | 2 | 6 | .0438251419 | 43 | 3 | 26 | .0123650103 | 43 | 5 | 12 | .0298175043 |

TABLE I-COVARIANCES OF NORMAL ORDER STATISTICS

| N | I | J | COVARIANCE | N | I | J | COVARIANCE | N | I | J | COVARIANCE |
|---|---|---|---|---|---|---|---|---|---|---|---|
| 43 | 5 | 13 | .0277578654 | 43 | 7 | 7 | .0533155730 | 43 | 9 | 9 | .0470244354 |
| 43 | 5 | 14 | .0259641416 | 43 | 7 | 8 | .0474333803 | 43 | 9 | 10 | .0428815589 |
| 43 | 5 | 15 | .0243845056 | 43 | 7 | 9 | .0427849981 | 43 | 9 | 11 | .0394346733 |
| 43 | 5 | 16 | .0229798499 | 43 | 7 | 10 | .0390058886 | 43 | 9 | 12 | .0365139596 |
| 43 | 5 | 17 | .0217200811 | 43 | 7 | 11 | .0358634725 | 43 | 9 | 13 | .0340011881 |
| 43 | 5 | 18 | .0205816487 | 43 | 7 | 12 | .0332020808 | 43 | 9 | 14 | .0318113840 |
| 43 | 5 | 19 | .0195458353 | 43 | 7 | 13 | .0309134137 | 43 | 9 | 15 | .0298818110 |
| 43 | 5 | 20 | .0185975407 | 43 | 7 | 14 | .0289196888 | 43 | 9 | 16 | .0281650754 |
| 43 | 5 | 21 | .0177244146 | 43 | 7 | 15 | .0271635303 | 43 | 9 | 17 | .0266246669 |
| 43 | 5 | 22 | .0169162294 | 43 | 7 | 16 | .0256016372 | 43 | 9 | 18 | .0252319899 |
| 43 | 5 | 23 | .0161644153 | 43 | 7 | 17 | .0242006656 | 43 | 9 | 19 | .0239643198 |
| 43 | 5 | 24 | .0154617371 | 43 | 7 | 18 | .0229344701 | 43 | 9 | 20 | .0228033553 |
| 43 | 5 | 25 | .0148020871 | 43 | 7 | 19 | .0217822163 | 43 | 9 | 21 | .0217341202 |
| 43 | 5 | 26 | .0141803017 | 43 | 7 | 20 | .0207270986 | 43 | 9 | 22 | .0207441031 |
| 43 | 5 | 27 | .0135918957 | 43 | 7 | 21 | .0197554524 | 43 | 9 | 23 | .0198227939 |
| 43 | 5 | 28 | .0130327466 | 43 | 7 | 22 | .0188560326 | 43 | 9 | 24 | .0189614204 |
| 43 | 5 | 29 | .0124989469 | 43 | 7 | 23 | .0180194243 | 43 | 9 | 25 | .0181525181 |
| 43 | 5 | 30 | .0119869786 | 43 | 7 | 24 | .0172376611 | 43 | 9 | 26 | .0173896251 |
| 43 | 5 | 31 | .0114939347 | 43 | 7 | 25 | .0165039350 | 43 | 9 | 27 | .0166673415 |
| 43 | 5 | 32 | .0110172803 | 43 | 7 | 26 | .0158122553 | 43 | 9 | 28 | .0159811493 |
| 43 | 5 | 33 | .0105542129 | 43 | 7 | 27 | .0151572165 | 43 | 9 | 29 | .0153265265 |
| 43 | 5 | 34 | .0101014627 | 43 | 7 | 28 | .0145341430 | 43 | 9 | 30 | .0146984231 |
| 43 | 5 | 35 | .0096558097 | 43 | 7 | 29 | .0139393669 | 43 | 9 | 31 | .0140923681 |
| 43 | 5 | 36 | .0092141321 | 43 | 7 | 30 | .0133698835 | 43 | 9 | 32 | .0135057516 |
| 43 | 5 | 37 | .0087723920 | 43 | 7 | 31 | .0128222963 | 43 | 9 | 33 | .0129365166 |
| 43 | 5 | 38 | .0083251326 | 43 | 7 | 32 | .0122923973 | 43 | 9 | 34 | .0123806416 |
| 43 | 5 | 39 | .0078653654 | 43 | 7 | 33 | .0117763510 | 43 | 9 | 35 | .0118328492 |
| 43 | 6 | 6 | .0580313971 | 43 | 7 | 34 | .0112717240 | 43 | 10 | 10 | .0447936076 |
| 43 | 6 | 7 | .0506756390 | 43 | 7 | 35 | .0107760832 | 43 | 10 | 11 | .0411979319 |
| 43 | 6 | 8 | .0450758952 | 43 | 7 | 36 | .0102851025 | 43 | 10 | 12 | .0381507489 |
| 43 | 6 | 9 | .0406522547 | 43 | 7 | 37 | .0097935342 | 43 | 10 | 13 | .0355289211 |
| 43 | 6 | 10 | .0370568444 | 43 | 8 | 8 | .0497754861 | 43 | 10 | 14 | .0332439145 |
| 43 | 6 | 11 | .0340678380 | 43 | 8 | 9 | .0449046750 | 43 | 10 | 15 | .0312303499 |
| 43 | 6 | 12 | .0315368313 | 43 | 8 | 10 | .0409434204 | 43 | 10 | 16 | .0294388406 |
| 43 | 6 | 13 | .0293606161 | 43 | 8 | 11 | .0376486236 | 43 | 10 | 17 | .0278313541 |
| 43 | 6 | 14 | .0274650826 | 43 | 8 | 12 | .0348575025 | 43 | 10 | 18 | .0263781043 |
| 43 | 6 | 15 | .0257955704 | 43 | 8 | 13 | .0324567709 | 43 | 10 | 19 | .0250553913 |
| 43 | 6 | 16 | .0243108392 | 43 | 8 | 14 | .0303650405 | 43 | 10 | 20 | .0238440453 |
| 43 | 6 | 17 | .0229791575 | 43 | 8 | 15 | .0285222518 | 43 | 10 | 21 | .0227283449 |
| 43 | 6 | 18 | .0217756740 | 43 | 8 | 16 | .0268830483 | 43 | 10 | 22 | .0216952780 |
| 43 | 6 | 19 | .0206806071 | 43 | 8 | 17 | .0254124790 | 43 | 10 | 23 | .0207339177 |
| 43 | 6 | 20 | .0196779716 | 43 | 8 | 18 | .0240831430 | 43 | 10 | 24 | .0198351151 |
| 43 | 6 | 21 | .0187546720 | 43 | 8 | 19 | .0228732580 | 43 | 10 | 25 | .0189913916 |
| 43 | 6 | 22 | .0178998968 | 43 | 8 | 20 | .0217652879 | 43 | 10 | 26 | .0181963323 |
| 43 | 6 | 23 | .0171047061 | 43 | 8 | 21 | .0207449321 | 43 | 10 | 27 | .0174438330 |
| 43 | 6 | 24 | .0163616354 | 43 | 8 | 22 | .0198004086 | 43 | 10 | 28 | .0167282070 |
| 43 | 6 | 25 | .0156642687 | 43 | 8 | 23 | .0189218273 | 43 | 10 | 29 | .0160447769 |
| 43 | 6 | 26 | .0150069275 | 43 | 8 | 24 | .0181006293 | 43 | 10 | 30 | .0153895537 |
| 43 | 6 | 27 | .0143845918 | 43 | 8 | 25 | .0173293981 | 43 | 10 | 31 | .0147584144 |
| 43 | 6 | 28 | .0137929165 | 43 | 8 | 26 | .0166019789 | 43 | 10 | 32 | .0141476231 |
| 43 | 6 | 29 | .0132280865 | 43 | 8 | 27 | .0159133443 | 43 | 10 | 33 | .0135546278 |
| 43 | 6 | 30 | .0126865671 | 43 | 8 | 28 | .0152589921 | 43 | 10 | 34 | .0129761615 |
| 43 | 6 | 31 | .0121651081 | 43 | 8 | 29 | .0146345166 | 43 | 11 | 11 | .0429810763 |
| 43 | 6 | 32 | .0116609259 | 43 | 8 | 30 | .0140358704 | 43 | 11 | 12 | .0398065428 |
| 43 | 6 | 33 | .0111713285 | 43 | 8 | 31 | .0134595833 | 43 | 11 | 13 | .0370748994 |
| 43 | 6 | 34 | .0106928162 | 43 | 8 | 32 | .0129021098 | 43 | 11 | 14 | .0346940453 |
| 43 | 6 | 35 | .0102212460 | 43 | 8 | 33 | .0123594493 | 43 | 11 | 15 | .0325959567 |
| 43 | 6 | 36 | .0097533114 | 43 | 8 | 34 | .0118283541 | 43 | 11 | 16 | .0307292483 |
| 43 | 6 | 37 | .0092860627 | 43 | 8 | 35 | .0113067602 | 43 | 11 | 17 | .0290543381 |
| 43 | 6 | 38 | .0088137190 | 43 | 8 | 36 | .0107909052 | 43 | 11 | 18 | .0275401931 |

TABLE I-COVARIANCES OF NORMAL ORDER STATISTICS

| N | I | J | COVARIANCE | N | I | J | COVARIANCE | N | I | J | COVARIANCE |
|---|---|---|---|---|---|---|---|---|---|---|---|
| 43 | 11 | 19 | .0261620788 | 43 | 14 | 19 | .0296933131 | 43 | 19 | 19 | .0365777859 |
| 43 | 11 | 20 | .0248999907 | 43 | 14 | 20 | .0282591062 | 43 | 19 | 20 | .0348181649 |
| 43 | 11 | 21 | .0237375389 | 43 | 14 | 21 | .0269375955 | 43 | 19 | 21 | .0331958380 |
| 43 | 11 | 22 | .0226612033 | 43 | 14 | 22 | .0257135752 | 43 | 19 | 22 | .0316923600 |
| 43 | 11 | 23 | .0216598556 | 43 | 14 | 23 | .0245742604 | 43 | 19 | 23 | .0302922906 |
| 43 | 11 | 24 | .0207241190 | 43 | 14 | 24 | .0235087850 | 43 | 19 | 24 | .0289825408 |
| 43 | 11 | 25 | .0198458447 | 43 | 14 | 25 | .0225077747 | 43 | 19 | 25 | .0277518616 |
| 43 | 11 | 26 | .0190180770 | 43 | 14 | 26 | .0215633872 | 43 | 20 | 20 | .0363287623 |
| 43 | 11 | 27 | .0182346847 | 43 | 14 | 27 | .0206691012 | 43 | 20 | 21 | .0346470555 |
| 43 | 11 | 28 | .0174897130 | 43 | 14 | 28 | .0198188177 | 43 | 20 | 22 | .0330891305 |
| 43 | 11 | 29 | .0167778107 | 43 | 14 | 29 | .0190069599 | 43 | 20 | 23 | .0316388924 |
| 43 | 11 | 30 | .0160953119 | 43 | 14 | 30 | .0182287173 | 43 | 20 | 24 | .0302827190 |
| 43 | 11 | 31 | .0154394545 | 43 | 15 | 15 | .0384960212 | 43 | 21 | 21 | .0361995034 |
| 43 | 11 | 32 | .0148060854 | 43 | 15 | 16 | .0363008503 | 43 | 21 | 22 | .0345755071 |
| 43 | 11 | 33 | .0141898585 | 43 | 15 | 17 | .0343303411 | 43 | 21 | 23 | .0330637167 |
| 43 | 12 | 12 | .0415514777 | 43 | 15 | 18 | .0325482505 | 43 | 22 | 22 | .0361640777 |
| 43 | 12 | 13 | .0387019695 | 43 | 15 | 19 | .0309257131 | 44 | 1 | 1 | .2225931020 |
| 43 | 12 | 14 | .0362176675 | 43 | 15 | 20 | .0294394235 | 44 | 1 | 2 | .1059704544 |
| 43 | 12 | 15 | .0340278274 | 43 | 15 | 21 | .0280702995 | 44 | 1 | 3 | .0713415888 |
| 43 | 12 | 16 | .0320789885 | 43 | 15 | 22 | .0268025038 | 44 | 1 | 4 | .0545443583 |
| 43 | 12 | 17 | .0303299410 | 43 | 15 | 23 | .0256227324 | 44 | 1 | 5 | .0445187628 |
| 43 | 12 | 18 | .0287483470 | 43 | 15 | 24 | .0245197621 | 44 | 1 | 6 | .0378031172 |
| 43 | 12 | 19 | .0273084253 | 43 | 15 | 25 | .0234841436 | 44 | 1 | 7 | .0329608091 |
| 43 | 12 | 20 | .0259893501 | 43 | 15 | 26 | .0225077345 | 44 | 1 | 8 | .0292859723 |
| 43 | 12 | 21 | .0247741432 | 43 | 15 | 27 | .0215835747 | 44 | 1 | 9 | .0263901226 |
| 43 | 12 | 22 | .0236488257 | 43 | 15 | 28 | .0207055699 | 44 | 1 | 10 | .0240412229 |
| 43 | 12 | 23 | .0226017297 | 43 | 15 | 29 | .0198675060 | 44 | 1 | 11 | .0220918251 |
| 43 | 12 | 24 | .0216230722 | 43 | 16 | 16 | .0378217825 | 44 | 1 | 12 | .0204435674 |
| 43 | 12 | 25 | .0207043370 | 43 | 16 | 17 | .0357715056 | 44 | 1 | 13 | .0190282106 |
| 43 | 12 | 26 | .0198377631 | 43 | 16 | 18 | .0339171587 | 44 | 1 | 14 | .0177968599 |
| 43 | 12 | 27 | .0190167588 | 43 | 16 | 19 | .0322287347 | 44 | 1 | 15 | .0167135198 |
| 43 | 12 | 28 | .0182360569 | 43 | 16 | 20 | .0306819895 | 44 | 1 | 16 | .0157510667 |
| 43 | 12 | 29 | .0174908102 | 43 | 16 | 21 | .0292570589 | 44 | 1 | 17 | .0148886510 |
| 43 | 12 | 30 | .0167763087 | 43 | 16 | 22 | .0279374757 | 44 | 1 | 18 | .0141099699 |
| 43 | 12 | 31 | .0160887543 | 43 | 16 | 23 | .0267094781 | 44 | 1 | 19 | .0134020715 |
| 43 | 12 | 32 | .0154244769 | 43 | 16 | 24 | .0255614910 | 44 | 1 | 20 | .0127545135 |
| 43 | 13 | 13 | .0403888765 | 43 | 16 | 25 | .0244837232 | 44 | 1 | 21 | .0121587832 |
| 43 | 13 | 14 | .0377966748 | 43 | 16 | 26 | .0234678060 | 44 | 1 | 22 | .0116078682 |
| 43 | 13 | 15 | .0355109437 | 43 | 16 | 27 | .0225061938 | 44 | 1 | 23 | .0110959085 |
| 43 | 13 | 16 | .0334760756 | 43 | 16 | 28 | .0215920916 | 44 | 1 | 24 | .0106179516 |
| 43 | 13 | 17 | .0316491877 | 43 | 17 | 17 | .0373252772 | 44 | 1 | 25 | .0101697957 |
| 43 | 13 | 18 | .0299966237 | 43 | 17 | 18 | .0353861156 | 44 | 1 | 26 | .0097478380 |
| 43 | 13 | 19 | .0284915732 | 43 | 17 | 19 | .0336196239 | 44 | 1 | 27 | .0093489304 |
| 43 | 13 | 20 | .0271124138 | 43 | 17 | 20 | .0320005489 | 44 | 1 | 28 | .0089703020 |
| 43 | 13 | 21 | .0258415102 | 43 | 17 | 21 | .0305082167 | 44 | 1 | 29 | .0086095097 |
| 43 | 13 | 22 | .0246643185 | 43 | 17 | 22 | .0291255194 | 44 | 1 | 30 | .0082643503 |
| 43 | 13 | 23 | .0235686388 | 43 | 17 | 23 | .0278381817 | 44 | 1 | 31 | .0079327829 |
| 43 | 13 | 24 | .0225439790 | 43 | 17 | 24 | .0266341932 | 44 | 1 | 32 | .0076129020 |
| 43 | 13 | 25 | .0215813829 | 43 | 17 | 25 | .0255032995 | 44 | 1 | 33 | .0073028967 |
| 43 | 13 | 26 | .0206731609 | 43 | 17 | 26 | .0244365716 | 44 | 1 | 34 | .0070009784 |
| 43 | 13 | 27 | .0198124319 | 43 | 17 | 27 | .0234261321 | 44 | 1 | 35 | .0067053148 |
| 43 | 13 | 28 | .0189935427 | 43 | 18 | 18 | .0369219675 | 44 | 1 | 36 | .0064139446 |
| 43 | 13 | 29 | .0182120772 | 43 | 18 | 19 | .0350755152 | 44 | 1 | 37 | .0061246493 |
| 43 | 13 | 30 | .0174632651 | 43 | 18 | 20 | .0333824521 | 44 | 1 | 38 | .0058347787 |
| 43 | 13 | 31 | .0167414359 | 43 | 18 | 21 | .0318213118 | 44 | 1 | 39 | .0055409354 |
| 43 | 14 | 14 | .0393705143 | 43 | 18 | 22 | .0303743234 | 44 | 1 | 40 | .0052383541 |
| 43 | 14 | 15 | .0369934930 | 43 | 18 | 23 | .0290266449 | 44 | 1 | 41 | .0049196536 |
| 43 | 14 | 16 | .0348772530 | 43 | 18 | 24 | .0277657302 | 44 | 1 | 42 | .0045717868 |
| 43 | 14 | 17 | .0329772550 | 43 | 18 | 25 | .0265808134 | 44 | 1 | 43 | .0041664760 |
| 43 | 14 | 18 | .0312585495 | 43 | 18 | 26 | .0254624885 | 44 | 1 | 44 | .0036135387 |

TABLE I-COVARIANCES OF NORMAL ORDER STATISTICS

| N | I | J | COVARIANCE | N | I | J | COVARIANCE | N | I | J | COVARIANCE |
|---|---|---|---|---|---|---|---|---|---|---|---|
| 44 | 2 | 2 | .1212023380 | 44 | 3 | 21 | .0153708128 | 44 | 5 | 5 | .0639246447 |
| 44 | 2 | 3 | .0818255702 | 44 | 3 | 22 | .0146752689 | 44 | 5 | 6 | .0543786563 |
| 44 | 2 | 4 | .0626489218 | 44 | 3 | 23 | .0140288069 | 44 | 5 | 7 | .0474737419 |
| 44 | 2 | 5 | .0511773283 | 44 | 3 | 24 | .0134252312 | 44 | 5 | 8 | .0422212503 |
| 44 | 2 | 6 | .0434819129 | 44 | 3 | 25 | .0128592532 | 44 | 5 | 9 | .0380746343 |
| 44 | 2 | 7 | .0379275057 | 44 | 3 | 26 | .0123263016 | 44 | 5 | 10 | .0347063194 |
| 44 | 2 | 8 | .0337090911 | 44 | 3 | 27 | .0118223915 | 44 | 5 | 11 | .0319075724 |
| 44 | 2 | 9 | .0303829700 | 44 | 3 | 28 | .0113440326 | 44 | 5 | 12 | .0295388351 |
| 44 | 2 | 10 | .0276838290 | 44 | 3 | 29 | .0108881435 | 44 | 5 | 13 | .0275031086 |
| 44 | 2 | 11 | .0254429239 | 44 | 3 | 30 | .0104519551 | 44 | 5 | 14 | .0257307790 |
| 44 | 2 | 12 | .0235476046 | 44 | 3 | 31 | .0100329202 | 44 | 5 | 15 | .0241705185 |
| 44 | 2 | 13 | .0219196730 | 44 | 3 | 32 | .0096286574 | 44 | 5 | 16 | .0227835986 |
| 44 | 2 | 14 | .0205030689 | 44 | 3 | 33 | .0092368938 | 44 | 5 | 17 | .0215402176 |
| 44 | 2 | 15 | .0192565012 | 44 | 3 | 34 | .0088553478 | 44 | 5 | 18 | .0204170566 |
| 44 | 2 | 16 | .0181488469 | 44 | 3 | 35 | .0084816112 | 44 | 5 | 19 | .0193955961 |
| 44 | 2 | 17 | .0171561758 | 44 | 3 | 36 | .0081131563 | 44 | 5 | 20 | .0184609140 |
| 44 | 2 | 18 | .0162597658 | 44 | 3 | 37 | .0077473315 | 44 | 5 | 21 | .0176008215 |
| 44 | 2 | 19 | .0154447419 | 44 | 3 | 38 | .0073809378 | 44 | 5 | 22 | .0168052300 |
| 44 | 2 | 20 | .0146991184 | 44 | 3 | 39 | .0070094987 | 44 | 5 | 23 | .0160656748 |
| 44 | 2 | 21 | .0140131135 | 44 | 3 | 40 | .0066268133 | 44 | 5 | 24 | .0153749821 |
| 44 | 2 | 22 | .0133786518 | 44 | 3 | 41 | .0062237608 | 44 | 5 | 25 | .0147270785 |
| 44 | 2 | 23 | .0127889936 | 44 | 3 | 42 | .0057838773 | 44 | 5 | 26 | .0141168631 |
| 44 | 2 | 24 | .0122384575 | 44 | 4 | 4 | .0735463865 | 44 | 5 | 27 | .0135400095 |
| 44 | 2 | 25 | .0117222129 | 44 | 4 | 5 | .0601548582 | 44 | 5 | 28 | .0129926451 |
| 44 | 2 | 26 | .0112361141 | 44 | 4 | 6 | .0511525624 | 44 | 5 | 29 | .0124710707 |
| 44 | 2 | 27 | .0107765555 | 44 | 4 | 7 | .0446452062 | 44 | 5 | 30 | .0119717791 |
| 44 | 2 | 28 | .0103403560 | 44 | 4 | 8 | .0396975834 | 44 | 5 | 31 | .0114917471 |
| 44 | 2 | 29 | .0099246764 | 44 | 4 | 9 | .0357931593 | 44 | 5 | 32 | .0110285348 |
| 44 | 2 | 30 | .0095269571 | 44 | 4 | 10 | .0326225660 | 44 | 5 | 33 | .0105798374 |
| 44 | 2 | 31 | .0091448668 | 44 | 4 | 11 | .0299887828 | 44 | 5 | 34 | .0101429455 |
| 44 | 2 | 32 | .0087762443 | 44 | 4 | 12 | .0277601422 | 44 | 5 | 35 | .0097148932 |
| 44 | 2 | 33 | .0084190129 | 44 | 4 | 13 | .0258451634 | 44 | 5 | 36 | .0092929303 |
| 44 | 2 | 34 | .0080710918 | 44 | 4 | 14 | .0241782180 | 44 | 5 | 37 | .0088740752 |
| 44 | 2 | 35 | .0077303447 | 44 | 4 | 15 | .0227109395 | 44 | 5 | 38 | .0084541999 |
| 44 | 2 | 36 | .0073945209 | 44 | 4 | 16 | .0214068507 | 44 | 5 | 39 | .0080282589 |
| 44 | 2 | 37 | .0070610934 | 44 | 4 | 17 | .0202378825 | 44 | 5 | 40 | .0075898907 |
| 44 | 2 | 38 | .0067269942 | 44 | 4 | 18 | .0191820532 | 44 | 6 | 6 | .0574261059 |
| 44 | 2 | 39 | .0063882844 | 44 | 4 | 19 | .0182218890 | 44 | 6 | 7 | .0501472771 |
| 44 | 2 | 40 | .0060395035 | 44 | 4 | 20 | .0173433226 | 44 | 6 | 8 | .0446077476 |
| 44 | 2 | 41 | .0056721409 | 44 | 4 | 21 | .0165348921 | 44 | 6 | 9 | .0402329247 |
| 44 | 2 | 42 | .0052711235 | 44 | 4 | 22 | .0157871363 | 44 | 6 | 10 | .0366781935 |
| 44 | 2 | 43 | .0048038724 | 44 | 4 | 23 | .0150921429 | 44 | 6 | 11 | .0337238442 |
| 44 | 3 | 3 | .0894136000 | 44 | 4 | 24 | .0144432190 | 44 | 6 | 12 | .0312229132 |
| 44 | 3 | 4 | .0685339079 | 44 | 4 | 25 | .0138346496 | 44 | 6 | 13 | .0290732107 |
| 44 | 3 | 5 | .0560220146 | 44 | 4 | 26 | .0132615101 | 44 | 6 | 14 | .0272013675 |
| 44 | 3 | 6 | .0476193256 | 44 | 4 | 27 | .0127195262 | 44 | 6 | 15 | .0255532734 |
| 44 | 3 | 7 | .0415496313 | 44 | 4 | 28 | .0122049816 | 44 | 6 | 16 | .0240881072 |
| 44 | 3 | 8 | .0369371661 | 44 | 4 | 29 | .0117146475 | 44 | 6 | 17 | .0227744697 |
| 44 | 3 | 9 | .0332986968 | 44 | 4 | 30 | .0112456506 | 44 | 6 | 18 | .0215877877 |
| 44 | 3 | 10 | .0303450207 | 44 | 4 | 31 | .0107952410 | 44 | 6 | 19 | .0205085192 |
| 44 | 3 | 11 | .0278920724 | 44 | 4 | 32 | .0103606168 | 44 | 6 | 20 | .0195208817 |
| 44 | 3 | 12 | .0258169037 | 44 | 4 | 33 | .0099390542 | 44 | 6 | 21 | .0186119308 |
| 44 | 3 | 13 | .0240341265 | 44 | 4 | 34 | .0095282320 | 44 | 6 | 22 | .0177709454 |
| 44 | 3 | 14 | .0224825012 | 44 | 4 | 35 | .0091261345 | 44 | 6 | 23 | .0169890479 |
| 44 | 3 | 15 | .0211169076 | 44 | 4 | 36 | .0087302553 | 44 | 6 | 24 | .0162588723 |
| 44 | 3 | 16 | .0199033302 | 44 | 4 | 37 | .0083370650 | 44 | 6 | 25 | .0155741718 |
| 44 | 3 | 17 | .0188156102 | 44 | 4 | 38 | .0079426267 | 44 | 6 | 26 | .0149294672 |
| 44 | 3 | 18 | .0178332738 | 44 | 4 | 39 | .0075428147 | 44 | 6 | 27 | .0143198877 |
| 44 | 3 | 19 | .0169400432 | 44 | 4 | 40 | .0071313909 | 44 | 6 | 28 | .0137411628 |
| 44 | 3 | 20 | .0161227933 | 44 | 4 | 41 | .0066978887 | 44 | 6 | 29 | .0131895326 |

TABLE I-COVARIANCES OF NORMAL ORDER STATISTICS

| N | I | J | COVARIANCE | N | I | J | COVARIANCE | N | I | J | COVARIANCE |
|---|---|---|---|---|---|---|---|---|---|---|---|
| 44 | 6 | 30 | .0126615188 | 44 | 8 | 26 | .0164964377 | 44 | 10 | 30 | .0153118340 |
| 44 | 6 | 31 | .0121538817 | 44 | 8 | 27 | .0158230468 | 44 | 10 | 31 | .0146987868 |
| 44 | 6 | 32 | .0116639373 | 44 | 8 | 28 | .0151838463 | 44 | 10 | 32 | .0141062025 |
| 44 | 6 | 33 | .0111895773 | 44 | 8 | 29 | .0145747550 | 44 | 10 | 33 | .0135316189 |
| 44 | 6 | 34 | .0107282927 | 44 | 8 | 30 | .0139919786 | 44 | 10 | 34 | .0129737054 |
| 44 | 6 | 35 | .0102762602 | 44 | 8 | 31 | .0134323411 | 44 | 10 | 35 | .0124285315 |
| 44 | 6 | 36 | .0098294612 | 44 | 8 | 32 | .0128927750 | 44 | 11 | 11 | .0423961364 |
| 44 | 6 | 37 | .0093854988 | 44 | 8 | 33 | .0123693716 | 44 | 11 | 12 | .0392701117 |
| 44 | 6 | 38 | .0089418929 | 44 | 8 | 34 | .0118581339 | 44 | 11 | 13 | .0365811461 |
| 44 | 6 | 39 | .0084924822 | 44 | 8 | 35 | .0113570307 | 44 | 11 | 14 | .0342383340 |
| 44 | 7 | 7 | .0527230432 | 44 | 8 | 36 | .0108649702 | 44 | 11 | 15 | .0321745426 |
| 44 | 7 | 8 | .0469078581 | 44 | 8 | 37 | .0103775091 | 44 | 11 | 16 | .0303390685 |
| 44 | 7 | 9 | .0423138143 | 44 | 9 | 9 | .0464514976 | 44 | 11 | 17 | .0286928714 |
| 44 | 7 | 10 | .0385799893 | 44 | 9 | 10 | .0423629019 | 44 | 11 | 18 | .0272053688 |
| 44 | 7 | 11 | .0354761541 | 44 | 9 | 11 | .0389621794 | 44 | 11 | 19 | .0258522087 |
| 44 | 7 | 12 | .0328482389 | 44 | 9 | 12 | .0360814413 | 44 | 11 | 20 | .0246136996 |
| 44 | 7 | 13 | .0305890722 | 44 | 9 | 13 | .0336038237 | 44 | 11 | 21 | .0234736800 |
| 44 | 7 | 14 | .0286216911 | 44 | 9 | 14 | .0314453520 | 44 | 11 | 22 | .0224187571 |
| 44 | 7 | 15 | .0268893342 | 44 | 9 | 15 | .0295440424 | 44 | 11 | 23 | .0214379151 |
| 44 | 7 | 16 | .0253491783 | 44 | 9 | 16 | .0278530743 | 44 | 11 | 24 | .0205219671 |
| 44 | 7 | 17 | .0239682655 | 44 | 9 | 17 | .0263363738 | 44 | 11 | 25 | .0196629805 |
| 44 | 7 | 18 | .0227207654 | 44 | 9 | 18 | .0249656814 | 44 | 11 | 26 | .0188542728 |
| 44 | 7 | 19 | .0215860889 | 44 | 9 | 19 | .0237185466 | 44 | 11 | 27 | .0180901235 |
| 44 | 7 | 20 | .0205476036 | 44 | 9 | 20 | .0225769237 | 44 | 11 | 28 | .0173648551 |
| 44 | 7 | 21 | .0195917747 | 44 | 9 | 21 | .0215261054 | 44 | 11 | 29 | .0166727821 |
| 44 | 7 | 22 | .0187074876 | 44 | 9 | 22 | .0205538227 | 44 | 11 | 30 | .0160096286 |
| 44 | 7 | 23 | .0178854766 | 44 | 9 | 23 | .0196497191 | 44 | 11 | 31 | .0153729846 |
| 44 | 7 | 24 | .0171179483 | 44 | 9 | 24 | .0188051111 | 44 | 11 | 32 | .0147601682 |
| 44 | 7 | 25 | .0163982993 | 44 | 9 | 25 | .0180125869 | 44 | 11 | 33 | .0141666709 |
| 44 | 7 | 26 | .0157207433 | 44 | 9 | 26 | .0172656805 | 44 | 11 | 34 | .0135880598 |
| 44 | 7 | 27 | .0150799612 | 44 | 9 | 27 | .0165590094 | 44 | 12 | 12 | .0409482521 |
| 44 | 7 | 28 | .0144711382 | 44 | 9 | 28 | .0158883698 | 44 | 12 | 13 | .0381467344 |
| 44 | 7 | 29 | .0138904031 | 44 | 9 | 29 | .0152499497 | 44 | 12 | 14 | .0357052910 |
| 44 | 7 | 30 | .0133349541 | 44 | 9 | 30 | .0146392083 | 44 | 12 | 15 | .0335541672 |
| 44 | 7 | 31 | .0128022044 | 44 | 9 | 31 | .0140512162 | 44 | 12 | 16 | .0316406654 |
| 44 | 7 | 32 | .0122886822 | 44 | 9 | 32 | .0134824856 | 44 | 12 | 17 | .0299241854 |
| 44 | 7 | 33 | .0117903298 | 44 | 9 | 33 | .0129313621 | 44 | 12 | 18 | .0283728854 |
| 44 | 7 | 34 | .0113040690 | 44 | 9 | 34 | .0123956033 | 44 | 12 | 19 | .0269613805 |
| 44 | 7 | 35 | .0108280196 | 44 | 9 | 35 | .0118708717 | 44 | 12 | 20 | .0256691426 |
| 44 | 7 | 36 | .0103594982 | 44 | 9 | 36 | .0113528517 | 44 | 12 | 21 | .0244794140 |
| 44 | 7 | 37 | .0098941825 | 44 | 10 | 10 | .0442273270 | 44 | 12 | 22 | .0233784104 |
| 44 | 7 | 38 | .0094277514 | 44 | 10 | 11 | .0406814875 | 44 | 12 | 23 | .0223546598 |
| 44 | 8 | 8 | .0491888318 | 44 | 10 | 12 | .0376773881 | 44 | 12 | 24 | .0213986564 |
| 44 | 8 | 9 | .0443782425 | 44 | 10 | 13 | .0350933809 | 44 | 12 | 25 | .0205022808 |
| 44 | 8 | 10 | .0404671921 | 44 | 10 | 14 | .0328420200 | 44 | 12 | 26 | .0196580192 |
| 44 | 8 | 11 | .0372151758 | 44 | 10 | 15 | .0308587373 | 44 | 12 | 27 | .0188592204 |
| 44 | 8 | 12 | .0344611769 | 44 | 10 | 16 | .0290947596 | 44 | 12 | 28 | .0181005505 |
| 44 | 8 | 13 | .0320931557 | 44 | 10 | 17 | .0275125281 | 44 | 12 | 29 | .0173772952 |
| 44 | 8 | 14 | .0300306355 | 44 | 10 | 18 | .0260826421 | 44 | 12 | 30 | .0166848138 |
| 44 | 8 | 15 | .0282142455 | 44 | 10 | 19 | .0247817417 | 44 | 12 | 31 | .0160195276 |
| 44 | 8 | 16 | .0265991650 | 44 | 10 | 20 | .0235909647 | 44 | 12 | 32 | .0153789934 |
| 44 | 8 | 17 | .0251508618 | 44 | 10 | 21 | .0224948448 | 44 | 12 | 33 | .0147592375 |
| 44 | 8 | 18 | .0238422545 | 44 | 10 | 22 | .0214805685 | 44 | 13 | 13 | .0397950416 |
| 44 | 8 | 19 | .0226518013 | 44 | 10 | 23 | .0205373165 | 44 | 13 | 14 | .0372477262 |
| 44 | 8 | 20 | .0215621612 | 44 | 10 | 24 | .0196559330 | 44 | 13 | 15 | .0350024223 |
| 44 | 8 | 21 | .0205592120 | 44 | 10 | 25 | .0188289865 | 44 | 13 | 16 | .0330043469 |
| 44 | 8 | 22 | .0196313728 | 44 | 10 | 26 | .0180503631 | 44 | 13 | 17 | .0312112685 |
| 44 | 8 | 23 | .0187690034 | 44 | 10 | 27 | .0173143997 | 44 | 13 | 18 | .0295900445 |
| 44 | 8 | 24 | .0179637694 | 44 | 10 | 28 | .0166156478 | 44 | 13 | 19 | .0281142625 |
| 44 | 8 | 25 | .0172083255 | 44 | 10 | 29 | .0159494307 | 44 | 13 | 20 | .0267626110 |

TABLE I-COVARIANCES OF NORMAL ORDER STATISTICS

| N | I | J | COVARIANCE | N | I | J | COVARIANCE | N | I | J | COVARIANCE |
|---|---|---|---|---|---|---|---|---|---|---|---|
| 44 | 13 | 21 | .0255177182 | 44 | 17 | 17 | .0366503894 | 45 | 1 | 19 | .0133392614 |
| 44 | 13 | 22 | .0243653084 | 44 | 17 | 18 | .0347585553 | 45 | 1 | 20 | .0126987871 |
| 44 | 13 | 23 | .0232934999 | 44 | 17 | 19 | .0330363656 | 45 | 1 | 21 | .0121098669 |
| 44 | 13 | 24 | .0222920904 | 44 | 17 | 20 | .0314590629 | 45 | 1 | 22 | .0115655484 |
| 44 | 13 | 25 | .0213523243 | 44 | 17 | 21 | .0300063915 | 45 | 1 | 23 | .0110600383 |
| 44 | 13 | 26 | .0204666794 | 44 | 17 | 22 | .0286615744 | 45 | 1 | 24 | .0105884350 |
| 44 | 13 | 27 | .0196282310 | 44 | 17 | 23 | .0274106041 | 45 | 1 | 25 | .0101465711 |
| 44 | 13 | 28 | .0188310882 | 44 | 17 | 24 | .0262417281 | 45 | 1 | 26 | .0097308952 |
| 44 | 13 | 29 | .0180710736 | 44 | 17 | 25 | .0251450000 | 45 | 1 | 27 | .0093383239 |
| 44 | 13 | 30 | .0173444306 | 44 | 17 | 26 | .0241118458 | 45 | 1 | 28 | .0089661318 |
| 44 | 13 | 31 | .0166462190 | 44 | 17 | 27 | .0231347978 | 45 | 1 | 29 | .0086119182 |
| 44 | 13 | 32 | .0159714813 | 44 | 17 | 28 | .0222069319 | 45 | 1 | 30 | .0082735570 |
| 44 | 14 | 14 | .0387878042 | 44 | 18 | 18 | .0362482728 | 45 | 1 | 31 | .0079490991 |
| 44 | 14 | 15 | .0364518480 | 44 | 18 | 19 | .0344460082 | 45 | 1 | 32 | .0076367210 |
| 44 | 14 | 16 | .0343727827 | 44 | 18 | 20 | .0327943967 | 45 | 1 | 33 | .0073347186 |
| 44 | 14 | 17 | .0325067454 | 44 | 18 | 21 | .0312723908 | 45 | 1 | 34 | .0070414572 |
| 44 | 14 | 18 | .0308193197 | 44 | 18 | 22 | .0298625724 | 45 | 1 | 35 | .0067553009 |
| 44 | 14 | 19 | .0292831033 | 44 | 18 | 23 | .0285504178 | 45 | 1 | 36 | .0064745596 |
| 44 | 14 | 20 | .0278760099 | 44 | 18 | 24 | .0273237076 | 45 | 1 | 37 | .0061974050 |
| 44 | 14 | 21 | .0265800349 | 44 | 18 | 25 | .0261720302 | 45 | 1 | 38 | .0059217492 |
| 44 | 14 | 22 | .0253802951 | 44 | 18 | 26 | .0250863314 | 45 | 1 | 39 | .0056450825 |
| 44 | 14 | 23 | .0242643010 | 44 | 18 | 27 | .0240585541 | 45 | 1 | 40 | .0053641613 |
| 44 | 14 | 24 | .0232214340 | 44 | 19 | 19 | .0358933855 | 45 | 1 | 41 | .0050744059 |
| 44 | 14 | 25 | .0222424127 | 44 | 19 | 20 | .0341740403 | 45 | 1 | 42 | .0047687016 |
| 44 | 14 | 26 | .0213193402 | 44 | 19 | 21 | .0325894296 | 45 | 1 | 43 | .0044344377 |
| 44 | 14 | 27 | .0204458248 | 44 | 19 | 22 | .0311215248 | 45 | 1 | 44 | .0040442305 |
| 44 | 14 | 28 | .0196160731 | 44 | 19 | 23 | .0297552500 | 45 | 1 | 45 | .0035106586 |
| 44 | 14 | 29 | .0188247941 | 44 | 19 | 24 | .0284778735 | 45 | 2 | 2 | .1203479881 |
| 44 | 14 | 30 | .0180677687 | 44 | 19 | 25 | .0272784800 | 45 | 2 | 3 | .0812226469 |
| 44 | 14 | 31 | .0173405780 | 44 | 19 | 26 | .0261475741 | 45 | 2 | 4 | .0621772185 |
| 44 | 15 | 15 | .0378890189 | 44 | 20 | 20 | .0356040115 | 45 | 2 | 5 | .0507880716 |
| 44 | 15 | 16 | .0357351502 | 44 | 20 | 21 | .0339644190 | 45 | 2 | 6 | .0431502089 |
| 44 | 15 | 17 | .0338023774 | 44 | 20 | 22 | .0324462407 | 45 | 2 | 7 | .0376388146 |
| 44 | 15 | 18 | .0320550512 | 44 | 20 | 23 | .0310337940 | 45 | 2 | 8 | .0334541298 |
| 44 | 15 | 19 | .0304647864 | 44 | 20 | 24 | .0297138249 | 45 | 2 | 9 | .0301554256 |
| 44 | 15 | 20 | .0290086938 | 44 | 20 | 25 | .0284749684 | 45 | 2 | 10 | .0274792024 |
| 44 | 15 | 21 | .0276680680 | 44 | 21 | 21 | .0354271282 | 45 | 2 | 11 | .0252578909 |
| 44 | 15 | 22 | .0264274042 | 44 | 21 | 22 | .0338528350 | 45 | 2 | 12 | .0233796395 |
| 44 | 15 | 23 | .0252736550 | 44 | 21 | 23 | .0323886013 | 45 | 2 | 13 | .0217668134 |
| 44 | 15 | 24 | .0241957437 | 44 | 21 | 24 | .0310206032 | 45 | 2 | 14 | .0203637634 |
| 44 | 15 | 25 | .0231843313 | 44 | 22 | 22 | .0353575851 | 45 | 2 | 15 | .0191295061 |
| 44 | 15 | 26 | .0222313814 | 44 | 22 | 23 | .0338248647 | 45 | 2 | 16 | .0180331550 |
| 44 | 15 | 27 | .0213301008 | 45 | 1 | 1 | .2213543125 | 45 | 2 | 17 | .0170509648 |
| 44 | 15 | 28 | .0204749412 | 45 | 1 | 2 | .1053090873 | 45 | 2 | 18 | .0161643612 |
| 44 | 15 | 29 | .0196603055 | 45 | 1 | 3 | .0708756107 | 45 | 2 | 19 | .0153585918 |
| 44 | 15 | 30 | .0188803822 | 45 | 1 | 4 | .0541801441 | 45 | 2 | 20 | .0146217751 |
| 44 | 16 | 16 | .0371700672 | 45 | 1 | 5 | .0442184928 | 45 | 2 | 21 | .0139442184 |
| 44 | 16 | 17 | .0351651093 | 45 | 1 | 6 | .0375475303 | 45 | 2 | 22 | .0133179233 |
| 44 | 16 | 18 | .0333527438 | 45 | 1 | 7 | .0327386682 | 45 | 2 | 23 | .0127362190 |
| 44 | 16 | 19 | .0317035276 | 45 | 1 | 8 | .0290901076 | 45 | 2 | 24 | .0121934848 |
| 44 | 16 | 20 | .0301936928 | 45 | 1 | 9 | .0262156598 | 45 | 2 | 25 | .0116849443 |
| 44 | 16 | 21 | .0288037809 | 45 | 1 | 10 | .0238846894 | 45 | 2 | 26 | .0112065089 |
| 44 | 16 | 22 | .0275176495 | 45 | 1 | 11 | .0219506558 | 45 | 2 | 27 | .0107546409 |
| 44 | 16 | 23 | .0263217766 | 45 | 1 | 12 | .0203158145 | 45 | 2 | 28 | .0103262320 |
| 44 | 16 | 24 | .0252047660 | 45 | 1 | 13 | .0189123615 | 45 | 2 | 29 | .0099185160 |
| 44 | 16 | 25 | .0241569966 | 45 | 1 | 14 | .0176917188 | 45 | 2 | 30 | .0095290070 |
| 44 | 16 | 26 | .0231704043 | 45 | 1 | 15 | .0166181298 | 45 | 2 | 31 | .0091554508 |
| 44 | 16 | 27 | .0222378554 | 45 | 1 | 16 | .0156646560 | 45 | 2 | 32 | .0087957823 |
| 44 | 16 | 28 | .0213528762 | 45 | 1 | 17 | .0148105896 | 45 | 2 | 33 | .0084480685 |
| 44 | 16 | 29 | .0205098614 | 45 | 1 | 18 | .0140397420 | 45 | 2 | 34 | .0081104252 |

TABLE I-COVARIANCES OF NORMAL ORDER STATISTICS

| N | I | J | COVARIANCE | N | I | J | COVARIANCE | N | I | J | COVARIANCE |
|---|---|---|---|---|---|---|---|---|---|---|---|
| 45 | 2 | 35 | .0077809481 | 45 | 4 | 13 | .0256295710 | 45 | 5 | 35 | .0097587819 |
| 45 | 2 | 36 | .0074576776 | 45 | 4 | 14 | .0239808096 | 45 | 5 | 36 | .0093534135 |
| 45 | 2 | 37 | .0071385265 | 45 | 4 | 15 | .0225299936 | 45 | 5 | 37 | .0089533792 |
| 45 | 2 | 38 | .0068211043 | 45 | 4 | 16 | .0212409735 | 45 | 5 | 38 | .0085555818 |
| 45 | 2 | 39 | .0065024911 | 45 | 4 | 17 | .0200859387 | 45 | 5 | 39 | .0081557971 |
| 45 | 2 | 40 | .0061789520 | 45 | 4 | 18 | .0190431109 | 45 | 5 | 40 | .0077496048 |
| 45 | 2 | 41 | .0058452607 | 45 | 4 | 19 | .0180951742 | 45 | 5 | 41 | .0073311909 |
| 45 | 2 | 42 | .0054931918 | 45 | 4 | 20 | .0172281872 | 45 | 6 | 6 | .0568452434 |
| 45 | 2 | 43 | .0051081902 | 45 | 4 | 21 | .0164307997 | 45 | 6 | 7 | .0496400516 |
| 45 | 2 | 44 | .0046587506 | 45 | 4 | 22 | .0156936603 | 45 | 6 | 8 | .0441581193 |
| 45 | 3 | 3 | .0886951534 | 45 | 4 | 23 | .0150089666 | 45 | 6 | 9 | .0398299584 |
| 45 | 3 | 4 | .0679712457 | 45 | 4 | 24 | .0143701297 | 45 | 6 | 10 | .0363140800 |
| 45 | 3 | 5 | .0555572905 | 45 | 4 | 25 | .0137715230 | 45 | 6 | 11 | .0333928044 |
| 45 | 3 | 6 | .0472229512 | 45 | 4 | 26 | .0132082875 | 45 | 6 | 12 | .0309205492 |
| 45 | 3 | 7 | .0412043032 | 45 | 4 | 27 | .0126761897 | 45 | 6 | 13 | .0287961013 |
| 45 | 3 | 8 | .0366318277 | 45 | 4 | 28 | .0121715414 | 45 | 6 | 14 | .0269468030 |
| 45 | 3 | 9 | .0330258250 | 45 | 4 | 29 | .0116911657 | 45 | 6 | 15 | .0253190716 |
| 45 | 3 | 10 | .0300992515 | 45 | 4 | 30 | .0112323284 | 45 | 6 | 16 | .0238724840 |
| 45 | 3 | 11 | .0276694413 | 45 | 4 | 31 | .0107925342 | 45 | 6 | 17 | .0225759520 |
| 45 | 3 | 12 | .0256143977 | 45 | 4 | 32 | .0103692453 | 45 | 6 | 18 | .0214051600 |
| 45 | 3 | 13 | .0238494033 | 45 | 4 | 33 | .0099598048 | 45 | 6 | 19 | .0203407935 |
| 45 | 3 | 14 | .0223137069 | 45 | 4 | 34 | .0095617562 | 45 | 6 | 20 | .0193672745 |
| 45 | 3 | 15 | .0209625533 | 45 | 4 | 35 | .0091731829 | 45 | 6 | 21 | .0184718235 |
| 45 | 3 | 16 | .0197622070 | 45 | 4 | 36 | .0087923486 | 45 | 6 | 22 | .0176438190 |
| 45 | 3 | 17 | .0186867305 | 45 | 4 | 37 | .0084167463 | 45 | 6 | 23 | .0168744283 |
| 45 | 3 | 18 | .0177158305 | 45 | 4 | 38 | .0080428816 | 45 | 6 | 24 | .0161563325 |
| 45 | 3 | 19 | .0168333777 | 45 | 4 | 39 | .0076671884 | 45 | 6 | 25 | .0154833918 |
| 45 | 3 | 20 | .0160263689 | 45 | 4 | 40 | .0072858955 | 45 | 6 | 26 | .0148502968 |
| 45 | 3 | 21 | .0152841906 | 45 | 4 | 41 | .0068928508 | 45 | 6 | 27 | .0142523551 |
| 45 | 3 | 22 | .0145980910 | 45 | 4 | 42 | .0064779497 | 45 | 6 | 28 | .0136854361 |
| 45 | 3 | 23 | .0139607888 | 45 | 5 | 5 | .0633159315 | 45 | 6 | 29 | .0131458886 |
| 45 | 3 | 24 | .0133661662 | 45 | 5 | 6 | .0538586098 | 45 | 6 | 30 | .0126303027 |
| 45 | 3 | 25 | .0128090182 | 45 | 5 | 7 | .0470198909 | 45 | 6 | 31 | .0121353741 |
| 45 | 3 | 26 | .0122848569 | 45 | 5 | 8 | .0418192098 | 45 | 6 | 32 | .0116582435 |
| 45 | 3 | 27 | .0117897712 | 45 | 5 | 9 | .0377146036 | 45 | 6 | 33 | .0111969600 |
| 45 | 3 | 28 | .0113203301 | 45 | 5 | 10 | .0343813030 | 45 | 6 | 34 | .0107499759 |
| 45 | 3 | 29 | .0108735076 | 45 | 5 | 11 | .0316123943 | 45 | 6 | 35 | .0103146032 |
| 45 | 3 | 30 | .0104466069 | 45 | 5 | 12 | .0292695562 | 45 | 6 | 36 | .0098866059 |
| 45 | 3 | 31 | .0100371828 | 45 | 5 | 13 | .0272566636 | 45 | 6 | 37 | .0094624145 |
| 45 | 3 | 32 | .0096429859 | 45 | 5 | 14 | .0255047422 | 45 | 6 | 38 | .0090406432 |
| 45 | 3 | 33 | .0092619199 | 45 | 5 | 15 | .0239629449 | 45 | 6 | 39 | .0086189069 |
| 45 | 3 | 34 | .0088919446 | 45 | 5 | 16 | .0225929094 | 45 | 6 | 40 | .0081905563 |
| 45 | 3 | 35 | .0085309226 | 45 | 5 | 17 | .0213651131 | 45 | 7 | 7 | .0521556638 |
| 45 | 3 | 36 | .0081765652 | 45 | 5 | 18 | .0202564543 | 45 | 7 | 8 | .0464044334 |
| 45 | 3 | 37 | .0078265603 | 45 | 5 | 19 | .0192485929 | 45 | 7 | 9 | .0418622088 |
| 45 | 3 | 38 | .0074785369 | 45 | 5 | 20 | .0183267692 | 45 | 7 | 10 | .0381715309 |
| 45 | 3 | 39 | .0071294552 | 45 | 5 | 21 | .0174789476 | 45 | 7 | 11 | .0351044216 |
| 45 | 3 | 40 | .0067749103 | 45 | 5 | 22 | .0166951840 | 45 | 7 | 12 | .0325083429 |
| 45 | 3 | 41 | .0064089765 | 45 | 5 | 23 | .0159671339 | 45 | 7 | 13 | .0302772036 |
| 45 | 3 | 42 | .0060229810 | 45 | 5 | 24 | .0152876959 | 45 | 7 | 14 | .0283348239 |
| 45 | 3 | 43 | .0056009774 | 45 | 5 | 25 | .0146508162 | 45 | 7 | 15 | .0266250255 |
| 45 | 4 | 4 | .0728968102 | 45 | 5 | 26 | .0140514014 | 45 | 7 | 16 | .0251054360 |
| 45 | 4 | 5 | .0596178539 | 45 | 5 | 27 | .0134852022 | 45 | 7 | 17 | .0237434631 |
| 45 | 4 | 6 | .0506941348 | 45 | 5 | 28 | .0129485493 | 45 | 7 | 18 | .0225135830 |
| 45 | 4 | 7 | .0442454478 | 45 | 5 | 29 | .0124380103 | 45 | 7 | 19 | .0213954559 |
| 45 | 4 | 8 | .0393437624 | 45 | 5 | 30 | .0119502268 | 45 | 7 | 20 | .0203726275 |
| 45 | 4 | 9 | .0354766068 | 45 | 5 | 31 | .0114821038 | 45 | 7 | 21 | .0194316811 |
| 45 | 4 | 10 | .0323370982 | 45 | 5 | 32 | .0110310942 | 45 | 7 | 22 | .0185615983 |
| 45 | 4 | 11 | .0297298276 | 45 | 5 | 33 | .0105950794 | 45 | 7 | 23 | .0177532129 |
| 45 | 4 | 12 | .0275242215 | 45 | 5 | 34 | .0101718332 | 45 | 7 | 24 | .0169988580 |

TABLE I-COVARIANCES OF NORMAL ORDER STATISTICS

| N | I | J | COVARIANCE | N | I | J | COVARIANCE | N | I | J | COVARIANCE |
|---|---|---|---|---|---|---|---|---|---|---|---|
| 45 | 7 | 25 | .0162921203 | 45 | 9 | 23 | .0194802340 | 45 | 11 | 29 | .0165615497 |
| 45 | 7 | 26 | .0156274692 | 45 | 9 | 24 | .0186514593 | 45 | 11 | 30 | .0159162219 |
| 45 | 7 | 27 | .0149997928 | 45 | 9 | 25 | .0178744796 | 45 | 11 | 31 | .0152967458 |
| 45 | 7 | 28 | .0144042400 | 45 | 9 | 26 | .0171427852 | 45 | 11 | 32 | .0147012541 |
| 45 | 7 | 29 | .0138366342 | 45 | 9 | 27 | .0164508439 | 45 | 11 | 33 | .0141265018 |
| 45 | 7 | 30 | .0132939822 | 45 | 9 | 28 | .0157944923 | 45 | 11 | 34 | .0135683108 |
| 45 | 7 | 31 | .0127741391 | 45 | 9 | 29 | .0151705026 | 45 | 11 | 35 | .0130238547 |
| 45 | 7 | 32 | .0122745743 | 45 | 9 | 30 | .0145752088 | 45 | 12 | 12 | .0403697620 |
| 45 | 7 | 33 | .0117917083 | 45 | 9 | 31 | .0140038865 | 45 | 12 | 13 | .0376138776 |
| 45 | 7 | 34 | .0113219694 | 45 | 9 | 32 | .0134521883 | 45 | 12 | 14 | .0352131695 |
| 45 | 7 | 35 | .0108630917 | 45 | 9 | 33 | .0129177938 | 45 | 12 | 15 | .0330988214 |
| 45 | 7 | 36 | .0104132582 | 45 | 9 | 34 | .0123993426 | 45 | 12 | 16 | .0312188554 |
| 45 | 7 | 37 | .0099692926 | 45 | 9 | 35 | .0118939378 | 45 | 12 | 17 | .0295332469 |
| 45 | 7 | 38 | .0095273460 | 45 | 9 | 36 | .0113975640 | 45 | 12 | 18 | .0280106305 |
| 45 | 7 | 39 | .0090842280 | 45 | 9 | 37 | .0109073966 | 45 | 12 | 19 | .0266260152 |
| 45 | 8 | 8 | .0486251264 | 45 | 10 | 10 | .0436894317 | 45 | 12 | 20 | .0253591766 |
| 45 | 8 | 9 | .0438721108 | 45 | 10 | 11 | .0401906874 | 45 | 12 | 21 | .0241935672 |
| 45 | 8 | 10 | .0400090372 | 45 | 10 | 12 | .0372272497 | 45 | 12 | 22 | .0231155564 |
| 45 | 8 | 11 | .0367978850 | 45 | 10 | 13 | .0346788893 | 45 | 12 | 23 | .0221137880 |
| 45 | 8 | 12 | .0340793272 | 45 | 10 | 14 | .0324591979 | 45 | 12 | 24 | .0211789324 |
| 45 | 8 | 13 | .0317425162 | 45 | 10 | 15 | .0305043825 | 45 | 12 | 25 | .0203032496 |
| 45 | 8 | 14 | .0297078440 | 45 | 10 | 16 | .0287662546 | 45 | 12 | 26 | .0194796000 |
| 45 | 8 | 15 | .0279165958 | 45 | 10 | 17 | .0272076996 | 45 | 12 | 27 | .0187014377 |
| 45 | 8 | 16 | .0263244642 | 45 | 10 | 18 | .0257996664 | 45 | 12 | 28 | .0179634034 |
| 45 | 8 | 17 | .0248973252 | 45 | 10 | 19 | .0245190996 | 45 | 12 | 29 | .0172608303 |
| 45 | 8 | 18 | .0236084135 | 45 | 10 | 20 | .0233474257 | 45 | 12 | 30 | .0165889245 |
| 45 | 8 | 19 | .0224364205 | 45 | 10 | 21 | .0222694412 | 45 | 12 | 31 | .0159437473 |
| 45 | 8 | 20 | .0213641781 | 45 | 10 | 22 | .0212725654 | 45 | 12 | 32 | .0153234823 |
| 45 | 8 | 21 | .0203777006 | 45 | 10 | 23 | .0203461331 | 45 | 12 | 33 | .0147262794 |
| 45 | 8 | 22 | .0194655571 | 45 | 10 | 24 | .0194809714 | 45 | 12 | 34 | .0141469090 |
| 45 | 8 | 23 | .0186183342 | 45 | 10 | 25 | .0186695756 | 45 | 13 | 13 | .0392203071 |
| 45 | 8 | 24 | .0178279637 | 45 | 10 | 26 | .0179059551 | 45 | 13 | 14 | .0367163963 |
| 45 | 8 | 25 | .0170872655 | 45 | 10 | 27 | .0171848078 | 45 | 13 | 15 | .0345102133 |
| 45 | 8 | 26 | .0163899869 | 45 | 10 | 28 | .0165010260 | 45 | 13 | 16 | .0325477643 |
| 45 | 8 | 27 | .0157309723 | 45 | 10 | 29 | .0158501108 | 45 | 13 | 17 | .0307874446 |
| 45 | 8 | 28 | .0151059338 | 45 | 10 | 30 | .0152283681 | 45 | 13 | 18 | .0291966136 |
| 45 | 8 | 29 | .0145109516 | 45 | 10 | 31 | .0146319382 | 45 | 13 | 19 | .0277492504 |
| 45 | 8 | 30 | .0139424064 | 45 | 10 | 32 | .0140562749 | 45 | 13 | 20 | .0264243366 |
| 45 | 8 | 31 | .0133973988 | 45 | 10 | 33 | .0134979728 | 45 | 13 | 21 | .0252047225 |
| 45 | 8 | 32 | .0128735205 | 45 | 10 | 34 | .0129563695 | 45 | 13 | 22 | .0240763353 |
| 45 | 8 | 33 | .0123675158 | 45 | 10 | 35 | .0124305967 | 45 | 13 | 23 | .0230275614 |
| 45 | 8 | 34 | .0118748665 | 45 | 10 | 36 | .0119152728 | 45 | 13 | 24 | .0220484958 |
| 45 | 8 | 35 | .0113921204 | 45 | 11 | 11 | .0418432189 | 45 | 13 | 25 | .0211306434 |
| 45 | 8 | 36 | .0109188306 | 45 | 11 | 12 | .0387626298 | 45 | 13 | 26 | .0202667714 |
| 45 | 8 | 37 | .0104542855 | 45 | 11 | 13 | .0361135674 | 45 | 13 | 27 | .0194500364 |
| 45 | 8 | 38 | .0099928713 | 45 | 11 | 14 | .0338062593 | 45 | 13 | 28 | .0186741920 |
| 45 | 9 | 9 | .0459008011 | 45 | 11 | 15 | .0317744171 | 45 | 13 | 29 | .0179348074 |
| 45 | 9 | 10 | .0418641214 | 45 | 11 | 16 | .0299679773 | 45 | 13 | 30 | .0172287374 |
| 45 | 9 | 11 | .0385075414 | 45 | 11 | 17 | .0283484012 | 45 | 13 | 31 | .0165520420 |
| 45 | 9 | 12 | .0356650183 | 45 | 11 | 18 | .0268855230 | 45 | 13 | 32 | .0159000635 |
| 45 | 9 | 13 | .0332209931 | 45 | 11 | 19 | .0255553544 | 45 | 13 | 33 | .0152693572 |
| 45 | 9 | 14 | .0310924467 | 45 | 11 | 20 | .0243385222 | 45 | 14 | 14 | .0382339423 |
| 45 | 9 | 15 | .0292181170 | 45 | 11 | 21 | .0232191167 | 45 | 14 | 15 | .0359368890 |
| 45 | 9 | 16 | .0275517375 | 45 | 11 | 22 | .0221838798 | 45 | 14 | 16 | .0338930317 |
| 45 | 9 | 17 | .0260576572 | 45 | 11 | 23 | .0212218617 | 45 | 14 | 17 | .0320591514 |
| 45 | 9 | 18 | .0247079344 | 45 | 11 | 24 | .0203239786 | 45 | 14 | 18 | .0304013414 |
| 45 | 9 | 19 | .0234803623 | 45 | 11 | 25 | .0194824098 | 45 | 14 | 19 | .0288925929 |
| 45 | 9 | 20 | .0223571087 | 45 | 11 | 26 | .0186906555 | 45 | 14 | 20 | .0275111313 |
| 45 | 9 | 21 | .0213237053 | 45 | 11 | 27 | .0179434341 | 45 | 14 | 21 | .0262392306 |
| 45 | 9 | 22 | .0203681342 | 45 | 11 | 28 | .0172356183 | 45 | 14 | 22 | .0250622941 |

TABLE I-COVARIANCES OF NORMAL ORDER STATISTICS

| N | I | J | COVARIANCE | N | I | J | COVARIANCE | N | I | J | COVARIANCE |
|---|---|---|---|---|---|---|---|---|---|---|---|
| 45 | 14 | 23 | .0239681283 | 45 | 18 | 23 | .0280983207 | 46 | 1 | 30 | .0082782850 |
| 45 | 14 | 24 | .0229464230 | 45 | 18 | 24 | .0269054644 | 46 | 1 | 31 | .0079602514 |
| 45 | 14 | 25 | .0219880857 | 45 | 18 | 25 | .0257865928 | 46 | 1 | 32 | .0076546115 |
| 45 | 14 | 26 | .0210851626 | 45 | 18 | 26 | .0247330212 | 46 | 1 | 33 | .0073597304 |
| 45 | 14 | 27 | .0202312474 | 45 | 18 | 27 | .0237370240 | 46 | 1 | 34 | .0070740809 |
| 45 | 14 | 28 | .0194206560 | 45 | 18 | 28 | .0227917681 | 46 | 1 | 35 | .0067961784 |
| 45 | 14 | 29 | .0186481492 | 45 | 19 | 19 | .0352498414 | 46 | 1 | 36 | .0065245183 |
| 45 | 14 | 30 | .0179099109 | 45 | 19 | 20 | .0335687948 | 46 | 1 | 37 | .0062575355 |
| 45 | 14 | 31 | .0172027114 | 45 | 19 | 21 | .0320200684 | 46 | 1 | 38 | .0059935196 |
| 45 | 14 | 32 | .0165214318 | 45 | 19 | 22 | .0305859984 | 46 | 1 | 39 | .0057305004 |
| 45 | 15 | 15 | .0373249167 | 45 | 19 | 23 | .0292518209 | 46 | 1 | 40 | .0054660960 |
| 45 | 15 | 16 | .0352084577 | 45 | 19 | 24 | .0280051211 | 46 | 1 | 41 | .0051972015 |
| 45 | 15 | 17 | .0333097574 | 45 | 19 | 25 | .0268353183 | 46 | 1 | 42 | .0049194090 |
| 45 | 15 | 18 | .0315937107 | 45 | 19 | 26 | .0257332345 | 46 | 1 | 43 | .0046258546 |
| 45 | 15 | 19 | .0300323651 | 45 | 19 | 27 | .0246907678 | 46 | 1 | 44 | .0043043370 |
| 45 | 15 | 20 | .0286031997 | 45 | 20 | 20 | .0349337762 | 46 | 1 | 45 | .0039283168 |
| 45 | 15 | 21 | .0272878503 | 45 | 20 | 21 | .0333309166 | 46 | 1 | 46 | .0034129885 |
| 45 | 15 | 22 | .0260711403 | 45 | 20 | 22 | .0318471964 | 46 | 2 | 2 | .1195226189 |
| 45 | 15 | 23 | .0249403115 | 45 | 20 | 23 | .0304673253 | 46 | 2 | 3 | .0806401681 |
| 45 | 15 | 24 | .0238844683 | 45 | 20 | 24 | .0291783969 | 46 | 2 | 4 | .0617214013 |
| 45 | 15 | 25 | .0228943707 | 45 | 20 | 25 | .0279693656 | 46 | 2 | 5 | .0504117824 |
| 45 | 15 | 26 | .0219620033 | 45 | 20 | 26 | .0268306383 | 46 | 2 | 6 | .0428293991 |
| 45 | 15 | 27 | .0210805095 | 45 | 21 | 21 | .0347081910 | 46 | 2 | 7 | .0373594412 |
| 45 | 15 | 28 | .0202446567 | 45 | 21 | 22 | .0331765362 | 46 | 2 | 8 | .0332072280 |
| 45 | 15 | 29 | .0194495703 | 45 | 21 | 23 | .0317528927 | 46 | 2 | 9 | .0299348988 |
| 45 | 15 | 30 | .0186898108 | 45 | 21 | 24 | .0304237779 | 46 | 2 | 10 | .0272807051 |
| 45 | 15 | 31 | .0179607124 | 45 | 21 | 25 | .0291777052 | 46 | 2 | 11 | .0250782121 |
| 45 | 16 | 16 | .0365600358 | 45 | 22 | 22 | .0345993421 | 46 | 2 | 12 | .0232163379 |
| 45 | 16 | 17 | .0345961833 | 45 | 22 | 23 | .0331177172 | 46 | 2 | 13 | .0216179913 |
| 45 | 16 | 18 | .0328217627 | 45 | 22 | 24 | .0317343955 | 46 | 2 | 14 | .0202279190 |
| 45 | 16 | 19 | .0312078550 | 45 | 23 | 23 | .0345723782 | 46 | 2 | 15 | .0190054344 |
| 45 | 16 | 20 | .0297311313 | 46 | 1 | 1 | .2201549964 | 46 | 2 | 16 | .0179198782 |
| 45 | 16 | 21 | .0283725236 | 46 | 1 | 2 | .1046690782 | 46 | 2 | 17 | .0169476831 |
| 45 | 16 | 22 | .0271162217 | 46 | 1 | 3 | .0704246395 | 46 | 2 | 18 | .0160704163 |
| 45 | 16 | 23 | .0259489523 | 46 | 1 | 4 | .0538275504 | 46 | 2 | 19 | .0152734409 |
| 45 | 16 | 24 | .0248594763 | 46 | 1 | 5 | .0439276720 | 46 | 2 | 20 | .0145449745 |
| 45 | 16 | 25 | .0238382445 | 46 | 1 | 6 | .0372998460 | 46 | 2 | 21 | .0138754097 |
| 45 | 16 | 26 | .0228773422 | 46 | 1 | 7 | .0325232500 | 46 | 2 | 22 | .0132568210 |
| 45 | 16 | 27 | .0219699530 | 46 | 1 | 8 | .0289000199 | 46 | 2 | 23 | .0126826028 |
| 45 | 16 | 28 | .0211099141 | 46 | 1 | 9 | .0260461880 | 46 | 2 | 24 | .0121471923 |
| 45 | 16 | 29 | .0202921354 | 46 | 1 | 10 | .0237324736 | 46 | 2 | 25 | .0116458623 |
| 45 | 16 | 30 | .0195114393 | 46 | 1 | 11 | .0218132138 | 46 | 2 | 26 | .0111745696 |
| 45 | 17 | 17 | .0360044699 | 46 | 1 | 12 | .0201912603 | 46 | 2 | 27 | .0107298313 |
| 45 | 17 | 18 | .0341576273 | 46 | 1 | 13 | .0187992298 | 46 | 2 | 28 | .0103086049 |
| 45 | 17 | 19 | .0324775244 | 46 | 1 | 14 | .0175888499 | 46 | 2 | 29 | .0099081907 |
| 45 | 17 | 20 | .0309398859 | 46 | 1 | 15 | .0165245935 | 46 | 2 | 30 | .0095261688 |
| 45 | 17 | 21 | .0295248686 | 46 | 1 | 16 | .0155797011 | 46 | 2 | 31 | .0091603525 |
| 45 | 17 | 22 | .0282160226 | 46 | 1 | 17 | .0147336030 | 46 | 2 | 32 | .0088087465 |
| 45 | 17 | 23 | .0269995807 | 46 | 1 | 18 | .0139702173 | 46 | 2 | 33 | .0084695108 |
| 45 | 17 | 24 | .0258639776 | 46 | 1 | 19 | .0132767863 | 46 | 2 | 34 | .0081409033 |
| 45 | 17 | 25 | .0247994768 | 46 | 1 | 20 | .0126430323 | 46 | 2 | 35 | .0078212054 |
| 45 | 17 | 26 | .0237977724 | 46 | 1 | 21 | .0120605618 | 46 | 2 | 36 | .0075086737 |
| 45 | 17 | 27 | .0228518103 | 46 | 1 | 22 | .0115224694 | 46 | 2 | 37 | .0072015098 |
| 45 | 17 | 28 | .0219551404 | 46 | 1 | 23 | .0110230236 | 46 | 2 | 38 | .0068977616 |
| 45 | 17 | 29 | .0211014734 | 46 | 1 | 24 | .0105573817 | 46 | 2 | 39 | .0065951490 |
| 45 | 18 | 18 | .0356010775 | 46 | 1 | 25 | .0101214073 | 46 | 2 | 40 | .0062908948 |
| 45 | 18 | 19 | .0338423043 | 46 | 1 | 26 | .0097115792 | 46 | 2 | 41 | .0059814651 |
| 45 | 18 | 20 | .0322315899 | 46 | 1 | 27 | .0093248743 | 46 | 2 | 42 | .0056618404 |
| 45 | 18 | 21 | .0307482759 | 46 | 1 | 28 | .0089586239 | 46 | 2 | 43 | .0053240515 |
| 45 | 18 | 22 | .0293752711 | 46 | 1 | 29 | .0086104547 | 46 | 2 | 44 | .0049540408 |

TABLE I-COVARIANCES OF NORMAL ORDER STATISTICS

| N | I | J | COVARIANCE | N | I | J | COVARIANCE | N | I | J | COVARIANCE |
|---|---|---|---|---|---|---|---|---|---|---|---|
| 46 | 2 | 45 | .0045213154 | 46 | 4 | 21 | .0163279402 | 46 | 5 | 42 | .0070879578 |
| 46 | 3 | 3 | .0880021308 | 46 | 4 | 22 | .0156008113 | 46 | 6 | 6 | .0562870889 |
| 46 | 3 | 4 | .0674283992 | 46 | 4 | 23 | .0149257870 | 46 | 6 | 7 | .0491524814 |
| 46 | 3 | 5 | .0551087906 | 46 | 4 | 24 | .0142963797 | 46 | 6 | 8 | .0437257234 |
| 46 | 3 | 6 | .0468402524 | 46 | 4 | 25 | .0137070599 | 46 | 6 | 9 | .0394422324 |
| 46 | 3 | 7 | .0408707160 | 46 | 4 | 26 | .0131530487 | 46 | 6 | 10 | .0359635224 |
| 46 | 3 | 8 | .0363366894 | 46 | 4 | 27 | .0126301613 | 46 | 6 | 11 | .0330738625 |
| 46 | 3 | 9 | .0327618807 | 46 | 4 | 28 | .0121347217 | 46 | 6 | 12 | .0306289959 |
| 46 | 3 | 10 | .0298613278 | 46 | 4 | 29 | .0116635508 | 46 | 6 | 13 | .0285286478 |
| 46 | 3 | 11 | .0274537142 | 46 | 4 | 30 | .0112139648 | 46 | 6 | 14 | .0267008444 |
| 46 | 3 | 12 | .0254179594 | 46 | 4 | 31 | .0107836555 | 46 | 6 | 15 | .0250925119 |
| 46 | 3 | 13 | .0236699927 | 46 | 4 | 32 | .0103703946 | 46 | 6 | 16 | .0236636079 |
| 46 | 3 | 14 | .0221495334 | 46 | 4 | 33 | .0099717602 | 46 | 6 | 17 | .0223833358 |
| 46 | 3 | 15 | .0208121768 | 46 | 4 | 34 | .0095852349 | 46 | 6 | 18 | .0212276164 |
| 46 | 3 | 16 | .0196244556 | 46 | 4 | 35 | .0092086866 | 46 | 6 | 19 | .0201773446 |
| 46 | 3 | 17 | .0185606438 | 46 | 4 | 36 | .0088405877 | 46 | 6 | 20 | .0192171413 |
| 46 | 3 | 18 | .0176006198 | 46 | 4 | 37 | .0084793526 | 46 | 6 | 21 | .0183344056 |
| 46 | 3 | 19 | .0167283977 | 46 | 4 | 38 | .0081223950 | 46 | 6 | 22 | .0175186368 |
| 46 | 3 | 20 | .0159310932 | 46 | 4 | 39 | .0077663400 | 46 | 6 | 23 | .0167610528 |
| 46 | 3 | 21 | .0151981888 | 46 | 4 | 40 | .0074080508 | 46 | 6 | 24 | .0160543525 |
| 46 | 3 | 22 | .0145210082 | 46 | 4 | 41 | .0070439832 | 46 | 6 | 25 | .0153924415 |
| 46 | 3 | 23 | .0138923340 | 46 | 4 | 42 | .0066679983 | 46 | 6 | 26 | .0147701172 |
| 46 | 3 | 24 | .0133061125 | 46 | 4 | 43 | .0062704300 | 46 | 6 | 27 | .0141828421 |
| 46 | 3 | 25 | .0127572105 | 46 | 5 | 5 | .0627297965 | 46 | 6 | 28 | .0136266564 |
| 46 | 3 | 26 | .0122412179 | 46 | 5 | 6 | .0533576637 | 46 | 6 | 29 | .0130980897 |
| 46 | 3 | 27 | .0117542993 | 46 | 5 | 7 | .0465825089 | 46 | 6 | 30 | .0125938979 |
| 46 | 3 | 28 | .0112930890 | 46 | 5 | 8 | .0414315514 | 46 | 6 | 31 | .0121107855 |
| 46 | 3 | 29 | .0108546126 | 46 | 5 | 9 | .0373672394 | 46 | 6 | 32 | .0116456038 |
| 46 | 3 | 30 | .0104362220 | 46 | 5 | 10 | .0340675023 | 46 | 6 | 33 | .0111960829 |
| 46 | 3 | 31 | .0100355339 | 46 | 5 | 11 | .0313271767 | 46 | 6 | 34 | .0107610881 |
| 46 | 3 | 32 | .0096503834 | 46 | 5 | 12 | .0290091294 | 46 | 6 | 35 | .0103393348 |
| 46 | 3 | 33 | .0092787988 | 46 | 5 | 13 | .0270180742 | 46 | 6 | 36 | .0099276200 |
| 46 | 3 | 34 | .0089189412 | 46 | 5 | 14 | .0252856507 | 46 | 6 | 37 | .0095214371 |
| 46 | 3 | 35 | .0085689439 | 46 | 5 | 15 | .0237614747 | 46 | 6 | 38 | .0091180428 |
| 46 | 3 | 36 | .0082267537 | 46 | 5 | 16 | .0224075389 | 46 | 6 | 39 | .0087169426 |
| 46 | 3 | 37 | .0078901957 | 46 | 5 | 17 | .0211945913 | 46 | 6 | 40 | .0083154127 |
| 46 | 3 | 38 | .0075572058 | 46 | 5 | 18 | .0200997360 | 46 | 6 | 41 | .0079064154 |
| 46 | 3 | 39 | .0072256775 | 46 | 5 | 19 | .0191047968 | 46 | 7 | 7 | .0516121769 |
| 46 | 3 | 40 | .0068926545 | 46 | 5 | 20 | .0181951569 | 46 | 7 | 8 | .0459220356 |
| 46 | 3 | 41 | .0065537789 | 46 | 5 | 21 | .0173589193 | 46 | 7 | 9 | .0414292652 |
| 46 | 3 | 42 | .0062034348 | 46 | 5 | 22 | .0165862822 | 46 | 7 | 10 | .0377797277 |
| 46 | 3 | 43 | .0058333647 | 46 | 5 | 23 | .0158690355 | 46 | 7 | 11 | .0347476030 |
| 46 | 3 | 44 | .0054280762 | 46 | 5 | 24 | .0152001742 | 46 | 7 | 12 | .0321818216 |
| 46 | 4 | 4 | .0722709949 | 46 | 5 | 25 | .0145736752 | 46 | 7 | 13 | .0299773282 |
| 46 | 4 | 5 | .0591003511 | 46 | 5 | 26 | .0139844247 | 46 | 7 | 14 | .0280586937 |
| 46 | 4 | 6 | .0502521873 | 46 | 5 | 27 | .0134281732 | 46 | 7 | 15 | .0263702947 |
| 46 | 4 | 7 | .0438598784 | 46 | 5 | 28 | .0129013649 | 46 | 7 | 16 | .0248701832 |
| 46 | 4 | 8 | .0390023082 | 46 | 5 | 29 | .0124008122 | 46 | 7 | 17 | .0235261089 |
| 46 | 4 | 9 | .0351709189 | 46 | 5 | 30 | .0119234062 | 46 | 7 | 18 | .0223128414 |
| 46 | 4 | 10 | .0320612206 | 46 | 5 | 31 | .0114661243 | 46 | 7 | 19 | .0212102925 |
| 46 | 4 | 11 | .0294793560 | 46 | 5 | 32 | .0110263097 | 46 | 7 | 20 | .0202021984 |
| 46 | 4 | 12 | .0272958052 | 46 | 5 | 33 | .0106018033 | 46 | 7 | 21 | .0192752656 |
| 46 | 4 | 13 | .0254206010 | 46 | 5 | 34 | .0101906175 | 46 | 7 | 22 | .0184185564 |
| 46 | 4 | 14 | .0237892166 | 46 | 5 | 35 | .0097905325 | 46 | 7 | 23 | .0176229618 |
| 46 | 4 | 15 | .0223541128 | 46 | 5 | 36 | .0093992651 | 46 | 7 | 24 | .0168808771 |
| 46 | 4 | 16 | .0210794544 | 46 | 5 | 37 | .0090148749 | 46 | 7 | 25 | .0161860222 |
| 46 | 4 | 17 | .0199376805 | 46 | 5 | 38 | .0086352163 | 46 | 7 | 26 | .0155331206 |
| 46 | 4 | 18 | .0189072114 | 46 | 5 | 39 | .0082568268 | 46 | 7 | 27 | .0149173676 |
| 46 | 4 | 19 | .0179708853 | 46 | 5 | 40 | .0078755200 | 46 | 7 | 28 | .0143340536 |
| 46 | 4 | 20 | .0171148790 | 46 | 5 | 41 | .0074877145 | 46 | 7 | 29 | .0137787858 |

TABLE I-COVARIANCES OF NORMAL ORDER STATISTICS

| N | I | J | COVARIANCE | N | I | J | COVARIANCE | N | I | J | COVARIANCE |
|---|---|---|---|---|---|---|---|---|---|---|---|
| 46 | 7 | 30 | .0132481600 | 46 | 9 | 26 | .0170210258 | 46 | 11 | 30 | .0158160737 |
| 46 | 7 | 31 | .0127399855 | 46 | 9 | 27 | .0163432770 | 46 | 11 | 31 | .0152124287 |
| 46 | 7 | 32 | .0122524049 | 46 | 9 | 28 | .0157004795 | 46 | 11 | 32 | .0146322759 |
| 46 | 7 | 33 | .0117827096 | 46 | 9 | 29 | .0150896445 | 46 | 11 | 33 | .0140732649 |
| 46 | 7 | 34 | .0113274699 | 46 | 9 | 30 | .0145079824 | 46 | 11 | 34 | .0135317560 |
| 46 | 7 | 35 | .0108839356 | 46 | 9 | 31 | .0139515651 | 46 | 11 | 35 | .0130051044 |
| 46 | 7 | 36 | .0104503818 | 46 | 9 | 32 | .0134158146 | 46 | 11 | 36 | .0124919918 |
| 46 | 7 | 37 | .0100244632 | 46 | 9 | 33 | .0128974687 | 46 | 12 | 12 | .0398171295 |
| 46 | 7 | 38 | .0096026064 | 46 | 9 | 34 | .0123950101 | 46 | 12 | 13 | .0371044046 |
| 46 | 7 | 39 | .0091820733 | 46 | 9 | 35 | .0119065047 | 46 | 12 | 14 | .0347421898 |
| 46 | 7 | 40 | .0087607040 | 46 | 9 | 36 | .0114285119 | 46 | 12 | 15 | .0326625584 |
| 46 | 8 | 8 | .0480833058 | 46 | 9 | 37 | .0109583465 | 46 | 12 | 16 | .0308142089 |
| 46 | 8 | 9 | .0433853611 | 46 | 9 | 38 | .0104946744 | 46 | 12 | 17 | .0291576542 |
| 46 | 8 | 10 | .0395681568 | 46 | 10 | 10 | .0431766656 | 46 | 12 | 18 | .0276619786 |
| 46 | 8 | 11 | .0363960536 | 46 | 10 | 11 | .0397226221 | 46 | 12 | 19 | .0263025784 |
| 46 | 8 | 12 | .0337113446 | 46 | 10 | 12 | .0367977412 | 46 | 12 | 20 | .0250595451 |
| 46 | 8 | 13 | .0314043233 | 46 | 10 | 13 | .0342831471 | 46 | 12 | 21 | .0239165551 |
| 46 | 8 | 14 | .0293962106 | 46 | 10 | 14 | .0320934224 | 46 | 12 | 22 | .0228601229 |
| 46 | 8 | 15 | .0276289147 | 46 | 10 | 15 | .0301655177 | 46 | 12 | 23 | .0218789495 |
| 46 | 8 | 16 | .0260586168 | 46 | 10 | 16 | .0284518073 | 46 | 12 | 24 | .0209637569 |
| 46 | 8 | 17 | .0246515892 | 46 | 10 | 17 | .0269155996 | 46 | 12 | 25 | .0201070836 |
| 46 | 8 | 18 | .0233813825 | 46 | 10 | 18 | .0255281637 | 46 | 12 | 26 | .0193021920 |
| 46 | 8 | 19 | .0222269242 | 46 | 10 | 19 | .0242667098 | 46 | 12 | 27 | .0185427868 |
| 46 | 8 | 20 | .0211712118 | 46 | 10 | 20 | .0231129192 | 46 | 12 | 28 | .0178236384 |
| 46 | 8 | 21 | .0202003654 | 46 | 10 | 21 | .0220518396 | 46 | 12 | 29 | .0171402535 |
| 46 | 8 | 22 | .0193030452 | 46 | 10 | 22 | .0210711484 | 46 | 12 | 30 | .0164877312 |
| 46 | 8 | 23 | .0184700087 | 46 | 10 | 23 | .0201604025 | 46 | 12 | 31 | .0158613596 |
| 46 | 8 | 24 | .0176934523 | 46 | 10 | 24 | .0193104660 | 46 | 12 | 32 | .0152588771 |
| 46 | 8 | 25 | .0169664356 | 46 | 10 | 25 | .0185137137 | 46 | 12 | 33 | .0146800291 |
| 46 | 8 | 26 | .0162827939 | 46 | 10 | 26 | .0177641188 | 46 | 12 | 34 | .0141222159 |
| 46 | 8 | 27 | .0156373193 | 46 | 10 | 27 | .0170565674 | 46 | 12 | 35 | .0135789013 |
| 46 | 8 | 28 | .0150256662 | 46 | 10 | 28 | .0163862332 | 46 | 13 | 13 | .0386617368 |
| 46 | 8 | 29 | .0144439180 | 46 | 10 | 29 | .0157488555 | 46 | 13 | 14 | .0361999243 |
| 46 | 8 | 30 | .0138884455 | 46 | 10 | 30 | .0151411139 | 46 | 13 | 15 | .0340317134 |
| 46 | 8 | 31 | .0133564087 | 46 | 10 | 31 | .0145597164 | 46 | 13 | 16 | .0321038655 |
| 46 | 8 | 32 | .0128459667 | 46 | 10 | 32 | .0140000992 | 46 | 13 | 17 | .0303753768 |
| 46 | 8 | 33 | .0123550023 | 46 | 10 | 33 | .0134575325 | 46 | 13 | 18 | .0288140975 |
| 46 | 8 | 34 | .0118794776 | 46 | 10 | 34 | .0129302233 | 46 | 13 | 19 | .0273943998 |
| 46 | 8 | 35 | .0114144744 | 46 | 10 | 35 | .0124191599 | 46 | 13 | 20 | .0260955600 |
| 46 | 8 | 36 | .0109578074 | 46 | 10 | 36 | .0119227491 | 46 | 13 | 21 | .0249006305 |
| 46 | 8 | 37 | .0105105475 | 46 | 10 | 37 | .0114340386 | 46 | 13 | 22 | .0237956816 |
| 46 | 8 | 38 | .0100715096 | 46 | 11 | 11 | .0413218340 | 46 | 13 | 23 | .0227692896 |
| 46 | 8 | 39 | .0096338329 | 46 | 11 | 12 | .0382837314 | 46 | 13 | 24 | .0218118006 |
| 46 | 9 | 9 | .0453697309 | 46 | 11 | 13 | .0356719059 | 46 | 13 | 25 | .0209149778 |
| 46 | 9 | 10 | .0413828598 | 46 | 11 | 14 | .0333976615 | 46 | 13 | 26 | .0200719852 |
| 46 | 9 | 11 | .0380686309 | 46 | 11 | 15 | .0313955125 | 46 | 13 | 27 | .0192762869 |
| 46 | 9 | 12 | .0352627699 | 46 | 11 | 16 | .0296159992 | 46 | 13 | 28 | .0185214092 |
| 46 | 9 | 13 | .0328509662 | 46 | 11 | 17 | .0280210466 | 46 | 13 | 29 | .0178023846 |
| 46 | 9 | 14 | .0307511140 | 46 | 11 | 18 | .0265808672 | 46 | 13 | 30 | .0171161076 |
| 46 | 9 | 15 | .0289026429 | 46 | 11 | 19 | .0252718068 | 46 | 13 | 31 | .0164593927 |
| 46 | 9 | 16 | .0272598232 | 46 | 11 | 20 | .0240748042 | 46 | 13 | 32 | .0158281109 |
| 46 | 9 | 17 | .0257874190 | 46 | 11 | 21 | .0229742281 | 46 | 13 | 33 | .0152190871 |
| 46 | 9 | 18 | .0244577988 | 46 | 11 | 22 | .0219569912 | 46 | 13 | 34 | .0146302670 |
| 46 | 9 | 19 | .0232489792 | 46 | 11 | 23 | .0210122087 | 46 | 14 | 14 | .0377015414 |
| 46 | 9 | 20 | .0221433010 | 46 | 11 | 24 | .0201308517 | 46 | 14 | 15 | .0354419720 |
| 46 | 9 | 21 | .0211264884 | 46 | 11 | 25 | .0193051050 | 46 | 14 | 16 | .0334320612 |
| 46 | 9 | 22 | .0201867551 | 46 | 11 | 26 | .0185284888 | 46 | 14 | 17 | .0316292104 |
| 46 | 9 | 23 | .0193141545 | 46 | 11 | 27 | .0177960403 | 46 | 14 | 18 | .0300000094 |
| 46 | 9 | 24 | .0185003352 | 46 | 11 | 28 | .0171033136 | 46 | 14 | 19 | .0285178305 |
| 46 | 9 | 25 | .0177381263 | 46 | 11 | 29 | .0164451239 | 46 | 14 | 20 | .0271611838 |

TABLE I-COVARIANCES OF NORMAL ORDER STATISTICS

| N | I | J | COVARIANCE | N | I | J | COVARIANCE | N | I | J | COVARIANCE |
|---|---|---|---|---|---|---|---|---|---|---|---|
| 46 | 14 | 21 | .0259125733 | 46 | 17 | 30 | .0200919299 | 47 | 1 | 18 | .0139014695 |
| 46 | 14 | 22 | .0247576326 | 46 | 18 | 18 | .0349749520 | 47 | 1 | 19 | .0132147507 |
| 46 | 14 | 23 | .0236844258 | 46 | 18 | 19 | .0332590299 | 47 | 1 | 20 | .0125873828 |
| 46 | 14 | 24 | .0226829753 | 46 | 18 | 20 | .0316886856 | 47 | 1 | 21 | .0120110341 |
| 46 | 14 | 25 | .0217444946 | 46 | 18 | 21 | .0302436370 | 47 | 1 | 22 | .0114788374 |
| 46 | 14 | 26 | .0208610700 | 46 | 18 | 22 | .0289071167 | 47 | 1 | 23 | .0109851089 |
| 46 | 14 | 27 | .0200262728 | 46 | 18 | 23 | .0276651108 | 47 | 1 | 24 | .0105250707 |
| 46 | 14 | 28 | .0192344223 | 46 | 18 | 24 | .0265058533 | 47 | 1 | 25 | .0100946272 |
| 46 | 14 | 29 | .0184800652 | 46 | 18 | 25 | .0254194459 | 47 | 1 | 26 | .0096902710 |
| 46 | 14 | 30 | .0177593058 | 46 | 18 | 26 | .0243975165 | 47 | 1 | 27 | .0093090186 |
| 46 | 14 | 31 | .0170699269 | 46 | 18 | 27 | .0234327028 | 47 | 1 | 28 | .0089482687 |
| 46 | 14 | 32 | .0164084030 | 46 | 18 | 28 | .0225185423 | 47 | 1 | 29 | .0086056855 |
| 46 | 14 | 33 | .0157686745 | 46 | 18 | 29 | .0216491076 | 47 | 1 | 30 | .0082792009 |
| 46 | 15 | 15 | .0367991551 | 46 | 19 | 19 | .0346363645 | 47 | 1 | 31 | .0079670085 |
| 46 | 15 | 16 | .0347168951 | 46 | 19 | 20 | .0329931012 | 47 | 1 | 32 | .0076674566 |
| 46 | 15 | 17 | .0328492696 | 46 | 19 | 21 | .0314798930 | 47 | 1 | 33 | .0073789735 |
| 46 | 15 | 18 | .0311616758 | 46 | 19 | 22 | .0300793900 | 47 | 1 | 34 | .0071000931 |
| 46 | 15 | 19 | .0296265561 | 46 | 19 | 23 | .0287770921 | 47 | 1 | 35 | .0068294436 |
| 46 | 15 | 20 | .0282217101 | 46 | 19 | 24 | .0275608410 | 47 | 1 | 36 | .0065656660 |
| 46 | 15 | 21 | .0269290658 | 46 | 19 | 25 | .0264203507 | 47 | 1 | 37 | .0063073699 |
| 46 | 15 | 22 | .0257337466 | 46 | 19 | 26 | .0253467790 | 47 | 1 | 38 | .0060531018 |
| 46 | 15 | 23 | .0246233028 | 46 | 19 | 27 | .0243323672 | 47 | 1 | 39 | .0058012556 |
| 46 | 15 | 24 | .0235870770 | 46 | 19 | 28 | .0233700180 | 47 | 1 | 40 | .0055499686 |
| 46 | 15 | 25 | .0226159863 | 46 | 20 | 20 | .0343115998 | 47 | 1 | 41 | .0052969747 |
| 46 | 15 | 26 | .0217020572 | 46 | 20 | 21 | .0327418962 | 47 | 1 | 42 | .0050392942 |
| 46 | 15 | 27 | .0208382111 | 46 | 20 | 22 | .0312891552 | 47 | 1 | 43 | .0047726792 |
| 46 | 15 | 28 | .0200192034 | 46 | 20 | 23 | .0299384453 | 47 | 1 | 44 | .0044905026 |
| 46 | 15 | 29 | .0192407067 | 46 | 20 | 24 | .0286771683 | 47 | 1 | 45 | .0041809475 |
| 46 | 15 | 30 | .0184978275 | 46 | 20 | 25 | .0274945973 | 47 | 1 | 46 | .0038182733 |
| 46 | 15 | 31 | .0177861146 | 46 | 20 | 26 | .0263814354 | 47 | 1 | 47 | .0033201558 |
| 46 | 15 | 32 | .0171023100 | 46 | 20 | 27 | .0253294892 | 47 | 2 | 2 | .1187246059 |
| 46 | 16 | 16 | .0359938878 | 46 | 21 | 21 | .0340430219 | 47 | 2 | 3 | .0800769926 |
| 46 | 16 | 17 | .0340667712 | 46 | 21 | 22 | .0325472338 | 47 | 2 | 4 | .0612805921 |
| 46 | 16 | 18 | .0323261155 | 46 | 21 | 23 | .0311574702 | 47 | 2 | 5 | .0500477537 |
| 46 | 16 | 19 | .0307434792 | 46 | 21 | 24 | .0298605968 | 47 | 2 | 6 | .0425188999 |
| 46 | 16 | 20 | .0292959256 | 46 | 21 | 25 | .0286454411 | 47 | 2 | 7 | .0370888972 |
| 46 | 16 | 21 | .0279647477 | 46 | 21 | 26 | .0275023254 | 47 | 2 | 8 | .0329679746 |
| 46 | 16 | 22 | .0267344686 | 46 | 22 | 22 | .0338874854 | 47 | 2 | 9 | .0297210438 |
| 46 | 16 | 23 | .0255920924 | 46 | 22 | 23 | .0324506885 | 47 | 2 | 10 | .0270880484 |
| 46 | 16 | 24 | .0245265661 | 46 | 22 | 24 | .0311104379 | 47 | 2 | 11 | .0249036492 |
| 46 | 16 | 25 | .0235283814 | 46 | 22 | 25 | .0298550541 | 47 | 2 | 12 | .0230575080 |
| 46 | 16 | 26 | .0225896363 | 46 | 23 | 23 | .0338330407 | 47 | 2 | 13 | .0214730578 |
| 46 | 16 | 27 | .0217036729 | 46 | 23 | 24 | .0324298993 | 47 | 2 | 14 | .0200954275 |
| 46 | 16 | 28 | .0208644753 | 47 | 1 | 1 | .2189930364 | 47 | 2 | 15 | .0188842165 |
| 46 | 16 | 29 | .0200674168 | 47 | 1 | 2 | .1040492774 | 47 | 2 | 16 | .0178089850 |
| 46 | 16 | 30 | .0193081947 | 47 | 1 | 3 | .0699878739 | 47 | 2 | 17 | .0168463370 |
| 46 | 16 | 31 | .0185811389 | 47 | 1 | 4 | .0534859646 | 47 | 2 | 18 | .0159779754 |
| 46 | 17 | 17 | .0353899288 | 47 | 1 | 5 | .0436458112 | 47 | 2 | 19 | .0151893729 |
| 46 | 17 | 18 | .0335850578 | 47 | 1 | 6 | .0370596647 | 47 | 2 | 20 | .0144688404 |
| 46 | 17 | 19 | .0319441313 | 47 | 1 | 7 | .0323142242 | 47 | 2 | 21 | .0138068522 |
| 46 | 17 | 20 | .0304433435 | 47 | 1 | 8 | .0287154360 | 47 | 2 | 22 | .0131955527 |
| 46 | 17 | 21 | .0290632557 | 47 | 1 | 9 | .0258814818 | 47 | 2 | 23 | .0126283971 |
| 46 | 17 | 22 | .0277877540 | 47 | 1 | 10 | .0235843925 | 47 | 2 | 24 | .0120998792 |
| 46 | 17 | 23 | .0266033172 | 47 | 1 | 11 | .0216793541 | 47 | 2 | 25 | .0116053192 |
| 46 | 17 | 24 | .0254985326 | 47 | 1 | 12 | .0200697945 | 47 | 2 | 26 | .0111407122 |
| 46 | 17 | 25 | .0244637837 | 47 | 1 | 13 | .0186887378 | 47 | 2 | 27 | .0107026163 |
| 46 | 17 | 26 | .0234908950 | 47 | 1 | 14 | .0174882069 | 47 | 2 | 28 | .0102880442 |
| 46 | 17 | 27 | .0225731023 | 47 | 1 | 15 | .0164328948 | 47 | 2 | 29 | .0098943580 |
| 46 | 17 | 28 | .0217045061 | 47 | 1 | 16 | .0154962152 | 47 | 2 | 30 | .0095191975 |
| 46 | 17 | 29 | .0208791295 | 47 | 1 | 17 | .0146577337 | 47 | 2 | 31 | .0091604371 |

TABLE I-COVARIANCES OF NORMAL ORDER STATISTICS

| N | I | J | COVARIANCE | N | I | J | COVARIANCE | N | I | J | COVARIANCE |
|---|---|---|---|---|---|---|---|---|---|---|---|
| 47 | 2 | 32 | .0088161440 | 47 | 4 | 6 | .0498257958 | 47 | 5 | 26 | .0139162931 |
| 47 | 2 | 33 | .0084845419 | 47 | 4 | 7 | .0434877162 | 47 | 5 | 27 | .0133694273 |
| 47 | 2 | 34 | .0081639797 | 47 | 4 | 8 | .0386725537 | 47 | 5 | 28 | .0128518015 |
| 47 | 2 | 35 | .0078528765 | 47 | 4 | 9 | .0348755242 | 47 | 5 | 29 | .0123603861 |
| 47 | 2 | 36 | .0075496602 | 47 | 4 | 10 | .0317944444 | 47 | 5 | 30 | .0118923339 |
| 47 | 2 | 37 | .0072527394 | 47 | 4 | 11 | .0292369520 | 47 | 5 | 31 | .0114448232 |
| 47 | 2 | 38 | .0069604585 | 47 | 4 | 12 | .0270745427 | 47 | 5 | 32 | .0110152125 |
| 47 | 2 | 39 | .0066709701 | 47 | 4 | 13 | .0252179637 | 47 | 5 | 33 | .0106012639 |
| 47 | 2 | 40 | .0063820846 | 47 | 4 | 14 | .0236032065 | 47 | 5 | 34 | .0102010353 |
| 47 | 2 | 41 | .0060911713 | 47 | 4 | 15 | .0221831197 | 47 | 5 | 35 | .0098124948 |
| 47 | 2 | 42 | .0057948929 | 47 | 4 | 16 | .0209221683 | 47 | 5 | 36 | .0094334841 |
| 47 | 2 | 43 | .0054884042 | 47 | 4 | 17 | .0197930319 | 47 | 5 | 37 | .0090622345 |
| 47 | 2 | 44 | .0051639701 | 47 | 4 | 18 | .0187743254 | 47 | 5 | 38 | .0086973883 |
| 47 | 2 | 45 | .0048080144 | 47 | 4 | 19 | .0178490393 | 47 | 5 | 39 | .0083366873 |
| 47 | 2 | 46 | .0043909979 | 47 | 4 | 20 | .0170034653 | 47 | 5 | 40 | .0079761404 |
| 47 | 3 | 3 | .0873330610 | 47 | 4 | 21 | .0162264398 | 47 | 5 | 41 | .0076118697 |
| 47 | 3 | 4 | .0669042298 | 47 | 4 | 22 | .0155087828 | 47 | 5 | 42 | .0072412783 |
| 47 | 3 | 5 | .0546755928 | 47 | 4 | 23 | .0148428694 | 47 | 5 | 43 | .0068589919 |
| 47 | 3 | 6 | .0464704629 | 47 | 4 | 24 | .0142223020 | 47 | 6 | 6 | .0557500066 |
| 47 | 3 | 7 | .0405482234 | 47 | 4 | 25 | .0136416497 | 47 | 6 | 7 | .0486831525 |
| 47 | 3 | 8 | .0360512020 | 47 | 4 | 26 | .0130962290 | 47 | 6 | 8 | .0433093263 |
| 47 | 3 | 9 | .0325063961 | 47 | 4 | 27 | .0125819256 | 47 | 6 | 9 | .0390686648 |
| 47 | 3 | 10 | .0296308527 | 47 | 4 | 28 | .0120950866 | 47 | 6 | 10 | .0356255702 |
| 47 | 3 | 11 | .0272445567 | 47 | 4 | 29 | .0116325076 | 47 | 6 | 11 | .0327661846 |
| 47 | 3 | 12 | .0252273113 | 47 | 4 | 30 | .0111914778 | 47 | 6 | 12 | .0303475248 |
| 47 | 3 | 13 | .0234956697 | 47 | 4 | 31 | .0107697642 | 47 | 6 | 13 | .0282702169 |
| 47 | 3 | 14 | .0219898047 | 47 | 4 | 32 | .0103653949 | 47 | 6 | 14 | .0264629466 |
| 47 | 3 | 15 | .0206656496 | 47 | 4 | 33 | .0099762941 | 47 | 6 | 15 | .0248731318 |
| 47 | 3 | 16 | .0194899946 | 47 | 4 | 34 | .0096001016 | 47 | 6 | 16 | .0234610981 |
| 47 | 3 | 17 | .0184373148 | 47 | 4 | 35 | .0092344858 | 47 | 6 | 17 | .0221963252 |
| 47 | 3 | 18 | .0174876524 | 47 | 4 | 36 | .0088776886 | 47 | 6 | 18 | .0210549496 |
| 47 | 3 | 19 | .0166251586 | 47 | 4 | 37 | .0085284979 | 47 | 6 | 19 | .0200180529 |
| 47 | 3 | 20 | .0158370655 | 47 | 4 | 38 | .0081853351 | 47 | 6 | 20 | .0190704412 |
| 47 | 3 | 21 | .0151129505 | 47 | 4 | 39 | .0078455226 | 47 | 6 | 21 | .0181996976 |
| 47 | 3 | 22 | .0144442116 | 47 | 4 | 40 | .0075059357 | 47 | 6 | 22 | .0173954672 |
| 47 | 3 | 23 | .0138236892 | 47 | 4 | 41 | .0071638693 | 47 | 6 | 23 | .0166490423 |
| 47 | 3 | 24 | .0132453818 | 47 | 4 | 42 | .0068158359 | 47 | 6 | 24 | .0159531385 |
| 47 | 3 | 25 | .0127042142 | 47 | 4 | 43 | .0064556943 | 47 | 6 | 25 | .0153016665 |
| 47 | 3 | 26 | .0121958435 | 47 | 4 | 44 | .0060743135 | 47 | 6 | 26 | .0146894618 |
| 47 | 3 | 27 | .0117165069 | 47 | 5 | 5 | .0621648351 | 47 | 6 | 27 | .0141120754 |
| 47 | 3 | 28 | .0112629090 | 47 | 5 | 6 | .0528746375 | 47 | 6 | 28 | .0135656896 |
| 47 | 3 | 29 | .0108321354 | 47 | 5 | 7 | .0461605875 | 47 | 6 | 29 | .0130470435 |
| 47 | 3 | 30 | .0104215837 | 47 | 5 | 8 | .0410574048 | 47 | 6 | 30 | .0125531641 |
| 47 | 3 | 31 | .0100289098 | 47 | 5 | 9 | .0370317868 | 47 | 6 | 31 | .0120809456 |
| 47 | 3 | 32 | .0096519921 | 47 | 5 | 10 | .0337642619 | 47 | 6 | 32 | .0116270690 |
| 47 | 3 | 33 | .0092889245 | 47 | 5 | 11 | .0310513516 | 47 | 6 | 33 | .0111886970 |
| 47 | 3 | 34 | .0089380015 | 47 | 5 | 12 | .0287570659 | 47 | 6 | 34 | .0107643789 |
| 47 | 3 | 35 | .0085976023 | 47 | 5 | 13 | .0267869245 | 47 | 6 | 35 | .0103536917 |
| 47 | 3 | 36 | .0082659656 | 47 | 5 | 14 | .0250731560 | 47 | 6 | 36 | .0099551920 |
| 47 | 3 | 37 | .0079410787 | 47 | 5 | 15 | .0235658215 | 47 | 6 | 37 | .0095649916 |
| 47 | 3 | 38 | .0076209028 | 47 | 5 | 16 | .0222272581 | 47 | 6 | 38 | .0091786803 |
| 47 | 3 | 39 | .0073036563 | 47 | 5 | 17 | .0210284789 | 47 | 6 | 39 | .0087946234 |
| 47 | 3 | 40 | .0069874545 | 47 | 5 | 18 | .0199467872 | 47 | 6 | 40 | .0084128496 |
| 47 | 3 | 41 | .0066693460 | 47 | 5 | 19 | .0189641583 | 47 | 6 | 41 | .0080299301 |
| 47 | 3 | 42 | .0063450247 | 47 | 5 | 20 | .0180660999 | 47 | 6 | 42 | .0076387668 |
| 47 | 3 | 43 | .0060092238 | 47 | 5 | 21 | .0172408334 | 47 | 7 | 7 | .0510913502 |
| 47 | 3 | 44 | .0056540497 | 47 | 5 | 22 | .0164786863 | 47 | 7 | 8 | .0454596110 |
| 47 | 3 | 45 | .0052644103 | 47 | 5 | 23 | .0157715899 | 47 | 7 | 9 | .0410140776 |
| 47 | 4 | 4 | .0716675716 | 47 | 5 | 24 | .0151126613 | 47 | 7 | 10 | .0374037997 |
| 47 | 4 | 5 | .0586012399 | 47 | 5 | 25 | .0144959387 | 47 | 7 | 11 | .0344050279 |

TABLE I-COVARIANCES OF NORMAL ORDER STATISTICS

| N | I | J | COVARIANCE | N | I | J | COVARIANCE | N | I | J | COVARIANCE |
|---|---|---|---|---|---|---|---|---|---|---|---|
| 47 | 7 | 12 | .0318681016 | 47 | 8 | 38 | .0101291159 | 47 | 10 | 36 | .0119181291 |
| 47 | 7 | 13 | .0296889610 | 47 | 8 | 39 | .0097134434 | 47 | 10 | 37 | .0114482613 |
| 47 | 7 | 14 | .0277928977 | 47 | 8 | 40 | .0092973049 | 47 | 10 | 38 | .0109833728 |
| 47 | 7 | 15 | .0261248197 | 47 | 9 | 9 | .0448561693 | 47 | 11 | 11 | .0408307953 |
| 47 | 7 | 16 | .0246431786 | 47 | 9 | 10 | .0409172026 | 47 | 11 | 12 | .0378324217 |
| 47 | 7 | 17 | .0233160411 | 47 | 9 | 11 | .0376437122 | 47 | 11 | 13 | .0352553393 |
| 47 | 7 | 18 | .0221184508 | 47 | 9 | 12 | .0348731215 | 47 | 11 | 14 | .0330118767 |
| 47 | 7 | 19 | .0210305672 | 47 | 9 | 13 | .0324923159 | 47 | 11 | 15 | .0310373144 |
| 47 | 7 | 20 | .0200363301 | 47 | 9 | 14 | .0304200626 | 47 | 11 | 16 | .0292827671 |
| 47 | 7 | 21 | .0191225874 | 47 | 9 | 15 | .0285964526 | 47 | 11 | 17 | .0277105923 |
| 47 | 7 | 22 | .0182784881 | 47 | 9 | 16 | .0269762751 | 47 | 11 | 18 | .0262913419 |
| 47 | 7 | 23 | .0174949586 | 47 | 9 | 17 | .0255247038 | 47 | 11 | 19 | .0250016569 |
| 47 | 7 | 24 | .0167643972 | 47 | 9 | 18 | .0242144188 | 47 | 11 | 20 | .0238227642 |
| 47 | 7 | 25 | .0160805723 | 47 | 9 | 19 | .0230236540 | 47 | 11 | 21 | .0227393246 |
| 47 | 7 | 26 | .0154383991 | 47 | 9 | 20 | .0219348927 | 47 | 11 | 22 | .0217384693 |
| 47 | 7 | 27 | .0148334125 | 47 | 9 | 21 | .0209340005 | 47 | 11 | 23 | .0208094168 |
| 47 | 7 | 28 | .0142612121 | 47 | 9 | 22 | .0200093781 | 47 | 11 | 24 | .0199432028 |
| 47 | 7 | 29 | .0137174036 | 47 | 9 | 23 | .0191512775 | 47 | 11 | 25 | .0191319608 |
| 47 | 7 | 30 | .0131981962 | 47 | 9 | 24 | .0183515715 | 47 | 11 | 26 | .0183690614 |
| 47 | 7 | 31 | .0127010100 | 47 | 9 | 25 | .0176033385 | 47 | 11 | 27 | .0176496261 |
| 47 | 7 | 32 | .0122241628 | 47 | 9 | 26 | .0169002071 | 47 | 11 | 28 | .0169698057 |
| 47 | 7 | 33 | .0117657042 | 47 | 9 | 27 | .0162362486 | 47 | 11 | 29 | .0163251874 |
| 47 | 7 | 34 | .0113228040 | 47 | 9 | 28 | .0156066621 | 47 | 11 | 30 | .0157105212 |
| 47 | 7 | 35 | .0108926414 | 47 | 9 | 29 | .0150082985 | 47 | 11 | 31 | .0151214115 |
| 47 | 7 | 36 | .0104733948 | 47 | 9 | 30 | .0144389138 | 47 | 11 | 32 | .0145554144 |
| 47 | 7 | 37 | .0100632762 | 47 | 9 | 31 | .0138955745 | 47 | 11 | 33 | .0140104596 |
| 47 | 7 | 38 | .0096589532 | 47 | 9 | 32 | .0133741723 | 47 | 11 | 34 | .0134831774 |
| 47 | 7 | 39 | .0092569579 | 47 | 9 | 33 | .0128709329 | 47 | 11 | 35 | .0129706935 |
| 47 | 7 | 40 | .0088562605 | 47 | 9 | 34 | .0123837687 | 47 | 11 | 36 | .0124728250 |
| 47 | 7 | 41 | .0084552502 | 47 | 9 | 35 | .0119110496 | 47 | 11 | 37 | .0119885636 |
| 47 | 8 | 8 | .0475626599 | 47 | 9 | 36 | .0114496749 | 47 | 12 | 12 | .0392921738 |
| 47 | 8 | 9 | .0429173967 | 47 | 9 | 37 | .0109964972 | 47 | 12 | 13 | .0366199911 |
| 47 | 8 | 10 | .0391440456 | 47 | 9 | 38 | .0105510933 | 47 | 12 | 14 | .0342938850 |
| 47 | 8 | 11 | .0360092523 | 47 | 9 | 39 | .0101128738 | 47 | 12 | 15 | .0322467708 |
| 47 | 8 | 12 | .0333568642 | 47 | 10 | 10 | .0426854855 | 47 | 12 | 16 | .0304279831 |
| 47 | 8 | 13 | .0310782693 | 47 | 10 | 11 | .0392740984 | 47 | 12 | 17 | .0287985324 |
| 47 | 8 | 14 | .0290954804 | 47 | 10 | 12 | .0363859899 | 47 | 12 | 18 | .0273279176 |
| 47 | 8 | 15 | .0273509971 | 47 | 10 | 13 | .0339035788 | 47 | 12 | 19 | .0259919042 |
| 47 | 8 | 16 | .0258014622 | 47 | 10 | 14 | .0317423950 | 47 | 12 | 20 | .0247709083 |
| 47 | 8 | 17 | .0244135275 | 47 | 10 | 15 | .0298401041 | 47 | 12 | 21 | .0236488567 |
| 47 | 8 | 18 | .0231610581 | 47 | 10 | 16 | .0281496266 | 47 | 12 | 22 | .0226124345 |
| 47 | 8 | 19 | .0220232269 | 47 | 10 | 17 | .0266346815 | 47 | 12 | 23 | .0216503866 |
| 47 | 8 | 20 | .0209832053 | 47 | 10 | 18 | .0252668388 | 47 | 12 | 24 | .0207533805 |
| 47 | 8 | 21 | .0200272075 | 47 | 10 | 19 | .0240235393 | 47 | 12 | 25 | .0199140862 |
| 47 | 8 | 22 | .0191439344 | 47 | 10 | 20 | .0228866767 | 47 | 12 | 26 | .0191260890 |
| 47 | 8 | 23 | .0183242399 | 47 | 10 | 21 | .0218415149 | 47 | 12 | 27 | .0183833803 |
| 47 | 8 | 24 | .0175605374 | 47 | 10 | 22 | .0208759884 | 47 | 12 | 28 | .0176809874 |
| 47 | 8 | 25 | .0168461515 | 47 | 10 | 23 | .0199799390 | 47 | 12 | 29 | .0170148172 |
| 47 | 8 | 26 | .0161751021 | 47 | 10 | 24 | .0191443611 | 47 | 12 | 30 | .0163801959 |
| 47 | 8 | 27 | .0155422369 | 47 | 10 | 25 | .0183615460 | 47 | 12 | 31 | .0157716997 |
| 47 | 8 | 28 | .0149431957 | 47 | 10 | 26 | .0176253491 | 47 | 12 | 32 | .0151857398 |
| 47 | 8 | 29 | .0143740158 | 47 | 10 | 27 | .0169306761 | 47 | 12 | 33 | .0146222065 |
| 47 | 8 | 30 | .0138309068 | 47 | 10 | 28 | .0162728146 | 47 | 12 | 34 | .0140811380 |
| 47 | 8 | 31 | .0133107787 | 47 | 10 | 29 | .0156476286 | 47 | 12 | 35 | .0135582374 |
| 47 | 8 | 32 | .0128119441 | 47 | 10 | 30 | .0150521461 | 47 | 12 | 36 | .0130468562 |
| 47 | 8 | 33 | .0123334147 | 47 | 10 | 31 | .0144839106 | 47 | 13 | 13 | .0381183952 |
| 47 | 8 | 34 | .0118726328 | 47 | 10 | 32 | .0139390850 | 47 | 13 | 14 | .0356973540 |
| 47 | 8 | 35 | .0114245106 | 47 | 10 | 33 | .0134121817 | 47 | 13 | 15 | .0335659388 |
| 47 | 8 | 36 | .0109844042 | 47 | 10 | 34 | .0128992483 | 47 | 13 | 16 | .0316716324 |
| 47 | 8 | 37 | .0105519482 | 47 | 10 | 35 | .0124006816 | 47 | 13 | 17 | .0299740006 |

TABLE I-COVARIANCES OF NORMAL ORDER STATISTICS

| N | I | J | COVARIANCE | N | I | J | COVARIANCE | N | I | J | COVARIANCE |
|---|---|---|---|---|---|---|---|---|---|---|---|
| 47 | 13 | 18 | .0284413685 | 47 | 16 | 18 | .0318666760 | 47 | 20 | 26 | .0259618916 |
| 47 | 13 | 19 | .0270485044 | 47 | 16 | 19 | .0303114470 | 47 | 20 | 27 | .0249334743 |
| 47 | 13 | 20 | .0257749941 | 47 | 16 | 20 | .0288892990 | 47 | 20 | 28 | .0239586330 |
| 47 | 13 | 21 | .0246040985 | 47 | 16 | 21 | .0275818402 | 47 | 21 | 21 | .0334300334 |
| 47 | 13 | 22 | .0235219968 | 47 | 16 | 22 | .0263739081 | 47 | 21 | 22 | .0319645555 |
| 47 | 13 | 23 | .0225173770 | 47 | 16 | 23 | .0252528061 | 47 | 21 | 23 | .0306031526 |
| 47 | 13 | 24 | .0215807506 | 47 | 16 | 24 | .0242077280 | 47 | 21 | 24 | .0293330474 |
| 47 | 13 | 25 | .0207040593 | 47 | 16 | 25 | .0232292507 | 47 | 21 | 25 | .0281433769 |
| 47 | 13 | 26 | .0198808819 | 47 | 16 | 26 | .0223094293 | 47 | 21 | 26 | .0270247452 |
| 47 | 13 | 27 | .0191052052 | 47 | 16 | 27 | .0214416368 | 47 | 21 | 27 | .0259688956 |
| 47 | 13 | 28 | .0183706263 | 47 | 16 | 28 | .0206197665 | 47 | 22 | 22 | .0332236993 |
| 47 | 13 | 29 | .0176716740 | 47 | 16 | 29 | .0198393640 | 47 | 22 | 23 | .0318245457 |
| 47 | 13 | 30 | .0170048841 | 47 | 16 | 30 | .0190970608 | 47 | 22 | 24 | .0305202587 |
| 47 | 13 | 31 | .0163673567 | 47 | 16 | 31 | .0183880261 | 47 | 22 | 25 | .0292994159 |
| 47 | 13 | 32 | .0157552782 | 47 | 16 | 32 | .0177066713 | 47 | 22 | 26 | .0281522471 |
| 47 | 13 | 33 | .0151655443 | 47 | 17 | 17 | .0348113066 | 47 | 23 | 23 | .0331337632 |
| 47 | 13 | 34 | .0145972562 | 47 | 17 | 18 | .0330446720 | 47 | 23 | 24 | .0317778940 |
| 47 | 13 | 35 | .0140482669 | 47 | 17 | 19 | .0314393007 | 47 | 23 | 25 | .0305086209 |
| 47 | 14 | 14 | .0371837242 | 47 | 17 | 20 | .0299718362 | 47 | 24 | 24 | .0331148603 |
| 47 | 14 | 15 | .0349608234 | 47 | 17 | 21 | .0286232231 | 48 | 1 | 1 | .2178664723 |
| 47 | 14 | 16 | .0329841654 | 47 | 17 | 22 | .0273776913 | 48 | 1 | 2 | .1034486204 |
| 47 | 14 | 17 | .0312117540 | 47 | 17 | 23 | .0262219931 | 48 | 1 | 3 | .0695645710 |
| 47 | 14 | 18 | .0296106691 | 47 | 17 | 24 | .0251448815 | 48 | 1 | 4 | .0531548185 |
| 47 | 14 | 19 | .0281546647 | 47 | 17 | 25 | .0241368207 | 48 | 1 | 5 | .0433724560 |
| 47 | 14 | 20 | .0268225284 | 47 | 17 | 26 | .0231896412 | 48 | 1 | 6 | .0368266141 |
| 47 | 14 | 21 | .0255969611 | 47 | 17 | 27 | .0222966596 | 48 | 1 | 7 | .0321112825 |
| 47 | 14 | 22 | .0244637665 | 47 | 17 | 28 | .0214524352 | 48 | 1 | 8 | .0285360995 |
| 47 | 14 | 23 | .0234111933 | 47 | 17 | 29 | .0206513959 | 48 | 1 | 9 | .0257213292 |
| 47 | 14 | 24 | .0224295619 | 47 | 17 | 30 | .0198888147 | 48 | 1 | 10 | .0234402726 |
| 47 | 14 | 25 | .0215104693 | 47 | 17 | 31 | .0191607517 | 48 | 1 | 11 | .0215489377 |
| 47 | 14 | 26 | .0206461718 | 47 | 18 | 18 | .0343684792 | 48 | 1 | 12 | .0199513100 |
| 47 | 14 | 27 | .0198303284 | 47 | 18 | 19 | .0326942760 | 48 | 1 | 13 | .0185808077 |
| 47 | 14 | 28 | .0190573273 | 47 | 18 | 20 | .0311632493 | 48 | 1 | 14 | .0173897400 |
| 47 | 14 | 29 | .0183213977 | 47 | 18 | 21 | .0297554978 | 48 | 1 | 15 | .0163430119 |
| 47 | 14 | 30 | .0176180482 | 47 | 18 | 22 | .0284545898 | 48 | 1 | 16 | .0154142026 |
| 47 | 14 | 31 | .0169453747 | 47 | 18 | 23 | .0272467554 | 48 | 1 | 17 | .0145830112 |
| 47 | 14 | 32 | .0163015868 | 47 | 18 | 24 | .0261203868 | 48 | 1 | 18 | .0138335555 |
| 47 | 14 | 33 | .0156818981 | 47 | 18 | 25 | .0250657045 | 48 | 1 | 19 | .0131532405 |
| 47 | 14 | 34 | .0150805564 | 47 | 18 | 26 | .0240745324 | 48 | 1 | 20 | .0125319499 |
| 47 | 15 | 15 | .0363050354 | 47 | 18 | 27 | .0231398009 | 48 | 1 | 21 | .0119614213 |
| 47 | 15 | 16 | .0342546108 | 47 | 18 | 28 | .0222554342 | 48 | 1 | 22 | .0114348238 |
| 47 | 15 | 17 | .0324158821 | 47 | 18 | 29 | .0214161048 | 48 | 1 | 23 | .0109465033 |
| 47 | 15 | 18 | .0307547179 | 47 | 18 | 30 | .0206158571 | 48 | 1 | 24 | .0104917419 |
| 47 | 15 | 19 | .0292439320 | 47 | 19 | 19 | .0340430340 | 48 | 1 | 25 | .0100665017 |
| 47 | 15 | 20 | .0278616071 | 47 | 19 | 20 | .0324379366 | 48 | 1 | 26 | .0096672893 |
| 47 | 15 | 21 | .0265899097 | 47 | 19 | 21 | .0309607698 | 48 | 1 | 27 | .0092911312 |
| 47 | 15 | 22 | .0254142105 | 47 | 19 | 22 | .0295944582 | 48 | 1 | 28 | .0089354853 |
| 47 | 15 | 23 | .0243223539 | 47 | 19 | 23 | .0283247365 | 48 | 1 | 29 | .0085980789 |
| 47 | 15 | 24 | .0233039614 | 47 | 19 | 24 | .0271396419 | 48 | 1 | 30 | .0082768539 |
| 47 | 15 | 25 | .0223501937 | 47 | 19 | 25 | .0260291071 | 48 | 1 | 31 | .0079700159 |
| 47 | 15 | 26 | .0214532432 | 47 | 19 | 26 | .0249845780 | 48 | 1 | 32 | .0076759907 |
| 47 | 15 | 27 | .0206057996 | 47 | 19 | 27 | .0239986849 | 48 | 1 | 33 | .0073932921 |
| 47 | 15 | 28 | .0198023446 | 47 | 19 | 28 | .0230646727 | 48 | 1 | 34 | .0071204950 |
| 47 | 15 | 29 | .0190387695 | 47 | 19 | 29 | .0221764032 | 48 | 1 | 35 | .0068562886 |
| 47 | 15 | 30 | .0183106006 | 47 | 20 | 20 | .0337278538 | 48 | 1 | 36 | .0065994348 |
| 47 | 15 | 31 | .0176136189 | 47 | 20 | 21 | .0321896993 | 48 | 1 | 37 | .0063486786 |
| 47 | 15 | 32 | .0169452851 | 47 | 20 | 22 | .0307664534 | 48 | 1 | 38 | .0061027306 |
| 47 | 15 | 33 | .0163024610 | 47 | 20 | 23 | .0294434886 | 48 | 1 | 39 | .0058602378 |
| 47 | 16 | 16 | .0354721828 | 47 | 20 | 24 | .0282084578 | 48 | 1 | 40 | .0056196873 |
| 47 | 16 | 17 | .0335775830 | 47 | 20 | 25 | .0270509179 | 48 | 1 | 41 | .0053793148 |

TABLE I-COVARIANCES OF NORMAL ORDER STATISTICS

| N | I | J | COVARIANCE | N | I | J | COVARIANCE | N | I | J | COVARIANCE |
|---|---|---|---|---|---|---|---|---|---|---|---|
| 48 | 1 | 42 | .0051369588 | 48 | 3 | 10 | .0294074540 | 48 | 4 | 27 | .0125319026 |
| 48 | 1 | 43 | .0048897523 | 48 | 3 | 11 | .0270416538 | 48 | 4 | 28 | .0120530925 |
| 48 | 1 | 44 | .0046335984 | 48 | 3 | 12 | .0250421902 | 48 | 4 | 29 | .0115985689 |
| 48 | 1 | 45 | .0043620940 | 48 | 3 | 13 | .0233262184 | 48 | 4 | 30 | .0111655547 |
| 48 | 1 | 46 | .0040637829 | 48 | 3 | 14 | .0218343494 | 48 | 4 | 31 | .0107517835 |
| 48 | 1 | 47 | .0037136813 | 48 | 3 | 15 | .0205228429 | 48 | 4 | 32 | .0103554128 |
| 48 | 1 | 48 | .0032318217 | 48 | 3 | 16 | .0193587366 | 48 | 4 | 33 | .0099746978 |
| 48 | 2 | 2 | .1179524493 | 48 | 3 | 17 | .0183166981 | 48 | 4 | 34 | .0096076206 |
| 48 | 2 | 3 | .0795320668 | 48 | 3 | 18 | .0173769245 | 48 | 4 | 35 | .0092518870 |
| 48 | 2 | 4 | .0608539790 | 48 | 3 | 19 | .0165236976 | 48 | 4 | 36 | .0089054277 |
| 48 | 2 | 5 | .0496953308 | 48 | 3 | 20 | .0157443620 | 48 | 4 | 37 | .0085668660 |
| 48 | 2 | 6 | .0422181699 | 48 | 3 | 21 | .0150285901 | 48 | 4 | 38 | .0082351736 |
| 48 | 2 | 7 | .0368267288 | 48 | 3 | 22 | .0143678549 | 48 | 4 | 39 | .0079086576 |
| 48 | 2 | 8 | .0327359867 | 48 | 3 | 23 | .0137550527 | 48 | 4 | 40 | .0075846220 |
| 48 | 2 | 9 | .0295135375 | 48 | 3 | 24 | .0131842261 | 48 | 4 | 41 | .0072603338 |
| 48 | 2 | 10 | .0269009609 | 48 | 3 | 25 | .0126503445 | 48 | 4 | 42 | .0069334410 |
| 48 | 2 | 11 | .0247339771 | 48 | 3 | 26 | .0121491188 | 48 | 4 | 43 | .0066003285 |
| 48 | 2 | 12 | .0229029666 | 48 | 3 | 27 | .0116768492 | 48 | 4 | 44 | .0062549103 |
| 48 | 2 | 13 | .0213318687 | 48 | 3 | 28 | .0112303093 | 48 | 4 | 45 | .0058886852 |
| 48 | 2 | 14 | .0199661814 | 48 | 3 | 29 | .0108066535 | 48 | 5 | 5 | .0616197894 |
| 48 | 2 | 15 | .0187657802 | 48 | 3 | 30 | .0104033368 | 48 | 5 | 6 | .0524084754 |
| 48 | 2 | 16 | .0177004369 | 48 | 3 | 31 | .0100180540 | 48 | 5 | 7 | .0457532253 |
| 48 | 2 | 17 | .0167469211 | 48 | 3 | 32 | .0096487056 | 48 | 5 | 8 | .0406959926 |
| 48 | 2 | 18 | .0158870667 | 48 | 3 | 33 | .0092934026 | 48 | 5 | 9 | .0367075715 |
| 48 | 2 | 19 | .0151064509 | 48 | 3 | 34 | .0089504922 | 48 | 5 | 10 | .0334709957 |
| 48 | 2 | 20 | .0143934717 | 48 | 3 | 35 | .0086185200 | 48 | 5 | 11 | .0307844105 |
| 48 | 2 | 21 | .0137386811 | 48 | 3 | 36 | .0082960283 | 48 | 5 | 12 | .0285129272 |
| 48 | 2 | 22 | .0131342910 | 48 | 3 | 37 | .0079812921 | 48 | 5 | 13 | .0265628405 |
| 48 | 2 | 23 | .0125738143 | 48 | 3 | 38 | .0076723227 | 48 | 5 | 14 | .0248669438 |
| 48 | 2 | 24 | .0120517976 | 48 | 3 | 39 | .0073672637 | 48 | 5 | 15 | .0233757261 |
| 48 | 2 | 25 | .0115636099 | 48 | 3 | 40 | .0070646355 | 48 | 5 | 16 | .0220518580 |
| 48 | 2 | 26 | .0111052826 | 48 | 3 | 41 | .0067627058 | 48 | 5 | 17 | .0208666132 |
| 48 | 2 | 27 | .0106734036 | 48 | 3 | 42 | .0064584595 | 48 | 5 | 18 | .0197974920 |
| 48 | 2 | 28 | .0102650269 | 48 | 3 | 43 | .0061476796 | 48 | 5 | 19 | .0188266145 |
| 48 | 2 | 29 | .0098775702 | 48 | 3 | 44 | .0058254799 | 48 | 5 | 20 | .0179395978 |
| 48 | 2 | 30 | .0095087288 | 48 | 3 | 45 | .0054842613 | 48 | 5 | 21 | .0171247587 |
| 48 | 2 | 31 | .0091564305 | 48 | 3 | 46 | .0051092936 | 48 | 5 | 22 | .0163725332 |
| 48 | 2 | 32 | .0088188017 | 48 | 4 | 4 | .0710852838 | 48 | 5 | 23 | .0156749869 |
| 48 | 2 | 33 | .0084941246 | 48 | 4 | 5 | .0581195010 | 48 | 5 | 24 | .0150253781 |
| 48 | 2 | 34 | .0081808064 | 48 | 4 | 6 | .0494141108 | 48 | 5 | 25 | .0144178434 |
| 48 | 2 | 35 | .0078773542 | 48 | 4 | 7 | .0431282420 | 48 | 5 | 26 | .0138472710 |
| 48 | 2 | 36 | .0075823250 | 48 | 4 | 8 | .0383538845 | 48 | 5 | 27 | .0133093091 |
| 48 | 2 | 37 | .0072942840 | 48 | 4 | 9 | .0345898957 | 48 | 5 | 28 | .0128003638 |
| 48 | 2 | 38 | .0070117839 | 48 | 4 | 10 | .0315363173 | 48 | 5 | 29 | .0123174537 |
| 48 | 2 | 39 | .0067332892 | 48 | 4 | 11 | .0290022295 | 48 | 5 | 30 | .0118579213 |
| 48 | 2 | 40 | .0064570255 | 48 | 4 | 12 | .0268601071 | 48 | 5 | 31 | .0114191880 |
| 48 | 2 | 41 | .0061808845 | 48 | 4 | 13 | .0250213868 | 48 | 5 | 32 | .0109987592 |
| 48 | 2 | 42 | .0059023923 | 48 | 4 | 14 | .0234225589 | 48 | 5 | 33 | .0105944139 |
| 48 | 2 | 43 | .0056183977 | 48 | 4 | 15 | .0220168437 | 48 | 5 | 34 | .0102042239 |
| 48 | 2 | 44 | .0053241955 | 48 | 4 | 16 | .0207689925 | 48 | 5 | 35 | .0098262401 |
| 48 | 2 | 45 | .0050122699 | 48 | 4 | 17 | .0196519155 | 48 | 5 | 36 | .0094582993 |
| 48 | 2 | 46 | .0046695134 | 48 | 4 | 18 | .0186444169 | 48 | 5 | 37 | .0090984985 |
| 48 | 2 | 47 | .0042672827 | 48 | 4 | 19 | .0177296394 | 48 | 5 | 38 | .0087457978 |
| 48 | 3 | 3 | .0866865861 | 48 | 4 | 20 | .0168939901 | 48 | 5 | 39 | .0083992713 |
| 48 | 3 | 4 | .0663976854 | 48 | 4 | 21 | .0161263894 | 48 | 5 | 40 | .0080561766 |
| 48 | 3 | 5 | .0542568437 | 48 | 4 | 22 | .0154177233 | 48 | 5 | 41 | .0077119855 |
| 48 | 3 | 6 | .0461128722 | 48 | 4 | 23 | .0147604291 | 48 | 5 | 42 | .0073634944 |
| 48 | 3 | 7 | .0402362251 | 48 | 4 | 24 | .0141481809 | 48 | 5 | 43 | .0070091046 |
| 48 | 3 | 8 | .0357748540 | 48 | 4 | 25 | .0135756374 | 48 | 5 | 44 | .0066431773 |
| 48 | 3 | 9 | .0322589346 | 48 | 4 | 26 | .0130382178 | 48 | 6 | 6 | .0552324517 |

TABLE I-COVARIANCES OF NORMAL ORDER STATISTICS

| N | I | J | COVARIANCE | N | I | J | COVARIANCE | N | I | J | COVARIANCE |
|---|---|---|---|---|---|---|---|---|---|---|---|
| 48 | 6 | 7 | .0482307248 | 48 | 7 | 30 | .0131445752 | 48 | 9 | 22 | .0198357101 |
| 48 | 6 | 8 | .0429077542 | 48 | 7 | 31 | .0126579170 | 48 | 9 | 23 | .0189914262 |
| 48 | 6 | 9 | .0387082221 | 48 | 7 | 32 | .0121911770 | 48 | 9 | 24 | .0182050453 |
| 48 | 6 | 10 | .0352993111 | 48 | 7 | 33 | .0117427454 | 48 | 9 | 25 | .0174699550 |
| 48 | 6 | 11 | .0324689661 | 48 | 7 | 34 | .0113103964 | 48 | 9 | 26 | .0167800606 |
| 48 | 6 | 12 | .0300754282 | 48 | 7 | 35 | .0108915668 | 48 | 9 | 27 | .0161294200 |
| 48 | 6 | 13 | .0280201894 | 48 | 7 | 36 | .0104844888 | 48 | 9 | 28 | .0155128386 |
| 48 | 6 | 14 | .0262325709 | 48 | 7 | 37 | .0100879809 | 48 | 9 | 29 | .0149267197 |
| 48 | 6 | 15 | .0246604671 | 48 | 7 | 38 | .0096994052 | 48 | 9 | 30 | .0143688950 |
| 48 | 6 | 16 | .0232645623 | 48 | 7 | 39 | .0093143491 | 48 | 9 | 31 | .0138372042 |
| 48 | 6 | 17 | .0220146020 | 48 | 7 | 40 | .0089302860 | 48 | 9 | 32 | .0133283632 |
| 48 | 6 | 18 | .0208869219 | 48 | 7 | 41 | .0085482134 | 48 | 9 | 33 | .0128387008 |
| 48 | 6 | 19 | .0198627661 | 48 | 7 | 42 | .0081662460 | 48 | 9 | 34 | .0123657736 |
| 48 | 6 | 20 | .0189271030 | 48 | 8 | 8 | .0470627417 | 48 | 9 | 35 | .0119080612 |
| 48 | 6 | 21 | .0180676942 | 48 | 8 | 9 | .0424678616 | 48 | 9 | 36 | .0114627196 |
| 48 | 6 | 22 | .0172743488 | 48 | 8 | 10 | .0387364199 | 48 | 9 | 37 | .0110256095 |
| 48 | 6 | 23 | .0165384663 | 48 | 8 | 11 | .0356372554 | 48 | 9 | 38 | .0105952226 |
| 48 | 6 | 24 | .0158528027 | 48 | 8 | 12 | .0330157085 | 48 | 9 | 39 | .0101737459 |
| 48 | 6 | 25 | .0152112661 | 48 | 8 | 13 | .0307642184 | 48 | 9 | 40 | .0097599346 |
| 48 | 6 | 26 | .0146086806 | 48 | 8 | 14 | .0288055567 | 48 | 10 | 10 | .0422122636 |
| 48 | 6 | 27 | .0140406073 | 48 | 8 | 15 | .0270827842 | 48 | 10 | 11 | .0388418220 |
| 48 | 6 | 28 | .0135032920 | 48 | 8 | 16 | .0255529773 | 48 | 10 | 12 | .0359890084 |
| 48 | 6 | 29 | .0129936388 | 48 | 8 | 17 | .0241831452 | 48 | 10 | 13 | .0335374817 |
| 48 | 6 | 30 | .0125089858 | 48 | 8 | 18 | .0229474605 | 48 | 10 | 14 | .0314036784 |
| 48 | 6 | 31 | .0120465990 | 48 | 8 | 19 | .0218253510 | 48 | 10 | 15 | .0295259516 |
| 48 | 6 | 32 | .0116032748 | 48 | 8 | 20 | .0208001817 | 48 | 10 | 16 | .0278577540 |
| 48 | 6 | 33 | .0111757094 | 48 | 8 | 21 | .0198582716 | 48 | 10 | 17 | .0263632108 |
| 48 | 6 | 34 | .0107616532 | 48 | 8 | 22 | .0189883333 | 48 | 10 | 18 | .0250141852 |
| 48 | 6 | 35 | .0103605895 | 48 | 8 | 23 | .0181812444 | 48 | 10 | 19 | .0237883248 |
| 48 | 6 | 36 | .0099724841 | 48 | 8 | 24 | .0174295566 | 48 | 10 | 20 | .0226676932 |
| 48 | 6 | 37 | .0095953220 | 48 | 8 | 25 | .0167268271 | 48 | 10 | 21 | .0216377193 |
| 48 | 6 | 38 | .0092246293 | 48 | 8 | 26 | .0160673045 | 48 | 10 | 22 | .0206865589 |
| 48 | 6 | 39 | .0088565866 | 48 | 8 | 27 | .0154459990 | 48 | 10 | 23 | .0198043710 |
| 48 | 6 | 40 | .0084907151 | 48 | 8 | 28 | .0148586540 | 48 | 10 | 24 | .0189823789 |
| 48 | 6 | 41 | .0081270019 | 48 | 8 | 29 | .0143013419 | 48 | 10 | 25 | .0182128854 |
| 48 | 6 | 42 | .0077611515 | 48 | 8 | 30 | .0137701055 | 48 | 10 | 26 | .0174896469 |
| 48 | 6 | 43 | .0073865196 | 48 | 8 | 31 | .0132613856 | 48 | 10 | 27 | .0168075011 |
| 48 | 7 | 7 | .0505919285 | 48 | 8 | 32 | .0127731027 | 48 | 10 | 28 | .0161616830 |
| 48 | 7 | 8 | .0450160780 | 48 | 8 | 33 | .0123048372 | 48 | 10 | 29 | .0155479596 |
| 48 | 7 | 9 | .0406157090 | 48 | 8 | 34 | .0118558239 | 48 | 10 | 30 | .0149634589 |
| 48 | 7 | 10 | .0370429342 | 48 | 8 | 35 | .0114223984 | 48 | 10 | 31 | .0144065143 |
| 48 | 7 | 11 | .0340759921 | 48 | 8 | 36 | .0109987086 | 48 | 10 | 32 | .0138746121 |
| 48 | 7 | 12 | .0315665754 | 48 | 8 | 37 | .0105812803 | 48 | 10 | 33 | .0133626607 |
| 48 | 7 | 13 | .0294115812 | 48 | 8 | 38 | .0101716130 | 48 | 10 | 34 | .0128648883 |
| 48 | 7 | 14 | .0275369961 | 48 | 8 | 39 | .0097715391 | 48 | 10 | 35 | .0123792233 |
| 48 | 7 | 15 | .0258882390 | 48 | 8 | 40 | .0093769212 | 48 | 10 | 36 | .0119079271 |
| 48 | 7 | 16 | .0244241405 | 48 | 8 | 41 | .0089803589 | 48 | 10 | 37 | .0114518262 |
| 48 | 7 | 17 | .0231130600 | 48 | 9 | 9 | .0443585836 | 48 | 10 | 38 | .0110056741 |
| 48 | 7 | 18 | .0219302903 | 48 | 9 | 10 | .0404657639 | 48 | 10 | 39 | .0105625760 |
| 48 | 7 | 19 | .0208562258 | 48 | 9 | 11 | .0372315264 | 48 | 11 | 11 | .0403680315 |
| 48 | 7 | 20 | .0198750172 | 48 | 9 | 12 | .0344949264 | 48 | 11 | 12 | .0374068920 |
| 48 | 7 | 21 | .0189736757 | 48 | 9 | 13 | .0321439975 | 48 | 11 | 13 | .0348623000 |
| 48 | 7 | 22 | .0181414620 | 48 | 9 | 14 | .0300983417 | 48 | 11 | 14 | .0326475607 |
| 48 | 7 | 23 | .0173693446 | 48 | 9 | 15 | .0282986804 | 48 | 11 | 15 | .0306986888 |
| 48 | 7 | 24 | .0166496888 | 48 | 9 | 16 | .0267003009 | 48 | 11 | 16 | .0289673511 |
| 48 | 7 | 25 | .0159762245 | 48 | 9 | 17 | .0252687799 | 48 | 11 | 17 | .0274163149 |
| 48 | 7 | 26 | .0153439532 | 48 | 9 | 18 | .0239771154 | 48 | 11 | 18 | .0260164389 |
| 48 | 7 | 27 | .0147486996 | 48 | 9 | 19 | .0228037662 | 48 | 11 | 19 | .0247446162 |
| 48 | 7 | 28 | .0141864614 | 48 | 9 | 20 | .0217313457 | 48 | 11 | 20 | .0235823205 |
| 48 | 7 | 29 | .0136530758 | 48 | 9 | 21 | .0207458166 | 48 | 11 | 21 | .0225144957 |

TABLE I-COVARIANCES OF NORMAL ORDER STATISTICS

| N | I | J | COVARIANCE | N | I | J | COVARIANCE | N | I | J | COVARIANCE |
|---|---|---|---|---|---|---|---|---|---|---|---|
| 48 | 11 | 22 | .0215285279 | 48 | 13 | 30 | .0168924584 | 48 | 16 | 27 | .0211869835 |
| 48 | 11 | 23 | .0206137956 | 48 | 13 | 31 | .0162737132 | 48 | 16 | 28 | .0203798220 |
| 48 | 11 | 24 | .0197614624 | 48 | 13 | 32 | .0156801299 | 48 | 16 | 29 | .0196130438 |
| 48 | 11 | 25 | .0189636326 | 48 | 13 | 33 | .0151080992 | 48 | 16 | 30 | .0188839034 |
| 48 | 11 | 26 | .0182134187 | 48 | 13 | 34 | .0145570869 | 48 | 16 | 31 | .0181887110 |
| 48 | 11 | 27 | .0175057643 | 48 | 13 | 35 | .0140277662 | 48 | 16 | 32 | .0175221730 |
| 48 | 11 | 28 | .0168371534 | 48 | 13 | 36 | .0135157970 | 48 | 16 | 33 | .0168804647 |
| 48 | 11 | 29 | .0162039728 | 48 | 14 | 14 | .0366750043 | 48 | 17 | 17 | .0342737156 |
| 48 | 11 | 30 | .0156015539 | 48 | 14 | 15 | .0344883551 | 48 | 17 | 18 | .0325410333 |
| 48 | 11 | 31 | .0150253219 | 48 | 14 | 16 | .0325446253 | 48 | 17 | 19 | .0309670678 |
| 48 | 11 | 32 | .0144723468 | 48 | 14 | 17 | .0308024071 | 48 | 17 | 20 | .0295288789 |
| 48 | 11 | 33 | .0139405334 | 48 | 14 | 18 | .0292292631 | 48 | 17 | 21 | .0282077545 |
| 48 | 11 | 34 | .0134264947 | 48 | 14 | 19 | .0277993371 | 48 | 17 | 22 | .0269882581 |
| 48 | 11 | 35 | .0129264327 | 48 | 14 | 20 | .0264917061 | 48 | 17 | 23 | .0258574446 |
| 48 | 11 | 36 | .0124398514 | 48 | 14 | 21 | .0252892596 | 48 | 17 | 24 | .0248042809 |
| 48 | 11 | 37 | .0119688881 | 48 | 14 | 22 | .0241779277 | 48 | 17 | 25 | .0238193399 |
| 48 | 11 | 38 | .0115108888 | 48 | 14 | 23 | .0231460585 | 48 | 17 | 26 | .0228944087 |
| 48 | 12 | 12 | .0387968513 | 48 | 14 | 24 | .0221841782 | 48 | 17 | 27 | .0220226788 |
| 48 | 12 | 13 | .0361624755 | 48 | 14 | 25 | .0212842713 | 48 | 17 | 28 | .0211989609 |
| 48 | 12 | 14 | .0338699748 | 48 | 14 | 26 | .0204388503 | 48 | 17 | 29 | .0204179831 |
| 48 | 12 | 15 | .0318530620 | 48 | 14 | 27 | .0196418067 | 48 | 17 | 30 | .0196753285 |
| 48 | 12 | 16 | .0300616719 | 48 | 14 | 28 | .0188878059 | 48 | 17 | 31 | .0189679221 |
| 48 | 12 | 17 | .0284572739 | 48 | 14 | 29 | .0181709122 | 48 | 17 | 32 | .0182911458 |
| 48 | 12 | 18 | .0270097410 | 48 | 14 | 30 | .0174857710 | 48 | 18 | 18 | .0337843265 |
| 48 | 12 | 19 | .0256951758 | 48 | 14 | 31 | .0168299474 | 48 | 18 | 19 | .0321498433 |
| 48 | 12 | 20 | .0244943137 | 48 | 14 | 32 | .0162027378 | 48 | 18 | 20 | .0306561873 |
| 48 | 12 | 21 | .0233913567 | 48 | 14 | 33 | .0156012169 | 48 | 18 | 21 | .0292838521 |
| 48 | 12 | 22 | .0223732083 | 48 | 14 | 34 | .0150200137 | 48 | 18 | 22 | .0280167564 |
| 48 | 12 | 23 | .0214286957 | 48 | 14 | 35 | .0144555273 | 48 | 18 | 23 | .0268414009 |
| 48 | 12 | 24 | .0205483779 | 48 | 15 | 15 | .0358342728 | 48 | 18 | 24 | .0257463442 |
| 48 | 12 | 25 | .0197249191 | 48 | 15 | 16 | .0338142309 | 48 | 18 | 25 | .0247218718 |
| 48 | 12 | 26 | .0189520841 | 48 | 15 | 17 | .0320030933 | 48 | 18 | 26 | .0237598803 |
| 48 | 12 | 27 | .0182240609 | 48 | 15 | 18 | .0303671813 | 48 | 18 | 27 | .0228534461 |
| 48 | 12 | 28 | .0175361267 | 48 | 15 | 19 | .0288796734 | 48 | 18 | 28 | .0219967382 |
| 48 | 12 | 29 | .0168847449 | 48 | 15 | 20 | .0275189156 | 48 | 18 | 29 | .0211851320 |
| 48 | 12 | 30 | .0162659142 | 48 | 15 | 21 | .0262672689 | 48 | 18 | 30 | .0204131576 |
| 48 | 12 | 31 | .0156740356 | 48 | 15 | 22 | .0251102856 | 48 | 18 | 31 | .0196755559 |
| 48 | 12 | 32 | .0151039953 | 48 | 15 | 23 | .0240360510 | 48 | 19 | 19 | .0334627571 |
| 48 | 12 | 33 | .0145543782 | 48 | 15 | 24 | .0230344617 | 48 | 19 | 20 | .0318964229 |
| 48 | 12 | 34 | .0140264463 | 48 | 15 | 25 | .0220969820 | 48 | 19 | 21 | .0304560047 |
| 48 | 12 | 35 | .0135189198 | 48 | 15 | 26 | .0212161439 | 48 | 19 | 22 | .0291246860 |
| 48 | 12 | 36 | .0130261807 | 48 | 15 | 27 | .0203845430 | 48 | 19 | 23 | .0278884369 |
| 48 | 12 | 37 | .0125433636 | 48 | 15 | 28 | .0195963025 | 48 | 19 | 24 | .0267354524 |
| 48 | 13 | 13 | .0375913308 | 48 | 15 | 29 | .0188472542 | 48 | 19 | 25 | .0256558009 |
| 48 | 13 | 14 | .0352095490 | 48 | 15 | 30 | .0181330622 | 48 | 19 | 26 | .0246411108 |
| 48 | 13 | 15 | .0331135721 | 48 | 15 | 31 | .0174494475 | 48 | 19 | 27 | .0236843472 |
| 48 | 13 | 16 | .0312515679 | 48 | 15 | 32 | .0167943326 | 48 | 19 | 28 | .0227791847 |
| 48 | 13 | 17 | .0295836367 | 48 | 15 | 33 | .0161664008 | 48 | 19 | 29 | .0219197703 |
| 48 | 13 | 18 | .0280785536 | 48 | 15 | 34 | .0155606408 | 48 | 19 | 30 | .0211011828 |
| 48 | 13 | 19 | .0267114784 | 48 | 16 | 16 | .0349928634 | 48 | 20 | 20 | .0331709213 |
| 48 | 13 | 20 | .0254623250 | 48 | 16 | 17 | .0331270444 | 48 | 20 | 21 | .0316644432 |
| 48 | 13 | 21 | .0243145915 | 48 | 16 | 18 | .0314423587 | 48 | 20 | 22 | .0302709707 |
| 48 | 13 | 22 | .0232545673 | 48 | 16 | 19 | .0299111678 | 48 | 20 | 23 | .0289761122 |
| 48 | 13 | 23 | .0222710010 | 48 | 16 | 20 | .0285111668 | 48 | 20 | 24 | .0277677117 |
| 48 | 13 | 24 | .0213544813 | 48 | 16 | 21 | .0272242223 | 48 | 20 | 25 | .0266355416 |
| 48 | 13 | 25 | .0204969896 | 48 | 16 | 22 | .0260354414 | 48 | 20 | 26 | .0255709115 |
| 48 | 13 | 26 | .0196924230 | 48 | 16 | 23 | .0249324206 | 48 | 20 | 27 | .0245662511 |
| 48 | 13 | 27 | .0189353936 | 48 | 16 | 24 | .0239046567 | 48 | 20 | 28 | .0236148665 |
| 48 | 13 | 28 | .0182198875 | 48 | 16 | 25 | .0229429056 | 48 | 20 | 29 | .0227104674 |
| 48 | 13 | 29 | .0175402045 | 48 | 16 | 26 | .0220392186 | 48 | 21 | 21 | .0328639426 |

TABLE I-COVARIANCES OF NORMAL ORDER STATISTICS

| N | I | J | COVARIANCE | N | I | J | COVARIANCE | N | I | J | COVARIANCE |
|---|---|---|---|---|---|---|---|---|---|---|---|
| 48 | 21 | 22 | .0314251203 | 49 | 1 | 42 | .0052177836 | 49 | 3 | 8 | .0355071684 |
| 48 | 21 | 23 | .0300884717 | 49 | 1 | 43 | .0049853627 | 49 | 3 | 9 | .0320190874 |
| 48 | 21 | 24 | .0288415608 | 49 | 1 | 44 | .0047479548 | 49 | 3 | 10 | .0291907821 |
| 48 | 21 | 25 | .0276738007 | 49 | 1 | 45 | .0045016075 | 49 | 3 | 11 | .0268447081 |
| 48 | 21 | 26 | .0265760962 | 49 | 1 | 46 | .0042401289 | 49 | 3 | 12 | .0248623458 |
| 48 | 21 | 27 | .0255404661 | 49 | 1 | 47 | .0039524014 | 49 | 3 | 13 | .0231614319 |
| 48 | 21 | 28 | .0245597474 | 49 | 1 | 48 | .0036141599 | 49 | 3 | 14 | .0216830007 |
| 48 | 22 | 22 | .0326104464 | 49 | 1 | 49 | .0031476778 | 49 | 3 | 15 | .0203836280 |
| 48 | 22 | 23 | .0312418467 | 49 | 2 | 2 | .1172047627 | 49 | 3 | 16 | .0192305904 |
| 48 | 22 | 24 | .0299664452 | 49 | 2 | 3 | .0790044159 | 49 | 3 | 17 | .0181987398 |
| 48 | 22 | 25 | .0287731199 | 49 | 2 | 4 | .0604408107 | 49 | 3 | 18 | .0172684200 |
| 48 | 22 | 26 | .0276523781 | 49 | 2 | 5 | .0493539072 | 49 | 3 | 19 | .0164240363 |
| 48 | 22 | 27 | .0265959017 | 49 | 2 | 6 | .0419267070 | 49 | 3 | 20 | .0156530406 |
| 48 | 23 | 23 | .0324742698 | 49 | 2 | 7 | .0365725140 | 49 | 3 | 21 | .0149451993 |
| 48 | 23 | 24 | .0311594631 | 49 | 2 | 8 | .0325109076 | 49 | 3 | 22 | .0142920626 |
| 48 | 23 | 25 | .0299298504 | 49 | 2 | 9 | .0293120776 | 49 | 3 | 23 | .0136865841 |
| 48 | 23 | 26 | .0287754536 | 49 | 2 | 10 | .0267191877 | 49 | 3 | 24 | .0131228474 |
| 48 | 24 | 24 | .0324342610 | 49 | 2 | 11 | .0245689830 | 49 | 3 | 25 | .0125958563 |
| 48 | 24 | 25 | .0311457132 | 49 | 2 | 12 | .0227525391 | 49 | 3 | 26 | .0121013611 |
| 49 | 1 | 1 | .2167734861 | 49 | 2 | 13 | .0211942849 | 49 | 3 | 27 | .0116357115 |
| 49 | 1 | 2 | .1028661202 | 49 | 2 | 14 | .0198400748 | 49 | 3 | 28 | .0111957410 |
| 49 | 1 | 3 | .0691540415 | 49 | 2 | 15 | .0186500511 | 49 | 3 | 29 | .0107786736 |
| 49 | 1 | 4 | .0528335843 | 49 | 2 | 16 | .0175941901 | 49 | 3 | 30 | .0103820357 |
| 49 | 1 | 5 | .0431071834 | 49 | 2 | 17 | .0166494215 | 49 | 3 | 31 | .0100035771 |
| 49 | 1 | 6 | .0366003471 | 49 | 2 | 18 | .0157977061 | 49 | 3 | 32 | .0096412247 |
| 49 | 1 | 7 | .0319141366 | 49 | 2 | 19 | .0150247213 | 49 | 3 | 33 | .0092930848 |
| 49 | 1 | 8 | .0283617700 | 49 | 2 | 20 | .0143189468 | 49 | 3 | 34 | .0089574923 |
| 49 | 1 | 9 | .0255655301 | 49 | 2 | 21 | .0136710072 | 49 | 3 | 35 | .0086330494 |
| 49 | 1 | 10 | .0232999491 | 49 | 2 | 22 | .0130731800 | 49 | 3 | 36 | .0083185237 |
| 49 | 1 | 11 | .0214218313 | 49 | 2 | 23 | .0125190328 | 49 | 3 | 37 | .0080125505 |
| 49 | 1 | 12 | .0198357024 | 49 | 2 | 24 | .0120031616 | 49 | 3 | 38 | .0077133843 |
| 49 | 1 | 13 | .0184753620 | 49 | 2 | 25 | .0115209836 | 49 | 3 | 39 | .0074190775 |
| 49 | 1 | 14 | .0172933973 | 49 | 2 | 26 | .0110685698 | 49 | 3 | 40 | .0071280133 |
| 49 | 1 | 15 | .0162549176 | 49 | 2 | 27 | .0106425326 | 49 | 3 | 41 | .0068390070 |
| 49 | 1 | 16 | .0153336602 | 49 | 2 | 28 | .0102399523 | 49 | 3 | 42 | .0065503860 |
| 49 | 1 | 17 | .0145094550 | 49 | 2 | 29 | .0098582914 | 49 | 3 | 43 | .0062590295 |
| 49 | 1 | 18 | .0137665185 | 49 | 2 | 30 | .0094952982 | 49 | 3 | 44 | .0059608751 |
| 49 | 1 | 19 | .0130923251 | 49 | 2 | 31 | .0091489477 | 49 | 3 | 45 | .0056514282 |
| 49 | 1 | 20 | .0124768277 | 49 | 2 | 32 | .0088174172 | 49 | 3 | 46 | .0053233007 |
| 49 | 1 | 21 | .0119118382 | 49 | 2 | 33 | .0084990488 | 49 | 3 | 47 | .0049621077 |
| 49 | 1 | 22 | .0113905699 | 49 | 2 | 34 | .0081923037 | 49 | 4 | 4 | .0705229728 |
| 49 | 1 | 23 | .0109073842 | 49 | 2 | 35 | .0078957416 | 49 | 4 | 5 | .0576541938 |
| 49 | 1 | 24 | .0104576041 | 49 | 2 | 36 | .0076080025 | 49 | 4 | 6 | .0490163478 |
| 49 | 1 | 25 | .0100372621 | 49 | 2 | 37 | .0073277672 | 49 | 4 | 7 | .0427807908 |
| 49 | 1 | 26 | .0096428982 | 49 | 2 | 38 | .0070537346 | 49 | 4 | 8 | .0380457314 |
| 49 | 1 | 27 | .0092715288 | 49 | 2 | 39 | .0067845922 | 49 | 4 | 9 | .0343135440 |
| 49 | 1 | 28 | .0089206385 | 49 | 2 | 40 | .0065188951 | 49 | 4 | 10 | .0312864187 |
| 49 | 1 | 29 | .0085880327 | 49 | 2 | 41 | .0062549146 | 49 | 4 | 11 | .0287748280 |
| 49 | 1 | 30 | .0082716937 | 49 | 2 | 42 | .0059906246 | 49 | 4 | 12 | .0266521920 |
| 49 | 1 | 31 | .0079698106 | 49 | 2 | 43 | .0057237222 | 49 | 4 | 13 | .0248306132 |
| 49 | 1 | 32 | .0076808373 | 49 | 2 | 44 | .0054512243 | 49 | 4 | 14 | .0232470633 |
| 49 | 1 | 33 | .0074033852 | 49 | 2 | 45 | .0051685299 | 49 | 4 | 15 | .0218551195 |
| 49 | 1 | 34 | .0071360957 | 49 | 2 | 46 | .0048683375 | 49 | 4 | 16 | .0206198061 |
| 49 | 1 | 35 | .0068776784 | 49 | 2 | 47 | .0045379959 | 49 | 4 | 17 | .0195142521 |
| 49 | 1 | 36 | .0066269667 | 49 | 2 | 48 | .0041497013 | 49 | 4 | 18 | .0185174450 |
| 49 | 1 | 37 | .0063828326 | 49 | 3 | 3 | .0860614494 | 49 | 4 | 19 | .0176126793 |
| 49 | 1 | 38 | .0061441070 | 49 | 3 | 4 | .0659077916 | 49 | 4 | 20 | .0167864798 |
| 49 | 1 | 39 | .0059095929 | 49 | 3 | 5 | .0538517518 | 49 | 4 | 21 | .0160278523 |
| 49 | 1 | 40 | .0056780277 | 49 | 3 | 6 | .0457668204 | 49 | 4 | 22 | .0153277421 |
| 49 | 1 | 41 | .0054479825 | 49 | 3 | 7 | .0399341622 | 49 | 4 | 23 | .0146786335 |

TABLE I-COVARIANCES OF NORMAL ORDER STATISTICS

| N | I | J | COVARIANCE | N | I | J | COVARIANCE | N | I | J | COVARIANCE |
|---|---|---|---|---|---|---|---|---|---|---|---|
| 49 | 4 | 24 | .0140742512 | 49 | 5 | 42 | .0074629472 | 49 | 7 | 24 | .0165368974 |
| 49 | 4 | 25 | .0135093242 | 49 | 5 | 43 | .0071291740 | 49 | 7 | 25 | .0158732808 |
| 49 | 4 | 26 | .0129793690 | 49 | 5 | 44 | .0067900951 | 49 | 7 | 26 | .0152503012 |
| 49 | 4 | 27 | .0124804769 | 49 | 5 | 45 | .0064394635 | 49 | 7 | 27 | .0146639575 |
| 49 | 4 | 28 | .0120091378 | 49 | 6 | 6 | .0547329816 | 49 | 7 | 28 | .0141106298 |
| 49 | 4 | 29 | .0115621572 | 49 | 6 | 7 | .0477939433 | 49 | 7 | 29 | .0135865434 |
| 49 | 4 | 30 | .0111366976 | 49 | 6 | 8 | .0425199045 | 49 | 7 | 30 | .0130878253 |
| 49 | 4 | 31 | .0107303927 | 49 | 6 | 9 | .0383599296 | 49 | 7 | 31 | .0126111318 |
| 49 | 4 | 32 | .0103413881 | 49 | 6 | 10 | .0349838815 | 49 | 7 | 32 | .0121541392 |
| 49 | 4 | 33 | .0099681448 | 49 | 6 | 11 | .0321814417 | 49 | 7 | 33 | .0117151779 |
| 49 | 4 | 34 | .0096090318 | 49 | 6 | 12 | .0298120296 | 49 | 7 | 34 | .0112923281 |
| 49 | 4 | 35 | .0092620420 | 49 | 6 | 13 | .0277779702 | 49 | 7 | 35 | .0108832601 |
| 49 | 4 | 36 | .0089250078 | 49 | 6 | 14 | .0260091955 | 49 | 7 | 36 | .0104862765 |
| 49 | 4 | 37 | .0085961935 | 49 | 6 | 15 | .0244540629 | 49 | 7 | 37 | .0101008323 |
| 49 | 4 | 38 | .0082745432 | 49 | 6 | 16 | .0230736078 | 49 | 7 | 38 | .0097256825 |
| 49 | 4 | 39 | .0079590447 | 49 | 6 | 17 | .0218378367 | 49 | 7 | 39 | .0093566354 |
| 49 | 4 | 40 | .0076478329 | 49 | 6 | 18 | .0207232733 | 49 | 7 | 40 | .0089886096 |
| 49 | 4 | 41 | .0073383465 | 49 | 6 | 19 | .0197113024 | 49 | 7 | 41 | .0086209747 |
| 49 | 4 | 42 | .0070283528 | 49 | 6 | 20 | .0187870267 | 49 | 7 | 42 | .0082566111 |
| 49 | 4 | 43 | .0067156851 | 49 | 6 | 21 | .0179383674 | 49 | 7 | 43 | .0078923508 |
| 49 | 4 | 44 | .0063964392 | 49 | 6 | 22 | .0171553036 | 49 | 8 | 8 | .0465832610 |
| 49 | 4 | 45 | .0060647199 | 49 | 6 | 23 | .0164293715 | 49 | 8 | 9 | .0420365444 |
| 49 | 4 | 46 | .0057127225 | 49 | 6 | 24 | .0157534053 | 49 | 8 | 10 | .0383451295 |
| 49 | 5 | 5 | .0610935379 | 49 | 6 | 25 | .0151213344 | 49 | 8 | 11 | .0352799617 |
| 49 | 5 | 6 | .0519582379 | 49 | 6 | 26 | .0145279582 | 49 | 8 | 12 | .0326878160 |
| 49 | 5 | 7 | .0453596216 | 49 | 6 | 27 | .0139687887 | 49 | 8 | 13 | .0304621427 |
| 49 | 5 | 8 | .0403466249 | 49 | 6 | 28 | .0134400385 | 49 | 8 | 14 | .0285264426 |
| 49 | 5 | 9 | .0363939953 | 49 | 6 | 29 | .0129386714 | 49 | 8 | 15 | .0268243110 |
| 49 | 5 | 10 | .0331871837 | 49 | 6 | 30 | .0124622778 | 49 | 8 | 16 | .0253132296 |
| 49 | 5 | 11 | .0305259022 | 49 | 6 | 31 | .0120085830 | 49 | 8 | 17 | .0239605393 |
| 49 | 5 | 12 | .0282763239 | 49 | 6 | 32 | .0115748010 | 49 | 8 | 18 | .0227407045 |
| 49 | 5 | 13 | .0263454896 | 49 | 6 | 33 | .0111575318 | 49 | 8 | 19 | .0216334114 |
| 49 | 5 | 14 | .0246667349 | 49 | 6 | 34 | .0107537104 | 49 | 8 | 20 | .0206222422 |
| 49 | 5 | 15 | .0231909585 | 49 | 6 | 35 | .0103619481 | 49 | 8 | 21 | .0196936508 |
| 49 | 5 | 16 | .0218811529 | 49 | 6 | 36 | .0099825532 | 49 | 8 | 22 | .0188363614 |
| 49 | 5 | 17 | .0207088477 | 49 | 6 | 37 | .0096154066 | 49 | 8 | 23 | .0180412230 |
| 49 | 5 | 18 | .0196517409 | 49 | 6 | 38 | .0092576801 | 49 | 8 | 24 | .0173008391 |
| 49 | 5 | 19 | .0186920964 | 49 | 6 | 39 | .0089047014 | 49 | 8 | 25 | .0166089229 |
| 49 | 5 | 20 | .0178156318 | 49 | 6 | 40 | .0085537029 | 49 | 8 | 26 | .0159599285 |
| 49 | 5 | 21 | .0170107388 | 49 | 6 | 41 | .0082050838 | 49 | 8 | 27 | .0153490808 |
| 49 | 5 | 22 | .0162679343 | 49 | 6 | 42 | .0078581592 | 49 | 8 | 28 | .0147723470 |
| 49 | 5 | 23 | .0155793989 | 49 | 6 | 43 | .0075079119 | 49 | 8 | 29 | .0142259976 |
| 49 | 5 | 24 | .0149385355 | 49 | 6 | 44 | .0071487489 | 49 | 8 | 30 | .0137060650 |
| 49 | 5 | 25 | .0143396118 | 49 | 7 | 7 | .0501126011 | 49 | 8 | 31 | .0132085213 |
| 49 | 5 | 26 | .0137775793 | 49 | 7 | 8 | .0445902973 | 49 | 8 | 32 | .0127304871 |
| 49 | 5 | 27 | .0132480604 | 49 | 7 | 9 | .0402331639 | 49 | 8 | 33 | .0122712824 |
| 49 | 5 | 28 | .0127473828 | 49 | 7 | 10 | .0366962594 | 49 | 8 | 34 | .0118314016 |
| 49 | 5 | 29 | .0122725236 | 49 | 7 | 11 | .0337597331 | 49 | 8 | 35 | .0114094558 |
| 49 | 5 | 30 | .0118209031 | 49 | 7 | 12 | .0312765780 | 49 | 8 | 36 | .0110004370 |
| 49 | 5 | 31 | .0113901335 | 49 | 7 | 13 | .0291446117 | 49 | 8 | 37 | .0105985482 |
| 49 | 5 | 32 | .0109779200 | 49 | 7 | 14 | .0272904921 | 49 | 8 | 38 | .0102021997 |
| 49 | 5 | 33 | .0105821700 | 49 | 7 | 15 | .0256601328 | 49 | 8 | 39 | .0098141191 |
| 49 | 5 | 34 | .0102010542 | 49 | 7 | 16 | .0242127278 | 49 | 8 | 40 | .0094348642 |
| 49 | 5 | 35 | .0098327475 | 49 | 7 | 17 | .0229169073 | 49 | 8 | 41 | .0090589105 |
| 49 | 5 | 36 | .0094750870 | 49 | 7 | 18 | .0217481863 | 49 | 8 | 42 | .0086803339 |
| 49 | 5 | 37 | .0091258360 | 49 | 7 | 19 | .0206871725 | 49 | 9 | 9 | .0438760649 |
| 49 | 5 | 38 | .0087836388 | 49 | 7 | 20 | .0197182234 | 49 | 9 | 10 | .0400277269 |
| 49 | 5 | 39 | .0084482963 | 49 | 7 | 21 | .0188285313 | 49 | 9 | 11 | .0368313335 |
| 49 | 5 | 40 | .0081189965 | 49 | 7 | 22 | .0180075024 | 49 | 9 | 12 | .0341275100 |
| 49 | 5 | 41 | .0077921728 | 49 | 7 | 23 | .0172461808 | 49 | 9 | 13 | .0318053946 |

TABLE I-COVARIANCES OF NORMAL ORDER STATISTICS

| N | I | J | COVARIANCE | N | I | J | COVARIANCE | N | I | J | COVARIANCE |
|---|---|---|---|---|---|---|---|---|---|---|---|
| 49 | 9 | 14 | .0297853880 | 49 | 11 | 12 | .0370044694 | 49 | 13 | 18 | .0277269800 |
| 49 | 9 | 15 | .0280088113 | 49 | 11 | 13 | .0344904172 | 49 | 13 | 19 | .0263843595 |
| 49 | 9 | 16 | .0264314252 | 49 | 11 | 14 | .0323026241 | 49 | 13 | 20 | .0251582762 |
| 49 | 9 | 17 | .0250191982 | 49 | 11 | 15 | .0303778105 | 49 | 13 | 21 | .0240325041 |
| 49 | 9 | 18 | .0237454524 | 49 | 11 | 16 | .0286681786 | 49 | 13 | 22 | .0229934779 |
| 49 | 9 | 19 | .0225888890 | 49 | 11 | 17 | .0271368925 | 49 | 13 | 23 | .0220299990 |
| 49 | 9 | 20 | .0215322583 | 49 | 11 | 18 | .0257550963 | 49 | 13 | 24 | .0211326807 |
| 49 | 9 | 21 | .0205615949 | 49 | 11 | 19 | .0244998956 | 49 | 13 | 25 | .0202933935 |
| 49 | 9 | 22 | .0196655081 | 49 | 11 | 20 | .0233529592 | 49 | 13 | 26 | .0195061633 |
| 49 | 9 | 23 | .0188344745 | 49 | 11 | 21 | .0222994778 | 49 | 13 | 27 | .0187661770 |
| 49 | 9 | 24 | .0180607171 | 49 | 11 | 22 | .0213271035 | 49 | 13 | 28 | .0180680099 |
| 49 | 9 | 25 | .0173379334 | 49 | 11 | 23 | .0204254207 | 49 | 13 | 29 | .0174060715 |
| 49 | 9 | 26 | .0166604208 | 49 | 11 | 24 | .0195858127 | 49 | 13 | 30 | .0167762645 |
| 49 | 9 | 27 | .0160224289 | 49 | 11 | 25 | .0188004534 | 49 | 13 | 31 | .0161756203 |
| 49 | 9 | 28 | .0154185289 | 49 | 11 | 26 | .0180621922 | 49 | 13 | 32 | .0156001784 |
| 49 | 9 | 29 | .0148445929 | 49 | 11 | 27 | .0173655969 | 49 | 13 | 33 | .0150453314 |
| 49 | 9 | 30 | .0142981348 | 49 | 11 | 28 | .0167071512 | 49 | 13 | 34 | .0145095682 |
| 49 | 9 | 31 | .0137773195 | 49 | 11 | 29 | .0160838219 | 49 | 13 | 35 | .0139953239 |
| 49 | 9 | 32 | .0132796005 | 49 | 11 | 30 | .0154916711 | 49 | 13 | 36 | .0135026609 |
| 49 | 9 | 33 | .0128017582 | 49 | 11 | 31 | .0149263826 | 49 | 13 | 37 | .0130238061 |
| 49 | 9 | 34 | .0123413748 | 49 | 11 | 32 | .0143848879 | 49 | 14 | 14 | .0361719996 |
| 49 | 9 | 35 | .0118972559 | 49 | 11 | 33 | .0138652050 | 49 | 14 | 15 | .0340213426 |
| 49 | 9 | 36 | .0114672458 | 49 | 11 | 34 | .0133640833 | 49 | 14 | 16 | .0321103573 |
| 49 | 9 | 37 | .0110465657 | 49 | 11 | 35 | .0128765576 | 49 | 14 | 17 | .0303982095 |
| 49 | 9 | 38 | .0106310447 | 49 | 11 | 36 | .0124000653 | 49 | 14 | 18 | .0288529312 |
| 49 | 9 | 39 | .0102221544 | 49 | 11 | 37 | .0119375749 | 49 | 14 | 19 | .0274490636 |
| 49 | 9 | 40 | .0098241195 | 49 | 11 | 38 | .0114920510 | 49 | 14 | 20 | .0261659965 |
| 49 | 9 | 41 | .0094333791 | 49 | 11 | 39 | .0110573587 | 49 | 14 | 21 | .0249868263 |
| 49 | 10 | 10 | .0417535299 | 49 | 12 | 12 | .0383326899 | 49 | 14 | 22 | .0238975973 |
| 49 | 10 | 11 | .0384226214 | 49 | 12 | 13 | .0357333365 | 49 | 14 | 23 | .0228866884 |
| 49 | 10 | 12 | .0356038991 | 49 | 12 | 14 | .0334718845 | 49 | 14 | 24 | .0219447086 |
| 49 | 10 | 13 | .0331822132 | 49 | 12 | 15 | .0314828031 | 49 | 14 | 25 | .0210639601 |
| 49 | 10 | 14 | .0310748679 | 49 | 12 | 16 | .0297165970 | 49 | 14 | 26 | .0202372070 |
| 49 | 10 | 15 | .0292208762 | 49 | 12 | 17 | .0281351618 | 49 | 14 | 27 | .0194586558 |
| 49 | 10 | 16 | .0275742091 | 49 | 12 | 18 | .0267087027 | 49 | 14 | 28 | .0187234928 |
| 49 | 10 | 17 | .0260993971 | 49 | 12 | 19 | .0254136163 | 49 | 14 | 29 | .0180259757 |
| 49 | 10 | 18 | .0247686010 | 49 | 12 | 20 | .0242309323 | 49 | 14 | 30 | .0173600024 |
| 49 | 10 | 19 | .0235596667 | 49 | 12 | 21 | .0231451359 | 49 | 14 | 31 | .0167220308 |
| 49 | 10 | 20 | .0224547960 | 49 | 12 | 22 | .0221433996 | 49 | 14 | 32 | .0161114493 |
| 49 | 10 | 21 | .0214395257 | 49 | 12 | 23 | .0212147104 | 49 | 14 | 33 | .0155268616 |
| 49 | 10 | 22 | .0205021676 | 49 | 12 | 24 | .0203495323 | 49 | 14 | 34 | .0149640035 |
| 49 | 10 | 23 | .0196331851 | 49 | 12 | 25 | .0195404414 | 49 | 14 | 35 | .0144191016 |
| 49 | 10 | 24 | .0188240999 | 49 | 12 | 26 | .0187812422 | 49 | 14 | 36 | .0138913667 |
| 49 | 10 | 25 | .0180673395 | 49 | 12 | 27 | .0180661510 | 49 | 15 | 15 | .0353779244 |
| 49 | 10 | 26 | .0173566551 | 49 | 12 | 28 | .0173905340 | 49 | 15 | 16 | .0333876750 |
| 49 | 10 | 27 | .0166868431 | 49 | 12 | 29 | .0167513727 | 49 | 15 | 17 | .0316036388 |
| 49 | 10 | 28 | .0160530131 | 49 | 12 | 30 | .0161456603 | 49 | 15 | 18 | .0299925845 |
| 49 | 10 | 29 | .0154506304 | 49 | 12 | 31 | .0155683689 | 49 | 15 | 19 | .0285280614 |
| 49 | 10 | 30 | .0148765517 | 49 | 12 | 32 | .0150133993 | 49 | 15 | 20 | .0271886805 |
| 49 | 10 | 31 | .0143295102 | 49 | 12 | 33 | .0144773398 | 49 | 15 | 21 | .0259569747 |
| 49 | 10 | 32 | .0138084328 | 49 | 12 | 34 | .0139608966 | 49 | 15 | 22 | .0248186182 |
| 49 | 10 | 33 | .0133097048 | 49 | 12 | 35 | .0134648855 | 49 | 15 | 23 | .0237618600 |
| 49 | 10 | 34 | .0128270887 | 49 | 12 | 36 | .0129859800 | 49 | 15 | 24 | .0227768141 |
| 49 | 10 | 35 | .0123557893 | 49 | 12 | 37 | .0125190824 | 49 | 15 | 25 | .0218552405 |
| 49 | 10 | 36 | .0118961134 | 49 | 12 | 38 | .0120625290 | 49 | 15 | 26 | .0209901599 |
| 49 | 10 | 37 | .0114508735 | 49 | 13 | 13 | .0370832743 | 49 | 15 | 27 | .0201742927 |
| 49 | 10 | 38 | .0110188404 | 49 | 13 | 14 | .0347389430 | 49 | 15 | 28 | .0194015033 |
| 49 | 10 | 39 | .0105938431 | 49 | 13 | 15 | .0326767609 | 49 | 15 | 29 | .0186674768 |
| 49 | 10 | 40 | .0101714996 | 49 | 13 | 16 | .0308455429 | 49 | 15 | 30 | .0179678096 |
| 49 | 11 | 11 | .0399305427 | 49 | 13 | 17 | .0292058856 | 49 | 15 | 31 | .0172977671 |

TABLE I-COVARIANCES OF NORMAL ORDER STATISTICS

| N | I | J | COVARIANCE | N | I | J | COVARIANCE | N | I | J | COVARIANCE |
|---|---|---|---|---|---|---|---|---|---|---|---|
| 49 | 15 | 32 | .0166550358 | 49 | 19 | 24 | .0263430583 | 50 | 1 | 17 | .0144370764 |
| 49 | 15 | 33 | .0160398370 | 49 | 19 | 25 | .0252948396 | 50 | 1 | 18 | .0137003898 |
| 49 | 15 | 34 | .0154498072 | 49 | 19 | 26 | .0243104686 | 50 | 1 | 19 | .0130320597 |
| 49 | 15 | 35 | .0148774450 | 49 | 19 | 27 | .0233831032 | 50 | 1 | 20 | .0124220953 |
| 49 | 16 | 16 | .0345508343 | 49 | 19 | 28 | .0225068073 | 50 | 1 | 21 | .0118623821 |
| 49 | 16 | 17 | .0327108372 | 49 | 19 | 29 | .0216760600 | 50 | 1 | 22 | .0113461920 |
| 49 | 16 | 18 | .0310496127 | 49 | 19 | 30 | .0208866015 | 50 | 1 | 23 | .0108678986 |
| 49 | 16 | 19 | .0295398558 | 49 | 19 | 31 | .0201330811 | 50 | 1 | 24 | .0104228390 |
| 49 | 16 | 20 | .0281595199 | 49 | 20 | 20 | .0326292544 | 50 | 1 | 25 | .0100071116 |
| 49 | 16 | 21 | .0268906754 | 49 | 20 | 21 | .0311557603 | 50 | 1 | 26 | .0096173193 |
| 49 | 16 | 22 | .0257186350 | 49 | 20 | 22 | .0297935313 | 50 | 1 | 27 | .0092504737 |
| 49 | 16 | 23 | .0246312439 | 49 | 20 | 23 | .0285283462 | 50 | 1 | 28 | .0089040442 |
| 49 | 16 | 24 | .0236183196 | 49 | 20 | 24 | .0273482006 | 50 | 1 | 29 | .0085758968 |
| 49 | 16 | 25 | .0226708951 | 49 | 20 | 25 | .0262430034 | 50 | 1 | 30 | .0082640951 |
| 49 | 16 | 26 | .0217811164 | 49 | 20 | 26 | .0252042648 | 50 | 1 | 31 | .0079668289 |
| 49 | 16 | 27 | .0209424992 | 49 | 20 | 27 | .0242247099 | 50 | 1 | 32 | .0076825235 |
| 49 | 16 | 28 | .0201483747 | 49 | 20 | 28 | .0232980551 | 50 | 1 | 33 | .0074098509 |
| 49 | 16 | 29 | .0193934760 | 49 | 20 | 29 | .0224184249 | 50 | 1 | 34 | .0071475593 |
| 49 | 16 | 30 | .0186751583 | 49 | 20 | 30 | .0215799539 | 50 | 1 | 35 | .0068943926 |
| 49 | 16 | 31 | .0179907074 | 49 | 21 | 21 | .0323354051 | 50 | 1 | 36 | .0066491931 |
| 49 | 16 | 32 | .0173355188 | 49 | 21 | 22 | .0309217657 | 50 | 1 | 37 | .0064109247 |
| 49 | 16 | 33 | .0167057874 | 49 | 21 | 23 | .0296084941 | 50 | 1 | 38 | .0061785450 |
| 49 | 16 | 34 | .0160998515 | 49 | 21 | 24 | .0283834407 | 50 | 1 | 39 | .0059509575 |
| 49 | 17 | 17 | .0337811538 | 49 | 21 | 25 | .0272362296 | 50 | 1 | 40 | .0057270521 |
| 49 | 17 | 18 | .0320779045 | 49 | 21 | 26 | .0261580381 | 50 | 1 | 41 | .0055056477 |
| 49 | 17 | 19 | .0305309933 | 49 | 21 | 27 | .0251412101 | 50 | 1 | 42 | .0052853917 |
| 49 | 17 | 20 | .0291178462 | 49 | 21 | 28 | .0241788555 | 50 | 1 | 43 | .0050646933 |
| 49 | 17 | 21 | .0278200394 | 49 | 21 | 29 | .0232647879 | 50 | 1 | 44 | .0048415664 |
| 49 | 17 | 22 | .0266224336 | 49 | 22 | 22 | .0320480291 | 50 | 1 | 45 | .0046133392 |
| 49 | 17 | 23 | .0255123938 | 49 | 22 | 23 | .0307040595 | 50 | 1 | 46 | .0043761992 |
| 49 | 17 | 24 | .0244791544 | 49 | 22 | 24 | .0294515872 | 50 | 1 | 47 | .0041241534 |
| 49 | 17 | 25 | .0235134769 | 49 | 22 | 25 | .0282798416 | 50 | 1 | 48 | .0038464013 |
| 49 | 17 | 26 | .0226071555 | 49 | 22 | 26 | .0271796217 | 50 | 1 | 49 | .0035193623 |
| 49 | 17 | 27 | .0217531379 | 49 | 22 | 27 | .0261428668 | 50 | 1 | 50 | .0030674426 |
| 49 | 17 | 28 | .0209462788 | 49 | 22 | 28 | .0251624026 | 50 | 2 | 2 | .1164802619 |
| 49 | 17 | 29 | .0201813744 | 49 | 23 | 23 | .0318581984 | 50 | 2 | 3 | .0784931367 |
| 49 | 17 | 30 | .0194540574 | 49 | 23 | 24 | .0305769982 | 50 | 2 | 4 | .0600403904 |
| 49 | 17 | 31 | .0187620832 | 49 | 23 | 25 | .0293796525 | 50 | 2 | 5 | .0490229200 |
| 49 | 17 | 32 | .0181022503 | 49 | 23 | 26 | .0282563502 | 50 | 2 | 6 | .0416440443 |
| 49 | 17 | 33 | .0174682507 | 49 | 23 | 27 | .0271987047 | 50 | 2 | 7 | .0363258600 |
| 49 | 18 | 18 | .0332284203 | 49 | 24 | 24 | .0317863193 | 50 | 2 | 8 | .0322924041 |
| 49 | 18 | 19 | .0316306408 | 49 | 24 | 25 | .0305422939 | 50 | 2 | 9 | .0291163813 |
| 49 | 18 | 20 | .0301713826 | 49 | 24 | 26 | .0293749238 | 50 | 2 | 10 | .0265424893 |
| 49 | 18 | 21 | .0288315453 | 49 | 25 | 25 | .0317752383 | 50 | 2 | 11 | .0244084659 |
| 49 | 18 | 22 | .0275953987 | 50 | 1 | 1 | .2157123890 | 50 | 2 | 12 | .0226060595 |
| 49 | 18 | 23 | .0264497477 | 50 | 1 | 2 | .1023008594 | 50 | 2 | 13 | .0210601725 |
| 49 | 18 | 24 | .0253833628 | 50 | 1 | 3 | .0687556445 | 50 | 2 | 14 | .0197170040 |
| 49 | 18 | 25 | .0243866017 | 50 | 1 | 4 | .0525217706 | 50 | 2 | 15 | .0185369544 |
| 49 | 18 | 26 | .0234513605 | 50 | 1 | 5 | .0428495992 | 50 | 2 | 16 | .0174901974 |
| 49 | 18 | 27 | .0225707133 | 50 | 1 | 6 | .0363805398 | 50 | 2 | 17 | .0165538174 |
| 49 | 18 | 28 | .0217388154 | 50 | 1 | 7 | .0317225166 | 50 | 2 | 18 | .0157098988 |
| 49 | 18 | 29 | .0209516258 | 50 | 1 | 8 | .0281922216 | 50 | 2 | 19 | .0149442163 |
| 49 | 18 | 30 | .0202043715 | 50 | 1 | 9 | .0254138960 | 50 | 2 | 20 | .0142453265 |
| 49 | 18 | 31 | .0194919504 | 50 | 1 | 10 | .0231632658 | 50 | 2 | 21 | .0136039206 |
| 49 | 18 | 32 | .0188110892 | 50 | 1 | 11 | .0212979079 | 50 | 2 | 22 | .0130123390 |
| 49 | 19 | 19 | .0328927158 | 50 | 1 | 12 | .0197228709 | 50 | 2 | 23 | .0124642031 |
| 49 | 19 | 20 | .0313653013 | 50 | 1 | 13 | .0183723243 | 50 | 2 | 24 | .0119541538 |
| 49 | 19 | 21 | .0299618489 | 50 | 1 | 14 | .0171991252 | 50 | 2 | 25 | .0114776539 |
| 49 | 19 | 22 | .0286658157 | 50 | 1 | 15 | .0161685811 | 50 | 2 | 26 | .0110308181 |
| 49 | 19 | 23 | .0274634340 | 50 | 1 | 16 | .0152545795 | 50 | 2 | 27 | .0106102861 |

TABLE I-COVARIANCES OF NORMAL ORDER STATISTICS

| N | I | J | COVARIANCE | N | I | J | COVARIANCE | N | I | J | COVARIANCE |
|---|---|---|---|---|---|---|---|---|---|---|---|
| 50 | 2 | 28 | .0102131534 | 50 | 3 | 41 | .0069020279 | 50 | 5 | 13 | .0261345793 |
| 50 | 2 | 29 | .0098369116 | 50 | 3 | 42 | .0066257542 | 50 | 5 | 14 | .0244722841 |
| 50 | 2 | 30 | .0094793568 | 50 | 3 | 43 | .0063495494 | 50 | 5 | 15 | .0230113178 |
| 50 | 2 | 31 | .0091385093 | 50 | 3 | 44 | .0060701834 | 50 | 5 | 16 | .0217149817 |
| 50 | 2 | 32 | .0088125864 | 50 | 3 | 45 | .0057838316 | 50 | 5 | 17 | .0205550556 |
| 50 | 2 | 33 | .0084999840 | 50 | 3 | 46 | .0054863714 | 50 | 5 | 18 | .0195094366 |
| 50 | 2 | 34 | .0081992257 | 50 | 3 | 47 | .0051705366 | 50 | 5 | 19 | .0185605365 |
| 50 | 2 | 35 | .0079089183 | 50 | 3 | 48 | .0048222946 | 50 | 5 | 20 | .0176941714 |
| 50 | 2 | 36 | .0076277478 | 50 | 4 | 4 | .0699795644 | 50 | 5 | 21 | .0168987939 |
| 50 | 2 | 37 | .0073544690 | 50 | 4 | 5 | .0572044451 | 50 | 5 | 22 | .0161649738 |
| 50 | 2 | 38 | .0070878802 | 50 | 4 | 6 | .0486317776 | 50 | 5 | 23 | .0154849767 |
| 50 | 2 | 39 | .0068268137 | 50 | 4 | 7 | .0424447433 | 50 | 5 | 24 | .0148523367 |
| 50 | 2 | 40 | .0065700765 | 50 | 4 | 8 | .0377475637 | 50 | 5 | 25 | .0142614694 |
| 50 | 2 | 41 | .0063162806 | 50 | 4 | 9 | .0340460122 | 50 | 5 | 26 | .0137074349 |
| 50 | 2 | 42 | .0060637223 | 50 | 4 | 10 | .0310443547 | 50 | 5 | 27 | .0131858803 |
| 50 | 2 | 43 | .0058104713 | 50 | 4 | 11 | .0285544087 | 50 | 5 | 28 | .0126930757 |
| 50 | 2 | 44 | .0055544078 | 50 | 4 | 12 | .0264505080 | 50 | 5 | 29 | .0122259144 |
| 50 | 2 | 45 | .0052926889 | 50 | 4 | 13 | .0246453984 | 50 | 5 | 30 | .0117817928 |
| 50 | 2 | 46 | .0050207862 | 50 | 4 | 14 | .0230765175 | 50 | 5 | 31 | .0113584075 |
| 50 | 2 | 47 | .0047316160 | 50 | 4 | 15 | .0216977858 | 50 | 5 | 32 | .0109536214 |
| 50 | 2 | 48 | .0044129694 | 50 | 4 | 16 | .0204744881 | 50 | 5 | 33 | .0105655068 |
| 50 | 2 | 49 | .0040378273 | 50 | 4 | 17 | .0193799605 | 50 | 5 | 34 | .0101924190 |
| 50 | 3 | 3 | .0854564866 | 50 | 4 | 18 | .0183933652 | 50 | 5 | 35 | .0098327824 |
| 50 | 3 | 4 | .0654336443 | 50 | 4 | 19 | .0174981461 | 50 | 5 | 36 | .0094846125 |
| 50 | 3 | 5 | .0534595821 | 50 | 4 | 20 | .0166809496 | 50 | 5 | 37 | .0091454223 |
| 50 | 3 | 6 | .0454316937 | 50 | 4 | 21 | .0159308715 | 50 | 5 | 38 | .0088131222 |
| 50 | 3 | 7 | .0396415141 | 50 | 4 | 22 | .0152389168 | 50 | 5 | 39 | .0084872098 |
| 50 | 3 | 8 | .0352476995 | 50 | 4 | 23 | .0145976070 | 50 | 5 | 40 | .0081682641 |
| 50 | 3 | 9 | .0317864719 | 50 | 4 | 24 | .0140006978 | 50 | 5 | 41 | .0078551434 |
| 50 | 3 | 10 | .0289805083 | 50 | 4 | 25 | .0134429638 | 50 | 5 | 42 | .0075432498 |
| 50 | 3 | 11 | .0266534390 | 50 | 4 | 26 | .0129199996 | 50 | 5 | 43 | .0072277309 |
| 50 | 3 | 12 | .0246875408 | 50 | 4 | 27 | .0124280081 | 50 | 5 | 44 | .0069078065 |
| 50 | 3 | 13 | .0230011121 | 50 | 4 | 28 | .0119635963 | 50 | 5 | 45 | .0065832234 |
| 50 | 3 | 14 | .0215355972 | 50 | 4 | 29 | .0115236424 | 50 | 5 | 46 | .0062468540 |
| 50 | 3 | 15 | .0202478776 | 50 | 4 | 30 | .0111052909 | 50 | 6 | 6 | .0542502698 |
| 50 | 3 | 16 | .0191054620 | 50 | 4 | 31 | .0107060669 | 50 | 6 | 7 | .0473716522 |
| 50 | 3 | 17 | .0180833794 | 50 | 4 | 32 | .0103240028 | 50 | 6 | 8 | .0421447594 |
| 50 | 3 | 18 | .0171621129 | 50 | 4 | 33 | .0099575984 | 50 | 6 | 9 | .0380228853 |
| 50 | 3 | 19 | .0163261831 | 50 | 4 | 34 | .0096055087 | 50 | 6 | 10 | .0346784792 |
| 50 | 3 | 20 | .0155631436 | 50 | 4 | 35 | .0092661284 | 50 | 6 | 11 | .0319028979 |
| 50 | 3 | 21 | .0148628520 | 50 | 4 | 36 | .0089374778 | 50 | 6 | 12 | .0295566961 |
| 50 | 3 | 22 | .0142169371 | 50 | 4 | 37 | .0086176100 | 50 | 6 | 13 | .0275429992 |
| 50 | 3 | 23 | .0136184142 | 50 | 4 | 38 | .0083051493 | 50 | 6 | 14 | .0257923274 |
| 50 | 3 | 24 | .0130614089 | 50 | 4 | 39 | .0079992264 | 50 | 6 | 15 | .0242534855 |
| 50 | 3 | 25 | .0125409540 | 50 | 4 | 40 | .0076986808 | 50 | 6 | 16 | .0228878548 |
| 50 | 3 | 26 | .0120528272 | 50 | 4 | 41 | .0074015130 | 50 | 6 | 17 | .0216657010 |
| 50 | 3 | 27 | .0115934129 | 50 | 4 | 42 | .0071055026 | 50 | 6 | 18 | .0205637320 |
| 50 | 3 | 28 | .0111595899 | 50 | 4 | 43 | .0068089487 | 50 | 6 | 19 | .0195634576 |
| 50 | 3 | 29 | .0107486426 | 50 | 4 | 44 | .0065096249 | 50 | 6 | 20 | .0186500853 |
| 50 | 3 | 30 | .0103581731 | 50 | 4 | 45 | .0062032398 | 50 | 6 | 21 | .0178116674 |
| 50 | 3 | 31 | .0099860073 | 50 | 4 | 46 | .0058842928 | 50 | 6 | 22 | .0170383425 |
| 50 | 3 | 32 | .0096301217 | 50 | 4 | 47 | .0055456874 | 50 | 6 | 23 | .0163218001 |
| 50 | 3 | 33 | .0092886266 | 50 | 5 | 5 | .0605850836 | 50 | 6 | 24 | .0156549914 |
| 50 | 3 | 34 | .0089598160 | 50 | 5 | 6 | .0515230920 | 50 | 6 | 25 | .0150319118 |
| 50 | 3 | 35 | .0086422560 | 50 | 5 | 7 | .0449790680 | 50 | 6 | 26 | .0144473617 |
| 50 | 3 | 36 | .0083348003 | 50 | 5 | 8 | .0400086924 | 50 | 6 | 27 | .0138967845 |
| 50 | 3 | 37 | .0080363887 | 50 | 5 | 9 | .0360905310 | 50 | 6 | 28 | .0133762822 |
| 50 | 3 | 38 | .0077456733 | 50 | 5 | 10 | .0329123685 | 50 | 6 | 29 | .0128827499 |
| 50 | 3 | 39 | .0074608578 | 50 | 5 | 11 | .0302754297 | 50 | 6 | 30 | .0124138899 |
| 50 | 3 | 40 | .0071800841 | 50 | 5 | 12 | .0280469132 | 50 | 6 | 31 | .0119678348 |

TABLE I-COVARIANCES OF NORMAL ORDER STATISTICS

| N | I | J | COVARIANCE | N | I | J | COVARIANCE | N | I | J | COVARIANCE |
|---|---|---|---|---|---|---|---|---|---|---|---|
| 50 | 6 | 32 | .0115423995 | 50 | 8 | 16 | .0250823212 | 50 | 9 | 41 | .0094997277 |
| 50 | 6 | 33 | .0111345304 | 50 | 8 | 17 | .0237458469 | 50 | 9 | 42 | .0091302559 |
| 50 | 6 | 34 | .0107407352 | 50 | 8 | 18 | .0225409550 | 50 | 10 | 10 | .0413062203 |
| 50 | 6 | 35 | .0103585169 | 50 | 8 | 19 | .0214475831 | 50 | 10 | 11 | .0380136787 |
| 50 | 6 | 36 | .0099874693 | 50 | 8 | 20 | .0204495494 | 50 | 10 | 12 | .0352280702 |
| 50 | 6 | 37 | .0096283691 | 50 | 8 | 21 | .0195334837 | 50 | 10 | 13 | .0328353914 |
| 50 | 6 | 38 | .0092805345 | 50 | 8 | 22 | .0186881500 | 50 | 10 | 14 | .0307537781 |
| 50 | 6 | 39 | .0089403829 | 50 | 8 | 23 | .0179043487 | 50 | 10 | 15 | .0289228746 |
| 50 | 6 | 40 | .0086036361 | 50 | 8 | 24 | .0171746659 | 50 | 10 | 16 | .0272971533 |
| 50 | 6 | 41 | .0082688240 | 50 | 8 | 25 | .0164928765 | 50 | 10 | 17 | .0258415500 |
| 50 | 6 | 42 | .0079366234 | 50 | 8 | 26 | .0158535603 | 50 | 10 | 18 | .0245285343 |
| 50 | 6 | 43 | .0076051539 | 50 | 8 | 27 | .0152521310 | 50 | 10 | 19 | .0233361585 |
| 50 | 6 | 44 | .0072691660 | 50 | 8 | 28 | .0146848385 | 50 | 10 | 20 | .0222467497 |
| 50 | 6 | 45 | .0069246589 | 50 | 8 | 29 | .0141483118 | 50 | 10 | 21 | .0212459049 |
| 50 | 7 | 7 | .0496519878 | 50 | 8 | 30 | .0136388269 | 50 | 10 | 22 | .0203220090 |
| 50 | 7 | 8 | .0441810584 | 50 | 8 | 31 | .0131521192 | 50 | 10 | 23 | .0194657734 |
| 50 | 7 | 9 | .0398653750 | 50 | 8 | 32 | .0126844187 | 50 | 10 | 24 | .0186690372 |
| 50 | 7 | 10 | .0363628324 | 50 | 8 | 33 | .0122340800 | 50 | 10 | 25 | .0179244388 |
| 50 | 7 | 11 | .0334554184 | 50 | 8 | 34 | .0118018143 | 50 | 10 | 26 | .0172258473 |
| 50 | 7 | 12 | .0309973757 | 50 | 8 | 35 | .0113882450 | 50 | 10 | 27 | .0165681365 |
| 50 | 7 | 13 | .0288874080 | 50 | 8 | 36 | .0109906768 | 50 | 10 | 28 | .0159463524 |
| 50 | 7 | 14 | .0270528222 | 50 | 8 | 37 | .0106031665 | 50 | 10 | 29 | .0153555798 |
| 50 | 7 | 15 | .0254400133 | 50 | 8 | 38 | .0102210381 | 50 | 10 | 30 | .0147920723 |
| 50 | 7 | 16 | .0240085283 | 50 | 8 | 39 | .0098447227 | 50 | 10 | 31 | .0142543804 |
| 50 | 7 | 17 | .0227272522 | 50 | 8 | 40 | .0094769516 | 50 | 10 | 32 | .0137424553 |
| 50 | 7 | 18 | .0215718961 | 50 | 8 | 41 | .0091162195 | 50 | 10 | 33 | .0132546501 |
| 50 | 7 | 19 | .0205232520 | 50 | 8 | 42 | .0087566517 | 50 | 10 | 34 | .0127857889 |
| 50 | 7 | 20 | .0195658683 | 50 | 8 | 43 | .0083949290 | 50 | 10 | 35 | .0123295365 |
| 50 | 7 | 21 | .0186871230 | 50 | 9 | 9 | .0434083115 | 50 | 10 | 36 | .0118832434 |
| 50 | 7 | 22 | .0178765999 | 50 | 9 | 10 | .0396028346 | 50 | 10 | 37 | .0114489676 |
| 50 | 7 | 23 | .0171254689 | 50 | 9 | 11 | .0364429087 | 50 | 10 | 38 | .0110282567 |
| 50 | 7 | 24 | .0164260622 | 50 | 9 | 12 | .0337706712 | 50 | 10 | 39 | .0106175897 |
| 50 | 7 | 25 | .0157718848 | 50 | 9 | 13 | .0314763245 | 50 | 10 | 40 | .0102118593 |
| 50 | 7 | 26 | .0151577779 | 50 | 9 | 14 | .0294810358 | 50 | 10 | 41 | .0098102488 |
| 50 | 7 | 27 | .0145797711 | 50 | 9 | 15 | .0277266953 | 50 | 11 | 11 | .0395145023 |
| 50 | 7 | 28 | .0140345199 | 50 | 9 | 16 | .0261695096 | 50 | 11 | 12 | .0366216997 |
| 50 | 7 | 29 | .0135186729 | 50 | 9 | 17 | .0247758208 | 50 | 11 | 13 | .0341365829 |
| 50 | 7 | 30 | .0130286644 | 50 | 9 | 18 | .0235192755 | 50 | 11 | 14 | .0319742832 |
| 50 | 7 | 31 | .0125611176 | 50 | 9 | 19 | .0223788386 | 50 | 11 | 15 | .0300721991 |
| 50 | 7 | 32 | .0121134092 | 50 | 9 | 20 | .0213374208 | 50 | 11 | 16 | .0283830554 |
| 50 | 7 | 33 | .0116836365 | 50 | 9 | 21 | .0203811274 | 50 | 11 | 17 | .0268704092 |
| 50 | 7 | 34 | .0112698876 | 50 | 9 | 22 | .0194986135 | 50 | 11 | 18 | .0255056815 |
| 50 | 7 | 35 | .0108698266 | 50 | 9 | 23 | .0186803607 | 50 | 11 | 19 | .0242661634 |
| 50 | 7 | 36 | .0104815558 | 50 | 9 | 24 | .0179186350 | 50 | 11 | 20 | .0231336659 |
| 50 | 7 | 37 | .0101048000 | 50 | 9 | 25 | .0172073778 | 50 | 11 | 21 | .0220935701 |
| 50 | 7 | 38 | .0097399467 | 50 | 9 | 26 | .0165413240 | 50 | 11 | 22 | .0211337701 |
| 50 | 7 | 39 | .0093847254 | 50 | 9 | 27 | .0159150978 | 50 | 11 | 23 | .0202440708 |
| 50 | 7 | 40 | .0090330537 | 50 | 9 | 28 | .0153232774 | 50 | 11 | 24 | .0194161584 |
| 50 | 7 | 41 | .0086800758 | 50 | 9 | 29 | .0147613199 | 50 | 11 | 25 | .0186424235 |
| 50 | 7 | 42 | .0083278448 | 50 | 9 | 30 | .0142262404 | 50 | 11 | 26 | .0179155541 |
| 50 | 7 | 43 | .0079804453 | 50 | 9 | 31 | .0137161027 | 50 | 11 | 27 | .0172296682 |
| 50 | 7 | 44 | .0076324738 | 50 | 9 | 32 | .0132287398 | 50 | 11 | 28 | .0165809674 |
| 50 | 8 | 8 | .0461239738 | 50 | 9 | 33 | .0127613339 | 50 | 11 | 29 | .0159666547 |
| 50 | 8 | 9 | .0416232761 | 50 | 9 | 34 | .0123116264 | 50 | 11 | 30 | .0153833824 |
| 50 | 8 | 10 | .0379700647 | 50 | 9 | 35 | .0118789245 | 50 | 11 | 31 | .0148272657 |
| 50 | 8 | 11 | .0349373092 | 50 | 9 | 36 | .0114623891 | 50 | 11 | 32 | .0142954042 |
| 50 | 8 | 12 | .0323731632 | 50 | 9 | 37 | .0110579056 | 50 | 11 | 33 | .0137862609 |
| 50 | 8 | 13 | .0301720508 | 50 | 9 | 38 | .0106589043 | 50 | 11 | 34 | .0132973053 |
| 50 | 8 | 14 | .0282581757 | 50 | 9 | 39 | .0102628230 | 50 | 11 | 35 | .0128230334 |
| 50 | 8 | 15 | .0265756453 | 50 | 9 | 40 | .0098748404 | 50 | 11 | 36 | .0123580120 |

TABLE I-COVARIANCES OF NORMAL ORDER STATISTICS

| N | I | J | COVARIANCE | N | I | J | COVARIANCE | N | I | J | COVARIANCE |
|---|---|---|---|---|---|---|---|---|---|---|---|
| 50 | 11 | 37 | .0119028505 | 50 | 14 | 16 | .0316803405 | 50 | 16 | 32 | .0171550850 |
| 50 | 11 | 38 | .0114636980 | 50 | 14 | 17 | .0299980473 | 50 | 16 | 33 | .0165345766 |
| 50 | 11 | 39 | .0110419687 | 50 | 14 | 18 | .0284804513 | 50 | 16 | 34 | .0159387486 |
| 50 | 11 | 40 | .0106273176 | 50 | 14 | 19 | .0271024894 | 50 | 16 | 35 | .0153658579 |
| 50 | 12 | 12 | .0379002950 | 50 | 14 | 20 | .0258438881 | 50 | 17 | 17 | .0333350409 |
| 50 | 12 | 13 | .0353332302 | 50 | 14 | 21 | .0246879885 | 50 | 17 | 18 | .0316568630 |
| 50 | 12 | 14 | .0331003107 | 50 | 14 | 22 | .0236209738 | 50 | 17 | 19 | .0301328452 |
| 50 | 12 | 15 | .0311367249 | 50 | 14 | 23 | .0226312244 | 50 | 17 | 20 | .0287407180 |
| 50 | 12 | 16 | .0293935241 | 50 | 14 | 24 | .0217093169 | 50 | 17 | 21 | .0274622841 |
| 50 | 12 | 17 | .0278330052 | 50 | 14 | 25 | .0208477586 | 50 | 17 | 22 | .0262826409 |
| 50 | 12 | 18 | .0264256696 | 50 | 14 | 26 | .0200394657 | 50 | 17 | 23 | .0251894346 |
| 50 | 12 | 19 | .0251481596 | 50 | 14 | 27 | .0192789207 | 50 | 17 | 24 | .0241721957 |
| 50 | 12 | 20 | .0239817551 | 50 | 14 | 28 | .0185619856 | 50 | 17 | 25 | .0232219646 |
| 50 | 12 | 21 | .0229112066 | 50 | 14 | 29 | .0178835294 | 50 | 17 | 26 | .0223306756 |
| 50 | 12 | 22 | .0219239870 | 50 | 14 | 30 | .0172371365 | 50 | 17 | 27 | .0214910358 |
| 50 | 12 | 23 | .0210093374 | 50 | 14 | 31 | .0166180667 | 50 | 17 | 28 | .0206978442 |
| 50 | 12 | 24 | .0201576955 | 50 | 14 | 32 | .0160249893 | 50 | 17 | 29 | .0199457516 |
| 50 | 12 | 25 | .0193615457 | 50 | 14 | 33 | .0154572098 | 50 | 17 | 30 | .0192300332 |
| 50 | 12 | 26 | .0186146554 | 50 | 14 | 34 | .0149114755 | 50 | 17 | 31 | .0185488446 |
| 50 | 12 | 27 | .0179111042 | 50 | 14 | 35 | .0143840801 | 50 | 17 | 32 | .0179006139 |
| 50 | 12 | 28 | .0172460903 | 50 | 14 | 36 | .0138748436 | 50 | 17 | 33 | .0172801481 |
| 50 | 12 | 29 | .0166168629 | 50 | 14 | 37 | .0133841301 | 50 | 17 | 34 | .0166812504 |
| 50 | 12 | 30 | .0160214457 | 50 | 15 | 15 | .0349275095 | 50 | 18 | 18 | .0327084229 |
| 50 | 12 | 31 | .0154560001 | 50 | 15 | 16 | .0329671710 | 50 | 18 | 19 | .0311433912 |
| 50 | 12 | 32 | .0149143923 | 50 | 15 | 17 | .0312104098 | 50 | 18 | 20 | .0297146367 |
| 50 | 12 | 33 | .0143914766 | 50 | 15 | 18 | .0296244457 | 50 | 18 | 21 | .0284034590 |
| 50 | 12 | 34 | .0138862532 | 50 | 15 | 19 | .0281832111 | 50 | 18 | 22 | .0271944550 |
| 50 | 12 | 35 | .0134000128 | 50 | 15 | 20 | .0268656027 | 50 | 18 | 23 | .0260747512 |
| 50 | 12 | 36 | .0129314156 | 50 | 15 | 21 | .0256543310 | 50 | 18 | 24 | .0250333897 |
| 50 | 12 | 37 | .0124760850 | 50 | 15 | 22 | .0245351566 | 50 | 18 | 25 | .0240608639 |
| 50 | 12 | 38 | .0120319010 | 50 | 15 | 23 | .0234964146 | 50 | 18 | 26 | .0231490760 |
| 50 | 12 | 39 | .0116002234 | 50 | 15 | 24 | .0225283423 | 50 | 18 | 27 | .0222910263 |
| 50 | 13 | 13 | .0365981043 | 50 | 15 | 25 | .0216229051 | 50 | 18 | 28 | .0214805833 |
| 50 | 13 | 14 | .0342890678 | 50 | 15 | 26 | .0207736786 | 50 | 18 | 29 | .0207140037 |
| 50 | 13 | 15 | .0322587049 | 50 | 15 | 27 | .0199737124 | 50 | 18 | 30 | .0199871569 |
| 50 | 13 | 16 | .0304564383 | 50 | 15 | 28 | .0192168180 | 50 | 18 | 31 | .0192952531 |
| 50 | 13 | 17 | .0288433240 | 50 | 15 | 29 | .0184986542 | 50 | 18 | 32 | .0186355023 |
| 50 | 13 | 18 | .0273889273 | 50 | 15 | 30 | .0178147774 | 50 | 18 | 33 | .0180051959 |
| 50 | 13 | 19 | .0260691221 | 50 | 15 | 31 | .0171597797 | 50 | 19 | 19 | .0323349267 |
| 50 | 13 | 20 | .0248644910 | 50 | 15 | 32 | .0165303158 | 50 | 19 | 20 | .0308456404 |
| 50 | 13 | 21 | .0237591164 | 50 | 15 | 33 | .0159270914 | 50 | 19 | 21 | .0294783689 |
| 50 | 13 | 22 | .0227396332 | 50 | 15 | 34 | .0153505806 | 50 | 19 | 22 | .0282168844 |
| 50 | 13 | 23 | .0217949319 | 50 | 15 | 35 | .0147955814 | 50 | 19 | 23 | .0270477204 |
| 50 | 13 | 24 | .0209156585 | 50 | 15 | 36 | .0142539958 | 50 | 19 | 24 | .0259594305 |
| 50 | 13 | 25 | .0200934613 | 50 | 16 | 16 | .0341381561 | 50 | 19 | 25 | .0249422488 |
| 50 | 13 | 26 | .0193222716 | 50 | 16 | 17 | .0323220036 | 50 | 19 | 26 | .0239878474 |
| 50 | 13 | 27 | .0185976696 | 50 | 16 | 18 | .0306824323 | 50 | 19 | 27 | .0230894062 |
| 50 | 13 | 28 | .0179148374 | 50 | 16 | 19 | .0291924421 | 50 | 19 | 28 | .0222412322 |
| 50 | 13 | 29 | .0172685250 | 50 | 16 | 20 | .0278302321 | 50 | 19 | 29 | .0214379991 |
| 50 | 13 | 30 | .0166547157 | 50 | 16 | 21 | .0265780369 | 50 | 19 | 30 | .0206762101 |
| 50 | 13 | 31 | .0160706685 | 50 | 16 | 22 | .0254213084 | 50 | 19 | 31 | .0199513399 |
| 50 | 13 | 32 | .0155125903 | 50 | 16 | 23 | .0243480632 | 50 | 19 | 32 | .0192573787 |
| 50 | 13 | 33 | .0149748946 | 50 | 16 | 24 | .0233484036 | 50 | 20 | 20 | .0320936705 |
| 50 | 13 | 34 | .0144539774 | 50 | 16 | 25 | .0224136952 | 50 | 20 | 21 | .0306549222 |
| 50 | 13 | 35 | .0139518817 | 50 | 16 | 26 | .0215362768 | 50 | 20 | 22 | .0293258360 |
| 50 | 13 | 36 | .0134724782 | 50 | 16 | 27 | .0207099885 | 50 | 20 | 23 | .0280923232 |
| 50 | 13 | 37 | .0130128277 | 50 | 16 | 28 | .0199280367 | 50 | 20 | 24 | .0269425276 |
| 50 | 13 | 38 | .0125625015 | 50 | 16 | 29 | .0191844471 | 50 | 20 | 25 | .0258664379 |
| 50 | 14 | 14 | .0356738617 | 50 | 16 | 30 | .0184761905 | 50 | 20 | 26 | .0248556582 |
| 50 | 14 | 15 | .0335588517 | 50 | 16 | 31 | .0178010656 | 50 | 20 | 27 | .0239030915 |

TABLE I-COVARIANCES OF NORMAL ORDER STATISTICS

| N | I | J | COVARIANCE | N | I | J | COVARIANCE | N | I | J | COVARIANCE |
|---|---|---|---|---|---|---|---|---|---|---|---|
| 50 | 20 | 28 | .0230028080 | 50 | 21 | 28 | .0238277493 | 50 | 23 | 23 | .0312901039 |
| 50 | 20 | 29 | .0221495029 | 50 | 21 | 29 | .0229324259 | 50 | 23 | 24 | .0300349585 |
| 50 | 20 | 30 | .0213375022 | 50 | 21 | 30 | .0220797530 | 50 | 23 | 25 | .0288622822 |
| 50 | 20 | 31 | .0205626704 | 50 | 22 | 22 | .0315325004 | 50 | 23 | 26 | .0277625206 |
| 50 | 21 | 21 | .0318320887 | 50 | 22 | 23 | .0302092256 | 50 | 23 | 27 | .0267275414 |
| 50 | 21 | 22 | .0304441799 | 50 | 22 | 24 | .0289757264 | 50 | 23 | 28 | .0257501194 |
| 50 | 21 | 23 | .0291549930 | 50 | 22 | 25 | .0278216011 | 50 | 24 | 24 | .0311721915 |
| 50 | 21 | 24 | .0279525704 | 50 | 22 | 26 | .0267379142 | 50 | 24 | 25 | .0299664019 |
| 50 | 21 | 25 | .0268266717 | 50 | 22 | 27 | .0257169051 | 50 | 24 | 26 | .0288362122 |
| 50 | 21 | 26 | .0257686683 | 50 | 22 | 28 | .0247516666 | 50 | 24 | 27 | .0277729877 |
| 50 | 21 | 27 | .0247712063 | 50 | 22 | 29 | .0238358480 | 50 | 25 | 25 | .0311462614 |
| 50 | 25 | 26 | .0299596731 | | | | | | | | |

TABLE II-EXPECTED VALUES OF NORMAL ORDER STATISTICS

| N | I | EXPECTATION | N | I | EXPECTATION | N | I | EXPECTATION |
|---|---|---|---|---|---|---|---|---|
| 2 | 1 | .5641895835 | 16 | 5 | .5700093557 | 22 | 11 | .0564154119 |
| 3 | 1 | .8462843753 | 16 | 6 | .3962227552 | 23 | 1 | 1.9291617120 |
| 4 | 1 | 1.0293753730 | 16 | 7 | .2337515783 | 23 | 2 | 1.4813657430 |
| 4 | 2 | .2970113823 | 16 | 8 | .0772874596 | 23 | 3 | 1.2144464930 |
| 5 | 1 | 1.1629644740 | 17 | 1 | 1.7939419810 | 23 | 4 | 1.0135591890 |
| 5 | 2 | .4950189705 | 17 | 2 | 1.3187819880 | 23 | 5 | .8469688035 |
| 6 | 1 | 1.2672063610 | 17 | 3 | 1.0294609890 | 23 | 6 | .7011503264 |
| 6 | 2 | .6417550388 | 17 | 4 | .8073849287 | 23 | 7 | .5689637234 |
| 6 | 3 | .2015468338 | 17 | 5 | .6194576510 | 23 | 8 | .4460926961 |
| 7 | 1 | 1.3521783760 | 17 | 6 | .4513334471 | 23 | 9 | .3296525009 |
| 7 | 2 | .7573742706 | 17 | 7 | .2951864865 | 23 | 10 | .2175470436 |
| 7 | 3 | .3527069592 | 17 | 8 | .1459874237 | 23 | 11 | .1081295394 |
| 8 | 1 | 1.4236003060 | 18 | 1 | 1.8200318790 | 24 | 1 | 1.9476740740 |
| 8 | 2 | .8522248625 | 18 | 2 | 1.3504137130 | 24 | 2 | 1.5033773720 |
| 8 | 3 | .4728224949 | 18 | 3 | 1.0657281830 | 24 | 3 | 1.2392378170 |
| 8 | 4 | .1525143995 | 18 | 4 | .8481250192 | 24 | 4 | 1.0409072230 |
| 9 | 1 | 1.4850131620 | 18 | 5 | .6647946122 | 24 | 5 | .8768190218 |
| 9 | 2 | .9322974567 | 18 | 6 | .5015815518 | 24 | 6 | .7335379741 |
| 9 | 3 | .5719707829 | 18 | 7 | .3508372378 | 24 | 7 | .6039873831 |
| 9 | 4 | .2745259191 | 18 | 8 | .2077353060 | 24 | 8 | .4839062642 |
| 10 | 1 | 1.5387527310 | 18 | 9 | .0688025709 | 24 | 9 | .3704655600 |
| 10 | 2 | 1.0013570450 | 19 | 1 | 1.8444815120 | 24 | 10 | .2616307358 |
| 10 | 3 | .6560591054 | 19 | 2 | 1.3799384920 | 24 | 11 | .1558298746 |
| 10 | 4 | .3757646970 | 19 | 3 | 1.0994530990 | 24 | 12 | .0517564160 |
| 10 | 5 | .1226677523 | 19 | 4 | .8858619619 | 25 | 1 | 1.9653146100 |
| 11 | 1 | 1.5864363520 | 19 | 5 | .7066114841 | 25 | 2 | 1.5243012210 |
| 11 | 2 | 1.0619165200 | 19 | 6 | .5477073711 | 25 | 3 | 1.2627531070 |
| 11 | 3 | .7288394047 | 19 | 7 | .4016422768 | 25 | 4 | 1.0667923590 |
| 11 | 4 | .4619783072 | 19 | 8 | .2637428852 | 25 | 5 | .9050102574 |
| 11 | 5 | .2248908792 | 19 | 9 | .1307248847 | 25 | 6 | .7640540796 |
| 12 | 1 | 1.6292276400 | 20 | 1 | 1.8674750600 | 25 | 7 | .6369036401 |
| 12 | 2 | 1.1157321840 | 20 | 2 | 1.4076040960 | 25 | 8 | .5193455795 |
| 12 | 3 | .7928381991 | 20 | 3 | 1.1309480520 | 25 | 9 | .4085977192 |
| 12 | 4 | .5368430214 | 20 | 4 | .9209817008 | 25 | 10 | .3026750547 |
| 12 | 5 | .3122488787 | 20 | 5 | .7453830060 | 25 | 11 | .2000642574 |
| 12 | 6 | .1025896798 | 20 | 6 | .5902969182 | 25 | 12 | .0995315692 |
| 13 | 1 | 1.6679901770 | 20 | 7 | .4483317610 | 26 | 1 | 1.9821578400 |
| 13 | 2 | 1.1640771940 | 20 | 8 | .3149332346 | 26 | 2 | 1.5442338590 |
| 13 | 3 | .8498346324 | 20 | 9 | .1869573609 | 26 | 3 | 1.2851095680 |
| 13 | 4 | .6028500882 | 20 | 10 | .0619963025 | 26 | 4 | 1.0913535750 |
| 13 | 5 | .3883271210 | 21 | 1 | 1.8891679150 | 26 | 5 | .9317056722 |
| 13 | 6 | .1905236911 | 21 | 2 | 1.4336179570 | 26 | 6 | .7928895152 |
| 14 | 1 | 1.7033815540 | 21 | 3 | 1.1604724120 | 26 | 7 | .6679359610 |
| 14 | 2 | 1.2079022750 | 21 | 4 | .9538018913 | 26 | 8 | .5526730547 |
| 14 | 3 | .9011267039 | 21 | 5 | .7814958915 | 26 | 9 | .4443587605 |
| 14 | 4 | .6617637035 | 21 | 6 | .6298217726 | 26 | 10 | .3410490856 |
| 14 | 5 | .4555660500 | 21 | 7 | .4914847823 | 26 | 11 | .2412766052 |
| 14 | 6 | .2672970489 | 21 | 8 | .3620257184 | 26 | 12 | .1438656014 |
| 14 | 7 | .0881592141 | 21 | 9 | .2384079485 | 26 | 13 | .0478085317 |
| 15 | 1 | 1.7359134450 | 21 | 10 | .1183565776 | 27 | 1 | 1.9982693020 |
| 15 | 2 | 1.2479350820 | 22 | 1 | 1.9096923220 | 27 | 2 | 1.5632598210 |
| 15 | 3 | .9476890303 | 22 | 2 | 1.4581553730 | 27 | 3 | 1.3064093380 |
| 15 | 4 | .7148773982 | 22 | 3 | 1.1882438010 | 27 | 4 | 1.1147114050 |
| 15 | 5 | .5157010431 | 22 | 4 | .9845869483 | 27 | 5 | .9570460498 |
| 15 | 6 | .3352960638 | 22 | 5 | .8152691346 | 27 | 6 | .8202080104 |
| 15 | 7 | .1652985264 | 22 | 6 | .6666668647 | 27 | 7 | .6972747819 |
| 16 | 1 | 1.7659913930 | 22 | 7 | .5315681934 | 27 | 8 | .5841107583 |
| 16 | 2 | 1.2847442230 | 22 | 8 | .4055917587 | 27 | 9 | .4780085086 |
| 16 | 3 | .9902710960 | 22 | 9 | .2857851480 | 27 | 10 | .3770592642 |
| 16 | 4 | .7631667457 | 22 | 10 | .1699742157 | 27 | 11 | .2798317821 |

TABLE II-EXPECTED VALUES OF NORMAL ORDER STATISTICS

| N | I | EXPECTATION | N | I | EXPECTATION | N | I | EXPECTATION |
|---|---|---|---|---|---|---|---|---|
| 27 | 12 | .1851963479 | 32 | 1 | 2.0696688280 | 35 | 12 | .4409172882 |
| 27 | 13 | .0922021682 | 32 | 2 | 1.6471174610 | 35 | 13 | .3637257685 |
| 28 | 1 | 2.0137069240 | 32 | 3 | 1.3998499460 | 35 | 14 | .2886207228 |
| 28 | 2 | 1.5814535040 | 32 | 4 | 1.2167194290 | 35 | 15 | .2151268958 |
| 28 | 3 | 1.3267419350 | 32 | 5 | 1.0672100660 | 35 | 16 | .1428211365 |
| 28 | 4 | 1.1369710290 | 32 | 6 | .9384119914 | 35 | 17 | .0712503714 |
| 28 | 5 | .9811536644 | 32 | 7 | .8235899583 | 36 | 1 | 2.1181232860 |
| 28 | 6 | .8461510226 | 32 | 8 | .7187478454 | 36 | 2 | 1.7036247940 |
| 28 | 7 | .7250836325 | 32 | 9 | .6212889319 | 36 | 3 | 1.4624376960 |
| 28 | 8 | .6138482304 | 32 | 10 | .5294239722 | 36 | 4 | 1.2846566380 |
| 28 | 9 | .5097670781 | 32 | 11 | .4418508675 | 36 | 5 | 1.1401631500 |
| 28 | 10 | .4109626396 | 32 | 12 | .3575522203 | 36 | 6 | 1.0162363410 |
| 28 | 11 | .3160331884 | 32 | 13 | .2757199120 | 36 | 7 | .9062477277 |
| 28 | 12 | .2238841541 | 32 | 14 | .1957114560 | 36 | 8 | .8062868965 |
| 28 | 13 | .1336126063 | 32 | 15 | .1169567118 | 36 | 9 | .7138265190 |
| 28 | 14 | .0444208934 | 32 | 16 | .0389112739 | 36 | 10 | .6271043962 |
| 29 | 1 | 2.0285221460 | 33 | 1 | 2.0824083360 | 36 | 11 | .5448693893 |
| 29 | 2 | 1.5988807100 | 33 | 2 | 1.6620045750 | 36 | 12 | .4662044873 |
| 29 | 3 | 1.3461862270 | 33 | 3 | 1.4163672060 | 36 | 13 | .3903428898 |
| 29 | 4 | 1.1582247370 | 33 | 4 | 1.2346773450 | 36 | 14 | .3166339386 |
| 29 | 5 | 1.0041353500 | 33 | 5 | 1.0865245390 | 36 | 15 | .2445999551 |
| 29 | 6 | .8708415720 | 33 | 6 | .9590490184 | 36 | 16 | .1738646126 |
| 29 | 7 | .7515039168 | 33 | 7 | .8455453701 | 36 | 17 | .1040167913 |
| 29 | 8 | .6420484531 | 33 | 8 | .7420412857 | 36 | 18 | .0346290786 |
| 29 | 9 | .5398226457 | 33 | 9 | .6459558446 | 37 | 1 | 2.1292770250 |
| 29 | 10 | .4429769278 | 33 | 10 | .5555104982 | 37 | 2 | 1.7165886990 |
| 29 | 11 | .3501354920 | 33 | 11 | .4694249625 | 37 | 3 | 1.4767564490 |
| 29 | 12 | .2602294189 | 33 | 12 | .3867026776 | 37 | 4 | 1.3001584880 |
| 29 | 13 | .1723950289 | 33 | 13 | .3065389200 | 37 | 5 | 1.1567663750 |
| 29 | 14 | .0858803939 | 33 | 14 | .2283060537 | 37 | 6 | 1.0339025150 |
| 30 | 1 | 2.0427608440 | 33 | 15 | .1514759305 | 37 | 7 | .9249611111 |
| 30 | 2 | 1.6155999010 | 33 | 16 | .0755336493 | 37 | 8 | .8260475131 |
| 30 | 3 | 1.3648120370 | 34 | 1 | 2.0947127550 | 37 | 9 | .7346546611 |
| 30 | 4 | 1.1785539340 | 34 | 2 | 1.6763624880 | 37 | 10 | .6490278547 |
| 30 | 5 | 1.0260849620 | 34 | 3 | 1.4322779670 | 37 | 11 | .5679110582 |
| 30 | 6 | .8943872897 | 34 | 4 | 1.2519560100 | 37 | 12 | .4904072629 |
| 30 | 7 | .7766587008 | 34 | 5 | 1.1050873520 | 37 | 13 | .4157820383 |
| 30 | 8 | .6688524835 | 34 | 6 | .9788602190 | 37 | 14 | .3433783080 |
| 30 | 9 | .5683373696 | 34 | 7 | .8665967487 | 37 | 15 | .2726967605 |
| 30 | 10 | .4732882901 | 34 | 8 | .7643471956 | 37 | 16 | .2033913073 |
| 30 | 11 | .3823542033 | 34 | 9 | .6695470784 | 37 | 17 | .1351108260 |
| 30 | 12 | .2944849907 | 34 | 10 | .5804246395 | 37 | 18 | .0674355741 |
| 30 | 13 | .2088460613 | 34 | 11 | .4957165590 | 38 | 1 | 2.1400914550 |
| 30 | 14 | .1247282942 | 34 | 12 | .4144516243 | 38 | 2 | 1.7291431200 |
| 30 | 15 | .0414827937 | 34 | 13 | .3358296087 | 38 | 3 | 1.4906091260 |
| 31 | 1 | 2.0564640980 | 34 | 14 | .2592231920 | 38 | 4 | 1.3151418820 |
| 31 | 2 | 1.6316632420 | 34 | 15 | .1841387132 | 38 | 5 | 1.1727996420 |
| 31 | 3 | 1.3826814600 | 34 | 16 | .1101030725 | 38 | 6 | 1.0509468070 |
| 31 | 4 | 1.1980307580 | 34 | 17 | .0366430481 | 38 | 7 | .9429996257 |
| 31 | 5 | 1.0470853670 | 35 | 1 | 2.1066094390 | 38 | 8 | .8450762607 |
| 31 | 6 | .9168828602 | 35 | 2 | 1.6902255110 | 38 | 9 | .7546897096 |
| 31 | 7 | .8006557461 | 35 | 3 | 1.4476226070 | 38 | 10 | .6700972826 |
| 31 | 8 | .6943831171 | 35 | 4 | 1.2686018060 | 38 | 11 | .5900334567 |
| 31 | 9 | .5954519120 | 35 | 5 | 1.1229510940 | 38 | 12 | .5136106256 |
| 31 | 10 | .5020573770 | 35 | 6 | .9979049058 | 38 | 13 | .4401333104 |
| 31 | 11 | .4128732075 | 35 | 7 | .8868108994 | 38 | 14 | .3689526687 |
| 31 | 12 | .3268651047 | 35 | 8 | .7857401459 | 38 | 15 | .2995365467 |
| 31 | 13 | .2432164768 | 35 | 9 | .6921459883 | 38 | 16 | .2315424216 |
| 31 | 14 | .1612562554 | 35 | 10 | .6042613387 | 38 | 17 | .1646835251 |
| 31 | 15 | .0803729127 | 35 | 11 | .5208328915 | 38 | 18 | .0985798447 |

TABLE II-EXPECTED VALUES OF NORMAL ORDER STATISTICS

| N | I | EXPECTATION | N | I | EXPECTATION | N | I | EXPECTATION |
|---|---|---|---|---|---|---|---|---|
| 38 | 19 | .0328308289 | 42 | 1 | 2.1803156080 | 44 | 19 | .1995710871 |
| 39 | 1 | 2.1505856570 | 42 | 2 | 1.7757118290 | 44 | 20 | .1422292935 |
| 39 | 2 | 1.7413117550 | 42 | 3 | 1.5418782930 | 44 | 21 | .0853162880 |
| 39 | 3 | 1.5040233630 | 42 | 4 | 1.3704813120 | 44 | 22 | .0284518812 |
| 39 | 4 | 1.3296382870 | 42 | 5 | 1.2318963300 | 45 | 1 | 2.2077195670 |
| 39 | 5 | 1.1882983310 | 42 | 6 | 1.1136413940 | 45 | 2 | 1.8073268440 |
| 39 | 6 | 1.0674085600 | 42 | 7 | 1.0092234250 | 45 | 3 | 1.5765845400 |
| 39 | 7 | .9604071636 | 42 | 8 | .9147905622 | 45 | 4 | 1.4078447250 |
| 39 | 8 | .8634223095 | 42 | 9 | .8279041251 | 45 | 5 | 1.2716955100 |
| 39 | 9 | .7739853217 | 42 | 10 | .7469163600 | 45 | 6 | 1.1557517610 |
| 39 | 10 | .6903710029 | 42 | 11 | .6705559856 | 45 | 7 | 1.0535932620 |
| 39 | 11 | .6113034939 | 42 | 12 | .5978566972 | 45 | 8 | .9613955412 |
| 39 | 12 | .5358915436 | 42 | 13 | .5282049258 | 45 | 9 | .8767021316 |
| 39 | 13 | .4634785599 | 42 | 14 | .4611413656 | 45 | 10 | .7979271325 |
| 39 | 14 | .3934428115 | 42 | 15 | .3961252450 | 45 | 11 | .7238982683 |
| 39 | 15 | .3252202708 | 42 | 16 | .3326256515 | 45 | 12 | .6535889104 |
| 39 | 16 | .2584425881 | 42 | 17 | .2703515710 | 45 | 13 | .5862586732 |
| 39 | 17 | .1928734321 | 42 | 18 | .2092023580 | 45 | 14 | .5215366752 |
| 39 | 18 | .1282024690 | 42 | 19 | .1489830642 | 45 | 15 | .4591157553 |
| 39 | 19 | .0640201164 | 42 | 20 | .0892977511 | 45 | 16 | .3984827785 |
| 40 | 1 | 2.1607771780 | 42 | 21 | .0297646992 | 45 | 17 | .3390957375 |
| 40 | 2 | 1.7531163500 | 43 | 1 | 2.1896912620 | 45 | 18 | .2807139129 |
| 40 | 3 | 1.5170244630 | 43 | 2 | 1.7865381580 | 45 | 19 | .2233693792 |
| 40 | 4 | 1.3436764610 | 43 | 3 | 1.5537720840 | 45 | 20 | .1670050031 |
| 40 | 5 | 1.2032947240 | 43 | 4 | 1.3832944180 | 45 | 21 | .1112596564 |
| 40 | 6 | 1.0833235760 | 43 | 5 | 1.2455535250 | 45 | 22 | .0556667242 |
| 40 | 7 | .9772234692 | 43 | 6 | 1.1281016410 | 46 | 1 | 2.2163951720 |
| 40 | 8 | .8811302943 | 43 | 7 | 1.0244698670 | 46 | 2 | 1.8173173530 |
| 40 | 9 | .7925903706 | 43 | 8 | .9308131526 | 46 | 3 | 1.5875356650 |
| 40 | 10 | .7099012642 | 43 | 9 | .8446917290 | 46 | 4 | 1.4196184160 |
| 40 | 11 | .6317802189 | 43 | 10 | .7644842881 | 46 | 5 | 1.2842209740 |
| 40 | 12 | .5573194007 | 43 | 11 | .6889421974 | 46 | 6 | 1.1689867060 |
| 40 | 13 | .4858932105 | 43 | 12 | .6170688241 | 46 | 7 | 1.0675187980 |
| 40 | 14 | .4169250548 | 43 | 13 | .5482253694 | 46 | 8 | .9760081311 |
| 40 | 15 | .3498329310 | 43 | 14 | .4820039019 | 46 | 9 | .8919857392 |
| 40 | 16 | .2841991705 | 43 | 15 | .4179261118 | 46 | 10 | .8138695227 |
| 40 | 17 | .2198077145 | 43 | 16 | .3554302936 | 46 | 11 | .7405345281 |
| 40 | 18 | .1564329324 | 43 | 17 | .2941428180 | 46 | 12 | .6709647146 |
| 40 | 19 | .0936985692 | 43 | 18 | .2339649580 | 46 | 13 | .6043574652 |
| 40 | 20 | .0312176160 | 43 | 19 | .1748098580 | 46 | 14 | .5403155859 |
| 41 | 1 | 2.1706821850 | 43 | 20 | .1163597456 | 46 | 15 | .4786134505 |
| 41 | 2 | 1.7645768980 | 43 | 21 | .0581764575 | 46 | 16 | .4188205185 |
| 41 | 3 | 1.5296356520 | 44 | 1 | 2.1988219510 | 46 | 17 | .3603495161 |
| 41 | 4 | 1.3572827420 | 44 | 2 | 1.7970716310 | 46 | 18 | .3028392916 |
| 41 | 5 | 1.2178183610 | 44 | 3 | 1.5653352190 | 46 | 19 | .2462966572 |
| 41 | 6 | 1.0987245410 | 44 | 4 | 1.3957425730 | 46 | 20 | .1907885105 |
| 41 | 7 | .9934846143 | 44 | 5 | 1.2588128710 | 46 | 21 | .1360864435 |
| 41 | 8 | .8982407646 | 44 | 6 | 1.1421306280 | 46 | 22 | .0817039575 |
| 41 | 9 | .8105496040 | 44 | 7 | 1.0392513940 | 46 | 23 | .0272624696 |
| 41 | 10 | .7287353185 | 44 | 8 | .9463389350 | 47 | 1 | 2.2248590720 |
| 41 | 11 | .6515156958 | 44 | 9 | .8609471318 | 47 | 2 | 1.8270557640 |
| 41 | 12 | .5779561911 | 44 | 10 | .7814762738 | 47 | 3 | 1.5982031020 |
| 41 | 13 | .5074471572 | 44 | 11 | .7067115364 | 47 | 4 | 1.4310799120 |
| 41 | 14 | .4394693254 | 44 | 12 | .6356341805 | 47 | 5 | 1.2964073350 |
| 41 | 15 | .3734468187 | 44 | 13 | .5675612071 | 47 | 6 | 1.1818555440 |
| 41 | 16 | .3089021923 | 44 | 14 | .5021168334 | 47 | 7 | 1.0810496430 |
| 41 | 17 | .2456006991 | 44 | 15 | .4389047631 | 47 | 8 | .9901996847 |
| 41 | 18 | .1833940892 | 44 | 16 | .3773673862 | 47 | 9 | .9068243068 |
| 41 | 19 | .1219825654 | 44 | 17 | .3170403815 | 47 | 10 | .8293340089 |
| 41 | 20 | .0609486788 | 44 | 18 | .2577760994 | 47 | 11 | .7566509235 |

TABLE II-EXPECTED VALUES OF NORMAL ORDER STATISTICS

| N | I | EXPECTATION | N | I | EXPECTATION | N | I | EXPECTATION |
|---|---|---|---|---|---|---|---|---|
| 47 | 12 | .6877899613 | 48 | 17 | .4005870562 | 49 | 21 | .2047870507 |
| 47 | 13 | .6218910785 | 48 | 18 | .3448191245 | 49 | 22 | .1533736319 |
| 47 | 14 | .5585003227 | 48 | 19 | .2898269250 | 49 | 23 | .1023932022 |
| 47 | 15 | .4974515637 | 48 | 20 | .2357704499 | 49 | 24 | .0513176987 |
| 47 | 16 | .4384254758 | 48 | 21 | .1827527283 | 50 | 1 | 2.2490736340 |
| 47 | 17 | .3808359137 | 48 | 22 | .1304844594 | 50 | 2 | 1.8548720280 |
| 47 | 18 | .3241970497 | 48 | 23 | .0784189863 | 50 | 3 | 1.6286337960 |
| 47 | 19 | .2684295703 | 48 | 24 | .0261824993 | 50 | 4 | 1.4637372470 |
| 47 | 20 | .2136797326 | 49 | 1 | 2.2411896020 | 50 | 5 | 1.3310933240 |
| 47 | 21 | .1598853607 | 49 | 2 | 1.8458224980 | 50 | 6 | 1.2184452760 |
| 47 | 22 | .1066211176 | 49 | 3 | 1.6187400030 | 50 | 7 | 1.1194690560 |
| 47 | 23 | .0533890029 | 49 | 4 | 1.4531257330 | 50 | 8 | 1.0304510830 |
| 48 | 1 | 2.2331208860 | 49 | 5 | 1.3198285200 | 50 | 9 | .9488967683 |
| 48 | 2 | 1.8365538250 | 49 | 6 | 1.2065681300 | 50 | 10 | .8731334166 |
| 48 | 3 | 1.6086003540 | 49 | 7 | 1.1070065400 | 50 | 11 | .8021705560 |
| 48 | 4 | 1.4422443280 | 49 | 8 | 1.0174023930 | 50 | 12 | .7352179030 |
| 48 | 5 | 1.3082713370 | 49 | 9 | .9352593650 | 50 | 13 | .6713599159 |
| 48 | 6 | 1.1943769150 | 49 | 10 | .8589408445 | 50 | 14 | .6098616464 |
| 48 | 7 | 1.0942059480 | 49 | 11 | .7874409724 | 50 | 15 | .5505084058 |
| 48 | 8 | 1.0039912860 | 49 | 12 | .7198919861 | 50 | 16 | .4933240350 |
| 48 | 9 | .9212416775 | 49 | 13 | .6553703658 | 50 | 17 | .4380220426 |
| 48 | 10 | .8443490338 | 49 | 14 | .5932427390 | 50 | 18 | .3839471900 |
| 48 | 11 | .7722769142 | 49 | 15 | .5333530945 | 50 | 19 | .3305636209 |
| 48 | 12 | .7040907730 | 49 | 16 | .4756273974 | 50 | 20 | .2778428120 |
| 48 | 13 | .6388875261 | 49 | 17 | .4196365927 | 50 | 21 | .2260815607 |
| 48 | 14 | .5761314120 | 49 | 18 | .3647291051 | 50 | 22 | .1753803463 |
| 48 | 15 | .5156819628 | 49 | 19 | .3105297135 | 50 | 23 | .1253650863 |
| 48 | 16 | .4573446857 | 49 | 20 | .2571383115 | 50 | 24 | .0754262079 |
|  |  |  |  |  |  | 50 | 25 | .0252001470 |

Selected Tables in Mathematical Statistics
Volume V, 1977

# MEANS, VARIANCES, AND COVARIANCES OF NORMAL ORDER STATISTICS IN THE PRESENCE OF AN OUTLIER

H. A. David, W. J. Kennedy, and R. D. Knight
Iowa State University

## SUMMARY

Let $X_1, \ldots, X_{n-1}$ be independent standard normal variates and Y a further independent variate representing the outlier. Tables of the means, variances, and covariances of the order statistics in the combined sample of size n are given for $n \leq 20$ in the following two cases:

(a) $Y \frown N(\lambda, 1)$ for $\lambda = 0(0.5)\ 3, 4$;

(b) $Y \frown N(0, \tau^2)$ for $\tau = 0.5, 2, 3, 4$.

For (b) additional values of the means and variances are provided for $\tau = 6, 8, 10$. Worked examples illustrate some uses of the tables in robustness studies.

## 1. INTRODUCTION

Consider the n independent absolutely continuous random variables $X_j$ $(j = 1, 2, \ldots, n-1)$ and Y, where

$X_j$ has cdf $F(x)$ and pdf $f(x)$,
Y has cdf $G(x)$ and pdf $g(x)$.

Arrange the n variates in combined ascending order to obtain

$$Z_{1:n} \leq Z_{2:n} \leq \cdots \leq Z_{n:n} .$$

Tables of the means, variances, and covariances of the $Z_{r:n}$ $(r = 1, 2, \ldots, n)$ for $n \leq 20$ are given to 4D in the following two cases:

(a) $X \frown N(0, 1)$, $Y \frown N(\lambda, 1)$ for $\lambda = 0(0.5)\ 3, 4$;

(b) $X \frown N(0, 1)$, $Y \frown N(0, \tau^2)$ for $\tau = 0.5, 2, 3, 4$.

For (b) additional values of the means and variances are provided for $\tau = 6, 8, 10$.

Received by the editors April 1976 and in revised form February 1977.
AMS(MOS) Subject Classifications (1970): Primary 62Q05; Secondary 62G30, 62G35, 62E99.
This work was supported by the U. S. Army Research Office.

The null case, corresponding to $\lambda = 0$ in (a) and $\tau = 1$ in (b), has been tabulated to 10D by Teichroew (1956) and Sarhan and Greenberg (1956). The tabulated values may be supplemented by results for $\lambda = \infty$ in (a) and $\tau = \infty$ in (b) which are obtainable from the null case for sample size n-1 by relations given in Section 3. Cases (a) and (b) constitute two well-known models for studying the effect of an outlier in normal populations (e.g., Dixon, 1950). With the help of the present tables it is easy to obtain the mean, variance, and mean square error of statistics expressible as linear functions of the order statistics. For example, in (a) the bias and mean square error of such estimators of location as the median, trimmed mean, and Winsorized mean may be found as a function of $\lambda$. These and similar investigations of robustness based on the present tables are published in detail elsewhere (David and Shu, 1977). Two examples of the use of the Tables are, however, given in Section 4.

## 2. BASIC FORMULAE

By distinguishing the cases $Y \leq x$, $x < Y \leq x + dx$, and $Y > x + dx$ we may write down the pdf $f_{r:n}(x)$ of $Z_{r:n}$ $(r = 1, 2, \ldots, n)$:

$$
\begin{aligned}
f_{r:n}(x) = \; & \frac{(n-1)!}{(r-2)!(n-r)!} F^{r-2}(x)\, G(x)\, [1 - F(x)]^{n-r} f(x) \\
& + \frac{(n-1)!}{(r-1)!(n-r)!} F^{r-1}(x)\, [1 - F(x)]^{n-r} g(x) \\
& + \frac{(n-1)!}{(r-1)!(n-r-1)!} F^{r-1}(x)\, [1 - F(x)]^{n-r-1} [1 - G(x)]\, f(x).
\end{aligned} \tag{1}
$$

The first term drops out if $r = 1$, the last if $r = n$.

A similar argument gives the joint pdf $f_{r,s:n}(x,y)$ of $Z_{r:n}$ and $Z_{s:n}$ for $x < y$ and $r < s$ as

$$
\begin{aligned}
& f_{r,s:n}(x, y) = \\
& \frac{(n-1)!}{(r-2)!(s-r-1)!(n-s)!} [F(x)]^{r-2} G(x)\, [F(y) - F(x)]^{s-r-1} [1 - F(y)]^{n-s} f(x) f(y) \\
& + \frac{(n-1)!}{(r-1)!(s-r-1)!(n-s)!} [F(x)]^{r-1} [F(y) - F(x)]^{s-r-1} [1 - F(y)]^{n-s} g(x) f(y) \\
& + \frac{(n-1)!}{(r-1)!(s-r-2)!(n-s)!} [F(x)]^{r-1} [F(y) - F(x)]^{s-r-2} [G(y) - G(x)][1 - F(y)]^{n-s} \cdot f(x) f(y) \\
& + \frac{(n-1)!}{(r-1)!(s-r-1)!(n-s)!} [F(x)]^{r-1} [F(y) - F(x)]^{s-r-1} [1 - F(y)]^{n-s} f(x) g(y) \\
& \frac{(n-1)!}{(r-1)!(s-r-1)!(n-s-1)!} [F(x)]^{r-1} [F(y) - F(x)]^{s-r-1} [1 - F(y)]^{n-s-1} \cdot [1 - G(y)]\, f(x) f(y)\,.
\end{aligned} \tag{2}
$$

Here the first term drops out if $r = 1$, the last if $s = n$, and the middle term if $s = r+1$.

The means, variances, and covariances of the $Z_{r:n}$ are now obtained in the usual manner:

$$E(Z_{r:n}) = \int_{-\infty}^{\infty} x \, f_{r:n}(x) \, dx \qquad r = 1, 2, \ldots, n \tag{3}$$

$$E(Z_{r:n} Z_{s:n}) = \int_{-\infty}^{\infty} \int_{-\infty}^{y} xy \, f_{r,s:n}(x,y) \, dx \, dy \qquad r < s \tag{4}$$

and, for $r \leq s$,

$$\mathrm{cov}(Z_{r:n}, Z_{s:n}) = E(Z_{r:n} Z_{s:n}) - E(Z_{r:n}) E(Z_{s:n}) . \tag{5}$$

It may be noted that eqs. (1) and (2) can be regarded formally as a special case of a result due to Vaughan and Venables (1972).

## 3. FURTHER RESULTS AND CHECKS

We now apply the general formulae of Section 2 to the cases (a) and (b) of Section 1. Many of the results obtained will, however, hold more widely. Let $\mu_{r:n}(\lambda)$, $\sigma_{r,s:n}(\lambda)$ denote the mean of $Z_{r:n}$ and the covariance of $Z_{r:n}$, $Z_{s:n}$ $(r \leq s)$ in case (a), with $\mu^*_{r:n}(\tau)$, $\sigma^*_{r,s:n}(\tau)$ having a corresponding meaning in case (b).

Under (a) we may set $Y = X_n + \lambda$, where $X_n$ is unit normal, independent of $X_1, X_2, \ldots, X_{n-1}$. It follows that for any set $X_1, X_2, \ldots, X_n$ the variate $Z_{r:n}$ is a non-decreasing function of $\lambda$, so that, in particular,

$$\mu_{r:n}(-\infty) \leq \mu_{r:n}(\lambda) \leq \mu_{r:n}(\infty) . \tag{6}$$

Clearly $\mu_{n:n}(\infty) = \infty$ and $\mu_{1:n}(-\infty) = -\infty$. In all other cases one has (David and Shu, 1977)

$$\mu_{r:n}(\infty) = \mu_{r:n-1}(0), \quad \mu_{r:n}(-\infty) = \mu_{r-1:n-1}(0) . \tag{7}$$

Likewise it can be shown that

$$\begin{aligned} \sigma_{r,s:n}(\infty) &= \sigma_{r,s:n-1}(0) \qquad && r,s = 1, \ldots, n-1, \\ \sigma_{r,s:n}(-\infty) &= \sigma_{r-1,s-1:n-1}(0) \qquad && r,s = 2, \ldots, n . \end{aligned} \tag{8}$$

Thus the moments for $\lambda = \pm \infty$ can be obtained from the null case for sample size $n-1$.

Also, for $\lambda \geq 0$,

$$Z_{r:n}(X_1, \ldots, X_{n-1}, X_n - \lambda) = -Z_{n+1-r:n}(-X_1, \ldots, -X_{n-1}, -X_n + \lambda).$$

But $(X_1, \ldots, X_n)$ and $(-X_1, \ldots, -X_n)$ have the same distribution, so that in generalization of well-known results for $\lambda = 0$, we have

$$\mu_{r:n}(-\lambda) = -\mu_{n+1-r:n}(\lambda),$$

$$\sigma_{r,s:n}(-\lambda) = \sigma_{n+1-s,n+1-r:n}(\lambda) \qquad r \leq s .$$

For case (b) results corresponding to (7) and (8) are:

$$\mu^*_{r:n}(\infty) = \tfrac{1}{2}[\mu_{r-1:n-1}(0) + \mu_{r:n-1}(0)] \qquad r = 2, \ldots, n-1,$$

$$\sigma^*_{r,s:n}(\infty) = \tfrac{1}{2}[\sigma_{r-1,s-1:n-1}(0) + \sigma_{r,s:n-1}(0)] + \tfrac{1}{4}[\mu_{r:n-1}(0) - \mu_{r-1:n-1}(0)]$$

$$\cdot [\mu_{s:n-1}(0) - \mu_{s-1:n-1}(0)] \qquad r,s = 2, \ldots, n-1.$$

Symmetry considerations allow major reductions in tabulation since

$$\mu^*_{n+1-r:n}(\tau) = -\mu^*_{r:n}(\tau), \tag{9}$$

$$\sigma^*_{n+1-s,\, n+1-r:n}(\tau) = \sigma^*_{r,s:n}(\tau) . \tag{10}$$

Sum checks. From the identity

$$\left(\sum_{r=1}^{n} Z^k_{r:n}\right)^m = \left(\sum_{j=1}^{n-1} X^k_j + Y^k\right)^m$$

it is easy to establish the following relations which do not depend on normality (cf. David, 1970, p. 30):

(a)

$$\sum_{r=1}^{n} \mu_{r:n}(\lambda) = \lambda$$

$$\sum_{r=1}^{n} E\, Z^2_{r:n} = n + \lambda^2$$

$$\sum_{r=1}^{n-1} \sum_{s=r+1}^{n} E(Z_{r:n}\, Z_{s:n}) = 0$$

$$\sum_{r=1}^{n} \sum_{s=1}^{n} \sigma_{r,s:n}(\lambda) = n$$

(b)

$$\sum_{r=1}^{n} \mu^*_{r:n}(\tau) = 0$$

$$\sum_{r=1}^{n} E\, Z^2_{r:n} = n - 1 + \tau^2$$

$$\sum_{r=1}^{n-1} \sum_{s=r+1}^{n} E(Z_{r:n}\, Z_{s:n}) = 0$$

$$\sum_{r=1}^{n} \sum_{s=1}^{n} \sigma^*_{r,s:n}(\tau) = n - 1 + \tau^2$$

## 4. EXAMPLES OF USE OF TABLES

Example 1. Consider the median in samples of 10, viz.

$$M = \tfrac{1}{2}(Z_{5:10} + Z_{6:10}) .$$

For case (a) of Section 3 we find the following results for bias and mean square error (MSE) as functions of $\lambda$:

| $\lambda$ | 0.0 | 1.0 | 2.0 | 3.0 | 4.0 | $\infty$ |
|---|---|---|---|---|---|---|
| Bias | 0 | 0.0877 | 0.1280 | 0.1364 | 0.1372 | 0.1373 |
| MSE | 0.1383 | 0.1496 | 0.1652 | 0.1707 | 0.1715 | 0.1715 |

Corresponding results for case (b) are:

| $\tau$ | 0.5 | 1.0 | 2.0 | 3.0 | 4.0 | $\infty$ |
|---|---|---|---|---|---|---|
| Var | 0.1173 | 0.1383 | 0.1538 | 0.1595 | 0.1625 | 0.1715 |

For example, in case (a), if $\lambda = 3$, Tables 1 and 3 give

$$\text{Bias} = \frac{1}{2}[\mu_{5:10}(3) + \mu_{6:10}(3)]$$

$$= \frac{1}{2}(-0.0005 + 0.2733) = 0.1364$$

$$\text{MSE} = \frac{1}{4}[\sigma_{5,5:10}(3) + 2\sigma_{5,6:10}(3) + \sigma_{6,6:10}(3)] + (\text{Bias})^2$$

$$= \frac{1}{4}[0.1657 + 2(0.1365) + 0.1697] + (0.1364)^2$$

$$= 0.1707$$

For $\lambda = \infty$, we have from (7) and (8)

$$\text{Bias} = \frac{1}{2}[\mu_{5:9}(0) + \mu_{6:9}(0)]$$

$$= \frac{1}{2}(0 + 0.2745) = 0.1373$$

$$\text{MSE} = \frac{1}{4}[\sigma_{5,5:9}(0) + 2\sigma_{5,6:9}(0) + \sigma_{6,6:9}(0)] + (\text{Bias})^2$$

$$= \frac{1}{4}[0.1661 + 2(0.1370) + 0.1706] + (0.1373)^2$$

$$= 0.1715$$

Example 2. In a random sample of size $n = 5$ from a normal population with variance $\sigma^2$ a good unbiased estimator of $\sigma$ is $cW_5$, where $c = 0.42994$ and $W_5 = X_{5:5} - X_{1:5}$. The bias and variance of this estimator under case (b) are as follows:

| $\tau$ | 0.5 | 1.0 | 2.0 | 3.0 | 4.0 | 6.0 | 8.0 | 10.0 | $\infty$ |
|---|---|---|---|---|---|---|---|---|---|
| $\frac{1}{\sigma}$ Bias | -0.0769 | 0 | 0.2532 | 0.5577 | 0.8801 | 1.5448 | 2.2200 | 2.8996 | $\infty$ |
| $\frac{1}{\sigma^2}$ Var | 0.1290 | 0.1380 | 0.2808 | 0.5867 | 0.9433 | - | - | - | $\infty$ |

For example, if $\tau = 2$, Tables 2 and 4 give

$$\frac{1}{\sigma}\text{ Bias} = c[\mu^*_{5:5}(2) - \mu^*_{1:5}(2)] - 1$$

$$= 2c\mu^*_{5:5}(2) - 1 \qquad \text{by (9)}$$

$$= 2c(1.4574 - 1) - 1 = 0.2532$$

$$\frac{1}{\sigma^2}\text{ Var} = c^2\,[\sigma^*_{5,5:5}(2) - 2\sigma^*_{1,5:5}(2) + \sigma^*_{1,1:5}(2)]$$

$$= 2c^2\,[\sigma^*_{5,5:5}(2) - \sigma^*_{1,5:5}(2)] \qquad \text{by (10)}$$

$$= 2c^2\,[0.9769 - 0.2173] = 0.2808$$

## 5. NUMERICAL ANALYSIS

Tabular values were calculated using double precision FORTRAN IV language programs in an IBM 360-65 computer. Checks used to verify the accuracy of computed values were:

1) Comparison with the Teichroew (1956) tables for the null cases $\lambda = 0$ in (a) and $\tau = 1$ in (b).
2) The sum checks listed in Section 3.

Computed values in the null case agreed completely with the ten-decimal-place Teichroew results. Additionally, the sum checks used in other cases indicated that results obtained were accurate to more than six decimal places. Values printed in the tables are rounded to four decimal places and should be accurate to within one part in the fourth decimal place.

The normal cdf, which appears in the integrand of integrals defined in equations (3) and (4), was evaluated by using the double precision error function routine DERF and the complementary error function routine DERFC. The basic relationship between the normal cdf and the error function is:

$$F(x) = \frac{1}{2}\,(1 + \text{erf}(x/\sqrt{2})) \qquad x \geq 0.$$

The integrals $E(Z_{r:n})$ and $E(Z^2_{r:n})$ were evaluated using the trapezoidal rule over a finite length interval. The interval length and associated grid size were chosen such that all computational checks were satisfied. In both cases (a) and (b) the actual interval of integration always contained the interval (-10, 10). Case (b), for $\tau \geq 2$, required the largest regions of integration.

Values of $E(Z_{r:n}\,Z_{s:n})$ were obtained using Legendre-Gauss six-point composite quadrature on the inner integral and the trapezoidal rule on the outer integral. The overall region of integration and subregion definitions were adjusted, for differing $\lambda$ and $\tau$, to achieve accuracy within the six-decimal-place limit previously specified.

Four separate tables were prepared which give values obtained for mean, variance, and covariances in cases (a) and (b). Reduction in overall size of Tables 2 and 4, which are specific to case (b), was made possible by taking advantage of the symmetry prescribed by equations (9) and (10).

## REFERENCES

David, H. A. (1970). Order Statistics. Wiley, New York.

David, H. A. and Shu, V. S. (1977). Robustness of location estimators in the presence of an outlier. In: David, H. A. (Ed.), Contributions to Survey Sampling and Applied Statistics - Statistical Papers in Honor of H. O. Hartley. Academic Press, New York.

Dixon, W. J. (1950). Analysis of extreme values. Ann. Math. Statist. 21, 488-506.

Sarhan, A. E. and Greenberg, B. G. (1956). Estimation of location and scale parameters by order statistics from singly and doubly censored samples. Part I. The normal distribution up to samples of size 10. Ann. Math. Statist. 27, 427-51.

Teichroew, D. (1956). Tables of expected values of order statistics and products of order statistics for samples of size twenty and less from the normal distribution. Ann. Math. Statist. 27, 410-26.

Vaughan, R. J. and Venables, W. N. (1972). Permanent expressions for order statistic densities. J. Roy. Statist. Soc. B 34, 308-10.

## TABLES 1, 2, 3 and 4

Table 1. Means and Variances of Order Statistics in the Presence of One Normal Outlier Differing in Location

Table 2. Means and Variances of Order Statistics in the Presence of One Normal Outlier Differing in Standard Deviation

Table 3. Covariances of Order Statistics in the Presence of One Normal Outlier Differing in Location

Table 4. Covariances of Order Statistics in the Presence of One Normal Outlier Differing in Standard Deviation

### Notation in Tables

N = sample size including the outlier

R = rank of order statistic

S = rank of second order statistic

N-1 of the N variates are a random sample from a normal N(0, 1) population

TABLE 1

OUTLIER MEAN 0.5 OUTLIER STANDARD DEVIATION 1.0

| N | R | MEAN | VAR | N | R | MEAN | VAR | N | R | MEAN | VAR |
|---|---|---|---|---|---|---|---|---|---|---|---|
| 2 | 1 | -0.3491 | 0.7036 | 9 | 1 | -1.4467 | 0.3616 | 12 | 6 | -0.0634 | 0.1275 |
| 2 | 2 | 0.8491 | 0.7036 | 9 | 2 | -0.8891 | 0.2282 | 12 | 7 | 0.1438 | 0.1275 |
| 3 | 1 | -0.7126 | 0.5752 | 9 | 3 | -0.5251 | 0.1882 | 12 | 8 | 0.3557 | 0.1316 |
| 3 | 2 | 0.1629 | 0.4563 | 9 | 4 | -0.2243 | 0.1721 | 12 | 9 | 0.5829 | 0.1410 |
| 3 | 3 | 1.0497 | 0.5802 | 9 | 5 | 0.0537 | 0.1676 | 12 | 10 | 0.8421 | 0.1597 |
| 4 | 1 | -0.9335 | 0.5032 | 9 | 6 | 0.3320 | 0.1722 | 12 | 11 | 1.1695 | 0.2002 |
| 4 | 2 | -0.1838 | 0.3671 | 9 | 7 | 0.6340 | 0.1885 | 12 | 12 | 1.6915 | 0.3316 |
| 4 | 3 | 0.4281 | 0.3673 | 9 | 8 | 1.0006 | 0.2293 | 13 | 1 | -1.6425 | 0.3178 |
| 4 | 4 | 1.1891 | 0.5098 | 9 | 9 | 1.5649 | 0.3675 | 13 | 2 | -1.1357 | 0.1920 |
| 5 | 1 | -1.0886 | 0.4564 | 10 | 1 | -1.5046 | 0.3481 | 13 | 3 | -0.8195 | 0.1525 |
| 5 | 2 | -0.4088 | 0.3167 | 10 | 2 | -0.9631 | 0.2167 | 13 | 4 | -0.5707 | 0.1340 |
| 5 | 3 | 0.0971 | 0.2911 | 10 | 3 | -0.6147 | 0.1766 | 13 | 5 | -0.3545 | 0.1241 |
| 5 | 4 | 0.6051 | 0.3172 | 10 | 4 | -0.3317 | 0.1593 | 13 | 6 | -0.1551 | 0.1191 |
| 5 | 5 | 1.2952 | 0.4633 | 10 | 5 | -0.0758 | 0.1523 | 13 | 7 | 0.0371 | 0.1175 |
| 6 | 1 | -1.2067 | 0.4230 | 10 | 6 | 0.1724 | 0.1523 | 13 | 8 | 0.2294 | 0.1191 |
| 6 | 2 | -0.5724 | 0.2837 | 10 | 7 | 0.4288 | 0.1594 | 13 | 9 | 0.4292 | 0.1241 |
| 6 | 3 | -0.1246 | 0.2494 | 10 | 8 | 0.7131 | 0.1770 | 13 | 10 | 0.6461 | 0.1341 |
| 6 | 4 | 0.2865 | 0.2495 | 10 | 9 | 1.0639 | 0.2179 | 13 | 11 | 0.8959 | 0.1530 |
| 6 | 5 | 0.7369 | 0.2845 | 10 | 10 | 1.6117 | 0.3536 | 13 | 12 | 1.2143 | 0.1932 |
| 6 | 6 | 1.3804 | 0.4299 | 11 | 1 | -1.5558 | 0.3365 | 13 | 13 | 1.7261 | 0.3226 |
| 7 | 1 | -1.3014 | 0.3978 | 11 | 2 | -1.0276 | 0.2071 | 14 | 1 | -1.6799 | 0.3101 |
| 7 | 2 | -0.6995 | 0.2601 | 11 | 3 | -0.6919 | 0.1672 | 14 | 2 | -1.1819 | 0.1858 |
| 7 | 3 | -0.2892 | 0.2223 | 11 | 4 | -0.4227 | 0.1491 | 14 | 3 | -0.8733 | 0.1468 |
| 7 | 4 | 0.0691 | 0.2128 | 11 | 5 | -0.1834 | 0.1407 | 14 | 4 | -0.6323 | 0.1281 |
| 7 | 5 | 0.4282 | 0.2225 | 11 | 6 | 0.0439 | 0.1382 | 14 | 5 | -0.4247 | 0.1178 |
| 7 | 6 | 0.8414 | 0.2611 | 11 | 7 | 0.2713 | 0.1407 | 14 | 6 | -0.2350 | 0.1122 |
| 7 | 7 | 1.4514 | 0.4043 | 11 | 8 | 0.5113 | 0.1493 | 14 | 7 | -0.0545 | 0.1097 |
| 8 | 1 | -1.3798 | 0.3778 | 11 | 9 | 0.7817 | 0.1675 | 14 | 8 | 0.1233 | 0.1097 |
| 8 | 2 | -0.8027 | 0.2423 | 11 | 10 | 1.1197 | 0.2083 | 14 | 9 | 0.3041 | 0.1122 |
| 8 | 3 | -0.4189 | 0.2029 | 11 | 11 | 1.6536 | 0.3418 | 14 | 10 | 0.4942 | 0.1179 |
| 8 | 4 | -0.0943 | 0.1891 | 12 | 1 | -1.6014 | 0.3266 | 14 | 11 | 0.7025 | 0.1283 |
| 8 | 5 | 0.2153 | 0.1891 | 12 | 2 | -1.0847 | 0.1990 | 14 | 12 | 0.9445 | 0.1472 |
| 8 | 6 | 0.5409 | 0.2032 | 12 | 3 | -0.7595 | 0.1593 | 14 | 13 | 1.2551 | 0.1870 |
| 8 | 7 | 0.9275 | 0.2434 | 12 | 4 | -0.5015 | 0.1409 | 14 | 14 | 1.7578 | 0.3146 |
| 8 | 8 | 1.5120 | 0.3841 | 12 | 5 | -0.2750 | 0.1315 | 15 | 1 | -1.7141 | 0.3032 |

TABLE 1

OUTLIER MEAN 0.5 OUTLIER STANDARD DEVIATION 1.0

| N | R | MEAN | VAR | N | R | MEAN | VAR | N | R | MEAN | VAR |
|---|---|---|---|---|---|---|---|---|---|---|---|
| 15 | 2 | -1.2238 | 0.1804 | 17 | 6 | -0.4259 | 0.0974 | 19 | 6 | -0.5255 | 0.0905 |
| 15 | 3 | -0.9220 | 0.1417 | 17 | 7 | -0.2688 | 0.0934 | 19 | 7 | -0.3786 | 0.0860 |
| 15 | 4 | -0.6878 | 0.1230 | 17 | 8 | -0.1186 | 0.0912 | 19 | 8 | -0.2400 | 0.0832 |
| 15 | 5 | -0.4873 | 0.1125 | 17 | 9 | 0.0283 | 0.0905 | 19 | 9 | -0.1062 | 0.0817 |
| 15 | 6 | -0.3057 | 0.1065 | 17 | 10 | 0.1754 | 0.0912 | 19 | 10 | 0.0253 | 0.0812 |
| 15 | 7 | -0.1344 | 0.1033 | 17 | 11 | 0.3257 | 0.0934 | 19 | 11 | 0.1569 | 0.0817 |
| 15 | 8 | 0.0321 | 0.1023 | 17 | 12 | 0.4831 | 0.0974 | 19 | 12 | 0.2908 | 0.0832 |
| 15 | 9 | 0.1988 | 0.1033 | 17 | 13 | 0.6526 | 0.1041 | 19 | 13 | 0.4297 | 0.0861 |
| 15 | 10 | 0.3702 | 0.1065 | 17 | 14 | 0.8422 | 0.1149 | 19 | 14 | 0.5768 | 0.0905 |
| 15 | 11 | 0.5523 | 0.1126 | 17 | 15 | 1.0664 | 0.1337 | 19 | 15 | 0.7369 | 0.0974 |
| 15 | 12 | 0.7534 | 0.1232 | 17 | 16 | 1.3588 | 0.1723 | 19 | 16 | 0.9176 | 0.1083 |
| 15 | 13 | 0.9887 | 0.1421 | 17 | 17 | 1.8399 | 0.2953 | 19 | 17 | 1.1331 | 0.1269 |
| 15 | 14 | 1.2924 | 0.1816 | 18 | 1 | -1.8022 | 0.2863 | 19 | 18 | 1.4163 | 0.1648 |
| 15 | 15 | 1.7872 | 0.3075 | 18 | 2 | -1.3308 | 0.1673 | 19 | 19 | 1.8862 | 0.2852 |
| 16 | 1 | -1.7457 | 0.2970 | 18 | 3 | -1.0449 | 0.1297 | 20 | 1 | -1.8516 | 0.2772 |
| 16 | 2 | -1.2623 | 0.1756 | 18 | 4 | -0.8262 | 0.1112 | 20 | 2 | -1.3902 | 0.1605 |
| 16 | 3 | -0.9664 | 0.1372 | 18 | 5 | -0.6419 | 0.1004 | 20 | 3 | -1.1124 | 0.1235 |
| 16 | 4 | -0.7380 | 0.1186 | 18 | 6 | -0.4779 | 0.0937 | 20 | 4 | -0.9016 | 0.1052 |
| 16 | 5 | -0.5437 | 0.1080 | 18 | 7 | -0.3263 | 0.0895 | 20 | 5 | -0.7252 | 0.0945 |
| 16 | 6 | -0.3689 | 0.1016 | 18 | 8 | -0.1823 | 0.0869 | 20 | 6 | -0.5694 | 0.0876 |
| 16 | 7 | -0.2053 | 0.0979 | 18 | 9 | -0.0425 | 0.0857 | 20 | 7 | -0.4267 | 0.0830 |
| 16 | 8 | -0.0477 | 0.0962 | 18 | 10 | 0.0960 | 0.0857 | 20 | 8 | -0.2926 | 0.0800 |
| 16 | 9 | 0.1080 | 0.0962 | 18 | 11 | 0.2359 | 0.0869 | 20 | 9 | -0.1640 | 0.0781 |
| 16 | 10 | 0.2657 | 0.0979 | 18 | 12 | 0.3800 | 0.0895 | 20 | 10 | -0.0383 | 0.0773 |
| 16 | 11 | 0.4295 | 0.1016 | 18 | 13 | 0.5319 | 0.0938 | 20 | 11 | 0.0864 | 0.0773 |
| 16 | 12 | 0.6048 | 0.1081 | 18 | 14 | 0.6965 | 0.1006 | 20 | 12 | 0.2122 | 0.0782 |
| 16 | 13 | 0.7997 | 0.1188 | 18 | 15 | 0.8814 | 0.1114 | 20 | 13 | 0.3410 | 0.0800 |
| 16 | 14 | 1.0291 | 0.1377 | 18 | 16 | 1.1010 | 0.1301 | 20 | 14 | 0.4753 | 0.0830 |
| 16 | 15 | 1.3269 | 0.1767 | 18 | 17 | 1.3885 | 0.1684 | 20 | 15 | 0.6182 | 0.0876 |
| 16 | 16 | 1.8145 | 0.3011 | 18 | 18 | 1.8637 | 0.2900 | 20 | 16 | 0.7745 | 0.0946 |
| 17 | 1 | -1.7750 | 0.2914 | 19 | 1 | -1.8277 | 0.2815 | 20 | 17 | 0.9515 | 0.1054 |
| 17 | 2 | -1.2979 | 0.1712 | 19 | 2 | -1.3615 | 0.1637 | 20 | 18 | 1.1632 | 0.1239 |
| 17 | 3 | -1.0072 | 0.1332 | 19 | 3 | -1.0798 | 0.1264 | 20 | 19 | 1.4424 | 0.1615 |
| 17 | 4 | -0.7840 | 0.1147 | 19 | 4 | -0.8653 | 0.1080 | 20 | 20 | 1.9073 | 0.2807 |
| 17 | 5 | -0.5950 | 0.1040 | 19 | 5 | -0.6852 | 0.0973 | | | | |

TABLE 1

OUTLIER MEAN 1.0 OUTLIER STANDARD DEVIATION 1.0

| N | R | MEAN | VAR | N | R | MEAN | VAR | N | R | MEAN | VAR |
|---|---|---|---|---|---|---|---|---|---|---|---|
| 2 | 1 | -0.1996 | 0.7605 | 9 | 1 | -1.4310 | 0.3675 | 12 | 6 | -0.0333 | 0.1298 |
| 2 | 2 | 1.1996 | 0.7605 | 9 | 2 | -0.8666 | 0.2328 | 12 | 7 | 0.1788 | 0.1300 |
| 3 | 1 | -0.6343 | 0.6083 | 9 | 3 | -0.4962 | 0.1924 | 12 | 8 | 0.3966 | 0.1344 |
| 3 | 2 | 0.3051 | 0.4865 | 9 | 4 | -0.1885 | 0.1761 | 12 | 9 | 0.6314 | 0.1445 |
| 3 | 3 | 1.3292 | 0.6431 | 9 | 5 | 0.0974 | 0.1717 | 12 | 10 | 0.9012 | 0.1648 |
| 4 | 1 | -0.8829 | 0.5246 | 9 | 6 | 0.3857 | 0.1769 | 12 | 11 | 1.2459 | 0.2101 |
| 4 | 2 | -0.1003 | 0.3848 | 9 | 7 | 0.7014 | 0.1948 | 12 | 12 | 1.8100 | 0.3689 |
| 4 | 3 | 0.5580 | 0.3875 | 9 | 8 | 1.0901 | 0.2409 | 13 | 1 | -1.6331 | 0.3211 |
| 4 | 4 | 1.4253 | 0.5707 | 9 | 9 | 1.7078 | 0.4115 | 13 | 2 | -1.1227 | 0.1945 |
| 5 | 1 | -1.0523 | 0.4714 | 10 | 1 | -1.4911 | 0.3530 | 13 | 3 | -0.8032 | 0.1548 |
| 5 | 2 | -0.3518 | 0.3289 | 10 | 2 | -0.9439 | 0.2206 | 13 | 4 | -0.5514 | 0.1360 |
| 5 | 3 | 0.1786 | 0.3032 | 10 | 3 | -0.5903 | 0.1801 | 13 | 5 | -0.3320 | 0.1261 |
| 5 | 4 | 0.7239 | 0.3338 | 10 | 4 | -0.3019 | 0.1626 | 13 | 6 | -0.1291 | 0.1211 |
| 5 | 5 | 1.5017 | 0.5205 | 10 | 5 | -0.0400 | 0.1556 | 13 | 7 | 0.0670 | 0.1196 |
| 6 | 1 | -1.1789 | 0.4343 | 10 | 6 | 0.2154 | 0.1558 | 13 | 8 | 0.2638 | 0.1213 |
| 6 | 2 | -0.5301 | 0.2928 | 10 | 7 | 0.4807 | 0.1635 | 13 | 9 | 0.4691 | 0.1267 |
| 6 | 3 | -0.0668 | 0.2580 | 10 | 8 | 0.7774 | 0.1828 | 13 | 10 | 0.6930 | 0.1374 |
| 6 | 4 | 0.3644 | 0.2589 | 10 | 9 | 1.1485 | 0.2288 | 13 | 11 | 0.9528 | 0.1578 |
| 6 | 5 | 0.8464 | 0.2992 | 10 | 10 | 1.7452 | 0.3951 | 13 | 12 | 1.2874 | 0.2026 |
| 6 | 6 | 1.5650 | 0.4833 | 11 | 1 | -1.5439 | 0.3408 | 13 | 13 | 1.8385 | 0.3581 |
| 7 | 1 | -1.2790 | 0.4066 | 11 | 2 | -1.0109 | 0.2104 | 14 | 1 | -1.6714 | 0.3130 |
| 7 | 2 | -0.6664 | 0.2672 | 11 | 3 | -0.6709 | 0.1701 | 14 | 2 | -1.1701 | 0.1881 |
| 7 | 3 | -0.2452 | 0.2288 | 11 | 4 | -0.3973 | 0.1519 | 14 | 3 | -0.8588 | 0.1487 |
| 7 | 4 | 0.1261 | 0.2194 | 11 | 5 | -0.1532 | 0.1434 | 14 | 4 | -0.6152 | 0.1299 |
| 7 | 5 | 0.5025 | 0.2304 | 11 | 6 | 0.0794 | 0.1410 | 14 | 5 | -0.4048 | 0.1196 |
| 7 | 6 | 0.9431 | 0.2745 | 11 | 7 | 0.3132 | 0.1438 | 14 | 6 | -0.2123 | 0.1139 |
| 7 | 7 | 1.6190 | 0.4542 | 11 | 8 | 0.5614 | 0.1530 | 14 | 7 | -0.0285 | 0.1114 |
| 8 | 1 | -1.3614 | 0.3849 | 11 | 9 | 0.8433 | 0.1729 | 14 | 8 | 0.1529 | 0.1115 |
| 8 | 2 | -0.7758 | 0.2479 | 11 | 10 | 1.2000 | 0.2187 | 14 | 9 | 0.3378 | 0.1142 |
| 8 | 3 | -0.3837 | 0.2080 | 11 | 11 | 1.7790 | 0.3811 | 14 | 10 | 0.5330 | 0.1203 |
| 8 | 4 | -0.0500 | 0.1941 | 12 | 1 | -1.5909 | 0.3303 | 14 | 11 | 0.7480 | 0.1313 |
| 8 | 5 | 0.2707 | 0.1945 | 12 | 2 | -1.0700 | 0.2019 | 14 | 12 | 0.9993 | 0.1518 |
| 8 | 6 | 0.6116 | 0.2101 | 12 | 3 | -0.7411 | 0.1618 | 14 | 13 | 1.3251 | 0.1961 |
| 8 | 7 | 1.0226 | 0.2557 | 12 | 4 | -0.4795 | 0.1432 | 14 | 14 | 1.8649 | 0.3485 |
| 8 | 8 | 1.6661 | 0.4308 | 12 | 5 | -0.2491 | 0.1338 | 15 | 1 | -1.7064 | 0.3058 |

TABLE 1

OUTLIER MEAN 1.0 OUTLIER STANDARD DEVIATION 1.0

| N | R | MEAN | VAR | N | R | MEAN | VAR | N | R | MEAN | VAR |
|---|---|---|---|---|---|---|---|---|---|---|---|
| 15 | 2 | -1.2132 | 0.1824 | 17 | 6 | -0.4096 | 0.0986 | 19 | 6 | -0.5119 | 0.0914 |
| 15 | 3 | -0.9089 | 0.1434 | 17 | 7 | -0.2505 | 0.0945 | 19 | 7 | -0.3635 | 0.0870 |
| 15 | 4 | -0.6724 | 0.1246 | 17 | 8 | -0.0963 | 0.0924 | 19 | 8 | -0.2233 | 0.0842 |
| 15 | 5 | -0.4696 | 0.1141 | 17 | 9 | 0.0510 | 0.0917 | 19 | 9 | -0.0878 | 0.0826 |
| 15 | 6 | -0.2855 | 0.1080 | 17 | 10 | 0.2006 | 0.0925 | 19 | 10 | 0.0456 | 0.0821 |
| 15 | 7 | -0.1116 | 0.1048 | 17 | 11 | 0.3539 | 0.0948 | 19 | 11 | 0.1792 | 0.0827 |
| 15 | 8 | 0.0579 | 0.1038 | 17 | 12 | 0.5148 | 0.0990 | 19 | 12 | 0.3154 | 0.0843 |
| 15 | 9 | 0.2279 | 0.1049 | 17 | 13 | 0.6886 | 0.1060 | 19 | 13 | 0.4569 | 0.0873 |
| 15 | 10 | 0.4033 | 0.1083 | 17 | 14 | 0.8638 | 0.1175 | 19 | 14 | 0.6072 | 0.0920 |
| 15 | 11 | 0.5902 | 0.1148 | 17 | 15 | 1.1159 | 0.1378 | 19 | 15 | 0.7713 | 0.0992 |
| 15 | 12 | 0.7975 | 0.1261 | 17 | 16 | 1.4213 | 0.1805 | 19 | 16 | 0.9571 | 0.1107 |
| 15 | 13 | 1.0416 | 0.1466 | 17 | 17 | 1.9339 | 0.3252 | 19 | 17 | 1.1798 | 0.1307 |
| 15 | 14 | 1.3597 | 0.1903 | 18 | 1 | -1.7961 | 0.2882 | 19 | 18 | 1.4748 | 0.1724 |
| 15 | 15 | 1.8894 | 0.3399 | 18 | 2 | -1.3226 | 0.1688 | 19 | 19 | 1.9734 | 0.3128 |
| 16 | 1 | -1.7386 | 0.2994 | 18 | 3 | -1.0349 | 0.1309 | 20 | 1 | -1.8463 | 0.2789 |
| 16 | 2 | -1.2526 | 0.1774 | 18 | 4 | -0.8146 | 0.1124 | 20 | 2 | -1.3831 | 0.1617 |
| 16 | 3 | -0.9545 | 0.1388 | 18 | 5 | -0.6287 | 0.1015 | 20 | 3 | -1.1039 | 0.1245 |
| 16 | 4 | -0.7241 | 0.1200 | 18 | 6 | -0.4630 | 0.0948 | 20 | 4 | -0.8917 | 0.1062 |
| 16 | 5 | -0.5278 | 0.1093 | 18 | 7 | -0.3097 | 0.0905 | 20 | 5 | -0.7140 | 0.0954 |
| 16 | 6 | -0.3508 | 0.1029 | 18 | 8 | -0.1639 | 0.0880 | 20 | 6 | -0.5569 | 0.0884 |
| 16 | 7 | -0.1850 | 0.0992 | 18 | 9 | -0.0222 | 0.0868 | 20 | 7 | -0.4129 | 0.0838 |
| 16 | 8 | -0.0250 | 0.0976 | 18 | 10 | 0.1185 | 0.0868 | 20 | 8 | -0.2774 | 0.0808 |
| 16 | 9 | 0.1335 | 0.0976 | 18 | 11 | 0.2608 | 0.0881 | 20 | 9 | -0.1472 | 0.0790 |
| 16 | 10 | 0.2943 | 0.0994 | 18 | 12 | 0.4078 | 0.0908 | 20 | 10 | -0.0199 | 0.0781 |
| 16 | 11 | 0.4619 | 0.1034 | 18 | 13 | 0.5630 | 0.0953 | 20 | 11 | 0.1065 | 0.0782 |
| 16 | 12 | 0.6417 | 0.1101 | 18 | 14 | 0.7316 | 0.1024 | 20 | 12 | 0.2342 | 0.0791 |
| 16 | 13 | 0.8426 | 0.1216 | 18 | 15 | 0.9219 | 0.1140 | 20 | 13 | 0.3652 | 0.0810 |
| 16 | 14 | 1.0803 | 0.1419 | 18 | 16 | 1.1490 | 0.1341 | 20 | 14 | 0.5021 | 0.0842 |
| 16 | 15 | 1.3917 | 0.1851 | 18 | 17 | 1.4489 | 0.1763 | 20 | 15 | 0.6481 | 0.0890 |
| 16 | 16 | 1.9123 | 0.3322 | 18 | 18 | 1.9542 | 0.3187 | 20 | 16 | 0.8081 | 0.0963 |
| 17 | 1 | -1.7684 | 0.2935 | 19 | 1 | -1.8220 | 0.2833 | 20 | 17 | 0.9899 | 0.1078 |
| 17 | 2 | -1.2889 | 0.1729 | 19 | 2 | -1.3538 | 0.1651 | 20 | 18 | 1.2085 | 0.1277 |
| 17 | 3 | -0.9963 | 0.1346 | 19 | 3 | -1.0706 | 0.1276 | 20 | 19 | 1.4991 | 0.1689 |
| 17 | 4 | -0.7713 | 0.1160 | 19 | 4 | -0.8546 | 0.1091 | 20 | 20 | 1.9915 | 0.3074 |
| 17 | 5 | -0.5805 | 0.1052 | 19 | 5 | -0.6730 | 0.0983 | | | | |

TABLE 1

OUTLIER MEAN 1.5    OUTLIER STANDARD DEVIATION 1.0

| N | R | MEAN | VAR | N | R | MEAN | VAR | N | R | MEAN | VAR |
|---|---|---|---|---|---|---|---|---|---|---|---|
| 2 | 1 | -0.1048 | 0.8318 | 9 | 1 | -1.4256 | 0.3709 | 12 | 6 | -0.0147 | 0.1327 |
| 2 | 2 | 1.6048 | 0.8318 | 9 | 2 | -0.8569 | 0.2365 | 12 | 7 | 0.2028 | 0.1332 |
| 3 | 1 | -0.5936 | 0.6403 | 9 | 3 | -0.4816 | 0.1964 | 12 | 8 | 0.4278 | 0.1383 |
| 3 | 2 | 0.4134 | 0.5293 | 9 | 4 | -0.1677 | 0.1807 | 12 | 9 | 0.6728 | 0.1496 |
| 3 | 3 | 1.6802 | 0.7340 | 9 | 5 | 0.1265 | 0.1771 | 12 | 10 | 0.9591 | 0.1725 |
| 4 | 1 | -0.8598 | 0.5423 | 9 | 6 | 0.4266 | 0.1835 | 12 | 11 | 1.3357 | 0.2263 |
| 4 | 2 | -0.0479 | 0.4069 | 9 | 7 | 0.7613 | 0.2043 | 12 | 12 | 2.0045 | 0.4518 |
| 4 | 3 | 0.6680 | 0.4191 | 9 | 8 | 1.1874 | 0.2593 | 13 | 1 | -1.6302 | 0.3228 |
| 4 | 4 | 1.7396 | 0.6676 | 9 | 9 | 1.9301 | 0.5017 | 13 | 2 | -1.1177 | 0.1962 |
| 5 | 1 | -1.0371 | 0.4826 | 10 | 1 | -1.4866 | 0.3558 | 13 | 3 | -0.7962 | 0.1565 |
| 5 | 2 | -0.3201 | 0.3422 | 10 | 2 | -0.9359 | 0.2235 | 13 | 4 | -0.5420 | 0.1379 |
| 5 | 3 | 0.2359 | 0.3201 | 10 | 3 | -0.5785 | 0.1833 | 13 | 5 | -0.3200 | 0.1281 |
| 5 | 4 | 0.8324 | 0.3600 | 10 | 4 | -0.2855 | 0.1661 | 13 | 6 | -0.1138 | 0.1234 |
| 5 | 5 | 1.7889 | 0.6185 | 10 | 5 | -0.0178 | 0.1596 | 13 | 7 | 0.0863 | 0.1222 |
| 6 | 1 | -1.1680 | 0.4420 | 10 | 6 | 0.2454 | 0.1606 | 13 | 8 | 0.2883 | 0.1243 |
| 6 | 2 | -0.5086 | 0.3016 | 10 | 7 | 0.5220 | 0.1696 | 13 | 9 | 0.5005 | 0.1303 |
| 6 | 3 | -0.0307 | 0.2685 | 10 | 8 | 0.8368 | 0.1915 | 13 | 10 | 0.7343 | 0.1422 |
| 6 | 4 | 0.4239 | 0.2727 | 10 | 9 | 1.2431 | 0.2464 | 13 | 11 | 1.0099 | 0.1652 |
| 6 | 5 | 0.9524 | 0.3221 | 10 | 10 | 1.9570 | 0.4828 | 13 | 12 | 1.3749 | 0.2184 |
| 6 | 6 | 1.8310 | 0.5802 | 11 | 1 | -1.5401 | 0.3431 | 13 | 13 | 2.0257 | 0.4389 |
| 7 | 1 | -1.2707 | 0.4122 | 11 | 2 | -1.0043 | 0.2128 | 14 | 1 | -1.6688 | 0.3145 |
| 7 | 2 | -0.6507 | 0.2734 | 11 | 3 | -0.6612 | 0.1727 | 14 | 2 | -1.1658 | 0.1895 |
| 7 | 3 | -0.2200 | 0.2360 | 11 | 4 | -0.3840 | 0.1547 | 14 | 3 | -0.8526 | 0.1502 |
| 7 | 4 | 0.1647 | 0.2282 | 11 | 5 | -0.1356 | 0.1466 | 14 | 4 | -0.6071 | 0.1315 |
| 7 | 5 | 0.5627 | 0.2422 | 11 | 6 | 0.1026 | 0.1446 | 14 | 5 | -0.3945 | 0.1213 |
| 7 | 6 | 1.0461 | 0.2954 | 11 | 7 | 0.3440 | 0.1481 | 14 | 6 | -0.1994 | 0.1158 |
| 7 | 7 | 1.8678 | 0.5492 | 11 | 8 | 0.6029 | 0.1585 | 14 | 7 | -0.0125 | 0.1135 |
| 8 | 1 | -1.3548 | 0.3893 | 11 | 9 | 0.9019 | 0.1811 | 14 | 8 | 0.1728 | 0.1139 |
| 8 | 2 | -0.7637 | 0.2526 | 11 | 10 | 1.2921 | 0.2355 | 14 | 9 | 0.3627 | 0.1170 |
| 8 | 3 | -0.3650 | 0.2133 | 11 | 11 | 1.9817 | 0.4664 | 14 | 10 | 0.5646 | 0.1237 |
| 8 | 4 | -0.0225 | 0.2002 | 12 | 1 | -1.5875 | 0.3323 | 14 | 11 | 0.7891 | 0.1359 |
| 8 | 5 | 0.3108 | 0.2021 | 12 | 2 | -1.0643 | 0.2039 | 14 | 12 | 1.0555 | 0.1589 |
| 8 | 6 | 0.6718 | 0.2206 | 12 | 3 | -0.7329 | 0.1639 | 14 | 13 | 1.4106 | 0.2114 |
| 8 | 7 | 1.1227 | 0.2752 | 12 | 4 | -0.4684 | 0.1455 | 14 | 14 | 2.0455 | 0.4272 |
| 8 | 8 | 1.9006 | 0.5235 | 12 | 5 | -0.2347 | 0.1364 | 15 | 1 | -1.7041 | 0.3072 |

TABLE 1

OUTLIER MEAN 1.5 OUTLIER STANDARD DEVIATION 1.0

| N | R | MEAN | VAR | N | R | MEAN | VAR | N | R | MEAN | VAR |
|---|---|---|---|---|---|---|---|---|---|---|---|
| 15 | 2 | -1.2094 | 0.1837 | 17 | 6 | -0.4012 | 0.0998 | 19 | 6 | -0.5052 | 0.0923 |
| 15 | 3 | -0.9035 | 0.1447 | 17 | 7 | -0.2404 | 0.0958 | 19 | 7 | -0.3556 | 0.0879 |
| 15 | 4 | -0.6653 | 0.1260 | 17 | 8 | -0.0861 | 0.0937 | 19 | 8 | -0.2139 | 0.0852 |
| 15 | 5 | -0.4607 | 0.1156 | 17 | 9 | 0.0655 | 0.0932 | 19 | 9 | -0.0768 | 0.0837 |
| 15 | 6 | -0.2745 | 0.1096 | 17 | 10 | 0.2180 | 0.0942 | 19 | 10 | 0.0585 | 0.0833 |
| 15 | 7 | -0.0981 | 0.1065 | 17 | 11 | 0.3748 | 0.0967 | 19 | 11 | 0.1943 | 0.0840 |
| 15 | 8 | 0.0745 | 0.1057 | 17 | 12 | 0.5402 | 0.1013 | 19 | 12 | 0.3332 | 0.0858 |
| 15 | 9 | 0.2482 | 0.1071 | 17 | 13 | 0.7201 | 0.1089 | 19 | 13 | 0.4781 | 0.0890 |
| 15 | 10 | 0.4284 | 0.1109 | 17 | 14 | 0.9240 | 0.1215 | 19 | 14 | 0.6327 | 0.0940 |
| 15 | 11 | 0.6218 | 0.1180 | 17 | 15 | 1.1697 | 0.1442 | 19 | 15 | 0.8025 | 0.1019 |
| 15 | 12 | 0.8383 | 0.1304 | 17 | 16 | 1.5012 | 0.1947 | 19 | 16 | 0.9965 | 0.1144 |
| 15 | 13 | 1.0970 | 0.1534 | 17 | 17 | 2.0981 | 0.3982 | 19 | 17 | 1.2320 | 0.1368 |
| 15 | 14 | 1.4432 | 0.2052 | 18 | 1 | -1.7944 | 0.2891 | 19 | 18 | 1.5516 | 0.1860 |
| 15 | 15 | 2.0641 | 0.4166 | 18 | 2 | -1.3198 | 0.1697 | 19 | 19 | 2.1287 | 0.3825 |
| 16 | 1 | -1.7365 | 0.3005 | 18 | 3 | -1.0310 | 0.1318 | 20 | 1 | -1.8449 | 0.2796 |
| 16 | 2 | -1.2492 | 0.1785 | 18 | 4 | -0.8097 | 0.1133 | 20 | 2 | -1.3807 | 0.1624 |
| 16 | 3 | -0.9497 | 0.1399 | 18 | 5 | -0.6226 | 0.1025 | 20 | 3 | -1.1007 | 0.1253 |
| 16 | 4 | -0.7179 | 0.1212 | 18 | 6 | -0.4556 | 0.0958 | 20 | 4 | -0.8876 | 0.1069 |
| 16 | 5 | -0.5200 | 0.1106 | 18 | 7 | -0.3008 | 0.0916 | 20 | 5 | -0.7090 | 0.0962 |
| 16 | 6 | -0.3412 | 0.1043 | 18 | 8 | -0.1533 | 0.0891 | 20 | 6 | -0.5509 | 0.0893 |
| 16 | 7 | -0.1733 | 0.1007 | 18 | 9 | -0.0096 | 0.0881 | 20 | 7 | -0.4058 | 0.0847 |
| 16 | 8 | -0.0109 | 0.0992 | 18 | 10 | 0.1334 | 0.0882 | 20 | 8 | -0.2691 | 0.0817 |
| 16 | 9 | 0.1505 | 0.0994 | 18 | 11 | 0.2785 | 0.0897 | 20 | 9 | -0.1375 | 0.0800 |
| 16 | 10 | 0.3150 | 0.1015 | 18 | 12 | 0.4289 | 0.0926 | 20 | 10 | -0.0086 | 0.0792 |
| 16 | 11 | 0.4872 | 0.1058 | 18 | 13 | 0.5885 | 0.0975 | 20 | 11 | 0.1197 | 0.0793 |
| 16 | 12 | 0.6733 | 0.1132 | 18 | 14 | 0.7630 | 0.1052 | 20 | 12 | 0.2496 | 0.0804 |
| 16 | 13 | 0.8831 | 0.1257 | 18 | 15 | 0.9616 | 0.1178 | 20 | 13 | 0.3832 | 0.0825 |
| 16 | 14 | 1.1348 | 0.1485 | 18 | 16 | 1.2020 | 0.1403 | 20 | 14 | 0.5234 | 0.0858 |
| 16 | 15 | 1.4733 | 0.1997 | 18 | 17 | 1.5272 | 0.1901 | 20 | 15 | 0.6735 | 0.0910 |
| 16 | 16 | 2.0816 | 0.4070 | 18 | 18 | 2.1138 | 0.3900 | 20 | 16 | 0.8391 | 0.0989 |
| 17 | 1 | -1.7665 | 0.2946 | 19 | 1 | -1.8205 | 0.2842 | 20 | 17 | 1.0289 | 0.1114 |
| 17 | 2 | -1.2859 | 0.1738 | 19 | 2 | -1.3513 | 0.1659 | 20 | 18 | 1.2600 | 0.1336 |
| 17 | 3 | -0.9921 | 0.1357 | 19 | 3 | -1.0671 | 0.1284 | 20 | 19 | 1.5744 | 0.1822 |
| 17 | 4 | -0.7658 | 0.1170 | 19 | 4 | -0.8501 | 0.1099 | 20 | 20 | 2.1429 | 0.3756 |
| 17 | 5 | -0.5737 | 0.1063 | 19 | 5 | -0.6675 | 0.0992 | | | | |

TABLE 1

OUTLIER MEAN 2.0 OUTLIER STANDARD DEVIATION 1.0

| N | R | MEAN | VAR | N | R | MEAN | VAR | N | R | MEAN | VAR |
|---|---|---|---|---|---|---|---|---|---|---|---|
| 2 | 1 | -0.0503 | 0.8970 | 9 | 1 | -1.4241 | 0.3723 | 12 | 6 | -0.0054 | 0.1350 |
| 2 | 2 | 2.0503 | 0.8970 | 9 | 2 | -0.8535 | 0.2384 | 12 | 7 | 0.2160 | 0.1362 |
| 3 | 1 | -0.5751 | 0.6622 | 9 | 3 | -0.4756 | 0.1990 | 12 | 8 | 0.4469 | 0.1423 |
| 3 | 2 | 0.4855 | 0.5773 | 9 | 4 | -0.1579 | 0.1842 | 12 | 9 | 0.7015 | 0.1553 |
| 3 | 3 | 2.0896 | 0.8277 | 9 | 5 | 0.1422 | 0.1818 | 12 | 10 | 1.0052 | 0.1818 |
| 4 | 1 | -0.8506 | 0.5526 | 9 | 6 | 0.4520 | 0.1904 | 12 | 11 | 1.4227 | 0.2469 |
| 4 | 2 | -0.0199 | 0.4260 | 9 | 7 | 0.8051 | 0.2154 | 12 | 12 | 2.2841 | 0.5773 |
| 4 | 3 | 0.7482 | 0.4577 | 9 | 8 | 1.2755 | 0.2826 | 13 | 1 | -1.6294 | 0.3234 |
| 4 | 4 | 2.1223 | 0.7760 | 9 | 9 | 2.2362 | 0.6280 | 13 | 2 | -1.1162 | 0.1969 |
| 5 | 1 | -1.0316 | 0.4884 | 10 | 1 | -1.4854 | 0.3569 | 13 | 3 | -0.7937 | 0.1575 |
| 5 | 2 | -0.3054 | 0.3519 | 10 | 2 | -0.9333 | 0.2250 | 13 | 4 | -0.5384 | 0.1391 |
| 5 | 3 | 0.2698 | 0.3366 | 10 | 3 | -0.5739 | 0.1852 | 13 | 5 | -0.3147 | 0.1296 |
| 5 | 4 | 0.9167 | 0.3929 | 10 | 4 | -0.2782 | 0.1686 | 13 | 6 | -0.1065 | 0.1251 |
| 5 | 5 | 2.1504 | 0.7352 | 10 | 5 | -0.0066 | 0.1629 | 13 | 7 | 0.0964 | 0.1244 |
| 6 | 1 | -1.1643 | 0.4457 | 10 | 6 | 0.2625 | 0.1651 | 13 | 8 | 0.3024 | 0.1271 |
| 6 | 2 | -0.4995 | 0.3073 | 10 | 7 | 0.5488 | 0.1760 | 13 | 9 | 0.5205 | 0.1341 |
| 6 | 3 | -0.0120 | 0.2774 | 10 | 8 | 0.8817 | 0.2019 | 13 | 10 | 0.7637 | 0.1476 |
| 6 | 4 | 0.4617 | 0.2873 | 10 | 9 | 1.3310 | 0.2686 | 13 | 11 | 1.0565 | 0.1741 |
| 6 | 5 | 1.0388 | 0.3513 | 10 | 10 | 2.2533 | 0.6093 | 13 | 12 | 1.4614 | 0.2383 |
| 6 | 6 | 2.1753 | 0.7017 | 11 | 1 | -1.5390 | 0.3440 | 13 | 13 | 2.2981 | 0.5636 |
| 7 | 1 | -1.2681 | 0.4148 | 11 | 2 | -1.0021 | 0.2140 | 14 | 1 | -1.6681 | 0.3150 |
| 7 | 2 | -0.6445 | 0.2772 | 11 | 3 | -0.6575 | 0.1741 | 14 | 2 | -1.1645 | 0.1902 |
| 7 | 3 | -0.2081 | 0.2415 | 11 | 4 | -0.3784 | 0.1566 | 14 | 3 | -0.8506 | 0.1510 |
| 7 | 4 | 0.1863 | 0.2363 | 11 | 5 | -0.1272 | 0.1490 | 14 | 4 | -0.6041 | 0.1325 |
| 7 | 5 | 0.6032 | 0.2553 | 11 | 6 | 0.1149 | 0.1477 | 14 | 5 | -0.3903 | 0.1225 |
| 7 | 6 | 1.1337 | 0.3220 | 11 | 7 | 0.3622 | 0.1523 | 14 | 6 | -0.1935 | 0.1172 |
| 7 | 7 | 2.1975 | 0.6735 | 11 | 8 | 0.6307 | 0.1646 | 14 | 7 | -0.0045 | 0.1152 |
| 8 | 1 | -1.3528 | 0.3911 | 11 | 9 | 0.9475 | 0.1909 | 14 | 8 | 0.1836 | 0.1160 |
| 8 | 2 | -0.7592 | 0.2552 | 11 | 10 | 1.3796 | 0.2568 | 14 | 9 | 0.3775 | 0.1197 |
| 8 | 3 | -0.3568 | 0.2170 | 11 | 11 | 2.2693 | 0.5925 | 14 | 10 | 0.5852 | 0.1273 |
| 8 | 4 | -0.0085 | 0.2053 | 12 | 1 | -1.5867 | 0.3330 | 14 | 11 | 0.8190 | 0.1410 |
| 8 | 5 | 0.3345 | 0.2095 | 12 | 2 | -1.0625 | 0.2048 | 14 | 12 | 1.1024 | 0.1674 |
| 8 | 6 | 0.7142 | 0.2325 | 12 | 3 | -0.7300 | 0.1651 | 14 | 13 | 1.4965 | 0.2308 |
| 8 | 7 | 1.2108 | 0.2999 | 12 | 4 | -0.4639 | 0.1470 | 14 | 14 | 2.3113 | 0.5510 |
| 8 | 8 | 2.2177 | 0.6492 | 12 | 5 | -0.2282 | 0.1382 | 15 | 1 | -1.7035 | 0.3076 |

TABLE 1

OUTLIER MEAN 2.0 OUTLIER STANDARD DEVIATION 1.0

| N | R | MEAN | VAR | N | R | MEAN | VAR | N | R | MEAN | VAR |
|---|---|---|---|---|---|---|---|---|---|---|---|
| 15 | 2 | -1.2082 | 0.1842 | 17 | 6 | -0.3977 | 0.1005 | 19 | 6 | -0.5026 | 0.0929 |
| 15 | 3 | -0.9017 | 0.1454 | 17 | 7 | -0.2359 | 0.0967 | 19 | 7 | -0.3523 | 0.0886 |
| 15 | 4 | -0.6628 | 0.1268 | 17 | 8 | -0.0804 | 0.0948 | 19 | 8 | -0.2098 | 0.0859 |
| 15 | 5 | -0.4571 | 0.1165 | 17 | 9 | 0.0729 | 0.0945 | 19 | 9 | -0.0716 | 0.0846 |
| 15 | 6 | -0.2696 | 0.1107 | 17 | 10 | 0.2275 | 0.0957 | 19 | 10 | 0.0650 | 0.0843 |
| 15 | 7 | -0.0916 | 0.1079 | 17 | 11 | 0.3872 | 0.0986 | 19 | 11 | 0.2025 | 0.0852 |
| 15 | 8 | 0.0830 | 0.1074 | 17 | 12 | 0.5566 | 0.1037 | 19 | 12 | 0.3436 | 0.0873 |
| 15 | 9 | 0.2596 | 0.1092 | 17 | 13 | 0.7422 | 0.1121 | 19 | 13 | 0.4913 | 0.0908 |
| 15 | 10 | 0.4433 | 0.1135 | 17 | 14 | 0.9551 | 0.1261 | 19 | 14 | 0.6498 | 0.0963 |
| 15 | 11 | 0.6430 | 0.1215 | 17 | 15 | 1.2170 | 0.1519 | 19 | 15 | 0.8253 | 0.1048 |
| 15 | 12 | 0.8688 | 0.1354 | 17 | 16 | 1.5852 | 0.2129 | 19 | 16 | 1.0281 | 0.1188 |
| 15 | 13 | 1.1441 | 0.1616 | 17 | 17 | 2.3470 | 0.5189 | 19 | 17 | 1.2793 | 0.1441 |
| 15 | 14 | 1.5285 | 0.2241 | 18 | 1 | -1.7940 | 0.2894 | 19 | 18 | 1.6342 | 0.2036 |
| 15 | 15 | 2.3238 | 0.5395 | 18 | 2 | -1.3190 | 0.1700 | 19 | 19 | 2.3681 | 0.5010 |
| 16 | 1 | -1.7360 | 0.3009 | 18 | 3 | -1.0298 | 0.1322 | 20 | 1 | -1.8446 | 0.2799 |
| 16 | 2 | -1.2482 | 0.1789 | 18 | 4 | -0.8080 | 0.1138 | 20 | 2 | -1.3801 | 0.1627 |
| 16 | 3 | -0.9482 | 0.1405 | 18 | 5 | -0.6203 | 0.1031 | 20 | 3 | -1.0997 | 0.1256 |
| 16 | 4 | -0.7157 | 0.1219 | 18 | 6 | -0.4526 | 0.0965 | 20 | 4 | -0.8863 | 0.1073 |
| 16 | 5 | -0.5170 | 0.1114 | 18 | 7 | -0.2970 | 0.0924 | 20 | 5 | -0.7073 | 0.0966 |
| 16 | 6 | -0.3372 | 0.1052 | 18 | 8 | -0.1485 | 0.0900 | 20 | 6 | -0.5486 | 0.0897 |
| 16 | 7 | -0.1680 | 0.1018 | 18 | 9 | -0.0034 | 0.0891 | 20 | 7 | -0.4029 | 0.0853 |
| 16 | 8 | -0.0039 | 0.1005 | 18 | 10 | 0.1412 | 0.0895 | 20 | 8 | -0.2654 | 0.0824 |
| 16 | 9 | 0.1596 | 0.1010 | 18 | 11 | 0.2884 | 0.0912 | 20 | 9 | -0.1330 | 0.0807 |
| 16 | 10 | 0.3269 | 0.1035 | 18 | 12 | 0.4417 | 0.0944 | 20 | 10 | -0.0031 | 0.0800 |
| 16 | 11 | 0.5031 | 0.1083 | 18 | 13 | 0.6052 | 0.0998 | 20 | 11 | 0.1266 | 0.0803 |
| 16 | 12 | 0.6950 | 0.1165 | 18 | 14 | 0.7855 | 0.1083 | 20 | 12 | 0.2581 | 0.0815 |
| 16 | 13 | 0.9139 | 0.1304 | 18 | 15 | 0.9930 | 0.1222 | 20 | 13 | 0.3939 | 0.0838 |
| 16 | 14 | 1.1821 | 0.1565 | 18 | 16 | 1.2493 | 0.1478 | 20 | 14 | 0.5369 | 0.0875 |
| 16 | 15 | 1.5579 | 0.2182 | 18 | 17 | 1.6105 | 0.2080 | 20 | 15 | 0.6910 | 0.0932 |
| 16 | 16 | 2.3357 | 0.5288 | 18 | 18 | 2.3578 | 0.5096 | 20 | 16 | 0.8621 | 0.1017 |
| 17 | 1 | -1.7661 | 0.2949 | 19 | 1 | -1.8201 | 0.2845 | 20 | 17 | 1.0607 | 0.1156 |
| 17 | 2 | -1.2850 | 0.1742 | 19 | 2 | -1.3506 | 0.1662 | 20 | 18 | 1.3073 | 0.1408 |
| 17 | 3 | -0.9907 | 0.1361 | 19 | 3 | -1.0660 | 0.1287 | 20 | 19 | 1.6564 | 0.1995 |
| 17 | 4 | -0.7639 | 0.1176 | 19 | 4 | -0.8486 | 0.1104 | 20 | 20 | 2.3780 | 0.4929 |
| 17 | 5 | -0.5710 | 0.1070 | 19 | 5 | -0.6655 | 0.0996 | | | | |

TABLE 1

OUTLIER MEAN 2.5 OUTLIER STANDARD DEVIATION 1.0

| N | R | MEAN | VAR | N | R | MEAN | VAR | N | R | MEAN | VAR |
|---|---|---|---|---|---|---|---|---|---|---|---|
| 2 | 1 | -0.0219 | 0.9448 | 9 | 1 | -1.4237 | 0.3728 | 12 | 6 | -0.0016 | 0.1363 |
| 2 | 2 | 2.5219 | 0.9448 | 9 | 2 | -0.8525 | 0.2391 | 12 | 7 | 0.2220 | 0.1382 |
| 3 | 1 | -0.5678 | 0.6740 | 9 | 3 | -0.4735 | 0.2002 | 12 | 8 | 0.4565 | 0.1453 |
| 3 | 2 | 0.5275 | 0.6196 | 9 | 4 | -0.1541 | 0.1861 | 12 | 9 | 0.7177 | 0.1602 |
| 3 | 3 | 2.5402 | 0.9031 | 9 | 5 | 0.1491 | 0.1849 | 12 | 10 | 1.0354 | 0.1909 |
| 4 | 1 | -0.8475 | 0.5572 | 9 | 6 | 0.4650 | 0.1957 | 12 | 11 | 1.4929 | 0.2698 |
| 4 | 2 | -0.0071 | 0.4384 | 9 | 7 | 0.8314 | 0.2255 | 12 | 12 | 2.6432 | 0.7179 |
| 4 | 3 | 0.7985 | 0.4952 | 9 | 8 | 1.3422 | 0.3084 | 13 | 1 | -1.6293 | 0.3236 |
| 4 | 4 | 2.5562 | 0.8695 | 9 | 9 | 2.6161 | 0.7602 | 13 | 2 | -1.1158 | 0.1972 |
| 5 | 1 | -1.0299 | 0.4908 | 10 | 1 | -1.4851 | 0.3573 | 13 | 3 | -0.7930 | 0.1578 |
| 5 | 2 | -0.2996 | 0.3572 | 10 | 2 | -0.9325 | 0.2255 | 13 | 4 | -0.5372 | 0.1396 |
| 5 | 3 | 0.2866 | 0.3488 | 10 | 3 | -0.5725 | 0.1660 | 13 | 5 | -0.3129 | 0.1303 |
| 5 | 4 | 0.9725 | 0.4267 | 10 | 4 | -0.2755 | 0.1699 | 13 | 6 | -0.1037 | 0.1261 |
| 5 | 5 | 2.5704 | 0.8414 | 10 | 5 | -0.0020 | 0.1649 | 13 | 7 | 0.1007 | 0.1257 |
| 6 | 1 | -1.1633 | 0.4470 | 10 | 6 | 0.2704 | 0.1681 | 13 | 8 | 0.3089 | 0.1291 |
| 6 | 2 | -0.4962 | 0.3101 | 10 | 7 | 0.5630 | 0.1811 | 13 | 9 | 0.5307 | 0.1370 |
| 6 | 3 | -0.0040 | 0.2829 | 10 | 8 | 0.9094 | 0.2116 | 13 | 10 | 0.7807 | 0.1523 |
| 6 | 4 | 0.4816 | 0.2990 | 10 | 9 | 1.3991 | 0.2932 | 13 | 11 | 1.0875 | 0.1826 |
| 6 | 5 | 1.0986 | 0.3824 | 10 | 10 | 2.6257 | 0.7448 | 13 | 12 | 1.5323 | 0.2605 |
| 6 | 6 | 2.5832 | 0.8172 | 11 | 1 | -1.5388 | 0.3443 | 13 | 13 | 2.6512 | 0.7059 |
| 7 | 1 | -1.2674 | 0.4156 | 11 | 2 | -1.0015 | 0.2144 | 14 | 1 | -1.6680 | 0.3152 |
| 7 | 2 | -0.6424 | 0.2788 | 11 | 3 | -0.6564 | 0.1748 | 14 | 2 | -1.1642 | 0.1904 |
| 7 | 3 | -0.2035 | 0.2444 | 11 | 4 | -0.3765 | 0.1575 | 14 | 3 | -0.8500 | 0.1513 |
| 7 | 4 | 0.1962 | 0.2418 | 11 | 5 | -0.1240 | 0.1503 | 14 | 4 | -0.6031 | 0.1329 |
| 7 | 5 | 0.6257 | 0.2664 | 11 | 6 | 0.1202 | 0.1497 | 14 | 5 | -0.3888 | 0.1230 |
| 7 | 6 | 1.1964 | 0.3509 | 11 | 7 | 0.3709 | 0.1554 | 14 | 6 | -0.1913 | 0.1179 |
| 7 | 7 | 2.5950 | 0.7960 | 11 | 8 | 0.6459 | 0.1696 | 14 | 7 | -0.0013 | 0.1162 |
| 8 | 1 | -1.3523 | 0.3917 | 11 | 9 | 0.9766 | 0.2003 | 14 | 8 | 0.1883 | 0.1174 |
| 8 | 2 | -0.7578 | 0.2563 | 11 | 10 | 1.4489 | 0.2806 | 14 | 9 | 0.3845 | 0.1217 |
| 8 | 3 | -0.3538 | 0.2188 | 11 | 11 | 2.6347 | 0.7308 | 14 | 10 | 0.5961 | 0.1301 |
| 8 | 4 | -0.0027 | 0.2084 | 12 | 1 | -1.5865 | 0.3332 | 14 | 11 | 0.8367 | 0.1456 |
| 8 | 5 | 0.3461 | 0.2149 | 12 | 2 | -1.0620 | 0.2051 | 14 | 12 | 1.1343 | 0.1759 |
| 8 | 6 | 0.7388 | 0.2432 | 12 | 3 | -0.7291 | 0.1655 | 14 | 13 | 1.5679 | 0.2524 |
| 8 | 7 | 1.2758 | 0.3271 | 12 | 4 | -0.4625 | 0.1477 | 14 | 14 | 2.6589 | 0.6947 |
| 8 | 8 | 2.6060 | 0.7772 | 12 | 5 | -0.2258 | 0.1391 | 15 | 1 | -1.7034 | 0.3077 |

TABLE 1

OUTLIER MEAN 2.5 OUTLIER STANDARD DEVIATION 1.0

| N | R | MEAN | VAR | N | R | MEAN | VAR | N | R | MEAN | VAR |
|---|---|---|---|---|---|---|---|---|---|---|---|
| 15 | 2 | -1.2080 | 0.1844 | 17 | 6 | -0.3966 | 0.1009 | 19 | 6 | -0.5018 | 0.0931 |
| 15 | 3 | -0.9013 | 0.1456 | 17 | 7 | -0.2343 | 0.0972 | 19 | 7 | -0.3512 | 0.0889 |
| 15 | 4 | -0.6620 | 0.1271 | 17 | 8 | -0.0782 | 0.0954 | 19 | 8 | -0.2083 | 0.0863 |
| 15 | 5 | -0.4559 | 0.1169 | 17 | 9 | 0.0760 | 0.0952 | 19 | 9 | -0.0696 | 0.0850 |
| 15 | 6 | -0.2679 | 0.1113 | 17 | 10 | 0.2317 | 0.0967 | 19 | 10 | 0.0677 | 0.0849 |
| 15 | 7 | -0.0891 | 0.1086 | 17 | 11 | 0.3931 | 0.1000 | 19 | 11 | 0.2061 | 0.0860 |
| 15 | 8 | 0.0866 | 0.1084 | 17 | 12 | 0.5650 | 0.1056 | 19 | 12 | 0.3484 | 0.0883 |
| 15 | 9 | 0.2648 | 0.1105 | 17 | 13 | 0.7547 | 0.1148 | 19 | 13 | 0.4979 | 0.0921 |
| 15 | 10 | 0.4513 | 0.1154 | 17 | 14 | 0.9746 | 0.1304 | 19 | 14 | 0.6590 | 0.0981 |
| 15 | 11 | 0.6544 | 0.1243 | 17 | 15 | 1.2508 | 0.1598 | 19 | 15 | 0.8386 | 0.1074 |
| 15 | 12 | 0.8871 | 0.1399 | 17 | 16 | 1.6578 | 0.2331 | 19 | 16 | 1.0486 | 0.1229 |
| 15 | 13 | 1.1766 | 0.1699 | 17 | 17 | 2.6801 | 0.6654 | 19 | 17 | 1.3140 | 0.1517 |
| 15 | 14 | 1.6004 | 0.2453 | 18 | 1 | -1.7940 | 0.2895 | 19 | 18 | 1.7073 | 0.2231 |
| 15 | 15 | 2.6663 | 0.6843 | 18 | 2 | -1.3188 | 0.1701 | 19 | 19 | 2.6929 | 0.6486 |
| 16 | 1 | -1.7359 | 0.3010 | 18 | 3 | -1.0295 | 0.1324 | 20 | 1 | -1.8445 | 0.2799 |
| 16 | 2 | -1.2480 | 0.1791 | 18 | 4 | -0.8075 | 0.1139 | 20 | 2 | -1.3800 | 0.1628 |
| 16 | 3 | -0.9478 | 0.1407 | 18 | 5 | -0.6197 | 0.1033 | 20 | 3 | -1.0995 | 0.1257 |
| 16 | 4 | -0.7151 | 0.1221 | 18 | 6 | -0.4516 | 0.0968 | 20 | 4 | -0.8859 | 0.1074 |
| 16 | 5 | -0.5160 | 0.1117 | 18 | 7 | -0.2956 | 0.0927 | 20 | 5 | -0.7067 | 0.0967 |
| 16 | 6 | -0.3358 | 0.1057 | 18 | 8 | -0.1467 | 0.0905 | 20 | 6 | -0.5479 | 0.0899 |
| 16 | 7 | -0.1660 | 0.1024 | 18 | 9 | -0.0010 | 0.0897 | 20 | 7 | -0.4019 | 0.0855 |
| 16 | 8 | -0.0012 | 0.1012 | 18 | 10 | 0.1445 | 0.0902 | 20 | 8 | -0.2642 | 0.0827 |
| 16 | 9 | 0.1635 | 0.1020 | 18 | 11 | 0.2930 | 0.0922 | 20 | 9 | -0.1314 | 0.0811 |
| 16 | 10 | 0.3325 | 0.1048 | 18 | 12 | 0.4479 | 0.0958 | 20 | 10 | -0.0009 | 0.0805 |
| 16 | 11 | 0.5111 | 0.1102 | 18 | 13 | 0.6141 | 0.1016 | 20 | 11 | 0.1295 | 0.0809 |
| 16 | 12 | 0.7069 | 0.1192 | 18 | 14 | 0.7984 | 0.1109 | 20 | 12 | 0.2619 | 0.0823 |
| 16 | 13 | 0.9328 | 0.1348 | 18 | 15 | 1.0130 | 0.1264 | 20 | 13 | 0.3990 | 0.0848 |
| 16 | 14 | 1.2153 | 0.1646 | 18 | 16 | 1.2836 | 0.1555 | 20 | 14 | 0.5437 | 0.0888 |
| 16 | 15 | 1.6303 | 0.2389 | 18 | 17 | 1.6834 | 0.2279 | 20 | 15 | 0.7005 | 0.0949 |
| 16 | 16 | 2.6733 | 0.6746 | 18 | 18 | 2.6866 | 0.6568 | 20 | 16 | 0.8758 | 0.1043 |
| 17 | 1 | -1.7660 | 0.2950 | 19 | 1 | -1.8200 | 0.2845 | 20 | 17 | 1.0816 | 0.1197 |
| 17 | 2 | -1.2848 | 0.1744 | 19 | 2 | -1.3504 | 0.1663 | 20 | 18 | 1.3424 | 0.1482 |
| 17 | 3 | -0.9904 | 0.1363 | 19 | 3 | -1.0658 | 0.1289 | 20 | 19 | 1.7296 | 0.2188 |
| 17 | 4 | -0.7633 | 0.1178 | 19 | 4 | -0.8482 | 0.1105 | 20 | 20 | 2.6989 | 0.6408 |
| 17 | 5 | -0.5702 | 0.1072 | 19 | 5 | -0.6650 | 0.0998 | | | | |

TABLE 1

OUTLIER MEAN 3.0 OUTLIER STANDARD DEVIATION 1.0

| N | R | MEAN | VAR | N | R | MEAN | VAR | N | R | MEAN | VAR |
|---|---|---|---|---|---|---|---|---|---|---|---|
| 2 | 1 | -0.0086 | 0.9741 | 9 | 1 | -1.4236 | 0.3729 | 12 | 6 | -0.0004 | 0.1369 |
| 2 | 2 | 3.0086 | 0.9741 | 9 | 2 | -0.8523 | 0.2393 | 12 | 7 | 0.2241 | 0.1392 |
| 3 | 1 | -0.5652 | 0.6791 | 9 | 3 | -0.4730 | 0.2006 | 12 | 8 | 0.4603 | 0.1469 |
| 3 | 2 | 0.5490 | 0.6497 | 9 | 4 | -0.1529 | 0.1869 | 12 | 9 | 0.7250 | 0.1633 |
| 3 | 3 | 3.0162 | 0.9527 | 9 | 5 | 0.1516 | 0.1864 | 12 | 10 | 1.0514 | 0.1979 |
| 4 | 1 | -0.8466 | 0.5588 | 9 | 6 | 0.4703 | 0.1988 | 12 | 11 | 1.5400 | 0.2922 |
| 4 | 2 | -0.0022 | 0.4448 | 9 | 7 | 0.8445 | 0.2328 | 12 | 12 | 3.0640 | 0.8391 |
| 4 | 3 | 0.8257 | 0.5244 | 9 | 8 | 1.3846 | 0.3328 | 13 | 1 | -1.6292 | 0.3236 |
| 4 | 4 | 3.0231 | 0.9343 | 9 | 9 | 3.0507 | 0.8678 | 13 | 2 | -1.1157 | 0.1973 |
| 5 | 1 | -1.0295 | 0.4915 | 10 | 1 | -1.4850 | 0.3573 | 13 | 3 | -0.7929 | 0.1580 |
| 5 | 2 | -0.2977 | 0.3594 | 10 | 2 | -0.9323 | 0.2257 | 13 | 4 | -0.5369 | 0.1398 |
| 5 | 3 | 0.2936 | 0.3557 | 10 | 3 | -0.5721 | 0.1863 | 13 | 5 | -0.3124 | 0.1305 |
| 5 | 4 | 1.0042 | 0.4548 | 10 | 4 | -0.2748 | 0.1704 | 13 | 6 | -0.1029 | 0.1265 |
| 5 | 5 | 3.0294 | 0.9182 | 10 | 5 | -0.0005 | 0.1657 | 13 | 7 | 0.1021 | 0.1264 |
| 6 | 1 | -1.1630 | 0.4474 | 10 | 6 | 0.2733 | 0.1697 | 13 | 8 | 0.3113 | 0.1301 |
| 6 | 2 | -0.4953 | 0.3111 | 10 | 7 | 0.5690 | 0.1842 | 13 | 9 | 0.5349 | 0.1387 |
| 6 | 3 | -0.0011 | 0.2855 | 10 | 8 | 0.9236 | 0.2188 | 13 | 10 | 0.7886 | 0.1555 |
| 6 | 4 | 0.4905 | 0.3062 | 10 | 9 | 1.4433 | 0.3169 | 13 | 11 | 1.1044 | 0.1898 |
| 6 | 5 | 1.1337 | 0.4094 | 10 | 10 | 3.0554 | 0.8576 | 13 | 12 | 1.5806 | 0.2825 |
| 6 | 6 | 3.0353 | 0.9038 | 11 | 1 | -1.5388 | 0.3443 | 13 | 13 | 3.0681 | 0.8307 |
| 7 | 1 | -1.2672 | 0.4159 | 11 | 2 | -1.0014 | 0.2145 | 14 | 1 | -1.6680 | 0.3152 |
| 7 | 2 | -0.6419 | 0.2794 | 11 | 3 | -0.6561 | 0.1749 | 14 | 2 | -1.1641 | 0.1904 |
| 7 | 3 | -0.2020 | 0.2457 | 11 | 4 | -0.3759 | 0.1578 | 14 | 3 | -0.8499 | 0.1514 |
| 7 | 4 | 0.2000 | 0.2446 | 11 | 5 | -0.1230 | 0.1508 | 14 | 4 | -0.6029 | 0.1330 |
| 7 | 5 | 0.6361 | 0.2737 | 11 | 6 | 0.1220 | 0.1506 | 14 | 5 | -0.3884 | 0.1232 |
| 7 | 6 | 1.2344 | 0.3769 | 11 | 7 | 0.3743 | 0.1570 | 14 | 6 | -0.1907 | 0.1182 |
| 7 | 7 | 3.0407 | 0.8907 | 11 | 8 | 0.6527 | 0.1727 | 14 | 7 | -0.0003 | 0.1166 |
| 8 | 1 | -1.3522 | 0.3919 | 11 | 9 | 0.9917 | 0.2074 | 14 | 8 | 0.1899 | 0.1180 |
| 8 | 2 | -0.7575 | 0.2566 | 11 | 10 | 1.4946 | 0.3036 | 14 | 9 | 0.3872 | 0.1227 |
| 8 | 3 | -0.3530 | 0.2195 | 11 | 11 | 3.0598 | 0.8480 | 14 | 10 | 0.6007 | 0.1319 |
| 8 | 4 | -0.0007 | 0.2098 | 12 | 1 | -1.5864 | 0.3332 | 14 | 11 | 0.8452 | 0.1488 |
| 8 | 5 | 0.3507 | 0.2179 | 12 | 2 | -1.0619 | 0.2052 | 14 | 12 | 1.1519 | 0.1828 |
| 8 | 6 | 0.7506 | 0.2505 | 12 | 3 | -0.7289 | 0.1657 | 14 | 13 | 1.6173 | 0.2739 |
| 8 | 7 | 1.3161 | 0.3523 | 12 | 4 | -0.4621 | 0.1479 | 14 | 14 | 3.0720 | 0.8227 |
| 8 | 8 | 3.0459 | 0.8788 | 12 | 5 | -0.2251 | 0.1395 | 15 | 1 | -1.7034 | 0.3077 |

TABLE 1

OUTLIER MEAN 3.0 OUTLIER STANDARD DEVIATION 1.0

| N | R | MEAN | VAR | N | R | MEAN | VAR | N | R | MEAN | VAR |
|---|---|---|---|---|---|---|---|---|---|---|---|
| 15 | 2 | -1.2079 | 0.1844 | 17 | 6 | -0.3963 | 0.1010 | 19 | 6 | -0.5016 | 0.0932 |
| 15 | 3 | -0.9011 | 0.1457 | 17 | 7 | -0.2339 | 0.0973 | 19 | 7 | -0.3509 | 0.0890 |
| 15 | 4 | -0.6618 | 0.1272 | 17 | 8 | -0.0775 | 0.0956 | 19 | 8 | -0.2079 | 0.0864 |
| 15 | 5 | -0.4556 | 0.1171 | 17 | 9 | 0.0770 | 0.0956 | 19 | 9 | -0.0690 | 0.0852 |
| 15 | 6 | -0.2674 | 0.1115 | 17 | 10 | 0.2332 | 0.0972 | 19 | 10 | 0.0685 | 0.0852 |
| 15 | 7 | -0.0884 | 0.1089 | 17 | 11 | 0.3953 | 0.1007 | 19 | 11 | 0.2073 | 0.0863 |
| 15 | 8 | 0.0878 | 0.1088 | 17 | 12 | 0.5685 | 0.1067 | 19 | 12 | 0.3502 | 0.0868 |
| 15 | 9 | 0.2666 | 0.1112 | 17 | 13 | 0.7603 | 0.1166 | 19 | 13 | 0.5005 | 0.0928 |
| 15 | 10 | 0.4543 | 0.1165 | 17 | 14 | 0.9845 | 0.1335 | 19 | 14 | 0.6630 | 0.0992 |
| 15 | 11 | 0.6594 | 0.1260 | 17 | 15 | 1.2704 | 0.1664 | 19 | 15 | 0.8448 | 0.1092 |
| 15 | 12 | 0.8961 | 0.1430 | 17 | 16 | 1.7099 | 0.2534 | 19 | 16 | 1.0592 | 0.1260 |
| 15 | 13 | 1.1950 | 0.1767 | 17 | 17 | 3.0829 | 0.8012 | 19 | 17 | 1.3347 | 0.1582 |
| 15 | 14 | 1.6508 | 0.2663 | 18 | 1 | -1.7939 | 0.2895 | 19 | 18 | 1.7607 | 0.2428 |
| 15 | 15 | 3.0757 | 0.8152 | 18 | 2 | -1.3188 | 0.1701 | 19 | 19 | 3.0895 | 0.7885 |
| 16 | 1 | -1.7359 | 0.3010 | 18 | 3 | -1.0295 | 0.1324 | 20 | 1 | -1.8445 | 0.2799 |
| 16 | 2 | -1.2479 | 0.1791 | 18 | 4 | -0.8074 | 0.1140 | 20 | 2 | -1.3799 | 0.1628 |
| 16 | 3 | -0.9477 | 0.1407 | 18 | 5 | -0.6195 | 0.1034 | 20 | 3 | -1.0995 | 0.1257 |
| 16 | 4 | -0.7149 | 0.1222 | 18 | 6 | -0.4514 | 0.0969 | 20 | 4 | -0.8859 | 0.1075 |
| 16 | 5 | -0.5158 | 0.1118 | 18 | 7 | -0.2953 | 0.0929 | 20 | 5 | -0.7066 | 0.0968 |
| 16 | 6 | -0.3354 | 0.1058 | 18 | 8 | -0.1461 | 0.0907 | 20 | 6 | -0.5477 | 0.0900 |
| 16 | 7 | -0.1655 | 0.1026 | 18 | 9 | -0.0002 | 0.0899 | 20 | 7 | -0.4017 | 0.0856 |
| 16 | 8 | -0.0003 | 0.1016 | 18 | 10 | 0.1456 | 0.0906 | 20 | 8 | -0.2638 | 0.0828 |
| 16 | 9 | 0.1648 | 0.1025 | 18 | 11 | 0.2946 | 0.0927 | 20 | 9 | -0.1309 | 0.0812 |
| 16 | 10 | 0.3345 | 0.1055 | 18 | 12 | 0.4503 | 0.0965 | 20 | 10 | -0.0002 | 0.0807 |
| 16 | 11 | 0.5143 | 0.1112 | 18 | 13 | 0.6178 | 0.1027 | 20 | 11 | 0.1304 | 0.0812 |
| 16 | 12 | 0.7123 | 0.1210 | 18 | 14 | 0.8043 | 0.1127 | 20 | 12 | 0.2633 | 0.0827 |
| 16 | 13 | 0.9423 | 0.1380 | 18 | 15 | 1.0233 | 0.1295 | 20 | 13 | 0.4009 | 0.0853 |
| 16 | 14 | 1.2343 | 0.1713 | 18 | 16 | 1.3038 | 0.1621 | 20 | 14 | 0.5465 | 0.0896 |
| 16 | 15 | 1.6815 | 0.2595 | 18 | 17 | 1.7362 | 0.2478 | 20 | 15 | 0.7046 | 0.0960 |
| 16 | 16 | 3.0794 | 0.8081 | 18 | 18 | 3.0863 | 0.7947 | 20 | 16 | 0.8824 | 0.1061 |
| 17 | 1 | -1.7660 | 0.2950 | 19 | 1 | -1.8200 | 0.2845 | 20 | 17 | 1.0926 | 0.1227 |
| 17 | 2 | -1.2847 | 0.1744 | 19 | 2 | -1.3504 | 0.1663 | 20 | 18 | 1.3636 | 0.1546 |
| 17 | 3 | -0.9903 | 0.1363 | 19 | 3 | -1.0657 | 0.1289 | 20 | 19 | 1.7837 | 0.2382 |
| 17 | 4 | -0.7632 | 0.1179 | 19 | 4 | -0.8481 | 0.1106 | 20 | 20 | 3.0927 | 0.7825 |
| 17 | 5 | -0.5701 | 0.1073 | 19 | 5 | -0.6648 | 0.0999 | | | | |

TABLE 1

OUTLIER MEAN 4.0 OUTLIER STANDARD DEVIATION 1.0

| N | R | MEAN | VAR | N | R | MEAN | VAR | N | R | MEAN | VAR |
|---|---|---|---|---|---|---|---|---|---|---|---|
| 2 | 1 | -0.0010 | 0.9961 | 9 | 1 | -1.4236 | 0.3729 | 12 | 6 | -0.0000 | 0.1372 |
| 2 | 2 | 4.0010 | 0.9961 | 9 | 2 | -0.8522 | 0.2394 | 12 | 7 | 0.2249 | 0.1396 |
| 3 | 1 | -0.5642 | 0.6815 | 9 | 3 | -0.4728 | 0.2008 | 12 | 8 | 0.4619 | 0.1479 |
| 3 | 2 | 0.5623 | 0.6762 | 9 | 4 | -0.1525 | 0.1872 | 12 | 9 | 0.7286 | 0.1655 |
| 3 | 3 | 4.0019 | 0.9925 | 9 | 5 | 0.1525 | 0.1871 | 12 | 10 | 1.0609 | 0.2042 |
| 4 | 1 | -0.8463 | 0.5594 | 9 | 6 | 0.4727 | 0.2006 | 12 | 11 | 1.5790 | 0.3226 |
| 4 | 2 | -0.0001 | 0.4484 | 9 | 7 | 0.8516 | 0.2386 | 12 | 12 | 4.0088 | 0.9677 |
| 4 | 3 | 0.8436 | 0.5529 | 9 | 8 | 1.4177 | 0.3633 | 13 | 1 | -1.6292 | 0.3236 |
| 4 | 4 | 4.0028 | 0.9891 | 9 | 9 | 4.0067 | 0.9749 | 13 | 2 | -1.1157 | 0.1973 |
| 5 | 1 | -1.0294 | 0.4917 | 10 | 1 | -1.4850 | 0.3574 | 13 | 3 | -0.7928 | 0.1580 |
| 5 | 2 | -0.2970 | 0.3604 | 10 | 2 | -0.9323 | 0.2257 | 13 | 4 | -0.5368 | 0.1398 |
| 5 | 3 | 0.2968 | 0.3600 | 10 | 3 | -0.5720 | 0.1864 | 13 | 5 | -0.3123 | 0.1306 |
| 5 | 4 | 1.0260 | 0.4843 | 10 | 4 | -0.2745 | 0.1706 | 13 | 6 | -0.1026 | 0.1266 |
| 5 | 5 | 4.0036 | 0.9860 | 10 | 5 | -0.0000 | 0.1661 | 13 | 7 | 0.1026 | 0.1266 |
| 6 | 1 | -1.1630 | 0.4475 | 10 | 6 | 0.2745 | 0.1705 | 13 | 8 | 0.3122 | 0.1306 |
| 6 | 2 | -0.4950 | 0.3115 | 10 | 7 | 0.5718 | 0.1862 | 13 | 9 | 0.5367 | 0.1397 |
| 6 | 3 | -0.0001 | 0.2868 | 10 | 8 | 0.9315 | 0.2248 | 13 | 10 | 0.7925 | 0.1577 |
| 6 | 4 | 0.4947 | 0.3110 | 10 | 9 | 1.4786 | 0.3473 | 13 | 11 | 1.1147 | 0.1962 |
| 6 | 5 | 1.1589 | 0.4394 | 10 | 10 | 4.0074 | 0.9724 | 13 | 12 | 1.6213 | 0.3127 |
| 6 | 6 | 4.0044 | 0.9830 | 11 | 1 | -1.5388 | 0.3443 | 13 | 13 | 4.0094 | 0.9655 |
| 7 | 1 | -1.2672 | 0.4159 | 11 | 2 | -1.0014 | 0.2145 | 14 | 1 | -1.6680 | 0.3152 |
| 7 | 2 | -0.6418 | 0.2796 | 11 | 3 | -0.6561 | 0.1750 | 14 | 2 | -1.1641 | 0.1904 |
| 7 | 3 | -0.2016 | 0.2462 | 11 | 4 | -0.3758 | 0.1579 | 14 | 3 | -0.8498 | 0.1514 |
| 7 | 4 | 0.2015 | 0.2461 | 11 | 5 | -0.1227 | 0.1510 | 14 | 4 | -0.6029 | 0.1330 |
| 7 | 5 | 0.6413 | 0.2789 | 11 | 6 | 0.1226 | 0.1510 | 14 | 5 | -0.3883 | 0.1232 |
| 7 | 6 | 1.2625 | 0.4072 | 11 | 7 | 0.3757 | 0.1579 | 14 | 6 | -0.1905 | 0.1183 |
| 7 | 7 | 4.0052 | 0.9802 | 11 | 8 | 0.6558 | 0.1748 | 14 | 7 | -0.0000 | 0.1168 |
| 8 | 1 | -1.3522 | 0.3919 | 11 | 9 | 1.0005 | 0.2136 | 14 | 8 | 0.1905 | 0.1183 |
| 8 | 2 | -0.7574 | 0.2567 | 11 | 10 | 1.5318 | 0.3340 | 14 | 9 | 0.3883 | 0.1232 |
| 8 | 3 | -0.3527 | 0.2197 | 11 | 11 | 4.0081 | 0.9700 | 14 | 10 | 0.6027 | 0.1329 |
| 8 | 4 | -0.0000 | 0.2104 | 12 | 1 | -1.5864 | 0.3332 | 14 | 11 | 0.8495 | 0.1511 |
| 8 | 5 | 0.3526 | 0.2196 | 12 | 2 | -1.0619 | 0.2052 | 14 | 12 | 1.1629 | 0.1893 |
| 8 | 6 | 0.7568 | 0.2560 | 12 | 3 | -0.7288 | 0.1657 | 14 | 13 | 1.6596 | 0.3040 |
| 8 | 7 | 1.3469 | 0.3827 | 12 | 4 | -0.4620 | 0.1480 | 14 | 14 | 4.0101 | 0.9634 |
| 8 | 8 | 4.0060 | 0.9775 | 12 | 5 | -0.2249 | 0.1396 | 15 | 1 | -1.7034 | 0.3077 |

TABLE 1

OUTLIER MEAN 4.0 OUTLIER STANDARD DEVIATION 1.0

| N | R | MEAN | VAR | N | R | MEAN | VAR | N | R | MEAN | VAR |
|---|---|---|---|---|---|---|---|---|---|---|---|
| 15 | 2 | -1.2079 | 0.1844 | 17 | 6 | -0.3962 | 0.1010 | 19 | 6 | -0.5016 | 0.0932 |
| 15 | 3 | -0.9011 | 0.1457 | 17 | 7 | -0.2338 | 0.0974 | 19 | 7 | -0.3508 | 0.0890 |
| 15 | 4 | -0.6618 | 0.1272 | 17 | 8 | -0.0773 | 0.0957 | 19 | 8 | -0.2077 | 0.0865 |
| 15 | 5 | -0.4556 | 0.1171 | 17 | 9 | 0.0773 | 0.0957 | 19 | 9 | -0.0688 | 0.0853 |
| 15 | 6 | -0.2673 | 0.1115 | 17 | 10 | 0.2337 | 0.0974 | 19 | 10 | 0.0688 | 0.0853 |
| 15 | 7 | -0.0882 | 0.1090 | 17 | 11 | 0.3962 | 0.1010 | 19 | 11 | 0.2077 | 0.0865 |
| 15 | 8 | 0.0881 | 0.1090 | 17 | 12 | 0.5699 | 0.1073 | 19 | 12 | 0.3508 | 0.0890 |
| 15 | 9 | 0.2673 | 0.1115 | 17 | 13 | 0.7630 | 0.1177 | 19 | 13 | 0.5015 | 0.0932 |
| 15 | 10 | 0.4555 | 0.1171 | 17 | 14 | 0.9898 | 0.1360 | 19 | 14 | 0.6647 | 0.0999 |
| 15 | 11 | 0.6616 | 0.1271 | 17 | 15 | 1.2833 | 0.1731 | 19 | 15 | 0.8479 | 0.1104 |
| 15 | 12 | 0.9008 | 0.1454 | 17 | 16 | 1.7563 | 0.2831 | 19 | 16 | 1.0652 | 0.1285 |
| 15 | 13 | 1.2066 | 0.1832 | 17 | 17 | 4.0119 | 0.9573 | 19 | 17 | 1.3487 | 0.1649 |
| 15 | 14 | 1.6946 | 0.2963 | 18 | 1 | -1.7939 | 0.2895 | 19 | 18 | 1.8096 | 0.2723 |
| 15 | 15 | 4.0107 | 0.9613 | 18 | 2 | -1.3188 | 0.1701 | 19 | 19 | 4.0131 | 0.9535 |
| 16 | 1 | -1.7359 | 0.3010 | 18 | 3 | -1.0295 | 0.1324 | 20 | 1 | -1.8445 | 0.2799 |
| 16 | 2 | -1.2479 | 0.1791 | 18 | 4 | -0.8074 | 0.1140 | 20 | 2 | -1.3799 | 0.1628 |
| 16 | 3 | -0.9477 | 0.1407 | 18 | 5 | -0.6195 | 0.1034 | 20 | 3 | -1.0995 | 0.1257 |
| 16 | 4 | -0.7149 | 0.1222 | 18 | 6 | -0.4513 | 0.0969 | 20 | 4 | -0.8859 | 0.1075 |
| 16 | 5 | -0.5157 | 0.1119 | 18 | 7 | -0.2952 | 0.0929 | 20 | 5 | -0.7066 | 0.0968 |
| 16 | 6 | -0.3353 | 0.1059 | 18 | 8 | -0.1460 | 0.0907 | 20 | 6 | -0.5477 | 0.0900 |
| 16 | 7 | -0.1653 | 0.1027 | 18 | 9 | -0.0000 | 0.0900 | 20 | 7 | -0.4016 | 0.0856 |
| 16 | 8 | -0.0000 | 0.1017 | 18 | 10 | 0.1460 | 0.0907 | 20 | 8 | -0.2637 | 0.0828 |
| 16 | 9 | 0.1653 | 0.1027 | 18 | 11 | 0.2952 | 0.0929 | 20 | 9 | -0.1307 | 0.0813 |
| 16 | 10 | 0.3353 | 0.1058 | 18 | 12 | 0.4513 | 0.0969 | 20 | 10 | -0.0000 | 0.0806 |
| 16 | 11 | 0.5156 | 0.1118 | 18 | 13 | 0.6194 | 0.1033 | 20 | 11 | 0.1307 | 0.0813 |
| 16 | 12 | 0.7147 | 0.1221 | 18 | 14 | 0.8072 | 0.1139 | 20 | 12 | 0.2637 | 0.0828 |
| 16 | 13 | 0.9473 | 0.1404 | 18 | 15 | 1.0290 | 0.1321 | 20 | 13 | 0.4016 | 0.0856 |
| 16 | 14 | 1.2465 | 0.1779 | 18 | 16 | 1.3172 | 0.1688 | 20 | 14 | 0.5476 | 0.0900 |
| 16 | 15 | 1.7267 | 0.2894 | 18 | 17 | 1.7839 | 0.2775 | 20 | 15 | 0.7065 | 0.0967 |
| 16 | 16 | 4.0113 | 0.9592 | 18 | 18 | 4.0125 | 0.9553 | 20 | 16 | 0.8856 | 0.1073 |
| 17 | 1 | -1.7660 | 0.2950 | 19 | 1 | -1.8200 | 0.2845 | 20 | 17 | 1.0989 | 0.1253 |
| 17 | 2 | -1.2847 | 0.1744 | 19 | 2 | -1.3504 | 0.1663 | 20 | 18 | 1.3781 | 0.1614 |
| 17 | 3 | -0.9903 | 0.1363 | 19 | 3 | -1.0657 | 0.1289 | 20 | 19 | 1.8336 | 0.2675 |
| 17 | 4 | -0.7632 | 0.1179 | 19 | 4 | -0.8481 | 0.1106 | 20 | 20 | 4.0137 | 0.9516 |
| 17 | 5 | -0.5700 | 0.1074 | 19 | 5 | -0.6648 | 0.0999 | | | | |

TABLE 2

OUTLIER MEAN 0.0 OUTLIER STANDARD DEVIATION 0.5

| N | R | MEAN | VAR | N | R | MEAN | VAR | N | R | MEAN | VAR |
|---|---|---|---|---|---|---|---|---|---|---|---|
| 2 | 2 | 0.4460 | 0.4261 | 12 | 9 | 0.5060 | 0.1273 | 17 | 12 | 0.4312 | 0.0897 |
| 3 | 2 | 0.0000 | 0.2832 | 12 | 10 | 0.7543 | 0.1496 | 17 | 13 | 0.5944 | 0.0976 |
| 3 | 3 | 0.7281 | 0.4532 | 12 | 11 | 1.0738 | 0.1944 | 17 | 14 | 0.7788 | 0.1100 |
| 4 | 3 | 0.2540 | 0.2601 | 12 | 12 | 1.5900 | 0.3278 | 17 | 15 | 0.9991 | 0.1305 |
| 4 | 4 | 0.9256 | 0.4438 | 13 | 7 | 0.0000 | 0.1021 | 17 | 16 | 1.2888 | 0.1707 |
| 5 | 3 | 0.0000 | 0.2120 | 13 | 8 | 0.1788 | 0.1042 | 17 | 17 | 1.7672 | 0.2932 |
| 5 | 4 | 0.4376 | 0.2495 | 13 | 9 | 0.3659 | 0.1107 | 18 | 10 | 0.0655 | 0.0772 |
| 5 | 5 | 1.0735 | 0.4256 | 13 | 10 | 0.5717 | 0.1231 | 18 | 11 | 0.1981 | 0.0788 |
| 6 | 4 | 0.1791 | 0.1924 | 13 | 11 | 0.8129 | 0.1451 | 18 | 12 | 0.3353 | 0.0819 |
| 6 | 5 | 0.5813 | 0.2403 | 13 | 12 | 1.1250 | 0.1887 | 18 | 13 | 0.4809 | 0.0871 |
| 6 | 6 | 1.1898 | 0.4067 | 13 | 13 | 1.6320 | 0.3194 | 18 | 14 | 0.6400 | 0.0951 |
| 7 | 4 | 0.0000 | 0.1674 | 14 | 8 | 0.0830 | 0.0963 | 18 | 15 | 0.8205 | 0.1074 |
| 7 | 5 | 0.3190 | 0.1811 | 14 | 9 | 0.2522 | 0.0996 | 18 | 16 | 1.0368 | 0.1275 |
| 7 | 6 | 0.6983 | 0.2313 | 14 | 10 | 0.4316 | 0.1068 | 18 | 17 | 1.3221 | 0.1670 |
| 7 | 7 | 1.2846 | 0.3892 | 14 | 11 | 0.6309 | 0.1194 | 18 | 18 | 1.7949 | 0.2881 |
| 8 | 5 | 0.1386 | 0.1536 | 14 | 12 | 0.8659 | 0.1410 | 19 | 10 | 0.0000 | 0.0734 |
| 8 | 6 | 0.4339 | 0.1728 | 14 | 13 | 1.1715 | 0.1836 | 19 | 11 | 0.1248 | 0.0741 |
| 8 | 7 | 0.7963 | 0.2228 | 14 | 14 | 1.6702 | 0.3119 | 19 | 12 | 0.2522 | 0.0760 |
| 8 | 8 | 1.3641 | 0.3736 | 15 | 8 | 0.0000 | 0.0904 | 19 | 13 | 0.3850 | 0.0795 |
| 9 | 5 | 0.0000 | 0.1381 | 15 | 9 | 0.1563 | 0.0917 | 19 | 14 | 0.5267 | 0.0848 |
| 9 | 6 | 0.2523 | 0.1444 | 15 | 10 | 0.3178 | 0.0958 | 19 | 15 | 0.6822 | 0.0928 |
| 9 | 7 | 0.5312 | 0.1660 | 15 | 11 | 0.4909 | 0.1034 | 19 | 16 | 0.8592 | 0.1049 |
| 9 | 8 | 0.8800 | 0.2147 | 15 | 12 | 0.6846 | 0.1160 | 19 | 17 | 1.0720 | 0.1248 |
| 9 | 9 | 1.4320 | 0.3599 | 15 | 13 | 0.9141 | 0.1372 | 19 | 18 | 1.3532 | 0.1637 |
| 10 | 6 | 0.1132 | 0.1281 | 15 | 14 | 1.2139 | 0.1789 | 19 | 19 | 1.8208 | 0.2833 |
| 10 | 7 | 0.3488 | 0.1376 | 15 | 15 | 1.7051 | 0.3051 | 20 | 11 | 0.0593 | 0.0703 |
| 10 | 8 | 0.6152 | 0.1600 | 16 | 9 | 0.0732 | 0.0857 | 20 | 12 | 0.1790 | 0.0714 |
| 10 | 9 | 0.9527 | 0.2073 | 16 | 10 | 0.2218 | 0.0879 | 20 | 13 | 0.3020 | 0.0737 |
| 10 | 10 | 1.4912 | 0.3479 | 16 | 11 | 0.3771 | 0.0926 | 20 | 14 | 0.4310 | 0.0773 |
| 11 | 6 | 0.0000 | 0.1174 | 16 | 12 | 0.5449 | 0.1004 | 20 | 15 | 0.5692 | 0.0827 |
| 11 | 7 | 0.2092 | 0.1209 | 16 | 13 | 0.7337 | 0.1129 | 20 | 16 | 0.7215 | 0.0907 |
| 11 | 8 | 0.4324 | 0.1320 | 16 | 14 | 0.9583 | 0.1337 | 20 | 17 | 0.8953 | 0.1026 |
| 11 | 9 | 0.6889 | 0.1546 | 16 | 15 | 1.2528 | 0.1746 | 20 | 18 | 1.1047 | 0.1222 |
| 11 | 10 | 1.0168 | 0.2005 | 16 | 16 | 1.7373 | 0.2989 | 20 | 19 | 1.3823 | 0.1606 |
| 11 | 11 | 1.5434 | 0.3372 | 17 | 9 | 0.0000 | 0.0810 | 20 | 20 | 1.8452 | 0.2789 |
| 12 | 7 | 0.0958 | 0.1099 | 17 | 10 | 0.1388 | 0.0819 | | | | |
| 12 | 8 | 0.2925 | 0.1153 | 17 | 11 | 0.2812 | 0.0847 | | | | |

TABLE 2

OUTLIER MEAN 0.0 OUTLIER STANDARD DEVIATION 1.0

| N | R | MEAN | VAR | N | R | MEAN | VAR | N | R | MEAN | VAR |
|---|---|---|---|---|---|---|---|---|---|---|---|
| 2 | 2 | 0.5642 | 0.6817 | 12 | 9 | 0.5368 | 0.1398 | 17 | 12 | 0.4513 | 0.0969 |
| 3 | 2 | 0.0000 | 0.4487 | 12 | 10 | 0.7928 | 0.1580 | 17 | 13 | 0.6195 | 0.1034 |
| 3 | 3 | 0.8463 | 0.5595 | 12 | 11 | 1.1157 | 0.1973 | 17 | 14 | 0.8074 | 0.1140 |
| 4 | 3 | 0.2970 | 0.3605 | 12 | 12 | 1.6292 | 0.3236 | 17 | 15 | 1.0295 | 0.1324 |
| 4 | 4 | 1.0294 | 0.4917 | 13 | 7 | 0.0000 | 0.1168 | 17 | 16 | 1.3188 | 0.1701 |
| 5 | 3 | 0.0000 | 0.2868 | 13 | 8 | 0.1905 | 0.1183 | 17 | 17 | 1.7939 | 0.2895 |
| 5 | 4 | 0.4950 | 0.3115 | 13 | 9 | 0.3883 | 0.1233 | 18 | 10 | 0.0688 | 0.0853 |
| 5 | 5 | 1.1630 | 0.4475 | 13 | 10 | 0.6029 | 0.1330 | 18 | 11 | 0.2077 | 0.0865 |
| 6 | 4 | 0.2015 | 0.2462 | 13 | 11 | 0.8498 | 0.1514 | 18 | 12 | 0.3508 | 0.0890 |
| 6 | 5 | 0.6418 | 0.2796 | 13 | 12 | 1.1641 | 0.1904 | 18 | 13 | 0.5016 | 0.0932 |
| 6 | 6 | 1.2672 | 0.4159 | 13 | 13 | 1.6680 | 0.3152 | 18 | 14 | 0.6648 | 0.0999 |
| 7 | 4 | 0.0000 | 0.2104 | 14 | 8 | 0.0882 | 0.1090 | 18 | 15 | 0.8481 | 0.1106 |
| 7 | 5 | 0.3527 | 0.2197 | 14 | 9 | 0.2673 | 0.1115 | 18 | 16 | 1.0657 | 0.1289 |
| 7 | 6 | 0.7574 | 0.2567 | 14 | 10 | 0.4556 | 0.1171 | 18 | 17 | 1.3504 | 0.1663 |
| 7 | 7 | 1.3522 | 0.3919 | 14 | 11 | 0.6618 | 0.1272 | 18 | 18 | 1.8200 | 0.2845 |
| 8 | 5 | 0.1525 | 0.1872 | 14 | 12 | 0.9011 | 0.1457 | 19 | 10 | 0.0000 | 0.0808 |
| 8 | 6 | 0.4728 | 0.2008 | 14 | 13 | 1.2079 | 0.1844 | 19 | 11 | 0.1307 | 0.0813 |
| 8 | 7 | 0.8522 | 0.2394 | 14 | 14 | 1.7034 | 0.3077 | 19 | 12 | 0.2637 | 0.0828 |
| 8 | 8 | 1.4236 | 0.3729 | 15 | 8 | 0.0000 | 0.1017 | 19 | 13 | 0.4016 | 0.0856 |
| 9 | 5 | 0.0000 | 0.1661 | 15 | 9 | 0.1653 | 0.1027 | 19 | 14 | 0.5477 | 0.0900 |
| 9 | 6 | 0.2745 | 0.1706 | 15 | 10 | 0.3353 | 0.1059 | 19 | 15 | 0.7066 | 0.0968 |
| 9 | 7 | 0.5720 | 0.1864 | 15 | 11 | 0.5157 | 0.1119 | 19 | 16 | 0.8859 | 0.1075 |
| 9 | 8 | 0.9323 | 0.2257 | 15 | 12 | 0.7149 | 0.1222 | 19 | 17 | 1.0995 | 0.1257 |
| 9 | 9 | 1.4850 | 0.3574 | 15 | 13 | 0.9477 | 0.1407 | 19 | 18 | 1.3799 | 0.1628 |
| 10 | 6 | 0.1227 | 0.1511 | 15 | 14 | 1.2479 | 0.1791 | 19 | 19 | 1.8445 | 0.2799 |
| 10 | 7 | 0.3758 | 0.1579 | 15 | 15 | 1.7359 | 0.3010 | 20 | 11 | 0.0620 | 0.0769 |
| 10 | 8 | 0.6561 | 0.1750 | 16 | 9 | 0.0773 | 0.0957 | 20 | 12 | 0.1870 | 0.0778 |
| 10 | 9 | 1.0014 | 0.2145 | 16 | 10 | 0.2338 | 0.0974 | 20 | 13 | 0.3149 | 0.0796 |
| 10 | 10 | 1.5388 | 0.3443 | 16 | 11 | 0.3962 | 0.1010 | 20 | 14 | 0.4483 | 0.0826 |
| 11 | 6 | 0.0000 | 0.1372 | 16 | 12 | 0.5700 | 0.1074 | 20 | 15 | 0.5903 | 0.0872 |
| 11 | 7 | 0.2249 | 0.1396 | 16 | 13 | 0.7632 | 0.1179 | 20 | 16 | 0.7454 | 0.0940 |
| 11 | 8 | 0.4620 | 0.1480 | 16 | 14 | 0.9903 | 0.1363 | 20 | 17 | 0.9210 | 0.1047 |
| 11 | 9 | 0.7288 | 0.1657 | 16 | 15 | 1.2847 | 0.1744 | 20 | 18 | 1.1309 | 0.1228 |
| 11 | 10 | 1.0619 | 0.2052 | 16 | 16 | 1.7660 | 0.2950 | 20 | 19 | 1.4076 | 0.1596 |
| 11 | 11 | 1.5864 | 0.3332 | 17 | 9 | 0.0000 | 0.0900 | 20 | 20 | 1.8675 | 0.2757 |
| 12 | 7 | 0.1026 | 0.1266 | 17 | 10 | 0.1460 | 0.0907 | | | | |
| 12 | 8 | 0.3122 | 0.1306 | 17 | 11 | 0.2952 | 0.0929 | | | | |

TABLE 2

OUTLIER MEAN 0.0 OUTLIER STANDARD DEVIATION 2.0

| N | R | MEAN | VAR | N | R | MEAN | VAR | N | R | MEAN | VAR |
|---|---|---|---|---|---|---|---|---|---|---|---|
| 2 | 2 | 0.8921 | 1.7042 | 12 | 9 | 0.5631 | 0.1549 | 17 | 12 | 0.4661 | 0.1036 |
| 3 | 2 | 0.0000 | 0.6596 | 12 | 10 | 0.8357 | 0.1792 | 17 | 13 | 0.6408 | 0.1114 |
| 3 | 3 | 1.1742 | 1.2916 | 12 | 11 | 1.1896 | 0.2390 | 17 | 14 | 0.8372 | 0.1244 |
| 4 | 3 | 0.3472 | 0.4863 | 12 | 12 | 1.8531 | 0.6628 | 17 | 15 | 1.0722 | 0.1481 |
| 4 | 4 | 1.3405 | 1.0962 | 13 | 7 | 0.0000 | 0.1265 | 17 | 16 | 1.3870 | 0.2031 |
| 5 | 3 | 0.0000 | 0.3571 | 13 | 8 | 0.1985 | 0.1284 | 17 | 17 | 1.9923 | 0.5792 |
| 5 | 4 | 0.5619 | 0.4048 | 13 | 9 | 0.4051 | 0.1344 | 18 | 10 | 0.0708 | 0.0903 |
| 5 | 5 | 1.4574 | 0.9769 | 13 | 10 | 0.6305 | 0.1467 | 18 | 11 | 0.2139 | 0.0917 |
| 6 | 4 | 0.2218 | 0.2962 | 13 | 11 | 0.8930 | 0.1711 | 18 | 12 | 0.3615 | 0.0946 |
| 6 | 5 | 0.7153 | 0.3556 | 13 | 12 | 1.2368 | 0.2298 | 18 | 13 | 0.5173 | 0.0995 |
| 6 | 6 | 1.5469 | 0.8945 | 13 | 13 | 1.8858 | 0.6421 | 18 | 14 | 0.6866 | 0.1074 |
| 7 | 4 | 0.0000 | 0.2452 | 14 | 8 | 0.0915 | 0.1174 | 18 | 15 | 0.8781 | 0.1204 |
| 7 | 5 | 0.3831 | 0.2592 | 14 | 9 | 0.2777 | 0.1204 | 18 | 16 | 1.1081 | 0.1439 |
| 7 | 6 | 0.8335 | 0.3219 | 14 | 10 | 0.4740 | 0.1272 | 18 | 17 | 1.4176 | 0.1981 |
| 7 | 7 | 1.6192 | 0.8331 | 14 | 11 | 0.6903 | 0.1399 | 18 | 18 | 2.0145 | 0.5670 |
| 8 | 5 | 0.1634 | 0.2141 | 14 | 12 | 0.9444 | 0.1641 | 19 | 10 | 0.0000 | 0.0853 |
| 8 | 6 | 0.5088 | 0.2338 | 14 | 13 | 1.2794 | 0.2219 | 19 | 11 | 0.1344 | 0.0858 |
| 8 | 7 | 0.9290 | 0.2970 | 14 | 14 | 1.9157 | 0.6238 | 19 | 12 | 0.2712 | 0.0876 |
| 8 | 8 | 1.6797 | 0.7851 | 15 | 8 | 0.0000 | 0.1089 | 19 | 13 | 0.4132 | 0.0908 |
| 9 | 5 | 0.0000 | 0.1867 | 15 | 9 | 0.1712 | 0.1101 | 19 | 14 | 0.5641 | 0.0959 |
| 9 | 6 | 0.2920 | 0.1928 | 15 | 10 | 0.3476 | 0.1139 | 19 | 15 | 0.7288 | 0.1039 |
| 9 | 7 | 0.6112 | 0.2151 | 15 | 11 | 0.5354 | 0.1211 | 19 | 16 | 0.9159 | 0.1168 |
| 9 | 8 | 1.0089 | 0.2778 | 15 | 12 | 0.7440 | 0.1340 | 19 | 17 | 1.1415 | 0.1401 |
| 9 | 9 | 1.7315 | 0.7463 | 15 | 13 | 0.9908 | 0.1581 | 19 | 18 | 1.4461 | 0.1935 |
| 10 | 6 | 0.1295 | 0.1679 | 15 | 14 | 1.3183 | 0.2149 | 19 | 19 | 2.0352 | 0.5558 |
| 10 | 7 | 0.3974 | 0.1770 | 15 | 15 | 1.9432 | 0.6074 | 20 | 11 | 0.0636 | 0.0810 |
| 10 | 8 | 0.6972 | 0.2005 | 16 | 9 | 0.0799 | 0.1021 | 20 | 12 | 0.1919 | 0.0820 |
| 10 | 9 | 1.0772 | 0.2624 | 16 | 10 | 0.2416 | 0.1041 | 20 | 13 | 0.3234 | 0.0840 |
| 10 | 10 | 1.7768 | 0.7140 | 16 | 11 | 0.4099 | 0.1084 | 20 | 14 | 0.4607 | 0.0875 |
| 11 | 6 | 0.0000 | 0.1508 | 16 | 12 | 0.5906 | 0.1159 | 20 | 15 | 0.6072 | 0.0927 |
| 11 | 7 | 0.2363 | 0.1540 | 16 | 13 | 0.7927 | 0.1289 | 20 | 16 | 0.7679 | 0.1007 |
| 11 | 8 | 0.4864 | 0.1647 | 16 | 14 | 1.0332 | 0.1528 | 20 | 17 | 0.9510 | 0.1136 |
| 11 | 9 | 0.7711 | 0.1888 | 16 | 15 | 1.3540 | 0.2086 | 20 | 18 | 1.1727 | 0.1366 |
| 11 | 10 | 1.1368 | 0.2497 | 16 | 16 | 1.9686 | 0.5926 | 20 | 19 | 1.4728 | 0.1893 |
| 11 | 11 | 1.8170 | 0.6865 | 17 | 9 | 0.0000 | 0.0957 | 20 | 20 | 2.0548 | 0.5454 |
| 12 | 7 | 0.1073 | 0.1382 | 17 | 10 | 0.1506 | 0.0965 | | | | |
| 12 | 8 | 0.3268 | 0.1431 | 17 | 11 | 0.3046 | 0.0990 | | | | |

TABLE 2

OUTLIER MEAN 0.0 OUTLIER STANDARD DEVIATION 3.0

| N | R | MEAN | VAR | N | R | MEAN | VAR | N | R | MEAN | VAR |
|---|---|---|---|---|---|---|---|---|---|---|---|
| 2 | 2 | 1.2616 | 3.4085 | 12 | 9 | 0.5735 | 0.1612 | 17 | 12 | 0.4717 | 0.1063 |
| 3 | 2 | 0.0000 | 0.7610 | 12 | 10 | 0.8543 | 0.1898 | 17 | 13 | 0.6490 | 0.1148 |
| 3 | 3 | 1.5437 | 2.7366 | 12 | 11 | 1.2287 | 0.2676 | 17 | 14 | 0.8496 | 0.1293 |
| 4 | 3 | 0.3702 | 0.5459 | 12 | 12 | 2.1725 | 1.6811 | 17 | 15 | 1.0921 | 0.1570 |
| 4 | 4 | 1.7023 | 2.4191 | 13 | 7 | 0.0000 | 0.1300 | 17 | 16 | 1.4262 | 0.2284 |
| 5 | 3 | 0.0000 | 0.3860 | 13 | 8 | 0.2014 | 0.1320 | 17 | 17 | 2.2975 | 1.5266 |
| 5 | 4 | 0.5926 | 0.4516 | 13 | 9 | 0.4114 | 0.1388 | 18 | 10 | 0.0715 | 0.0921 |
| 5 | 5 | 1.8115 | 2.2225 | 13 | 10 | 0.6414 | 0.1527 | 18 | 11 | 0.2161 | 0.0936 |
| 6 | 4 | 0.2299 | 0.3166 | 13 | 11 | 0.9120 | 0.1812 | 18 | 12 | 0.3653 | 0.0967 |
| 6 | 5 | 0.7496 | 0.3961 | 13 | 12 | 1.2760 | 0.2576 | 18 | 13 | 0.5231 | 0.1020 |
| 6 | 6 | 1.8942 | 2.0846 | 13 | 13 | 2.2019 | 1.6433 | 18 | 14 | 0.6952 | 0.1107 |
| 7 | 4 | 0.0000 | 0.2586 | 14 | 8 | 0.0928 | 0.1204 | 18 | 15 | 0.8908 | 0.1252 |
| 7 | 5 | 0.3953 | 0.2757 | 14 | 9 | 0.2816 | 0.1237 | 18 | 16 | 1.1282 | 0.1525 |
| 7 | 6 | 0.8697 | 0.3586 | 14 | 10 | 0.4810 | 0.1312 | 18 | 17 | 1.4568 | 0.2230 |
| 7 | 7 | 1.9604 | 1.9804 | 14 | 11 | 0.7017 | 0.1455 | 18 | 18 | 2.3174 | 1.5034 |
| 8 | 5 | 0.1676 | 0.2245 | 14 | 12 | 0.9637 | 0.1739 | 19 | 10 | 0.0000 | 0.0868 |
| 8 | 6 | 0.5234 | 0.2481 | 14 | 13 | 1.3187 | 0.2489 | 19 | 11 | 0.1356 | 0.0874 |
| 8 | 7 | 0.9665 | 0.3312 | 14 | 14 | 2.2288 | 1.6097 | 19 | 12 | 0.2738 | 0.0893 |
| 8 | 8 | 2.0155 | 1.8977 | 15 | 8 | 0.0000 | 0.1115 | 19 | 13 | 0.4175 | 0.0928 |
| 9 | 5 | 0.0000 | 0.1944 | 15 | 9 | 0.1734 | 0.1128 | 19 | 14 | 0.5702 | 0.0983 |
| 9 | 6 | 0.2986 | 0.2014 | 15 | 10 | 0.3521 | 0.1169 | 19 | 15 | 0.7376 | 0.1070 |
| 9 | 7 | 0.6273 | 0.2280 | 15 | 11 | 0.5429 | 0.1249 | 19 | 16 | 0.9287 | 0.1215 |
| 9 | 8 | 1.0470 | 0.3101 | 15 | 12 | 0.7559 | 0.1393 | 19 | 17 | 1.1617 | 0.1486 |
| 9 | 9 | 2.0626 | 1.8299 | 15 | 13 | 1.0104 | 0.1675 | 19 | 18 | 1.4852 | 0.2180 |
| 10 | 6 | 0.1320 | 0.1742 | 15 | 14 | 1.3576 | 0.2413 | 19 | 19 | 2.3360 | 1.4821 |
| 10 | 7 | 0.4057 | 0.1846 | 15 | 15 | 2.2535 | 1.5793 | 20 | 11 | 0.0642 | 0.0824 |
| 10 | 8 | 0.7144 | 0.2124 | 16 | 9 | 0.0808 | 0.1044 | 20 | 12 | 0.1937 | 0.0835 |
| 10 | 9 | 1.1159 | 0.2932 | 16 | 10 | 0.2445 | 0.1065 | 20 | 13 | 0.3265 | 0.0857 |
| 10 | 10 | 2.1036 | 1.7728 | 16 | 11 | 0.4150 | 0.1112 | 20 | 14 | 0.4653 | 0.0893 |
| 11 | 6 | 0.0000 | 0.1558 | 16 | 12 | 0.5985 | 0.1195 | 20 | 15 | 0.6136 | 0.0950 |
| 11 | 7 | 0.2405 | 0.1593 | 16 | 13 | 0.8049 | 0.1340 | 20 | 16 | 0.7769 | 0.1038 |
| 11 | 8 | 0.4959 | 0.1716 | 16 | 14 | 1.0530 | 0.1619 | 20 | 17 | 0.9640 | 0.1181 |
| 11 | 9 | 0.7891 | 0.2000 | 16 | 15 | 1.3933 | 0.2345 | 20 | 18 | 1.1929 | 0.1449 |
| 11 | 10 | 1.1758 | 0.2793 | 16 | 16 | 2.2763 | 1.5517 | 20 | 19 | 1.5119 | 0.2135 |
| 11 | 11 | 2.1399 | 1.7238 | 17 | 9 | 0.0000 | 0.0976 | 20 | 20 | 2.3535 | 1.4622 |
| 12 | 7 | 0.1089 | 0.1424 | 17 | 10 | 0.1522 | 0.0985 | | | | |
| 12 | 8 | 0.3322 | 0.1479 | 17 | 11 | 0.3080 | 0.1012 | | | | |

TABLE 2

OUTLIER MEAN 0.0 OUTLIER STANDARD DEVIATION 4.0

| N | R | MEAN | VAR | N | R | MEAN | VAR | N | R | MEAN | VAR |
|---|---|---|---|---|---|---|---|---|---|---|---|
| 2 | 2 | 1.6449 | 5.7944 | 12 | 9 | 0.5788 | 0.1645 | 17 | 12 | 0.4745 | 0.1076 |
| 3 | 2 | 0.0000 | 0.8173 | 12 | 10 | 0.8642 | 0.1955 | 17 | 13 | 0.6533 | 0.1166 |
| 3 | 3 | 1.9270 | 4.8781 | 12 | 11 | 1.2509 | 0.2844 | 17 | 14 | 0.8562 | 0.1320 |
| 4 | 3 | 0.3828 | 0.5786 | 12 | 12 | 2.5280 | 3.3990 | 17 | 15 | 1.1029 | 0.1619 |
| 4 | 4 | 2.0815 | 4.4424 | 13 | 7 | 0.0000 | 0.1318 | 17 | 16 | 1.4490 | 0.2437 |
| 5 | 3 | 0.0000 | 0.4012 | 13 | 8 | 0.2029 | 0.1339 | 17 | 17 | 2.6450 | 3.1708 |
| 5 | 4 | 0.6094 | 0.4775 | 13 | 9 | 0.4146 | 0.1410 | 18 | 10 | 0.0719 | 0.0930 |
| 5 | 5 | 2.1865 | 4.1699 | 13 | 10 | 0.6471 | 0.1558 | 18 | 11 | 0.2172 | 0.0945 |
| 6 | 4 | 0.2342 | 0.3273 | 13 | 11 | 0.9221 | 0.1867 | 18 | 12 | 0.3673 | 0.0978 |
| 6 | 5 | 0.7684 | 0.4188 | 13 | 12 | 1.2983 | 0.2740 | 18 | 13 | 0.5262 | 0.1033 |
| 6 | 6 | 2.2654 | 3.9767 | 13 | 13 | 2.5556 | 3.3437 | 18 | 14 | 0.6996 | 0.1124 |
| 7 | 4 | 0.0000 | 0.2655 | 14 | 8 | 0.0934 | 0.1220 | 18 | 15 | 0.8975 | 0.1277 |
| 7 | 5 | 0.4017 | 0.2844 | 14 | 9 | 0.2835 | 0.1254 | 18 | 16 | 1.1391 | 0.1574 |
| 7 | 6 | 0.8898 | 0.3794 | 14 | 10 | 0.4846 | 0.1333 | 18 | 17 | 1.4796 | 0.2380 |
| 7 | 7 | 2.3283 | 3.8295 | 14 | 11 | 0.7077 | 0.1484 | 18 | 18 | 2.6635 | 3.1362 |
| 8 | 5 | 0.1697 | 0.2298 | 14 | 12 | 0.9740 | 0.1792 | 19 | 10 | 0.0000 | 0.0876 |
| 8 | 6 | 0.5311 | 0.2557 | 14 | 13 | 1.3412 | 0.2649 | 19 | 11 | 0.1363 | 0.0883 |
| 8 | 7 | 0.9872 | 0.3507 | 14 | 14 | 2.5808 | 3.2941 | 19 | 12 | 0.2752 | 0.0902 |
| 8 | 8 | 2.3804 | 3.7119 | 15 | 8 | 0.0000 | 0.1128 | 19 | 13 | 0.4196 | 0.0938 |
| 9 | 5 | 0.0000 | 0.1984 | 15 | 9 | 0.1744 | 0.1142 | 19 | 14 | 0.5734 | 0.0995 |
| 9 | 6 | 0.3021 | 0.2059 | 15 | 10 | 0.3544 | 0.1185 | 19 | 15 | 0.7422 | 0.1087 |
| 9 | 7 | 0.6359 | 0.2349 | 15 | 11 | 0.5468 | 0.1269 | 19 | 16 | 0.9356 | 0.1240 |
| 9 | 8 | 1.0683 | 0.3286 | 15 | 12 | 0.7621 | 0.1422 | 19 | 17 | 1.1727 | 0.1533 |
| 9 | 9 | 2.4248 | 3.6147 | 15 | 13 | 1.0209 | 0.1727 | 19 | 18 | 1.5081 | 0.2328 |
| 10 | 6 | 0.1333 | 0.1774 | 15 | 14 | 1.3802 | 0.2570 | 19 | 19 | 2.6808 | 3.1041 |
| 10 | 7 | 0.4100 | 0.1885 | 15 | 15 | 2.6038 | 3.2492 | 20 | 11 | 0.0645 | 0.0831 |
| 10 | 8 | 0.7235 | 0.2188 | 16 | 9 | 0.0812 | 0.1055 | 20 | 12 | 0.1945 | 0.0842 |
| 10 | 9 | 1.1375 | 0.3110 | 16 | 10 | 0.2459 | 0.1077 | 20 | 13 | 0.3280 | 0.0865 |
| 10 | 10 | 2.4634 | 3.5324 | 16 | 11 | 0.4176 | 0.1126 | 20 | 14 | 0.4676 | 0.0903 |
| 11 | 6 | 0.0000 | 0.1583 | 16 | 12 | 0.6027 | 0.1213 | 20 | 15 | 0.6170 | 0.0961 |
| 11 | 7 | 0.2426 | 0.1621 | 16 | 13 | 0.8113 | 0.1367 | 20 | 16 | 0.7816 | 0.1053 |
| 11 | 8 | 0.5008 | 0.1752 | 16 | 14 | 1.0637 | 0.1670 | 20 | 17 | 0.9709 | 0.1206 |
| 11 | 9 | 0.7986 | 0.2060 | 16 | 15 | 1.4160 | 0.2500 | 20 | 18 | 1.2040 | 0.1496 |
| 11 | 10 | 1.1977 | 0.2965 | 16 | 16 | 2.6252 | 3.2083 | 20 | 19 | 1.5348 | 0.2281 |
| 11 | 11 | 2.4975 | 3.4613 | 17 | 9 | 0.0000 | 0.0987 | 20 | 20 | 2.6971 | 3.0743 |
| 12 | 7 | 0.1098 | 0.1445 | 17 | 10 | 0.1530 | 0.0995 | | | | |
| 12 | 8 | 0.3350 | 0.1504 | 17 | 11 | 0.3097 | 0.1024 | | | | |

TABLE 2

OUTLIER MEAN 0.0 OUTLIER STANDARD DEVIATION 6.0

| N | R | MEAN | VAR | N | R | MEAN | VAR | N | R | MEAN | VAR |
|---|---|---|---|---|---|---|---|---|---|---|---|
| 2 | 2 | 2.4267 | 12.6113 | 12 | 9 | 0.5843 | 0.1679 | 17 | 12 | 0.4773 | 0.1090 |
| 3 | 2 | 0.0000 | 0.8764 | 12 | 10 | 0.8744 | 0.2014 | 17 | 13 | 0.6577 | 0.1184 |
| 3 | 3 | 2.7088 | 11.2244 | 12 | 11 | 1.2744 | 0.3022 | 17 | 14 | 0.8629 | 0.1347 |
| 4 | 3 | 0.3959 | 0.6127 | 12 | 12 | 3.2807 | 8.9095 | 17 | 15 | 1.1142 | 0.1670 |
| 4 | 4 | 2.8589 | 10.5573 | 13 | 7 | 0.0000 | 0.1336 | 17 | 16 | 1.4734 | 0.2600 |
| 5 | 3 | 0.0000 | 0.4168 | 13 | 8 | 0.2044 | 0.1358 | 17 | 17 | 3.3889 | 8.5342 |
| 5 | 4 | 0.6269 | 0.5047 | 13 | 9 | 0.4179 | 0.1433 | 18 | 10 | 0.0723 | 0.0939 |
| 5 | 5 | 2.9595 | 10.1352 | 13 | 10 | 0.6530 | 0.1589 | 18 | 11 | 0.2183 | 0.0955 |
| 6 | 4 | 0.2386 | 0.3382 | 13 | 11 | 0.9327 | 0.1924 | 18 | 12 | 0.3693 | 0.0988 |
| 6 | 5 | 0.7881 | 0.4426 | 13 | 12 | 1.3222 | 0.2914 | 18 | 13 | 0.5292 | 0.1046 |
| 6 | 6 | 3.0345 | 9.8331 | 13 | 13 | 3.3063 | 8.8192 | 18 | 14 | 0.7042 | 0.1141 |
| 7 | 4 | 0.0000 | 0.2725 | 14 | 8 | 0.0940 | 0.1235 | 18 | 15 | 0.9043 | 0.1304 |
| 7 | 5 | 0.4083 | 0.2933 | 14 | 9 | 0.2855 | 0.1271 | 18 | 16 | 1.1505 | 0.1623 |
| 7 | 6 | 0.9107 | 0.4012 | 14 | 10 | 0.4882 | 0.1354 | 18 | 17 | 1.5041 | 0.2541 |
| 7 | 7 | 3.0939 | 9.6009 | 14 | 11 | 0.7139 | 0.1514 | 18 | 18 | 3.4060 | 8.4767 |
| 8 | 5 | 0.1719 | 0.2352 | 14 | 12 | 0.9848 | 0.1847 | 19 | 10 | 0.0000 | 0.0884 |
| 8 | 6 | 0.5390 | 0.2635 | 14 | 13 | 1.3652 | 0.2820 | 19 | 11 | 0.1370 | 0.0891 |
| 8 | 7 | 1.0091 | 0.3712 | 14 | 14 | 3.3296 | 8.7378 | 19 | 12 | 0.2765 | 0.0911 |
| 8 | 8 | 3.1429 | 9.4139 | 15 | 8 | 0.0000 | 0.1141 | 19 | 13 | 0.4218 | 0.0948 |
| 9 | 5 | 0.0000 | 0.2024 | 15 | 9 | 0.1755 | 0.1155 | 19 | 14 | 0.5767 | 0.1008 |
| 9 | 6 | 0.3056 | 0.2105 | 15 | 10 | 0.3567 | 0.1200 | 19 | 15 | 0.7469 | 0.1103 |
| 9 | 7 | 0.6446 | 0.2419 | 15 | 11 | 0.5507 | 0.1288 | 19 | 16 | 0.9426 | 0.1265 |
| 9 | 8 | 1.0908 | 0.3482 | 15 | 12 | 0.7685 | 0.1450 | 19 | 17 | 1.1842 | 0.1582 |
| 9 | 9 | 3.1845 | 9.2584 | 15 | 13 | 1.0319 | 0.1780 | 19 | 18 | 1.5327 | 0.2487 |
| 10 | 6 | 0.1346 | 0.1806 | 15 | 14 | 1.4044 | 0.2738 | 19 | 19 | 3.4219 | 8.4232 |
| 10 | 7 | 0.4144 | 0.1925 | 15 | 15 | 3.3509 | 8.6639 | 20 | 11 | 0.0648 | 0.0839 |
| 10 | 8 | 0.7329 | 0.2254 | 16 | 9 | 0.0817 | 0.1067 | 20 | 12 | 0.1954 | 0.0850 |
| 10 | 9 | 1.1605 | 0.3299 | 16 | 10 | 0.2474 | 0.1090 | 20 | 13 | 0.3296 | 0.0873 |
| 10 | 10 | 3.2206 | 9.1259 | 16 | 11 | 0.4202 | 0.1141 | 20 | 14 | 0.4699 | 0.0912 |
| 11 | 6 | 0.0000 | 0.1609 | 16 | 12 | 0.6068 | 0.1232 | 20 | 15 | 0.6203 | 0.0973 |
| 11 | 7 | 0.2448 | 0.1649 | 16 | 13 | 0.8179 | 0.1395 | 20 | 16 | 0.7864 | 0.1070 |
| 11 | 8 | 0.5058 | 0.1788 | 16 | 14 | 1.0748 | 0.1721 | 20 | 17 | 0.9781 | 0.1231 |
| 11 | 9 | 0.8085 | 0.2122 | 16 | 15 | 1.4403 | 0.2665 | 20 | 18 | 1.2156 | 0.1544 |
| 11 | 10 | 1.2210 | 0.3148 | 16 | 16 | 3.3706 | 8.5964 | 20 | 19 | 1.5595 | 0.2438 |
| 11 | 11 | 3.2523 | 9.0108 | 17 | 9 | 0.0000 | 0.0997 | 20 | 20 | 3.4369 | 8.3732 |
| 12 | 7 | 0.1107 | 0.1467 | 17 | 10 | 0.1538 | 0.1006 | | | | |
| 12 | 8 | 0.3378 | 0.1529 | 17 | 11 | 0.3115 | 0.1035 | | | | |

TABLE 2

OUTLIER MEAN 0.0 OUTLIER STANDARD DEVIATION 8.0

| N | R | MEAN | VAR | N | R | MEAN | VAR | N | R | MEAN | VAR |
|---|---|---|---|---|---|---|---|---|---|---|---|
| 2 | 2 | 3.2164 | 22.1549 | 12 | 9 | 0.5871 | 0.1696 | 17 | 12 | 0.4788 | 0.1097 |
| 3 | 2 | 0.0000 | 0.9068 | 12 | 10 | 0.8796 | 0.2044 | 17 | 13 | 0.6599 | 0.1193 |
| 3 | 3 | 3.4985 | 20.3073 | 12 | 11 | 1.2866 | 0.3113 | 17 | 14 | 0.8664 | 0.1360 |
| 4 | 3 | 0.4026 | 0.6302 | 12 | 12 | 4.0554 | 17.1652 | 17 | 15 | 1.1199 | 0.1695 |
| 4 | 4 | 3.6464 | 19.4118 | 13 | 7 | 0.0000 | 0.1345 | 17 | 16 | 1.4861 | 0.2683 |
| 5 | 3 | 0.0000 | 0.4247 | 13 | 8 | 0.2052 | 0.1368 | 17 | 17 | 4.1591 | 16.6441 |
| 5 | 4 | 0.6359 | 0.5185 | 13 | 9 | 0.4196 | 0.1444 | 18 | 10 | 0.0724 | 0.0944 |
| 5 | 5 | 3.7447 | 18.8418 | 13 | 10 | 0.6559 | 0.1605 | 18 | 11 | 0.2189 | 0.0959 |
| 6 | 4 | 0.2408 | 0.3437 | 13 | 11 | 0.9380 | 0.1952 | 18 | 12 | 0.3703 | 0.0994 |
| 6 | 5 | 0.7982 | 0.4547 | 13 | 12 | 1.3345 | 0.3003 | 18 | 13 | 0.5308 | 0.1053 |
| 6 | 6 | 3.8177 | 18.4317 | 13 | 13 | 4.0800 | 17.0402 | 18 | 14 | 0.7065 | 0.1150 |
| 7 | 4 | 0.0000 | 0.2761 | 14 | 8 | 0.0943 | 0.1243 | 18 | 15 | 0.9078 | 0.1317 |
| 7 | 5 | 0.4116 | 0.2978 | 14 | 9 | 0.2865 | 0.1280 | 18 | 16 | 1.1563 | 0.1648 |
| 7 | 6 | 0.9215 | 0.4124 | 14 | 10 | 0.4901 | 0.1365 | 18 | 17 | 1.5169 | 0.2623 |
| 7 | 7 | 3.8753 | 18.1152 | 14 | 11 | 0.7170 | 0.1530 | 18 | 18 | 4.1754 | 16.5638 |
| 8 | 5 | 0.1730 | 0.2379 | 14 | 12 | 0.9903 | 0.1874 | 19 | 10 | 0.0000 | 0.0888 |
| 8 | 6 | 0.5430 | 0.2674 | 14 | 13 | 1.3777 | 0.2907 | 19 | 11 | 0.1373 | 0.0895 |
| 8 | 7 | 1.0203 | 0.3817 | 14 | 14 | 4.1023 | 16.9273 | 19 | 12 | 0.2772 | 0.0915 |
| 8 | 8 | 3.9227 | 17.8594 | 15 | 8 | 0.0000 | 0.1148 | 19 | 13 | 0.4229 | 0.0953 |
| 9 | 5 | 0.0000 | 0.2044 | 15 | 9 | 0.1761 | 0.1162 | 19 | 14 | 0.5783 | 0.1014 |
| 9 | 6 | 0.3073 | 0.2127 | 15 | 10 | 0.3579 | 0.1208 | 19 | 15 | 0.7493 | 0.1112 |
| 9 | 7 | 0.6491 | 0.2455 | 15 | 11 | 0.5527 | 0.1298 | 19 | 16 | 0.9461 | 0.1278 |
| 9 | 8 | 1.1024 | 0.3582 | 15 | 12 | 0.7717 | 0.1465 | 19 | 17 | 1.1901 | 0.1606 |
| 9 | 9 | 3.9629 | 17.6461 | 15 | 13 | 1.0375 | 0.1807 | 19 | 18 | 1.5455 | 0.2568 |
| 10 | 6 | 0.1353 | 0.1823 | 15 | 14 | 1.4169 | 0.2823 | 19 | 19 | 4.1907 | 16.4891 |
| 10 | 7 | 0.4166 | 0.1945 | 15 | 15 | 4.1228 | 16.8247 | 20 | 11 | 0.0649 | 0.0842 |
| 10 | 8 | 0.7377 | 0.2287 | 16 | 9 | 0.0819 | 0.1073 | 20 | 12 | 0.1959 | 0.0853 |
| 10 | 9 | 1.1723 | 0.3395 | 16 | 10 | 0.2481 | 0.1096 | 20 | 13 | 0.3303 | 0.0877 |
| 10 | 10 | 3.9976 | 17.4638 | 16 | 11 | 0.4215 | 0.1148 | 20 | 14 | 0.4711 | 0.0917 |
| 11 | 6 | 0.0000 | 0.1622 | 16 | 12 | 0.6089 | 0.1242 | 20 | 15 | 0.6220 | 0.0979 |
| 11 | 7 | 0.2459 | 0.1663 | 16 | 13 | 0.8212 | 0.1409 | 20 | 16 | 0.7889 | 0.1078 |
| 11 | 8 | 0.5083 | 0.1806 | 16 | 14 | 1.0805 | 0.1748 | 20 | 17 | 0.9817 | 0.1243 |
| 11 | 9 | 0.8135 | 0.2153 | 16 | 15 | 1.4529 | 0.2749 | 20 | 18 | 1.2215 | 0.1568 |
| 11 | 10 | 1.2331 | 0.3241 | 16 | 16 | 4.1416 | 16.7307 | 20 | 19 | 1.5723 | 0.2519 |
| 11 | 11 | 4.0282 | 17.3053 | 17 | 9 | 0.0000 | 0.1002 | 20 | 20 | 4.2050 | 16.4192 |
| 12 | 7 | 0.1111 | 0.1478 | 17 | 10 | 0.1543 | 0.1011 | | | | |
| 12 | 8 | 0.3392 | 0.1541 | 17 | 11 | 0.3123 | 0.1041 | | | | |

TABLE 2

OUTLIER MEAN 0.0 OUTLIER STANDARD DEVIATION 10.0

| N | R | MEAN | VAR | N | R | MEAN | VAR | N | R | MEAN | VAR |
|---|---|---|---|---|---|---|---|---|---|---|---|
| 2 | 2 | 4.0093 | 34.4253 | 12 | 9 | 0.5887 | 0.1706 | 17 | 12 | 0.4796 | 0.1101 |
| 3 | 2 | 0.0000 | 0.9253 | 12 | 10 | 0.8827 | 0.2062 | 17 | 13 | 0.6612 | 0.1198 |
| 3 | 3 | 4.2914 | 32.1211 | 12 | 11 | 1.2941 | 0.3167 | 17 | 14 | 0.8684 | 0.1368 |
| 4 | 3 | 0.4067 | 0.6407 | 12 | 12 | 4.8393 | 28.1561 | 17 | 15 | 1.1234 | 0.1711 |
| 4 | 4 | 4.4379 | 30.9985 | 13 | 7 | 0.0000 | 0.1350 | 17 | 16 | 1.4939 | 0.2733 |
| 5 | 3 | 0.0000 | 0.4295 | 13 | 8 | 0.2056 | 0.1373 | 17 | 17 | 4.9402 | 27.4899 |
| 5 | 4 | 0.6413 | 0.5269 | 13 | 9 | 0.4206 | 0.1451 | 18 | 10 | 0.0726 | 0.0946 |
| 5 | 5 | 4.5350 | 30.2813 | 13 | 10 | 0.6577 | 0.1615 | 18 | 11 | 0.2192 | 0.0962 |
| 6 | 4 | 0.2421 | 0.3471 | 13 | 11 | 0.9412 | 0.1970 | 18 | 12 | 0.3709 | 0.0997 |
| 6 | 5 | 0.8043 | 0.4620 | 13 | 12 | 1.3420 | 0.3056 | 18 | 13 | 0.5317 | 0.1057 |
| 6 | 6 | 4.6067 | 29.7637 | 13 | 13 | 4.8632 | 27.9965 | 18 | 14 | 0.7079 | 0.1155 |
| 7 | 4 | 0.0000 | 0.2782 | 14 | 8 | 0.0945 | 0.1248 | 18 | 15 | 0.9099 | 0.1325 |
| 7 | 5 | 0.4136 | 0.3005 | 14 | 9 | 0.2870 | 0.1285 | 18 | 16 | 1.1599 | 0.1663 |
| 7 | 6 | 0.9281 | 0.4191 | 14 | 10 | 0.4912 | 0.1371 | 18 | 17 | 1.5247 | 0.2672 |
| 7 | 7 | 4.6632 | 29.3633 | 14 | 11 | 0.7188 | 0.1539 | 18 | 18 | 4.9561 | 27.3870 |
| 8 | 5 | 0.1737 | 0.2396 | 14 | 12 | 0.9936 | 0.1891 | 19 | 10 | 0.0000 | 0.0891 |
| 8 | 6 | 0.5454 | 0.2698 | 14 | 13 | 1.3852 | 0.2959 | 19 | 11 | 0.1375 | 0.0897 |
| 8 | 7 | 1.0272 | 0.3880 | 14 | 14 | 4.8850 | 27.8523 | 19 | 12 | 0.2776 | 0.0918 |
| 8 | 8 | 4.7097 | 29.0391 | 15 | 8 | 0.0000 | 0.1152 | 19 | 13 | 0.4236 | 0.0956 |
| 9 | 5 | 0.0000 | 0.2056 | 15 | 9 | 0.1764 | 0.1166 | 19 | 14 | 0.5793 | 0.1018 |
| 9 | 6 | 0.3084 | 0.2141 | 15 | 10 | 0.3586 | 0.1213 | 19 | 15 | 0.7507 | 0.1117 |
| 9 | 7 | 0.6517 | 0.2476 | 15 | 11 | 0.5539 | 0.1304 | 19 | 16 | 0.9483 | 0.1286 |
| 9 | 8 | 1.1094 | 0.3642 | 15 | 12 | 0.7736 | 0.1474 | 19 | 17 | 1.1937 | 0.1621 |
| 9 | 9 | 4.7489 | 28.7682 | 15 | 13 | 1.0409 | 0.1823 | 19 | 18 | 1.5534 | 0.2617 |
| 10 | 6 | 0.1357 | 0.1832 | 15 | 14 | 1.4246 | 0.2875 | 19 | 19 | 4.9709 | 27.2912 |
| 10 | 7 | 0.4179 | 0.1958 | 15 | 15 | 4.9049 | 27.7211 | 20 | 11 | 0.0650 | 0.0844 |
| 10 | 8 | 0.7406 | 0.2306 | 16 | 9 | 0.0821 | 0.1076 | 20 | 12 | 0.1962 | 0.0856 |
| 10 | 9 | 1.1795 | 0.3453 | 16 | 10 | 0.2485 | 0.1100 | 20 | 13 | 0.3308 | 0.0880 |
| 10 | 10 | 4.7829 | 28.5365 | 16 | 11 | 0.4223 | 0.1153 | 20 | 14 | 0.4718 | 0.0920 |
| 11 | 6 | 0.0000 | 0.1630 | 16 | 12 | 0.6102 | 0.1247 | 20 | 15 | 0.6231 | 0.0983 |
| 11 | 7 | 0.2466 | 0.1671 | 16 | 13 | 0.8232 | 0.1417 | 20 | 16 | 0.7903 | 0.1083 |
| 11 | 8 | 0.5098 | 0.1817 | 16 | 14 | 1.0839 | 0.1763 | 20 | 17 | 0.9839 | 0.1251 |
| 11 | 9 | 0.8165 | 0.2172 | 16 | 15 | 1.4606 | 0.2800 | 20 | 18 | 1.2251 | 0.1583 |
| 11 | 10 | 1.2404 | 0.3297 | 16 | 16 | 4.9232 | 27.6008 | 20 | 19 | 1.5802 | 0.2567 |
| 11 | 11 | 4.8127 | 28.3345 | 17 | 9 | 0.0000 | 0.1005 | 20 | 20 | 4.9848 | 27.2016 |
| 12 | 7 | 0.1114 | 0.1484 | 17 | 10 | 0.1545 | 0.1014 | | | | |
| 12 | 8 | 0.3400 | 0.1549 | 17 | 11 | 0.3129 | 0.1044 | | | | |

TABLE 3

OUTLIER MEAN 0.0 OUTLIER STANDARD DEVIATION 1.0

| R | S | N | COVAR | R | S | N | COVAR | R | S | N | COVAR |
|---|---|---|---|---|---|---|---|---|---|---|---|
| 1 | 2 | 2 | 0.3183 | 3 | 8 | 8 | 0.0602 | 5 | 10 | 10 | 0.0584 |
| 1 | 3 | 3 | 0.1649 | 4 | 8 | 8 | 0.0748 | 6 | 10 | 10 | 0.0707 |
| 2 | 3 | 3 | 0.2757 | 5 | 8 | 8 | 0.0947 | 7 | 10 | 10 | 0.0882 |
| 2 | 3 | 4 | 0.2359 | 6 | 8 | 8 | 0.1260 | 8 | 10 | 10 | 0.1163 |
| 1 | 4 | 4 | 0.1047 | 7 | 8 | 8 | 0.1863 | 9 | 10 | 10 | 0.1713 |
| 2 | 4 | 4 | 0.1580 | 4 | 6 | 9 | 0.1127 | 5 | 7 | 11 | 0.0992 |
| 3 | 4 | 4 | 0.2456 | 5 | 6 | 9 | 0.1370 | 6 | 7 | 11 | 0.1167 |
| 2 | 4 | 5 | 0.1499 | 3 | 7 | 9 | 0.0772 | 4 | 8 | 11 | 0.0725 |
| 3 | 4 | 5 | 0.2084 | 4 | 7 | 9 | 0.0934 | 5 | 8 | 11 | 0.0849 |
| 1 | 5 | 5 | 0.0742 | 5 | 7 | 9 | 0.1138 | 6 | 8 | 11 | 0.1000 |
| 2 | 5 | 5 | 0.1058 | 6 | 7 | 9 | 0.1421 | 7 | 8 | 11 | 0.1199 |
| 3 | 5 | 5 | 0.1481 | 2 | 8 | 9 | 0.0517 | 3 | 9 | 11 | 0.0528 |
| 4 | 5 | 5 | 0.2243 | 3 | 8 | 9 | 0.0632 | 4 | 9 | 11 | 0.0619 |
| 3 | 4 | 6 | 0.1833 | 4 | 8 | 9 | 0.0765 | 5 | 9 | 11 | 0.0725 |
| 2 | 5 | 6 | 0.1059 | 5 | 8 | 9 | 0.0934 | 6 | 9 | 11 | 0.0855 |
| 3 | 5 | 6 | 0.1397 | 6 | 8 | 9 | 0.1170 | 7 | 9 | 11 | 0.1026 |
| 4 | 5 | 6 | 0.1890 | 7 | 8 | 9 | 0.1541 | 8 | 9 | 11 | 0.1270 |
| 1 | 6 | 6 | 0.0563 | 1 | 9 | 9 | 0.0311 | 2 | 10 | 11 | 0.0371 |
| 2 | 6 | 6 | 0.0774 | 2 | 9 | 9 | 0.0401 | 3 | 10 | 11 | 0.0443 |
| 3 | 6 | 6 | 0.1024 | 3 | 9 | 9 | 0.0491 | 4 | 10 | 11 | 0.0520 |
| 4 | 6 | 6 | 0.1394 | 4 | 9 | 9 | 0.0595 | 5 | 10 | 11 | 0.0609 |
| 5 | 6 | 6 | 0.2085 | 5 | 9 | 9 | 0.0727 | 6 | 10 | 11 | 0.0719 |
| 3 | 5 | 7 | 0.1296 | 6 | 9 | 9 | 0.0913 | 7 | 10 | 11 | 0.0864 |
| 4 | 5 | 7 | 0.1656 | 7 | 9 | 9 | 0.1207 | 8 | 10 | 11 | 0.1071 |
| 2 | 6 | 7 | 0.0800 | 8 | 9 | 9 | 0.1781 | 9 | 10 | 11 | 0.1403 |
| 3 | 6 | 7 | 0.1020 | 5 | 6 | 10 | 0.1256 | 1 | 11 | 11 | 0.0233 |
| 4 | 6 | 7 | 0.1307 | 4 | 7 | 10 | 0.0889 | 2 | 11 | 11 | 0.0294 |
| 5 | 6 | 7 | 0.1745 | 5 | 7 | 10 | 0.1058 | 3 | 11 | 11 | 0.0351 |
| 1 | 7 | 7 | 0.0448 | 6 | 7 | 10 | 0.1275 | 4 | 11 | 11 | 0.0412 |
| 2 | 7 | 7 | 0.0599 | 3 | 8 | 10 | 0.0630 | 5 | 11 | 11 | 0.0484 |
| 3 | 7 | 7 | 0.0766 | 4 | 8 | 10 | 0.0749 | 6 | 11 | 11 | 0.0572 |
| 4 | 7 | 7 | 0.0985 | 5 | 8 | 10 | 0.0892 | 7 | 11 | 11 | 0.0688 |
| 5 | 7 | 7 | 0.1321 | 6 | 8 | 10 | 0.1077 | 8 | 11 | 11 | 0.0855 |
| 6 | 7 | 7 | 0.1962 | 7 | 8 | 10 | 0.1338 | 9 | 11 | 11 | 0.1124 |
| 4 | 5 | 8 | 0.1492 | 2 | 9 | 10 | 0.0434 | 10 | 11 | 11 | 0.1654 |
| 3 | 6 | 8 | 0.0978 | 3 | 9 | 10 | 0.0523 | 6 | 7 | 12 | 0.1084 |
| 4 | 6 | 8 | 0.1210 | 4 | 9 | 10 | 0.0622 | 5 | 8 | 12 | 0.0809 |
| 5 | 6 | 8 | 0.1524 | 5 | 9 | 10 | 0.0742 | 6 | 8 | 12 | 0.0937 |
| 2 | 7 | 8 | 0.0632 | 6 | 9 | 10 | 0.0897 | 7 | 8 | 12 | 0.1096 |
| 3 | 7 | 8 | 0.0787 | 7 | 9 | 10 | 0.1117 | 4 | 9 | 12 | 0.0607 |
| 4 | 7 | 8 | 0.0976 | 8 | 9 | 10 | 0.1466 | 5 | 9 | 12 | 0.0701 |
| 5 | 7 | 8 | 0.1233 | 1 | 10 | 10 | 0.0267 | 6 | 9 | 12 | 0.0812 |
| 6 | 7 | 8 | 0.1632 | 2 | 10 | 10 | 0.0340 | 7 | 9 | 12 | 0.0952 |
| 1 | 8 | 8 | 0.0368 | 3 | 10 | 10 | 0.0411 | 8 | 9 | 12 | 0.1136 |
| 2 | 8 | 8 | 0.0483 | 4 | 10 | 10 | 0.0489 | 3 | 10 | 12 | 0.0450 |

TABLE 3

OUTLIER MEAN 0.0 OUTLIER STANDARD DEVIATION 1.0

| R | S | N | COVAR | R | S | N | COVAR | R | S | N | COVAR |
|---|---|---|---|---|---|---|---|---|---|---|---|
| 4 | 10 | 12 | 0.0523 | 10 | 11 | 13 | 0.1163 | 9 | 12 | 14 | 0.0767 |
| 5 | 10 | 12 | 0.0604 | 2 | 12 | 13 | 0.0284 | 10 | 12 | 14 | 0.0911 |
| 6 | 10 | 12 | 0.0701 | 3 | 12 | 13 | 0.0333 | 11 | 12 | 14 | 0.1120 |
| 7 | 10 | 12 | 0.0822 | 4 | 12 | 13 | 0.0384 | 2 | 13 | 14 | 0.0253 |
| 8 | 10 | 12 | 0.0983 | 5 | 12 | 13 | 0.0439 | 3 | 13 | 14 | 0.0295 |
| 9 | 10 | 12 | 0.1212 | 6 | 12 | 13 | 0.0503 | 4 | 13 | 14 | 0.0337 |
| 2 | 11 | 12 | 0.0323 | 7 | 12 | 13 | 0.0580 | 5 | 13 | 14 | 0.0382 |
| 3 | 11 | 12 | 0.0381 | 8 | 12 | 13 | 0.0678 | 6 | 13 | 14 | 0.0434 |
| 4 | 11 | 12 | 0.0443 | 9 | 12 | 13 | 0.0809 | 7 | 13 | 14 | 0.0494 |
| 5 | 11 | 12 | 0.0512 | 10 | 12 | 13 | 0.0997 | 8 | 13 | 14 | 0.0567 |
| 6 | 11 | 12 | 0.0595 | 11 | 12 | 13 | 0.1302 | 9 | 13 | 14 | 0.0660 |
| 7 | 11 | 12 | 0.0698 | 1 | 13 | 13 | 0.0184 | 10 | 13 | 14 | 0.0785 |
| 8 | 11 | 12 | 0.0835 | 2 | 13 | 13 | 0.0229 | 11 | 13 | 14 | 0.0967 |
| 9 | 11 | 12 | 0.1032 | 3 | 13 | 13 | 0.0269 | 12 | 13 | 14 | 0.1261 |
| 10 | 11 | 12 | 0.1349 | 4 | 13 | 13 | 0.0309 | 1 | 14 | 14 | 0.0166 |
| 1 | 12 | 12 | 0.0206 | 5 | 13 | 13 | 0.0354 | 2 | 14 | 14 | 0.0205 |
| 2 | 12 | 12 | 0.0258 | 6 | 13 | 13 | 0.0406 | 3 | 14 | 14 | 0.0239 |
| 3 | 12 | 12 | 0.0305 | 7 | 13 | 13 | 0.0469 | 4 | 14 | 14 | 0.0273 |
| 4 | 12 | 12 | 0.0354 | 8 | 13 | 13 | 0.0548 | 5 | 14 | 14 | 0.0310 |
| 5 | 12 | 12 | 0.0410 | 9 | 13 | 13 | 0.0655 | 6 | 14 | 14 | 0.0352 |
| 6 | 12 | 12 | 0.0477 | 10 | 13 | 13 | 0.0809 | 7 | 14 | 14 | 0.0401 |
| 7 | 12 | 12 | 0.0560 | 11 | 13 | 13 | 0.1059 | 8 | 14 | 14 | 0.0461 |
| 8 | 12 | 12 | 0.0671 | 12 | 13 | 13 | 0.1557 | 9 | 14 | 14 | 0.0537 |
| 9 | 12 | 12 | 0.0831 | 7 | 8 | 14 | 0.0953 | 10 | 14 | 14 | 0.0640 |
| 10 | 12 | 12 | 0.1089 | 6 | 9 | 14 | 0.0739 | 11 | 14 | 14 | 0.0789 |
| 11 | 12 | 12 | 0.1602 | 7 | 9 | 14 | 0.0840 | 12 | 14 | 14 | 0.1032 |
| 6 | 8 | 13 | 0.0884 | 8 | 9 | 14 | 0.0961 | 13 | 14 | 14 | 0.1517 |
| 7 | 8 | 13 | 0.1017 | 5 | 10 | 14 | 0.0576 | 7 | 9 | 15 | 0.0797 |
| 5 | 9 | 13 | 0.0676 | 6 | 10 | 14 | 0.0653 | 8 | 9 | 15 | 0.0900 |
| 6 | 9 | 13 | 0.0774 | 7 | 10 | 14 | 0.0742 | 6 | 10 | 15 | 0.0630 |
| 7 | 9 | 13 | 0.0890 | 8 | 10 | 14 | 0.0851 | 7 | 10 | 15 | 0.0709 |
| 8 | 9 | 13 | 0.1037 | 9 | 10 | 14 | 0.0988 | 8 | 10 | 15 | 0.0801 |
| 4 | 10 | 13 | 0.0517 | 4 | 11 | 14 | 0.0448 | 9 | 10 | 15 | 0.0915 |
| 5 | 10 | 13 | 0.0592 | 5 | 11 | 14 | 0.0508 | 5 | 11 | 15 | 0.0499 |
| 6 | 10 | 13 | 0.0677 | 6 | 11 | 14 | 0.0576 | 6 | 11 | 15 | 0.0561 |
| 7 | 10 | 13 | 0.0780 | 7 | 11 | 14 | 0.0655 | 7 | 11 | 15 | 0.0631 |
| 8 | 10 | 13 | 0.0910 | 8 | 11 | 14 | 0.0752 | 8 | 11 | 15 | 0.0714 |
| 9 | 10 | 13 | 0.1083 | 9 | 11 | 14 | 0.0874 | 9 | 11 | 15 | 0.0816 |
| 3 | 11 | 13 | 0.0391 | 10 | 11 | 14 | 0.1037 | 10 | 11 | 15 | 0.0945 |
| 4 | 11 | 13 | 0.0450 | 3 | 12 | 14 | 0.0343 | 4 | 12 | 15 | 0.0394 |
| 5 | 11 | 13 | 0.0514 | 4 | 12 | 14 | 0.0392 | 5 | 12 | 15 | 0.0443 |
| 6 | 11 | 13 | 0.0589 | 5 | 12 | 14 | 0.0445 | 6 | 12 | 15 | 0.0498 |
| 7 | 11 | 13 | 0.0679 | 6 | 12 | 14 | 0.0505 | 7 | 12 | 15 | 0.0561 |
| 8 | 11 | 13 | 0.0793 | 7 | 12 | 14 | 0.0574 | 8 | 12 | 15 | 0.0635 |
| 9 | 11 | 13 | 0.0945 | 8 | 12 | 14 | 0.0659 | 9 | 12 | 15 | 0.0726 |

TABLE 3

OUTLIER MEAN 0.0 OUTLIER STANDARD DEVIATION 1.0

| R | S | N | COVAR | R | S | N | COVAR | R | S | N | COVAR |
|---|---|---|---|---|---|---|---|---|---|---|---|
| 10 | 12 | 15 | 0.0842 | 9 | 11 | 16 | 0.0768 | 4 | 16 | 16 | 0.0220 |
| 11 | 12 | 15 | 0.0997 | 10 | 11 | 16 | 0.0875 | 5 | 16 | 16 | 0.0246 |
| 3 | 13 | 15 | 0.0305 | 5 | 12 | 16 | 0.0438 | 6 | 16 | 16 | 0.0275 |
| 4 | 13 | 15 | 0.0347 | 6 | 12 | 16 | 0.0489 | 7 | 16 | 16 | 0.0308 |
| 5 | 13 | 15 | 0.0390 | 7 | 12 | 16 | 0.0546 | 8 | 16 | 16 | 0.0345 |
| 6 | 13 | 15 | 0.0439 | 8 | 12 | 16 | 0.0611 | 9 | 16 | 16 | 0.0390 |
| 7 | 13 | 15 | 0.0494 | 9 | 12 | 16 | 0.0689 | 10 | 16 | 16 | 0.0446 |
| 8 | 13 | 15 | 0.0560 | 10 | 12 | 16 | 0.0785 | 11 | 16 | 16 | 0.0517 |
| 9 | 13 | 15 | 0.0641 | 11 | 12 | 16 | 0.0908 | 12 | 16 | 16 | 0.0613 |
| 10 | 13 | 15 | 0.0743 | 4 | 13 | 16 | 0.0349 | 13 | 16 | 16 | 0.0754 |
| 11 | 13 | 15 | 0.0882 | 5 | 13 | 16 | 0.0391 | 14 | 16 | 16 | 0.0985 |
| 12 | 13 | 15 | 0.1082 | 6 | 13 | 16 | 0.0437 | 15 | 16 | 16 | 0.1449 |
| 2 | 14 | 15 | 0.0227 | 7 | 13 | 16 | 0.0488 | 8 | 10 | 17 | 0.0725 |
| 3 | 14 | 15 | 0.0263 | 8 | 13 | 16 | 0.0547 | 9 | 10 | 17 | 0.0808 |
| 4 | 14 | 15 | 0.0299 | 9 | 13 | 16 | 0.0617 | 7 | 11 | 17 | 0.0588 |
| 5 | 14 | 15 | 0.0337 | 10 | 13 | 16 | 0.0703 | 8 | 11 | 17 | 0.0653 |
| 6 | 14 | 15 | 0.0379 | 11 | 13 | 16 | 0.0813 | 9 | 11 | 17 | 0.0728 |
| 7 | 14 | 15 | 0.0427 | 12 | 13 | 16 | 0.0963 | 10 | 11 | 17 | 0.0818 |
| 8 | 14 | 15 | 0.0484 | 3 | 14 | 16 | 0.0274 | 6 | 12 | 17 | 0.0478 |
| 9 | 14 | 15 | 0.0554 | 4 | 14 | 16 | 0.0309 | 7 | 12 | 17 | 0.0530 |
| 10 | 14 | 15 | 0.0643 | 5 | 14 | 16 | 0.0346 | 8 | 12 | 17 | 0.0589 |
| 11 | 14 | 15 | 0.0764 | 6 | 14 | 16 | 0.0387 | 9 | 12 | 17 | 0.0657 |
| 12 | 14 | 15 | 0.0939 | 7 | 14 | 16 | 0.0432 | 10 | 12 | 17 | 0.0739 |
| 13 | 14 | 15 | 0.1224 | 8 | 14 | 16 | 0.0484 | 11 | 12 | 17 | 0.0840 |
| 1 | 15 | 15 | 0.0151 | 9 | 14 | 16 | 0.0547 | 5 | 13 | 17 | 0.0388 |
| 2 | 15 | 15 | 0.0185 | 10 | 14 | 16 | 0.0624 | 6 | 13 | 17 | 0.0431 |
| 3 | 15 | 15 | 0.0215 | 11 | 14 | 16 | 0.0722 | 7 | 13 | 17 | 0.0478 |
| 4 | 15 | 15 | 0.0244 | 12 | 14 | 16 | 0.0855 | 8 | 13 | 17 | 0.0531 |
| 5 | 15 | 15 | 0.0275 | 13 | 14 | 16 | 0.1049 | 9 | 13 | 17 | 0.0593 |
| 6 | 15 | 15 | 0.0310 | 2 | 15 | 16 | 0.0206 | 10 | 13 | 17 | 0.0667 |
| 7 | 15 | 15 | 0.0349 | 3 | 15 | 16 | 0.0237 | 11 | 13 | 17 | 0.0759 |
| 8 | 15 | 15 | 0.0396 | 4 | 15 | 16 | 0.0268 | 12 | 13 | 17 | 0.0876 |
| 9 | 15 | 15 | 0.0453 | 5 | 15 | 16 | 0.0300 | 4 | 14 | 17 | 0.0313 |
| 10 | 15 | 15 | 0.0527 | 6 | 15 | 16 | 0.0336 | 5 | 14 | 17 | 0.0349 |
| 11 | 15 | 15 | 0.0626 | 7 | 15 | 16 | 0.0375 | 6 | 14 | 17 | 0.0387 |
| 12 | 15 | 15 | 0.0771 | 8 | 15 | 16 | 0.0421 | 7 | 14 | 17 | 0.0429 |
| 13 | 15 | 15 | 0.1007 | 9 | 15 | 16 | 0.0475 | 8 | 14 | 17 | 0.0477 |
| 14 | 15 | 15 | 0.1481 | 10 | 15 | 16 | 0.0542 | 9 | 14 | 17 | 0.0533 |
| 8 | 9 | 16 | 0.0850 | 11 | 15 | 16 | 0.0628 | 10 | 14 | 17 | 0.0600 |
| 7 | 10 | 16 | 0.0679 | 12 | 15 | 16 | 0.0745 | 11 | 14 | 17 | 0.0682 |
| 8 | 10 | 16 | 0.0760 | 13 | 15 | 16 | 0.0914 | 12 | 14 | 17 | 0.0788 |
| 9 | 10 | 16 | 0.0856 | 14 | 15 | 16 | 0.1191 | 13 | 14 | 17 | 0.0932 |
| 6 | 11 | 16 | 0.0545 | 1 | 16 | 16 | 0.0138 | 3 | 15 | 17 | 0.0247 |
| 7 | 11 | 16 | 0.0609 | 2 | 16 | 16 | 0.0169 | 4 | 15 | 17 | 0.0278 |
| 8 | 11 | 16 | 0.0682 | 3 | 16 | 16 | 0.0195 | 5 | 15 | 17 | 0.0310 |

TABLE 3

OUTLIER MEAN 0.0 OUTLIER STANDARD DEVIATION 1.0

| R | S | N | COVAR | R | S | N | COVAR | R | S | N | COVAR |
|---|---|---|---|---|---|---|---|---|---|---|---|
| 6 | 15 | 17 | 0.0344 | 9 | 12 | 18 | 0.0629 | 4 | 17 | 18 | 0.0221 |
| 7 | 15 | 17 | 0.0382 | 10 | 12 | 18 | 0.0700 | 5 | 17 | 18 | 0.0245 |
| 8 | 15 | 17 | 0.0425 | 11 | 12 | 18 | 0.0785 | 6 | 17 | 18 | 0.0270 |
| 9 | 15 | 17 | 0.0475 | 6 | 13 | 18 | 0.0424 | 7 | 17 | 18 | 0.0298 |
| 10 | 15 | 17 | 0.0534 | 7 | 13 | 18 | 0.0467 | 8 | 17 | 18 | 0.0330 |
| 11 | 15 | 17 | 0.0608 | 8 | 13 | 18 | 0.0516 | 9 | 17 | 18 | 0.0365 |
| 12 | 15 | 17 | 0.0703 | 9 | 13 | 18 | 0.0571 | 10 | 17 | 18 | 0.0407 |
| 13 | 15 | 17 | 0.0831 | 10 | 13 | 18 | 0.0636 | 11 | 17 | 18 | 0.0458 |
| 14 | 15 | 17 | 0.1019 | 11 | 13 | 18 | 0.0713 | 12 | 17 | 18 | 0.0520 |
| 2 | 16 | 17 | 0.0188 | 12 | 13 | 18 | 0.0809 | 13 | 17 | 18 | 0.0601 |
| 3 | 16 | 17 | 0.0215 | 5 | 14 | 18 | 0.0348 | 14 | 17 | 18 | 0.0711 |
| 4 | 16 | 17 | 0.0242 | 6 | 14 | 18 | 0.0384 | 15 | 17 | 18 | 0.0872 |
| 5 | 16 | 17 | 0.0270 | 7 | 14 | 18 | 0.0423 | 16 | 17 | 18 | 0.1135 |
| 6 | 16 | 17 | 0.0300 | 8 | 14 | 18 | 0.0467 | 1 | 18 | 18 | 0.0118 |
| 7 | 16 | 17 | 0.0333 | 9 | 14 | 18 | 0.0518 | 2 | 18 | 18 | 0.0142 |
| 8 | 16 | 17 | 0.0370 | 10 | 14 | 18 | 0.0577 | 3 | 18 | 18 | 0.0163 |
| 9 | 16 | 17 | 0.0414 | 11 | 14 | 18 | 0.0647 | 4 | 18 | 18 | 0.0182 |
| 10 | 16 | 17 | 0.0466 | 12 | 14 | 18 | 0.0734 | 5 | 18 | 18 | 0.0203 |
| 11 | 16 | 17 | 0.0531 | 13 | 14 | 18 | 0.0847 | 6 | 18 | 18 | 0.0224 |
| 12 | 16 | 17 | 0.0614 | 4 | 15 | 18 | 0.0283 | 7 | 18 | 18 | 0.0247 |
| 13 | 16 | 17 | 0.0727 | 5 | 15 | 18 | 0.0313 | 8 | 18 | 18 | 0.0273 |
| 14 | 16 | 17 | 0.0892 | 6 | 15 | 18 | 0.0346 | 9 | 18 | 18 | 0.0303 |
| 15 | 16 | 17 | 0.1162 | 7 | 15 | 18 | 0.0382 | 10 | 18 | 18 | 0.0337 |
| 1 | 17 | 17 | 0.0127 | 8 | 15 | 18 | 0.0422 | 11 | 18 | 18 | 0.0379 |
| 2 | 17 | 17 | 0.0155 | 9 | 15 | 18 | 0.0467 | 12 | 18 | 18 | 0.0431 |
| 3 | 17 | 17 | 0.0177 | 10 | 15 | 18 | 0.0520 | 13 | 18 | 18 | 0.0499 |
| 4 | 17 | 17 | 0.0200 | 11 | 15 | 18 | 0.0584 | 14 | 18 | 18 | 0.0590 |
| 5 | 17 | 17 | 0.0223 | 12 | 15 | 18 | 0.0664 | 15 | 18 | 18 | 0.0725 |
| 6 | 17 | 17 | 0.0247 | 13 | 15 | 18 | 0.0766 | 16 | 18 | 18 | 0.0946 |
| 7 | 17 | 17 | 0.0274 | 14 | 15 | 18 | 0.0904 | 17 | 18 | 18 | 0.1393 |
| 8 | 17 | 17 | 0.0305 | 3 | 16 | 18 | 0.0225 | 9 | 11 | 19 | 0.0664 |
| 9 | 17 | 17 | 0.0341 | 4 | 16 | 18 | 0.0252 | 10 | 11 | 19 | 0.0733 |
| 10 | 17 | 17 | 0.0385 | 5 | 16 | 18 | 0.0280 | 8 | 12 | 19 | 0.0550 |
| 11 | 17 | 17 | 0.0438 | 6 | 16 | 18 | 0.0309 | 9 | 12 | 19 | 0.0604 |
| 12 | 17 | 17 | 0.0507 | 7 | 16 | 18 | 0.0341 | 10 | 12 | 19 | 0.0667 |
| 13 | 17 | 17 | 0.0601 | 8 | 16 | 18 | 0.0377 | 11 | 12 | 19 | 0.0740 |
| 14 | 17 | 17 | 0.0739 | 9 | 16 | 18 | 0.0417 | 7 | 13 | 19 | 0.0457 |
| 15 | 17 | 17 | 0.0965 | 10 | 16 | 18 | 0.0465 | 8 | 13 | 19 | 0.0501 |
| 16 | 17 | 17 | 0.1419 | 11 | 16 | 18 | 0.0522 | 9 | 13 | 19 | 0.0551 |
| 9 | 10 | 18 | 0.0767 | 12 | 16 | 18 | 0.0594 | 10 | 13 | 19 | 0.0608 |
| 8 | 11 | 18 | 0.0627 | 13 | 16 | 18 | 0.0685 | 11 | 13 | 19 | 0.0675 |
| 9 | 11 | 18 | 0.0694 | 14 | 16 | 18 | 0.0810 | 12 | 13 | 19 | 0.0756 |
| 10 | 11 | 18 | 0.0772 | 15 | 16 | 18 | 0.0992 | 6 | 14 | 19 | 0.0379 |
| 7 | 12 | 18 | 0.0515 | 2 | 17 | 18 | 0.0172 | 7 | 14 | 19 | 0.0416 |
| 8 | 12 | 18 | 0.0569 | 3 | 17 | 18 | 0.0197 | 8 | 14 | 19 | 0.0457 |

TABLE 3

OUTLIER MEAN 0.0 OUTLIER STANDARD DEVIATION 1.0

| R | S | N | COVAR | R | S | N | COVAR | R | S | N | COVAR |
|---|---|---|---|---|---|---|---|---|---|---|---|
| 9 | 14 | 19 | 0.0502 | 6 | 18 | 19 | 0.0246 | 13 | 14 | 20 | 0.0730 |
| 10 | 14 | 19 | 0.0555 | 7 | 18 | 19 | 0.0270 | 6 | 15 | 20 | 0.0342 |
| 11 | 14 | 19 | 0.0616 | 8 | 18 | 19 | 0.0296 | 7 | 15 | 20 | 0.0374 |
| 12 | 14 | 19 | 0.0690 | 9 | 18 | 19 | 0.0326 | 8 | 15 | 20 | 0.0408 |
| 13 | 14 | 19 | 0.0782 | 10 | 18 | 19 | 0.0360 | 9 | 15 | 20 | 0.0447 |
| 5 | 15 | 19 | 0.0314 | 11 | 18 | 19 | 0.0401 | 10 | 15 | 20 | 0.0490 |
| 6 | 15 | 19 | 0.0345 | 12 | 18 | 19 | 0.0450 | 11 | 15 | 20 | 0.0540 |
| 7 | 15 | 19 | 0.0379 | 13 | 18 | 19 | 0.0510 | 12 | 15 | 20 | 0.0599 |
| 8 | 15 | 19 | 0.0416 | 14 | 18 | 19 | 0.0589 | 13 | 15 | 20 | 0.0670 |
| 9 | 15 | 19 | 0.0457 | 15 | 18 | 19 | 0.0696 | 14 | 15 | 20 | 0.0758 |
| 10 | 15 | 19 | 0.0505 | 16 | 18 | 19 | 0.0853 | 5 | 16 | 20 | 0.0285 |
| 11 | 15 | 19 | 0.0561 | 17 | 18 | 19 | 0.1111 | 6 | 16 | 20 | 0.0312 |
| 12 | 15 | 19 | 0.0629 | 1 | 19 | 19 | 0.0109 | 7 | 16 | 20 | 0.0341 |
| 13 | 15 | 19 | 0.0713 | 2 | 19 | 19 | 0.0132 | 8 | 16 | 20 | 0.0373 |
| 14 | 15 | 19 | 0.0821 | 3 | 19 | 19 | 0.0150 | 9 | 16 | 20 | 0.0408 |
| 4 | 16 | 19 | 0.0257 | 4 | 19 | 19 | 0.0168 | 10 | 16 | 20 | 0.0448 |
| 5 | 16 | 19 | 0.0284 | 5 | 19 | 19 | 0.0185 | 11 | 16 | 20 | 0.0493 |
| 6 | 16 | 19 | 0.0312 | 6 | 19 | 19 | 0.0204 | 12 | 16 | 20 | 0.0547 |
| 7 | 16 | 19 | 0.0343 | 7 | 19 | 19 | 0.0224 | 13 | 16 | 20 | 0.0612 |
| 8 | 16 | 19 | 0.0376 | 8 | 19 | 19 | 0.0246 | 14 | 16 | 20 | 0.0693 |
| 9 | 16 | 19 | 0.0414 | 9 | 19 | 19 | 0.0271 | 15 | 16 | 20 | 0.0798 |
| 10 | 16 | 19 | 0.0458 | 10 | 19 | 19 | 0.0300 | 4 | 17 | 20 | 0.0235 |
| 11 | 16 | 19 | 0.0509 | 11 | 19 | 19 | 0.0333 | 5 | 17 | 20 | 0.0259 |
| 12 | 16 | 19 | 0.0570 | 12 | 19 | 19 | 0.0374 | 6 | 17 | 20 | 0.0284 |
| 13 | 16 | 19 | 0.0646 | 13 | 19 | 19 | 0.0425 | 7 | 17 | 20 | 0.0310 |
| 14 | 16 | 19 | 0.0745 | 14 | 19 | 19 | 0.0490 | 8 | 17 | 20 | 0.0339 |
| 15 | 16 | 19 | 0.0879 | 15 | 19 | 19 | 0.0580 | 9 | 17 | 20 | 0.0371 |
| 3 | 17 | 19 | 0.0206 | 16 | 19 | 19 | 0.0712 | 10 | 17 | 20 | 0.0407 |
| 4 | 17 | 19 | 0.0230 | 17 | 19 | 19 | 0.0929 | 11 | 17 | 20 | 0.0448 |
| 5 | 17 | 19 | 0.0254 | 18 | 19 | 19 | 0.1368 | 12 | 17 | 20 | 0.0498 |
| 6 | 17 | 19 | 0.0280 | 10 | 11 | 20 | 0.0699 | 13 | 17 | 20 | 0.0557 |
| 7 | 17 | 19 | 0.0307 | 9 | 12 | 20 | 0.0582 | 14 | 17 | 20 | 0.0631 |
| 8 | 17 | 19 | 0.0337 | 10 | 12 | 20 | 0.0638 | 15 | 17 | 20 | 0.0726 |
| 9 | 17 | 19 | 0.0371 | 11 | 12 | 20 | 0.0703 | 16 | 17 | 20 | 0.0856 |
| 10 | 17 | 19 | 0.0410 | 8 | 13 | 20 | 0.0487 | 3 | 18 | 20 | 0.0190 |
| 11 | 17 | 19 | 0.0456 | 9 | 13 | 20 | 0.0533 | 4 | 18 | 20 | 0.0211 |
| 12 | 17 | 19 | 0.0512 | 10 | 13 | 20 | 0.0584 | 5 | 18 | 20 | 0.0233 |
| 13 | 17 | 19 | 0.0580 | 11 | 13 | 20 | 0.0643 | 6 | 18 | 20 | 0.0255 |
| 14 | 17 | 19 | 0.0669 | 12 | 13 | 20 | 0.0713 | 7 | 18 | 20 | 0.0279 |
| 15 | 17 | 19 | 0.0790 | 7 | 14 | 20 | 0.0408 | 8 | 18 | 20 | 0.0305 |
| 16 | 17 | 19 | 0.0967 | 8 | 14 | 20 | 0.0446 | 9 | 18 | 20 | 0.0333 |
| 2 | 18 | 19 | 0.0159 | 9 | 14 | 20 | 0.0488 | 10 | 18 | 20 | 0.0366 |
| 3 | 18 | 19 | 0.0181 | 10 | 14 | 20 | 0.0535 | 11 | 18 | 20 | 0.0403 |
| 4 | 18 | 19 | 0.0202 | 11 | 14 | 20 | 0.0589 | 12 | 18 | 20 | 0.0448 |
| 5 | 18 | 19 | 0.0223 | 12 | 14 | 20 | 0.0653 | 13 | 18 | 20 | 0.0501 |

TABLE 3

OUTLIER MEAN 0.0 OUTLIER STANDARD DEVIATION 1.0

| R | S | N | COVAR | R | S | N | COVAR | R | S | N | COVAR |
|---|---|---|---|---|---|---|---|---|---|---|---|
| 14 | 18 | 20 | 0.0568 | 12 | 19 | 20 | 0.0395 | 8 | 20 | 20 | 0.0224 |
| 15 | 18 | 20 | 0.0655 | 13 | 19 | 20 | 0.0442 | 9 | 20 | 20 | 0.0245 |
| 16 | 18 | 20 | 0.0772 | 14 | 19 | 20 | 0.0501 | 10 | 20 | 20 | 0.0269 |
| 17 | 18 | 20 | 0.0945 | 15 | 19 | 20 | 0.0578 | 11 | 20 | 20 | 0.0297 |
| 2 | 19 | 20 | 0.0147 | 16 | 19 | 20 | 0.0682 | 12 | 20 | 20 | 0.0329 |
| 3 | 19 | 20 | 0.0167 | 17 | 19 | 20 | 0.0836 | 13 | 20 | 20 | 0.0369 |
| 4 | 19 | 20 | 0.0186 | 18 | 19 | 20 | 0.1088 | 14 | 20 | 20 | 0.0418 |
| 5 | 19 | 20 | 0.0205 | 1 | 20 | 20 | 0.0102 | 15 | 20 | 20 | 0.0483 |
| 6 | 19 | 20 | 0.0225 | 2 | 20 | 20 | 0.0123 | 16 | 20 | 20 | 0.0571 |
| 7 | 19 | 20 | 0.0245 | 3 | 20 | 20 | 0.0139 | 17 | 20 | 20 | 0.0700 |
| 8 | 19 | 20 | 0.0268 | 4 | 20 | 20 | 0.0155 | 18 | 20 | 20 | 0.0913 |
| 9 | 19 | 20 | 0.0294 | 5 | 20 | 20 | 0.0171 | 19 | 20 | 20 | 0.1345 |
| 10 | 19 | 20 | 0.0322 | 6 | 20 | 20 | 0.0187 | | | | |
| 11 | 19 | 20 | 0.0356 | 7 | 20 | 20 | 0.0205 | | | | |

TABLE 3

OUTLIER MEAN 0.5 OUTLIER STANDARD DEVIATION 1.0

| R | S | N | COVAR | R | S | N | COVAR | R | S | N | COVAR |
|---|---|---|---|---|---|---|---|---|---|---|---|
| 1 | 2 | 2 | 0.2964 | 1 | 6 | 7 | 0.0576 | 1 | 5 | 9 | 0.0722 |
| 1 | 2 | 3 | 0.2734 | 2 | 6 | 7 | 0.0779 | 2 | 5 | 9 | 0.0933 |
| 1 | 3 | 3 | 0.1514 | 3 | 6 | 7 | 0.1004 | 3 | 5 | 9 | 0.1140 |
| 2 | 3 | 3 | 0.2684 | 4 | 6 | 7 | 0.1300 | 4 | 5 | 9 | 0.1377 |
| 1 | 2 | 4 | 0.2467 | 5 | 6 | 7 | 0.1753 | 1 | 6 | 9 | 0.0586 |
| 1 | 3 | 4 | 0.1541 | 1 | 7 | 7 | 0.0414 | 2 | 6 | 9 | 0.0759 |
| 2 | 3 | 4 | 0.2353 | 2 | 7 | 7 | 0.0566 | 3 | 6 | 9 | 0.0930 |
| 1 | 4 | 4 | 0.0960 | 3 | 7 | 7 | 0.0735 | 4 | 6 | 9 | 0.1128 |
| 2 | 4 | 4 | 0.1509 | 4 | 7 | 7 | 0.0959 | 5 | 6 | 9 | 0.1377 |
| 3 | 4 | 4 | 0.2433 | 5 | 7 | 7 | 0.1307 | 1 | 7 | 9 | 0.0479 |
| 1 | 2 | 5 | 0.2262 | 6 | 7 | 7 | 0.1976 | 2 | 7 | 9 | 0.0622 |
| 1 | 3 | 5 | 0.1469 | 1 | 2 | 8 | 0.1880 | 3 | 7 | 9 | 0.0764 |
| 2 | 3 | 5 | 0.2093 | 1 | 3 | 8 | 0.1264 | 4 | 7 | 9 | 0.0929 |
| 1 | 4 | 5 | 0.1023 | 2 | 3 | 8 | 0.1644 | 5 | 7 | 9 | 0.1138 |
| 2 | 4 | 5 | 0.1477 | 1 | 4 | 8 | 0.0944 | 6 | 7 | 9 | 0.1428 |
| 3 | 4 | 5 | 0.2086 | 2 | 4 | 8 | 0.1235 | 1 | 8 | 9 | 0.0385 |
| 1 | 5 | 5 | 0.0682 | 3 | 4 | 8 | 0.1533 | 2 | 8 | 9 | 0.0502 |
| 2 | 5 | 5 | 0.1002 | 1 | 5 | 8 | 0.0739 | 3 | 8 | 9 | 0.0619 |
| 3 | 5 | 5 | 0.1441 | 2 | 5 | 8 | 0.0971 | 4 | 8 | 9 | 0.0754 |
| 4 | 5 | 5 | 0.2242 | 3 | 5 | 8 | 0.1210 | 5 | 8 | 9 | 0.0927 |
| 1 | 2 | 6 | 0.2104 | 4 | 5 | 8 | 0.1500 | 6 | 8 | 9 | 0.1169 |
| 1 | 3 | 6 | 0.1392 | 1 | 6 | 8 | 0.0588 | 7 | 8 | 9 | 0.1552 |
| 2 | 3 | 6 | 0.1902 | 2 | 6 | 8 | 0.0776 | 1 | 9 | 9 | 0.0289 |
| 1 | 4 | 6 | 0.1008 | 3 | 6 | 8 | 0.0971 | 2 | 9 | 9 | 0.0379 |
| 2 | 4 | 6 | 0.1390 | 4 | 6 | 8 | 0.1208 | 3 | 9 | 9 | 0.0469 |
| 3 | 4 | 6 | 0.1840 | 5 | 6 | 8 | 0.1531 | 4 | 9 | 9 | 0.0575 |
| 1 | 5 | 6 | 0.0745 | 1 | 7 | 8 | 0.0464 | 5 | 9 | 9 | 0.0710 |
| 2 | 5 | 6 | 0.1036 | 2 | 7 | 8 | 0.0615 | 6 | 9 | 9 | 0.0900 |
| 3 | 5 | 6 | 0.1384 | 3 | 7 | 8 | 0.0772 | 7 | 9 | 9 | 0.1204 |
| 4 | 5 | 6 | 0.1896 | 4 | 7 | 8 | 0.0965 | 8 | 9 | 9 | 0.1799 |
| 1 | 6 | 6 | 0.0519 | 5 | 7 | 8 | 0.1229 | 1 | 2 | 10 | 0.1727 |
| 2 | 6 | 6 | 0.0731 | 6 | 7 | 8 | 0.1642 | 1 | 3 | 10 | 0.1168 |
| 3 | 6 | 6 | 0.0987 | 1 | 8 | 8 | 0.0342 | 2 | 3 | 10 | 0.1477 |
| 4 | 6 | 6 | 0.1370 | 2 | 8 | 8 | 0.0456 | 1 | 4 | 10 | 0.0883 |
| 5 | 6 | 6 | 0.2094 | 3 | 8 | 8 | 0.0576 | 2 | 4 | 10 | 0.1121 |
| 1 | 2 | 7 | 0.1980 | 4 | 8 | 8 | 0.0725 | 3 | 4 | 10 | 0.1347 |
| 1 | 3 | 7 | 0.1324 | 5 | 8 | 8 | 0.0929 | 1 | 5 | 10 | 0.0705 |
| 2 | 3 | 7 | 0.1758 | 6 | 8 | 8 | 0.1252 | 2 | 5 | 10 | 0.0897 |
| 1 | 4 | 7 | 0.0977 | 7 | 8 | 8 | 0.1879 | 3 | 5 | 10 | 0.1081 |
| 2 | 4 | 7 | 0.1307 | 1 | 2 | 9 | 0.1797 | 4 | 5 | 10 | 0.1282 |
| 3 | 4 | 7 | 0.1665 | 1 | 3 | 9 | 0.1213 | 1 | 6 | 10 | 0.0579 |
| 1 | 5 | 7 | 0.0750 | 2 | 3 | 9 | 0.1553 | 2 | 6 | 10 | 0.0739 |
| 2 | 5 | 7 | 0.1009 | 1 | 4 | 9 | 0.0912 | 3 | 6 | 10 | 0.0891 |
| 3 | 5 | 7 | 0.1292 | 2 | 4 | 9 | 0.1174 | 4 | 6 | 10 | 0.1060 |
| 4 | 5 | 7 | 0.1663 | 3 | 4 | 9 | 0.1430 | 5 | 6 | 10 | 0.1262 |

TABLE 3

OUTLIER MEAN 0.5 OUTLIER STANDARD DEVIATION 1.0

| R | S | N | COVAR | R | S | N | COVAR | R | S | N | COVAR |
|---|---|---|---|---|---|---|---|---|---|---|---|
| 1 | 7 | 10 | 0.0481 | 1 | 7 | 11 | 0.0478 | 3 | 4 | 12 | 0.1220 |
| 2 | 7 | 10 | 0.0616 | 2 | 7 | 11 | 0.0605 | 1 | 5 | 12 | 0.0670 |
| 3 | 7 | 10 | 0.0744 | 3 | 7 | 11 | 0.0722 | 2 | 5 | 12 | 0.0837 |
| 4 | 7 | 10 | 0.0887 | 4 | 7 | 11 | 0.0848 | 3 | 5 | 12 | 0.0986 |
| 5 | 7 | 10 | 0.1059 | 5 | 7 | 11 | 0.0994 | 4 | 5 | 12 | 0.1142 |
| 6 | 7 | 10 | 0.1282 | 6 | 7 | 11 | 0.1173 | 1 | 6 | 12 | 0.0558 |
| 1 | 8 | 10 | 0.0400 | 1 | 8 | 11 | 0.0405 | 2 | 6 | 12 | 0.0697 |
| 2 | 8 | 10 | 0.0513 | 2 | 8 | 11 | 0.0513 | 3 | 6 | 12 | 0.0823 |
| 3 | 8 | 10 | 0.0622 | 3 | 8 | 11 | 0.0614 | 4 | 6 | 12 | 0.0955 |
| 4 | 8 | 10 | 0.0743 | 4 | 8 | 11 | 0.0722 | 5 | 6 | 12 | 0.1102 |
| 5 | 8 | 10 | 0.0889 | 5 | 8 | 11 | 0.0847 | 1 | 7 | 12 | 0.0473 |
| 6 | 8 | 10 | 0.1078 | 6 | 8 | 11 | 0.1002 | 2 | 7 | 12 | 0.0592 |
| 7 | 8 | 10 | 0.1346 | 7 | 8 | 11 | 0.1205 | 3 | 7 | 12 | 0.0700 |
| 1 | 9 | 10 | 0.0327 | 1 | 9 | 11 | 0.0342 | 4 | 7 | 12 | 0.0813 |
| 2 | 9 | 10 | 0.0421 | 2 | 9 | 11 | 0.0434 | 5 | 7 | 12 | 0.0939 |
| 3 | 9 | 10 | 0.0511 | 3 | 9 | 11 | 0.0520 | 6 | 7 | 12 | 0.1089 |
| 4 | 9 | 10 | 0.0611 | 4 | 9 | 11 | 0.0612 | 1 | 8 | 12 | 0.0405 |
| 5 | 9 | 10 | 0.0733 | 5 | 9 | 11 | 0.0720 | 2 | 8 | 12 | 0.0508 |
| 6 | 9 | 10 | 0.0892 | 6 | 9 | 11 | 0.0853 | 3 | 8 | 12 | 0.0601 |
| 7 | 9 | 10 | 0.1117 | 7 | 9 | 11 | 0.1028 | 4 | 8 | 12 | 0.0699 |
| 8 | 9 | 10 | 0.1477 | 8 | 9 | 11 | 0.1277 | 5 | 8 | 12 | 0.0809 |
| 1 | 10 | 10 | 0.0249 | 1 | 10 | 11 | 0.0283 | 6 | 8 | 12 | 0.0939 |
| 2 | 10 | 10 | 0.0322 | 2 | 10 | 11 | 0.0360 | 7 | 8 | 12 | 0.1102 |
| 3 | 10 | 10 | 0.0393 | 3 | 10 | 11 | 0.0432 | 1 | 9 | 12 | 0.0348 |
| 4 | 10 | 10 | 0.0472 | 4 | 10 | 11 | 0.0510 | 2 | 9 | 12 | 0.0437 |
| 5 | 10 | 10 | 0.0568 | 5 | 10 | 11 | 0.0600 | 3 | 9 | 12 | 0.0518 |
| 6 | 10 | 10 | 0.0694 | 6 | 10 | 11 | 0.0713 | 4 | 9 | 12 | 0.0603 |
| 7 | 10 | 10 | 0.0873 | 7 | 10 | 11 | 0.0861 | 5 | 9 | 12 | 0.0698 |
| 8 | 10 | 10 | 0.1162 | 8 | 10 | 11 | 0.1073 | 6 | 9 | 12 | 0.0812 |
| 9 | 10 | 10 | 0.1731 | 9 | 10 | 11 | 0.1414 | 7 | 9 | 12 | 0.0954 |
| 1 | 2 | 11 | 0.1666 | 1 | 11 | 11 | 0.0218 | 8 | 9 | 12 | 0.1142 |
| 1 | 3 | 11 | 0.1129 | 2 | 11 | 11 | 0.0279 | 1 | 10 | 12 | 0.0297 |
| 2 | 3 | 11 | 0.1413 | 3 | 11 | 11 | 0.0336 | 2 | 10 | 12 | 0.0373 |
| 1 | 4 | 11 | 0.0857 | 4 | 11 | 11 | 0.0397 | 3 | 10 | 12 | 0.0443 |
| 2 | 4 | 11 | 0.1076 | 5 | 11 | 11 | 0.0470 | 4 | 10 | 12 | 0.0516 |
| 3 | 4 | 11 | 0.1278 | 6 | 11 | 11 | 0.0559 | 5 | 10 | 12 | 0.0599 |
| 1 | 5 | 11 | 0.0687 | 7 | 11 | 11 | 0.0678 | 6 | 10 | 12 | 0.0698 |
| 2 | 5 | 11 | 0.0865 | 8 | 11 | 11 | 0.0849 | 7 | 10 | 12 | 0.0821 |
| 3 | 5 | 11 | 0.1030 | 9 | 11 | 11 | 0.1125 | 8 | 10 | 12 | 0.0985 |
| 4 | 5 | 11 | 0.1205 | 10 | 11 | 11 | 0.1672 | 9 | 10 | 12 | 0.1219 |
| 1 | 6 | 11 | 0.0568 | 1 | 2 | 12 | 0.1614 | 1 | 11 | 12 | 0.0248 |
| 2 | 6 | 11 | 0.0717 | 1 | 3 | 12 | 0.1095 | 2 | 11 | 12 | 0.0313 |
| 3 | 6 | 11 | 0.0855 | 2 | 3 | 12 | 0.1358 | 3 | 11 | 12 | 0.0372 |
| 4 | 6 | 11 | 0.1003 | 1 | 4 | 12 | 0.0833 | 4 | 11 | 12 | 0.0434 |
| 5 | 6 | 11 | 0.1173 | 2 | 4 | 12 | 0.1037 | 5 | 11 | 12 | 0.0504 |

TABLE 3

OUTLIER MEAN 0.5 OUTLIER STANDARD DEVIATION 1.0

| R | S | N | COVAR | R | S | N | COVAR | R | S | N | COVAR |
|---|---|---|---|---|---|---|---|---|---|---|---|
| 6 | 11 | 12 | 0.0588 | 2 | 9 | 13 | 0.0435 | 9 | 13 | 13 | 0.0648 |
| 7 | 11 | 12 | 0.0693 | 3 | 9 | 13 | 0.0511 | 10 | 13 | 13 | 0.0806 |
| 8 | 11 | 12 | 0.0833 | 4 | 9 | 13 | 0.0589 | 11 | 13 | 13 | 0.1062 |
| 9 | 11 | 12 | 0.1034 | 5 | 9 | 13 | 0.0675 | 12 | 13 | 13 | 0.1575 |
| 10 | 11 | 12 | 0.1360 | 6 | 9 | 13 | 0.0774 | 1 | 2 | 14 | 0.1527 |
| 1 | 12 | 12 | 0.0194 | 7 | 9 | 13 | 0.0893 | 1 | 3 | 14 | 0.1037 |
| 2 | 12 | 12 | 0.0245 | 8 | 9 | 13 | 0.1042 | 2 | 3 | 14 | 0.1269 |
| 3 | 12 | 12 | 0.0292 | 1 | 10 | 13 | 0.0303 | 1 | 4 | 14 | 0.0791 |
| 4 | 12 | 12 | 0.0341 | 2 | 10 | 13 | 0.0378 | 2 | 4 | 14 | 0.0971 |
| 5 | 12 | 12 | 0.0398 | 3 | 10 | 13 | 0.0445 | 3 | 4 | 14 | 0.1126 |
| 6 | 12 | 12 | 0.0465 | 4 | 10 | 13 | 0.0513 | 1 | 5 | 14 | 0.0640 |
| 7 | 12 | 12 | 0.0550 | 5 | 10 | 13 | 0.0588 | 2 | 5 | 14 | 0.0787 |
| 8 | 12 | 12 | 0.0663 | 6 | 10 | 13 | 0.0675 | 3 | 5 | 14 | 0.0915 |
| 9 | 12 | 12 | 0.0826 | 7 | 10 | 13 | 0.0780 | 4 | 5 | 14 | 0.1042 |
| 10 | 12 | 12 | 0.1092 | 8 | 10 | 13 | 0.0912 | 1 | 6 | 14 | 0.0536 |
| 11 | 12 | 12 | 0.1621 | 9 | 10 | 13 | 0.1088 | 2 | 6 | 14 | 0.0661 |
| 1 | 2 | 13 | 0.1568 | 1 | 11 | 13 | 0.0261 | 3 | 6 | 14 | 0.0769 |
| 1 | 3 | 13 | 0.1064 | 2 | 11 | 13 | 0.0326 | 4 | 6 | 14 | 0.0877 |
| 2 | 3 | 13 | 0.1311 | 3 | 11 | 13 | 0.0384 | 5 | 6 | 14 | 0.0993 |
| 1 | 4 | 13 | 0.0811 | 4 | 11 | 13 | 0.0444 | 1 | 7 | 14 | 0.0459 |
| 2 | 4 | 13 | 0.1002 | 5 | 11 | 13 | 0.0509 | 2 | 7 | 14 | 0.0566 |
| 3 | 4 | 13 | 0.1170 | 6 | 11 | 13 | 0.0585 | 3 | 7 | 14 | 0.0659 |
| 1 | 5 | 13 | 0.0655 | 7 | 11 | 13 | 0.0677 | 4 | 7 | 14 | 0.0753 |
| 2 | 5 | 13 | 0.0811 | 8 | 11 | 13 | 0.0792 | 5 | 7 | 14 | 0.0853 |
| 3 | 5 | 13 | 0.0948 | 9 | 11 | 13 | 0.0947 | 6 | 7 | 14 | 0.0966 |
| 4 | 5 | 13 | 0.1088 | 10 | 11 | 13 | 0.1170 | 1 | 8 | 14 | 0.0398 |
| 1 | 6 | 13 | 0.0547 | 1 | 12 | 13 | 0.0220 | 2 | 8 | 14 | 0.0492 |
| 2 | 6 | 13 | 0.0678 | 2 | 12 | 13 | 0.0275 | 3 | 8 | 14 | 0.0573 |
| 3 | 6 | 13 | 0.0794 | 3 | 12 | 13 | 0.0325 | 4 | 8 | 14 | 0.0655 |
| 4 | 6 | 13 | 0.0913 | 4 | 12 | 13 | 0.0375 | 5 | 8 | 14 | 0.0743 |
| 5 | 6 | 13 | 0.1043 | 5 | 12 | 13 | 0.0432 | 6 | 8 | 14 | 0.0842 |
| 1 | 7 | 13 | 0.0466 | 6 | 12 | 13 | 0.0496 | 7 | 8 | 14 | 0.0957 |
| 2 | 7 | 13 | 0.0579 | 7 | 12 | 13 | 0.0575 | 1 | 9 | 14 | 0.0349 |
| 3 | 7 | 13 | 0.0679 | 8 | 12 | 13 | 0.0674 | 2 | 9 | 14 | 0.0431 |
| 4 | 7 | 13 | 0.0781 | 9 | 12 | 13 | 0.0808 | 3 | 9 | 14 | 0.0502 |
| 5 | 7 | 13 | 0.0893 | 10 | 12 | 13 | 0.1000 | 4 | 9 | 14 | 0.0575 |
| 6 | 7 | 13 | 0.1022 | 11 | 12 | 13 | 0.1313 | 5 | 9 | 14 | 0.0652 |
| 1 | 8 | 13 | 0.0402 | 1 | 13 | 13 | 0.0173 | 6 | 9 | 14 | 0.0740 |
| 2 | 8 | 13 | 0.0500 | 2 | 13 | 13 | 0.0217 | 7 | 9 | 14 | 0.0842 |
| 3 | 8 | 13 | 0.0587 | 3 | 13 | 13 | 0.0257 | 8 | 9 | 14 | 0.0966 |
| 4 | 8 | 13 | 0.0676 | 4 | 13 | 13 | 0.0298 | 1 | 10 | 14 | 0.0306 |
| 5 | 8 | 13 | 0.0774 | 5 | 13 | 13 | 0.0343 | 2 | 10 | 14 | 0.0378 |
| 6 | 8 | 13 | 0.0887 | 6 | 13 | 13 | 0.0395 | 3 | 10 | 14 | 0.0442 |
| 7 | 8 | 13 | 0.1022 | 7 | 13 | 13 | 0.0459 | 4 | 10 | 14 | 0.0505 |
| 1 | 9 | 13 | 0.0349 | 8 | 13 | 13 | 0.0540 | 5 | 10 | 14 | 0.0574 |

TABLE 3

OUTLIER MEAN 0.5 OUTLIER STANDARD DEVIATION 1.0

| R | S | N | COVAR | R | S | N | COVAR | R | S | N | COVAR |
|---|---|---|---|---|---|---|---|---|---|---|---|
| 6 | 10 | 14 | 0.0652 | 9 | 14 | 14 | 0.0530 | 5 | 10 | 15 | 0.0560 |
| 7 | 10 | 14 | 0.0743 | 10 | 14 | 14 | 0.0635 | 6 | 10 | 15 | 0.0629 |
| 8 | 10 | 14 | 0.0853 | 11 | 14 | 14 | 0.0787 | 7 | 10 | 15 | 0.0709 |
| 9 | 10 | 14 | 0.0992 | 12 | 14 | 14 | 0.1036 | 8 | 10 | 15 | 0.0804 |
| 1 | 11 | 14 | 0.0268 | 13 | 14 | 14 | 0.1535 | 9 | 10 | 15 | 0.0919 |
| 2 | 11 | 14 | 0.0332 | 1 | 2 | 15 | 0.1491 | 1 | 11 | 15 | 0.0271 |
| 3 | 11 | 14 | 0.0388 | 1 | 3 | 15 | 0.1012 | 2 | 11 | 15 | 0.0333 |
| 4 | 11 | 14 | 0.0444 | 2 | 3 | 15 | 0.1231 | 3 | 11 | 15 | 0.0387 |
| 5 | 11 | 14 | 0.0505 | 1 | 4 | 15 | 0.0773 | 4 | 11 | 15 | 0.0440 |
| 6 | 11 | 14 | 0.0573 | 2 | 4 | 15 | 0.0943 | 5 | 11 | 15 | 0.0497 |
| 7 | 11 | 14 | 0.0654 | 3 | 4 | 15 | 0.1088 | 6 | 11 | 15 | 0.0559 |
| 8 | 11 | 14 | 0.0752 | 1 | 5 | 15 | 0.0627 | 7 | 11 | 15 | 0.0631 |
| 9 | 11 | 14 | 0.0876 | 2 | 5 | 15 | 0.0766 | 8 | 11 | 15 | 0.0715 |
| 10 | 11 | 14 | 0.1042 | 3 | 5 | 15 | 0.0885 | 9 | 11 | 15 | 0.0818 |
| 1 | 12 | 14 | 0.0233 | 4 | 5 | 15 | 0.1002 | 10 | 11 | 15 | 0.0950 |
| 2 | 12 | 14 | 0.0288 | 1 | 6 | 15 | 0.0526 | 1 | 12 | 15 | 0.0239 |
| 3 | 12 | 14 | 0.0337 | 2 | 6 | 15 | 0.0644 | 2 | 12 | 15 | 0.0294 |
| 4 | 12 | 14 | 0.0387 | 3 | 6 | 15 | 0.0745 | 3 | 12 | 15 | 0.0342 |
| 5 | 12 | 14 | 0.0440 | 4 | 6 | 15 | 0.0845 | 4 | 12 | 15 | 0.0389 |
| 6 | 12 | 14 | 0.0500 | 5 | 6 | 15 | 0.0950 | 5 | 12 | 15 | 0.0440 |
| 7 | 12 | 14 | 0.0571 | 1 | 7 | 15 | 0.0452 | 6 | 12 | 15 | 0.0495 |
| 8 | 12 | 14 | 0.0657 | 2 | 7 | 15 | 0.0554 | 7 | 12 | 15 | 0.0559 |
| 9 | 12 | 14 | 0.0767 | 3 | 7 | 15 | 0.0641 | 8 | 12 | 15 | 0.0634 |
| 10 | 12 | 14 | 0.0914 | 4 | 7 | 15 | 0.0728 | 9 | 12 | 15 | 0.0726 |
| 11 | 12 | 14 | 0.1127 | 5 | 7 | 15 | 0.0819 | 10 | 12 | 15 | 0.0844 |
| 1 | 13 | 14 | 0.0198 | 6 | 7 | 15 | 0.0919 | 11 | 12 | 15 | 0.1003 |
| 2 | 13 | 14 | 0.0245 | 1 | 8 | 15 | 0.0394 | 1 | 13 | 15 | 0.0209 |
| 3 | 13 | 14 | 0.0287 | 2 | 8 | 15 | 0.0483 | 2 | 13 | 15 | 0.0258 |
| 4 | 13 | 14 | 0.0330 | 3 | 8 | 15 | 0.0560 | 3 | 13 | 15 | 0.0300 |
| 5 | 13 | 14 | 0.0375 | 4 | 8 | 15 | 0.0635 | 4 | 13 | 15 | 0.0341 |
| 6 | 13 | 14 | 0.0427 | 5 | 8 | 15 | 0.0716 | 5 | 13 | 15 | 0.0386 |
| 7 | 13 | 14 | 0.0488 | 6 | 8 | 15 | 0.0804 | 6 | 13 | 15 | 0.0435 |
| 8 | 13 | 14 | 0.0562 | 7 | 8 | 15 | 0.0904 | 7 | 13 | 15 | 0.0491 |
| 9 | 13 | 14 | 0.0657 | 1 | 9 | 15 | 0.0346 | 8 | 13 | 15 | 0.0557 |
| 10 | 13 | 14 | 0.0785 | 2 | 9 | 15 | 0.0425 | 9 | 13 | 15 | 0.0639 |
| 11 | 13 | 14 | 0.0970 | 3 | 9 | 15 | 0.0493 | 10 | 13 | 15 | 0.0744 |
| 12 | 13 | 14 | 0.1271 | 4 | 9 | 15 | 0.0560 | 11 | 13 | 15 | 0.0884 |
| 1 | 14 | 14 | 0.0157 | 5 | 9 | 15 | 0.0631 | 12 | 13 | 15 | 0.1089 |
| 2 | 14 | 14 | 0.0195 | 6 | 9 | 15 | 0.0710 | 1 | 14 | 15 | 0.0179 |
| 3 | 14 | 14 | 0.0229 | 7 | 9 | 15 | 0.0799 | 2 | 14 | 15 | 0.0220 |
| 4 | 14 | 14 | 0.0263 | 8 | 9 | 15 | 0.0904 | 3 | 14 | 15 | 0.0257 |
| 5 | 14 | 14 | 0.0300 | 1 | 10 | 15 | 0.0306 | 4 | 14 | 15 | 0.0293 |
| 6 | 14 | 14 | 0.0343 | 2 | 10 | 15 | 0.0376 | 5 | 14 | 15 | 0.0331 |
| 7 | 14 | 14 | 0.0392 | 3 | 10 | 15 | 0.0436 | 6 | 14 | 15 | 0.0373 |
| 8 | 14 | 14 | 0.0453 | 4 | 10 | 15 | 0.0496 | 7 | 14 | 15 | 0.0422 |

TABLE 3

OUTLIER MEAN 0.5 OUTLIER STANDARD DEVIATION 1.0

| R | S | N | COVAR | R | S | N | COVAR | R | S | N | COVAR |
|---|---|---|---|---|---|---|---|---|---|---|---|
| 8 | 14 | 15 | 0.0480 | 5 | 8 | 16 | 0.0691 | 5 | 13 | 16 | 0.0388 |
| 9 | 14 | 15 | 0.0550 | 6 | 8 | 16 | 0.0771 | 6 | 13 | 16 | 0.0434 |
| 10 | 14 | 15 | 0.0641 | 7 | 8 | 16 | 0.0860 | 7 | 13 | 16 | 0.0485 |
| 11 | 14 | 15 | 0.0764 | 1 | 9 | 16 | 0.0343 | 8 | 13 | 16 | 0.0545 |
| 12 | 14 | 15 | 0.0943 | 2 | 9 | 16 | 0.0419 | 9 | 13 | 16 | 0.0616 |
| 13 | 14 | 15 | 0.1234 | 3 | 9 | 16 | 0.0483 | 10 | 13 | 16 | 0.0704 |
| 1 | 15 | 15 | 0.0143 | 4 | 9 | 16 | 0.0546 | 11 | 13 | 16 | 0.0816 |
| 2 | 15 | 15 | 0.0177 | 5 | 9 | 16 | 0.0612 | 12 | 13 | 16 | 0.0968 |
| 3 | 15 | 15 | 0.0206 | 6 | 9 | 16 | 0.0683 | 1 | 14 | 16 | 0.0190 |
| 4 | 15 | 15 | 0.0235 | 7 | 9 | 16 | 0.0762 | 2 | 14 | 16 | 0.0232 |
| 5 | 15 | 15 | 0.0266 | 8 | 9 | 16 | 0.0854 | 3 | 14 | 16 | 0.0269 |
| 6 | 15 | 15 | 0.0301 | 1 | 10 | 16 | 0.0305 | 4 | 14 | 16 | 0.0304 |
| 7 | 15 | 15 | 0.0341 | 2 | 10 | 16 | 0.0373 | 5 | 14 | 16 | 0.0342 |
| 8 | 15 | 15 | 0.0388 | 3 | 10 | 16 | 0.0430 | 6 | 14 | 16 | 0.0382 |
| 9 | 15 | 15 | 0.0446 | 4 | 10 | 16 | 0.0486 | 7 | 14 | 16 | 0.0428 |
| 10 | 15 | 15 | 0.0521 | 5 | 10 | 16 | 0.0545 | 8 | 14 | 16 | 0.0481 |
| 11 | 15 | 15 | 0.0622 | 6 | 10 | 16 | 0.0609 | 9 | 14 | 16 | 0.0545 |
| 12 | 15 | 15 | 0.0770 | 7 | 10 | 16 | 0.0680 | 10 | 14 | 16 | 0.0623 |
| 13 | 15 | 15 | 0.1012 | 8 | 10 | 16 | 0.0762 | 11 | 14 | 16 | 0.0723 |
| 14 | 15 | 15 | 0.1498 | 9 | 10 | 16 | 0.0860 | 12 | 14 | 16 | 0.0858 |
| 1 | 2 | 16 | 0.1458 | 1 | 11 | 16 | 0.0272 | 13 | 14 | 16 | 0.1055 |
| 1 | 3 | 16 | 0.0990 | 2 | 11 | 16 | 0.0333 | 1 | 15 | 16 | 0.0163 |
| 2 | 3 | 16 | 0.1198 | 3 | 11 | 16 | 0.0384 | 2 | 15 | 16 | 0.0200 |
| 1 | 4 | 16 | 0.0757 | 4 | 11 | 16 | 0.0434 | 3 | 15 | 16 | 0.0231 |
| 2 | 4 | 16 | 0.0918 | 5 | 11 | 16 | 0.0487 | 4 | 15 | 16 | 0.0262 |
| 3 | 4 | 16 | 0.1054 | 6 | 11 | 16 | 0.0544 | 5 | 15 | 16 | 0.0295 |
| 1 | 5 | 16 | 0.0614 | 7 | 11 | 16 | 0.0608 | 6 | 15 | 16 | 0.0330 |
| 2 | 5 | 16 | 0.0747 | 8 | 11 | 16 | 0.0682 | 7 | 15 | 16 | 0.0370 |
| 3 | 5 | 16 | 0.0859 | 9 | 11 | 16 | 0.0770 | 8 | 15 | 16 | 0.0416 |
| 4 | 5 | 16 | 0.0967 | 10 | 11 | 16 | 0.0878 | 9 | 15 | 16 | 0.0471 |
| 1 | 6 | 16 | 0.0517 | 1 | 12 | 16 | 0.0243 | 10 | 15 | 16 | 0.0539 |
| 2 | 6 | 16 | 0.0629 | 2 | 12 | 16 | 0.0297 | 11 | 15 | 16 | 0.0627 |
| 3 | 6 | 16 | 0.0724 | 3 | 12 | 16 | 0.0343 | 12 | 15 | 16 | 0.0745 |
| 4 | 6 | 16 | 0.0817 | 4 | 12 | 16 | 0.0388 | 13 | 15 | 16 | 0.0918 |
| 5 | 6 | 16 | 0.0912 | 5 | 12 | 16 | 0.0435 | 14 | 15 | 16 | 0.1201 |
| 1 | 7 | 16 | 0.0445 | 6 | 12 | 16 | 0.0487 | 1 | 16 | 16 | 0.0131 |
| 2 | 7 | 16 | 0.0542 | 7 | 12 | 16 | 0.0544 | 2 | 16 | 16 | 0.0161 |
| 3 | 7 | 16 | 0.0624 | 8 | 12 | 16 | 0.0611 | 3 | 16 | 16 | 0.0187 |
| 4 | 7 | 16 | 0.0705 | 9 | 12 | 16 | 0.0690 | 4 | 16 | 16 | 0.0212 |
| 5 | 7 | 16 | 0.0788 | 10 | 12 | 16 | 0.0788 | 5 | 16 | 16 | 0.0239 |
| 6 | 7 | 16 | 0.0878 | 11 | 12 | 16 | 0.0912 | 6 | 16 | 16 | 0.0268 |
| 1 | 8 | 16 | 0.0389 | 1 | 13 | 16 | 0.0216 | 7 | 16 | 16 | 0.0300 |
| 2 | 8 | 16 | 0.0474 | 2 | 13 | 16 | 0.0264 | 8 | 16 | 16 | 0.0338 |
| 3 | 8 | 16 | 0.0547 | 3 | 13 | 16 | 0.0305 | 9 | 16 | 16 | 0.0383 |
| 4 | 8 | 16 | 0.0617 | 4 | 13 | 16 | 0.0345 | 10 | 16 | 16 | 0.0440 |

TABLE 3

OUTLIER MEAN 0.5 OUTLIER STANDARD DEVIATION 1.0

| R | S | N | COVAR | R | S | N | COVAR | R | S | N | COVAR |
|---|---|---|---|---|---|---|---|---|---|---|---|
| 11 | 16 | 16 | 0.0512 | 5 | 10 | 17 | 0.0531 | 8 | 14 | 17 | 0.0475 |
| 12 | 16 | 16 | 0.0610 | 6 | 10 | 17 | 0.0589 | 9 | 14 | 17 | 0.0532 |
| 13 | 16 | 16 | 0.0754 | 7 | 10 | 17 | 0.0654 | 10 | 14 | 17 | 0.0599 |
| 14 | 16 | 16 | 0.0990 | 8 | 10 | 17 | 0.0727 | 11 | 14 | 17 | 0.0683 |
| 15 | 16 | 16 | 0.1465 | 9 | 10 | 17 | 0.0811 | 12 | 14 | 17 | 0.0791 |
| 1 | 2 | 17 | 0.1427 | 1 | 11 | 17 | 0.0272 | 13 | 14 | 17 | 0.0937 |
| 1 | 3 | 17 | 0.0969 | 2 | 11 | 17 | 0.0331 | 1 | 15 | 17 | 0.0173 |
| 2 | 3 | 17 | 0.1168 | 3 | 11 | 17 | 0.0380 | 2 | 15 | 17 | 0.0211 |
| 1 | 4 | 17 | 0.0741 | 4 | 11 | 17 | 0.0428 | 3 | 15 | 17 | 0.0243 |
| 2 | 4 | 17 | 0.0896 | 5 | 11 | 17 | 0.0477 | 4 | 15 | 17 | 0.0274 |
| 3 | 4 | 17 | 0.1024 | 6 | 11 | 17 | 0.0530 | 5 | 15 | 17 | 0.0306 |
| 1 | 5 | 17 | 0.0602 | 7 | 11 | 17 | 0.0588 | 6 | 15 | 17 | 0.0340 |
| 2 | 5 | 17 | 0.0729 | 8 | 11 | 17 | 0.0654 | 7 | 15 | 17 | 0.0378 |
| 3 | 5 | 17 | 0.0835 | 9 | 11 | 17 | 0.0730 | 8 | 15 | 17 | 0.0422 |
| 4 | 5 | 17 | 0.0936 | 10 | 11 | 17 | 0.0821 | 9 | 15 | 17 | 0.0472 |
| 1 | 6 | 17 | 0.0508 | 1 | 12 | 17 | 0.0244 | 10 | 15 | 17 | 0.0533 |
| 2 | 6 | 17 | 0.0615 | 2 | 12 | 17 | 0.0297 | 11 | 15 | 17 | 0.0607 |
| 3 | 6 | 17 | 0.0705 | 3 | 12 | 17 | 0.0342 | 12 | 15 | 17 | 0.0704 |
| 4 | 6 | 17 | 0.0791 | 4 | 12 | 17 | 0.0385 | 13 | 15 | 17 | 0.0835 |
| 5 | 6 | 17 | 0.0880 | 5 | 12 | 17 | 0.0429 | 14 | 15 | 17 | 0.1025 |
| 1 | 7 | 17 | 0.0438 | 6 | 12 | 17 | 0.0477 | 1 | 16 | 17 | 0.0149 |
| 2 | 7 | 17 | 0.0531 | 7 | 12 | 17 | 0.0529 | 2 | 16 | 17 | 0.0182 |
| 3 | 7 | 17 | 0.0609 | 8 | 12 | 17 | 0.0589 | 3 | 16 | 17 | 0.0210 |
| 4 | 7 | 17 | 0.0684 | 9 | 12 | 17 | 0.0658 | 4 | 16 | 17 | 0.0237 |
| 5 | 7 | 17 | 0.0761 | 10 | 12 | 17 | 0.0741 | 5 | 16 | 17 | 0.0265 |
| 6 | 7 | 17 | 0.0843 | 11 | 12 | 17 | 0.0843 | 6 | 16 | 17 | 0.0295 |
| 1 | 8 | 17 | 0.0384 | 1 | 13 | 17 | 0.0219 | 7 | 16 | 17 | 0.0328 |
| 2 | 8 | 17 | 0.0466 | 2 | 13 | 17 | 0.0267 | 8 | 16 | 17 | 0.0366 |
| 3 | 8 | 17 | 0.0534 | 3 | 13 | 17 | 0.0307 | 9 | 16 | 17 | 0.0410 |
| 4 | 8 | 17 | 0.0601 | 4 | 13 | 17 | 0.0346 | 10 | 16 | 17 | 0.0463 |
| 5 | 8 | 17 | 0.0669 | 5 | 13 | 17 | 0.0386 | 11 | 16 | 17 | 0.0528 |
| 6 | 8 | 17 | 0.0741 | 6 | 13 | 17 | 0.0429 | 12 | 16 | 17 | 0.0613 |
| 7 | 8 | 17 | 0.0822 | 7 | 13 | 17 | 0.0476 | 13 | 16 | 17 | 0.0728 |
| 1 | 9 | 17 | 0.0340 | 8 | 13 | 17 | 0.0530 | 14 | 16 | 17 | 0.0896 |
| 2 | 9 | 17 | 0.0413 | 9 | 13 | 17 | 0.0593 | 15 | 16 | 17 | 0.1171 |
| 3 | 9 | 17 | 0.0474 | 10 | 13 | 17 | 0.0668 | 1 | 17 | 17 | 0.0121 |
| 4 | 9 | 17 | 0.0533 | 11 | 13 | 17 | 0.0761 | 2 | 17 | 17 | 0.0148 |
| 5 | 9 | 17 | 0.0594 | 12 | 13 | 17 | 0.0880 | 3 | 17 | 17 | 0.0170 |
| 6 | 9 | 17 | 0.0659 | 1 | 14 | 17 | 0.0196 | 4 | 17 | 17 | 0.0193 |
| 7 | 9 | 17 | 0.0730 | 2 | 14 | 17 | 0.0238 | 5 | 17 | 17 | 0.0215 |
| 8 | 9 | 17 | 0.0811 | 3 | 14 | 17 | 0.0274 | 6 | 17 | 17 | 0.0240 |
| 1 | 10 | 17 | 0.0303 | 4 | 14 | 17 | 0.0309 | 7 | 17 | 17 | 0.0268 |
| 2 | 10 | 17 | 0.0369 | 5 | 14 | 17 | 0.0345 | 8 | 17 | 17 | 0.0299 |
| 3 | 10 | 17 | 0.0423 | 6 | 14 | 17 | 0.0384 | 9 | 17 | 17 | 0.0335 |
| 4 | 10 | 17 | 0.0476 | 7 | 14 | 17 | 0.0427 | 10 | 17 | 17 | 0.0379 |

TABLE 3

OUTLIER MEAN 0.5 OUTLIER STANDARD DEVIATION 1.0

| R | S | N | COVAR | R | S | N | COVAR | R | S | N | COVAR |
|---|---|---|---|---|---|---|---|---|---|---|---|
| 11 | 17 | 17 | 0.0433 | 4 | 10 | 18 | 0.0467 | 7 | 14 | 18 | 0.0422 |
| 12 | 17 | 17 | 0.0503 | 5 | 10 | 18 | 0.0518 | 8 | 14 | 18 | 0.0466 |
| 13 | 17 | 17 | 0.0599 | 6 | 10 | 18 | 0.0572 | 9 | 14 | 18 | 0.0517 |
| 14 | 17 | 17 | 0.0739 | 7 | 10 | 18 | 0.0630 | 10 | 14 | 18 | 0.0577 |
| 15 | 17 | 17 | 0.0970 | 8 | 10 | 18 | 0.0696 | 11 | 14 | 18 | 0.0648 |
| 16 | 17 | 17 | 0.1435 | 9 | 10 | 18 | 0.0770 | 12 | 14 | 18 | 0.0737 |
| 1 | 2 | 18 | 0.1400 | 1 | 11 | 18 | 0.0271 | 13 | 14 | 18 | 0.0851 |
| 1 | 3 | 18 | 0.0950 | 2 | 11 | 18 | 0.0328 | 1 | 15 | 18 | 0.0179 |
| 2 | 3 | 18 | 0.1141 | 3 | 11 | 18 | 0.0375 | 2 | 15 | 18 | 0.0217 |
| 1 | 4 | 18 | 0.0727 | 4 | 11 | 18 | 0.0421 | 3 | 15 | 18 | 0.0249 |
| 2 | 4 | 18 | 0.0875 | 5 | 11 | 18 | 0.0467 | 4 | 15 | 18 | 0.0279 |
| 3 | 4 | 18 | 0.0997 | 6 | 11 | 18 | 0.0516 | 5 | 15 | 18 | 0.0310 |
| 1 | 5 | 18 | 0.0592 | 7 | 11 | 18 | 0.0569 | 6 | 15 | 18 | 0.0343 |
| 2 | 5 | 18 | 0.0713 | 8 | 11 | 18 | 0.0628 | 7 | 15 | 18 | 0.0379 |
| 3 | 5 | 18 | 0.0813 | 9 | 11 | 18 | 0.0696 | 8 | 15 | 18 | 0.0419 |
| 4 | 5 | 18 | 0.0908 | 10 | 11 | 18 | 0.0775 | 9 | 15 | 18 | 0.0466 |
| 1 | 6 | 18 | 0.0499 | 1 | 12 | 18 | 0.0245 | 10 | 15 | 18 | 0.0519 |
| 2 | 6 | 18 | 0.0602 | 2 | 12 | 18 | 0.0296 | 11 | 15 | 18 | 0.0584 |
| 3 | 6 | 18 | 0.0687 | 3 | 12 | 18 | 0.0339 | 12 | 15 | 18 | 0.0665 |
| 4 | 6 | 18 | 0.0768 | 4 | 12 | 18 | 0.0380 | 13 | 15 | 18 | 0.0768 |
| 5 | 6 | 18 | 0.0851 | 5 | 12 | 18 | 0.0422 | 14 | 15 | 18 | 0.0909 |
| 1 | 7 | 18 | 0.0431 | 6 | 12 | 18 | 0.0466 | 1 | 16 | 18 | 0.0159 |
| 2 | 7 | 18 | 0.0521 | 7 | 12 | 18 | 0.0515 | 2 | 16 | 18 | 0.0193 |
| 3 | 7 | 18 | 0.0595 | 8 | 12 | 18 | 0.0569 | 3 | 16 | 18 | 0.0221 |
| 4 | 7 | 18 | 0.0665 | 9 | 12 | 18 | 0.0630 | 4 | 16 | 18 | 0.0248 |
| 5 | 7 | 18 | 0.0737 | 10 | 12 | 18 | 0.0702 | 5 | 16 | 18 | 0.0276 |
| 6 | 7 | 18 | 0.0813 | 11 | 12 | 18 | 0.0788 | 6 | 16 | 18 | 0.0305 |
| 1 | 8 | 18 | 0.0378 | 1 | 13 | 18 | 0.0221 | 7 | 16 | 18 | 0.0338 |
| 2 | 8 | 18 | 0.0457 | 2 | 13 | 18 | 0.0268 | 8 | 16 | 18 | 0.0374 |
| 3 | 8 | 18 | 0.0523 | 3 | 13 | 18 | 0.0307 | 9 | 16 | 18 | 0.0415 |
| 4 | 8 | 18 | 0.0585 | 4 | 13 | 18 | 0.0344 | 10 | 16 | 18 | 0.0463 |
| 5 | 8 | 18 | 0.0649 | 5 | 13 | 18 | 0.0382 | 11 | 16 | 18 | 0.0521 |
| 6 | 8 | 18 | 0.0716 | 6 | 13 | 18 | 0.0422 | 12 | 16 | 18 | 0.0593 |
| 7 | 8 | 18 | 0.0788 | 7 | 13 | 18 | 0.0466 | 13 | 16 | 18 | 0.0686 |
| 1 | 9 | 18 | 0.0336 | 8 | 13 | 18 | 0.0515 | 14 | 16 | 18 | 0.0813 |
| 2 | 9 | 18 | 0.0407 | 9 | 13 | 18 | 0.0571 | 15 | 16 | 18 | 0.0998 |
| 3 | 9 | 18 | 0.0465 | 10 | 13 | 18 | 0.0637 | 1 | 17 | 18 | 0.0138 |
| 4 | 9 | 18 | 0.0521 | 11 | 13 | 18 | 0.0715 | 2 | 17 | 18 | 0.0167 |
| 5 | 9 | 18 | 0.0577 | 12 | 13 | 18 | 0.0813 | 3 | 17 | 18 | 0.0192 |
| 6 | 9 | 18 | 0.0637 | 1 | 14 | 18 | 0.0199 | 4 | 17 | 18 | 0.0216 |
| 7 | 9 | 18 | 0.0702 | 2 | 14 | 18 | 0.0242 | 5 | 17 | 18 | 0.0240 |
| 8 | 9 | 18 | 0.0775 | 3 | 14 | 18 | 0.0277 | 6 | 17 | 18 | 0.0266 |
| 1 | 10 | 18 | 0.0301 | 4 | 14 | 18 | 0.0311 | 7 | 17 | 18 | 0.0294 |
| 2 | 10 | 18 | 0.0364 | 5 | 14 | 18 | 0.0345 | 8 | 17 | 18 | 0.0325 |
| 3 | 10 | 18 | 0.0417 | 6 | 14 | 18 | 0.0382 | 9 | 17 | 18 | 0.0362 |

TABLE 3

OUTLIER MEAN 0.5 OUTLIER STANDARD DEVIATION 1.0

| R | S | N | COVAR | R | S | N | COVAR | R | S | N | COVAR |
|---|---|---|---|---|---|---|---|---|---|---|---|
| 10 | 17 | 18 | 0.0404 | 1 | 8 | 19 | 0.0373 | 1 | 13 | 19 | 0.0222 |
| 11 | 17 | 18 | 0.0455 | 2 | 8 | 19 | 0.0450 | 2 | 13 | 19 | 0.0268 |
| 12 | 17 | 18 | 0.0518 | 3 | 8 | 19 | 0.0512 | 3 | 13 | 19 | 0.0305 |
| 13 | 17 | 18 | 0.0600 | 4 | 8 | 19 | 0.0571 | 4 | 13 | 19 | 0.0341 |
| 14 | 17 | 18 | 0.0712 | 5 | 8 | 19 | 0.0630 | 5 | 13 | 19 | 0.0377 |
| 15 | 17 | 18 | 0.0876 | 6 | 8 | 19 | 0.0693 | 6 | 13 | 19 | 0.0415 |
| 16 | 17 | 18 | 0.1144 | 7 | 8 | 19 | 0.0759 | 7 | 13 | 19 | 0.0456 |
| 1 | 18 | 18 | 0.0112 | 1 | 9 | 19 | 0.0332 | 8 | 13 | 19 | 0.0501 |
| 2 | 18 | 18 | 0.0136 | 2 | 9 | 19 | 0.0400 | 9 | 13 | 19 | 0.0551 |
| 3 | 18 | 18 | 0.0156 | 3 | 9 | 19 | 0.0456 | 10 | 13 | 19 | 0.0609 |
| 4 | 18 | 18 | 0.0176 | 4 | 9 | 19 | 0.0509 | 11 | 13 | 19 | 0.0677 |
| 5 | 18 | 18 | 0.0196 | 5 | 9 | 19 | 0.0562 | 12 | 13 | 19 | 0.0759 |
| 6 | 18 | 18 | 0.0217 | 6 | 9 | 19 | 0.0618 | 1 | 14 | 19 | 0.0201 |
| 7 | 18 | 18 | 0.0241 | 7 | 9 | 19 | 0.0677 | 2 | 14 | 19 | 0.0243 |
| 8 | 18 | 18 | 0.0267 | 8 | 9 | 19 | 0.0743 | 3 | 14 | 19 | 0.0277 |
| 9 | 18 | 18 | 0.0297 | 1 | 10 | 19 | 0.0298 | 4 | 14 | 19 | 0.0310 |
| 10 | 18 | 18 | 0.0332 | 2 | 10 | 19 | 0.0359 | 5 | 14 | 19 | 0.0343 |
| 11 | 18 | 18 | 0.0374 | 3 | 10 | 19 | 0.0410 | 6 | 14 | 19 | 0.0378 |
| 12 | 18 | 18 | 0.0427 | 4 | 10 | 19 | 0.0457 | 7 | 14 | 19 | 0.0415 |
| 13 | 18 | 18 | 0.0495 | 5 | 10 | 19 | 0.0505 | 8 | 14 | 19 | 0.0456 |
| 14 | 18 | 18 | 0.0589 | 6 | 10 | 19 | 0.0556 | 9 | 14 | 19 | 0.0502 |
| 15 | 18 | 18 | 0.0726 | 7 | 10 | 19 | 0.0610 | 10 | 14 | 19 | 0.0555 |
| 16 | 18 | 18 | 0.0951 | 8 | 10 | 19 | 0.0669 | 11 | 14 | 19 | 0.0617 |
| 17 | 18 | 18 | 0.1408 | 9 | 10 | 19 | 0.0735 | 12 | 14 | 19 | 0.0692 |
| 1 | 2 | 19 | 0.1375 | 1 | 11 | 19 | 0.0269 | 13 | 14 | 19 | 0.0785 |
| 1 | 3 | 19 | 0.0933 | 2 | 11 | 19 | 0.0325 | 1 | 15 | 19 | 0.0182 |
| 2 | 3 | 19 | 0.1116 | 3 | 11 | 19 | 0.0370 | 2 | 15 | 19 | 0.0220 |
| 1 | 4 | 19 | 0.0714 | 4 | 11 | 19 | 0.0413 | 3 | 15 | 19 | 0.0252 |
| 2 | 4 | 19 | 0.0857 | 5 | 11 | 19 | 0.0457 | 4 | 15 | 19 | 0.0281 |
| 3 | 4 | 19 | 0.0972 | 6 | 11 | 19 | 0.0503 | 5 | 15 | 19 | 0.0311 |
| 1 | 5 | 19 | 0.0581 | 7 | 11 | 19 | 0.0552 | 6 | 15 | 19 | 0.0343 |
| 2 | 5 | 19 | 0.0698 | 8 | 11 | 19 | 0.0606 | 7 | 15 | 19 | 0.0377 |
| 3 | 5 | 19 | 0.0793 | 9 | 11 | 19 | 0.0666 | 8 | 15 | 19 | 0.0414 |
| 4 | 5 | 19 | 0.0883 | 10 | 11 | 19 | 0.0735 | 9 | 15 | 19 | 0.0456 |
| 1 | 6 | 19 | 0.0491 | 1 | 12 | 19 | 0.0244 | 10 | 15 | 19 | 0.0505 |
| 2 | 6 | 19 | 0.0590 | 2 | 12 | 19 | 0.0294 | 11 | 15 | 19 | 0.0561 |
| 3 | 6 | 19 | 0.0671 | 3 | 12 | 19 | 0.0336 | 12 | 15 | 19 | 0.0630 |
| 4 | 6 | 19 | 0.0748 | 4 | 12 | 19 | 0.0375 | 13 | 15 | 19 | 0.0715 |
| 5 | 6 | 19 | 0.0825 | 5 | 12 | 19 | 0.0415 | 14 | 15 | 19 | 0.0825 |
| 1 | 7 | 19 | 0.0425 | 6 | 12 | 19 | 0.0456 | 1 | 16 | 19 | 0.0164 |
| 2 | 7 | 19 | 0.0511 | 7 | 12 | 19 | 0.0501 | 2 | 16 | 19 | 0.0199 |
| 3 | 7 | 19 | 0.0582 | 8 | 12 | 19 | 0.0550 | 3 | 16 | 19 | 0.0227 |
| 4 | 7 | 19 | 0.0648 | 9 | 12 | 19 | 0.0605 | 4 | 16 | 19 | 0.0254 |
| 5 | 7 | 19 | 0.0715 | 10 | 12 | 19 | 0.0669 | 5 | 16 | 19 | 0.0281 |
| 6 | 7 | 19 | 0.0785 | 11 | 12 | 19 | 0.0743 | 6 | 16 | 19 | 0.0309 |

TABLE 3

OUTLIER MEAN 0.5 OUTLIER STANDARD DEVIATION 1.0

| R | S | N | COVAR | R | S | N | COVAR | R | S | N | COVAR |
|---|---|---|---|---|---|---|---|---|---|---|---|
| 7 | 16 | 19 | 0.0340 | 4 | 19 | 19 | 0.0162 | 3 | 9 | 20 | 0.0448 |
| 8 | 16 | 19 | 0.0374 | 5 | 19 | 19 | 0.0180 | 4 | 9 | 20 | 0.0498 |
| 9 | 16 | 19 | 0.0412 | 6 | 19 | 19 | 0.0198 | 5 | 9 | 20 | 0.0548 |
| 10 | 16 | 19 | 0.0456 | 7 | 19 | 19 | 0.0218 | 6 | 9 | 20 | 0.0600 |
| 11 | 16 | 19 | 0.0508 | 8 | 19 | 19 | 0.0240 | 7 | 9 | 20 | 0.0655 |
| 12 | 16 | 19 | 0.0570 | 9 | 19 | 19 | 0.0266 | 8 | 9 | 20 | 0.0715 |
| 13 | 16 | 19 | 0.0648 | 10 | 19 | 19 | 0.0294 | 1 | 10 | 20 | 0.0296 |
| 14 | 16 | 19 | 0.0748 | 11 | 19 | 19 | 0.0328 | 2 | 10 | 20 | 0.0355 |
| 15 | 16 | 19 | 0.0884 | 12 | 19 | 19 | 0.0370 | 3 | 10 | 20 | 0.0403 |
| 1 | 17 | 19 | 0.0146 | 13 | 19 | 19 | 0.0421 | 4 | 10 | 20 | 0.0448 |
| 2 | 17 | 19 | 0.0177 | 14 | 19 | 19 | 0.0488 | 5 | 10 | 20 | 0.0494 |
| 3 | 17 | 19 | 0.0202 | 15 | 19 | 19 | 0.0579 | 6 | 10 | 20 | 0.0541 |
| 4 | 17 | 19 | 0.0226 | 16 | 19 | 19 | 0.0713 | 7 | 10 | 20 | 0.0591 |
| 5 | 17 | 19 | 0.0251 | 17 | 19 | 19 | 0.0934 | 8 | 10 | 20 | 0.0645 |
| 6 | 17 | 19 | 0.0276 | 18 | 19 | 19 | 0.1383 | 9 | 10 | 20 | 0.0705 |
| 7 | 17 | 19 | 0.0304 | 1 | 2 | 20 | 0.1352 | 1 | 11 | 20 | 0.0267 |
| 8 | 17 | 19 | 0.0334 | 1 | 3 | 20 | 0.0917 | 2 | 11 | 20 | 0.0321 |
| 9 | 17 | 19 | 0.0369 | 2 | 3 | 20 | 0.1093 | 3 | 11 | 20 | 0.0365 |
| 10 | 17 | 19 | 0.0408 | 1 | 4 | 20 | 0.0702 | 4 | 11 | 20 | 0.0406 |
| 11 | 17 | 19 | 0.0454 | 2 | 4 | 20 | 0.0839 | 5 | 11 | 20 | 0.0447 |
| 12 | 17 | 19 | 0.0511 | 3 | 4 | 20 | 0.0949 | 6 | 11 | 20 | 0.0490 |
| 13 | 17 | 19 | 0.0580 | 1 | 5 | 20 | 0.0572 | 7 | 11 | 20 | 0.0536 |
| 14 | 17 | 19 | 0.0671 | 2 | 5 | 20 | 0.0684 | 8 | 11 | 20 | 0.0585 |
| 15 | 17 | 19 | 0.0794 | 3 | 5 | 20 | 0.0775 | 9 | 11 | 20 | 0.0640 |
| 16 | 17 | 19 | 0.0974 | 4 | 5 | 20 | 0.0860 | 10 | 11 | 20 | 0.0702 |
| 1 | 18 | 19 | 0.0127 | 1 | 6 | 20 | 0.0483 | 1 | 12 | 20 | 0.0243 |
| 2 | 18 | 19 | 0.0154 | 2 | 6 | 20 | 0.0579 | 2 | 12 | 20 | 0.0292 |
| 3 | 18 | 19 | 0.0176 | 3 | 6 | 20 | 0.0657 | 3 | 12 | 20 | 0.0332 |
| 4 | 18 | 19 | 0.0197 | 4 | 6 | 20 | 0.0729 | 4 | 12 | 20 | 0.0370 |
| 5 | 18 | 19 | 0.0219 | 5 | 6 | 20 | 0.0801 | 5 | 12 | 20 | 0.0407 |
| 6 | 18 | 19 | 0.0241 | 1 | 7 | 20 | 0.0419 | 6 | 12 | 20 | 0.0446 |
| 7 | 18 | 19 | 0.0265 | 2 | 7 | 20 | 0.0502 | 7 | 12 | 20 | 0.0488 |
| 8 | 18 | 19 | 0.0292 | 3 | 7 | 20 | 0.0569 | 8 | 12 | 20 | 0.0533 |
| 9 | 18 | 19 | 0.0322 | 4 | 7 | 20 | 0.0633 | 9 | 12 | 20 | 0.0583 |
| 10 | 18 | 19 | 0.0357 | 5 | 7 | 20 | 0.0696 | 10 | 12 | 20 | 0.0640 |
| 11 | 18 | 19 | 0.0398 | 6 | 7 | 20 | 0.0761 | 11 | 12 | 20 | 0.0705 |
| 12 | 18 | 19 | 0.0447 | 1 | 8 | 20 | 0.0369 | 1 | 13 | 20 | 0.0222 |
| 13 | 18 | 19 | 0.0509 | 2 | 8 | 20 | 0.0442 | 2 | 13 | 20 | 0.0267 |
| 14 | 18 | 19 | 0.0589 | 3 | 8 | 20 | 0.0502 | 3 | 13 | 20 | 0.0303 |
| 15 | 18 | 19 | 0.0697 | 4 | 8 | 20 | 0.0558 | 4 | 13 | 20 | 0.0337 |
| 16 | 18 | 19 | 0.0857 | 5 | 8 | 20 | 0.0614 | 5 | 13 | 20 | 0.0372 |
| 17 | 18 | 19 | 0.1120 | 6 | 8 | 20 | 0.0672 | 6 | 13 | 20 | 0.0408 |
| 1 | 19 | 19 | 0.0104 | 7 | 8 | 20 | 0.0733 | 7 | 13 | 20 | 0.0446 |
| 2 | 19 | 19 | 0.0126 | 1 | 9 | 20 | 0.0329 | 8 | 13 | 20 | 0.0487 |
| 3 | 19 | 19 | 0.0144 | 2 | 9 | 20 | 0.0394 | 9 | 13 | 20 | 0.0533 |

TABLE 3

OUTLIER MEAN 0.5 OUTLIER STANDARD DEVIATION 1.0

| R | S | N | COVAR | R | S | N | COVAR | R | S | N | COVAR |
|---|---|---|---|---|---|---|---|---|---|---|---|
| 10 | 13 | 20 | 0.0585 | 10 | 16 | 20 | 0.0447 | 1 | 19 | 20 | 0.0119 |
| 11 | 13 | 20 | 0.0645 | 11 | 16 | 20 | 0.0493 | 2 | 19 | 20 | 0.0143 |
| 12 | 13 | 20 | 0.0715 | 12 | 16 | 20 | 0.0547 | 3 | 19 | 20 | 0.0163 |
| 1 | 14 | 20 | 0.0202 | 13 | 16 | 20 | 0.0613 | 4 | 19 | 20 | 0.0182 |
| 2 | 14 | 20 | 0.0243 | 14 | 16 | 20 | 0.0695 | 5 | 19 | 20 | 0.0201 |
| 3 | 14 | 20 | 0.0277 | 15 | 16 | 20 | 0.0801 | 6 | 19 | 20 | 0.0220 |
| 4 | 14 | 20 | 0.0308 | 1 | 17 | 20 | 0.0152 | 7 | 19 | 20 | 0.0242 |
| 5 | 14 | 20 | 0.0340 | 2 | 17 | 20 | 0.0183 | 8 | 19 | 20 | 0.0265 |
| 6 | 14 | 20 | 0.0373 | 3 | 17 | 20 | 0.0208 | 9 | 19 | 20 | 0.0290 |
| 7 | 14 | 20 | 0.0408 | 4 | 17 | 20 | 0.0232 | 10 | 19 | 20 | 0.0319 |
| 8 | 14 | 20 | 0.0446 | 5 | 17 | 20 | 0.0256 | 11 | 19 | 20 | 0.0353 |
| 9 | 14 | 20 | 0.0488 | 6 | 17 | 20 | 0.0281 | 12 | 19 | 20 | 0.0392 |
| 10 | 14 | 20 | 0.0535 | 7 | 17 | 20 | 0.0308 | 13 | 19 | 20 | 0.0440 |
| 11 | 14 | 20 | 0.0590 | 8 | 17 | 20 | 0.0337 | 14 | 19 | 20 | 0.0500 |
| 12 | 14 | 20 | 0.0655 | 9 | 17 | 20 | 0.0369 | 15 | 19 | 20 | 0.0578 |
| 13 | 14 | 20 | 0.0733 | 10 | 17 | 20 | 0.0405 | 16 | 19 | 20 | 0.0684 |
| 1 | 15 | 20 | 0.0185 | 11 | 17 | 20 | 0.0447 | 17 | 19 | 20 | 0.0840 |
| 2 | 15 | 20 | 0.0222 | 12 | 17 | 20 | 0.0497 | 18 | 19 | 20 | 0.1097 |
| 3 | 15 | 20 | 0.0253 | 13 | 17 | 20 | 0.0557 | 1 | 20 | 20 | 0.0097 |
| 4 | 15 | 20 | 0.0282 | 14 | 17 | 20 | 0.0632 | 2 | 20 | 20 | 0.0117 |
| 5 | 15 | 20 | 0.0310 | 15 | 17 | 20 | 0.0729 | 3 | 20 | 20 | 0.0134 |
| 6 | 15 | 20 | 0.0340 | 16 | 17 | 20 | 0.0861 | 4 | 20 | 20 | 0.0150 |
| 7 | 15 | 20 | 0.0372 | 1 | 18 | 20 | 0.0136 | 5 | 20 | 20 | 0.0165 |
| 8 | 15 | 20 | 0.0407 | 2 | 18 | 20 | 0.0164 | 6 | 20 | 20 | 0.0182 |
| 9 | 15 | 20 | 0.0446 | 3 | 18 | 20 | 0.0186 | 7 | 20 | 20 | 0.0199 |
| 10 | 15 | 20 | 0.0490 | 4 | 18 | 20 | 0.0208 | 8 | 20 | 20 | 0.0218 |
| 11 | 15 | 20 | 0.0540 | 5 | 18 | 20 | 0.0229 | 9 | 20 | 20 | 0.0240 |
| 12 | 15 | 20 | 0.0600 | 6 | 18 | 20 | 0.0252 | 10 | 20 | 20 | 0.0264 |
| 13 | 15 | 20 | 0.0672 | 7 | 18 | 20 | 0.0276 | 11 | 20 | 20 | 0.0292 |
| 14 | 15 | 20 | 0.0761 | 8 | 18 | 20 | 0.0302 | 12 | 20 | 20 | 0.0325 |
| 1 | 16 | 20 | 0.0168 | 9 | 18 | 20 | 0.0331 | 13 | 20 | 20 | 0.0365 |
| 2 | 16 | 20 | 0.0202 | 10 | 18 | 20 | 0.0364 | 14 | 20 | 20 | 0.0415 |
| 3 | 16 | 20 | 0.0230 | 11 | 18 | 20 | 0.0402 | 15 | 20 | 20 | 0.0480 |
| 4 | 16 | 20 | 0.0256 | 12 | 18 | 20 | 0.0446 | 16 | 20 | 20 | 0.0570 |
| 5 | 16 | 20 | 0.0283 | 13 | 18 | 20 | 0.0501 | 17 | 20 | 20 | 0.0701 |
| 6 | 16 | 20 | 0.0310 | 14 | 18 | 20 | 0.0568 | 18 | 20 | 20 | 0.0919 |
| 7 | 16 | 20 | 0.0339 | 15 | 18 | 20 | 0.0656 | 19 | 20 | 20 | 0.1360 |
| 8 | 16 | 20 | 0.0371 | 16 | 18 | 20 | 0.0776 | | | | |
| 9 | 16 | 20 | 0.0407 | 17 | 18 | 20 | 0.0951 | | | | |

TABLE 3

OUTLIER MEAN 1.0 OUTLIER STANDARD DEVIATION 1.0

| R | S | N | COVAR | R | S | N | COVAR | R | S | N | COVAR |
|---|---|---|---|---|---|---|---|---|---|---|---|
| 1 | 2 | 2 | 0.2395 | 1 | 6 | 7 | 0.0537 | 1 | 5 | 9 | 0.0722 |
| 1 | 2 | 3 | 0.2742 | 2 | 6 | 7 | 0.0737 | 2 | 5 | 9 | 0.0937 |
| 1 | 3 | 3 | 0.1175 | 3 | 6 | 7 | 0.0966 | 3 | 5 | 9 | 0.1151 |
| 2 | 3 | 3 | 0.2393 | 4 | 6 | 7 | 0.1278 | 4 | 5 | 9 | 0.1400 |
| 1 | 2 | 4 | 0.2523 | 5 | 6 | 7 | 0.1770 | 1 | 6 | 9 | 0.0578 |
| 1 | 3 | 4 | 0.1489 | 1 | 7 | 7 | 0.0329 | 2 | 6 | 9 | 0.0753 |
| 2 | 3 | 4 | 0.2358 | 2 | 7 | 7 | 0.0465 | 3 | 6 | 9 | 0.0929 |
| 1 | 4 | 4 | 0.0743 | 3 | 7 | 7 | 0.0626 | 4 | 6 | 9 | 0.1135 |
| 2 | 4 | 4 | 0.1272 | 4 | 7 | 7 | 0.0851 | 5 | 6 | 9 | 0.1398 |
| 3 | 4 | 4 | 0.2279 | 5 | 7 | 7 | 0.1218 | 1 | 7 | 9 | 0.0462 |
| 1 | 2 | 5 | 0.2316 | 6 | 7 | 7 | 0.1968 | 2 | 7 | 9 | 0.0605 |
| 1 | 3 | 5 | 0.1469 | 1 | 2 | 8 | 0.1913 | 3 | 7 | 9 | 0.0749 |
| 2 | 3 | 5 | 0.2131 | 1 | 3 | 8 | 0.1280 | 4 | 7 | 9 | 0.0919 |
| 1 | 4 | 5 | 0.0969 | 2 | 3 | 8 | 0.1675 | 5 | 7 | 9 | 0.1139 |
| 2 | 4 | 5 | 0.1436 | 1 | 4 | 8 | 0.0949 | 6 | 7 | 9 | 0.1449 |
| 3 | 4 | 5 | 0.2094 | 2 | 4 | 8 | 0.1249 | 1 | 8 | 9 | 0.0357 |
| 1 | 5 | 5 | 0.0531 | 3 | 4 | 8 | 0.1561 | 2 | 8 | 9 | 0.0470 |
| 2 | 5 | 5 | 0.0828 | 1 | 5 | 8 | 0.0733 | 3 | 8 | 9 | 0.0586 |
| 3 | 5 | 5 | 0.1273 | 2 | 5 | 8 | 0.0970 | 4 | 8 | 9 | 0.0723 |
| 4 | 5 | 5 | 0.2162 | 3 | 5 | 8 | 0.1219 | 5 | 8 | 9 | 0.0902 |
| 1 | 2 | 6 | 0.2151 | 4 | 5 | 8 | 0.1525 | 6 | 8 | 9 | 0.1158 |
| 1 | 3 | 6 | 0.1406 | 1 | 6 | 8 | 0.0571 | 7 | 8 | 9 | 0.1574 |
| 2 | 3 | 6 | 0.1942 | 2 | 6 | 8 | 0.0760 | 1 | 9 | 9 | 0.0234 |
| 1 | 4 | 6 | 0.0994 | 3 | 6 | 8 | 0.0960 | 2 | 9 | 9 | 0.0314 |
| 2 | 4 | 6 | 0.1389 | 4 | 6 | 8 | 0.1208 | 3 | 9 | 9 | 0.0398 |
| 3 | 4 | 6 | 0.1869 | 5 | 6 | 8 | 0.1553 | 4 | 9 | 9 | 0.0499 |
| 1 | 5 | 6 | 0.0698 | 1 | 7 | 8 | 0.0431 | 5 | 9 | 9 | 0.0634 |
| 2 | 5 | 6 | 0.0990 | 2 | 7 | 8 | 0.0578 | 6 | 9 | 9 | 0.0830 |
| 3 | 5 | 6 | 0.1353 | 3 | 7 | 8 | 0.0736 | 7 | 9 | 9 | 0.1154 |
| 4 | 5 | 6 | 0.1909 | 4 | 7 | 8 | 0.0934 | 8 | 9 | 9 | 0.1822 |
| 1 | 6 | 6 | 0.0408 | 5 | 7 | 8 | 0.1213 | 1 | 2 | 10 | 0.1751 |
| 2 | 6 | 6 | 0.0601 | 6 | 7 | 8 | 0.1662 | 1 | 3 | 10 | 0.1182 |
| 3 | 6 | 6 | 0.0851 | 1 | 8 | 8 | 0.0274 | 2 | 3 | 10 | 0.1500 |
| 4 | 6 | 6 | 0.1249 | 2 | 8 | 8 | 0.0376 | 1 | 4 | 10 | 0.0890 |
| 5 | 6 | 6 | 0.2058 | 3 | 8 | 8 | 0.0489 | 2 | 4 | 10 | 0.1135 |
| 1 | 2 | 7 | 0.2019 | 4 | 8 | 8 | 0.0634 | 3 | 4 | 10 | 0.1368 |
| 1 | 3 | 7 | 0.1340 | 5 | 8 | 8 | 0.0842 | 1 | 5 | 10 | 0.0706 |
| 2 | 3 | 7 | 0.1793 | 6 | 8 | 8 | 0.1186 | 2 | 5 | 10 | 0.0904 |
| 1 | 4 | 7 | 0.0977 | 7 | 8 | 8 | 0.1890 | 3 | 5 | 10 | 0.1093 |
| 2 | 4 | 7 | 0.1318 | 1 | 2 | 9 | 0.1825 | 4 | 5 | 10 | 0.1303 |
| 3 | 4 | 7 | 0.1695 | 1 | 3 | 9 | 0.1228 | 1 | 6 | 10 | 0.0575 |
| 1 | 5 | 7 | 0.0733 | 2 | 3 | 9 | 0.1580 | 2 | 6 | 10 | 0.0738 |
| 2 | 5 | 7 | 0.0996 | 1 | 4 | 9 | 0.0919 | 3 | 6 | 10 | 0.0895 |
| 3 | 5 | 7 | 0.1292 | 2 | 4 | 9 | 0.1188 | 4 | 6 | 10 | 0.1070 |
| 4 | 5 | 7 | 0.1688 | 3 | 4 | 9 | 0.1455 | 5 | 6 | 10 | 0.1282 |

TABLE 3

OUTLIER MEAN 1.0 OUTLIER STANDARD DEVIATION 1.0

| R | S | N | COVAR | R | S | N | COVAR | R | S | N | COVAR |
|---|---|---|---|---|---|---|---|---|---|---|---|
| 1 | 7 | 10 | 0.0473 | 1 | 7 | 11 | 0.0474 | 3 | 4 | 12 | 0.1237 |
| 2 | 7 | 10 | 0.0608 | 2 | 7 | 11 | 0.0602 | 1 | 5 | 12 | 0.0674 |
| 3 | 7 | 10 | 0.0739 | 3 | 7 | 11 | 0.0721 | 2 | 5 | 12 | 0.0844 |
| 4 | 7 | 10 | 0.0886 | 4 | 7 | 11 | 0.0851 | 3 | 5 | 12 | 0.0998 |
| 5 | 7 | 10 | 0.1066 | 5 | 7 | 11 | 0.1003 | 4 | 5 | 12 | 0.1158 |
| 6 | 7 | 10 | 0.1301 | 6 | 7 | 11 | 0.1190 | 1 | 6 | 12 | 0.0558 |
| 1 | 8 | 10 | 0.0385 | 1 | 8 | 11 | 0.0397 | 2 | 6 | 12 | 0.0700 |
| 2 | 8 | 10 | 0.0497 | 2 | 8 | 11 | 0.0505 | 3 | 6 | 12 | 0.0829 |
| 3 | 8 | 10 | 0.0607 | 3 | 8 | 11 | 0.0606 | 4 | 6 | 12 | 0.0965 |
| 4 | 8 | 10 | 0.0730 | 4 | 8 | 11 | 0.0717 | 5 | 6 | 12 | 0.1117 |
| 5 | 8 | 10 | 0.0881 | 5 | 8 | 11 | 0.0847 | 1 | 7 | 12 | 0.0471 |
| 6 | 8 | 10 | 0.1080 | 6 | 8 | 11 | 0.1008 | 2 | 7 | 12 | 0.0591 |
| 7 | 8 | 10 | 0.1366 | 7 | 8 | 11 | 0.1222 | 3 | 7 | 12 | 0.0702 |
| 1 | 9 | 10 | 0.0303 | 1 | 9 | 11 | 0.0328 | 4 | 7 | 12 | 0.0818 |
| 2 | 9 | 10 | 0.0393 | 2 | 9 | 11 | 0.0419 | 5 | 7 | 12 | 0.0949 |
| 3 | 9 | 10 | 0.0481 | 3 | 9 | 11 | 0.0505 | 6 | 7 | 12 | 0.1104 |
| 4 | 9 | 10 | 0.0582 | 4 | 9 | 11 | 0.0599 | 1 | 8 | 12 | 0.0400 |
| 5 | 9 | 10 | 0.0707 | 5 | 9 | 11 | 0.0709 | 2 | 8 | 12 | 0.0503 |
| 6 | 9 | 10 | 0.0872 | 6 | 9 | 11 | 0.0847 | 3 | 8 | 12 | 0.0598 |
| 7 | 9 | 10 | 0.1111 | 7 | 9 | 11 | 0.1031 | 4 | 8 | 12 | 0.0698 |
| 8 | 9 | 10 | 0.1501 | 8 | 9 | 11 | 0.1297 | 5 | 8 | 12 | 0.0811 |
| 1 | 10 | 10 | 0.0204 | 1 | 10 | 11 | 0.0262 | 6 | 8 | 12 | 0.0947 |
| 2 | 10 | 10 | 0.0268 | 2 | 10 | 11 | 0.0335 | 7 | 8 | 12 | 0.1117 |
| 3 | 10 | 10 | 0.0333 | 3 | 10 | 11 | 0.0406 | 1 | 9 | 12 | 0.0340 |
| 4 | 10 | 10 | 0.0409 | 4 | 10 | 11 | 0.0483 | 2 | 9 | 12 | 0.0428 |
| 5 | 10 | 10 | 0.0503 | 5 | 10 | 11 | 0.0575 | 3 | 9 | 12 | 0.0510 |
| 6 | 10 | 10 | 0.0629 | 6 | 10 | 11 | 0.0690 | 4 | 9 | 12 | 0.0596 |
| 7 | 10 | 10 | 0.0816 | 7 | 10 | 11 | 0.0845 | 5 | 9 | 12 | 0.0694 |
| 8 | 10 | 10 | 0.1125 | 8 | 10 | 11 | 0.1070 | 6 | 9 | 12 | 0.0811 |
| 9 | 10 | 10 | 0.1762 | 9 | 10 | 11 | 0.1439 | 7 | 9 | 12 | 0.0960 |
| 1 | 2 | 11 | 0.1688 | 1 | 11 | 11 | 0.0180 | 8 | 9 | 12 | 0.1158 |
| 1 | 3 | 11 | 0.1142 | 2 | 11 | 11 | 0.0233 | 1 | 10 | 12 | 0.0285 |
| 2 | 3 | 11 | 0.1434 | 3 | 11 | 11 | 0.0286 | 2 | 10 | 12 | 0.0360 |
| 1 | 4 | 11 | 0.0864 | 4 | 11 | 11 | 0.0344 | 3 | 10 | 12 | 0.0429 |
| 2 | 4 | 11 | 0.1089 | 5 | 11 | 11 | 0.0414 | 4 | 10 | 12 | 0.0503 |
| 3 | 4 | 11 | 0.1297 | 6 | 11 | 11 | 0.0502 | 5 | 10 | 12 | 0.0587 |
| 1 | 5 | 11 | 0.0690 | 7 | 11 | 11 | 0.0623 | 6 | 10 | 12 | 0.0688 |
| 2 | 5 | 11 | 0.0872 | 8 | 11 | 11 | 0.0801 | 7 | 10 | 12 | 0.0816 |
| 3 | 5 | 11 | 0.1042 | 9 | 11 | 11 | 0.1098 | 8 | 10 | 12 | 0.0988 |
| 4 | 5 | 11 | 0.1224 | 10 | 11 | 11 | 0.1710 | 9 | 10 | 12 | 0.1239 |
| 1 | 6 | 11 | 0.0568 | 1 | 2 | 12 | 0.1633 | 1 | 11 | 12 | 0.0229 |
| 2 | 6 | 11 | 0.0719 | 1 | 3 | 12 | 0.1106 | 2 | 11 | 12 | 0.0291 |
| 3 | 6 | 11 | 0.0861 | 2 | 3 | 12 | 0.1376 | 3 | 11 | 12 | 0.0349 |
| 4 | 6 | 11 | 0.1013 | 1 | 4 | 12 | 0.0840 | 4 | 11 | 12 | 0.0410 |
| 5 | 6 | 11 | 0.1191 | 2 | 4 | 12 | 0.1048 | 5 | 11 | 12 | 0.0480 |

TABLE 3

OUTLIER MEAN 1.0 OUTLIER STANDARD DEVIATION 1.0

| R | S | N | COVAR | R | S | N | COVAR | R | S | N | COVAR |
|---|---|---|---|---|---|---|---|---|---|---|---|
| 6 | 11 | 12 | 0.0565 | 2 | 9 | 13 | 0.0430 | 9 | 13 | 13 | 0.0607 |
| 7 | 11 | 12 | 0.0673 | 3 | 9 | 13 | 0.0507 | 10 | 13 | 13 | 0.0773 |
| 8 | 11 | 12 | 0.0820 | 4 | 9 | 13 | 0.0586 | 11 | 13 | 13 | 0.1049 |
| 9 | 11 | 12 | 0.1033 | 5 | 9 | 13 | 0.0674 | 12 | 13 | 13 | 0.1620 |
| 10 | 11 | 12 | 0.1386 | 6 | 9 | 13 | 0.0776 | 1 | 2 | 14 | 0.1542 |
| 1 | 12 | 12 | 0.0160 | 7 | 9 | 13 | 0.0900 | 1 | 3 | 14 | 0.1046 |
| 2 | 12 | 12 | 0.0206 | 8 | 9 | 13 | 0.1056 | 2 | 3 | 14 | 0.1283 |
| 3 | 12 | 12 | 0.0249 | 1 | 10 | 13 | 0.0296 | 1 | 4 | 14 | 0.0798 |
| 4 | 12 | 12 | 0.0295 | 2 | 10 | 13 | 0.0370 | 2 | 4 | 14 | 0.0981 |
| 5 | 12 | 12 | 0.0349 | 3 | 10 | 13 | 0.0436 | 3 | 4 | 14 | 0.1140 |
| 6 | 12 | 12 | 0.0415 | 4 | 10 | 13 | 0.0506 | 1 | 5 | 14 | 0.0644 |
| 7 | 12 | 12 | 0.0500 | 5 | 10 | 13 | 0.0582 | 2 | 5 | 14 | 0.0794 |
| 8 | 12 | 12 | 0.0615 | 6 | 10 | 13 | 0.0672 | 3 | 5 | 14 | 0.0925 |
| 9 | 12 | 12 | 0.0787 | 7 | 10 | 13 | 0.0780 | 4 | 5 | 14 | 0.1055 |
| 10 | 12 | 12 | 0.1072 | 8 | 10 | 13 | 0.0918 | 1 | 6 | 14 | 0.0538 |
| 11 | 12 | 12 | 0.1662 | 9 | 10 | 13 | 0.1104 | 2 | 6 | 14 | 0.0665 |
| 1 | 2 | 13 | 0.1585 | 1 | 11 | 13 | 0.0250 | 3 | 6 | 14 | 0.0775 |
| 1 | 3 | 13 | 0.1075 | 2 | 11 | 13 | 0.0314 | 4 | 6 | 14 | 0.0886 |
| 2 | 3 | 13 | 0.1327 | 3 | 11 | 13 | 0.0371 | 5 | 6 | 14 | 0.1005 |
| 1 | 4 | 13 | 0.0818 | 4 | 11 | 13 | 0.0431 | 1 | 7 | 14 | 0.0459 |
| 2 | 4 | 13 | 0.1013 | 5 | 11 | 13 | 0.0497 | 2 | 7 | 14 | 0.0568 |
| 3 | 4 | 13 | 0.1185 | 6 | 11 | 13 | 0.0574 | 3 | 7 | 14 | 0.0663 |
| 1 | 5 | 13 | 0.0659 | 7 | 11 | 13 | 0.0668 | 4 | 7 | 14 | 0.0759 |
| 2 | 5 | 13 | 0.0818 | 8 | 11 | 13 | 0.0789 | 5 | 7 | 14 | 0.0862 |
| 3 | 5 | 13 | 0.0959 | 9 | 11 | 13 | 0.0951 | 6 | 7 | 14 | 0.0978 |
| 4 | 5 | 13 | 0.1103 | 10 | 11 | 13 | 0.1189 | 1 | 8 | 14 | 0.0397 |
| 1 | 6 | 13 | 0.0548 | 1 | 12 | 13 | 0.0204 | 2 | 8 | 14 | 0.0491 |
| 2 | 6 | 13 | 0.0682 | 2 | 12 | 13 | 0.0256 | 3 | 8 | 14 | 0.0574 |
| 3 | 6 | 13 | 0.0801 | 3 | 12 | 13 | 0.0304 | 4 | 8 | 14 | 0.0658 |
| 4 | 6 | 13 | 0.0923 | 4 | 12 | 13 | 0.0354 | 5 | 8 | 14 | 0.0748 |
| 5 | 6 | 13 | 0.1057 | 5 | 12 | 13 | 0.0410 | 6 | 8 | 14 | 0.0850 |
| 1 | 7 | 13 | 0.0466 | 6 | 12 | 13 | 0.0475 | 7 | 8 | 14 | 0.0970 |
| 2 | 7 | 13 | 0.0580 | 7 | 12 | 13 | 0.0555 | 1 | 9 | 14 | 0.0345 |
| 3 | 7 | 13 | 0.0682 | 8 | 12 | 13 | 0.0658 | 2 | 9 | 14 | 0.0428 |
| 4 | 7 | 13 | 0.0787 | 9 | 12 | 13 | 0.0797 | 3 | 9 | 14 | 0.0500 |
| 5 | 7 | 13 | 0.0902 | 10 | 12 | 13 | 0.1001 | 4 | 9 | 14 | 0.0574 |
| 6 | 7 | 13 | 0.1035 | 11 | 12 | 13 | 0.1339 | 5 | 9 | 14 | 0.0654 |
| 1 | 8 | 13 | 0.0400 | 1 | 13 | 13 | 0.0145 | 6 | 9 | 14 | 0.0743 |
| 2 | 8 | 13 | 0.0498 | 2 | 13 | 13 | 0.0184 | 7 | 9 | 14 | 0.0849 |
| 3 | 8 | 13 | 0.0587 | 3 | 13 | 13 | 0.0220 | 8 | 9 | 14 | 0.0978 |
| 4 | 8 | 13 | 0.0678 | 4 | 13 | 13 | 0.0258 | 1 | 10 | 14 | 0.0301 |
| 5 | 8 | 13 | 0.0778 | 5 | 13 | 13 | 0.0301 | 2 | 10 | 14 | 0.0373 |
| 6 | 8 | 13 | 0.0895 | 6 | 13 | 13 | 0.0352 | 3 | 10 | 14 | 0.0437 |
| 7 | 8 | 13 | 0.1035 | 7 | 13 | 13 | 0.0415 | 4 | 10 | 14 | 0.0502 |
| 1 | 9 | 13 | 0.0344 | 8 | 13 | 13 | 0.0496 | 5 | 10 | 14 | 0.0572 |

TABLE 3

OUTLIER MEAN 1.0 OUTLIER STANDARD DEVIATION 1.0

| R | S | N | COVAR | R | S | N | COVAR | R | S | N | COVAR |
|---|---|---|---|---|---|---|---|---|---|---|---|
| 6 | 10 | 14 | 0.0651 | 9 | 14 | 14 | 0.0491 | 5 | 10 | 15 | 0.0559 |
| 7 | 10 | 14 | 0.0745 | 10 | 14 | 14 | 0.0599 | 6 | 10 | 15 | 0.0631 |
| 8 | 10 | 14 | 0.0859 | 11 | 14 | 14 | 0.0759 | 7 | 10 | 15 | 0.0713 |
| 9 | 10 | 14 | 0.1006 | 12 | 14 | 14 | 0.1027 | 8 | 10 | 15 | 0.0810 |
| 1 | 11 | 14 | 0.0261 | 13 | 14 | 14 | 0.1582 | 9 | 10 | 15 | 0.0930 |
| 2 | 11 | 14 | 0.0324 | 1 | 2 | 15 | 0.1504 | 1 | 11 | 15 | 0.0266 |
| 3 | 11 | 14 | 0.0380 | 1 | 3 | 15 | 0.1021 | 2 | 11 | 15 | 0.0328 |
| 4 | 11 | 14 | 0.0436 | 2 | 3 | 15 | 0.1245 | 3 | 11 | 15 | 0.0382 |
| 5 | 11 | 14 | 0.0498 | 1 | 4 | 15 | 0.0779 | 4 | 11 | 15 | 0.0436 |
| 6 | 11 | 14 | 0.0568 | 2 | 4 | 15 | 0.0952 | 5 | 11 | 15 | 0.0493 |
| 7 | 11 | 14 | 0.0651 | 3 | 4 | 15 | 0.1101 | 6 | 11 | 15 | 0.0557 |
| 8 | 11 | 14 | 0.0752 | 1 | 5 | 15 | 0.0631 | 7 | 11 | 15 | 0.0630 |
| 9 | 11 | 14 | 0.0882 | 2 | 5 | 15 | 0.0773 | 8 | 11 | 15 | 0.0717 |
| 10 | 11 | 14 | 0.1057 | 3 | 5 | 15 | 0.0894 | 9 | 11 | 15 | 0.0824 |
| 1 | 12 | 14 | 0.0223 | 4 | 5 | 15 | 0.1014 | 10 | 11 | 15 | 0.0962 |
| 2 | 12 | 14 | 0.0277 | 1 | 6 | 15 | 0.0529 | 1 | 12 | 15 | 0.0233 |
| 3 | 12 | 14 | 0.0325 | 2 | 6 | 15 | 0.0648 | 2 | 12 | 15 | 0.0287 |
| 4 | 12 | 14 | 0.0374 | 3 | 6 | 15 | 0.0751 | 3 | 12 | 15 | 0.0334 |
| 5 | 12 | 14 | 0.0428 | 4 | 6 | 15 | 0.0853 | 4 | 12 | 15 | 0.0382 |
| 6 | 12 | 14 | 0.0489 | 5 | 6 | 15 | 0.0961 | 5 | 12 | 15 | 0.0433 |
| 7 | 12 | 14 | 0.0561 | 1 | 7 | 15 | 0.0453 | 6 | 12 | 15 | 0.0489 |
| 8 | 12 | 14 | 0.0650 | 2 | 7 | 15 | 0.0556 | 7 | 12 | 15 | 0.0554 |
| 9 | 12 | 14 | 0.0764 | 3 | 7 | 15 | 0.0645 | 8 | 12 | 15 | 0.0632 |
| 10 | 12 | 14 | 0.0919 | 4 | 7 | 15 | 0.0733 | 9 | 12 | 15 | 0.0727 |
| 11 | 12 | 14 | 0.1146 | 5 | 7 | 15 | 0.0827 | 10 | 12 | 15 | 0.0850 |
| 1 | 13 | 14 | 0.0183 | 6 | 7 | 15 | 0.0930 | 11 | 12 | 15 | 0.1017 |
| 2 | 13 | 14 | 0.0228 | 1 | 8 | 15 | 0.0393 | 1 | 13 | 15 | 0.0200 |
| 3 | 13 | 14 | 0.0269 | 2 | 8 | 15 | 0.0483 | 2 | 13 | 15 | 0.0247 |
| 4 | 13 | 14 | 0.0310 | 3 | 8 | 15 | 0.0561 | 3 | 13 | 15 | 0.0288 |
| 5 | 13 | 14 | 0.0356 | 4 | 8 | 15 | 0.0639 | 4 | 13 | 15 | 0.0330 |
| 6 | 13 | 14 | 0.0408 | 5 | 8 | 15 | 0.0721 | 5 | 13 | 15 | 0.0374 |
| 7 | 13 | 14 | 0.0469 | 6 | 8 | 15 | 0.0811 | 6 | 13 | 15 | 0.0424 |
| 8 | 13 | 14 | 0.0545 | 7 | 8 | 15 | 0.0915 | 7 | 13 | 15 | 0.0481 |
| 9 | 13 | 14 | 0.0643 | 1 | 9 | 15 | 0.0344 | 8 | 13 | 15 | 0.0549 |
| 10 | 13 | 14 | 0.0776 | 2 | 9 | 15 | 0.0424 | 9 | 13 | 15 | 0.0633 |
| 11 | 13 | 14 | 0.0973 | 3 | 9 | 15 | 0.0492 | 10 | 13 | 15 | 0.0742 |
| 12 | 13 | 14 | 0.1298 | 4 | 9 | 15 | 0.0561 | 11 | 13 | 15 | 0.0890 |
| 1 | 14 | 14 | 0.0132 | 5 | 9 | 15 | 0.0633 | 12 | 13 | 15 | 0.1107 |
| 2 | 14 | 14 | 0.0166 | 6 | 9 | 15 | 0.0714 | 1 | 14 | 15 | 0.0166 |
| 3 | 14 | 14 | 0.0196 | 7 | 9 | 15 | 0.0806 | 2 | 14 | 15 | 0.0205 |
| 4 | 14 | 14 | 0.0228 | 8 | 9 | 15 | 0.0915 | 3 | 14 | 15 | 0.0240 |
| 5 | 14 | 14 | 0.0264 | 1 | 10 | 15 | 0.0303 | 4 | 14 | 15 | 0.0275 |
| 6 | 14 | 14 | 0.0304 | 2 | 10 | 15 | 0.0373 | 5 | 14 | 15 | 0.0313 |
| 7 | 14 | 14 | 0.0353 | 3 | 10 | 15 | 0.0434 | 6 | 14 | 15 | 0.0355 |
| 8 | 14 | 14 | 0.0413 | 4 | 10 | 15 | 0.0494 | 7 | 14 | 15 | 0.0404 |

TABLE 3

OUTLIER MEAN 1.0 OUTLIER STANDARD DEVIATION 1.0

| R | S | N | COVAR | R | S | N | COVAR | R | S | N | COVAR |
|---|---|---|---|---|---|---|---|---|---|---|---|
| 8 | 14 | 15 | 0.0462 | 5 | 8 | 16 | 0.0696 | 5 | 13 | 16 | 0.0381 |
| 9 | 14 | 15 | 0.0535 | 6 | 8 | 16 | 0.0778 | 6 | 13 | 16 | 0.0427 |
| 10 | 14 | 15 | 0.0629 | 7 | 8 | 16 | 0.0870 | 7 | 13 | 16 | 0.0480 |
| 11 | 14 | 15 | 0.0758 | 1 | 9 | 16 | 0.0342 | 8 | 13 | 16 | 0.0541 |
| 12 | 14 | 15 | 0.0947 | 2 | 9 | 16 | 0.0419 | 9 | 13 | 16 | 0.0614 |
| 13 | 14 | 15 | 0.1261 | 3 | 9 | 16 | 0.0484 | 10 | 13 | 16 | 0.0705 |
| 1 | 15 | 15 | 0.0120 | 4 | 9 | 16 | 0.0548 | 11 | 13 | 16 | 0.0822 |
| 2 | 15 | 15 | 0.0151 | 5 | 9 | 16 | 0.0615 | 12 | 13 | 16 | 0.0982 |
| 3 | 15 | 15 | 0.0177 | 6 | 9 | 16 | 0.0687 | 1 | 14 | 16 | 0.0181 |
| 4 | 15 | 15 | 0.0204 | 7 | 9 | 16 | 0.0769 | 2 | 14 | 16 | 0.0223 |
| 5 | 15 | 15 | 0.0234 | 8 | 9 | 16 | 0.0864 | 3 | 14 | 16 | 0.0258 |
| 6 | 15 | 15 | 0.0267 | 1 | 10 | 16 | 0.0303 | 4 | 14 | 16 | 0.0294 |
| 7 | 15 | 15 | 0.0306 | 2 | 10 | 16 | 0.0371 | 5 | 14 | 16 | 0.0331 |
| 8 | 15 | 15 | 0.0352 | 3 | 10 | 16 | 0.0429 | 6 | 14 | 16 | 0.0372 |
| 9 | 15 | 15 | 0.0411 | 4 | 10 | 16 | 0.0486 | 7 | 14 | 16 | 0.0418 |
| 10 | 15 | 15 | 0.0486 | 5 | 10 | 16 | 0.0546 | 8 | 14 | 16 | 0.0472 |
| 11 | 15 | 15 | 0.0591 | 6 | 10 | 16 | 0.0611 | 9 | 14 | 16 | 0.0537 |
| 12 | 15 | 15 | 0.0747 | 7 | 10 | 16 | 0.0684 | 10 | 14 | 16 | 0.0618 |
| 13 | 15 | 15 | 0.1007 | 8 | 10 | 16 | 0.0769 | 11 | 14 | 16 | 0.0722 |
| 14 | 15 | 15 | 0.1547 | 9 | 10 | 16 | 0.0870 | 12 | 14 | 16 | 0.0864 |
| 1 | 2 | 16 | 0.1470 | 1 | 11 | 16 | 0.0269 | 13 | 14 | 16 | 0.1074 |
| 1 | 3 | 16 | 0.0998 | 2 | 11 | 16 | 0.0329 | 1 | 15 | 16 | 0.0151 |
| 2 | 3 | 16 | 0.1210 | 3 | 11 | 16 | 0.0381 | 2 | 15 | 16 | 0.0186 |
| 1 | 4 | 16 | 0.0762 | 4 | 11 | 16 | 0.0432 | 3 | 15 | 16 | 0.0216 |
| 2 | 4 | 16 | 0.0927 | 5 | 11 | 16 | 0.0485 | 4 | 15 | 16 | 0.0246 |
| 3 | 4 | 16 | 0.1066 | 6 | 11 | 16 | 0.0544 | 5 | 15 | 16 | 0.0278 |
| 1 | 5 | 16 | 0.0618 | 7 | 11 | 16 | 0.0609 | 6 | 15 | 16 | 0.0313 |
| 2 | 5 | 16 | 0.0753 | 8 | 11 | 16 | 0.0686 | 7 | 15 | 16 | 0.0353 |
| 3 | 5 | 16 | 0.0867 | 9 | 11 | 16 | 0.0777 | 8 | 15 | 16 | 0.0400 |
| 4 | 5 | 16 | 0.0978 | 10 | 11 | 16 | 0.0889 | 9 | 15 | 16 | 0.0456 |
| 1 | 6 | 16 | 0.0519 | 1 | 12 | 16 | 0.0238 | 10 | 15 | 16 | 0.0525 |
| 2 | 6 | 16 | 0.0633 | 2 | 12 | 16 | 0.0292 | 11 | 15 | 16 | 0.0616 |
| 3 | 6 | 16 | 0.0730 | 3 | 12 | 16 | 0.0338 | 12 | 15 | 16 | 0.0740 |
| 4 | 6 | 16 | 0.0824 | 4 | 12 | 16 | 0.0383 | 13 | 15 | 16 | 0.0924 |
| 5 | 6 | 16 | 0.0923 | 5 | 12 | 16 | 0.0431 | 14 | 15 | 16 | 0.1228 |
| 1 | 7 | 16 | 0.0446 | 6 | 12 | 16 | 0.0483 | 1 | 16 | 16 | 0.0111 |
| 2 | 7 | 16 | 0.0544 | 7 | 12 | 16 | 0.0542 | 2 | 16 | 16 | 0.0138 |
| 3 | 7 | 16 | 0.0628 | 8 | 12 | 16 | 0.0611 | 3 | 16 | 16 | 0.0161 |
| 4 | 7 | 16 | 0.0710 | 9 | 12 | 16 | 0.0693 | 4 | 16 | 16 | 0.0185 |
| 5 | 7 | 16 | 0.0796 | 10 | 12 | 16 | 0.0794 | 5 | 16 | 16 | 0.0210 |
| 6 | 7 | 16 | 0.0889 | 11 | 12 | 16 | 0.0924 | 6 | 16 | 16 | 0.0237 |
| 1 | 8 | 16 | 0.0389 | 1 | 13 | 16 | 0.0209 | 7 | 16 | 16 | 0.0269 |
| 2 | 8 | 16 | 0.0475 | 2 | 13 | 16 | 0.0257 | 8 | 16 | 16 | 0.0306 |
| 3 | 8 | 16 | 0.0549 | 3 | 13 | 16 | 0.0298 | 9 | 16 | 16 | 0.0351 |
| 4 | 8 | 16 | 0.0621 | 4 | 13 | 16 | 0.0338 | 10 | 16 | 16 | 0.0408 |

TABLE 3

OUTLIER MEAN 1.0 OUTLIER STANDARD DEVIATION 1.0

| R | S | N | COVAR | R | S | N | COVAR | R | S | N | COVAR |
|---|---|---|---|---|---|---|---|---|---|---|---|
| 11 | 16 | 16 | 0.0481 | 5 | 10 | 17 | 0.0532 | 8 | 14 | 17 | 0.0470 |
| 12 | 16 | 16 | 0.0583 | 6 | 10 | 17 | 0.0592 | 9 | 14 | 17 | 0.0528 |
| 13 | 16 | 16 | 0.0735 | 7 | 10 | 17 | 0.0658 | 10 | 14 | 17 | 0.0598 |
| 14 | 16 | 16 | 0.0989 | 8 | 10 | 17 | 0.0733 | 11 | 14 | 17 | 0.0685 |
| 15 | 16 | 16 | 0.1515 | 9 | 10 | 17 | 0.0820 | 12 | 14 | 17 | 0.0797 |
| 1 | 2 | 17 | 0.1439 | 1 | 11 | 17 | 0.0270 | 13 | 14 | 17 | 0.0950 |
| 1 | 3 | 17 | 0.0977 | 2 | 11 | 17 | 0.0329 | 1 | 15 | 17 | 0.0165 |
| 2 | 3 | 17 | 0.1179 | 3 | 11 | 17 | 0.0378 | 2 | 15 | 17 | 0.0202 |
| 1 | 4 | 17 | 0.0747 | 4 | 11 | 17 | 0.0426 | 3 | 15 | 17 | 0.0233 |
| 2 | 4 | 17 | 0.0904 | 5 | 11 | 17 | 0.0476 | 4 | 15 | 17 | 0.0264 |
| 3 | 4 | 17 | 0.1034 | 6 | 11 | 17 | 0.0530 | 5 | 15 | 17 | 0.0296 |
| 1 | 5 | 17 | 0.0606 | 7 | 11 | 17 | 0.0590 | 6 | 15 | 17 | 0.0330 |
| 2 | 5 | 17 | 0.0735 | 8 | 11 | 17 | 0.0657 | 7 | 15 | 17 | 0.0369 |
| 3 | 5 | 17 | 0.0842 | 9 | 11 | 17 | 0.0736 | 8 | 15 | 17 | 0.0412 |
| 4 | 5 | 17 | 0.0946 | 10 | 11 | 17 | 0.0831 | 9 | 15 | 17 | 0.0464 |
| 1 | 6 | 17 | 0.0510 | 1 | 12 | 17 | 0.0241 | 10 | 15 | 17 | 0.0526 |
| 2 | 6 | 17 | 0.0619 | 2 | 12 | 17 | 0.0294 | 11 | 15 | 17 | 0.0603 |
| 3 | 6 | 17 | 0.0710 | 3 | 12 | 17 | 0.0338 | 12 | 15 | 17 | 0.0704 |
| 4 | 6 | 17 | 0.0799 | 4 | 12 | 17 | 0.0382 | 13 | 15 | 17 | 0.0841 |
| 5 | 6 | 17 | 0.0889 | 5 | 12 | 17 | 0.0427 | 14 | 15 | 17 | 0.1043 |
| 1 | 7 | 17 | 0.0439 | 6 | 12 | 17 | 0.0475 | 1 | 16 | 17 | 0.0138 |
| 2 | 7 | 17 | 0.0534 | 7 | 12 | 17 | 0.0529 | 2 | 16 | 17 | 0.0170 |
| 3 | 7 | 17 | 0.0613 | 8 | 12 | 17 | 0.0590 | 3 | 16 | 17 | 0.0196 |
| 4 | 7 | 17 | 0.0689 | 9 | 12 | 17 | 0.0661 | 4 | 16 | 17 | 0.0223 |
| 5 | 7 | 17 | 0.0768 | 10 | 12 | 17 | 0.0747 | 5 | 16 | 17 | 0.0250 |
| 6 | 7 | 17 | 0.0853 | 11 | 12 | 17 | 0.0853 | 6 | 16 | 17 | 0.0280 |
| 1 | 8 | 17 | 0.0384 | 1 | 13 | 17 | 0.0215 | 7 | 16 | 17 | 0.0313 |
| 2 | 8 | 17 | 0.0467 | 2 | 13 | 17 | 0.0262 | 8 | 16 | 17 | 0.0351 |
| 3 | 8 | 17 | 0.0537 | 3 | 13 | 17 | 0.0302 | 9 | 16 | 17 | 0.0395 |
| 4 | 8 | 17 | 0.0604 | 4 | 13 | 17 | 0.0341 | 10 | 16 | 17 | 0.0449 |
| 5 | 8 | 17 | 0.0674 | 5 | 13 | 17 | 0.0381 | 11 | 16 | 17 | 0.0516 |
| 6 | 8 | 17 | 0.0748 | 6 | 13 | 17 | 0.0425 | 12 | 16 | 17 | 0.0604 |
| 7 | 8 | 17 | 0.0831 | 7 | 13 | 17 | 0.0473 | 13 | 16 | 17 | 0.0724 |
| 1 | 9 | 17 | 0.0339 | 8 | 13 | 17 | 0.0528 | 14 | 16 | 17 | 0.0902 |
| 2 | 9 | 17 | 0.0413 | 9 | 13 | 17 | 0.0593 | 15 | 16 | 17 | 0.1199 |
| 3 | 9 | 17 | 0.0475 | 10 | 13 | 17 | 0.0670 | 1 | 17 | 17 | 0.0103 |
| 4 | 9 | 17 | 0.0535 | 11 | 13 | 17 | 0.0767 | 2 | 17 | 17 | 0.0127 |
| 5 | 9 | 17 | 0.0597 | 12 | 13 | 17 | 0.0891 | 3 | 17 | 17 | 0.0148 |
| 6 | 9 | 17 | 0.0663 | 1 | 14 | 17 | 0.0190 | 4 | 17 | 17 | 0.0168 |
| 7 | 9 | 17 | 0.0737 | 2 | 14 | 17 | 0.0232 | 5 | 17 | 17 | 0.0190 |
| 8 | 9 | 17 | 0.0820 | 3 | 14 | 17 | 0.0267 | 6 | 17 | 17 | 0.0213 |
| 1 | 10 | 17 | 0.0302 | 4 | 14 | 17 | 0.0302 | 7 | 17 | 17 | 0.0239 |
| 2 | 10 | 17 | 0.0368 | 5 | 14 | 17 | 0.0338 | 8 | 17 | 17 | 0.0270 |
| 3 | 10 | 17 | 0.0423 | 6 | 14 | 17 | 0.0377 | 9 | 17 | 17 | 0.0306 |
| 4 | 10 | 17 | 0.0477 | 7 | 14 | 17 | 0.0421 | 10 | 17 | 17 | 0.0349 |

TABLE 3

OUTLIER MEAN 1.0 OUTLIER STANDARD DEVIATION 1.0

| R | S | N | COVAR | R | S | N | COVAR | R | S | N | COVAR |
|---|---|---|---|---|---|---|---|---|---|---|---|
| 11 | 17 | 17 | 0.0404 | 4 | 10 | 18 | 0.0468 | 7 | 14 | 18 | 0.0418 |
| 12 | 17 | 17 | 0.0476 | 5 | 10 | 18 | 0.0520 | 8 | 14 | 18 | 0.0463 |
| 13 | 17 | 17 | 0.0575 | 6 | 10 | 18 | 0.0575 | 9 | 14 | 18 | 0.0515 |
| 14 | 17 | 17 | 0.0723 | 7 | 10 | 18 | 0.0635 | 10 | 14 | 18 | 0.0577 |
| 15 | 17 | 17 | 0.0971 | 8 | 10 | 18 | 0.0702 | 11 | 14 | 18 | 0.0650 |
| 16 | 17 | 17 | 0.1486 | 9 | 10 | 18 | 0.0779 | 12 | 14 | 18 | 0.0742 |
| 1 | 2 | 18 | 0.1410 | 1 | 11 | 18 | 0.0269 | 13 | 14 | 18 | 0.0862 |
| 1 | 3 | 18 | 0.0957 | 2 | 11 | 18 | 0.0327 | 1 | 15 | 18 | 0.0173 |
| 2 | 3 | 18 | 0.1151 | 3 | 11 | 18 | 0.0374 | 2 | 15 | 18 | 0.0211 |
| 1 | 4 | 18 | 0.0732 | 4 | 11 | 18 | 0.0420 | 3 | 15 | 18 | 0.0242 |
| 2 | 4 | 18 | 0.0883 | 5 | 11 | 18 | 0.0467 | 4 | 15 | 18 | 0.0273 |
| 3 | 4 | 18 | 0.1006 | 6 | 11 | 18 | 0.0517 | 5 | 15 | 18 | 0.0304 |
| 1 | 5 | 18 | 0.0595 | 7 | 11 | 18 | 0.0571 | 6 | 15 | 18 | 0.0337 |
| 2 | 5 | 18 | 0.0718 | 8 | 11 | 18 | 0.0632 | 7 | 15 | 18 | 0.0373 |
| 3 | 5 | 18 | 0.0820 | 9 | 11 | 18 | 0.0702 | 8 | 15 | 18 | 0.0414 |
| 4 | 5 | 18 | 0.0917 | 10 | 11 | 18 | 0.0783 | 9 | 15 | 18 | 0.0461 |
| 1 | 6 | 18 | 0.0502 | 1 | 12 | 18 | 0.0242 | 10 | 15 | 18 | 0.0516 |
| 2 | 6 | 18 | 0.0606 | 2 | 12 | 18 | 0.0294 | 11 | 15 | 18 | 0.0583 |
| 3 | 6 | 18 | 0.0693 | 3 | 12 | 18 | 0.0337 | 12 | 15 | 18 | 0.0666 |
| 4 | 6 | 18 | 0.0775 | 4 | 12 | 18 | 0.0378 | 13 | 15 | 18 | 0.0774 |
| 5 | 6 | 18 | 0.0859 | 5 | 12 | 18 | 0.0421 | 14 | 15 | 18 | 0.0922 |
| 1 | 7 | 18 | 0.0433 | 6 | 12 | 18 | 0.0466 | 1 | 16 | 18 | 0.0152 |
| 2 | 7 | 18 | 0.0523 | 7 | 12 | 18 | 0.0515 | 2 | 16 | 18 | 0.0184 |
| 3 | 7 | 18 | 0.0598 | 8 | 12 | 18 | 0.0571 | 3 | 16 | 18 | 0.0212 |
| 4 | 7 | 18 | 0.0670 | 9 | 12 | 18 | 0.0634 | 4 | 16 | 18 | 0.0239 |
| 5 | 7 | 18 | 0.0744 | 10 | 12 | 18 | 0.0708 | 5 | 16 | 18 | 0.0267 |
| 6 | 7 | 18 | 0.0821 | 11 | 12 | 18 | 0.0797 | 6 | 16 | 18 | 0.0296 |
| 1 | 8 | 18 | 0.0379 | 1 | 13 | 18 | 0.0218 | 7 | 16 | 18 | 0.0328 |
| 2 | 8 | 18 | 0.0459 | 2 | 13 | 18 | 0.0264 | 8 | 16 | 18 | 0.0365 |
| 3 | 8 | 18 | 0.0525 | 3 | 13 | 18 | 0.0303 | 9 | 16 | 18 | 0.0406 |
| 4 | 8 | 18 | 0.0589 | 4 | 13 | 18 | 0.0341 | 10 | 16 | 18 | 0.0456 |
| 5 | 8 | 18 | 0.0653 | 5 | 13 | 18 | 0.0379 | 11 | 16 | 18 | 0.0515 |
| 6 | 8 | 18 | 0.0722 | 6 | 13 | 18 | 0.0420 | 12 | 16 | 18 | 0.0590 |
| 7 | 8 | 18 | 0.0797 | 7 | 13 | 18 | 0.0465 | 13 | 16 | 18 | 0.0687 |
| 1 | 9 | 18 | 0.0336 | 8 | 13 | 18 | 0.0515 | 14 | 16 | 18 | 0.0820 |
| 2 | 9 | 18 | 0.0407 | 9 | 13 | 18 | 0.0572 | 15 | 16 | 18 | 0.1016 |
| 3 | 9 | 18 | 0.0466 | 10 | 13 | 18 | 0.0640 | 1 | 17 | 18 | 0.0128 |
| 4 | 9 | 18 | 0.0523 | 11 | 13 | 18 | 0.0721 | 2 | 17 | 18 | 0.0156 |
| 5 | 9 | 18 | 0.0581 | 12 | 13 | 18 | 0.0822 | 3 | 17 | 18 | 0.0180 |
| 6 | 9 | 18 | 0.0642 | 1 | 14 | 18 | 0.0195 | 4 | 17 | 18 | 0.0203 |
| 7 | 9 | 18 | 0.0708 | 2 | 14 | 18 | 0.0237 | 5 | 17 | 18 | 0.0226 |
| 8 | 9 | 18 | 0.0783 | 3 | 14 | 18 | 0.0272 | 6 | 17 | 18 | 0.0252 |
| 1 | 10 | 18 | 0.0300 | 4 | 14 | 18 | 0.0306 | 7 | 17 | 18 | 0.0280 |
| 2 | 10 | 18 | 0.0364 | 5 | 14 | 18 | 0.0341 | 8 | 17 | 18 | 0.0311 |
| 3 | 10 | 18 | 0.0417 | 6 | 14 | 18 | 0.0378 | 9 | 17 | 18 | 0.0347 |

TABLE 3

OUTLIER MEAN 1.0 OUTLIER STANDARD DEVIATION 1.0

| R | S | N | COVAR | R | S | N | COVAR | R | S | N | COVAR |
|---|---|---|---|---|---|---|---|---|---|---|---|
| 10 | 17 | 18 | 0.0390 | 1 | 8 | 19 | 0.0374 | 1 | 13 | 19 | 0.0219 |
| 11 | 17 | 18 | 0.0442 | 2 | 8 | 19 | 0.0451 | 2 | 13 | 19 | 0.0265 |
| 12 | 17 | 18 | 0.0508 | 3 | 8 | 19 | 0.0515 | 3 | 13 | 19 | 0.0303 |
| 13 | 17 | 18 | 0.0593 | 4 | 8 | 19 | 0.0575 | 4 | 13 | 19 | 0.0339 |
| 14 | 17 | 18 | 0.0710 | 5 | 8 | 19 | 0.0635 | 5 | 13 | 19 | 0.0376 |
| 15 | 17 | 18 | 0.0883 | 6 | 8 | 19 | 0.0698 | 6 | 13 | 19 | 0.0414 |
| 16 | 17 | 18 | 0.1171 | 7 | 8 | 19 | 0.0767 | 7 | 13 | 19 | 0.0456 |
| 1 | 18 | 18 | 0.0096 | 1 | 9 | 19 | 0.0333 | 8 | 13 | 19 | 0.0501 |
| 2 | 18 | 18 | 0.0117 | 2 | 9 | 19 | 0.0401 | 9 | 13 | 19 | 0.0553 |
| 3 | 18 | 18 | 0.0136 | 3 | 9 | 19 | 0.0458 | 10 | 13 | 19 | 0.0613 |
| 4 | 18 | 18 | 0.0154 | 4 | 9 | 19 | 0.0511 | 11 | 13 | 19 | 0.0683 |
| 5 | 18 | 18 | 0.0173 | 5 | 9 | 19 | 0.0565 | 12 | 13 | 19 | 0.0767 |
| 6 | 18 | 18 | 0.0193 | 6 | 9 | 19 | 0.0622 | 1 | 14 | 19 | 0.0198 |
| 7 | 18 | 18 | 0.0215 | 7 | 9 | 19 | 0.0683 | 2 | 14 | 19 | 0.0240 |
| 8 | 18 | 18 | 0.0241 | 8 | 9 | 19 | 0.0750 | 3 | 14 | 19 | 0.0274 |
| 9 | 18 | 18 | 0.0270 | 1 | 10 | 19 | 0.0298 | 4 | 14 | 19 | 0.0307 |
| 10 | 18 | 18 | 0.0305 | 2 | 10 | 19 | 0.0360 | 5 | 14 | 19 | 0.0340 |
| 11 | 18 | 18 | 0.0347 | 3 | 10 | 19 | 0.0410 | 6 | 14 | 19 | 0.0375 |
| 12 | 18 | 18 | 0.0401 | 4 | 10 | 19 | 0.0459 | 7 | 14 | 19 | 0.0413 |
| 13 | 18 | 18 | 0.0471 | 5 | 10 | 19 | 0.0507 | 8 | 14 | 19 | 0.0455 |
| 14 | 18 | 18 | 0.0568 | 6 | 10 | 19 | 0.0559 | 9 | 14 | 19 | 0.0502 |
| 15 | 18 | 18 | 0.0712 | 7 | 10 | 19 | 0.0614 | 10 | 14 | 19 | 0.0556 |
| 16 | 18 | 18 | 0.0955 | 8 | 10 | 19 | 0.0674 | 11 | 14 | 19 | 0.0620 |
| 17 | 18 | 18 | 0.1458 | 9 | 10 | 19 | 0.0743 | 12 | 14 | 19 | 0.0698 |
| 1 | 2 | 19 | 0.1385 | 1 | 11 | 19 | 0.0268 | 13 | 14 | 19 | 0.0794 |
| 1 | 3 | 19 | 0.0940 | 2 | 11 | 19 | 0.0324 | 1 | 15 | 19 | 0.0178 |
| 2 | 3 | 19 | 0.1125 | 3 | 11 | 19 | 0.0370 | 2 | 15 | 19 | 0.0216 |
| 1 | 4 | 19 | 0.0719 | 4 | 11 | 19 | 0.0414 | 3 | 15 | 19 | 0.0247 |
| 2 | 4 | 19 | 0.0863 | 5 | 11 | 19 | 0.0458 | 4 | 15 | 19 | 0.0277 |
| 3 | 4 | 19 | 0.0981 | 6 | 11 | 19 | 0.0504 | 5 | 15 | 19 | 0.0307 |
| 1 | 5 | 19 | 0.0585 | 7 | 11 | 19 | 0.0554 | 6 | 15 | 19 | 0.0339 |
| 2 | 5 | 19 | 0.0703 | 8 | 11 | 19 | 0.0609 | 7 | 15 | 19 | 0.0373 |
| 3 | 5 | 19 | 0.0800 | 9 | 11 | 19 | 0.0672 | 8 | 15 | 19 | 0.0411 |
| 4 | 5 | 19 | 0.0891 | 10 | 11 | 19 | 0.0743 | 9 | 15 | 19 | 0.0454 |
| 1 | 6 | 19 | 0.0493 | 1 | 12 | 19 | 0.0243 | 10 | 15 | 19 | 0.0503 |
| 2 | 6 | 19 | 0.0594 | 2 | 12 | 19 | 0.0293 | 11 | 15 | 19 | 0.0562 |
| 3 | 6 | 19 | 0.0676 | 3 | 12 | 19 | 0.0335 | 12 | 15 | 19 | 0.0632 |
| 4 | 6 | 19 | 0.0754 | 4 | 12 | 19 | 0.0374 | 13 | 15 | 19 | 0.0721 |
| 5 | 6 | 19 | 0.0832 | 5 | 12 | 19 | 0.0414 | 14 | 15 | 19 | 0.0835 |
| 1 | 7 | 19 | 0.0426 | 6 | 12 | 19 | 0.0457 | 1 | 16 | 19 | 0.0159 |
| 2 | 7 | 19 | 0.0514 | 7 | 12 | 19 | 0.0502 | 2 | 16 | 19 | 0.0193 |
| 3 | 7 | 19 | 0.0585 | 8 | 12 | 19 | 0.0552 | 3 | 16 | 19 | 0.0221 |
| 4 | 7 | 19 | 0.0653 | 9 | 12 | 19 | 0.0609 | 4 | 16 | 19 | 0.0247 |
| 5 | 7 | 19 | 0.0721 | 10 | 12 | 19 | 0.0674 | 5 | 16 | 19 | 0.0275 |
| 6 | 7 | 19 | 0.0793 | 11 | 12 | 19 | 0.0751 | 6 | 16 | 19 | 0.0303 |

TABLE 3

OUTLIER MEAN 1.0 OUTLIER STANDARD DEVIATION 1.0

| R | S | N | COVAR | R | S | N | COVAR | R | S | N | COVAR |
|---|---|---|---|---|---|---|---|---|---|---|---|
| 7 | 16 | 19 | 0.0334 | 4 | 19 | 19 | 0.0142 | 3 | 9 | 20 | 0.0450 |
| 8 | 16 | 19 | 0.0368 | 5 | 19 | 19 | 0.0159 | 4 | 9 | 20 | 0.0500 |
| 9 | 16 | 19 | 0.0407 | 6 | 19 | 19 | 0.0176 | 5 | 9 | 20 | 0.0551 |
| 10 | 16 | 19 | 0.0452 | 7 | 19 | 19 | 0.0195 | 6 | 9 | 20 | 0.0604 |
| 11 | 16 | 19 | 0.0505 | 8 | 19 | 19 | 0.0217 | 7 | 9 | 20 | 0.0661 |
| 12 | 16 | 19 | 0.0569 | 9 | 19 | 19 | 0.0241 | 8 | 9 | 20 | 0.0722 |
| 13 | 16 | 19 | 0.0649 | 10 | 19 | 19 | 0.0270 | 1 | 10 | 20 | 0.0295 |
| 14 | 16 | 19 | 0.0754 | 11 | 19 | 19 | 0.0304 | 2 | 10 | 20 | 0.0355 |
| 15 | 16 | 19 | 0.0897 | 12 | 19 | 19 | 0.0345 | 3 | 10 | 20 | 0.0404 |
| 1 | 17 | 19 | 0.0140 | 13 | 19 | 19 | 0.0397 | 4 | 10 | 20 | 0.0450 |
| 2 | 17 | 19 | 0.0170 | 14 | 19 | 19 | 0.0466 | 5 | 10 | 20 | 0.0496 |
| 3 | 17 | 19 | 0.0194 | 15 | 19 | 19 | 0.0561 | 6 | 10 | 20 | 0.0544 |
| 4 | 17 | 19 | 0.0218 | 16 | 19 | 19 | 0.0702 | 7 | 10 | 20 | 0.0595 |
| 5 | 17 | 19 | 0.0242 | 17 | 19 | 19 | 0.0940 | 8 | 10 | 20 | 0.0650 |
| 6 | 17 | 19 | 0.0267 | 18 | 19 | 19 | 0.1433 | 9 | 10 | 20 | 0.0712 |
| 7 | 17 | 19 | 0.0295 | 1 | 2 | 20 | 0.1361 | 1 | 11 | 20 | 0.0267 |
| 8 | 17 | 19 | 0.0326 | 1 | 3 | 20 | 0.0923 | 2 | 11 | 20 | 0.0321 |
| 9 | 17 | 19 | 0.0360 | 2 | 3 | 20 | 0.1102 | 3 | 11 | 20 | 0.0365 |
| 10 | 17 | 19 | 0.0400 | 1 | 4 | 20 | 0.0707 | 4 | 11 | 20 | 0.0407 |
| 11 | 17 | 19 | 0.0448 | 2 | 4 | 20 | 0.0845 | 5 | 11 | 20 | 0.0449 |
| 12 | 17 | 19 | 0.0505 | 3 | 4 | 20 | 0.0957 | 6 | 11 | 20 | 0.0492 |
| 13 | 17 | 19 | 0.0577 | 1 | 5 | 20 | 0.0575 | 7 | 11 | 20 | 0.0538 |
| 14 | 17 | 19 | 0.0671 | 2 | 5 | 20 | 0.0689 | 8 | 11 | 20 | 0.0589 |
| 15 | 17 | 19 | 0.0800 | 3 | 5 | 20 | 0.0781 | 9 | 11 | 20 | 0.0645 |
| 16 | 17 | 19 | 0.0991 | 4 | 5 | 20 | 0.0868 | 10 | 11 | 20 | 0.0709 |
| 1 | 18 | 19 | 0.0118 | 1 | 6 | 20 | 0.0486 | 1 | 12 | 20 | 0.0242 |
| 2 | 18 | 19 | 0.0144 | 2 | 6 | 20 | 0.0583 | 2 | 12 | 20 | 0.0291 |
| 3 | 18 | 19 | 0.0165 | 3 | 6 | 20 | 0.0661 | 3 | 12 | 20 | 0.0331 |
| 4 | 18 | 19 | 0.0186 | 4 | 6 | 20 | 0.0735 | 4 | 12 | 20 | 0.0369 |
| 5 | 18 | 19 | 0.0206 | 5 | 6 | 20 | 0.0808 | 5 | 12 | 20 | 0.0408 |
| 6 | 18 | 19 | 0.0228 | 1 | 7 | 20 | 0.0420 | 6 | 12 | 20 | 0.0447 |
| 7 | 18 | 19 | 0.0252 | 2 | 7 | 20 | 0.0504 | 7 | 12 | 20 | 0.0490 |
| 8 | 18 | 19 | 0.0279 | 3 | 7 | 20 | 0.0573 | 8 | 12 | 20 | 0.0536 |
| 9 | 18 | 19 | 0.0309 | 4 | 7 | 20 | 0.0637 | 9 | 12 | 20 | 0.0587 |
| 10 | 18 | 19 | 0.0344 | 5 | 7 | 20 | 0.0701 | 10 | 12 | 20 | 0.0645 |
| 11 | 18 | 19 | 0.0386 | 6 | 7 | 20 | 0.0768 | 11 | 12 | 20 | 0.0712 |
| 12 | 18 | 19 | 0.0436 | 1 | 8 | 20 | 0.0370 | 1 | 13 | 20 | 0.0220 |
| 13 | 18 | 19 | 0.0499 | 2 | 8 | 20 | 0.0444 | 2 | 13 | 20 | 0.0265 |
| 14 | 18 | 19 | 0.0582 | 3 | 8 | 20 | 0.0504 | 3 | 13 | 20 | 0.0302 |
| 15 | 18 | 19 | 0.0696 | 4 | 8 | 20 | 0.0561 | 4 | 13 | 20 | 0.0336 |
| 16 | 18 | 19 | 0.0865 | 5 | 8 | 20 | 0.0618 | 5 | 13 | 20 | 0.0371 |
| 17 | 18 | 19 | 0.1147 | 6 | 8 | 20 | 0.0677 | 6 | 13 | 20 | 0.0407 |
| 1 | 19 | 19 | 0.0089 | 7 | 8 | 20 | 0.0740 | 7 | 13 | 20 | 0.0446 |
| 2 | 19 | 19 | 0.0109 | 1 | 9 | 20 | 0.0329 | 8 | 13 | 20 | 0.0488 |
| 3 | 19 | 19 | 0.0126 | 2 | 9 | 20 | 0.0395 | 9 | 13 | 20 | 0.0535 |

TABLE 3

OUTLIER MEAN 1.0 OUTLIER STANDARD DEVIATION 1.0

| R | S | N | COVAR | R | S | N | COVAR | R | S | N | COVAR |
|---|---|---|---|---|---|---|---|---|---|---|---|
| 10 | 13 | 20 | 0.0588 | 10 | 16 | 20 | 0.0445 | 1 | 19 | 20 | 0.0110 |
| 11 | 13 | 20 | 0.0650 | 11 | 16 | 20 | 0.0492 | 2 | 19 | 20 | 0.0133 |
| 12 | 13 | 20 | 0.0722 | 12 | 16 | 20 | 0.0548 | 3 | 19 | 20 | 0.0153 |
| 1 | 14 | 20 | 0.0200 | 13 | 16 | 20 | 0.0616 | 4 | 19 | 20 | 0.0171 |
| 2 | 14 | 20 | 0.0241 | 14 | 16 | 20 | 0.0701 | 5 | 19 | 20 | 0.0189 |
| 3 | 14 | 20 | 0.0274 | 15 | 16 | 20 | 0.0812 | 6 | 19 | 20 | 0.0209 |
| 4 | 14 | 20 | 0.0306 | 1 | 17 | 20 | 0.0147 | 7 | 19 | 20 | 0.0229 |
| 5 | 14 | 20 | 0.0338 | 2 | 17 | 20 | 0.0177 | 8 | 19 | 20 | 0.0252 |
| 6 | 14 | 20 | 0.0371 | 3 | 17 | 20 | 0.0202 | 9 | 19 | 20 | 0.0278 |
| 7 | 14 | 20 | 0.0406 | 4 | 17 | 20 | 0.0226 | 10 | 19 | 20 | 0.0307 |
| 8 | 14 | 20 | 0.0445 | 5 | 17 | 20 | 0.0250 | 11 | 19 | 20 | 0.0341 |
| 9 | 14 | 20 | 0.0488 | 6 | 17 | 20 | 0.0275 | 12 | 19 | 20 | 0.0381 |
| 10 | 14 | 20 | 0.0537 | 7 | 17 | 20 | 0.0302 | 13 | 19 | 20 | 0.0430 |
| 11 | 14 | 20 | 0.0594 | 8 | 17 | 20 | 0.0331 | 14 | 19 | 20 | 0.0491 |
| 12 | 14 | 20 | 0.0660 | 9 | 17 | 20 | 0.0364 | 15 | 19 | 20 | 0.0572 |
| 13 | 14 | 20 | 0.0741 | 10 | 17 | 20 | 0.0401 | 16 | 19 | 20 | 0.0683 |
| 1 | 15 | 20 | 0.0182 | 11 | 17 | 20 | 0.0444 | 17 | 19 | 20 | 0.0848 |
| 2 | 15 | 20 | 0.0219 | 12 | 17 | 20 | 0.0495 | 18 | 19 | 20 | 0.1124 |
| 3 | 15 | 20 | 0.0249 | 13 | 17 | 20 | 0.0556 | 1 | 20 | 20 | 0.0084 |
| 4 | 15 | 20 | 0.0278 | 14 | 17 | 20 | 0.0634 | 2 | 20 | 20 | 0.0102 |
| 5 | 15 | 20 | 0.0307 | 15 | 17 | 20 | 0.0735 | 3 | 20 | 20 | 0.0117 |
| 6 | 15 | 20 | 0.0338 | 16 | 17 | 20 | 0.0874 | 4 | 20 | 20 | 0.0132 |
| 7 | 15 | 20 | 0.0370 | 1 | 18 | 20 | 0.0130 | 5 | 20 | 20 | 0.0146 |
| 8 | 15 | 20 | 0.0405 | 2 | 18 | 20 | 0.0157 | 6 | 20 | 20 | 0.0162 |
| 9 | 15 | 20 | 0.0445 | 3 | 18 | 20 | 0.0179 | 7 | 20 | 20 | 0.0178 |
| 10 | 15 | 20 | 0.0490 | 4 | 18 | 20 | 0.0200 | 8 | 20 | 20 | 0.0197 |
| 11 | 15 | 20 | 0.0541 | 5 | 18 | 20 | 0.0221 | 9 | 20 | 20 | 0.0218 |
| 12 | 15 | 20 | 0.0603 | 6 | 18 | 20 | 0.0243 | 10 | 20 | 20 | 0.0241 |
| 13 | 15 | 20 | 0.0677 | 7 | 18 | 20 | 0.0267 | 11 | 20 | 20 | 0.0269 |
| 14 | 15 | 20 | 0.0770 | 8 | 18 | 20 | 0.0293 | 12 | 20 | 20 | 0.0302 |
| 1 | 16 | 20 | 0.0164 | 9 | 18 | 20 | 0.0323 | 13 | 20 | 20 | 0.0342 |
| 2 | 16 | 20 | 0.0198 | 10 | 18 | 20 | 0.0356 | 14 | 20 | 20 | 0.0394 |
| 3 | 16 | 20 | 0.0226 | 11 | 18 | 20 | 0.0394 | 15 | 20 | 20 | 0.0461 |
| 4 | 16 | 20 | 0.0252 | 12 | 18 | 20 | 0.0440 | 16 | 20 | 20 | 0.0554 |
| 5 | 16 | 20 | 0.0278 | 13 | 18 | 20 | 0.0496 | 17 | 20 | 20 | 0.0693 |
| 6 | 16 | 20 | 0.0306 | 14 | 18 | 20 | 0.0566 | 18 | 20 | 20 | 0.0926 |
| 7 | 16 | 20 | 0.0335 | 15 | 18 | 20 | 0.0657 | 19 | 20 | 20 | 0.1410 |
| 8 | 16 | 20 | 0.0368 | 16 | 18 | 20 | 0.0783 | | | | |
| 9 | 16 | 20 | 0.0404 | 17 | 18 | 20 | 0.0968 | | | | |

TABLE 3

OUTLIER MEAN 1.5 OUTLIER STANDARD DEVIATION 1.0

| R | S | N | COVAR | R | S | N | COVAR | R | S | N | COVAR |
|---|---|---|---|---|---|---|---|---|---|---|---|
| 1 | 2 | 2 | 0.1682 | 1 | 6 | 7 | 0.0513 | 1 | 5 | 9 | 0.0729 |
| 1 | 2 | 3 | 0.2823 | 2 | 6 | 7 | 0.0708 | 2 | 5 | 9 | 0.0949 |
| 1 | 3 | 3 | 0.0775 | 3 | 6 | 7 | 0.0937 | 3 | 5 | 9 | 0.1171 |
| 2 | 3 | 3 | 0.1885 | 4 | 6 | 7 | 0.1260 | 4 | 5 | 9 | 0.1432 |
| 1 | 2 | 4 | 0.2605 | 5 | 6 | 7 | 0.1798 | 1 | 6 | 9 | 0.0580 |
| 1 | 3 | 4 | 0.1484 | 1 | 7 | 7 | 0.0227 | 2 | 6 | 9 | 0.0758 |
| 2 | 3 | 4 | 0.2407 | 2 | 7 | 7 | 0.0328 | 3 | 6 | 9 | 0.0939 |
| 1 | 4 | 4 | 0.0488 | 3 | 7 | 7 | 0.0458 | 4 | 6 | 9 | 0.1153 |
| 2 | 4 | 4 | 0.0918 | 4 | 7 | 7 | 0.0655 | 5 | 6 | 9 | 0.1432 |
| 3 | 4 | 4 | 0.1918 | 5 | 7 | 7 | 0.1009 | 1 | 7 | 9 | 0.0457 |
| 1 | 2 | 5 | 0.2377 | 6 | 7 | 7 | 0.1831 | 2 | 7 | 9 | 0.0599 |
| 1 | 3 | 5 | 0.1496 | 1 | 2 | 8 | 0.1940 | 3 | 7 | 9 | 0.0745 |
| 2 | 3 | 5 | 0.2197 | 1 | 3 | 8 | 0.1299 | 4 | 7 | 9 | 0.0919 |
| 1 | 4 | 5 | 0.0947 | 2 | 3 | 8 | 0.1708 | 5 | 7 | 9 | 0.1149 |
| 2 | 4 | 5 | 0.1421 | 1 | 4 | 8 | 0.0961 | 6 | 7 | 9 | 0.1484 |
| 3 | 4 | 5 | 0.2130 | 2 | 4 | 8 | 0.1271 | 1 | 8 | 9 | 0.0337 |
| 1 | 5 | 5 | 0.0354 | 3 | 4 | 8 | 0.1597 | 2 | 8 | 9 | 0.0445 |
| 2 | 5 | 5 | 0.0584 | 1 | 5 | 8 | 0.0739 | 3 | 8 | 9 | 0.0557 |
| 3 | 5 | 5 | 0.0977 | 2 | 5 | 8 | 0.0981 | 4 | 8 | 9 | 0.0693 |
| 4 | 5 | 5 | 0.1902 | 3 | 5 | 8 | 0.1239 | 5 | 8 | 9 | 0.0875 |
| 1 | 2 | 6 | 0.2196 | 4 | 5 | 8 | 0.1563 | 6 | 8 | 9 | 0.1143 |
| 1 | 3 | 6 | 0.1432 | 1 | 6 | 8 | 0.0567 | 7 | 8 | 9 | 0.1601 |
| 2 | 3 | 6 | 0.1995 | 2 | 6 | 8 | 0.0757 | 1 | 9 | 9 | 0.0166 |
| 1 | 4 | 6 | 0.1001 | 3 | 6 | 8 | 0.0961 | 2 | 9 | 9 | 0.0225 |
| 2 | 4 | 6 | 0.1410 | 4 | 6 | 8 | 0.1221 | 3 | 9 | 9 | 0.0291 |
| 3 | 4 | 6 | 0.1923 | 5 | 6 | 8 | 0.1592 | 4 | 9 | 9 | 0.0375 |
| 1 | 5 | 6 | 0.0674 | 1 | 7 | 8 | 0.0409 | 5 | 9 | 9 | 0.0493 |
| 2 | 5 | 6 | 0.0962 | 2 | 7 | 8 | 0.0550 | 6 | 9 | 9 | 0.0675 |
| 3 | 5 | 6 | 0.1335 | 3 | 7 | 8 | 0.0705 | 7 | 9 | 9 | 0.1003 |
| 4 | 5 | 6 | 0.1939 | 4 | 7 | 8 | 0.0905 | 8 | 9 | 9 | 0.1755 |
| 1 | 6 | 6 | 0.0276 | 5 | 7 | 8 | 0.1197 | 1 | 2 | 10 | 0.1769 |
| 2 | 6 | 6 | 0.0422 | 6 | 7 | 8 | 0.1689 | 1 | 3 | 10 | 0.1195 |
| 3 | 6 | 6 | 0.0630 | 1 | 8 | 8 | 0.0192 | 2 | 3 | 10 | 0.1522 |
| 4 | 6 | 6 | 0.1001 | 2 | 8 | 8 | 0.0268 | 1 | 4 | 10 | 0.0900 |
| 5 | 6 | 6 | 0.1869 | 3 | 8 | 8 | 0.0357 | 2 | 4 | 10 | 0.1151 |
| 1 | 2 | 7 | 0.2054 | 4 | 8 | 8 | 0.0479 | 3 | 4 | 10 | 0.1392 |
| 1 | 3 | 7 | 0.1363 | 5 | 8 | 8 | 0.0669 | 1 | 5 | 10 | 0.0714 |
| 2 | 3 | 7 | 0.1835 | 6 | 8 | 8 | 0.1008 | 2 | 5 | 10 | 0.0915 |
| 1 | 4 | 7 | 0.0989 | 7 | 8 | 8 | 0.1793 | 3 | 5 | 10 | 0.1110 |
| 2 | 4 | 7 | 0.1341 | 1 | 2 | 9 | 0.1847 | 4 | 5 | 10 | 0.1329 |
| 3 | 4 | 7 | 0.1740 | 1 | 3 | 9 | 0.1244 | 1 | 6 | 10 | 0.0579 |
| 1 | 5 | 7 | 0.0732 | 2 | 3 | 9 | 0.1606 | 2 | 6 | 10 | 0.0744 |
| 2 | 5 | 7 | 0.1000 | 1 | 4 | 9 | 0.0930 | 3 | 6 | 10 | 0.0905 |
| 3 | 5 | 7 | 0.1308 | 2 | 4 | 9 | 0.1207 | 4 | 6 | 10 | 0.1087 |
| 4 | 5 | 7 | 0.1732 | 3 | 4 | 9 | 0.1484 | 5 | 6 | 10 | 0.1310 |

TABLE 3

OUTLIER MEAN 1.5 OUTLIER STANDARD DEVIATION 1.0

| R | S | N | COVAR | R | S | N | COVAR | R | S | N | COVAR |
|---|---|---|---|---|---|---|---|---|---|---|---|
| 1 | 7 | 10 | 0.0473 | 1 | 7 | 11 | 0.0476 | 3 | 4 | 12 | 0.1253 |
| 2 | 7 | 10 | 0.0609 | 2 | 7 | 11 | 0.0605 | 1 | 5 | 12 | 0.0680 |
| 3 | 7 | 10 | 0.0742 | 3 | 7 | 11 | 0.0727 | 2 | 5 | 12 | 0.0853 |
| 4 | 7 | 10 | 0.0894 | 4 | 7 | 11 | 0.0861 | 3 | 5 | 12 | 0.1010 |
| 5 | 7 | 10 | 0.1081 | 5 | 7 | 11 | 0.1018 | 4 | 5 | 12 | 0.1176 |
| 6 | 7 | 10 | 0.1331 | 6 | 7 | 11 | 0.1216 | 1 | 6 | 12 | 0.0563 |
| 1 | 8 | 10 | 0.0380 | 1 | 8 | 11 | 0.0396 | 2 | 6 | 12 | 0.0707 |
| 2 | 8 | 10 | 0.0490 | 2 | 8 | 11 | 0.0504 | 3 | 6 | 12 | 0.0839 |
| 3 | 8 | 10 | 0.0600 | 3 | 8 | 11 | 0.0607 | 4 | 6 | 12 | 0.0979 |
| 4 | 8 | 10 | 0.0725 | 4 | 8 | 11 | 0.0720 | 5 | 6 | 12 | 0.1137 |
| 5 | 8 | 10 | 0.0880 | 5 | 8 | 11 | 0.0853 | 1 | 7 | 12 | 0.0474 |
| 6 | 8 | 10 | 0.1089 | 6 | 8 | 11 | 0.1022 | 2 | 7 | 12 | 0.0596 |
| 7 | 8 | 10 | 0.1397 | 7 | 8 | 11 | 0.1250 | 3 | 7 | 12 | 0.0708 |
| 1 | 9 | 10 | 0.0285 | 1 | 9 | 11 | 0.0322 | 4 | 7 | 12 | 0.0827 |
| 2 | 9 | 10 | 0.0370 | 2 | 9 | 11 | 0.0411 | 5 | 7 | 12 | 0.0963 |
| 3 | 9 | 10 | 0.0456 | 3 | 9 | 11 | 0.0497 | 6 | 7 | 12 | 0.1126 |
| 4 | 9 | 10 | 0.0554 | 4 | 9 | 11 | 0.0591 | 1 | 8 | 12 | 0.0401 |
| 5 | 9 | 10 | 0.0678 | 5 | 9 | 11 | 0.0703 | 2 | 8 | 12 | 0.0505 |
| 6 | 9 | 10 | 0.0846 | 6 | 9 | 11 | 0.0845 | 3 | 8 | 12 | 0.0601 |
| 7 | 9 | 10 | 0.1098 | 7 | 9 | 11 | 0.1039 | 4 | 8 | 12 | 0.0703 |
| 8 | 9 | 10 | 0.1529 | 8 | 9 | 11 | 0.1326 | 5 | 8 | 12 | 0.0820 |
| 1 | 10 | 10 | 0.0146 | 1 | 10 | 11 | 0.0246 | 6 | 8 | 12 | 0.0961 |
| 2 | 10 | 10 | 0.0194 | 2 | 10 | 11 | 0.0315 | 7 | 8 | 12 | 0.1140 |
| 3 | 10 | 10 | 0.0245 | 3 | 10 | 11 | 0.0382 | 1 | 9 | 12 | 0.0338 |
| 4 | 10 | 10 | 0.0306 | 4 | 10 | 11 | 0.0457 | 2 | 9 | 12 | 0.0426 |
| 5 | 10 | 10 | 0.0387 | 5 | 10 | 11 | 0.0547 | 3 | 9 | 12 | 0.0508 |
| 6 | 10 | 10 | 0.0501 | 6 | 10 | 11 | 0.0662 | 4 | 9 | 12 | 0.0595 |
| 7 | 10 | 10 | 0.0678 | 7 | 10 | 11 | 0.0821 | 5 | 9 | 12 | 0.0696 |
| 8 | 10 | 10 | 0.0995 | 8 | 10 | 11 | 0.1059 | 6 | 9 | 12 | 0.0817 |
| 9 | 10 | 10 | 0.1720 | 9 | 10 | 11 | 0.1468 | 7 | 9 | 12 | 0.0972 |
| 1 | 2 | 11 | 0.1703 | 1 | 11 | 11 | 0.0130 | 8 | 9 | 12 | 0.1184 |
| 1 | 3 | 11 | 0.1153 | 2 | 11 | 11 | 0.0171 | 1 | 10 | 12 | 0.0279 |
| 2 | 3 | 11 | 0.1451 | 3 | 11 | 11 | 0.0211 | 2 | 10 | 12 | 0.0352 |
| 1 | 4 | 11 | 0.0873 | 4 | 11 | 11 | 0.0258 | 3 | 10 | 12 | 0.0421 |
| 2 | 4 | 11 | 0.1102 | 5 | 11 | 11 | 0.0317 | 4 | 10 | 12 | 0.0494 |
| 3 | 4 | 11 | 0.1317 | 6 | 11 | 11 | 0.0395 | 5 | 10 | 12 | 0.0579 |
| 1 | 5 | 11 | 0.0697 | 7 | 11 | 11 | 0.0505 | 6 | 10 | 12 | 0.0682 |
| 2 | 5 | 11 | 0.0883 | 8 | 11 | 11 | 0.0678 | 7 | 10 | 12 | 0.0815 |
| 3 | 5 | 11 | 0.1057 | 9 | 11 | 11 | 0.0985 | 8 | 10 | 12 | 0.0996 |
| 4 | 5 | 11 | 0.1245 | 10 | 11 | 11 | 0.1686 | 9 | 10 | 12 | 0.1267 |
| 1 | 6 | 11 | 0.0572 | 1 | 2 | 12 | 0.1646 | 1 | 11 | 12 | 0.0215 |
| 2 | 6 | 11 | 0.0726 | 1 | 3 | 12 | 0.1116 | 2 | 11 | 12 | 0.0273 |
| 3 | 6 | 11 | 0.0871 | 2 | 3 | 12 | 0.1391 | 3 | 11 | 12 | 0.0328 |
| 4 | 6 | 11 | 0.1029 | 1 | 4 | 12 | 0.0847 | 4 | 11 | 12 | 0.0386 |
| 5 | 6 | 11 | 0.1214 | 2 | 4 | 12 | 0.1060 | 5 | 11 | 12 | 0.0455 |

TABLE 3

OUTLIER MEAN 1.5 OUTLIER STANDARD DEVIATION 1.0

| R | S | N | COVAR | R | S | N | COVAR | R | S | N | COVAR |
|---|---|---|---|---|---|---|---|---|---|---|---|
| 6 | 11 | 12 | 0.0539 | 2 | 9 | 13 | 0.0430 | 9 | 13 | 13 | 0.0508 |
| 7 | 11 | 12 | 0.0648 | 3 | 9 | 13 | 0.0508 | 10 | 13 | 13 | 0.0673 |
| 8 | 11 | 12 | 0.0798 | 4 | 9 | 13 | 0.0589 | 11 | 13 | 13 | 0.0965 |
| 9 | 11 | 12 | 0.1025 | 5 | 9 | 13 | 0.0679 | 12 | 13 | 13 | 0.1625 |
| 10 | 11 | 12 | 0.1415 | 6 | 9 | 13 | 0.0784 | 1 | 2 | 14 | 0.1552 |
| 1 | 12 | 12 | 0.0118 | 7 | 9 | 13 | 0.0913 | 1 | 3 | 14 | 0.1054 |
| 2 | 12 | 12 | 0.0152 | 8 | 9 | 13 | 0.1078 | 2 | 3 | 14 | 0.1294 |
| 3 | 12 | 12 | 0.0185 | 1 | 10 | 13 | 0.0294 | 1 | 4 | 14 | 0.0803 |
| 4 | 12 | 12 | 0.0223 | 2 | 10 | 13 | 0.0367 | 2 | 4 | 14 | 0.0990 |
| 5 | 12 | 12 | 0.0267 | 3 | 10 | 13 | 0.0434 | 3 | 4 | 14 | 0.1152 |
| 6 | 12 | 12 | 0.0324 | 4 | 10 | 13 | 0.0504 | 1 | 5 | 14 | 0.0649 |
| 7 | 12 | 12 | 0.0400 | 5 | 10 | 13 | 0.0581 | 2 | 5 | 14 | 0.0801 |
| 8 | 12 | 12 | 0.0508 | 6 | 10 | 13 | 0.0673 | 3 | 5 | 14 | 0.0934 |
| 9 | 12 | 12 | 0.0676 | 7 | 10 | 13 | 0.0785 | 4 | 5 | 14 | 0.1069 |
| 10 | 12 | 12 | 0.0975 | 8 | 10 | 13 | 0.0930 | 1 | 6 | 14 | 0.0542 |
| 11 | 12 | 12 | 0.1654 | 9 | 10 | 13 | 0.1128 | 2 | 6 | 14 | 0.0670 |
| 1 | 2 | 13 | 0.1596 | 1 | 11 | 13 | 0.0245 | 3 | 6 | 14 | 0.0783 |
| 1 | 3 | 13 | 0.1083 | 2 | 11 | 13 | 0.0306 | 4 | 6 | 14 | 0.0896 |
| 2 | 3 | 13 | 0.1339 | 3 | 11 | 13 | 0.0363 | 5 | 6 | 14 | 0.1019 |
| 1 | 4 | 13 | 0.0824 | 4 | 11 | 13 | 0.0422 | 1 | 7 | 14 | 0.0463 |
| 2 | 4 | 13 | 0.1023 | 5 | 11 | 13 | 0.0488 | 2 | 7 | 14 | 0.0572 |
| 3 | 4 | 13 | 0.1199 | 6 | 11 | 13 | 0.0566 | 3 | 7 | 14 | 0.0669 |
| 1 | 5 | 13 | 0.0664 | 7 | 11 | 13 | 0.0662 | 4 | 7 | 14 | 0.0767 |
| 2 | 5 | 13 | 0.0826 | 8 | 11 | 13 | 0.0787 | 5 | 7 | 14 | 0.0873 |
| 3 | 5 | 13 | 0.0970 | 9 | 11 | 13 | 0.0959 | 6 | 7 | 14 | 0.0993 |
| 4 | 5 | 13 | 0.1118 | 10 | 11 | 13 | 0.1216 | 1 | 8 | 14 | 0.0399 |
| 1 | 6 | 13 | 0.0553 | 1 | 12 | 13 | 0.0191 | 2 | 8 | 14 | 0.0494 |
| 2 | 6 | 13 | 0.0688 | 2 | 12 | 13 | 0.0240 | 3 | 8 | 14 | 0.0578 |
| 3 | 6 | 13 | 0.0810 | 3 | 12 | 13 | 0.0285 | 4 | 8 | 14 | 0.0664 |
| 4 | 6 | 13 | 0.0935 | 4 | 12 | 13 | 0.0333 | 5 | 8 | 14 | 0.0756 |
| 5 | 6 | 13 | 0.1073 | 5 | 12 | 13 | 0.0387 | 6 | 8 | 14 | 0.0862 |
| 1 | 7 | 13 | 0.0469 | 6 | 12 | 13 | 0.0450 | 7 | 8 | 14 | 0.0986 |
| 2 | 7 | 13 | 0.0584 | 7 | 12 | 13 | 0.0530 | 1 | 9 | 14 | 0.0346 |
| 3 | 7 | 13 | 0.0688 | 8 | 12 | 13 | 0.0633 | 2 | 9 | 14 | 0.0429 |
| 4 | 7 | 13 | 0.0796 | 9 | 12 | 13 | 0.0777 | 3 | 9 | 14 | 0.0503 |
| 5 | 7 | 13 | 0.0915 | 10 | 12 | 13 | 0.0995 | 4 | 9 | 14 | 0.0578 |
| 6 | 7 | 13 | 0.1053 | 11 | 12 | 13 | 0.1370 | 5 | 9 | 14 | 0.0659 |
| 1 | 8 | 13 | 0.0401 | 1 | 13 | 13 | 0.0107 | 6 | 9 | 14 | 0.0752 |
| 2 | 8 | 13 | 0.0501 | 2 | 13 | 13 | 0.0137 | 7 | 9 | 14 | 0.0861 |
| 3 | 8 | 13 | 0.0591 | 3 | 13 | 13 | 0.0165 | 8 | 9 | 14 | 0.0996 |
| 4 | 8 | 13 | 0.0684 | 4 | 13 | 13 | 0.0195 | 1 | 10 | 14 | 0.0301 |
| 5 | 8 | 13 | 0.0787 | 5 | 13 | 13 | 0.0231 | 2 | 10 | 14 | 0.0373 |
| 6 | 8 | 13 | 0.0908 | 6 | 13 | 13 | 0.0274 | 3 | 10 | 14 | 0.0437 |
| 7 | 8 | 13 | 0.1055 | 7 | 13 | 13 | 0.0329 | 4 | 10 | 14 | 0.0502 |
| 1 | 9 | 13 | 0.0345 | 8 | 13 | 13 | 0.0403 | 5 | 10 | 14 | 0.0574 |

TABLE 3

OUTLIER MEAN 1.5 OUTLIER STANDARD DEVIATION 1.0

| R | S | N | COVAR | R | S | N | COVAR | R | S | N | COVAR |
|---|---|---|---|---|---|---|---|---|---|---|---|
| 6 | 10 | 14 | 0.0655 | 9 | 14 | 14 | 0.0405 | 5 | 10 | 15 | 0.0562 |
| 7 | 10 | 14 | 0.0752 | 10 | 14 | 14 | 0.0508 | 6 | 10 | 15 | 0.0636 |
| 8 | 10 | 14 | 0.0871 | 11 | 14 | 14 | 0.0669 | 7 | 10 | 15 | 0.0720 |
| 9 | 10 | 14 | 0.1026 | 12 | 14 | 14 | 0.0954 | 8 | 10 | 15 | 0.0821 |
| 1 | 11 | 14 | 0.0258 | 13 | 14 | 14 | 0.1597 | 9 | 10 | 15 | 0.0947 |
| 2 | 11 | 14 | 0.0321 | 1 | 2 | 15 | 0.1513 | 1 | 11 | 15 | 0.0266 |
| 3 | 11 | 14 | 0.0377 | 1 | 3 | 15 | 0.1027 | 2 | 11 | 15 | 0.0327 |
| 4 | 11 | 14 | 0.0433 | 2 | 3 | 15 | 0.1254 | 3 | 11 | 15 | 0.0381 |
| 5 | 11 | 14 | 0.0496 | 1 | 4 | 15 | 0.0784 | 4 | 11 | 15 | 0.0436 |
| 6 | 11 | 14 | 0.0567 | 2 | 4 | 15 | 0.0960 | 5 | 11 | 15 | 0.0494 |
| 7 | 11 | 14 | 0.0652 | 3 | 4 | 15 | 0.1111 | 6 | 11 | 15 | 0.0558 |
| 8 | 11 | 14 | 0.0757 | 1 | 5 | 15 | 0.0635 | 7 | 11 | 15 | 0.0634 |
| 9 | 11 | 14 | 0.0893 | 2 | 5 | 15 | 0.0779 | 8 | 11 | 15 | 0.0724 |
| 10 | 11 | 14 | 0.1080 | 3 | 5 | 15 | 0.0903 | 9 | 11 | 15 | 0.0835 |
| 1 | 12 | 14 | 0.0217 | 4 | 5 | 15 | 0.1026 | 10 | 11 | 15 | 0.0981 |
| 2 | 12 | 14 | 0.0270 | 1 | 6 | 15 | 0.0532 | 1 | 12 | 15 | 0.0230 |
| 3 | 12 | 14 | 0.0317 | 2 | 6 | 15 | 0.0654 | 2 | 12 | 15 | 0.0284 |
| 4 | 12 | 14 | 0.0366 | 3 | 6 | 15 | 0.0758 | 3 | 12 | 15 | 0.0331 |
| 5 | 12 | 14 | 0.0419 | 4 | 6 | 15 | 0.0863 | 4 | 12 | 15 | 0.0378 |
| 6 | 12 | 14 | 0.0480 | 5 | 6 | 15 | 0.0973 | 5 | 12 | 15 | 0.0429 |
| 7 | 12 | 14 | 0.0553 | 1 | 7 | 15 | 0.0456 | 6 | 12 | 15 | 0.0486 |
| 8 | 12 | 14 | 0.0644 | 2 | 7 | 15 | 0.0560 | 7 | 12 | 15 | 0.0553 |
| 9 | 12 | 14 | 0.0763 | 3 | 7 | 15 | 0.0651 | 8 | 12 | 15 | 0.0632 |
| 10 | 12 | 14 | 0.0926 | 4 | 7 | 15 | 0.0741 | 9 | 12 | 15 | 0.0731 |
| 11 | 12 | 14 | 0.1172 | 5 | 7 | 15 | 0.0837 | 10 | 12 | 15 | 0.0860 |
| 1 | 13 | 14 | 0.0171 | 6 | 7 | 15 | 0.0943 | 11 | 12 | 15 | 0.1038 |
| 2 | 13 | 14 | 0.0214 | 1 | 8 | 15 | 0.0396 | 1 | 13 | 15 | 0.0195 |
| 3 | 13 | 14 | 0.0252 | 2 | 8 | 15 | 0.0487 | 2 | 13 | 15 | 0.0241 |
| 4 | 13 | 14 | 0.0291 | 3 | 8 | 15 | 0.0566 | 3 | 13 | 15 | 0.0281 |
| 5 | 13 | 14 | 0.0334 | 4 | 8 | 15 | 0.0644 | 4 | 13 | 15 | 0.0322 |
| 6 | 13 | 14 | 0.0385 | 5 | 8 | 15 | 0.0728 | 5 | 13 | 15 | 0.0365 |
| 7 | 13 | 14 | 0.0445 | 6 | 8 | 15 | 0.0822 | 6 | 13 | 15 | 0.0415 |
| 8 | 13 | 14 | 0.0521 | 7 | 8 | 15 | 0.0930 | 7 | 13 | 15 | 0.0472 |
| 9 | 13 | 14 | 0.0620 | 1 | 9 | 15 | 0.0346 | 8 | 13 | 15 | 0.0541 |
| 10 | 13 | 14 | 0.0759 | 2 | 9 | 15 | 0.0426 | 9 | 13 | 15 | 0.0627 |
| 11 | 13 | 14 | 0.0968 | 3 | 9 | 15 | 0.0495 | 10 | 13 | 15 | 0.0741 |
| 12 | 13 | 14 | 0.1329 | 4 | 9 | 15 | 0.0565 | 11 | 13 | 15 | 0.0897 |
| 1 | 14 | 14 | 0.0098 | 5 | 9 | 15 | 0.0639 | 12 | 13 | 15 | 0.1133 |
| 2 | 14 | 14 | 0.0124 | 6 | 9 | 15 | 0.0722 | 1 | 14 | 15 | 0.0155 |
| 3 | 14 | 14 | 0.0148 | 7 | 9 | 15 | 0.0817 | 2 | 14 | 15 | 0.0192 |
| 4 | 14 | 14 | 0.0174 | 8 | 9 | 15 | 0.0931 | 3 | 14 | 15 | 0.0224 |
| 5 | 14 | 14 | 0.0203 | 1 | 10 | 15 | 0.0303 | 4 | 14 | 15 | 0.0258 |
| 6 | 14 | 14 | 0.0237 | 2 | 10 | 15 | 0.0374 | 5 | 14 | 15 | 0.0294 |
| 7 | 14 | 14 | 0.0279 | 3 | 10 | 15 | 0.0435 | 6 | 14 | 15 | 0.0334 |
| 8 | 14 | 14 | 0.0332 | 4 | 10 | 15 | 0.0497 | 7 | 14 | 15 | 0.0382 |

TABLE 3

OUTLIER MEAN 1.5 OUTLIER STANDARD DEVIATION 1.0

| R | S | N | COVAR | R | S | N | COVAR | R | S | N | COVAR |
|---|---|---|---|---|---|---|---|---|---|---|---|
| 8 | 14 | 15 | 0.0440 | 5 | 8 | 16 | 0.0703 | 5 | 13 | 16 | 0.0377 |
| 9 | 14 | 15 | 0.0512 | 6 | 8 | 16 | 0.0787 | 6 | 13 | 16 | 0.0424 |
| 10 | 14 | 15 | 0.0608 | 7 | 8 | 16 | 0.0882 | 7 | 13 | 16 | 0.0477 |
| 11 | 14 | 15 | 0.0742 | 1 | 9 | 16 | 0.0344 | 8 | 13 | 16 | 0.0539 |
| 12 | 14 | 15 | 0.0944 | 2 | 9 | 16 | 0.0421 | 9 | 13 | 16 | 0.0614 |
| 13 | 14 | 15 | 0.1293 | 3 | 9 | 16 | 0.0487 | 10 | 13 | 16 | 0.0709 |
| 1 | 15 | 15 | 0.0091 | 4 | 9 | 16 | 0.0552 | 11 | 13 | 16 | 0.0832 |
| 2 | 15 | 15 | 0.0114 | 5 | 9 | 16 | 0.0620 | 12 | 13 | 16 | 0.1002 |
| 3 | 15 | 15 | 0.0135 | 6 | 9 | 16 | 0.0695 | 1 | 14 | 16 | 0.0176 |
| 4 | 15 | 15 | 0.0156 | 7 | 9 | 16 | 0.0779 | 2 | 14 | 16 | 0.0216 |
| 5 | 15 | 15 | 0.0180 | 8 | 9 | 16 | 0.0877 | 3 | 14 | 16 | 0.0251 |
| 6 | 15 | 15 | 0.0208 | 1 | 10 | 16 | 0.0304 | 4 | 14 | 16 | 0.0286 |
| 7 | 15 | 15 | 0.0241 | 2 | 10 | 16 | 0.0372 | 5 | 14 | 16 | 0.0322 |
| 8 | 15 | 15 | 0.0282 | 3 | 10 | 16 | 0.0431 | 6 | 14 | 16 | 0.0363 |
| 9 | 15 | 15 | 0.0335 | 4 | 10 | 16 | 0.0489 | 7 | 14 | 16 | 0.0409 |
| 10 | 15 | 15 | 0.0406 | 5 | 10 | 16 | 0.0549 | 8 | 14 | 16 | 0.0464 |
| 11 | 15 | 15 | 0.0507 | 6 | 10 | 16 | 0.0616 | 9 | 14 | 16 | 0.0529 |
| 12 | 15 | 15 | 0.0665 | 7 | 10 | 16 | 0.0691 | 10 | 14 | 16 | 0.0612 |
| 13 | 15 | 15 | 0.0944 | 8 | 10 | 16 | 0.0779 | 11 | 14 | 16 | 0.0721 |
| 14 | 15 | 15 | 0.1571 | 9 | 10 | 16 | 0.0884 | 12 | 14 | 16 | 0.0872 |
| 1 | 2 | 16 | 0.1477 | 1 | 11 | 16 | 0.0269 | 13 | 14 | 16 | 0.1098 |
| 1 | 3 | 16 | 0.1004 | 2 | 11 | 16 | 0.0330 | 1 | 15 | 16 | 0.0141 |
| 2 | 3 | 16 | 0.1219 | 3 | 11 | 16 | 0.0382 | 2 | 15 | 16 | 0.0174 |
| 1 | 4 | 16 | 0.0767 | 4 | 11 | 16 | 0.0433 | 3 | 15 | 16 | 0.0202 |
| 2 | 4 | 16 | 0.0934 | 5 | 11 | 16 | 0.0487 | 4 | 15 | 16 | 0.0230 |
| 3 | 4 | 16 | 0.1075 | 6 | 11 | 16 | 0.0547 | 5 | 15 | 16 | 0.0261 |
| 1 | 5 | 16 | 0.0622 | 7 | 11 | 16 | 0.0614 | 6 | 15 | 16 | 0.0294 |
| 2 | 5 | 16 | 0.0759 | 8 | 11 | 16 | 0.0693 | 7 | 15 | 16 | 0.0333 |
| 3 | 5 | 16 | 0.0875 | 9 | 11 | 16 | 0.0787 | 8 | 15 | 16 | 0.0378 |
| 4 | 5 | 16 | 0.0988 | 10 | 11 | 16 | 0.0905 | 9 | 15 | 16 | 0.0434 |
| 1 | 6 | 16 | 0.0523 | 1 | 12 | 16 | 0.0237 | 10 | 15 | 16 | 0.0504 |
| 2 | 6 | 16 | 0.0638 | 2 | 12 | 16 | 0.0291 | 11 | 15 | 16 | 0.0596 |
| 3 | 6 | 16 | 0.0736 | 3 | 12 | 16 | 0.0337 | 12 | 15 | 16 | 0.0726 |
| 4 | 6 | 16 | 0.0833 | 4 | 12 | 16 | 0.0382 | 13 | 15 | 16 | 0.0922 |
| 5 | 6 | 16 | 0.0933 | 5 | 12 | 16 | 0.0431 | 14 | 15 | 16 | 0.1261 |
| 1 | 7 | 16 | 0.0449 | 6 | 12 | 16 | 0.0484 | 1 | 16 | 16 | 0.0084 |
| 2 | 7 | 16 | 0.0548 | 7 | 12 | 16 | 0.0544 | 2 | 16 | 16 | 0.0105 |
| 3 | 7 | 16 | 0.0633 | 8 | 12 | 16 | 0.0614 | 3 | 16 | 16 | 0.0123 |
| 4 | 7 | 16 | 0.0717 | 9 | 12 | 16 | 0.0698 | 4 | 16 | 16 | 0.0142 |
| 5 | 7 | 16 | 0.0805 | 10 | 12 | 16 | 0.0804 | 5 | 16 | 16 | 0.0162 |
| 6 | 7 | 16 | 0.0900 | 11 | 12 | 16 | 0.0942 | 6 | 16 | 16 | 0.0185 |
| 1 | 8 | 16 | 0.0391 | 1 | 13 | 16 | 0.0207 | 7 | 16 | 16 | 0.0212 |
| 2 | 8 | 16 | 0.0478 | 2 | 13 | 16 | 0.0254 | 8 | 16 | 16 | 0.0244 |
| 3 | 8 | 16 | 0.0553 | 3 | 13 | 16 | 0.0294 | 9 | 16 | 16 | 0.0285 |
| 4 | 8 | 16 | 0.0626 | 4 | 13 | 16 | 0.0334 | 10 | 16 | 16 | 0.0336 |

TABLE 3

OUTLIER MEAN 1.5 OUTLIER STANDARD DEVIATION 1.0

| R | S | N | COVAR | R | S | N | COVAR | R | S | N | COVAR |
|---|---|---|---|---|---|---|---|---|---|---|---|
| 11 | 16 | 16 | 0.0406 | 5 | 10 | 17 | 0.0536 | 8 | 14 | 17 | 0.0467 |
| 12 | 16 | 16 | 0.0505 | 6 | 10 | 17 | 0.0597 | 9 | 14 | 17 | 0.0526 |
| 13 | 16 | 16 | 0.0660 | 7 | 10 | 17 | 0.0665 | 10 | 14 | 17 | 0.0598 |
| 14 | 16 | 16 | 0.0933 | 8 | 10 | 17 | 0.0742 | 11 | 14 | 17 | 0.0688 |
| 15 | 16 | 16 | 0.1546 | 9 | 10 | 17 | 0.0833 | 12 | 14 | 17 | 0.0806 |
| 1 | 2 | 17 | 0.1445 | 1 | 11 | 17 | 0.0270 | 13 | 14 | 17 | 0.0970 |
| 1 | 3 | 17 | 0.0982 | 2 | 11 | 17 | 0.0329 | 1 | 15 | 17 | 0.0160 |
| 2 | 3 | 17 | 0.1187 | 3 | 11 | 17 | 0.0379 | 2 | 15 | 17 | 0.0196 |
| 1 | 4 | 17 | 0.0751 | 4 | 11 | 17 | 0.0428 | 3 | 15 | 17 | 0.0227 |
| 2 | 4 | 17 | 0.0910 | 5 | 11 | 17 | 0.0479 | 4 | 15 | 17 | 0.0256 |
| 3 | 4 | 17 | 0.1042 | 6 | 11 | 17 | 0.0534 | 5 | 15 | 17 | 0.0288 |
| 1 | 5 | 17 | 0.0610 | 7 | 11 | 17 | 0.0595 | 6 | 15 | 17 | 0.0322 |
| 2 | 5 | 17 | 0.0740 | 8 | 11 | 17 | 0.0664 | 7 | 15 | 17 | 0.0360 |
| 3 | 5 | 17 | 0.0849 | 9 | 11 | 17 | 0.0746 | 8 | 15 | 17 | 0.0404 |
| 4 | 5 | 17 | 0.0955 | 10 | 11 | 17 | 0.0844 | 9 | 15 | 17 | 0.0455 |
| 1 | 6 | 17 | 0.0513 | 1 | 12 | 17 | 0.0241 | 10 | 15 | 17 | 0.0518 |
| 2 | 6 | 17 | 0.0624 | 2 | 12 | 17 | 0.0294 | 11 | 15 | 17 | 0.0598 |
| 3 | 6 | 17 | 0.0716 | 3 | 12 | 17 | 0.0338 | 12 | 15 | 17 | 0.0703 |
| 4 | 6 | 17 | 0.0806 | 4 | 12 | 17 | 0.0382 | 13 | 15 | 17 | 0.0848 |
| 5 | 6 | 17 | 0.0898 | 5 | 12 | 17 | 0.0428 | 14 | 15 | 17 | 0.1068 |
| 1 | 7 | 17 | 0.0442 | 6 | 12 | 17 | 0.0477 | 1 | 16 | 17 | 0.0129 |
| 2 | 7 | 17 | 0.0537 | 7 | 12 | 17 | 0.0532 | 2 | 16 | 17 | 0.0158 |
| 3 | 7 | 17 | 0.0618 | 8 | 12 | 17 | 0.0594 | 3 | 16 | 17 | 0.0183 |
| 4 | 7 | 17 | 0.0696 | 9 | 12 | 17 | 0.0668 | 4 | 16 | 17 | 0.0208 |
| 5 | 7 | 17 | 0.0776 | 10 | 12 | 17 | 0.0757 | 5 | 16 | 17 | 0.0234 |
| 6 | 7 | 17 | 0.0863 | 11 | 12 | 17 | 0.0868 | 6 | 16 | 17 | 0.0262 |
| 1 | 8 | 17 | 0.0386 | 1 | 13 | 17 | 0.0214 | 7 | 16 | 17 | 0.0294 |
| 2 | 8 | 17 | 0.0470 | 2 | 13 | 17 | 0.0261 | 8 | 16 | 17 | 0.0331 |
| 3 | 8 | 17 | 0.0541 | 3 | 13 | 17 | 0.0301 | 9 | 16 | 17 | 0.0374 |
| 4 | 8 | 17 | 0.0609 | 4 | 13 | 17 | 0.0340 | 10 | 16 | 17 | 0.0428 |
| 5 | 8 | 17 | 0.0680 | 5 | 13 | 17 | 0.0380 | 11 | 16 | 17 | 0.0496 |
| 6 | 8 | 17 | 0.0757 | 6 | 13 | 17 | 0.0424 | 12 | 16 | 17 | 0.0586 |
| 7 | 8 | 17 | 0.0841 | 7 | 13 | 17 | 0.0473 | 13 | 16 | 17 | 0.0712 |
| 1 | 9 | 17 | 0.0341 | 8 | 13 | 17 | 0.0530 | 14 | 16 | 17 | 0.0903 |
| 2 | 9 | 17 | 0.0415 | 9 | 13 | 17 | 0.0596 | 15 | 16 | 17 | 0.1232 |
| 3 | 9 | 17 | 0.0478 | 10 | 13 | 17 | 0.0676 | 1 | 17 | 17 | 0.0079 |
| 4 | 9 | 17 | 0.0539 | 11 | 13 | 17 | 0.0777 | 2 | 17 | 17 | 0.0097 |
| 5 | 9 | 17 | 0.0602 | 12 | 13 | 17 | 0.0908 | 3 | 17 | 17 | 0.0113 |
| 6 | 9 | 17 | 0.0670 | 1 | 14 | 17 | 0.0187 | 4 | 17 | 17 | 0.0130 |
| 7 | 9 | 17 | 0.0746 | 2 | 14 | 17 | 0.0229 | 5 | 17 | 17 | 0.0147 |
| 8 | 9 | 17 | 0.0832 | 3 | 14 | 17 | 0.0264 | 6 | 17 | 17 | 0.0167 |
| 1 | 10 | 17 | 0.0303 | 4 | 14 | 17 | 0.0298 | 7 | 17 | 17 | 0.0189 |
| 2 | 10 | 17 | 0.0369 | 5 | 14 | 17 | 0.0334 | 8 | 17 | 17 | 0.0215 |
| 3 | 10 | 17 | 0.0425 | 6 | 14 | 17 | 0.0373 | 9 | 17 | 17 | 0.0247 |
| 4 | 10 | 17 | 0.0480 | 7 | 14 | 17 | 0.0417 | 10 | 17 | 17 | 0.0286 |

TABLE 3

OUTLIER MEAN 1.5 OUTLIER STANDARD DEVIATION 1.0

| R | S | N | COVAR | R | S | N | COVAR | R | S | N | COVAR |
|---|---|---|---|---|---|---|---|---|---|---|---|
| 11 | 17 | 17 | 0.0337 | 4 | 10 | 18 | 0.0471 | 7 | 14 | 18 | 0.0417 |
| 12 | 17 | 17 | 0.0406 | 5 | 10 | 18 | 0.0523 | 8 | 14 | 18 | 0.0463 |
| 13 | 17 | 17 | 0.0503 | 6 | 10 | 18 | 0.0580 | 9 | 14 | 18 | 0.0516 |
| 14 | 17 | 17 | 0.0655 | 7 | 10 | 18 | 0.0641 | 10 | 14 | 18 | 0.0579 |
| 15 | 17 | 17 | 0.0923 | 8 | 10 | 18 | 0.0710 | 11 | 14 | 18 | 0.0656 |
| 16 | 17 | 17 | 0.1523 | 9 | 10 | 18 | 0.0789 | 12 | 14 | 18 | 0.0752 |
| 1 | 2 | 18 | 0.1416 | 1 | 11 | 18 | 0.0270 | 13 | 14 | 18 | 0.0878 |
| 1 | 3 | 18 | 0.0962 | 2 | 11 | 18 | 0.0328 | 1 | 15 | 18 | 0.0171 |
| 2 | 3 | 18 | 0.1158 | 3 | 11 | 18 | 0.0376 | 2 | 15 | 18 | 0.0208 |
| 1 | 4 | 18 | 0.0736 | 4 | 11 | 18 | 0.0422 | 3 | 15 | 18 | 0.0239 |
| 2 | 4 | 18 | 0.0888 | 5 | 11 | 18 | 0.0470 | 4 | 15 | 18 | 0.0269 |
| 3 | 4 | 18 | 0.1013 | 6 | 11 | 18 | 0.0521 | 5 | 15 | 18 | 0.0300 |
| 1 | 5 | 18 | 0.0598 | 7 | 11 | 18 | 0.0576 | 6 | 15 | 18 | 0.0333 |
| 2 | 5 | 18 | 0.0723 | 8 | 11 | 18 | 0.0639 | 7 | 15 | 18 | 0.0369 |
| 3 | 5 | 18 | 0.0826 | 9 | 11 | 18 | 0.0710 | 8 | 15 | 18 | 0.0410 |
| 4 | 5 | 18 | 0.0925 | 10 | 11 | 18 | 0.0795 | 9 | 15 | 18 | 0.0458 |
| 1 | 6 | 18 | 0.0504 | 1 | 12 | 18 | 0.0243 | 10 | 15 | 18 | 0.0514 |
| 2 | 6 | 18 | 0.0610 | 2 | 12 | 18 | 0.0294 | 11 | 15 | 18 | 0.0583 |
| 3 | 6 | 18 | 0.0698 | 3 | 12 | 18 | 0.0338 | 12 | 15 | 18 | 0.0670 |
| 4 | 6 | 18 | 0.0782 | 4 | 12 | 18 | 0.0380 | 13 | 15 | 18 | 0.0783 |
| 5 | 6 | 18 | 0.0868 | 5 | 12 | 18 | 0.0423 | 14 | 15 | 18 | 0.0941 |
| 1 | 7 | 18 | 0.0435 | 6 | 12 | 18 | 0.0468 | 1 | 16 | 18 | 0.0147 |
| 2 | 7 | 18 | 0.0527 | 7 | 12 | 18 | 0.0519 | 2 | 16 | 18 | 0.0179 |
| 3 | 7 | 18 | 0.0603 | 8 | 12 | 18 | 0.0575 | 3 | 16 | 18 | 0.0206 |
| 4 | 7 | 18 | 0.0676 | 9 | 12 | 18 | 0.0640 | 4 | 16 | 18 | 0.0232 |
| 5 | 7 | 18 | 0.0751 | 10 | 12 | 18 | 0.0717 | 5 | 16 | 18 | 0.0259 |
| 6 | 7 | 18 | 0.0830 | 11 | 12 | 18 | 0.0810 | 6 | 16 | 18 | 0.0288 |
| 1 | 8 | 18 | 0.0381 | 1 | 13 | 18 | 0.0217 | 7 | 16 | 18 | 0.0320 |
| 2 | 8 | 18 | 0.0462 | 2 | 13 | 18 | 0.0264 | 8 | 16 | 18 | 0.0356 |
| 3 | 8 | 18 | 0.0529 | 3 | 13 | 18 | 0.0303 | 9 | 16 | 18 | 0.0398 |
| 4 | 8 | 18 | 0.0594 | 4 | 13 | 18 | 0.0341 | 10 | 16 | 18 | 0.0447 |
| 5 | 8 | 18 | 0.0659 | 5 | 13 | 18 | 0.0379 | 11 | 16 | 18 | 0.0508 |
| 6 | 8 | 18 | 0.0729 | 6 | 13 | 18 | 0.0421 | 12 | 16 | 18 | 0.0585 |
| 7 | 8 | 18 | 0.0806 | 7 | 13 | 18 | 0.0466 | 13 | 16 | 18 | 0.0686 |
| 1 | 9 | 18 | 0.0338 | 8 | 13 | 18 | 0.0517 | 14 | 16 | 18 | 0.0827 |
| 2 | 9 | 18 | 0.0409 | 9 | 13 | 18 | 0.0576 | 15 | 16 | 18 | 0.1040 |
| 3 | 9 | 18 | 0.0469 | 10 | 13 | 18 | 0.0646 | 1 | 17 | 18 | 0.0119 |
| 4 | 9 | 18 | 0.0527 | 11 | 13 | 18 | 0.0730 | 2 | 17 | 18 | 0.0145 |
| 5 | 9 | 18 | 0.0585 | 12 | 13 | 18 | 0.0836 | 3 | 17 | 18 | 0.0168 |
| 6 | 9 | 18 | 0.0648 | 1 | 14 | 18 | 0.0194 | 4 | 17 | 18 | 0.0189 |
| 7 | 9 | 18 | 0.0716 | 2 | 14 | 18 | 0.0236 | 5 | 17 | 18 | 0.0212 |
| 8 | 9 | 18 | 0.0793 | 3 | 14 | 18 | 0.0270 | 6 | 17 | 18 | 0.0236 |
| 1 | 10 | 18 | 0.0302 | 4 | 14 | 18 | 0.0304 | 7 | 17 | 18 | 0.0262 |
| 2 | 10 | 18 | 0.0366 | 5 | 14 | 18 | 0.0339 | 8 | 17 | 18 | 0.0293 |
| 3 | 10 | 18 | 0.0419 | 6 | 14 | 18 | 0.0376 | 9 | 17 | 18 | 0.0328 |

TABLE 3

OUTLIER MEAN 1.5 OUTLIER STANDARD DEVIATION 1.0

| R | S | N | COVAR | R | S | N | COVAR | R | S | N | COVAR |
|---|---|---|---|---|---|---|---|---|---|---|---|
| 10 | 17 | 18 | 0.0371 | 1 | 8 | 19 | 0.0376 | 1 | 13 | 19 | 0.0220 |
| 11 | 17 | 18 | 0.0422 | 2 | 8 | 19 | 0.0454 | 2 | 13 | 19 | 0.0265 |
| 12 | 17 | 18 | 0.0488 | 3 | 8 | 19 | 0.0518 | 3 | 13 | 19 | 0.0303 |
| 13 | 17 | 18 | 0.0576 | 4 | 8 | 19 | 0.0579 | 4 | 13 | 19 | 0.0340 |
| 14 | 17 | 18 | 0.0698 | 5 | 8 | 19 | 0.0640 | 5 | 13 | 19 | 0.0376 |
| 15 | 17 | 18 | 0.0884 | 6 | 8 | 19 | 0.0705 | 6 | 13 | 19 | 0.0415 |
| 16 | 17 | 18 | 0.1205 | 7 | 8 | 19 | 0.0775 | 7 | 13 | 19 | 0.0458 |
| 1 | 18 | 18 | 0.0074 | 1 | 9 | 19 | 0.0334 | 8 | 13 | 19 | 0.0504 |
| 2 | 18 | 18 | 0.0090 | 2 | 9 | 19 | 0.0404 | 9 | 13 | 19 | 0.0557 |
| 3 | 18 | 18 | 0.0105 | 3 | 9 | 19 | 0.0461 | 10 | 13 | 19 | 0.0619 |
| 4 | 18 | 18 | 0.0119 | 4 | 9 | 19 | 0.0515 | 11 | 13 | 19 | 0.0691 |
| 5 | 18 | 18 | 0.0135 | 5 | 9 | 19 | 0.0570 | 12 | 13 | 19 | 0.0779 |
| 6 | 18 | 18 | 0.0151 | 6 | 9 | 19 | 0.0628 | 1 | 14 | 19 | 0.0198 |
| 7 | 18 | 18 | 0.0170 | 7 | 9 | 19 | 0.0690 | 2 | 14 | 19 | 0.0239 |
| 8 | 18 | 18 | 0.0192 | 8 | 9 | 19 | 0.0759 | 3 | 14 | 19 | 0.0274 |
| 9 | 18 | 18 | 0.0218 | 1 | 10 | 19 | 0.0299 | 4 | 14 | 19 | 0.0306 |
| 10 | 18 | 18 | 0.0249 | 2 | 10 | 19 | 0.0361 | 5 | 14 | 19 | 0.0340 |
| 11 | 18 | 18 | 0.0287 | 3 | 10 | 19 | 0.0413 | 6 | 14 | 19 | 0.0375 |
| 12 | 18 | 18 | 0.0337 | 4 | 10 | 19 | 0.0462 | 7 | 14 | 19 | 0.0413 |
| 13 | 18 | 18 | 0.0405 | 5 | 10 | 19 | 0.0511 | 8 | 14 | 19 | 0.0456 |
| 14 | 18 | 18 | 0.0501 | 6 | 10 | 19 | 0.0563 | 9 | 14 | 19 | 0.0504 |
| 15 | 18 | 18 | 0.0650 | 7 | 10 | 19 | 0.0620 | 10 | 14 | 19 | 0.0560 |
| 16 | 18 | 18 | 0.0914 | 8 | 10 | 19 | 0.0682 | 11 | 14 | 19 | 0.0626 |
| 17 | 18 | 18 | 0.1502 | 9 | 10 | 19 | 0.0752 | 12 | 14 | 19 | 0.0707 |
| 1 | 2 | 19 | 0.1390 | 1 | 11 | 19 | 0.0269 | 13 | 14 | 19 | 0.0808 |
| 1 | 3 | 19 | 0.0944 | 2 | 11 | 19 | 0.0325 | 1 | 15 | 19 | 0.0177 |
| 2 | 3 | 19 | 0.1131 | 3 | 11 | 19 | 0.0372 | 2 | 15 | 19 | 0.0214 |
| 1 | 4 | 19 | 0.0722 | 4 | 11 | 19 | 0.0416 | 3 | 15 | 19 | 0.0245 |
| 2 | 4 | 19 | 0.0868 | 5 | 11 | 19 | 0.0461 | 4 | 15 | 19 | 0.0275 |
| 3 | 4 | 19 | 0.0987 | 6 | 11 | 19 | 0.0508 | 5 | 15 | 19 | 0.0305 |
| 1 | 5 | 19 | 0.0588 | 7 | 11 | 19 | 0.0559 | 6 | 15 | 19 | 0.0337 |
| 2 | 5 | 19 | 0.0707 | 8 | 11 | 19 | 0.0615 | 7 | 15 | 19 | 0.0371 |
| 3 | 5 | 19 | 0.0805 | 9 | 11 | 19 | 0.0679 | 8 | 15 | 19 | 0.0410 |
| 4 | 5 | 19 | 0.0898 | 10 | 11 | 19 | 0.0753 | 9 | 15 | 19 | 0.0453 |
| 1 | 6 | 19 | 0.0496 | 1 | 12 | 19 | 0.0243 | 10 | 15 | 19 | 0.0504 |
| 2 | 6 | 19 | 0.0598 | 2 | 12 | 19 | 0.0294 | 11 | 15 | 19 | 0.0564 |
| 3 | 6 | 19 | 0.0681 | 3 | 12 | 19 | 0.0336 | 12 | 15 | 19 | 0.0637 |
| 4 | 6 | 19 | 0.0760 | 4 | 12 | 19 | 0.0376 | 13 | 15 | 19 | 0.0730 |
| 5 | 6 | 19 | 0.0840 | 5 | 12 | 19 | 0.0416 | 14 | 15 | 19 | 0.0851 |
| 1 | 7 | 19 | 0.0429 | 6 | 12 | 19 | 0.0459 | 1 | 16 | 19 | 0.0157 |
| 2 | 7 | 19 | 0.0517 | 7 | 12 | 19 | 0.0506 | 2 | 16 | 19 | 0.0190 |
| 3 | 7 | 19 | 0.0589 | 8 | 12 | 19 | 0.0557 | 3 | 16 | 19 | 0.0217 |
| 4 | 7 | 19 | 0.0658 | 9 | 12 | 19 | 0.0615 | 4 | 16 | 19 | 0.0244 |
| 5 | 7 | 19 | 0.0728 | 10 | 12 | 19 | 0.0682 | 5 | 16 | 19 | 0.0271 |
| 6 | 7 | 19 | 0.0801 | 11 | 12 | 19 | 0.0762 | 6 | 16 | 19 | 0.0299 |

TABLE 3

OUTLIER MEAN 1.5 OUTLIER STANDARD DEVIATION 1.0

| R | S | N | COVAR | R | S | N | COVAR | R | S | N | COVAR |
|---|---|---|---|---|---|---|---|---|---|---|---|
| 7 | 16 | 19 | 0.0330 | 4 | 19 | 19 | 0.0111 | 3 | 9 | 20 | 0.0452 |
| 8 | 16 | 19 | 0.0364 | 5 | 19 | 19 | 0.0124 | 4 | 9 | 20 | 0.0504 |
| 9 | 16 | 19 | 0.0404 | 6 | 19 | 19 | 0.0139 | 5 | 9 | 20 | 0.0556 |
| 10 | 16 | 19 | 0.0449 | 7 | 19 | 19 | 0.0155 | 6 | 9 | 20 | 0.0610 |
| 11 | 16 | 19 | 0.0503 | 8 | 19 | 19 | 0.0173 | 7 | 9 | 20 | 0.0667 |
| 12 | 16 | 19 | 0.0569 | 9 | 19 | 19 | 0.0194 | 8 | 9 | 20 | 0.0730 |
| 13 | 16 | 19 | 0.0653 | 10 | 19 | 19 | 0.0219 | 1 | 10 | 20 | 0.0297 |
| 14 | 16 | 19 | 0.0763 | 11 | 19 | 19 | 0.0250 | 2 | 10 | 20 | 0.0357 |
| 15 | 16 | 19 | 0.0915 | 12 | 19 | 19 | 0.0288 | 3 | 10 | 20 | 0.0406 |
| 1 | 17 | 19 | 0.0136 | 13 | 19 | 19 | 0.0337 | 4 | 10 | 20 | 0.0453 |
| 2 | 17 | 19 | 0.0164 | 14 | 19 | 19 | 0.0404 | 5 | 10 | 20 | 0.0499 |
| 3 | 17 | 19 | 0.0188 | 15 | 19 | 19 | 0.0499 | 6 | 10 | 20 | 0.0548 |
| 4 | 17 | 19 | 0.0211 | 16 | 19 | 19 | 0.0645 | 7 | 10 | 20 | 0.0600 |
| 5 | 17 | 19 | 0.0235 | 17 | 19 | 19 | 0.0904 | 8 | 10 | 20 | 0.0657 |
| 6 | 17 | 19 | 0.0260 | 18 | 19 | 19 | 0.1481 | 9 | 10 | 20 | 0.0720 |
| 7 | 17 | 19 | 0.0287 | 1 | 2 | 20 | 0.1365 | 1 | 11 | 20 | 0.0268 |
| 8 | 17 | 19 | 0.0317 | 1 | 3 | 20 | 0.0927 | 2 | 11 | 20 | 0.0322 |
| 9 | 17 | 19 | 0.0352 | 2 | 3 | 20 | 0.1107 | 3 | 11 | 20 | 0.0367 |
| 10 | 17 | 19 | 0.0392 | 1 | 4 | 20 | 0.0710 | 4 | 11 | 20 | 0.0409 |
| 11 | 17 | 19 | 0.0440 | 2 | 4 | 20 | 0.0850 | 5 | 11 | 20 | 0.0451 |
| 12 | 17 | 19 | 0.0499 | 3 | 4 | 20 | 0.0963 | 6 | 11 | 20 | 0.0496 |
| 13 | 17 | 19 | 0.0573 | 1 | 5 | 20 | 0.0578 | 7 | 11 | 20 | 0.0543 |
| 14 | 17 | 19 | 0.0671 | 2 | 5 | 20 | 0.0693 | 8 | 11 | 20 | 0.0595 |
| 15 | 17 | 19 | 0.0808 | 3 | 5 | 20 | 0.0786 | 9 | 11 | 20 | 0.0652 |
| 16 | 17 | 19 | 0.1015 | 4 | 5 | 20 | 0.0874 | 10 | 11 | 20 | 0.0717 |
| 1 | 18 | 19 | 0.0111 | 1 | 6 | 20 | 0.0488 | 1 | 12 | 20 | 0.0243 |
| 2 | 18 | 19 | 0.0134 | 2 | 6 | 20 | 0.0586 | 2 | 12 | 20 | 0.0292 |
| 3 | 18 | 19 | 0.0154 | 3 | 6 | 20 | 0.0665 | 3 | 12 | 20 | 0.0333 |
| 4 | 18 | 19 | 0.0173 | 4 | 6 | 20 | 0.0740 | 4 | 12 | 20 | 0.0371 |
| 5 | 18 | 19 | 0.0193 | 5 | 6 | 20 | 0.0815 | 5 | 12 | 20 | 0.0410 |
| 6 | 18 | 19 | 0.0214 | 1 | 7 | 20 | 0.0422 | 6 | 12 | 20 | 0.0450 |
| 7 | 18 | 19 | 0.0236 | 2 | 7 | 20 | 0.0507 | 7 | 12 | 20 | 0.0493 |
| 8 | 18 | 19 | 0.0262 | 3 | 7 | 20 | 0.0577 | 8 | 12 | 20 | 0.0540 |
| 9 | 18 | 19 | 0.0291 | 4 | 7 | 20 | 0.0642 | 9 | 12 | 20 | 0.0593 |
| 10 | 18 | 19 | 0.0326 | 5 | 7 | 20 | 0.0707 | 10 | 12 | 20 | 0.0652 |
| 11 | 18 | 19 | 0.0367 | 6 | 7 | 20 | 0.0775 | 11 | 12 | 20 | 0.0722 |
| 12 | 18 | 19 | 0.0417 | 1 | 8 | 20 | 0.0372 | 1 | 13 | 20 | 0.0220 |
| 13 | 18 | 19 | 0.0481 | 2 | 8 | 20 | 0.0447 | 2 | 13 | 20 | 0.0265 |
| 14 | 18 | 19 | 0.0566 | 3 | 8 | 20 | 0.0508 | 3 | 13 | 20 | 0.0302 |
| 15 | 18 | 19 | 0.0686 | 4 | 8 | 20 | 0.0565 | 4 | 13 | 20 | 0.0337 |
| 16 | 18 | 19 | 0.0868 | 5 | 8 | 20 | 0.0623 | 5 | 13 | 20 | 0.0372 |
| 17 | 18 | 19 | 0.1181 | 6 | 8 | 20 | 0.0683 | 6 | 13 | 20 | 0.0409 |
| 1 | 19 | 19 | 0.0069 | 7 | 8 | 20 | 0.0747 | 7 | 13 | 20 | 0.0448 |
| 2 | 19 | 19 | 0.0085 | 1 | 9 | 20 | 0.0331 | 8 | 13 | 20 | 0.0491 |
| 3 | 19 | 19 | 0.0098 | 2 | 9 | 20 | 0.0398 | 9 | 13 | 20 | 0.0539 |

TABLE 3

OUTLIER MEAN 1.5 OUTLIER STANDARD DEVIATION 1.0

| R | S | N | COVAR | R | S | N | COVAR | R | S | N | COVAR |
|---|---|---|---|---|---|---|---|---|---|---|---|
| 10 | 13 | 20 | 0.0594 | 10 | 16 | 20 | 0.0444 | 1 | 19 | 20 | 0.0103 |
| 11 | 13 | 20 | 0.0658 | 11 | 16 | 20 | 0.0492 | 2 | 19 | 20 | 0.0124 |
| 12 | 13 | 20 | 0.0733 | 12 | 16 | 20 | 0.0550 | 3 | 19 | 20 | 0.0142 |
| 1 | 14 | 20 | 0.0200 | 13 | 16 | 20 | 0.0621 | 4 | 19 | 20 | 0.0159 |
| 2 | 14 | 20 | 0.0241 | 14 | 16 | 20 | 0.0710 | 5 | 19 | 20 | 0.0177 |
| 3 | 14 | 20 | 0.0275 | 15 | 16 | 20 | 0.0827 | 6 | 19 | 20 | 0.0195 |
| 4 | 14 | 20 | 0.0306 | 1 | 17 | 20 | 0.0145 | 7 | 19 | 20 | 0.0215 |
| 5 | 14 | 20 | 0.0338 | 2 | 17 | 20 | 0.0175 | 8 | 19 | 20 | 0.0237 |
| 6 | 14 | 20 | 0.0372 | 3 | 17 | 20 | 0.0199 | 9 | 19 | 20 | 0.0261 |
| 7 | 14 | 20 | 0.0408 | 4 | 17 | 20 | 0.0222 | 10 | 19 | 20 | 0.0289 |
| 8 | 14 | 20 | 0.0447 | 5 | 17 | 20 | 0.0246 | 11 | 19 | 20 | 0.0323 |
| 9 | 14 | 20 | 0.0491 | 6 | 17 | 20 | 0.0271 | 12 | 19 | 20 | 0.0363 |
| 10 | 14 | 20 | 0.0541 | 7 | 17 | 20 | 0.0297 | 13 | 19 | 20 | 0.0412 |
| 11 | 14 | 20 | 0.0599 | 8 | 17 | 20 | 0.0327 | 14 | 19 | 20 | 0.0474 |
| 12 | 14 | 20 | 0.0668 | 9 | 17 | 20 | 0.0359 | 15 | 19 | 20 | 0.0558 |
| 13 | 14 | 20 | 0.0752 | 10 | 17 | 20 | 0.0397 | 16 | 19 | 20 | 0.0674 |
| 1 | 15 | 20 | 0.0181 | 11 | 17 | 20 | 0.0441 | 17 | 19 | 20 | 0.0852 |
| 2 | 15 | 20 | 0.0218 | 12 | 17 | 20 | 0.0493 | 18 | 19 | 20 | 0.1158 |
| 3 | 15 | 20 | 0.0249 | 13 | 17 | 20 | 0.0557 | 1 | 20 | 20 | 0.0065 |
| 4 | 15 | 20 | 0.0278 | 14 | 17 | 20 | 0.0637 | 2 | 20 | 20 | 0.0079 |
| 5 | 15 | 20 | 0.0307 | 15 | 17 | 20 | 0.0744 | 3 | 20 | 20 | 0.0091 |
| 6 | 15 | 20 | 0.0337 | 16 | 17 | 20 | 0.0892 | 4 | 20 | 20 | 0.0103 |
| 7 | 15 | 20 | 0.0370 | 1 | 18 | 20 | 0.0126 | 5 | 20 | 20 | 0.0115 |
| 8 | 15 | 20 | 0.0406 | 2 | 18 | 20 | 0.0152 | 6 | 20 | 20 | 0.0128 |
| 9 | 15 | 20 | 0.0446 | 3 | 18 | 20 | 0.0173 | 7 | 20 | 20 | 0.0142 |
| 10 | 15 | 20 | 0.0492 | 4 | 18 | 20 | 0.0194 | 8 | 20 | 20 | 0.0157 |
| 11 | 15 | 20 | 0.0545 | 5 | 18 | 20 | 0.0214 | 9 | 20 | 20 | 0.0175 |
| 12 | 15 | 20 | 0.0608 | 6 | 18 | 20 | 0.0236 | 10 | 20 | 20 | 0.0196 |
| 13 | 15 | 20 | 0.0685 | 7 | 18 | 20 | 0.0259 | 11 | 20 | 20 | 0.0221 |
| 14 | 15 | 20 | 0.0782 | 8 | 18 | 20 | 0.0285 | 12 | 20 | 20 | 0.0251 |
| 1 | 16 | 20 | 0.0163 | 9 | 18 | 20 | 0.0314 | 13 | 20 | 20 | 0.0288 |
| 2 | 16 | 20 | 0.0196 | 10 | 18 | 20 | 0.0347 | 14 | 20 | 20 | 0.0337 |
| 3 | 16 | 20 | 0.0224 | 11 | 18 | 20 | 0.0386 | 15 | 20 | 20 | 0.0402 |
| 4 | 16 | 20 | 0.0250 | 12 | 18 | 20 | 0.0433 | 16 | 20 | 20 | 0.0496 |
| 5 | 16 | 20 | 0.0276 | 13 | 18 | 20 | 0.0490 | 17 | 20 | 20 | 0.0641 |
| 6 | 16 | 20 | 0.0304 | 14 | 18 | 20 | 0.0562 | 18 | 20 | 20 | 0.0895 |
| 7 | 16 | 20 | 0.0334 | 15 | 18 | 20 | 0.0657 | 19 | 20 | 20 | 0.1462 |
| 8 | 16 | 20 | 0.0366 | 16 | 18 | 20 | 0.0791 | | | | |
| 9 | 16 | 20 | 0.0402 | 17 | 18 | 20 | 0.0992 | | | | |

TABLE 3

OUTLIER MEAN 2.0 OUTLIER STANDARD DEVIATION 1.0

| R | S | N | COVAR | R | S | N | COVAR | R | S | N | COVAR |
|---|---|---|---|---|---|---|---|---|---|---|---|
| 1 | 2 | 2 | 0.1030 | 1 | 6 | 7 | 0.0513 | 1 | 5 | 9 | 0.0738 |
| 1 | 2 | 3 | 0.2941 | 2 | 6 | 7 | 0.0707 | 2 | 5 | 9 | 0.0961 |
| 1 | 3 | 3 | 0.0437 | 3 | 6 | 7 | 0.0936 | 3 | 5 | 9 | 0.1190 |
| 2 | 3 | 3 | 0.1286 | 4 | 6 | 7 | 0.1269 | 4 | 5 | 9 | 0.1461 |
| 1 | 2 | 4 | 0.2678 | 5 | 6 | 7 | 0.1848 | 1 | 6 | 9 | 0.0588 |
| 1 | 3 | 4 | 0.1521 | 1 | 7 | 7 | 0.0136 | 2 | 6 | 9 | 0.0769 |
| 2 | 3 | 4 | 0.2500 | 2 | 7 | 7 | 0.0199 | 3 | 6 | 9 | 0.0953 |
| 1 | 4 | 4 | 0.0276 | 3 | 7 | 7 | 0.0284 | 4 | 6 | 9 | 0.1175 |
| 2 | 4 | 4 | 0.0562 | 4 | 7 | 7 | 0.0426 | 5 | 6 | 9 | 0.1469 |
| 3 | 4 | 4 | 0.1402 | 5 | 7 | 7 | 0.0712 | 1 | 7 | 9 | 0.0462 |
| 1 | 2 | 5 | 0.2421 | 6 | 7 | 7 | 0.1507 | 2 | 7 | 9 | 0.0606 |
| 1 | 3 | 5 | 0.1530 | 1 | 2 | 8 | 0.1954 | 3 | 7 | 9 | 0.0754 |
| 2 | 3 | 5 | 0.2266 | 1 | 3 | 8 | 0.1312 | 4 | 7 | 9 | 0.0932 |
| 1 | 4 | 5 | 0.0961 | 2 | 3 | 8 | 0.1729 | 5 | 7 | 9 | 0.1172 |
| 2 | 4 | 5 | 0.1446 | 1 | 4 | 8 | 0.0973 | 6 | 7 | 9 | 0.1529 |
| 3 | 4 | 5 | 0.2203 | 2 | 4 | 8 | 0.1289 | 1 | 8 | 9 | 0.0334 |
| 1 | 5 | 5 | 0.0204 | 3 | 4 | 8 | 0.1627 | 2 | 8 | 9 | 0.0440 |
| 2 | 5 | 5 | 0.0349 | 1 | 5 | 8 | 0.0749 | 3 | 8 | 9 | 0.0550 |
| 3 | 5 | 5 | 0.0635 | 2 | 5 | 8 | 0.0997 | 4 | 8 | 9 | 0.0684 |
| 4 | 5 | 5 | 0.1461 | 3 | 5 | 8 | 0.1263 | 5 | 8 | 9 | 0.0867 |
| 1 | 2 | 6 | 0.2224 | 4 | 5 | 8 | 0.1603 | 6 | 8 | 9 | 0.1145 |
| 1 | 3 | 6 | 0.1457 | 1 | 6 | 8 | 0.0575 | 7 | 8 | 9 | 0.1642 |
| 2 | 3 | 6 | 0.2040 | 2 | 6 | 8 | 0.0767 | 1 | 9 | 9 | 0.0103 |
| 1 | 4 | 6 | 0.1021 | 3 | 6 | 8 | 0.0977 | 2 | 9 | 9 | 0.0141 |
| 2 | 4 | 6 | 0.1442 | 4 | 6 | 8 | 0.1246 | 3 | 9 | 9 | 0.0183 |
| 3 | 4 | 6 | 0.1985 | 5 | 6 | 8 | 0.1642 | 4 | 9 | 9 | 0.0240 |
| 1 | 5 | 6 | 0.0678 | 1 | 7 | 8 | 0.0407 | 5 | 9 | 9 | 0.0324 |
| 2 | 5 | 6 | 0.0968 | 2 | 7 | 8 | 0.0546 | 6 | 9 | 9 | 0.0467 |
| 3 | 5 | 6 | 0.1351 | 3 | 7 | 8 | 0.0700 | 7 | 9 | 9 | 0.0750 |
| 4 | 5 | 6 | 0.1998 | 4 | 7 | 8 | 0.0900 | 8 | 9 | 9 | 0.1512 |
| 1 | 6 | 6 | 0.0163 | 5 | 7 | 8 | 0.1201 | 1 | 2 | 10 | 0.1778 |
| 2 | 6 | 6 | 0.0253 | 6 | 7 | 8 | 0.1733 | 1 | 3 | 10 | 0.1203 |
| 3 | 6 | 6 | 0.0395 | 1 | 8 | 8 | 0.0117 | 2 | 3 | 10 | 0.1534 |
| 4 | 6 | 6 | 0.0681 | 2 | 8 | 8 | 0.0164 | 1 | 4 | 10 | 0.0908 |
| 5 | 6 | 6 | 0.1492 | 3 | 8 | 8 | 0.0222 | 2 | 4 | 10 | 0.1162 |
| 1 | 2 | 7 | 0.2073 | 4 | 8 | 8 | 0.0307 | 3 | 4 | 10 | 0.1409 |
| 1 | 3 | 7 | 0.1380 | 5 | 8 | 8 | 0.0449 | 1 | 5 | 10 | 0.0721 |
| 2 | 3 | 7 | 0.1865 | 6 | 8 | 8 | 0.0734 | 2 | 5 | 10 | 0.0925 |
| 1 | 4 | 7 | 0.1005 | 7 | 8 | 8 | 0.1512 | 3 | 5 | 10 | 0.1124 |
| 2 | 4 | 7 | 0.1367 | 1 | 2 | 9 | 0.1858 | 4 | 5 | 10 | 0.1350 |
| 3 | 4 | 7 | 0.1783 | 1 | 3 | 9 | 0.1253 | 1 | 6 | 10 | 0.0586 |
| 1 | 5 | 7 | 0.0744 | 2 | 3 | 9 | 0.1622 | 2 | 6 | 10 | 0.0754 |
| 2 | 5 | 7 | 0.1018 | 1 | 4 | 9 | 0.0940 | 3 | 6 | 10 | 0.0918 |
| 3 | 5 | 7 | 0.1337 | 2 | 4 | 9 | 0.1221 | 4 | 6 | 10 | 0.1105 |
| 4 | 5 | 7 | 0.1788 | 3 | 4 | 9 | 0.1506 | 5 | 6 | 10 | 0.1338 |

TABLE 3

OUTLIER MEAN 2.0 OUTLIER STANDARD DEVIATION 1.0

| R | S | N | COVAR | R | S | N | COVAR | R | S | N | COVAR |
|---|---|---|---|---|---|---|---|---|---|---|---|
| 1 | 7 | 10 | 0.0479 | 1 | 7 | 11 | 0.0481 | 3 | 4 | 12 | 0.1263 |
| 2 | 7 | 10 | 0.0617 | 2 | 7 | 11 | 0.0612 | 1 | 5 | 12 | 0.0685 |
| 3 | 7 | 10 | 0.0752 | 3 | 7 | 11 | 0.0736 | 2 | 5 | 12 | 0.0859 |
| 4 | 7 | 10 | 0.0908 | 4 | 7 | 11 | 0.0873 | 3 | 5 | 12 | 0.1020 |
| 5 | 7 | 10 | 0.1102 | 5 | 7 | 11 | 0.1035 | 4 | 5 | 12 | 0.1189 |
| 6 | 7 | 10 | 0.1366 | 6 | 7 | 11 | 0.1242 | 1 | 6 | 12 | 0.0568 |
| 1 | 8 | 10 | 0.0383 | 1 | 8 | 11 | 0.0400 | 2 | 6 | 12 | 0.0713 |
| 2 | 8 | 10 | 0.0495 | 2 | 8 | 11 | 0.0509 | 3 | 6 | 12 | 0.0848 |
| 3 | 8 | 10 | 0.0605 | 3 | 8 | 11 | 0.0614 | 4 | 6 | 12 | 0.0990 |
| 4 | 8 | 10 | 0.0732 | 4 | 8 | 11 | 0.0729 | 5 | 6 | 12 | 0.1153 |
| 5 | 8 | 10 | 0.0891 | 5 | 8 | 11 | 0.0866 | 1 | 7 | 12 | 0.0478 |
| 6 | 8 | 10 | 0.1109 | 6 | 8 | 11 | 0.1041 | 2 | 7 | 12 | 0.0602 |
| 7 | 8 | 10 | 0.1439 | 7 | 8 | 11 | 0.1282 | 3 | 7 | 12 | 0.0716 |
| 1 | 9 | 10 | 0.0282 | 1 | 9 | 11 | 0.0325 | 4 | 7 | 12 | 0.0837 |
| 2 | 9 | 10 | 0.0365 | 2 | 9 | 11 | 0.0414 | 5 | 7 | 12 | 0.0977 |
| 3 | 9 | 10 | 0.0448 | 3 | 9 | 11 | 0.0500 | 6 | 7 | 12 | 0.1146 |
| 4 | 9 | 10 | 0.0544 | 4 | 9 | 11 | 0.0595 | 1 | 8 | 12 | 0.0405 |
| 5 | 9 | 10 | 0.0667 | 5 | 9 | 11 | 0.0709 | 2 | 8 | 12 | 0.0510 |
| 6 | 9 | 10 | 0.0837 | 6 | 9 | 11 | 0.0855 | 3 | 8 | 12 | 0.0608 |
| 7 | 9 | 10 | 0.1098 | 7 | 9 | 11 | 0.1057 | 4 | 8 | 12 | 0.0712 |
| 8 | 9 | 10 | 0.1567 | 8 | 9 | 11 | 0.1365 | 5 | 8 | 12 | 0.0831 |
| 1 | 10 | 10 | 0.0092 | 1 | 10 | 11 | 0.0242 | 6 | 8 | 12 | 0.0977 |
| 2 | 10 | 10 | 0.0123 | 2 | 10 | 11 | 0.0309 | 7 | 8 | 12 | 0.1165 |
| 3 | 10 | 10 | 0.0156 | 3 | 10 | 11 | 0.0374 | 1 | 9 | 12 | 0.0341 |
| 4 | 10 | 10 | 0.0197 | 4 | 10 | 11 | 0.0447 | 2 | 9 | 12 | 0.0430 |
| 5 | 10 | 10 | 0.0253 | 5 | 10 | 11 | 0.0535 | 3 | 9 | 12 | 0.0513 |
| 6 | 10 | 10 | 0.0338 | 6 | 10 | 11 | 0.0650 | 4 | 9 | 12 | 0.0602 |
| 7 | 10 | 10 | 0.0480 | 7 | 10 | 11 | 0.0810 | 5 | 9 | 12 | 0.0704 |
| 8 | 10 | 10 | 0.0760 | 8 | 10 | 11 | 0.1058 | 6 | 9 | 12 | 0.0829 |
| 9 | 10 | 10 | 0.1509 | 9 | 10 | 11 | 0.1504 | 7 | 9 | 12 | 0.0990 |
| 1 | 2 | 11 | 0.1710 | 1 | 11 | 11 | 0.0084 | 8 | 9 | 12 | 0.1214 |
| 1 | 3 | 11 | 0.1159 | 2 | 11 | 11 | 0.0109 | 1 | 10 | 12 | 0.0280 |
| 2 | 3 | 11 | 0.1461 | 3 | 11 | 11 | 0.0135 | 2 | 10 | 12 | 0.0354 |
| 1 | 4 | 11 | 0.0879 | 4 | 11 | 11 | 0.0167 | 3 | 10 | 12 | 0.0422 |
| 2 | 4 | 11 | 0.1111 | 5 | 11 | 11 | 0.0207 | 4 | 10 | 12 | 0.0496 |
| 3 | 4 | 11 | 0.1329 | 6 | 11 | 11 | 0.0264 | 5 | 10 | 12 | 0.0581 |
| 1 | 5 | 11 | 0.0703 | 7 | 11 | 11 | 0.0348 | 6 | 10 | 12 | 0.0686 |
| 2 | 5 | 11 | 0.0891 | 8 | 11 | 11 | 0.0490 | 7 | 10 | 12 | 0.0823 |
| 3 | 5 | 11 | 0.1068 | 9 | 11 | 11 | 0.0768 | 8 | 10 | 12 | 0.1013 |
| 4 | 5 | 11 | 0.1262 | 10 | 11 | 11 | 0.1503 | 9 | 10 | 12 | 0.1303 |
| 1 | 6 | 11 | 0.0578 | 1 | 2 | 12 | 0.1651 | 1 | 11 | 12 | 0.0211 |
| 2 | 6 | 11 | 0.0734 | 1 | 3 | 12 | 0.1121 | 2 | 11 | 12 | 0.0267 |
| 3 | 6 | 11 | 0.0882 | 2 | 3 | 12 | 0.1399 | 3 | 11 | 12 | 0.0320 |
| 4 | 6 | 11 | 0.1043 | 1 | 4 | 12 | 0.0852 | 4 | 11 | 12 | 0.0377 |
| 5 | 6 | 11 | 0.1235 | 2 | 4 | 12 | 0.1067 | 5 | 11 | 12 | 0.0443 |

TABLE 3

OUTLIER MEAN 2.0 OUTLIER STANDARD DEVIATION 1.0

| R | S | N | COVAR | R | S | N | COVAR | R | S | N | COVAR |
|---|---|---|---|---|---|---|---|---|---|---|---|
| 6 | 11 | 12 | 0.0525 | 2 | 9 | 13 | 0.0435 | 9 | 13 | 13 | 0.0364 |
| 7 | 11 | 12 | 0.0633 | 3 | 9 | 13 | 0.0513 | 10 | 13 | 13 | 0.0504 |
| 8 | 11 | 12 | 0.0786 | 4 | 9 | 13 | 0.0595 | 11 | 13 | 13 | 0.0777 |
| 9 | 11 | 12 | 0.1023 | 5 | 9 | 13 | 0.0687 | 12 | 13 | 13 | 0.1487 |
| 10 | 11 | 12 | 0.1450 | 6 | 9 | 13 | 0.0795 | 1 | 2 | 14 | 0.1556 |
| 1 | 12 | 12 | 0.0076 | 7 | 9 | 13 | 0.0928 | 1 | 3 | 14 | 0.1057 |
| 2 | 12 | 12 | 0.0098 | 8 | 9 | 13 | 0.1101 | 2 | 3 | 14 | 0.1300 |
| 3 | 12 | 12 | 0.0120 | 1 | 10 | 13 | 0.0296 | 1 | 4 | 14 | 0.0807 |
| 4 | 12 | 12 | 0.0145 | 2 | 10 | 13 | 0.0370 | 2 | 4 | 14 | 0.0995 |
| 5 | 12 | 12 | 0.0176 | 3 | 10 | 13 | 0.0438 | 3 | 4 | 14 | 0.1159 |
| 6 | 12 | 12 | 0.0216 | 4 | 10 | 13 | 0.0508 | 1 | 5 | 14 | 0.0652 |
| 7 | 12 | 12 | 0.0272 | 5 | 10 | 13 | 0.0587 | 2 | 5 | 14 | 0.0806 |
| 8 | 12 | 12 | 0.0357 | 6 | 10 | 13 | 0.0680 | 3 | 5 | 14 | 0.0941 |
| 9 | 12 | 12 | 0.0498 | 7 | 10 | 13 | 0.0796 | 4 | 5 | 14 | 0.1077 |
| 10 | 12 | 12 | 0.0774 | 8 | 10 | 13 | 0.0946 | 1 | 6 | 14 | 0.0546 |
| 11 | 12 | 12 | 0.1495 | 9 | 10 | 13 | 0.1156 | 2 | 6 | 14 | 0.0675 |
| 1 | 2 | 13 | 0.1601 | 1 | 11 | 13 | 0.0246 | 3 | 6 | 14 | 0.0789 |
| 1 | 3 | 13 | 0.1087 | 2 | 11 | 13 | 0.0307 | 4 | 6 | 14 | 0.0904 |
| 2 | 3 | 13 | 0.1346 | 3 | 11 | 13 | 0.0363 | 5 | 6 | 14 | 0.1030 |
| 1 | 4 | 13 | 0.0828 | 4 | 11 | 13 | 0.0422 | 1 | 7 | 14 | 0.0466 |
| 2 | 4 | 13 | 0.1029 | 5 | 11 | 13 | 0.0489 | 2 | 7 | 14 | 0.0576 |
| 3 | 4 | 13 | 0.1207 | 6 | 11 | 13 | 0.0568 | 3 | 7 | 14 | 0.0674 |
| 1 | 5 | 13 | 0.0668 | 7 | 11 | 13 | 0.0666 | 4 | 7 | 14 | 0.0774 |
| 2 | 5 | 13 | 0.0831 | 8 | 11 | 13 | 0.0794 | 5 | 7 | 14 | 0.0882 |
| 3 | 5 | 13 | 0.0977 | 9 | 11 | 13 | 0.0974 | 6 | 7 | 14 | 0.1006 |
| 4 | 5 | 13 | 0.1129 | 10 | 11 | 13 | 0.1250 | 1 | 8 | 14 | 0.0402 |
| 1 | 6 | 13 | 0.0557 | 1 | 12 | 13 | 0.0187 | 2 | 8 | 14 | 0.0498 |
| 2 | 6 | 13 | 0.0693 | 2 | 12 | 13 | 0.0234 | 3 | 8 | 14 | 0.0583 |
| 3 | 6 | 13 | 0.0817 | 3 | 12 | 13 | 0.0278 | 4 | 8 | 14 | 0.0670 |
| 4 | 6 | 13 | 0.0944 | 4 | 12 | 13 | 0.0324 | 5 | 8 | 14 | 0.0765 |
| 5 | 6 | 13 | 0.1086 | 5 | 12 | 13 | 0.0376 | 6 | 8 | 14 | 0.0873 |
| 1 | 7 | 13 | 0.0473 | 6 | 12 | 13 | 0.0438 | 7 | 8 | 14 | 0.1001 |
| 2 | 7 | 13 | 0.0589 | 7 | 12 | 13 | 0.0515 | 1 | 9 | 14 | 0.0350 |
| 3 | 7 | 13 | 0.0695 | 8 | 12 | 13 | 0.0619 | 2 | 9 | 14 | 0.0433 |
| 4 | 7 | 13 | 0.0804 | 9 | 12 | 13 | 0.0765 | 3 | 9 | 14 | 0.0507 |
| 5 | 7 | 13 | 0.0926 | 10 | 12 | 13 | 0.0992 | 4 | 9 | 14 | 0.0583 |
| 6 | 7 | 13 | 0.1069 | 11 | 12 | 13 | 0.1403 | 5 | 9 | 14 | 0.0666 |
| 1 | 8 | 13 | 0.0405 | 1 | 13 | 13 | 0.0070 | 6 | 9 | 14 | 0.0761 |
| 2 | 8 | 13 | 0.0506 | 2 | 13 | 13 | 0.0089 | 7 | 9 | 14 | 0.0874 |
| 3 | 8 | 13 | 0.0597 | 3 | 13 | 13 | 0.0108 | 8 | 9 | 14 | 0.1014 |
| 4 | 8 | 13 | 0.0691 | 4 | 13 | 13 | 0.0128 | 1 | 10 | 14 | 0.0303 |
| 5 | 8 | 13 | 0.0797 | 5 | 13 | 13 | 0.0152 | 2 | 10 | 14 | 0.0376 |
| 6 | 8 | 13 | 0.0921 | 6 | 13 | 13 | 0.0183 | 3 | 10 | 14 | 0.0441 |
| 7 | 8 | 13 | 0.1074 | 7 | 13 | 13 | 0.0223 | 4 | 10 | 14 | 0.0507 |
| 1 | 9 | 13 | 0.0348 | 8 | 13 | 13 | 0.0279 | 5 | 10 | 14 | 0.0580 |

TABLE 3

OUTLIER MEAN 2.0 OUTLIER STANDARD DEVIATION 1.0

| R | S | N | COVAR | R | S | N | COVAR | R | S | N | COVAR |
|---|---|---|---|---|---|---|---|---|---|---|---|
| 6 | 10 | 14 | 0.0663 | 9 | 14 | 14 | 0.0285 | 5 | 10 | 15 | 0.0568 |
| 7 | 10 | 14 | 0.0762 | 10 | 14 | 14 | 0.0369 | 6 | 10 | 15 | 0.0642 |
| 8 | 10 | 14 | 0.0886 | 11 | 14 | 14 | 0.0509 | 7 | 10 | 15 | 0.0729 |
| 9 | 10 | 14 | 0.1048 | 12 | 14 | 14 | 0.0779 | 8 | 10 | 15 | 0.0834 |
| 1 | 11 | 14 | 0.0261 | 13 | 14 | 14 | 0.1478 | 9 | 10 | 15 | 0.0964 |
| 2 | 11 | 14 | 0.0323 | 1 | 2 | 15 | 0.1516 | 1 | 11 | 15 | 0.0268 |
| 3 | 11 | 14 | 0.0379 | 1 | 3 | 15 | 0.1030 | 2 | 11 | 15 | 0.0330 |
| 4 | 11 | 14 | 0.0437 | 2 | 3 | 15 | 0.1259 | 3 | 11 | 15 | 0.0384 |
| 5 | 11 | 14 | 0.0499 | 1 | 4 | 15 | 0.0787 | 4 | 11 | 15 | 0.0439 |
| 6 | 11 | 14 | 0.0572 | 2 | 4 | 15 | 0.0964 | 5 | 11 | 15 | 0.0498 |
| 7 | 11 | 14 | 0.0658 | 3 | 4 | 15 | 0.1117 | 6 | 11 | 15 | 0.0564 |
| 8 | 11 | 14 | 0.0766 | 1 | 5 | 15 | 0.0638 | 7 | 11 | 15 | 0.0641 |
| 9 | 11 | 14 | 0.0908 | 2 | 5 | 15 | 0.0783 | 8 | 11 | 15 | 0.0733 |
| 10 | 11 | 14 | 0.1107 | 3 | 5 | 15 | 0.0908 | 9 | 11 | 15 | 0.0849 |
| 1 | 12 | 14 | 0.0218 | 4 | 5 | 15 | 0.1033 | 10 | 11 | 15 | 0.1002 |
| 2 | 12 | 14 | 0.0270 | 1 | 6 | 15 | 0.0535 | 1 | 12 | 15 | 0.0232 |
| 3 | 12 | 14 | 0.0317 | 2 | 6 | 15 | 0.0657 | 2 | 12 | 15 | 0.0286 |
| 4 | 12 | 14 | 0.0366 | 3 | 6 | 15 | 0.0763 | 3 | 12 | 15 | 0.0333 |
| 5 | 12 | 14 | 0.0419 | 4 | 6 | 15 | 0.0869 | 4 | 12 | 15 | 0.0381 |
| 6 | 12 | 14 | 0.0480 | 5 | 6 | 15 | 0.0982 | 5 | 12 | 15 | 0.0432 |
| 7 | 12 | 14 | 0.0554 | 1 | 7 | 15 | 0.0459 | 6 | 12 | 15 | 0.0490 |
| 8 | 12 | 14 | 0.0646 | 2 | 7 | 15 | 0.0564 | 7 | 12 | 15 | 0.0557 |
| 9 | 12 | 14 | 0.0769 | 3 | 7 | 15 | 0.0655 | 8 | 12 | 15 | 0.0638 |
| 10 | 12 | 14 | 0.0940 | 4 | 7 | 15 | 0.0747 | 9 | 12 | 15 | 0.0740 |
| 11 | 12 | 14 | 0.1205 | 5 | 7 | 15 | 0.0844 | 10 | 12 | 15 | 0.0875 |
| 1 | 13 | 14 | 0.0167 | 6 | 7 | 15 | 0.0953 | 11 | 12 | 15 | 0.1064 |
| 2 | 13 | 14 | 0.0208 | 1 | 8 | 15 | 0.0398 | 1 | 13 | 15 | 0.0195 |
| 3 | 13 | 14 | 0.0244 | 2 | 8 | 15 | 0.0490 | 2 | 13 | 15 | 0.0241 |
| 4 | 13 | 14 | 0.0282 | 3 | 8 | 15 | 0.0570 | 3 | 13 | 15 | 0.0281 |
| 5 | 13 | 14 | 0.0324 | 4 | 8 | 15 | 0.0650 | 4 | 13 | 15 | 0.0321 |
| 6 | 13 | 14 | 0.0373 | 5 | 8 | 15 | 0.0735 | 5 | 13 | 15 | 0.0365 |
| 7 | 13 | 14 | 0.0431 | 6 | 8 | 15 | 0.0831 | 6 | 13 | 15 | 0.0414 |
| 8 | 13 | 14 | 0.0506 | 7 | 8 | 15 | 0.0942 | 7 | 13 | 15 | 0.0471 |
| 9 | 13 | 14 | 0.0605 | 1 | 9 | 15 | 0.0349 | 8 | 13 | 15 | 0.0541 |
| 10 | 13 | 14 | 0.0746 | 2 | 9 | 15 | 0.0429 | 9 | 13 | 15 | 0.0629 |
| 11 | 13 | 14 | 0.0965 | 3 | 9 | 15 | 0.0499 | 10 | 13 | 15 | 0.0746 |
| 12 | 13 | 14 | 0.1362 | 4 | 9 | 15 | 0.0570 | 11 | 13 | 15 | 0.0910 |
| 1 | 14 | 14 | 0.0065 | 5 | 9 | 15 | 0.0645 | 12 | 13 | 15 | 0.1164 |
| 2 | 14 | 14 | 0.0082 | 6 | 9 | 15 | 0.0730 | 1 | 14 | 15 | 0.0151 |
| 3 | 14 | 14 | 0.0098 | 7 | 9 | 15 | 0.0828 | 2 | 14 | 15 | 0.0186 |
| 4 | 14 | 14 | 0.0115 | 8 | 9 | 15 | 0.0945 | 3 | 14 | 15 | 0.0218 |
| 5 | 14 | 14 | 0.0135 | 1 | 10 | 15 | 0.0306 | 4 | 14 | 15 | 0.0249 |
| 6 | 14 | 14 | 0.0158 | 2 | 10 | 15 | 0.0377 | 5 | 14 | 15 | 0.0284 |
| 7 | 14 | 14 | 0.0189 | 3 | 10 | 15 | 0.0439 | 6 | 14 | 15 | 0.0323 |
| 8 | 14 | 14 | 0.0229 | 4 | 10 | 15 | 0.0501 | 7 | 14 | 15 | 0.0369 |

TABLE 3

OUTLIER MEAN 2.0 OUTLIER STANDARD DEVIATION 1.0

| R | S | N | COVAR | R | S | N | COVAR | R | S | N | COVAR |
|---|---|---|---|---|---|---|---|---|---|---|---|
| 8 | 14 | 15 | 0.0425 | 5 | 8 | 16 | 0.0709 | 5 | 13 | 16 | 0.0379 |
| 9 | 14 | 15 | 0.0497 | 6 | 8 | 16 | 0.0795 | 6 | 13 | 16 | 0.0426 |
| 10 | 14 | 15 | 0.0592 | 7 | 8 | 16 | 0.0892 | 7 | 13 | 16 | 0.0480 |
| 11 | 14 | 15 | 0.0729 | 1 | 9 | 16 | 0.0346 | 8 | 13 | 16 | 0.0543 |
| 12 | 14 | 15 | 0.0941 | 2 | 9 | 16 | 0.0424 | 9 | 13 | 16 | 0.0620 |
| 13 | 14 | 15 | 0.1326 | 3 | 9 | 16 | 0.0490 | 10 | 13 | 16 | 0.0717 |
| 1 | 15 | 15 | 0.0061 | 4 | 9 | 16 | 0.0556 | 11 | 13 | 16 | 0.0846 |
| 2 | 15 | 15 | 0.0076 | 5 | 9 | 16 | 0.0625 | 12 | 13 | 16 | 0.1027 |
| 3 | 15 | 15 | 0.0090 | 6 | 9 | 16 | 0.0701 | 1 | 14 | 16 | 0.0176 |
| 4 | 15 | 15 | 0.0104 | 7 | 9 | 16 | 0.0788 | 2 | 14 | 16 | 0.0216 |
| 5 | 15 | 15 | 0.0121 | 8 | 9 | 16 | 0.0889 | 3 | 14 | 16 | 0.0251 |
| 6 | 15 | 15 | 0.0140 | 1 | 10 | 16 | 0.0306 | 4 | 14 | 16 | 0.0285 |
| 7 | 15 | 15 | 0.0163 | 2 | 10 | 16 | 0.0375 | 5 | 14 | 16 | 0.0321 |
| 8 | 15 | 15 | 0.0193 | 3 | 10 | 16 | 0.0434 | 6 | 14 | 16 | 0.0362 |
| 9 | 15 | 15 | 0.0233 | 4 | 10 | 16 | 0.0493 | 7 | 14 | 16 | 0.0408 |
| 10 | 15 | 15 | 0.0289 | 5 | 10 | 16 | 0.0554 | 8 | 14 | 16 | 0.0462 |
| 11 | 15 | 15 | 0.0374 | 6 | 10 | 16 | 0.0622 | 9 | 14 | 16 | 0.0529 |
| 12 | 15 | 15 | 0.0513 | 7 | 10 | 16 | 0.0699 | 10 | 14 | 16 | 0.0613 |
| 13 | 15 | 15 | 0.0781 | 8 | 10 | 16 | 0.0789 | 11 | 14 | 16 | 0.0725 |
| 14 | 15 | 15 | 0.1468 | 9 | 10 | 16 | 0.0898 | 12 | 14 | 16 | 0.0884 |
| 1 | 2 | 16 | 0.1480 | 1 | 11 | 16 | 0.0271 | 13 | 14 | 16 | 0.1129 |
| 1 | 3 | 16 | 0.1006 | 2 | 11 | 16 | 0.0332 | 1 | 15 | 16 | 0.0137 |
| 2 | 3 | 16 | 0.1222 | 3 | 11 | 16 | 0.0385 | 2 | 15 | 16 | 0.0168 |
| 1 | 4 | 16 | 0.0769 | 4 | 11 | 16 | 0.0437 | 3 | 15 | 16 | 0.0196 |
| 2 | 4 | 16 | 0.0937 | 5 | 11 | 16 | 0.0492 | 4 | 15 | 16 | 0.0223 |
| 3 | 4 | 16 | 0.1080 | 6 | 11 | 16 | 0.0552 | 5 | 15 | 16 | 0.0252 |
| 1 | 5 | 16 | 0.0624 | 7 | 11 | 16 | 0.0621 | 6 | 15 | 16 | 0.0284 |
| 2 | 5 | 16 | 0.0762 | 8 | 11 | 16 | 0.0701 | 7 | 15 | 16 | 0.0321 |
| 3 | 5 | 16 | 0.0879 | 9 | 11 | 16 | 0.0799 | 8 | 15 | 16 | 0.0365 |
| 4 | 5 | 16 | 0.0994 | 10 | 11 | 16 | 0.0921 | 9 | 15 | 16 | 0.0419 |
| 1 | 6 | 16 | 0.0525 | 1 | 12 | 16 | 0.0239 | 10 | 15 | 16 | 0.0488 |
| 2 | 6 | 16 | 0.0641 | 2 | 12 | 16 | 0.0293 | 11 | 15 | 16 | 0.0581 |
| 3 | 6 | 16 | 0.0740 | 3 | 12 | 16 | 0.0339 | 12 | 15 | 16 | 0.0713 |
| 4 | 6 | 16 | 0.0838 | 4 | 12 | 16 | 0.0385 | 13 | 15 | 16 | 0.0920 |
| 5 | 6 | 16 | 0.0940 | 5 | 12 | 16 | 0.0434 | 14 | 15 | 16 | 0.1294 |
| 1 | 7 | 16 | 0.0451 | 6 | 12 | 16 | 0.0488 | 1 | 16 | 16 | 0.0057 |
| 2 | 7 | 16 | 0.0552 | 7 | 12 | 16 | 0.0549 | 2 | 16 | 16 | 0.0070 |
| 3 | 7 | 16 | 0.0637 | 8 | 12 | 16 | 0.0620 | 3 | 16 | 16 | 0.0083 |
| 4 | 7 | 16 | 0.0722 | 9 | 12 | 16 | 0.0708 | 4 | 16 | 16 | 0.0095 |
| 5 | 7 | 16 | 0.0811 | 10 | 12 | 16 | 0.0817 | 5 | 16 | 16 | 0.0109 |
| 6 | 7 | 16 | 0.0908 | 11 | 12 | 16 | 0.0962 | 6 | 16 | 16 | 0.0125 |
| 1 | 8 | 16 | 0.0393 | 1 | 13 | 16 | 0.0208 | 7 | 16 | 16 | 0.0144 |
| 2 | 8 | 16 | 0.0481 | 2 | 13 | 16 | 0.0255 | 8 | 16 | 16 | 0.0168 |
| 3 | 8 | 16 | 0.0557 | 3 | 13 | 16 | 0.0296 | 9 | 16 | 16 | 0.0198 |
| 4 | 8 | 16 | 0.0631 | 4 | 13 | 16 | 0.0336 | 10 | 16 | 16 | 0.0238 |

TABLE 3

OUTLIER MEAN 2.0 OUTLIER STANDARD DEVIATION 1.0

| R | S | N | COVAR | R | S | N | COVAR | R | S | N | COVAR |
|---|---|---|---|---|---|---|---|---|---|---|---|
| 11 | 16 | 16 | 0.0293 | 5 | 10 | 17 | 0.0541 | 8 | 14 | 17 | 0.0470 |
| 12 | 16 | 16 | 0.0377 | 6 | 10 | 17 | 0.0603 | 9 | 14 | 17 | 0.0530 |
| 13 | 16 | 16 | 0.0516 | 7 | 10 | 17 | 0.0672 | 10 | 14 | 17 | 0.0603 |
| 14 | 16 | 16 | 0.0781 | 8 | 10 | 17 | 0.0751 | 11 | 14 | 17 | 0.0696 |
| 15 | 16 | 16 | 0.1458 | 9 | 10 | 17 | 0.0844 | 12 | 14 | 17 | 0.0820 |
| 1 | 2 | 17 | 0.1448 | 1 | 11 | 17 | 0.0272 | 13 | 14 | 17 | 0.0994 |
| 1 | 3 | 17 | 0.0984 | 2 | 11 | 17 | 0.0332 | 1 | 15 | 17 | 0.0160 |
| 2 | 3 | 17 | 0.1190 | 3 | 11 | 17 | 0.0382 | 2 | 15 | 17 | 0.0196 |
| 1 | 4 | 17 | 0.0753 | 4 | 11 | 17 | 0.0432 | 3 | 15 | 17 | 0.0226 |
| 2 | 4 | 17 | 0.0913 | 5 | 11 | 17 | 0.0483 | 4 | 15 | 17 | 0.0255 |
| 3 | 4 | 17 | 0.1047 | 6 | 11 | 17 | 0.0538 | 5 | 15 | 17 | 0.0286 |
| 1 | 5 | 17 | 0.0612 | 7 | 11 | 17 | 0.0600 | 6 | 15 | 17 | 0.0320 |
| 2 | 5 | 17 | 0.0743 | 8 | 11 | 17 | 0.0672 | 7 | 15 | 17 | 0.0358 |
| 3 | 5 | 17 | 0.0853 | 9 | 11 | 17 | 0.0756 | 8 | 15 | 17 | 0.0402 |
| 4 | 5 | 17 | 0.0960 | 10 | 11 | 17 | 0.0858 | 9 | 15 | 17 | 0.0454 |
| 1 | 6 | 17 | 0.0515 | 1 | 12 | 17 | 0.0243 | 10 | 15 | 17 | 0.0517 |
| 2 | 6 | 17 | 0.0626 | 2 | 12 | 17 | 0.0296 | 11 | 15 | 17 | 0.0599 |
| 3 | 6 | 17 | 0.0720 | 3 | 12 | 17 | 0.0341 | 12 | 15 | 17 | 0.0707 |
| 4 | 6 | 17 | 0.0811 | 4 | 12 | 17 | 0.0385 | 13 | 15 | 17 | 0.0860 |
| 5 | 6 | 17 | 0.0904 | 5 | 12 | 17 | 0.0431 | 14 | 15 | 17 | 0.1097 |
| 1 | 7 | 17 | 0.0444 | 6 | 12 | 17 | 0.0481 | 1 | 16 | 17 | 0.0126 |
| 2 | 7 | 17 | 0.0540 | 7 | 12 | 17 | 0.0537 | 2 | 16 | 17 | 0.0153 |
| 3 | 7 | 17 | 0.0621 | 8 | 12 | 17 | 0.0601 | 3 | 16 | 17 | 0.0177 |
| 4 | 7 | 17 | 0.0700 | 9 | 12 | 17 | 0.0676 | 4 | 16 | 17 | 0.0201 |
| 5 | 7 | 17 | 0.0782 | 10 | 12 | 17 | 0.0768 | 5 | 16 | 17 | 0.0225 |
| 6 | 7 | 17 | 0.0870 | 11 | 12 | 17 | 0.0884 | 6 | 16 | 17 | 0.0252 |
| 1 | 8 | 17 | 0.0388 | 1 | 13 | 17 | 0.0215 | 7 | 16 | 17 | 0.0283 |
| 2 | 8 | 17 | 0.0473 | 2 | 13 | 17 | 0.0262 | 8 | 16 | 17 | 0.0318 |
| 3 | 8 | 17 | 0.0544 | 3 | 13 | 17 | 0.0303 | 9 | 16 | 17 | 0.0361 |
| 4 | 8 | 17 | 0.0613 | 4 | 13 | 17 | 0.0342 | 10 | 16 | 17 | 0.0413 |
| 5 | 8 | 17 | 0.0685 | 5 | 13 | 17 | 0.0383 | 11 | 16 | 17 | 0.0480 |
| 6 | 8 | 17 | 0.0763 | 6 | 13 | 17 | 0.0427 | 12 | 16 | 17 | 0.0570 |
| 7 | 8 | 17 | 0.0849 | 7 | 13 | 17 | 0.0477 | 13 | 16 | 17 | 0.0699 |
| 1 | 9 | 17 | 0.0343 | 8 | 13 | 17 | 0.0534 | 14 | 16 | 17 | 0.0900 |
| 2 | 9 | 17 | 0.0418 | 9 | 13 | 17 | 0.0602 | 15 | 16 | 17 | 0.1264 |
| 3 | 9 | 17 | 0.0481 | 10 | 13 | 17 | 0.0685 | 1 | 17 | 17 | 0.0053 |
| 4 | 9 | 17 | 0.0543 | 11 | 13 | 17 | 0.0789 | 2 | 17 | 17 | 0.0066 |
| 5 | 9 | 17 | 0.0607 | 12 | 13 | 17 | 0.0927 | 3 | 17 | 17 | 0.0077 |
| 6 | 9 | 17 | 0.0676 | 1 | 14 | 17 | 0.0188 | 4 | 17 | 17 | 0.0088 |
| 7 | 9 | 17 | 0.0753 | 2 | 14 | 17 | 0.0230 | 5 | 17 | 17 | 0.0100 |
| 8 | 9 | 17 | 0.0841 | 3 | 14 | 17 | 0.0265 | 6 | 17 | 17 | 0.0113 |
| 1 | 10 | 17 | 0.0305 | 4 | 14 | 17 | 0.0300 | 7 | 17 | 17 | 0.0129 |
| 2 | 10 | 17 | 0.0372 | 5 | 14 | 17 | 0.0336 | 8 | 17 | 17 | 0.0148 |
| 3 | 10 | 17 | 0.0428 | 6 | 14 | 17 | 0.0375 | 9 | 17 | 17 | 0.0171 |
| 4 | 10 | 17 | 0.0483 | 7 | 14 | 17 | 0.0419 | 10 | 17 | 17 | 0.0201 |

TABLE 3

OUTLIER MEAN 2.0 OUTLIER STANDARD DEVIATION 1.0

| R | S | N | COVAR | R | S | N | COVAR | R | S | N | COVAR |
|---|---|---|---|---|---|---|---|---|---|---|---|
| 11 | 17 | 17 | 0.0241 | 4 | 10 | 18 | 0.0474 | 7 | 14 | 18 | 0.0420 |
| 12 | 17 | 17 | 0.0297 | 5 | 10 | 18 | 0.0527 | 8 | 14 | 18 | 0.0467 |
| 13 | 17 | 17 | 0.0380 | 6 | 10 | 18 | 0.0584 | 9 | 14 | 18 | 0.0521 |
| 14 | 17 | 17 | 0.0518 | 7 | 10 | 18 | 0.0647 | 10 | 14 | 18 | 0.0585 |
| 15 | 17 | 17 | 0.0781 | 8 | 10 | 18 | 0.0717 | 11 | 14 | 18 | 0.0664 |
| 16 | 17 | 17 | 0.1449 | 9 | 10 | 18 | 0.0799 | 12 | 14 | 18 | 0.0764 |
| 1 | 2 | 18 | 0.1419 | 1 | 11 | 18 | 0.0272 | 13 | 14 | 18 | 0.0896 |
| 1 | 3 | 18 | 0.0964 | 2 | 11 | 18 | 0.0330 | 1 | 15 | 18 | 0.0172 |
| 2 | 3 | 18 | 0.1161 | 3 | 11 | 18 | 0.0379 | 2 | 15 | 18 | 0.0209 |
| 1 | 4 | 18 | 0.0738 | 4 | 11 | 18 | 0.0425 | 3 | 15 | 18 | 0.0240 |
| 2 | 4 | 18 | 0.0891 | 5 | 11 | 18 | 0.0473 | 4 | 15 | 18 | 0.0270 |
| 3 | 4 | 18 | 0.1017 | 6 | 11 | 18 | 0.0525 | 5 | 15 | 18 | 0.0300 |
| 1 | 5 | 18 | 0.0600 | 7 | 11 | 18 | 0.0581 | 6 | 15 | 18 | 0.0334 |
| 2 | 5 | 18 | 0.0726 | 8 | 11 | 18 | 0.0645 | 7 | 15 | 18 | 0.0370 |
| 3 | 5 | 18 | 0.0830 | 9 | 11 | 18 | 0.0719 | 8 | 15 | 18 | 0.0412 |
| 4 | 5 | 18 | 0.0929 | 10 | 11 | 18 | 0.0806 | 9 | 15 | 18 | 0.0460 |
| 1 | 6 | 18 | 0.0506 | 1 | 12 | 18 | 0.0244 | 10 | 15 | 18 | 0.0517 |
| 2 | 6 | 18 | 0.0613 | 2 | 12 | 18 | 0.0296 | 11 | 15 | 18 | 0.0588 |
| 3 | 6 | 18 | 0.0701 | 3 | 12 | 18 | 0.0340 | 12 | 15 | 18 | 0.0677 |
| 4 | 6 | 18 | 0.0786 | 4 | 12 | 18 | 0.0382 | 13 | 15 | 18 | 0.0796 |
| 5 | 6 | 18 | 0.0873 | 5 | 12 | 18 | 0.0426 | 14 | 15 | 18 | 0.0964 |
| 1 | 7 | 18 | 0.0437 | 6 | 12 | 18 | 0.0472 | 1 | 16 | 18 | 0.0147 |
| 2 | 7 | 18 | 0.0529 | 7 | 12 | 18 | 0.0523 | 2 | 16 | 18 | 0.0178 |
| 3 | 7 | 18 | 0.0606 | 8 | 12 | 18 | 0.0581 | 3 | 16 | 18 | 0.0205 |
| 4 | 7 | 18 | 0.0680 | 9 | 12 | 18 | 0.0647 | 4 | 16 | 18 | 0.0231 |
| 5 | 7 | 18 | 0.0755 | 10 | 12 | 18 | 0.0726 | 5 | 16 | 18 | 0.0257 |
| 6 | 7 | 18 | 0.0836 | 11 | 12 | 18 | 0.0823 | 6 | 16 | 18 | 0.0286 |
| 1 | 8 | 18 | 0.0383 | 1 | 13 | 18 | 0.0219 | 7 | 16 | 18 | 0.0318 |
| 2 | 8 | 18 | 0.0464 | 2 | 13 | 18 | 0.0266 | 8 | 16 | 18 | 0.0353 |
| 3 | 8 | 18 | 0.0532 | 3 | 13 | 18 | 0.0305 | 9 | 16 | 18 | 0.0395 |
| 4 | 8 | 18 | 0.0597 | 4 | 13 | 18 | 0.0343 | 10 | 16 | 18 | 0.0445 |
| 5 | 8 | 18 | 0.0664 | 5 | 13 | 18 | 0.0382 | 11 | 16 | 18 | 0.0507 |
| 6 | 8 | 18 | 0.0735 | 6 | 13 | 18 | 0.0424 | 12 | 16 | 18 | 0.0585 |
| 7 | 8 | 18 | 0.0813 | 7 | 13 | 18 | 0.0470 | 13 | 16 | 18 | 0.0690 |
| 1 | 9 | 18 | 0.0340 | 8 | 13 | 18 | 0.0522 | 14 | 16 | 18 | 0.0838 |
| 2 | 9 | 18 | 0.0412 | 9 | 13 | 18 | 0.0582 | 15 | 16 | 18 | 0.1069 |
| 3 | 9 | 18 | 0.0472 | 10 | 13 | 18 | 0.0654 | 1 | 17 | 18 | 0.0116 |
| 4 | 9 | 18 | 0.0530 | 11 | 13 | 18 | 0.0741 | 2 | 17 | 18 | 0.0141 |
| 5 | 9 | 18 | 0.0590 | 12 | 13 | 18 | 0.0852 | 3 | 17 | 18 | 0.0162 |
| 6 | 9 | 18 | 0.0653 | 1 | 14 | 18 | 0.0195 | 4 | 17 | 18 | 0.0182 |
| 7 | 9 | 18 | 0.0723 | 2 | 14 | 18 | 0.0237 | 5 | 17 | 18 | 0.0204 |
| 8 | 9 | 18 | 0.0801 | 3 | 14 | 18 | 0.0272 | 6 | 17 | 18 | 0.0227 |
| 1 | 10 | 18 | 0.0303 | 4 | 14 | 18 | 0.0306 | 7 | 17 | 18 | 0.0252 |
| 2 | 10 | 18 | 0.0368 | 5 | 14 | 18 | 0.0341 | 8 | 17 | 18 | 0.0281 |
| 3 | 10 | 18 | 0.0422 | 6 | 14 | 18 | 0.0379 | 9 | 17 | 18 | 0.0315 |

TABLE 3

OUTLIER MEAN 2.0 OUTLIER STANDARD DEVIATION 1.0

| R | S | N | COVAR | R | S | N | COVAR | R | S | N | COVAR |
|---|---|---|---|---|---|---|---|---|---|---|---|
| 10 | 17 | 18 | 0.0356 | 1 | 8 | 19 | 0.0378 | 1 | 13 | 19 | 0.0221 |
| 11 | 17 | 18 | 0.0407 | 2 | 8 | 19 | 0.0456 | 2 | 13 | 19 | 0.0267 |
| 12 | 17 | 18 | 0.0472 | 3 | 8 | 19 | 0.0521 | 3 | 13 | 19 | 0.0305 |
| 13 | 17 | 18 | 0.0560 | 4 | 8 | 19 | 0.0582 | 4 | 13 | 19 | 0.0342 |
| 14 | 17 | 18 | 0.0686 | 5 | 8 | 19 | 0.0644 | 5 | 13 | 19 | 0.0379 |
| 15 | 17 | 18 | 0.0882 | 6 | 8 | 19 | 0.0710 | 6 | 13 | 19 | 0.0419 |
| 16 | 17 | 18 | 0.1238 | 7 | 8 | 19 | 0.0781 | 7 | 13 | 19 | 0.0461 |
| 1 | 18 | 18 | 0.0050 | 1 | 9 | 19 | 0.0336 | 8 | 13 | 19 | 0.0509 |
| 2 | 18 | 18 | 0.0062 | 2 | 9 | 19 | 0.0406 | 9 | 13 | 19 | 0.0563 |
| 3 | 18 | 18 | 0.0072 | 3 | 9 | 19 | 0.0463 | 10 | 13 | 19 | 0.0626 |
| 4 | 18 | 18 | 0.0081 | 4 | 9 | 19 | 0.0518 | 11 | 13 | 19 | 0.0700 |
| 5 | 18 | 18 | 0.0092 | 5 | 9 | 19 | 0.0574 | 12 | 13 | 19 | 0.0792 |
| 6 | 18 | 18 | 0.0104 | 6 | 9 | 19 | 0.0632 | 1 | 14 | 19 | 0.0199 |
| 7 | 18 | 18 | 0.0117 | 7 | 9 | 19 | 0.0696 | 2 | 14 | 19 | 0.0241 |
| 8 | 18 | 18 | 0.0132 | 8 | 9 | 19 | 0.0766 | 3 | 14 | 19 | 0.0275 |
| 9 | 18 | 18 | 0.0151 | 1 | 10 | 19 | 0.0301 | 4 | 14 | 19 | 0.0308 |
| 10 | 18 | 18 | 0.0174 | 2 | 10 | 19 | 0.0363 | 5 | 14 | 19 | 0.0342 |
| 11 | 18 | 18 | 0.0204 | 3 | 10 | 19 | 0.0415 | 6 | 14 | 19 | 0.0378 |
| 12 | 18 | 18 | 0.0244 | 4 | 10 | 19 | 0.0464 | 7 | 14 | 19 | 0.0416 |
| 13 | 18 | 18 | 0.0300 | 5 | 10 | 19 | 0.0515 | 8 | 14 | 19 | 0.0459 |
| 14 | 18 | 18 | 0.0383 | 6 | 10 | 19 | 0.0567 | 9 | 14 | 19 | 0.0508 |
| 15 | 18 | 18 | 0.0519 | 7 | 10 | 19 | 0.0625 | 10 | 14 | 19 | 0.0566 |
| 16 | 18 | 18 | 0.0780 | 8 | 10 | 19 | 0.0688 | 11 | 14 | 19 | 0.0634 |
| 17 | 18 | 18 | 0.1439 | 9 | 10 | 19 | 0.0760 | 12 | 14 | 19 | 0.0717 |
| 1 | 2 | 19 | 0.1392 | 1 | 11 | 19 | 0.0271 | 13 | 14 | 19 | 0.0823 |
| 1 | 3 | 19 | 0.0946 | 2 | 11 | 19 | 0.0327 | 1 | 15 | 19 | 0.0178 |
| 2 | 3 | 19 | 0.1134 | 3 | 11 | 19 | 0.0374 | 2 | 15 | 19 | 0.0216 |
| 1 | 4 | 19 | 0.0724 | 4 | 11 | 19 | 0.0419 | 3 | 15 | 19 | 0.0247 |
| 2 | 4 | 19 | 0.0871 | 5 | 11 | 19 | 0.0464 | 4 | 15 | 19 | 0.0276 |
| 3 | 4 | 19 | 0.0990 | 6 | 11 | 19 | 0.0512 | 5 | 15 | 19 | 0.0307 |
| 1 | 5 | 19 | 0.0589 | 7 | 11 | 19 | 0.0564 | 6 | 15 | 19 | 0.0339 |
| 2 | 5 | 19 | 0.0710 | 8 | 11 | 19 | 0.0621 | 7 | 15 | 19 | 0.0373 |
| 3 | 5 | 19 | 0.0808 | 9 | 11 | 19 | 0.0686 | 8 | 15 | 19 | 0.0412 |
| 4 | 5 | 19 | 0.0902 | 10 | 11 | 19 | 0.0762 | 9 | 15 | 19 | 0.0457 |
| 1 | 6 | 19 | 0.0498 | 1 | 12 | 19 | 0.0245 | 10 | 15 | 19 | 0.0508 |
| 2 | 6 | 19 | 0.0600 | 2 | 12 | 19 | 0.0296 | 11 | 15 | 19 | 0.0570 |
| 3 | 6 | 19 | 0.0684 | 3 | 12 | 19 | 0.0338 | 12 | 15 | 19 | 0.0645 |
| 4 | 6 | 19 | 0.0764 | 4 | 12 | 19 | 0.0378 | 13 | 15 | 19 | 0.0741 |
| 5 | 6 | 19 | 0.0844 | 5 | 12 | 19 | 0.0419 | 14 | 15 | 19 | 0.0869 |
| 1 | 7 | 19 | 0.0430 | 6 | 12 | 19 | 0.0463 | 1 | 16 | 19 | 0.0158 |
| 2 | 7 | 19 | 0.0519 | 7 | 12 | 19 | 0.0510 | 2 | 16 | 19 | 0.0191 |
| 3 | 7 | 19 | 0.0592 | 8 | 12 | 19 | 0.0562 | 3 | 16 | 19 | 0.0218 |
| 4 | 7 | 19 | 0.0661 | 9 | 12 | 19 | 0.0621 | 4 | 16 | 19 | 0.0244 |
| 5 | 7 | 19 | 0.0732 | 10 | 12 | 19 | 0.0690 | 5 | 16 | 19 | 0.0271 |
| 6 | 7 | 19 | 0.0806 | 11 | 12 | 19 | 0.0772 | 6 | 16 | 19 | 0.0300 |

TABLE 3

OUTLIER MEAN 2.0 OUTLIER STANDARD DEVIATION 1.0

| R | S | N | COVAR | R | S | N | COVAR | R | S | N | COVAR |
|---|---|---|---|---|---|---|---|---|---|---|---|
| 7 | 16 | 19 | 0.0331 | 4 | 19 | 19 | 0.0076 | 3 | 9 | 20 | 0.0454 |
| 8 | 16 | 19 | 0.0365 | 5 | 19 | 19 | 0.0085 | 4 | 9 | 20 | 0.0507 |
| 9 | 16 | 19 | 0.0405 | 6 | 19 | 19 | 0.0095 | 5 | 9 | 20 | 0.0559 |
| 10 | 16 | 19 | 0.0451 | 7 | 19 | 19 | 0.0107 | 6 | 9 | 20 | 0.0613 |
| 11 | 16 | 19 | 0.0506 | 8 | 19 | 19 | 0.0120 | 7 | 9 | 20 | 0.0672 |
| 12 | 16 | 19 | 0.0574 | 9 | 19 | 19 | 0.0135 | 8 | 9 | 20 | 0.0736 |
| 13 | 16 | 19 | 0.0660 | 10 | 19 | 19 | 0.0154 | 1 | 10 | 20 | 0.0298 |
| 14 | 16 | 19 | 0.0775 | 11 | 19 | 19 | 0.0177 | 2 | 10 | 20 | 0.0359 |
| 15 | 16 | 19 | 0.0937 | 12 | 19 | 19 | 0.0207 | 3 | 10 | 20 | 0.0408 |
| 1 | 17 | 19 | 0.0135 | 13 | 19 | 19 | 0.0247 | 4 | 10 | 20 | 0.0455 |
| 2 | 17 | 19 | 0.0163 | 14 | 19 | 19 | 0.0302 | 5 | 10 | 20 | 0.0502 |
| 3 | 17 | 19 | 0.0187 | 15 | 19 | 19 | 0.0385 | 6 | 10 | 20 | 0.0552 |
| 4 | 17 | 19 | 0.0210 | 16 | 19 | 19 | 0.0521 | 7 | 10 | 20 | 0.0605 |
| 5 | 17 | 19 | 0.0233 | 17 | 19 | 19 | 0.0779 | 8 | 10 | 20 | 0.0662 |
| 6 | 17 | 19 | 0.0258 | 18 | 19 | 19 | 0.1429 | 9 | 10 | 20 | 0.0727 |
| 7 | 17 | 19 | 0.0285 | 1 | 2 | 20 | 0.1367 | 1 | 11 | 20 | 0.0269 |
| 8 | 17 | 19 | 0.0315 | 1 | 3 | 20 | 0.0928 | 2 | 11 | 20 | 0.0324 |
| 9 | 17 | 19 | 0.0349 | 2 | 3 | 20 | 0.1110 | 3 | 11 | 20 | 0.0369 |
| 10 | 17 | 19 | 0.0389 | 1 | 4 | 20 | 0.0711 | 4 | 11 | 20 | 0.0412 |
| 11 | 17 | 19 | 0.0437 | 2 | 4 | 20 | 0.0852 | 5 | 11 | 20 | 0.0454 |
| 12 | 17 | 19 | 0.0497 | 3 | 4 | 20 | 0.0966 | 6 | 11 | 20 | 0.0499 |
| 13 | 17 | 19 | 0.0573 | 1 | 5 | 20 | 0.0579 | 7 | 11 | 20 | 0.0547 |
| 14 | 17 | 19 | 0.0675 | 2 | 5 | 20 | 0.0695 | 8 | 11 | 20 | 0.0600 |
| 15 | 17 | 19 | 0.0819 | 3 | 5 | 20 | 0.0789 | 9 | 11 | 20 | 0.0658 |
| 16 | 17 | 19 | 0.1043 | 4 | 5 | 20 | 0.0877 | 10 | 11 | 20 | 0.0725 |
| 1 | 18 | 19 | 0.0107 | 1 | 6 | 20 | 0.0490 | 1 | 12 | 20 | 0.0244 |
| 2 | 18 | 19 | 0.0130 | 2 | 6 | 20 | 0.0588 | 2 | 12 | 20 | 0.0294 |
| 3 | 18 | 19 | 0.0148 | 3 | 6 | 20 | 0.0668 | 3 | 12 | 20 | 0.0335 |
| 4 | 18 | 19 | 0.0167 | 4 | 6 | 20 | 0.0743 | 4 | 12 | 20 | 0.0373 |
| 5 | 18 | 19 | 0.0185 | 5 | 6 | 20 | 0.0819 | 5 | 12 | 20 | 0.0412 |
| 6 | 18 | 19 | 0.0205 | 1 | 7 | 20 | 0.0424 | 6 | 12 | 20 | 0.0453 |
| 7 | 18 | 19 | 0.0227 | 2 | 7 | 20 | 0.0509 | 7 | 12 | 20 | 0.0497 |
| 8 | 18 | 19 | 0.0251 | 3 | 7 | 20 | 0.0579 | 8 | 12 | 20 | 0.0545 |
| 9 | 18 | 19 | 0.0279 | 4 | 7 | 20 | 0.0645 | 9 | 12 | 20 | 0.0598 |
| 10 | 18 | 19 | 0.0312 | 5 | 7 | 20 | 0.0711 | 10 | 12 | 20 | 0.0659 |
| 11 | 18 | 19 | 0.0352 | 6 | 7 | 20 | 0.0779 | 11 | 12 | 20 | 0.0730 |
| 12 | 18 | 19 | 0.0402 | 1 | 8 | 20 | 0.0373 | 1 | 13 | 20 | 0.0222 |
| 13 | 18 | 19 | 0.0465 | 2 | 8 | 20 | 0.0448 | 2 | 13 | 20 | 0.0267 |
| 14 | 18 | 19 | 0.0551 | 3 | 8 | 20 | 0.0510 | 3 | 13 | 20 | 0.0304 |
| 15 | 18 | 19 | 0.0674 | 4 | 8 | 20 | 0.0568 | 4 | 13 | 20 | 0.0339 |
| 16 | 18 | 19 | 0.0866 | 5 | 8 | 20 | 0.0627 | 5 | 13 | 20 | 0.0375 |
| 17 | 18 | 19 | 0.1213 | 6 | 8 | 20 | 0.0687 | 6 | 13 | 20 | 0.0412 |
| 1 | 19 | 19 | 0.0048 | 7 | 8 | 20 | 0.0753 | 7 | 13 | 20 | 0.0452 |
| 2 | 19 | 19 | 0.0058 | 1 | 9 | 20 | 0.0332 | 8 | 13 | 20 | 0.0496 |
| 3 | 19 | 19 | 0.0067 | 2 | 9 | 20 | 0.0399 | 9 | 13 | 20 | 0.0544 |

TABLE 3

OUTLIER MEAN 2.0 OUTLIER STANDARD DEVIATION 1.0

| R | S | N | COVAR | R | S | N | COVAR | R | S | N | COVAR |
|---|---|---|---|---|---|---|---|---|---|---|---|
| 10 | 13 | 20 | 0.0600 | 10 | 16 | 20 | 0.0447 | 1 | 19 | 20 | 0.0099 |
| 11 | 13 | 20 | 0.0665 | 11 | 16 | 20 | 0.0496 | 2 | 19 | 20 | 0.0120 |
| 12 | 13 | 20 | 0.0743 | 12 | 16 | 20 | 0.0555 | 3 | 19 | 20 | 0.0137 |
| 1 | 14 | 20 | 0.0201 | 13 | 16 | 20 | 0.0628 | 4 | 19 | 20 | 0.0153 |
| 2 | 14 | 20 | 0.0243 | 14 | 16 | 20 | 0.0721 | 5 | 19 | 20 | 0.0170 |
| 3 | 14 | 20 | 0.0276 | 15 | 16 | 20 | 0.0844 | 6 | 19 | 20 | 0.0187 |
| 4 | 14 | 20 | 0.0308 | 1 | 17 | 20 | 0.0145 | 7 | 19 | 20 | 0.0206 |
| 5 | 14 | 20 | 0.0341 | 2 | 17 | 20 | 0.0175 | 8 | 19 | 20 | 0.0227 |
| 6 | 14 | 20 | 0.0374 | 3 | 17 | 20 | 0.0200 | 9 | 19 | 20 | 0.0250 |
| 7 | 14 | 20 | 0.0411 | 4 | 17 | 20 | 0.0223 | 10 | 19 | 20 | 0.0277 |
| 8 | 14 | 20 | 0.0451 | 5 | 17 | 20 | 0.0246 | 11 | 19 | 20 | 0.0309 |
| 9 | 14 | 20 | 0.0495 | 6 | 17 | 20 | 0.0271 | 12 | 19 | 20 | 0.0348 |
| 10 | 14 | 20 | 0.0546 | 7 | 17 | 20 | 0.0298 | 13 | 19 | 20 | 0.0396 |
| 11 | 14 | 20 | 0.0606 | 8 | 17 | 20 | 0.0327 | 14 | 19 | 20 | 0.0459 |
| 12 | 14 | 20 | 0.0677 | 9 | 17 | 20 | 0.0360 | 15 | 19 | 20 | 0.0542 |
| 13 | 14 | 20 | 0.0765 | 10 | 17 | 20 | 0.0398 | 16 | 19 | 20 | 0.0663 |
| 1 | 15 | 20 | 0.0182 | 11 | 17 | 20 | 0.0442 | 17 | 19 | 20 | 0.0851 |
| 2 | 15 | 20 | 0.0220 | 12 | 17 | 20 | 0.0495 | 18 | 19 | 20 | 0.1191 |
| 3 | 15 | 20 | 0.0250 | 13 | 17 | 20 | 0.0561 | 1 | 20 | 20 | 0.0045 |
| 4 | 15 | 20 | 0.0279 | 14 | 17 | 20 | 0.0644 | 2 | 20 | 20 | 0.0055 |
| 5 | 15 | 20 | 0.0309 | 15 | 17 | 20 | 0.0756 | 3 | 20 | 20 | 0.0063 |
| 6 | 15 | 20 | 0.0339 | 16 | 17 | 20 | 0.0913 | 4 | 20 | 20 | 0.0071 |
| 7 | 15 | 20 | 0.0372 | 1 | 18 | 20 | 0.0125 | 5 | 20 | 20 | 0.0079 |
| 8 | 15 | 20 | 0.0409 | 2 | 18 | 20 | 0.0151 | 6 | 20 | 20 | 0.0088 |
| 9 | 15 | 20 | 0.0449 | 3 | 18 | 20 | 0.0172 | 7 | 20 | 20 | 0.0098 |
| 10 | 15 | 20 | 0.0496 | 4 | 18 | 20 | 0.0192 | 8 | 20 | 20 | 0.0109 |
| 11 | 15 | 20 | 0.0550 | 5 | 18 | 20 | 0.0212 | 9 | 20 | 20 | 0.0122 |
| 12 | 15 | 20 | 0.0615 | 6 | 18 | 20 | 0.0234 | 10 | 20 | 20 | 0.0138 |
| 13 | 15 | 20 | 0.0695 | 7 | 18 | 20 | 0.0257 | 11 | 20 | 20 | 0.0156 |
| 14 | 15 | 20 | 0.0797 | 8 | 18 | 20 | 0.0283 | 12 | 20 | 20 | 0.0179 |
| 1 | 16 | 20 | 0.0164 | 9 | 18 | 20 | 0.0311 | 13 | 20 | 20 | 0.0209 |
| 2 | 16 | 20 | 0.0197 | 10 | 18 | 20 | 0.0344 | 14 | 20 | 20 | 0.0249 |
| 3 | 16 | 20 | 0.0225 | 11 | 18 | 20 | 0.0383 | 15 | 20 | 20 | 0.0304 |
| 4 | 16 | 20 | 0.0251 | 12 | 18 | 20 | 0.0430 | 16 | 20 | 20 | 0.0386 |
| 5 | 16 | 20 | 0.0278 | 13 | 18 | 20 | 0.0488 | 17 | 20 | 20 | 0.0521 |
| 6 | 16 | 20 | 0.0305 | 14 | 18 | 20 | 0.0561 | 18 | 20 | 20 | 0.0777 |
| 7 | 16 | 20 | 0.0335 | 15 | 18 | 20 | 0.0660 | 19 | 20 | 20 | 0.1419 |
| 8 | 16 | 20 | 0.0368 | 16 | 18 | 20 | 0.0801 | | | | |
| 9 | 16 | 20 | 0.0405 | 17 | 18 | 20 | 0.1019 | | | | |

TABLE 3

OUTLIER MEAN 2.5 OUTLIER STANDARD DEVIATION 1.0

| R | S | N | COVAR | R | S | N | COVAR | R | S | N | COVAR |
|---|---|---|---|---|---|---|---|---|---|---|---|
| 1 | 2 | 2 | 0.0552 | 1 | 6 | 7 | 0.0527 | 1 | 5 | 9 | 0.0743 |
| 1 | 2 | 3 | 0.3048 | 2 | 6 | 7 | 0.0724 | 2 | 5 | 9 | 0.0970 |
| 1 | 3 | 3 | 0.0212 | 3 | 6 | 7 | 0.0959 | 3 | 5 | 9 | 0.1201 |
| 2 | 3 | 3 | 0.0756 | 4 | 6 | 7 | 0.1302 | 4 | 5 | 9 | 0.1479 |
| 1 | 2 | 4 | 0.2723 | 5 | 6 | 7 | 0.1920 | 1 | 6 | 9 | 0.0595 |
| 1 | 3 | 4 | 0.1570 | 1 | 7 | 7 | 0.0072 | 2 | 6 | 9 | 0.0778 |
| 2 | 3 | 4 | 0.2601 | 2 | 7 | 7 | 0.0104 | 3 | 6 | 9 | 0.0966 |
| 1 | 4 | 4 | 0.0135 | 3 | 7 | 7 | 0.0151 | 4 | 6 | 9 | 0.1193 |
| 2 | 4 | 4 | 0.0292 | 4 | 7 | 7 | 0.0233 | 5 | 6 | 9 | 0.1498 |
| 3 | 4 | 4 | 0.0878 | 5 | 7 | 7 | 0.0421 | 1 | 7 | 9 | 0.0471 |
| 1 | 2 | 5 | 0.2443 | 6 | 7 | 7 | 0.1059 | 2 | 7 | 9 | 0.0617 |
| 1 | 3 | 5 | 0.1557 | 1 | 2 | 8 | 0.1960 | 3 | 7 | 9 | 0.0767 |
| 2 | 3 | 5 | 0.2316 | 1 | 3 | 8 | 0.1318 | 4 | 7 | 9 | 0.0950 |
| 1 | 4 | 5 | 0.0990 | 2 | 3 | 8 | 0.1740 | 5 | 7 | 9 | 0.1197 |
| 2 | 4 | 5 | 0.1491 | 1 | 4 | 8 | 0.0980 | 6 | 7 | 9 | 0.1574 |
| 3 | 4 | 5 | 0.2293 | 2 | 4 | 8 | 0.1300 | 1 | 8 | 9 | 0.0342 |
| 1 | 5 | 5 | 0.0102 | 3 | 4 | 8 | 0.1645 | 2 | 8 | 9 | 0.0449 |
| 2 | 5 | 5 | 0.0178 | 1 | 5 | 8 | 0.0758 | 3 | 8 | 9 | 0.0560 |
| 3 | 5 | 5 | 0.0347 | 2 | 5 | 8 | 0.1009 | 4 | 8 | 9 | 0.0696 |
| 4 | 5 | 5 | 0.0959 | 3 | 5 | 8 | 0.1281 | 5 | 8 | 9 | 0.0882 |
| 1 | 2 | 6 | 0.2237 | 4 | 5 | 8 | 0.1632 | 6 | 8 | 9 | 0.1169 |
| 1 | 3 | 6 | 0.1472 | 1 | 6 | 8 | 0.0585 | 7 | 8 | 9 | 0.1701 |
| 2 | 3 | 6 | 0.2067 | 2 | 6 | 8 | 0.0781 | 1 | 9 | 9 | 0.0057 |
| 1 | 4 | 6 | 0.1039 | 3 | 6 | 8 | 0.0995 | 2 | 9 | 9 | 0.0076 |
| 2 | 4 | 6 | 0.1471 | 4 | 6 | 8 | 0.1273 | 3 | 9 | 9 | 0.0099 |
| 3 | 4 | 6 | 0.2034 | 5 | 6 | 8 | 0.1688 | 4 | 9 | 9 | 0.0131 |
| 1 | 5 | 6 | 0.0698 | 1 | 7 | 8 | 0.0418 | 5 | 9 | 9 | 0.0180 |
| 2 | 5 | 6 | 0.0994 | 2 | 7 | 8 | 0.0559 | 6 | 9 | 9 | 0.0270 |
| 3 | 5 | 6 | 0.1390 | 3 | 7 | 8 | 0.0714 | 7 | 9 | 9 | 0.0469 |
| 4 | 5 | 6 | 0.2078 | 4 | 7 | 8 | 0.0919 | 8 | 9 | 9 | 0.1117 |
| 1 | 6 | 6 | 0.0084 | 5 | 7 | 8 | 0.1229 | 1 | 2 | 10 | 0.1780 |
| 2 | 6 | 6 | 0.0130 | 6 | 7 | 8 | 0.1798 | 1 | 3 | 10 | 0.1206 |
| 3 | 6 | 6 | 0.0209 | 1 | 8 | 8 | 0.0063 | 2 | 3 | 10 | 0.1539 |
| 4 | 6 | 6 | 0.0388 | 2 | 8 | 8 | 0.0088 | 1 | 4 | 10 | 0.0911 |
| 5 | 6 | 6 | 0.1017 | 3 | 8 | 8 | 0.0119 | 2 | 4 | 10 | 0.1167 |
| 1 | 2 | 7 | 0.2082 | 4 | 8 | 8 | 0.0167 | 3 | 4 | 10 | 0.1417 |
| 1 | 3 | 7 | 0.1389 | 5 | 8 | 8 | 0.0253 | 1 | 5 | 10 | 0.0725 |
| 2 | 3 | 7 | 0.1881 | 6 | 8 | 8 | 0.0447 | 2 | 5 | 10 | 0.0931 |
| 1 | 4 | 7 | 0.1016 | 7 | 8 | 8 | 0.1092 | 3 | 5 | 10 | 0.1133 |
| 2 | 4 | 7 | 0.1384 | 1 | 2 | 9 | 0.1862 | 4 | 5 | 10 | 0.1362 |
| 3 | 4 | 7 | 0.1811 | 1 | 3 | 9 | 0.1258 | 1 | 6 | 10 | 0.0591 |
| 1 | 5 | 7 | 0.0758 | 2 | 3 | 9 | 0.1629 | 2 | 6 | 10 | 0.0760 |
| 2 | 5 | 7 | 0.1037 | 1 | 4 | 9 | 0.0944 | 3 | 6 | 10 | 0.0927 |
| 3 | 5 | 7 | 0.1365 | 2 | 4 | 9 | 0.1229 | 4 | 6 | 10 | 0.1117 |
| 4 | 5 | 7 | 0.1836 | 3 | 4 | 9 | 0.1517 | 5 | 6 | 10 | 0.1356 |

TABLE 3

OUTLIER MEAN 2.5 OUTLIER STANDARD DEVIATION 1.C

| R | S | N | COVAR | R | S | N | COVAR | R | S | N | COVAR |
|---|---|---|---|---|---|---|---|---|---|---|---|
| 1 | 7 | 10 | 0.0485 | 1 | 7 | 11 | 0.0486 | 3 | 4 | 12 | 0.1268 |
| 2 | 7 | 10 | 0.0624 | 2 | 7 | 11 | 0.0618 | 1 | 5 | 12 | 0.0687 |
| 3 | 7 | 10 | 0.0762 | 3 | 7 | 11 | 0.0743 | 2 | 5 | 12 | 0.0863 |
| 4 | 7 | 10 | 0.0921 | 4 | 7 | 11 | 0.0882 | 3 | 5 | 12 | 0.1024 |
| 5 | 7 | 10 | 0.1120 | 5 | 7 | 11 | 0.1048 | 4 | 5 | 12 | 0.1196 |
| 6 | 7 | 10 | 0.1393 | 6 | 7 | 11 | 0.1260 | 1 | 6 | 12 | 0.0570 |
| 1 | 8 | 10 | 0.0390 | 1 | 8 | 11 | 0.0405 | 2 | 6 | 12 | 0.0717 |
| 2 | 8 | 10 | 0.0503 | 2 | 8 | 11 | 0.0516 | 3 | 6 | 12 | 0.0852 |
| 3 | 8 | 10 | 0.0615 | 3 | 8 | 11 | 0.0621 | 4 | 6 | 12 | 0.0997 |
| 4 | 8 | 10 | 0.0745 | 4 | 8 | 11 | 0.0738 | 5 | 6 | 12 | 0.1162 |
| 5 | 8 | 10 | 0.0908 | 5 | 8 | 11 | 0.0879 | 1 | 7 | 12 | 0.0481 |
| 6 | 8 | 10 | 0.1134 | 6 | 8 | 11 | 0.1059 | 2 | 7 | 12 | 0.0606 |
| 7 | 8 | 10 | 0.1481 | 7 | 8 | 11 | 0.1310 | 3 | 7 | 12 | 0.0721 |
| 1 | 9 | 10 | 0.0288 | 1 | 9 | 11 | 0.0331 | 4 | 7 | 12 | 0.0844 |
| 2 | 9 | 10 | 0.0372 | 2 | 9 | 11 | 0.0421 | 5 | 7 | 12 | 0.0986 |
| 3 | 9 | 10 | 0.0455 | 3 | 9 | 11 | 0.0508 | 6 | 7 | 12 | 0.1158 |
| 4 | 9 | 10 | 0.0552 | 4 | 9 | 11 | 0.0604 | 1 | 8 | 12 | 0.0409 |
| 5 | 9 | 10 | 0.0676 | 5 | 9 | 11 | 0.0720 | 2 | 8 | 12 | 0.0515 |
| 6 | 9 | 10 | 0.0849 | 6 | 9 | 11 | 0.0870 | 3 | 8 | 12 | 0.0614 |
| 7 | 9 | 10 | 0.1119 | 7 | 9 | 11 | 0.1080 | 4 | 8 | 12 | 0.0719 |
| 8 | 9 | 10 | 0.1621 | 8 | 9 | 11 | 0.1406 | 5 | 8 | 12 | 0.0841 |
| 1 | 10 | 10 | 0.0051 | 1 | 10 | 11 | 0.0247 | 6 | 8 | 12 | 0.0990 |
| 2 | 10 | 10 | 0.0068 | 2 | 10 | 11 | 0.0315 | 7 | 8 | 12 | 0.1183 |
| 3 | 10 | 10 | 0.0086 | 3 | 10 | 11 | 0.0380 | 1 | 9 | 12 | 0.0346 |
| 4 | 10 | 10 | 0.0108 | 4 | 10 | 11 | 0.0453 | 2 | 9 | 12 | 0.0436 |
| 5 | 10 | 10 | 0.0141 | 5 | 10 | 11 | 0.0542 | 3 | 9 | 12 | 0.0520 |
| 6 | 10 | 10 | 0.0191 | 6 | 10 | 11 | 0.0657 | 4 | 9 | 12 | 0.0609 |
| 7 | 10 | 10 | 0.0284 | 7 | 10 | 11 | 0.0819 | 5 | 9 | 12 | 0.0713 |
| 8 | 10 | 10 | 0.0487 | 8 | 10 | 11 | 0.1075 | 6 | 9 | 12 | 0.0841 |
| 9 | 10 | 10 | 0.1136 | 9 | 10 | 11 | 0.1554 | 7 | 9 | 12 | 0.1007 |
| 1 | 2 | 11 | 0.1712 | 1 | 11 | 11 | 0.0047 | 8 | 9 | 12 | 0.1240 |
| 1 | 3 | 11 | 0.1162 | 2 | 11 | 11 | 0.0061 | 1 | 10 | 12 | 0.0285 |
| 2 | 3 | 11 | 0.1465 | 3 | 11 | 11 | 0.0076 | 2 | 10 | 12 | 0.0360 |
| 1 | 4 | 11 | 0.0881 | 4 | 11 | 11 | 0.0093 | 3 | 10 | 12 | 0.0429 |
| 2 | 4 | 11 | 0.1115 | 5 | 11 | 11 | 0.0116 | 4 | 10 | 12 | 0.0504 |
| 3 | 4 | 11 | 0.1335 | 6 | 11 | 11 | 0.0149 | 5 | 10 | 12 | 0.0591 |
| 1 | 5 | 11 | 0.0706 | 7 | 11 | 11 | 0.0201 | 6 | 10 | 12 | 0.0697 |
| 2 | 5 | 11 | 0.0895 | 8 | 11 | 11 | 0.0296 | 7 | 10 | 12 | 0.0837 |
| 3 | 5 | 11 | 0.1074 | 9 | 11 | 11 | 0.0502 | 8 | 10 | 12 | 0.1034 |
| 4 | 5 | 11 | 0.1270 | 10 | 11 | 11 | 0.1152 | 9 | 10 | 12 | 0.1342 |
| 1 | 6 | 11 | 0.0582 | 1 | 2 | 12 | 0.1653 | 1 | 11 | 12 | 0.0215 |
| 2 | 6 | 11 | 0.0739 | 1 | 3 | 12 | 0.1123 | 2 | 11 | 12 | 0.0271 |
| 3 | 6 | 11 | 0.0888 | 2 | 3 | 12 | 0.1402 | 3 | 11 | 12 | 0.0324 |
| 4 | 6 | 11 | 0.1052 | 1 | 4 | 12 | 0.0854 | 4 | 11 | 12 | 0.0381 |
| 5 | 6 | 11 | 0.1248 | 2 | 4 | 12 | 0.1070 | 5 | 11 | 12 | 0.0448 |

TABLE 3

OUTLIER MEAN 2.5 OUTLIER STANDARD DEVIATION 1.0

| R | S | N | COVAR | R | S | N | COVAR | R | S | N | COVAR |
|---|---|---|---|---|---|---|---|---|---|---|---|
| 6 | 11 | 12 | 0.0530 | 2 | 9 | 13 | 0.0439 | 9 | 13 | 13 | 0.0217 |
| 7 | 11 | 12 | 0.0639 | 3 | 9 | 13 | 0.0518 | 10 | 13 | 13 | 0.0315 |
| 8 | 11 | 12 | 0.0793 | 4 | 9 | 13 | 0.0601 | 11 | 13 | 13 | 0.0526 |
| 9 | 11 | 12 | 0.1038 | 5 | 9 | 13 | 0.0694 | 12 | 13 | 13 | 0.1175 |
| 10 | 11 | 12 | 0.1497 | 6 | 9 | 13 | 0.0804 | 1 | 2 | 14 | 0.1557 |
| 1 | 12 | 12 | 0.0044 | 7 | 9 | 13 | 0.0940 | 1 | 3 | 14 | 0.1059 |
| 2 | 12 | 12 | 0.0056 | 8 | 9 | 13 | 0.1119 | 2 | 3 | 14 | 0.1301 |
| 3 | 12 | 12 | 0.0068 | 1 | 10 | 13 | 0.0300 | 1 | 4 | 14 | 0.0808 |
| 4 | 12 | 12 | 0.0082 | 2 | 10 | 13 | 0.0375 | 2 | 4 | 14 | 0.0997 |
| 5 | 12 | 12 | 0.0099 | 3 | 10 | 13 | 0.0443 | 3 | 4 | 14 | 0.1162 |
| 6 | 12 | 12 | 0.0122 | 4 | 10 | 13 | 0.0514 | 1 | 5 | 14 | 0.0654 |
| 7 | 12 | 12 | 0.0156 | 5 | 10 | 13 | 0.0594 | 2 | 5 | 14 | 0.0808 |
| 8 | 12 | 12 | 0.0210 | 6 | 10 | 13 | 0.0689 | 3 | 5 | 14 | 0.0943 |
| 9 | 12 | 12 | 0.0306 | 7 | 10 | 13 | 0.0807 | 4 | 5 | 14 | 0.1081 |
| 10 | 12 | 12 | 0.0515 | 8 | 10 | 13 | 0.0963 | 1 | 6 | 14 | 0.0547 |
| 11 | 12 | 12 | 0.1165 | 9 | 10 | 13 | 0.1182 | 2 | 6 | 14 | 0.0677 |
| 1 | 2 | 13 | 0.1602 | 1 | 11 | 13 | 0.0250 | 3 | 6 | 14 | 0.0791 |
| 1 | 3 | 13 | 0.1089 | 2 | 11 | 13 | 0.0312 | 4 | 6 | 14 | 0.0908 |
| 2 | 3 | 13 | 0.1348 | 3 | 11 | 13 | 0.0369 | 5 | 6 | 14 | 0.1035 |
| 1 | 4 | 13 | 0.0830 | 4 | 11 | 13 | 0.0429 | 1 | 7 | 14 | 0.0468 |
| 2 | 4 | 13 | 0.1031 | 5 | 11 | 13 | 0.0496 | 2 | 7 | 14 | 0.0579 |
| 3 | 4 | 13 | 0.1211 | 6 | 11 | 13 | 0.0576 | 3 | 7 | 14 | 0.0677 |
| 1 | 5 | 13 | 0.0670 | 7 | 11 | 13 | 0.0676 | 4 | 7 | 14 | 0.0778 |
| 2 | 5 | 13 | 0.0834 | 8 | 11 | 13 | 0.0808 | 5 | 7 | 14 | 0.0887 |
| 3 | 5 | 13 | 0.0981 | 9 | 11 | 13 | 0.0994 | 6 | 7 | 14 | 0.1013 |
| 4 | 5 | 13 | 0.1133 | 10 | 11 | 13 | 0.1287 | 1 | 8 | 14 | 0.0405 |
| 1 | 6 | 13 | 0.0559 | 1 | 12 | 13 | 0.0190 | 2 | 8 | 14 | 0.0501 |
| 2 | 6 | 13 | 0.0696 | 2 | 12 | 13 | 0.0238 | 3 | 8 | 14 | 0.0587 |
| 3 | 6 | 13 | 0.0820 | 3 | 12 | 13 | 0.0281 | 4 | 8 | 14 | 0.0674 |
| 4 | 6 | 13 | 0.0949 | 4 | 12 | 13 | 0.0327 | 5 | 8 | 14 | 0.0770 |
| 5 | 6 | 13 | 0.1093 | 5 | 12 | 13 | 0.0379 | 6 | 8 | 14 | 0.0880 |
| 1 | 7 | 13 | 0.0475 | 6 | 12 | 13 | 0.0441 | 7 | 8 | 14 | 0.1010 |
| 2 | 7 | 13 | 0.0593 | 7 | 12 | 13 | 0.0519 | 1 | 9 | 14 | 0.0352 |
| 3 | 7 | 13 | 0.0699 | 8 | 12 | 13 | 0.0622 | 2 | 9 | 14 | 0.0436 |
| 4 | 7 | 13 | 0.0809 | 9 | 12 | 13 | 0.0770 | 3 | 9 | 14 | 0.0511 |
| 5 | 7 | 13 | 0.0933 | 10 | 12 | 13 | 0.1006 | 4 | 9 | 14 | 0.0588 |
| 6 | 7 | 13 | 0.1078 | 11 | 12 | 13 | 0.1448 | 5 | 9 | 14 | 0.0672 |
| 1 | 8 | 13 | 0.0408 | 1 | 13 | 13 | 0.0041 | 6 | 9 | 14 | 0.0768 |
| 2 | 8 | 13 | 0.0509 | 2 | 13 | 13 | 0.0051 | 7 | 9 | 14 | 0.0883 |
| 3 | 8 | 13 | 0.0601 | 3 | 13 | 13 | 0.0062 | 8 | 9 | 14 | 0.1027 |
| 4 | 8 | 13 | 0.0697 | 4 | 13 | 13 | 0.0073 | 1 | 10 | 14 | 0.0306 |
| 5 | 8 | 13 | 0.0804 | 5 | 13 | 13 | 0.0087 | 2 | 10 | 14 | 0.0380 |
| 6 | 8 | 13 | 0.0930 | 6 | 13 | 13 | 0.0104 | 3 | 10 | 14 | 0.0445 |
| 7 | 8 | 13 | 0.1086 | 7 | 13 | 13 | 0.0128 | 4 | 10 | 14 | 0.0512 |
| 1 | 9 | 13 | 0.0351 | 8 | 13 | 13 | 0.0163 | 5 | 10 | 14 | 0.0586 |

TABLE 3

OUTLIER MEAN 2.5 OUTLIER STANDARD DEVIATION 1.0

| R | S | N | COVAR | R | S | N | COVAR | R | S | N | COVAR |
|---|---|---|---|---|---|---|---|---|---|---|---|
| 6 | 10 | 14 | 0.0670 | 9 | 14 | 14 | 0.0168 | 5 | 10 | 15 | 0.0572 |
| 7 | 10 | 14 | 0.0771 | 10 | 14 | 14 | 0.0224 | 6 | 10 | 15 | 0.0648 |
| 8 | 10 | 14 | 0.0898 | 11 | 14 | 14 | 0.0323 | 7 | 10 | 15 | 0.0736 |
| 9 | 10 | 14 | 0.1065 | 12 | 14 | 14 | 0.0535 | 8 | 10 | 15 | 0.0843 |
| 1 | 11 | 14 | 0.0264 | 13 | 14 | 14 | 0.1183 | 9 | 10 | 15 | 0.0977 |
| 2 | 11 | 14 | 0.0327 | 1 | 2 | 15 | 0.1517 | 1 | 11 | 15 | 0.0270 |
| 3 | 11 | 14 | 0.0384 | 1 | 3 | 15 | 0.1031 | 2 | 11 | 15 | 0.0333 |
| 4 | 11 | 14 | 0.0442 | 2 | 3 | 15 | 0.1260 | 3 | 11 | 15 | 0.0388 |
| 5 | 11 | 14 | 0.0506 | 1 | 4 | 15 | 0.0788 | 4 | 11 | 15 | 0.0443 |
| 6 | 11 | 14 | 0.0579 | 2 | 4 | 15 | 0.0966 | 5 | 11 | 15 | 0.0503 |
| 7 | 11 | 14 | 0.0667 | 3 | 4 | 15 | 0.1119 | 6 | 11 | 15 | 0.0570 |
| 8 | 11 | 14 | 0.0778 | 1 | 5 | 15 | 0.0639 | 7 | 11 | 15 | 0.0648 |
| 9 | 11 | 14 | 0.0924 | 2 | 5 | 15 | 0.0785 | 8 | 11 | 15 | 0.0742 |
| 10 | 11 | 14 | 0.1132 | 3 | 5 | 15 | 0.0910 | 9 | 11 | 15 | 0.0861 |
| 1 | 12 | 14 | 0.0221 | 4 | 5 | 15 | 0.1036 | 10 | 11 | 15 | 0.1019 |
| 2 | 12 | 14 | 0.0275 | 1 | 6 | 15 | 0.0536 | 1 | 12 | 15 | 0.0235 |
| 3 | 12 | 14 | 0.0322 | 2 | 6 | 15 | 0.0659 | 2 | 12 | 15 | 0.0289 |
| 4 | 12 | 14 | 0.0371 | 3 | 6 | 15 | 0.0766 | 3 | 12 | 15 | 0.0337 |
| 5 | 12 | 14 | 0.0425 | 4 | 6 | 15 | 0.0872 | 4 | 12 | 15 | 0.0385 |
| 6 | 12 | 14 | 0.0487 | 5 | 6 | 15 | 0.0986 | 5 | 12 | 15 | 0.0437 |
| 7 | 12 | 14 | 0.0562 | 1 | 7 | 15 | 0.0460 | 6 | 12 | 15 | 0.0496 |
| 8 | 12 | 14 | 0.0656 | 2 | 7 | 15 | 0.0566 | 7 | 12 | 15 | 0.0564 |
| 9 | 12 | 14 | 0.0781 | 3 | 7 | 15 | 0.0658 | 8 | 12 | 15 | 0.0647 |
| 10 | 12 | 14 | 0.0960 | 4 | 7 | 15 | 0.0750 | 9 | 12 | 15 | 0.0751 |
| 11 | 12 | 14 | 0.1240 | 5 | 7 | 15 | 0.0848 | 10 | 12 | 15 | 0.0891 |
| 1 | 13 | 14 | 0.0169 | 6 | 7 | 15 | 0.0958 | 11 | 12 | 15 | 0.1089 |
| 2 | 13 | 14 | 0.0210 | 1 | 8 | 15 | 0.0400 | 1 | 13 | 15 | 0.0198 |
| 3 | 13 | 14 | 0.0247 | 2 | 8 | 15 | 0.0492 | 2 | 13 | 15 | 0.0244 |
| 4 | 13 | 14 | 0.0285 | 3 | 8 | 15 | 0.0573 | 3 | 13 | 15 | 0.0285 |
| 5 | 13 | 14 | 0.0327 | 4 | 8 | 15 | 0.0653 | 4 | 13 | 15 | 0.0326 |
| 6 | 13 | 14 | 0.0375 | 5 | 8 | 15 | 0.0740 | 5 | 13 | 15 | 0.0370 |
| 7 | 13 | 14 | 0.0433 | 6 | 8 | 15 | 0.0836 | 6 | 13 | 15 | 0.0419 |
| 8 | 13 | 14 | 0.0508 | 7 | 8 | 15 | 0.0949 | 7 | 13 | 15 | 0.0477 |
| 9 | 13 | 14 | 0.0607 | 1 | 9 | 15 | 0.0351 | 8 | 13 | 15 | 0.0548 |
| 10 | 13 | 14 | 0.0750 | 2 | 9 | 15 | 0.0432 | 9 | 13 | 15 | 0.0638 |
| 11 | 13 | 14 | 0.0977 | 3 | 9 | 15 | 0.0502 | 10 | 13 | 15 | 0.0758 |
| 12 | 13 | 14 | 0.1405 | 4 | 9 | 15 | 0.0574 | 11 | 13 | 15 | 0.0929 |
| 1 | 14 | 14 | 0.0038 | 5 | 9 | 15 | 0.0650 | 12 | 13 | 15 | 0.1199 |
| 2 | 14 | 14 | 0.0048 | 6 | 9 | 15 | 0.0735 | 1 | 14 | 15 | 0.0153 |
| 3 | 14 | 14 | 0.0056 | 7 | 9 | 15 | 0.0835 | 2 | 14 | 15 | 0.0188 |
| 4 | 14 | 14 | 0.0066 | 8 | 9 | 15 | 0.0955 | 3 | 14 | 15 | 0.0220 |
| 5 | 14 | 14 | 0.0077 | 1 | 10 | 15 | 0.0308 | 4 | 14 | 15 | 0.0251 |
| 6 | 14 | 14 | 0.0091 | 2 | 10 | 15 | 0.0380 | 5 | 14 | 15 | 0.0286 |
| 7 | 14 | 14 | 0.0109 | 3 | 10 | 15 | 0.0442 | 6 | 14 | 15 | 0.0325 |
| 8 | 14 | 14 | 0.0133 | 4 | 10 | 15 | 0.0505 | 7 | 14 | 15 | 0.0370 |

TABLE 3

OUTLIER MEAN 2.5 OUTLIER STANDARD DEVIATION 1.0

| R | S | N | COVAR | R | S | N | COVAR | R | S | N | COVAR |
|---|---|---|---|---|---|---|---|---|---|---|---|
| 8 | 14 | 15 | 0.0426 | 5 | 8 | 16 | 0.0712 | 5 | 13 | 16 | 0.0383 |
| 9 | 14 | 15 | 0.0497 | 6 | 8 | 16 | 0.0799 | 6 | 13 | 16 | 0.0431 |
| 10 | 14 | 15 | 0.0593 | 7 | 8 | 16 | 0.0897 | 7 | 13 | 16 | 0.0485 |
| 11 | 14 | 15 | 0.0731 | 1 | 9 | 16 | 0.0348 | 8 | 13 | 16 | 0.0549 |
| 12 | 14 | 15 | 0.0951 | 2 | 9 | 16 | 0.0426 | 9 | 13 | 16 | 0.0628 |
| 13 | 14 | 15 | 0.1367 | 3 | 9 | 16 | 0.0493 | 10 | 13 | 16 | 0.0728 |
| 1 | 15 | 15 | 0.0036 | 4 | 9 | 16 | 0.0559 | 11 | 13 | 16 | 0.0861 |
| 2 | 15 | 15 | 0.0044 | 5 | 9 | 16 | 0.0629 | 12 | 13 | 16 | 0.1050 |
| 3 | 15 | 15 | 0.0052 | 6 | 9 | 16 | 0.0706 | 1 | 14 | 16 | 0.0179 |
| 4 | 15 | 15 | 0.0061 | 7 | 9 | 16 | 0.0793 | 2 | 14 | 16 | 0.0219 |
| 5 | 15 | 15 | 0.0070 | 8 | 9 | 16 | 0.0896 | 3 | 14 | 16 | 0.0254 |
| 6 | 15 | 15 | 0.0081 | 1 | 10 | 16 | 0.0308 | 4 | 14 | 16 | 0.0289 |
| 7 | 15 | 15 | 0.0095 | 2 | 10 | 16 | 0.0377 | 5 | 14 | 16 | 0.0326 |
| 8 | 15 | 15 | 0.0113 | 3 | 10 | 16 | 0.0437 | 6 | 14 | 16 | 0.0366 |
| 9 | 15 | 15 | 0.0138 | 4 | 10 | 16 | 0.0496 | 7 | 14 | 16 | 0.0413 |
| 10 | 15 | 15 | 0.0173 | 5 | 10 | 16 | 0.0558 | 8 | 14 | 16 | 0.0468 |
| 11 | 15 | 15 | 0.0230 | 6 | 10 | 16 | 0.0626 | 9 | 14 | 16 | 0.0535 |
| 12 | 15 | 15 | 0.0330 | 7 | 10 | 16 | 0.0704 | 10 | 14 | 16 | 0.0621 |
| 13 | 15 | 15 | 0.0544 | 8 | 10 | 16 | 0.0796 | 11 | 14 | 16 | 0.0736 |
| 14 | 15 | 15 | 0.1190 | 9 | 10 | 16 | 0.0908 | 12 | 14 | 16 | 0.0901 |
| 1 | 2 | 16 | 0.1481 | 1 | 11 | 16 | 0.0273 | 13 | 14 | 16 | 0.1162 |
| 1 | 3 | 16 | 0.1007 | 2 | 11 | 16 | 0.0335 | 1 | 15 | 16 | 0.0139 |
| 2 | 3 | 16 | 0.1224 | 3 | 11 | 16 | 0.0388 | 2 | 15 | 16 | 0.0170 |
| 1 | 4 | 16 | 0.0770 | 4 | 11 | 16 | 0.0440 | 3 | 15 | 16 | 0.0197 |
| 2 | 4 | 16 | 0.0939 | 5 | 11 | 16 | 0.0495 | 4 | 15 | 16 | 0.0224 |
| 3 | 4 | 16 | 0.1081 | 6 | 11 | 16 | 0.0557 | 5 | 15 | 16 | 0.0253 |
| 1 | 5 | 16 | 0.0625 | 7 | 11 | 16 | 0.0626 | 6 | 15 | 16 | 0.0285 |
| 2 | 5 | 16 | 0.0763 | 8 | 11 | 16 | 0.0708 | 7 | 15 | 16 | 0.0322 |
| 3 | 5 | 16 | 0.0881 | 9 | 11 | 16 | 0.0808 | 8 | 15 | 16 | 0.0365 |
| 4 | 5 | 16 | 0.0996 | 10 | 11 | 16 | 0.0934 | 9 | 15 | 16 | 0.0419 |
| 1 | 6 | 16 | 0.0526 | 1 | 12 | 16 | 0.0241 | 10 | 15 | 16 | 0.0487 |
| 2 | 6 | 16 | 0.0643 | 2 | 12 | 16 | 0.0296 | 11 | 15 | 16 | 0.0580 |
| 3 | 6 | 16 | 0.0742 | 3 | 12 | 16 | 0.0343 | 12 | 15 | 16 | 0.0714 |
| 4 | 6 | 16 | 0.0841 | 4 | 12 | 16 | 0.0389 | 13 | 15 | 16 | 0.0929 |
| 5 | 6 | 16 | 0.0944 | 5 | 12 | 16 | 0.0438 | 14 | 15 | 16 | 0.1333 |
| 1 | 7 | 16 | 0.0452 | 6 | 12 | 16 | 0.0492 | 1 | 16 | 16 | 0.0034 |
| 2 | 7 | 16 | 0.0553 | 7 | 12 | 16 | 0.0554 | 2 | 16 | 16 | 0.0042 |
| 3 | 7 | 16 | 0.0640 | 8 | 12 | 16 | 0.0627 | 3 | 16 | 16 | 0.0049 |
| 4 | 7 | 16 | 0.0725 | 9 | 12 | 16 | 0.0716 | 4 | 16 | 16 | 0.0056 |
| 5 | 7 | 16 | 0.0814 | 10 | 12 | 16 | 0.0829 | 5 | 16 | 16 | 0.0064 |
| 6 | 7 | 16 | 0.0913 | 11 | 12 | 16 | 0.0979 | 6 | 16 | 16 | 0.0073 |
| 1 | 8 | 16 | 0.0395 | 1 | 13 | 16 | 0.0211 | 7 | 16 | 16 | 0.0084 |
| 2 | 8 | 16 | 0.0483 | 2 | 13 | 16 | 0.0258 | 8 | 16 | 16 | 0.0098 |
| 3 | 8 | 16 | 0.0559 | 3 | 13 | 16 | 0.0299 | 9 | 16 | 16 | 0.0117 |
| 4 | 8 | 16 | 0.0634 | 4 | 13 | 16 | 0.0340 | 10 | 16 | 16 | 0.0142 |

TABLE 3

OUTLIER MEAN 2.5 OUTLIER STANDARD DEVIATION 1.0

| R | S | N | COVAR | R | S | N | COVAR | R | S | N | COVAR |
|---|---|---|---|---|---|---|---|---|---|---|---|
| 11 | 16 | 16 | 0.0178 | 5 | 10 | 17 | 0.0544 | 8 | 14 | 17 | 0.0475 |
| 12 | 16 | 16 | 0.0235 | 6 | 10 | 17 | 0.0606 | 9 | 14 | 17 | 0.0536 |
| 13 | 16 | 16 | 0.0337 | 7 | 10 | 17 | 0.0676 | 10 | 14 | 17 | 0.0611 |
| 14 | 16 | 16 | 0.0551 | 8 | 10 | 17 | 0.0756 | 11 | 14 | 17 | 0.0706 |
| 15 | 16 | 16 | 0.1195 | 9 | 10 | 17 | 0.0851 | 12 | 14 | 17 | 0.0834 |
| 1 | 2 | 17 | 0.1449 | 1 | 11 | 17 | 0.0274 | 13 | 14 | 17 | 0.1017 |
| 1 | 3 | 17 | 0.0985 | 2 | 11 | 17 | 0.0334 | 1 | 15 | 17 | 0.0163 |
| 2 | 3 | 17 | 0.1191 | 3 | 11 | 17 | 0.0385 | 2 | 15 | 17 | 0.0198 |
| 1 | 4 | 17 | 0.0754 | 4 | 11 | 17 | 0.0434 | 3 | 15 | 17 | 0.0229 |
| 2 | 4 | 17 | 0.0914 | 5 | 11 | 17 | 0.0486 | 4 | 15 | 17 | 0.0259 |
| 3 | 4 | 17 | 0.1048 | 6 | 11 | 17 | 0.0542 | 5 | 15 | 17 | 0.0290 |
| 1 | 5 | 17 | 0.0613 | 7 | 11 | 17 | 0.0605 | 6 | 15 | 17 | 0.0324 |
| 2 | 5 | 17 | 0.0744 | 8 | 11 | 17 | 0.0677 | 7 | 15 | 17 | 0.0362 |
| 3 | 5 | 17 | 0.0855 | 9 | 11 | 17 | 0.0763 | 8 | 15 | 17 | 0.0406 |
| 4 | 5 | 17 | 0.0962 | 10 | 11 | 17 | 0.0867 | 9 | 15 | 17 | 0.0459 |
| 1 | 6 | 17 | 0.0516 | 1 | 12 | 17 | 0.0245 | 10 | 15 | 17 | 0.0523 |
| 2 | 6 | 17 | 0.0628 | 2 | 12 | 17 | 0.0298 | 11 | 15 | 17 | 0.0606 |
| 3 | 6 | 17 | 0.0721 | 3 | 12 | 17 | 0.0344 | 12 | 15 | 17 | 0.0717 |
| 4 | 6 | 17 | 0.0813 | 4 | 12 | 17 | 0.0388 | 13 | 15 | 17 | 0.0877 |
| 5 | 6 | 17 | 0.0907 | 5 | 12 | 17 | 0.0434 | 14 | 15 | 17 | 0.1129 |
| 1 | 7 | 17 | 0.0445 | 6 | 12 | 17 | 0.0485 | 1 | 16 | 17 | 0.0127 |
| 2 | 7 | 17 | 0.0541 | 7 | 12 | 17 | 0.0541 | 2 | 16 | 17 | 0.0155 |
| 3 | 7 | 17 | 0.0623 | 8 | 12 | 17 | 0.0606 | 3 | 16 | 17 | 0.0178 |
| 4 | 7 | 17 | 0.0702 | 9 | 12 | 17 | 0.0683 | 4 | 16 | 17 | 0.0202 |
| 5 | 7 | 17 | 0.0784 | 10 | 12 | 17 | 0.0777 | 5 | 16 | 17 | 0.0226 |
| 6 | 7 | 17 | 0.0873 | 11 | 12 | 17 | 0.0897 | 6 | 16 | 17 | 0.0253 |
| 1 | 8 | 17 | 0.0390 | 1 | 13 | 17 | 0.0217 | 7 | 16 | 17 | 0.0283 |
| 2 | 8 | 17 | 0.0474 | 2 | 13 | 17 | 0.0265 | 8 | 16 | 17 | 0.0318 |
| 3 | 8 | 17 | 0.0546 | 3 | 13 | 17 | 0.0306 | 9 | 16 | 17 | 0.0360 |
| 4 | 8 | 17 | 0.0616 | 4 | 13 | 17 | 0.0345 | 10 | 16 | 17 | 0.0412 |
| 5 | 8 | 17 | 0.0688 | 5 | 13 | 17 | 0.0386 | 11 | 16 | 17 | 0.0478 |
| 6 | 8 | 17 | 0.0766 | 6 | 13 | 17 | 0.0431 | 12 | 16 | 17 | 0.0568 |
| 7 | 8 | 17 | 0.0854 | 7 | 13 | 17 | 0.0482 | 13 | 16 | 17 | 0.0699 |
| 1 | 9 | 17 | 0.0345 | 8 | 13 | 17 | 0.0540 | 14 | 16 | 17 | 0.0908 |
| 2 | 9 | 17 | 0.0420 | 9 | 13 | 17 | 0.0609 | 15 | 16 | 17 | 0.1302 |
| 3 | 9 | 17 | 0.0483 | 10 | 13 | 17 | 0.0693 | 1 | 17 | 17 | 0.0032 |
| 4 | 9 | 17 | 0.0545 | 11 | 13 | 17 | 0.0801 | 2 | 17 | 17 | 0.0039 |
| 5 | 9 | 17 | 0.0610 | 12 | 13 | 17 | 0.0944 | 3 | 17 | 17 | 0.0045 |
| 6 | 9 | 17 | 0.0679 | 1 | 14 | 17 | 0.0191 | 4 | 17 | 17 | 0.0052 |
| 7 | 9 | 17 | 0.0757 | 2 | 14 | 17 | 0.0233 | 5 | 17 | 17 | 0.0059 |
| 8 | 9 | 17 | 0.0847 | 3 | 14 | 17 | 0.0268 | 6 | 17 | 17 | 0.0067 |
| 1 | 10 | 17 | 0.0307 | 4 | 14 | 17 | 0.0303 | 7 | 17 | 17 | 0.0076 |
| 2 | 10 | 17 | 0.0374 | 5 | 14 | 17 | 0.0340 | 8 | 17 | 17 | 0.0087 |
| 3 | 10 | 17 | 0.0430 | 6 | 14 | 17 | 0.0379 | 9 | 17 | 17 | 0.0101 |
| 4 | 10 | 17 | 0.0486 | 7 | 14 | 17 | 0.0424 | 10 | 17 | 17 | 0.0120 |

TABLE 3

OUTLIER MEAN 2.5 OUTLIER STANDARD DEVIATION 1.0

| R | S | N | COVAR | R | S | N | COVAR | R | S | N | COVAR |
|---|---|---|---|---|---|---|---|---|---|---|---|
| 11 | 17 | 17 | 0.0145 | 4 | 10 | 18 | 0.0476 | 7 | 14 | 18 | 0.0424 |
| 12 | 17 | 17 | 0.0182 | 5 | 10 | 18 | 0.0530 | 8 | 14 | 18 | 0.0471 |
| 13 | 17 | 17 | 0.0240 | 6 | 10 | 18 | 0.0587 | 9 | 14 | 18 | 0.0526 |
| 14 | 17 | 17 | 0.0343 | 7 | 10 | 18 | 0.0650 | 10 | 14 | 18 | 0.0592 |
| 15 | 17 | 17 | 0.0558 | 8 | 10 | 18 | 0.0722 | 11 | 14 | 18 | 0.0672 |
| 16 | 17 | 17 | 0.1199 | 9 | 10 | 18 | 0.0804 | 12 | 14 | 18 | 0.0775 |
| 1 | 2 | 18 | 0.1419 | 1 | 11 | 18 | 0.0273 | 13 | 14 | 18 | 0.0913 |
| 1 | 3 | 18 | 0.0965 | 2 | 11 | 18 | 0.0332 | 1 | 15 | 18 | 0.0174 |
| 2 | 3 | 18 | 0.1162 | 3 | 11 | 18 | 0.0380 | 2 | 15 | 18 | 0.0211 |
| 1 | 4 | 18 | 0.0739 | 4 | 11 | 18 | 0.0428 | 3 | 15 | 18 | 0.0242 |
| 2 | 4 | 18 | 0.0892 | 5 | 11 | 18 | 0.0476 | 4 | 15 | 18 | 0.0273 |
| 3 | 4 | 18 | 0.1018 | 6 | 11 | 18 | 0.0528 | 5 | 15 | 18 | 0.0304 |
| 1 | 5 | 18 | 0.0601 | 7 | 11 | 18 | 0.0585 | 6 | 15 | 18 | 0.0337 |
| 2 | 5 | 18 | 0.0727 | 8 | 11 | 18 | 0.0649 | 7 | 15 | 18 | 0.0374 |
| 3 | 5 | 18 | 0.0831 | 9 | 11 | 18 | 0.0724 | 8 | 15 | 18 | 0.0416 |
| 4 | 5 | 18 | 0.0931 | 10 | 11 | 18 | 0.0813 | 9 | 15 | 18 | 0.0465 |
| 1 | 6 | 18 | 0.0507 | 1 | 12 | 18 | 0.0246 | 10 | 15 | 18 | 0.0523 |
| 2 | 6 | 18 | 0.0614 | 2 | 12 | 18 | 0.0298 | 11 | 15 | 18 | 0.0595 |
| 3 | 6 | 18 | 0.0702 | 3 | 12 | 18 | 0.0342 | 12 | 15 | 18 | 0.0687 |
| 4 | 6 | 18 | 0.0788 | 4 | 12 | 18 | 0.0385 | 13 | 15 | 18 | 0.0810 |
| 5 | 6 | 18 | 0.0875 | 5 | 12 | 18 | 0.0429 | 14 | 15 | 18 | 0.0987 |
| 1 | 7 | 18 | 0.0438 | 6 | 12 | 18 | 0.0475 | 1 | 16 | 18 | 0.0149 |
| 2 | 7 | 18 | 0.0530 | 7 | 12 | 18 | 0.0527 | 2 | 16 | 18 | 0.0181 |
| 3 | 7 | 18 | 0.0607 | 8 | 12 | 18 | 0.0585 | 3 | 16 | 18 | 0.0208 |
| 4 | 7 | 18 | 0.0682 | 9 | 12 | 18 | 0.0653 | 4 | 16 | 18 | 0.0234 |
| 5 | 7 | 18 | 0.0758 | 10 | 12 | 18 | 0.0733 | 5 | 16 | 18 | 0.0260 |
| 6 | 7 | 18 | 0.0839 | 11 | 12 | 18 | 0.0832 | 6 | 16 | 18 | 0.0289 |
| 1 | 8 | 18 | 0.0384 | 1 | 13 | 18 | 0.0221 | 7 | 16 | 18 | 0.0321 |
| 2 | 8 | 18 | 0.0466 | 2 | 13 | 18 | 0.0268 | 8 | 16 | 18 | 0.0357 |
| 3 | 8 | 18 | 0.0533 | 3 | 13 | 18 | 0.0308 | 9 | 16 | 18 | 0.0399 |
| 4 | 8 | 18 | 0.0599 | 4 | 13 | 18 | 0.0346 | 10 | 16 | 18 | 0.0450 |
| 5 | 8 | 18 | 0.0666 | 5 | 13 | 18 | 0.0385 | 11 | 16 | 18 | 0.0512 |
| 6 | 8 | 18 | 0.0738 | 6 | 13 | 18 | 0.0427 | 12 | 16 | 18 | 0.0592 |
| 7 | 8 | 18 | 0.0816 | 7 | 13 | 18 | 0.0474 | 13 | 16 | 18 | 0.0700 |
| 1 | 9 | 18 | 0.0341 | 8 | 13 | 18 | 0.0527 | 14 | 16 | 18 | 0.0855 |
| 2 | 9 | 18 | 0.0413 | 9 | 13 | 18 | 0.0588 | 15 | 16 | 18 | 0.1100 |
| 3 | 9 | 18 | 0.0474 | 10 | 13 | 18 | 0.0661 | 1 | 17 | 18 | 0.0116 |
| 4 | 9 | 18 | 0.0532 | 11 | 13 | 18 | 0.0750 | 2 | 17 | 18 | 0.0142 |
| 5 | 9 | 18 | 0.0592 | 12 | 13 | 18 | 0.0864 | 3 | 17 | 18 | 0.0163 |
| 6 | 9 | 18 | 0.0656 | 1 | 14 | 18 | 0.0197 | 4 | 17 | 18 | 0.0183 |
| 7 | 9 | 18 | 0.0726 | 2 | 14 | 18 | 0.0239 | 5 | 17 | 18 | 0.0204 |
| 8 | 9 | 18 | 0.0805 | 3 | 14 | 18 | 0.0275 | 6 | 17 | 18 | 0.0227 |
| 1 | 10 | 18 | 0.0305 | 4 | 14 | 18 | 0.0309 | 7 | 17 | 18 | 0.0252 |
| 2 | 10 | 18 | 0.0369 | 5 | 14 | 18 | 0.0344 | 8 | 17 | 18 | 0.0281 |
| 3 | 10 | 18 | 0.0424 | 6 | 14 | 18 | 0.0382 | 9 | 17 | 18 | 0.0315 |

TABLE 3

OUTLIER MEAN 2.5 OUTLIER STANDARD DEVIATION 1.0

| R | S | N | COVAR | R | S | N | COVAR | R | S | N | COVAR |
|---|---|---|---|---|---|---|---|---|---|---|---|
| 10 | 17 | 18 | 0.0355 | 1 | 8 | 19 | 0.0379 | 1 | 13 | 19 | 0.0223 |
| 11 | 17 | 18 | 0.0405 | 2 | 8 | 19 | 0.0457 | 2 | 13 | 19 | 0.0269 |
| 12 | 17 | 18 | 0.0470 | 3 | 8 | 19 | 0.0522 | 3 | 13 | 19 | 0.0307 |
| 13 | 17 | 18 | 0.0558 | 4 | 8 | 19 | 0.0584 | 4 | 13 | 19 | 0.0344 |
| 14 | 17 | 18 | 0.0685 | 5 | 8 | 19 | 0.0646 | 5 | 13 | 19 | 0.0382 |
| 15 | 17 | 18 | 0.0889 | 6 | 8 | 19 | 0.0712 | 6 | 13 | 19 | 0.0421 |
| 16 | 17 | 18 | 0.1274 | 7 | 8 | 19 | 0.0784 | 7 | 13 | 19 | 0.0464 |
| 1 | 18 | 18 | 0.0030 | 1 | 9 | 19 | 0.0337 | 8 | 13 | 19 | 0.0513 |
| 2 | 18 | 18 | 0.0037 | 2 | 9 | 19 | 0.0407 | 9 | 13 | 19 | 0.0567 |
| 3 | 18 | 18 | 0.0043 | 3 | 9 | 19 | 0.0464 | 10 | 13 | 19 | 0.0631 |
| 4 | 18 | 18 | 0.0049 | 4 | 9 | 19 | 0.0520 | 11 | 13 | 19 | 0.0707 |
| 5 | 18 | 18 | 0.0055 | 5 | 9 | 19 | 0.0576 | 12 | 13 | 19 | 0.0801 |
| 6 | 18 | 18 | 0.0061 | 6 | 9 | 19 | 0.0635 | 1 | 14 | 19 | 0.0201 |
| 7 | 18 | 18 | 0.0069 | 7 | 9 | 19 | 0.0699 | 2 | 14 | 19 | 0.0243 |
| 8 | 18 | 18 | 0.0079 | 8 | 9 | 19 | 0.0770 | 3 | 14 | 19 | 0.0278 |
| 9 | 18 | 18 | 0.0090 | 1 | 10 | 19 | 0.0302 | 4 | 14 | 19 | 0.0311 |
| 10 | 18 | 18 | 0.0104 | 2 | 10 | 19 | 0.0365 | 5 | 14 | 19 | 0.0345 |
| 11 | 18 | 18 | 0.0123 | 3 | 10 | 19 | 0.0417 | 6 | 14 | 19 | 0.0381 |
| 12 | 18 | 18 | 0.0149 | 4 | 10 | 19 | 0.0466 | 7 | 14 | 19 | 0.0420 |
| 13 | 18 | 18 | 0.0186 | 5 | 10 | 19 | 0.0517 | 8 | 14 | 19 | 0.0463 |
| 14 | 18 | 18 | 0.0245 | 6 | 10 | 19 | 0.0570 | 9 | 14 | 19 | 0.0513 |
| 15 | 18 | 18 | 0.0348 | 7 | 10 | 19 | 0.0628 | 10 | 14 | 19 | 0.0571 |
| 16 | 18 | 18 | 0.0564 | 8 | 10 | 19 | 0.0692 | 11 | 14 | 19 | 0.0640 |
| 17 | 18 | 18 | 0.1203 | 9 | 10 | 19 | 0.0765 | 12 | 14 | 19 | 0.0726 |
| 1 | 2 | 19 | 0.1392 | 1 | 11 | 19 | 0.0272 | 13 | 14 | 19 | 0.0835 |
| 1 | 3 | 19 | 0.0946 | 2 | 11 | 19 | 0.0329 | 1 | 15 | 19 | 0.0180 |
| 2 | 3 | 19 | 0.1135 | 3 | 11 | 19 | 0.0376 | 2 | 15 | 19 | 0.0218 |
| 1 | 4 | 19 | 0.0725 | 4 | 11 | 19 | 0.0420 | 3 | 15 | 19 | 0.0249 |
| 2 | 4 | 19 | 0.0871 | 5 | 11 | 19 | 0.0466 | 4 | 15 | 19 | 0.0279 |
| 3 | 4 | 19 | 0.0991 | 6 | 11 | 19 | 0.0514 | 5 | 15 | 19 | 0.0309 |
| 1 | 5 | 19 | 0.0590 | 7 | 11 | 19 | 0.0567 | 6 | 15 | 19 | 0.0342 |
| 2 | 5 | 19 | 0.0711 | 8 | 11 | 19 | 0.0625 | 7 | 15 | 19 | 0.0377 |
| 3 | 5 | 19 | 0.0810 | 9 | 11 | 19 | 0.0691 | 8 | 15 | 19 | 0.0416 |
| 4 | 5 | 19 | 0.0903 | 10 | 11 | 19 | 0.0768 | 9 | 15 | 19 | 0.0461 |
| 1 | 6 | 19 | 0.0498 | 1 | 12 | 19 | 0.0246 | 10 | 15 | 19 | 0.0513 |
| 2 | 6 | 19 | 0.0601 | 2 | 12 | 19 | 0.0297 | 11 | 15 | 19 | 0.0576 |
| 3 | 6 | 19 | 0.0685 | 3 | 12 | 19 | 0.0340 | 12 | 15 | 19 | 0.0653 |
| 4 | 6 | 19 | 0.0765 | 4 | 12 | 19 | 0.0380 | 13 | 15 | 19 | 0.0752 |
| 5 | 6 | 19 | 0.0846 | 5 | 12 | 19 | 0.0422 | 14 | 15 | 19 | 0.0885 |
| 1 | 7 | 19 | 0.0431 | 6 | 12 | 19 | 0.0465 | 1 | 16 | 19 | 0.0159 |
| 2 | 7 | 19 | 0.0520 | 7 | 12 | 19 | 0.0513 | 2 | 16 | 19 | 0.0193 |
| 3 | 7 | 19 | 0.0593 | 8 | 12 | 19 | 0.0566 | 3 | 16 | 19 | 0.0220 |
| 4 | 7 | 19 | 0.0663 | 9 | 12 | 19 | 0.0626 | 4 | 16 | 19 | 0.0247 |
| 5 | 7 | 19 | 0.0734 | 10 | 12 | 19 | 0.0696 | 5 | 16 | 19 | 0.0274 |
| 6 | 7 | 19 | 0.0808 | 11 | 12 | 19 | 0.0780 | 6 | 16 | 19 | 0.0303 |

TABLE 3

OUTLIER MEAN 2.5 OUTLIER STANDARD DEVIATION 1.0

| R | S | N | COVAR | R | S | N | COVAR | R | S | N | COVAR |
|---|---|---|---|---|---|---|---|---|---|---|---|
| 7 | 16 | 19 | 0.0334 | 4 | 19 | 19 | 0.0046 | 3 | 9 | 20 | 0.0456 |
| 8 | 16 | 19 | 0.0369 | 5 | 19 | 19 | 0.0051 | 4 | 9 | 20 | 0.0508 |
| 9 | 16 | 19 | 0.0409 | 6 | 19 | 19 | 0.0057 | 5 | 9 | 20 | 0.0560 |
| 10 | 16 | 19 | 0.0456 | 7 | 19 | 19 | 0.0064 | 6 | 9 | 20 | 0.0615 |
| 11 | 16 | 19 | 0.0512 | 8 | 19 | 19 | 0.0072 | 7 | 9 | 20 | 0.0674 |
| 12 | 16 | 19 | 0.0581 | 9 | 19 | 19 | 0.0081 | 8 | 9 | 20 | 0.0739 |
| 13 | 16 | 19 | 0.0669 | 10 | 19 | 19 | 0.0092 | 1 | 10 | 20 | 0.0299 |
| 14 | 16 | 19 | 0.0789 | 11 | 19 | 19 | 0.0107 | 2 | 10 | 20 | 0.0360 |
| 15 | 16 | 19 | 0.0959 | 12 | 19 | 19 | 0.0126 | 3 | 10 | 20 | 0.0410 |
| 1 | 17 | 19 | 0.0137 | 13 | 19 | 19 | 0.0152 | 4 | 10 | 20 | 0.0457 |
| 2 | 17 | 19 | 0.0166 | 14 | 19 | 19 | 0.0189 | 5 | 10 | 20 | 0.0504 |
| 3 | 17 | 19 | 0.0190 | 15 | 19 | 19 | 0.0249 | 6 | 10 | 20 | 0.0554 |
| 4 | 17 | 19 | 0.0212 | 16 | 19 | 19 | 0.0352 | 7 | 10 | 20 | 0.0607 |
| 5 | 17 | 19 | 0.0236 | 17 | 19 | 19 | 0.0569 | 8 | 10 | 20 | 0.0665 |
| 6 | 17 | 19 | 0.0261 | 18 | 19 | 19 | 0.1205 | 9 | 10 | 20 | 0.0731 |
| 7 | 17 | 19 | 0.0288 | 1 | 2 | 20 | 0.1368 | 1 | 11 | 20 | 0.0270 |
| 8 | 17 | 19 | 0.0318 | 1 | 3 | 20 | 0.0929 | 2 | 11 | 20 | 0.0325 |
| 9 | 17 | 19 | 0.0352 | 2 | 3 | 20 | 0.1110 | 3 | 11 | 20 | 0.0370 |
| 10 | 17 | 19 | 0.0393 | 1 | 4 | 20 | 0.0712 | 4 | 11 | 20 | 0.0413 |
| 11 | 17 | 19 | 0.0441 | 2 | 4 | 20 | 0.0853 | 5 | 11 | 20 | 0.0456 |
| 12 | 17 | 19 | 0.0502 | 3 | 4 | 20 | 0.0967 | 6 | 11 | 20 | 0.0501 |
| 13 | 17 | 19 | 0.0579 | 1 | 5 | 20 | 0.0580 | 7 | 11 | 20 | 0.0549 |
| 14 | 17 | 19 | 0.0684 | 2 | 5 | 20 | 0.0696 | 8 | 11 | 20 | 0.0603 |
| 15 | 17 | 19 | 0.0834 | 3 | 5 | 20 | 0.0790 | 9 | 11 | 20 | 0.0662 |
| 16 | 17 | 19 | 0.1073 | 4 | 5 | 20 | 0.0879 | 10 | 11 | 20 | 0.0730 |
| 1 | 18 | 19 | 0.0108 | 1 | 6 | 20 | 0.0490 | 1 | 12 | 20 | 0.0245 |
| 2 | 18 | 19 | 0.0130 | 2 | 6 | 20 | 0.0589 | 2 | 12 | 20 | 0.0295 |
| 3 | 18 | 19 | 0.0149 | 3 | 6 | 20 | 0.0669 | 3 | 12 | 20 | 0.0336 |
| 4 | 18 | 19 | 0.0167 | 4 | 6 | 20 | 0.0745 | 4 | 12 | 20 | 0.0375 |
| 5 | 18 | 19 | 0.0186 | 5 | 6 | 20 | 0.0820 | 5 | 12 | 20 | 0.0414 |
| 6 | 18 | 19 | 0.0205 | 1 | 7 | 20 | 0.0424 | 6 | 12 | 20 | 0.0455 |
| 7 | 18 | 19 | 0.0227 | 2 | 7 | 20 | 0.0510 | 7 | 12 | 20 | 0.0499 |
| 8 | 18 | 19 | 0.0251 | 3 | 7 | 20 | 0.0580 | 8 | 12 | 20 | 0.0548 |
| 9 | 18 | 19 | 0.0278 | 4 | 7 | 20 | 0.0646 | 9 | 12 | 20 | 0.0602 |
| 10 | 18 | 19 | 0.0311 | 5 | 7 | 20 | 0.0712 | 10 | 12 | 20 | 0.0664 |
| 11 | 18 | 19 | 0.0350 | 6 | 7 | 20 | 0.0781 | 11 | 12 | 20 | 0.0736 |
| 12 | 18 | 19 | 0.0399 | 1 | 8 | 20 | 0.0374 | 1 | 13 | 20 | 0.0223 |
| 13 | 18 | 19 | 0.0462 | 2 | 8 | 20 | 0.0449 | 2 | 13 | 20 | 0.0269 |
| 14 | 18 | 19 | 0.0548 | 3 | 8 | 20 | 0.0511 | 3 | 13 | 20 | 0.0306 |
| 15 | 18 | 19 | 0.0672 | 4 | 8 | 20 | 0.0569 | 4 | 13 | 20 | 0.0341 |
| 16 | 18 | 19 | 0.0872 | 5 | 8 | 20 | 0.0628 | 5 | 13 | 20 | 0.0377 |
| 17 | 18 | 19 | 0.1249 | 6 | 8 | 20 | 0.0689 | 6 | 13 | 20 | 0.0414 |
| 1 | 19 | 19 | 0.0029 | 7 | 8 | 20 | 0.0755 | 7 | 13 | 20 | 0.0454 |
| 2 | 19 | 19 | 0.0035 | 1 | 9 | 20 | 0.0333 | 8 | 13 | 20 | 0.0499 |
| 3 | 19 | 19 | 0.0040 | 2 | 9 | 20 | 0.0400 | 9 | 13 | 20 | 0.0548 |

TABLE 3

OUTLIER MEAN 2.5 OUTLIER STANDARD DEVIATION 1.0

| R | S | N | COVAR | R | S | N | COVAR | R | S | N | COVAR |
|---|---|---|---|---|---|---|---|---|---|---|---|
| 10 | 13 | 20 | 0.0605 | 10 | 16 | 20 | 0.0451 | 1 | 19 | 20 | 0.0100 |
| 11 | 13 | 20 | 0.0671 | 11 | 16 | 20 | 0.0501 | 2 | 19 | 20 | 0.0120 |
| 12 | 13 | 20 | 0.0750 | 12 | 16 | 20 | 0.0562 | 3 | 19 | 20 | 0.0137 |
| 1 | 14 | 20 | 0.0203 | 13 | 16 | 20 | 0.0636 | 4 | 19 | 20 | 0.0154 |
| 2 | 14 | 20 | 0.0244 | 14 | 16 | 20 | 0.0732 | 5 | 19 | 20 | 0.0170 |
| 3 | 14 | 20 | 0.0278 | 15 | 16 | 20 | 0.0860 | 6 | 19 | 20 | 0.0187 |
| 4 | 14 | 20 | 0.0310 | 1 | 17 | 20 | 0.0147 | 7 | 19 | 20 | 0.0205 |
| 5 | 14 | 20 | 0.0343 | 2 | 17 | 20 | 0.0177 | 8 | 19 | 20 | 0.0226 |
| 6 | 14 | 20 | 0.0377 | 3 | 17 | 20 | 0.0202 | 9 | 19 | 20 | 0.0249 |
| 7 | 14 | 20 | 0.0414 | 4 | 17 | 20 | 0.0225 | 10 | 19 | 20 | 0.0276 |
| 8 | 14 | 20 | 0.0454 | 5 | 17 | 20 | 0.0249 | 11 | 19 | 20 | 0.0307 |
| 9 | 14 | 20 | 0.0499 | 6 | 17 | 20 | 0.0274 | 12 | 19 | 20 | 0.0345 |
| 10 | 14 | 20 | 0.0551 | 7 | 17 | 20 | 0.0301 | 13 | 19 | 20 | 0.0393 |
| 11 | 14 | 20 | 0.0611 | 8 | 17 | 20 | 0.0330 | 14 | 19 | 20 | 0.0455 |
| 12 | 14 | 20 | 0.0684 | 9 | 17 | 20 | 0.0364 | 15 | 19 | 20 | 0.0538 |
| 13 | 14 | 20 | 0.0774 | 10 | 17 | 20 | 0.0402 | 16 | 19 | 20 | 0.0660 |
| 1 | 15 | 20 | 0.0184 | 11 | 17 | 20 | 0.0447 | 17 | 19 | 20 | 0.0856 |
| 2 | 15 | 20 | 0.0221 | 12 | 17 | 20 | 0.0501 | 18 | 19 | 20 | 0.1226 |
| 3 | 15 | 20 | 0.0252 | 13 | 17 | 20 | 0.0567 | 1 | 20 | 20 | 0.0028 |
| 4 | 15 | 20 | 0.0281 | 14 | 17 | 20 | 0.0653 | 2 | 20 | 20 | 0.0033 |
| 5 | 15 | 20 | 0.0311 | 15 | 17 | 20 | 0.0769 | 3 | 20 | 20 | 0.0038 |
| 6 | 15 | 20 | 0.0342 | 16 | 17 | 20 | 0.0935 | 4 | 20 | 20 | 0.0043 |
| 7 | 15 | 20 | 0.0375 | 1 | 18 | 20 | 0.0127 | 5 | 20 | 20 | 0.0048 |
| 8 | 15 | 20 | 0.0412 | 2 | 18 | 20 | 0.0153 | 6 | 20 | 20 | 0.0053 |
| 9 | 15 | 20 | 0.0453 | 3 | 18 | 20 | 0.0174 | 7 | 20 | 20 | 0.0059 |
| 10 | 15 | 20 | 0.0500 | 4 | 18 | 20 | 0.0194 | 8 | 20 | 20 | 0.0066 |
| 11 | 15 | 20 | 0.0556 | 5 | 18 | 20 | 0.0215 | 9 | 20 | 20 | 0.0074 |
| 12 | 15 | 20 | 0.0622 | 6 | 18 | 20 | 0.0236 | 10 | 20 | 20 | 0.0083 |
| 13 | 15 | 20 | 0.0704 | 7 | 18 | 20 | 0.0260 | 11 | 20 | 20 | 0.0094 |
| 14 | 15 | 20 | 0.0809 | 8 | 18 | 20 | 0.0285 | 12 | 20 | 20 | 0.0109 |
| 1 | 16 | 20 | 0.0165 | 9 | 18 | 20 | 0.0314 | 13 | 20 | 20 | 0.0128 |
| 2 | 16 | 20 | 0.0199 | 10 | 18 | 20 | 0.0347 | 14 | 20 | 20 | 0.0154 |
| 3 | 16 | 20 | 0.0227 | 11 | 18 | 20 | 0.0386 | 15 | 20 | 20 | 0.0193 |
| 4 | 16 | 20 | 0.0254 | 12 | 18 | 20 | 0.0434 | 16 | 20 | 20 | 0.0252 |
| 5 | 16 | 20 | 0.0280 | 13 | 18 | 20 | 0.0492 | 17 | 20 | 20 | 0.0357 |
| 6 | 16 | 20 | 0.0308 | 14 | 18 | 20 | 0.0567 | 18 | 20 | 20 | 0.0573 |
| 7 | 16 | 20 | 0.0338 | 15 | 18 | 20 | 0.0669 | 19 | 20 | 20 | 0.1207 |
| 8 | 16 | 20 | 0.0371 | 16 | 18 | 20 | 0.0816 | | | | |
| 9 | 16 | 20 | 0.0409 | 17 | 18 | 20 | 0.1049 | | | | |

TABLE 3

OUTLIER MEAN 3.0 OUTLIER STANDARD DEVIATION 1.0

| R | S | N | COVAR | R | S | N | COVAR | R | S | N | COVAR |
|---|---|---|---|---|---|---|---|---|---|---|---|
| 1 | 2 | 2 | 0.0259 | 1 | 6 | 7 | 0.0543 | 1 | 5 | 9 | 0.0746 |
| 1 | 2 | 3 | 0.3119 | 2 | 6 | 7 | 0.0745 | 2 | 5 | 9 | 0.0974 |
| 1 | 3 | 3 | 0.0090 | 3 | 6 | 7 | 0.0986 | 3 | 5 | 9 | 0.1207 |
| 2 | 3 | 3 | 0.0384 | 4 | 6 | 7 | 0.1340 | 4 | 5 | 9 | 0.1488 |
| 1 | 2 | 4 | 0.2745 | 5 | 6 | 7 | 0.1989 | 1 | 6 | 9 | 0.0599 |
| 1 | 3 | 4 | 0.1609 | 1 | 7 | 7 | 0.0033 | 2 | 6 | 9 | 0.0784 |
| 2 | 3 | 4 | 0.2677 | 2 | 7 | 7 | 0.0048 | 3 | 6 | 9 | 0.0974 |
| 1 | 4 | 4 | 0.0058 | 3 | 7 | 7 | 0.0069 | 4 | 6 | 9 | 0.1203 |
| 2 | 4 | 4 | 0.0129 | 4 | 7 | 7 | 0.0108 | 5 | 6 | 9 | 0.1514 |
| 3 | 4 | 4 | 0.0469 | 5 | 7 | 7 | 0.0207 | 1 | 7 | 9 | 0.0477 |
| 1 | 2 | 5 | 0.2452 | 6 | 7 | 7 | 0.0628 | 2 | 7 | 9 | 0.0625 |
| 1 | 3 | 5 | 0.1571 | 1 | 2 | 8 | 0.1962 | 3 | 7 | 9 | 0.0778 |
| 2 | 3 | 5 | 0.2343 | 1 | 3 | 8 | 0.1320 | 4 | 7 | 9 | 0.0964 |
| 1 | 4 | 5 | 0.1017 | 2 | 3 | 8 | 0.1743 | 5 | 7 | 9 | 0.1216 |
| 2 | 4 | 5 | 0.1532 | 1 | 4 | 8 | 0.0983 | 6 | 7 | 9 | 0.1605 |
| 3 | 4 | 5 | 0.2368 | 2 | 4 | 8 | 0.1305 | 1 | 8 | 9 | 0.0353 |
| 1 | 5 | 5 | 0.0045 | 3 | 4 | 8 | 0.1652 | 2 | 8 | 9 | 0.0462 |
| 2 | 5 | 5 | 0.0078 | 1 | 5 | 8 | 0.0763 | 3 | 8 | 9 | 0.0576 |
| 3 | 5 | 5 | 0.0160 | 2 | 5 | 8 | 0.1016 | 4 | 8 | 9 | 0.0716 |
| 4 | 5 | 5 | 0.0534 | 3 | 5 | 8 | 0.1291 | 5 | 8 | 9 | 0.0906 |
| 1 | 2 | 6 | 0.2242 | 4 | 5 | 8 | 0.1647 | 6 | 8 | 9 | 0.1203 |
| 1 | 3 | 6 | 0.1478 | 1 | 6 | 8 | 0.0593 | 7 | 8 | 9 | 0.1764 |
| 2 | 3 | 6 | 0.2079 | 2 | 6 | 8 | 0.0792 | 1 | 9 | 9 | 0.0027 |
| 1 | 4 | 6 | 0.1050 | 3 | 6 | 8 | 0.1009 | 2 | 9 | 9 | 0.0036 |
| 2 | 4 | 6 | 0.1488 | 4 | 6 | 8 | 0.1292 | 3 | 9 | 9 | 0.0047 |
| 3 | 4 | 6 | 0.2064 | 5 | 6 | 8 | 0.1720 | 4 | 9 | 9 | 0.0061 |
| 1 | 5 | 6 | 0.0718 | 1 | 7 | 8 | 0.0430 | 5 | 9 | 9 | 0.0085 |
| 2 | 5 | 6 | 0.1022 | 2 | 7 | 8 | 0.0575 | 6 | 9 | 9 | 0.0130 |
| 3 | 5 | 6 | 0.1430 | 3 | 7 | 8 | 0.0735 | 7 | 9 | 9 | 0.0243 |
| 4 | 5 | 6 | 0.2151 | 4 | 7 | 8 | 0.0945 | 8 | 9 | 9 | 0.0693 |
| 1 | 6 | 6 | 0.0038 | 5 | 7 | 8 | 0.1265 | 1 | 2 | 10 | 0.1781 |
| 2 | 6 | 6 | 0.0058 | 6 | 7 | 8 | 0.1864 | 1 | 3 | 10 | 0.1207 |
| 3 | 6 | 6 | 0.0094 | 1 | 8 | 8 | 0.0030 | 2 | 3 | 10 | 0.1541 |
| 4 | 6 | 6 | 0.0186 | 2 | 8 | 8 | 0.0041 | 1 | 4 | 10 | 0.0913 |
| 5 | 6 | 6 | 0.0586 | 3 | 8 | 8 | 0.0055 | 2 | 4 | 10 | 0.1169 |
| 1 | 2 | 7 | 0.2084 | 4 | 8 | 8 | 0.0077 | 3 | 4 | 10 | 0.1420 |
| 1 | 3 | 7 | 0.1393 | 5 | 8 | 8 | 0.0120 | 1 | 5 | 10 | 0.0727 |
| 2 | 3 | 7 | 0.1887 | 6 | 8 | 8 | 0.0226 | 2 | 5 | 10 | 0.0933 |
| 1 | 4 | 7 | 0.1021 | 7 | 8 | 8 | 0.0663 | 3 | 5 | 10 | 0.1136 |
| 2 | 4 | 7 | 0.1392 | 1 | 2 | 9 | 0.1863 | 4 | 5 | 10 | 0.1368 |
| 3 | 4 | 7 | 0.1825 | 1 | 3 | 9 | 0.1259 | 1 | 6 | 10 | 0.0593 |
| 1 | 5 | 7 | 0.0767 | 2 | 3 | 9 | 0.1631 | 2 | 6 | 10 | 0.0764 |
| 2 | 5 | 7 | 0.1049 | 1 | 4 | 9 | 0.0946 | 3 | 6 | 10 | 0.0931 |
| 3 | 5 | 7 | 0.1383 | 2 | 4 | 9 | 0.1231 | 4 | 6 | 10 | 0.1124 |
| 4 | 5 | 7 | 0.1867 | 3 | 4 | 9 | 0.1522 | 5 | 6 | 10 | 0.1365 |

TABLE 3

OUTLIER MEAN 3.0 OUTLIER STANDARD DEVIATION 1.0

| R | S | N | COVAR | R | S | N | COVAR | R | S | N | COVAR |
|---|---|---|---|---|---|---|---|---|---|---|---|
| 1 | 7 | 10 | 0.0488 | 1 | 7 | 11 | 0.0488 | 3 | 4 | 12 | 0.1269 |
| 2 | 7 | 10 | 0.0629 | 2 | 7 | 11 | 0.0621 | 1 | 5 | 12 | 0.0688 |
| 3 | 7 | 10 | 0.0768 | 3 | 7 | 11 | 0.0747 | 2 | 5 | 12 | 0.0864 |
| 4 | 7 | 10 | 0.0928 | 4 | 7 | 11 | 0.0887 | 3 | 5 | 12 | 0.1026 |
| 5 | 7 | 10 | 0.1131 | 5 | 7 | 11 | 0.1054 | 4 | 5 | 12 | 0.1198 |
| 6 | 7 | 10 | 0.1410 | 6 | 7 | 11 | 0.1270 | 1 | 6 | 12 | 0.0572 |
| 1 | 8 | 10 | 0.0396 | 1 | 8 | 11 | 0.0408 | 2 | 6 | 12 | 0.0719 |
| 2 | 8 | 10 | 0.0510 | 2 | 8 | 11 | 0.0520 | 3 | 6 | 12 | 0.0854 |
| 3 | 8 | 10 | 0.0624 | 3 | 8 | 11 | 0.0627 | 4 | 6 | 12 | 0.0999 |
| 4 | 8 | 10 | 0.0755 | 4 | 8 | 11 | 0.0745 | 5 | 6 | 12 | 0.1166 |
| 5 | 8 | 10 | 0.0922 | 5 | 8 | 11 | 0.0887 | 1 | 7 | 12 | 0.0483 |
| 6 | 8 | 10 | 0.1153 | 6 | 8 | 11 | 0.1070 | 2 | 7 | 12 | 0.0608 |
| 7 | 8 | 10 | 0.1513 | 7 | 8 | 11 | 0.1326 | 3 | 7 | 12 | 0.0724 |
| 1 | 9 | 10 | 0.0296 | 1 | 9 | 11 | 0.0336 | 4 | 7 | 12 | 0.0847 |
| 2 | 9 | 10 | 0.0383 | 2 | 9 | 11 | 0.0428 | 5 | 7 | 12 | 0.0990 |
| 3 | 9 | 10 | 0.0468 | 3 | 9 | 11 | 0.0516 | 6 | 7 | 12 | 0.1164 |
| 4 | 9 | 10 | 0.0568 | 4 | 9 | 11 | 0.0613 | 1 | 8 | 12 | 0.0411 |
| 5 | 9 | 10 | 0.0694 | 5 | 9 | 11 | 0.0731 | 2 | 8 | 12 | 0.0518 |
| 6 | 9 | 10 | 0.0871 | 6 | 9 | 11 | 0.0884 | 3 | 8 | 12 | 0.0617 |
| 7 | 9 | 10 | 0.1150 | 7 | 9 | 11 | 0.1099 | 4 | 8 | 12 | 0.0723 |
| 8 | 9 | 10 | 0.1681 | 8 | 9 | 11 | 0.1437 | 5 | 8 | 12 | 0.0846 |
| 1 | 10 | 10 | 0.0025 | 1 | 10 | 11 | 0.0254 | 6 | 8 | 12 | 0.0996 |
| 2 | 10 | 10 | 0.0033 | 2 | 10 | 11 | 0.0324 | 7 | 8 | 12 | 0.1193 |
| 3 | 10 | 10 | 0.0041 | 3 | 10 | 11 | 0.0391 | 1 | 9 | 12 | 0.0349 |
| 4 | 10 | 10 | 0.0052 | 4 | 10 | 11 | 0.0466 | 2 | 9 | 12 | 0.0440 |
| 5 | 10 | 10 | 0.0067 | 5 | 10 | 11 | 0.0556 | 3 | 9 | 12 | 0.0524 |
| 6 | 10 | 10 | 0.0092 | 6 | 10 | 11 | 0.0674 | 4 | 9 | 12 | 0.0615 |
| 7 | 10 | 10 | 0.0140 | 7 | 10 | 11 | 0.0841 | 5 | 9 | 12 | 0.0720 |
| 8 | 10 | 10 | 0.0257 | 8 | 10 | 11 | 0.1105 | 6 | 9 | 12 | 0.0849 |
| 9 | 10 | 10 | 0.0719 | 9 | 10 | 11 | 0.1612 | 7 | 9 | 12 | 0.1018 |
| 1 | 2 | 11 | 0.1713 | 1 | 11 | 11 | 0.0023 | 8 | 9 | 12 | 0.1257 |
| 1 | 3 | 11 | 0.1162 | 2 | 11 | 11 | 0.0030 | 1 | 10 | 12 | 0.0290 |
| 2 | 3 | 11 | 0.1466 | 3 | 11 | 11 | 0.0037 | 2 | 10 | 12 | 0.0365 |
| 1 | 4 | 11 | 0.0882 | 4 | 11 | 11 | 0.0045 | 3 | 10 | 12 | 0.0436 |
| 2 | 4 | 11 | 0.1117 | 5 | 11 | 11 | 0.0056 | 4 | 10 | 12 | 0.0512 |
| 3 | 4 | 11 | 0.1337 | 6 | 11 | 11 | 0.0072 | 5 | 10 | 12 | 0.0600 |
| 1 | 5 | 11 | 0.0707 | 7 | 11 | 11 | 0.0098 | 6 | 10 | 12 | 0.0708 |
| 2 | 5 | 11 | 0.0897 | 8 | 11 | 11 | 0.0148 | 7 | 10 | 12 | 0.0851 |
| 3 | 5 | 11 | 0.1077 | 9 | 11 | 11 | 0.0271 | 8 | 10 | 12 | 0.1053 |
| 4 | 5 | 11 | 0.1274 | 10 | 11 | 11 | 0.0741 | 9 | 10 | 12 | 0.1373 |
| 1 | 6 | 11 | 0.0583 | 1 | 2 | 12 | 0.1654 | 1 | 11 | 12 | 0.0222 |
| 2 | 6 | 11 | 0.0741 | 1 | 3 | 12 | 0.1123 | 2 | 11 | 12 | 0.0280 |
| 3 | 6 | 11 | 0.0891 | 2 | 3 | 12 | 0.1403 | 3 | 11 | 12 | 0.0334 |
| 4 | 6 | 11 | 0.1056 | 1 | 4 | 12 | 0.0855 | 4 | 11 | 12 | 0.0392 |
| 5 | 6 | 11 | 0.1253 | 2 | 4 | 12 | 0.1071 | 5 | 11 | 12 | 0.0460 |

TABLE 3

OUTLIER MEAN 3.0 OUTLIER STANDARD DEVIATION 1.0

| R | S | N | COVAR | R | S | N | COVAR | R | S | N | COVAR |
|---|---|---|---|---|---|---|---|---|---|---|---|
| 6 | 11 | 12 | 0.0544 | 2 | 9 | 13 | 0.0441 | 9 | 13 | 13 | 0.0109 |
| 7 | 11 | 12 | 0.0655 | 3 | 9 | 13 | 0.0521 | 10 | 13 | 13 | 0.0163 |
| 8 | 11 | 12 | 0.0813 | 4 | 9 | 13 | 0.0604 | 11 | 13 | 13 | 0.0293 |
| 9 | 11 | 12 | 0.1066 | 5 | 9 | 13 | 0.0698 | 12 | 13 | 13 | 0.0779 |
| 10 | 11 | 12 | 0.1552 | 6 | 9 | 13 | 0.0809 | 1 | 2 | 14 | 0.1557 |
| 1 | 12 | 12 | 0.0022 | 7 | 9 | 13 | 0.0947 | 1 | 3 | 14 | 0.1059 |
| 2 | 12 | 12 | 0.0028 | 8 | 9 | 13 | 0.1129 | 2 | 3 | 14 | 0.1302 |
| 3 | 12 | 12 | 0.0033 | 1 | 10 | 13 | 0.0303 | 1 | 4 | 14 | 0.0809 |
| 4 | 12 | 12 | 0.0040 | 2 | 10 | 13 | 0.0379 | 2 | 4 | 14 | 0.0997 |
| 5 | 12 | 12 | 0.0048 | 3 | 10 | 13 | 0.0447 | 3 | 4 | 14 | 0.1162 |
| 6 | 12 | 12 | 0.0059 | 4 | 10 | 13 | 0.0519 | 1 | 5 | 14 | 0.0654 |
| 7 | 12 | 12 | 0.0076 | 5 | 10 | 13 | 0.0600 | 2 | 5 | 14 | 0.0809 |
| 8 | 12 | 12 | 0.0103 | 6 | 10 | 13 | 0.0696 | 3 | 5 | 14 | 0.0944 |
| 9 | 12 | 12 | 0.0156 | 7 | 10 | 13 | 0.0816 | 4 | 5 | 14 | 0.1082 |
| 10 | 12 | 12 | 0.0282 | 8 | 10 | 13 | 0.0974 | 1 | 6 | 14 | 0.0548 |
| 11 | 12 | 12 | 0.0761 | 9 | 10 | 13 | 0.1199 | 2 | 6 | 14 | 0.0678 |
| 1 | 2 | 13 | 0.1602 | 1 | 11 | 13 | 0.0254 | 3 | 6 | 14 | 0.0793 |
| 1 | 3 | 13 | 0.1089 | 2 | 11 | 13 | 0.0317 | 4 | 6 | 14 | 0.0909 |
| 2 | 3 | 13 | 0.1349 | 3 | 11 | 13 | 0.0375 | 5 | 6 | 14 | 0.1037 |
| 1 | 4 | 13 | 0.0831 | 4 | 11 | 13 | 0.0436 | 1 | 7 | 14 | 0.0468 |
| 2 | 4 | 13 | 0.1032 | 5 | 11 | 13 | 0.0504 | 2 | 7 | 14 | 0.0580 |
| 3 | 4 | 13 | 0.1212 | 6 | 11 | 13 | 0.0585 | 3 | 7 | 14 | 0.0679 |
| 1 | 5 | 13 | 0.0671 | 7 | 11 | 13 | 0.0686 | 4 | 7 | 14 | 0.0780 |
| 2 | 5 | 13 | 0.0835 | 8 | 11 | 13 | 0.0821 | 5 | 7 | 14 | 0.0890 |
| 3 | 5 | 13 | 0.0982 | 9 | 11 | 13 | 0.1013 | 6 | 7 | 14 | 0.1016 |
| 4 | 5 | 13 | 0.1135 | 10 | 11 | 13 | 0.1318 | 1 | 8 | 14 | 0.0406 |
| 1 | 6 | 13 | 0.0560 | 1 | 12 | 13 | 0.0196 | 2 | 8 | 14 | 0.0503 |
| 2 | 6 | 13 | 0.0697 | 2 | 12 | 13 | 0.0245 | 3 | 8 | 14 | 0.0588 |
| 3 | 6 | 13 | 0.0822 | 3 | 12 | 13 | 0.0289 | 4 | 8 | 14 | 0.0676 |
| 4 | 6 | 13 | 0.0951 | 4 | 12 | 13 | 0.0336 | 5 | 8 | 14 | 0.0772 |
| 5 | 6 | 13 | 0.1095 | 5 | 12 | 13 | 0.0389 | 6 | 8 | 14 | 0.0883 |
| 1 | 7 | 13 | 0.0476 | 6 | 12 | 13 | 0.0452 | 7 | 8 | 14 | 0.1015 |
| 2 | 7 | 13 | 0.0594 | 7 | 12 | 13 | 0.0532 | 1 | 9 | 14 | 0.0353 |
| 3 | 7 | 13 | 0.0700 | 8 | 12 | 13 | 0.0637 | 2 | 9 | 14 | 0.0438 |
| 4 | 7 | 13 | 0.0811 | 9 | 12 | 13 | 0.0789 | 3 | 9 | 14 | 0.0513 |
| 5 | 7 | 13 | 0.0936 | 10 | 12 | 13 | 0.1032 | 4 | 9 | 14 | 0.0590 |
| 6 | 7 | 13 | 0.1082 | 11 | 12 | 13 | 0.1501 | 5 | 9 | 14 | 0.0675 |
| 1 | 8 | 13 | 0.0409 | 1 | 13 | 13 | 0.0020 | 6 | 9 | 14 | 0.0772 |
| 2 | 8 | 13 | 0.0511 | 2 | 13 | 13 | 0.0026 | 7 | 9 | 14 | 0.0888 |
| 3 | 8 | 13 | 0.0603 | 3 | 13 | 13 | 0.0031 | 8 | 9 | 14 | 0.1034 |
| 4 | 8 | 13 | 0.0699 | 4 | 13 | 13 | 0.0036 | 1 | 10 | 14 | 0.0308 |
| 5 | 8 | 13 | 0.0807 | 5 | 13 | 13 | 0.0043 | 2 | 10 | 14 | 0.0382 |
| 6 | 8 | 13 | 0.0935 | 6 | 13 | 13 | 0.0051 | 3 | 10 | 14 | 0.0448 |
| 7 | 8 | 13 | 0.1093 | 7 | 13 | 13 | 0.0063 | 4 | 10 | 14 | 0.0515 |
| 1 | 9 | 13 | 0.0353 | 8 | 13 | 13 | 0.0080 | 5 | 10 | 14 | 0.0589 |

TABLE 3

OUTLIER MEAN 3.0 OUTLIER STANDARD DEVIATION 1.0

| R | S | N | COVAR | R | S | N | COVAR | R | S | N | COVAR |
|---|---|---|---|---|---|---|---|---|---|---|---|
| 6 | 10 | 14 | 0.0674 | 9 | 14 | 14 | 0.0084 | 5 | 10 | 15 | 0.0575 |
| 7 | 10 | 14 | 0.0777 | 10 | 14 | 14 | 0.0113 | 6 | 10 | 15 | 0.0651 |
| 8 | 10 | 14 | 0.0905 | 11 | 14 | 14 | 0.0170 | 7 | 10 | 15 | 0.0740 |
| 9 | 10 | 14 | 0.1076 | 12 | 14 | 14 | 0.0303 | 8 | 10 | 15 | 0.0848 |
| 1 | 11 | 14 | 0.0267 | 13 | 14 | 14 | 0.0794 | 9 | 10 | 15 | 0.0984 |
| 2 | 11 | 14 | 0.0331 | 1 | 2 | 15 | 0.1517 | 1 | 11 | 15 | 0.0272 |
| 3 | 11 | 14 | 0.0388 | 1 | 3 | 15 | 0.1032 | 2 | 11 | 15 | 0.0335 |
| 4 | 11 | 14 | 0.0446 | 2 | 3 | 15 | 0.1261 | 3 | 11 | 15 | 0.0391 |
| 5 | 11 | 14 | 0.0511 | 1 | 4 | 15 | 0.0789 | 4 | 11 | 15 | 0.0446 |
| 6 | 11 | 14 | 0.0585 | 2 | 4 | 15 | 0.0966 | 5 | 11 | 15 | 0.0506 |
| 7 | 11 | 14 | 0.0674 | 3 | 4 | 15 | 0.1120 | 6 | 11 | 15 | 0.0574 |
| 8 | 11 | 14 | 0.0786 | 1 | 5 | 15 | 0.0640 | 7 | 11 | 15 | 0.0652 |
| 9 | 11 | 14 | 0.0936 | 2 | 5 | 15 | 0.0785 | 8 | 11 | 15 | 0.0748 |
| 10 | 11 | 14 | 0.1149 | 3 | 5 | 15 | 0.0911 | 9 | 11 | 15 | 0.0869 |
| 1 | 12 | 14 | 0.0225 | 4 | 5 | 15 | 0.1037 | 10 | 11 | 15 | 0.1030 |
| 2 | 12 | 14 | 0.0279 | 1 | 6 | 15 | 0.0537 | 1 | 12 | 15 | 0.0237 |
| 3 | 12 | 14 | 0.0327 | 2 | 6 | 15 | 0.0660 | 2 | 12 | 15 | 0.0292 |
| 4 | 12 | 14 | 0.0377 | 3 | 6 | 15 | 0.0766 | 3 | 12 | 15 | 0.0341 |
| 5 | 12 | 14 | 0.0432 | 4 | 6 | 15 | 0.0873 | 4 | 12 | 15 | 0.0389 |
| 6 | 12 | 14 | 0.0495 | 5 | 6 | 15 | 0.0987 | 5 | 12 | 15 | 0.0442 |
| 7 | 12 | 14 | 0.0570 | 1 | 7 | 15 | 0.0461 | 6 | 12 | 15 | 0.0501 |
| 8 | 12 | 14 | 0.0666 | 2 | 7 | 15 | 0.0567 | 7 | 12 | 15 | 0.0570 |
| 9 | 12 | 14 | 0.0794 | 3 | 7 | 15 | 0.0659 | 8 | 12 | 15 | 0.0653 |
| 10 | 12 | 14 | 0.0977 | 4 | 7 | 15 | 0.0751 | 9 | 12 | 15 | 0.0760 |
| 11 | 12 | 14 | 0.1270 | 5 | 7 | 15 | 0.0850 | 10 | 12 | 15 | 0.0902 |
| 1 | 13 | 14 | 0.0174 | 6 | 7 | 15 | 0.0961 | 11 | 12 | 15 | 0.1106 |
| 2 | 13 | 14 | 0.0217 | 1 | 8 | 15 | 0.0401 | 1 | 13 | 15 | 0.0201 |
| 3 | 13 | 14 | 0.0254 | 2 | 8 | 15 | 0.0493 | 2 | 13 | 15 | 0.0248 |
| 4 | 13 | 14 | 0.0293 | 3 | 8 | 15 | 0.0574 | 3 | 13 | 15 | 0.0289 |
| 5 | 13 | 14 | 0.0335 | 4 | 8 | 15 | 0.0655 | 4 | 13 | 15 | 0.0331 |
| 6 | 13 | 14 | 0.0385 | 5 | 8 | 15 | 0.0741 | 5 | 13 | 15 | 0.0375 |
| 7 | 13 | 14 | 0.0444 | 6 | 8 | 15 | 0.0839 | 6 | 13 | 15 | 0.0426 |
| 8 | 13 | 14 | 0.0520 | 7 | 8 | 15 | 0.0952 | 7 | 13 | 15 | 0.0485 |
| 9 | 13 | 14 | 0.0621 | 1 | 9 | 15 | 0.0352 | 8 | 13 | 15 | 0.0557 |
| 10 | 13 | 14 | 0.0767 | 2 | 9 | 15 | 0.0433 | 9 | 13 | 15 | 0.0648 |
| 11 | 13 | 14 | 0.1002 | 3 | 9 | 15 | 0.0504 | 10 | 13 | 15 | 0.0770 |
| 12 | 13 | 14 | 0.1456 | 4 | 9 | 15 | 0.0575 | 11 | 13 | 15 | 0.0946 |
| 1 | 14 | 14 | 0.0019 | 5 | 9 | 15 | 0.0652 | 12 | 13 | 15 | 0.1228 |
| 2 | 14 | 14 | 0.0024 | 6 | 9 | 15 | 0.0738 | 1 | 14 | 15 | 0.0157 |
| 3 | 14 | 14 | 0.0028 | 7 | 9 | 15 | 0.0838 | 2 | 14 | 15 | 0.0194 |
| 4 | 14 | 14 | 0.0033 | 8 | 9 | 15 | 0.0959 | 3 | 14 | 15 | 0.0226 |
| 5 | 14 | 14 | 0.0039 | 1 | 10 | 15 | 0.0310 | 4 | 14 | 15 | 0.0258 |
| 6 | 14 | 14 | 0.0045 | 2 | 10 | 15 | 0.0381 | 5 | 14 | 15 | 0.0293 |
| 7 | 14 | 14 | 0.0054 | 3 | 10 | 15 | 0.0444 | 6 | 14 | 15 | 0.0333 |
| 8 | 14 | 14 | 0.0066 | 4 | 10 | 15 | 0.0507 | 7 | 14 | 15 | 0.0379 |

TABLE 3

OUTLIER MEAN 3.0 OUTLIER STANDARD DEVIATION 1.0

| R | S | N | COVAR | R | S | N | COVAR | R | S | N | COVAR |
|---|---|---|---|---|---|---|---|---|---|---|---|
| 8 | 14 | 15 | 0.0436 | 5 | 8 | 16 | 0.0714 | 5 | 13 | 16 | 0.0387 |
| 9 | 14 | 15 | 0.0508 | 6 | 8 | 16 | 0.0801 | 6 | 13 | 16 | 0.0435 |
| 10 | 14 | 15 | 0.0606 | 7 | 8 | 16 | 0.0900 | 7 | 13 | 16 | 0.0490 |
| 11 | 14 | 15 | 0.0747 | 1 | 9 | 16 | 0.0349 | 8 | 13 | 16 | 0.0555 |
| 12 | 14 | 15 | 0.0975 | 2 | 9 | 16 | 0.0427 | 9 | 13 | 16 | 0.0635 |
| 13 | 14 | 15 | 0.1416 | 3 | 9 | 16 | 0.0494 | 10 | 13 | 16 | 0.0736 |
| 1 | 15 | 15 | 0.0018 | 4 | 9 | 16 | 0.0560 | 11 | 13 | 16 | 0.0872 |
| 2 | 15 | 15 | 0.0023 | 5 | 9 | 16 | 0.0631 | 12 | 13 | 16 | 0.1068 |
| 3 | 15 | 15 | 0.0027 | 6 | 9 | 16 | 0.0708 | 1 | 14 | 16 | 0.0182 |
| 4 | 15 | 15 | 0.0031 | 7 | 9 | 16 | 0.0796 | 2 | 14 | 16 | 0.0223 |
| 5 | 15 | 15 | 0.0035 | 8 | 9 | 16 | 0.0899 | 3 | 14 | 16 | 0.0258 |
| 6 | 15 | 15 | 0.0041 | 1 | 10 | 16 | 0.0309 | 4 | 14 | 16 | 0.0294 |
| 7 | 15 | 15 | 0.0047 | 2 | 10 | 16 | 0.0379 | 5 | 14 | 16 | 0.0331 |
| 8 | 15 | 15 | 0.0056 | 3 | 10 | 16 | 0.0438 | 6 | 14 | 16 | 0.0372 |
| 9 | 15 | 15 | 0.0069 | 4 | 10 | 16 | 0.0497 | 7 | 14 | 16 | 0.0419 |
| 10 | 15 | 15 | 0.0087 | 5 | 10 | 16 | 0.0560 | 8 | 14 | 16 | 0.0475 |
| 11 | 15 | 15 | 0.0118 | 6 | 10 | 16 | 0.0629 | 9 | 14 | 16 | 0.0544 |
| 12 | 15 | 15 | 0.0176 | 7 | 10 | 16 | 0.0707 | 10 | 14 | 16 | 0.0631 |
| 13 | 15 | 15 | 0.0312 | 8 | 10 | 16 | 0.0800 | 11 | 14 | 16 | 0.0749 |
| 14 | 15 | 15 | 0.0809 | 9 | 10 | 16 | 0.0912 | 12 | 14 | 16 | 0.0919 |
| 1 | 2 | 16 | 0.1481 | 1 | 11 | 16 | 0.0274 | 13 | 14 | 16 | 0.1191 |
| 1 | 3 | 16 | 0.1007 | 2 | 11 | 16 | 0.0336 | 1 | 15 | 16 | 0.0143 |
| 2 | 3 | 16 | 0.1224 | 3 | 11 | 16 | 0.0389 | 2 | 15 | 16 | 0.0175 |
| 1 | 4 | 16 | 0.0771 | 4 | 11 | 16 | 0.0442 | 3 | 15 | 16 | 0.0203 |
| 2 | 4 | 16 | 0.0939 | 5 | 11 | 16 | 0.0498 | 4 | 15 | 16 | 0.0230 |
| 3 | 4 | 16 | 0.1082 | 6 | 11 | 16 | 0.0559 | 5 | 15 | 16 | 0.0260 |
| 1 | 5 | 16 | 0.0626 | 7 | 11 | 16 | 0.0629 | 6 | 15 | 16 | 0.0292 |
| 2 | 5 | 16 | 0.0764 | 8 | 11 | 16 | 0.0712 | 7 | 15 | 16 | 0.0330 |
| 3 | 5 | 16 | 0.0881 | 9 | 11 | 16 | 0.0813 | 8 | 15 | 16 | 0.0374 |
| 4 | 5 | 16 | 0.0997 | 10 | 11 | 16 | 0.0941 | 9 | 15 | 16 | 0.0428 |
| 1 | 6 | 16 | 0.0526 | 1 | 12 | 16 | 0.0243 | 10 | 15 | 16 | 0.0498 |
| 2 | 6 | 16 | 0.0643 | 2 | 12 | 16 | 0.0298 | 11 | 15 | 16 | 0.0592 |
| 3 | 6 | 16 | 0.0743 | 3 | 12 | 16 | 0.0345 | 12 | 15 | 16 | 0.0729 |
| 4 | 6 | 16 | 0.0841 | 4 | 12 | 16 | 0.0392 | 13 | 15 | 16 | 0.0950 |
| 5 | 6 | 16 | 0.0945 | 5 | 12 | 16 | 0.0441 | 14 | 15 | 16 | 0.1380 |
| 1 | 7 | 16 | 0.0453 | 6 | 12 | 16 | 0.0496 | 1 | 16 | 16 | 0.0017 |
| 2 | 7 | 16 | 0.0554 | 7 | 12 | 16 | 0.0558 | 2 | 16 | 16 | 0.0021 |
| 3 | 7 | 16 | 0.0640 | 8 | 12 | 16 | 0.0632 | 3 | 16 | 16 | 0.0025 |
| 4 | 7 | 16 | 0.0726 | 9 | 12 | 16 | 0.0722 | 4 | 16 | 16 | 0.0029 |
| 5 | 7 | 16 | 0.0815 | 10 | 12 | 16 | 0.0837 | 5 | 16 | 16 | 0.0033 |
| 6 | 7 | 16 | 0.0914 | 11 | 12 | 16 | 0.0990 | 6 | 16 | 16 | 0.0037 |
| 1 | 8 | 16 | 0.0395 | 1 | 13 | 16 | 0.0213 | 7 | 16 | 16 | 0.0043 |
| 2 | 8 | 16 | 0.0484 | 2 | 13 | 16 | 0.0261 | 8 | 16 | 16 | 0.0050 |
| 3 | 8 | 16 | 0.0560 | 3 | 13 | 16 | 0.0302 | 9 | 16 | 16 | 0.0059 |
| 4 | 8 | 16 | 0.0635 | 4 | 13 | 16 | 0.0344 | 10 | 16 | 16 | 0.0071 |

TABLE 3

OUTLIER MEAN 3.0 OUTLIER STANDARD DEVIATION 1.0

| R | S | N | COVAR | R | S | N | COVAR | R | S | N | COVAR |
|---|---|---|---|---|---|---|---|---|---|---|---|
| 11 | 16 | 16 | 0.0090 | 5 | 10 | 17 | 0.0545 | 8 | 14 | 17 | 0.0480 |
| 12 | 16 | 16 | 0.0122 | 6 | 10 | 17 | 0.0608 | 9 | 14 | 17 | 0.0542 |
| 13 | 16 | 16 | 0.0181 | 7 | 10 | 17 | 0.0678 | 10 | 14 | 17 | 0.0617 |
| 14 | 16 | 16 | 0.0320 | 8 | 10 | 17 | 0.0759 | 11 | 14 | 17 | 0.0715 |
| 15 | 16 | 16 | 0.0821 | 9 | 10 | 17 | 0.0855 | 12 | 14 | 17 | 0.0845 |
| 1 | 2 | 17 | 0.1449 | 1 | 11 | 17 | 0.0275 | 13 | 14 | 17 | 0.1034 |
| 1 | 3 | 17 | 0.0985 | 2 | 11 | 17 | 0.0335 | 1 | 15 | 17 | 0.0165 |
| 2 | 3 | 17 | 0.1191 | 3 | 11 | 17 | 0.0386 | 2 | 15 | 17 | 0.0202 |
| 1 | 4 | 17 | 0.0754 | 4 | 11 | 17 | 0.0436 | 3 | 15 | 17 | 0.0233 |
| 2 | 4 | 17 | 0.0914 | 5 | 11 | 17 | 0.0488 | 4 | 15 | 17 | 0.0263 |
| 3 | 4 | 17 | 0.1049 | 6 | 11 | 17 | 0.0544 | 5 | 15 | 17 | 0.0295 |
| 1 | 5 | 17 | 0.0613 | 7 | 11 | 17 | 0.0607 | 6 | 15 | 17 | 0.0329 |
| 2 | 5 | 17 | 0.0744 | 8 | 11 | 17 | 0.0680 | 7 | 15 | 17 | 0.0368 |
| 3 | 5 | 17 | 0.0855 | 9 | 11 | 17 | 0.0766 | 8 | 15 | 17 | 0.0412 |
| 4 | 5 | 17 | 0.0962 | 10 | 11 | 17 | 0.0872 | 9 | 15 | 17 | 0.0466 |
| 1 | 6 | 17 | 0.0517 | 1 | 12 | 17 | 0.0246 | 10 | 15 | 17 | 0.0531 |
| 2 | 6 | 17 | 0.0628 | 2 | 12 | 17 | 0.0300 | 11 | 15 | 17 | 0.0616 |
| 3 | 6 | 17 | 0.0722 | 3 | 12 | 17 | 0.0345 | 12 | 15 | 17 | 0.0729 |
| 4 | 6 | 17 | 0.0813 | 4 | 12 | 17 | 0.0390 | 13 | 15 | 17 | 0.0894 |
| 5 | 6 | 17 | 0.0908 | 5 | 12 | 17 | 0.0437 | 14 | 15 | 17 | 0.1158 |
| 1 | 7 | 17 | 0.0445 | 6 | 12 | 17 | 0.0487 | 1 | 16 | 17 | 0.0130 |
| 2 | 7 | 17 | 0.0542 | 7 | 12 | 17 | 0.0544 | 2 | 16 | 17 | 0.0159 |
| 3 | 7 | 17 | 0.0623 | 8 | 12 | 17 | 0.0609 | 3 | 16 | 17 | 0.0183 |
| 4 | 7 | 17 | 0.0703 | 9 | 12 | 17 | 0.0687 | 4 | 16 | 17 | 0.0207 |
| 5 | 7 | 17 | 0.0785 | 10 | 12 | 17 | 0.0782 | 5 | 16 | 17 | 0.0232 |
| 6 | 7 | 17 | 0.0874 | 11 | 12 | 17 | 0.0904 | 6 | 16 | 17 | 0.0260 |
| 1 | 8 | 17 | 0.0390 | 1 | 13 | 17 | 0.0219 | 7 | 16 | 17 | 0.0290 |
| 2 | 8 | 17 | 0.0475 | 2 | 13 | 17 | 0.0267 | 8 | 16 | 17 | 0.0326 |
| 3 | 8 | 17 | 0.0546 | 3 | 13 | 17 | 0.0308 | 9 | 16 | 17 | 0.0368 |
| 4 | 8 | 17 | 0.0616 | 4 | 13 | 17 | 0.0348 | 10 | 16 | 17 | 0.0421 |
| 5 | 8 | 17 | 0.0689 | 5 | 13 | 17 | 0.0389 | 11 | 16 | 17 | 0.0488 |
| 6 | 8 | 17 | 0.0768 | 6 | 13 | 17 | 0.0434 | 12 | 16 | 17 | 0.0579 |
| 7 | 8 | 17 | 0.0856 | 7 | 13 | 17 | 0.0485 | 13 | 16 | 17 | 0.0713 |
| 1 | 9 | 17 | 0.0345 | 8 | 13 | 17 | 0.0544 | 14 | 16 | 17 | 0.0928 |
| 2 | 9 | 17 | 0.0420 | 9 | 13 | 17 | 0.0613 | 15 | 16 | 17 | 0.1348 |
| 3 | 9 | 17 | 0.0484 | 10 | 13 | 17 | 0.0699 | 1 | 17 | 17 | 0.0017 |
| 4 | 9 | 17 | 0.0546 | 11 | 13 | 17 | 0.0808 | 2 | 17 | 17 | 0.0020 |
| 5 | 9 | 17 | 0.0611 | 12 | 13 | 17 | 0.0955 | 3 | 17 | 17 | 0.0024 |
| 6 | 9 | 17 | 0.0681 | 1 | 14 | 17 | 0.0193 | 4 | 17 | 17 | 0.0027 |
| 7 | 9 | 17 | 0.0759 | 2 | 14 | 17 | 0.0235 | 5 | 17 | 17 | 0.0030 |
| 8 | 9 | 17 | 0.0849 | 3 | 14 | 17 | 0.0271 | 6 | 17 | 17 | 0.0034 |
| 1 | 10 | 17 | 0.0307 | 4 | 14 | 17 | 0.0306 | 7 | 17 | 17 | 0.0039 |
| 2 | 10 | 17 | 0.0375 | 5 | 14 | 17 | 0.0343 | 8 | 17 | 17 | 0.0044 |
| 3 | 10 | 17 | 0.0431 | 6 | 14 | 17 | 0.0383 | 9 | 17 | 17 | 0.0051 |
| 4 | 10 | 17 | 0.0487 | 7 | 14 | 17 | 0.0428 | 10 | 17 | 17 | 0.0061 |

TABLE 3

OUTLIER MEAN 3.0 OUTLIER STANDARD DEVIATION 1.0

| R | S | N | COVAR | R | S | N | COVAR | R | S | N | COVAR |
|---|---|---|---|---|---|---|---|---|---|---|---|
| 11 | 17 | 17 | 0.0074 | 4 | 10 | 18 | 0.0477 | 7 | 14 | 18 | 0.0427 |
| 12 | 17 | 17 | 0.0093 | 5 | 10 | 18 | 0.0531 | 8 | 14 | 18 | 0.0475 |
| 13 | 17 | 17 | 0.0126 | 6 | 10 | 18 | 0.0588 | 9 | 14 | 18 | 0.0530 |
| 14 | 17 | 17 | 0.0187 | 7 | 10 | 18 | 0.0652 | 10 | 14 | 18 | 0.0596 |
| 15 | 17 | 17 | 0.0328 | 8 | 10 | 18 | 0.0724 | 11 | 14 | 18 | 0.0678 |
| 16 | 17 | 17 | 0.0833 | 9 | 10 | 18 | 0.0807 | 12 | 14 | 18 | 0.0783 |
| 1 | 2 | 18 | 0.1419 | 1 | 11 | 18 | 0.0274 | 13 | 14 | 18 | 0.0924 |
| 1 | 3 | 18 | 0.0965 | 2 | 11 | 18 | 0.0333 | 1 | 15 | 18 | 0.0176 |
| 2 | 3 | 18 | 0.1162 | 3 | 11 | 18 | 0.0381 | 2 | 15 | 18 | 0.0213 |
| 1 | 4 | 18 | 0.0739 | 4 | 11 | 18 | 0.0429 | 3 | 15 | 18 | 0.0245 |
| 2 | 4 | 18 | 0.0892 | 5 | 11 | 18 | 0.0477 | 4 | 15 | 18 | 0.0276 |
| 3 | 4 | 18 | 0.1019 | 6 | 11 | 18 | 0.0529 | 5 | 15 | 18 | 0.0307 |
| 1 | 5 | 18 | 0.0601 | 7 | 11 | 18 | 0.0587 | 6 | 15 | 18 | 0.0341 |
| 2 | 5 | 18 | 0.0727 | 8 | 11 | 18 | 0.0652 | 7 | 15 | 18 | 0.0378 |
| 3 | 5 | 18 | 0.0831 | 9 | 11 | 18 | 0.0727 | 8 | 15 | 18 | 0.0421 |
| 4 | 5 | 18 | 0.0931 | 10 | 11 | 18 | 0.0816 | 9 | 15 | 18 | 0.0470 |
| 1 | 6 | 18 | 0.0507 | 1 | 12 | 18 | 0.0247 | 10 | 15 | 18 | 0.0529 |
| 2 | 6 | 18 | 0.0614 | 2 | 12 | 18 | 0.0299 | 11 | 15 | 18 | 0.0602 |
| 3 | 6 | 18 | 0.0703 | 3 | 12 | 18 | 0.0344 | 12 | 15 | 18 | 0.0695 |
| 4 | 6 | 18 | 0.0788 | 4 | 12 | 18 | 0.0386 | 13 | 15 | 18 | 0.0821 |
| 5 | 6 | 18 | 0.0876 | 5 | 12 | 18 | 0.0430 | 14 | 15 | 18 | 0.1004 |
| 1 | 7 | 18 | 0.0438 | 6 | 12 | 18 | 0.0477 | 1 | 16 | 18 | 0.0151 |
| 2 | 7 | 18 | 0.0531 | 7 | 12 | 18 | 0.0529 | 2 | 16 | 18 | 0.0184 |
| 3 | 7 | 18 | 0.0608 | 8 | 12 | 18 | 0.0588 | 3 | 16 | 18 | 0.0211 |
| 4 | 7 | 18 | 0.0682 | 9 | 12 | 18 | 0.0656 | 4 | 16 | 18 | 0.0237 |
| 5 | 7 | 18 | 0.0758 | 10 | 12 | 18 | 0.0737 | 5 | 16 | 18 | 0.0265 |
| 6 | 7 | 18 | 0.0839 | 11 | 12 | 18 | 0.0837 | 6 | 16 | 18 | 0.0294 |
| 1 | 8 | 18 | 0.0385 | 1 | 13 | 18 | 0.0222 | 7 | 16 | 18 | 0.0326 |
| 2 | 8 | 18 | 0.0466 | 2 | 13 | 18 | 0.0269 | 8 | 16 | 18 | 0.0363 |
| 3 | 8 | 18 | 0.0534 | 3 | 13 | 18 | 0.0309 | 9 | 16 | 18 | 0.0406 |
| 4 | 8 | 18 | 0.0600 | 4 | 13 | 18 | 0.0348 | 10 | 16 | 18 | 0.0457 |
| 5 | 8 | 18 | 0.0667 | 5 | 13 | 18 | 0.0387 | 11 | 16 | 18 | 0.0520 |
| 6 | 8 | 18 | 0.0739 | 6 | 13 | 18 | 0.0430 | 12 | 16 | 18 | 0.0601 |
| 7 | 8 | 18 | 0.0818 | 7 | 13 | 18 | 0.0476 | 13 | 16 | 18 | 0.0711 |
| 1 | 9 | 18 | 0.0341 | 8 | 13 | 18 | 0.0529 | 14 | 16 | 18 | 0.0871 |
| 2 | 9 | 18 | 0.0414 | 9 | 13 | 18 | 0.0591 | 15 | 16 | 18 | 0.1128 |
| 3 | 9 | 18 | 0.0474 | 10 | 13 | 18 | 0.0665 | 1 | 17 | 18 | 0.0120 |
| 4 | 9 | 18 | 0.0533 | 11 | 13 | 18 | 0.0755 | 2 | 17 | 18 | 0.0145 |
| 5 | 9 | 18 | 0.0593 | 12 | 13 | 18 | 0.0871 | 3 | 17 | 18 | 0.0167 |
| 6 | 9 | 18 | 0.0657 | 1 | 14 | 18 | 0.0199 | 4 | 17 | 18 | 0.0188 |
| 7 | 9 | 18 | 0.0728 | 2 | 14 | 18 | 0.0241 | 5 | 17 | 18 | 0.0210 |
| 8 | 9 | 18 | 0.0807 | 3 | 14 | 18 | 0.0277 | 6 | 17 | 18 | 0.0233 |
| 1 | 10 | 18 | 0.0305 | 4 | 14 | 18 | 0.0311 | 7 | 17 | 18 | 0.0258 |
| 2 | 10 | 18 | 0.0370 | 5 | 14 | 18 | 0.0347 | 8 | 17 | 18 | 0.0288 |
| 3 | 10 | 18 | 0.0424 | 6 | 14 | 18 | 0.0385 | 9 | 17 | 18 | 0.0322 |

TABLE 3

OUTLIER MEAN 3.0 OUTLIER STANDARD DEVIATION 1.0

| R | S | N | COVAR | R | S | N | COVAR | R | S | N | COVAR |
|---|---|---|---|---|---|---|---|---|---|---|---|
| 10 | 17 | 18 | 0.0363 | 1 | 8 | 19 | 0.0379 | 1 | 13 | 19 | 0.0223 |
| 11 | 17 | 18 | 0.0413 | 2 | 8 | 19 | 0.0458 | 2 | 13 | 19 | 0.0270 |
| 12 | 17 | 18 | 0.0479 | 3 | 8 | 19 | 0.0522 | 3 | 13 | 19 | 0.0308 |
| 13 | 17 | 18 | 0.0568 | 4 | 8 | 19 | 0.0584 | 4 | 13 | 19 | 0.0345 |
| 14 | 17 | 18 | 0.0698 | 5 | 8 | 19 | 0.0647 | 5 | 13 | 19 | 0.0383 |
| 15 | 17 | 18 | 0.0909 | 6 | 8 | 19 | 0.0713 | 6 | 13 | 19 | 0.0423 |
| 16 | 17 | 18 | 0.1319 | 7 | 8 | 19 | 0.0785 | 7 | 13 | 19 | 0.0466 |
| 1 | 18 | 18 | 0.0016 | 1 | 9 | 19 | 0.0337 | 8 | 13 | 19 | 0.0515 |
| 2 | 18 | 18 | 0.0019 | 2 | 9 | 19 | 0.0407 | 9 | 13 | 19 | 0.0570 |
| 3 | 18 | 18 | 0.0022 | 3 | 9 | 19 | 0.0465 | 10 | 13 | 19 | 0.0634 |
| 4 | 18 | 18 | 0.0025 | 4 | 9 | 19 | 0.0520 | 11 | 13 | 19 | 0.0711 |
| 5 | 18 | 18 | 0.0028 | 5 | 9 | 19 | 0.0576 | 12 | 13 | 19 | 0.0806 |
| 6 | 18 | 18 | 0.0032 | 6 | 9 | 19 | 0.0635 | 1 | 14 | 19 | 0.0202 |
| 7 | 18 | 18 | 0.0036 | 7 | 9 | 19 | 0.0700 | 2 | 14 | 19 | 0.0244 |
| 8 | 18 | 18 | 0.0040 | 8 | 9 | 19 | 0.0771 | 3 | 14 | 19 | 0.0279 |
| 9 | 18 | 18 | 0.0046 | 1 | 10 | 19 | 0.0302 | 4 | 14 | 19 | 0.0312 |
| 10 | 18 | 18 | 0.0053 | 2 | 10 | 19 | 0.0365 | 5 | 14 | 19 | 0.0347 |
| 11 | 18 | 18 | 0.0063 | 3 | 10 | 19 | 0.0417 | 6 | 14 | 19 | 0.0383 |
| 12 | 18 | 18 | 0.0076 | 4 | 10 | 19 | 0.0467 | 7 | 14 | 19 | 0.0422 |
| 13 | 18 | 18 | 0.0096 | 5 | 10 | 19 | 0.0517 | 8 | 14 | 19 | 0.0466 |
| 14 | 18 | 18 | 0.0129 | 6 | 10 | 19 | 0.0571 | 9 | 14 | 19 | 0.0516 |
| 15 | 18 | 18 | 0.0191 | 7 | 10 | 19 | 0.0629 | 10 | 14 | 19 | 0.0574 |
| 16 | 18 | 18 | 0.0335 | 8 | 10 | 19 | 0.0693 | 11 | 14 | 19 | 0.0644 |
| 17 | 18 | 18 | 0.0844 | 9 | 10 | 19 | 0.0767 | 12 | 14 | 19 | 0.0731 |
| 1 | 2 | 19 | 0.1392 | 1 | 11 | 19 | 0.0273 | 13 | 14 | 19 | 0.0842 |
| 1 | 3 | 19 | 0.0946 | 2 | 11 | 19 | 0.0329 | 1 | 15 | 19 | 0.0181 |
| 2 | 3 | 19 | 0.1135 | 3 | 11 | 19 | 0.0376 | 2 | 15 | 19 | 0.0219 |
| 1 | 4 | 19 | 0.0725 | 4 | 11 | 19 | 0.0421 | 3 | 15 | 19 | 0.0251 |
| 2 | 4 | 19 | 0.0872 | 5 | 11 | 19 | 0.0467 | 4 | 15 | 19 | 0.0281 |
| 3 | 4 | 19 | 0.0992 | 6 | 11 | 19 | 0.0515 | 5 | 15 | 19 | 0.0312 |
| 1 | 5 | 19 | 0.0590 | 7 | 11 | 19 | 0.0568 | 6 | 15 | 19 | 0.0344 |
| 2 | 5 | 19 | 0.0711 | 8 | 11 | 19 | 0.0626 | 7 | 15 | 19 | 0.0380 |
| 3 | 5 | 19 | 0.0810 | 9 | 11 | 19 | 0.0693 | 8 | 15 | 19 | 0.0419 |
| 4 | 5 | 19 | 0.0904 | 10 | 11 | 19 | 0.0771 | 9 | 15 | 19 | 0.0464 |
| 1 | 6 | 19 | 0.0499 | 1 | 12 | 19 | 0.0247 | 10 | 15 | 19 | 0.0517 |
| 2 | 6 | 19 | 0.0601 | 2 | 12 | 19 | 0.0298 | 11 | 15 | 19 | 0.0581 |
| 3 | 6 | 19 | 0.0685 | 3 | 12 | 19 | 0.0341 | 12 | 15 | 19 | 0.0659 |
| 4 | 6 | 19 | 0.0765 | 4 | 12 | 19 | 0.0381 | 13 | 15 | 19 | 0.0760 |
| 5 | 6 | 19 | 0.0847 | 5 | 12 | 19 | 0.0423 | 14 | 15 | 19 | 0.0896 |
| 1 | 7 | 19 | 0.0431 | 6 | 12 | 19 | 0.0467 | 1 | 16 | 19 | 0.0161 |
| 2 | 7 | 19 | 0.0520 | 7 | 12 | 19 | 0.0514 | 2 | 16 | 19 | 0.0195 |
| 3 | 7 | 19 | 0.0593 | 8 | 12 | 19 | 0.0568 | 3 | 16 | 19 | 0.0223 |
| 4 | 7 | 19 | 0.0663 | 9 | 12 | 19 | 0.0628 | 4 | 16 | 19 | 0.0250 |
| 5 | 7 | 19 | 0.0734 | 10 | 12 | 19 | 0.0699 | 5 | 16 | 19 | 0.0277 |
| 6 | 7 | 19 | 0.0809 | 11 | 12 | 19 | 0.0783 | 6 | 16 | 19 | 0.0306 |

TABLE 3

OUTLIER MEAN 3.0 OUTLIER STANDARD DEVIATION 1.0

| R | S | N | COVAR | R | S | N | COVAR | R | S | N | COVAR |
|---|---|---|---|---|---|---|---|---|---|---|---|
| 7 | 16 | 19 | 0.0338 | 4 | 19 | 19 | 0.0024 | 3 | 9 | 20 | 0.0456 |
| 8 | 16 | 19 | 0.0373 | 5 | 19 | 19 | 0.0027 | 4 | 9 | 20 | 0.0508 |
| 9 | 16 | 19 | 0.0413 | 6 | 19 | 19 | 0.0030 | 5 | 9 | 20 | 0.0561 |
| 10 | 16 | 19 | 0.0460 | 7 | 19 | 19 | 0.0033 | 6 | 9 | 20 | 0.0616 |
| 11 | 16 | 19 | 0.0517 | 8 | 19 | 19 | 0.0037 | 7 | 9 | 20 | 0.0675 |
| 12 | 16 | 19 | 0.0587 | 9 | 19 | 19 | 0.0042 | 8 | 9 | 20 | 0.0740 |
| 13 | 16 | 19 | 0.0677 | 10 | 19 | 19 | 0.0047 | 1 | 10 | 20 | 0.0299 |
| 14 | 16 | 19 | 0.0800 | 11 | 19 | 19 | 0.0055 | 2 | 10 | 20 | 0.0360 |
| 15 | 16 | 19 | 0.0976 | 12 | 19 | 19 | 0.0065 | 3 | 10 | 20 | 0.0410 |
| 1 | 17 | 19 | 0.0139 | 13 | 19 | 19 | 0.0078 | 4 | 10 | 20 | 0.0457 |
| 2 | 17 | 19 | 0.0169 | 14 | 19 | 19 | 0.0099 | 5 | 10 | 20 | 0.0505 |
| 3 | 17 | 19 | 0.0193 | 15 | 19 | 19 | 0.0133 | 6 | 10 | 20 | 0.0555 |
| 4 | 17 | 19 | 0.0216 | 16 | 19 | 19 | 0.0196 | 7 | 10 | 20 | 0.0608 |
| 5 | 17 | 19 | 0.0240 | 17 | 19 | 19 | 0.0342 | 8 | 10 | 20 | 0.0666 |
| 6 | 17 | 19 | 0.0265 | 18 | 19 | 19 | 0.0853 | 9 | 10 | 20 | 0.0732 |
| 7 | 17 | 19 | 0.0292 | 1 | 2 | 20 | 0.1368 | 1 | 11 | 20 | 0.0271 |
| 8 | 17 | 19 | 0.0323 | 1 | 3 | 20 | 0.0929 | 2 | 11 | 20 | 0.0326 |
| 9 | 17 | 19 | 0.0358 | 2 | 3 | 20 | 0.1111 | 3 | 11 | 20 | 0.0371 |
| 10 | 17 | 19 | 0.0399 | 1 | 4 | 20 | 0.0712 | 4 | 11 | 20 | 0.0414 |
| 11 | 17 | 19 | 0.0448 | 2 | 4 | 20 | 0.0853 | 5 | 11 | 20 | 0.0457 |
| 12 | 17 | 19 | 0.0509 | 3 | 4 | 20 | 0.0967 | 6 | 11 | 20 | 0.0502 |
| 13 | 17 | 19 | 0.0588 | 1 | 5 | 20 | 0.0580 | 7 | 11 | 20 | 0.0551 |
| 14 | 17 | 19 | 0.0695 | 2 | 5 | 20 | 0.0696 | 8 | 11 | 20 | 0.0604 |
| 15 | 17 | 19 | 0.0850 | 3 | 5 | 20 | 0.0790 | 9 | 11 | 20 | 0.0664 |
| 16 | 17 | 19 | 0.1101 | 4 | 5 | 20 | 0.0879 | 10 | 11 | 20 | 0.0732 |
| 1 | 18 | 19 | 0.0111 | 1 | 6 | 20 | 0.0490 | 1 | 12 | 20 | 0.0246 |
| 2 | 18 | 19 | 0.0134 | 2 | 6 | 20 | 0.0589 | 2 | 12 | 20 | 0.0296 |
| 3 | 18 | 19 | 0.0153 | 3 | 6 | 20 | 0.0669 | 3 | 12 | 20 | 0.0337 |
| 4 | 18 | 19 | 0.0172 | 4 | 6 | 20 | 0.0745 | 4 | 12 | 20 | 0.0376 |
| 5 | 18 | 19 | 0.0190 | 5 | 6 | 20 | 0.0821 | 5 | 12 | 20 | 0.0415 |
| 6 | 18 | 19 | 0.0210 | 1 | 7 | 20 | 0.0425 | 6 | 12 | 20 | 0.0456 |
| 7 | 18 | 19 | 0.0232 | 2 | 7 | 20 | 0.0510 | 7 | 12 | 20 | 0.0501 |
| 8 | 18 | 19 | 0.0257 | 3 | 7 | 20 | 0.0580 | 8 | 12 | 20 | 0.0549 |
| 9 | 18 | 19 | 0.0285 | 4 | 7 | 20 | 0.0646 | 9 | 12 | 20 | 0.0604 |
| 10 | 18 | 19 | 0.0318 | 5 | 7 | 20 | 0.0713 | 10 | 12 | 20 | 0.0666 |
| 11 | 18 | 19 | 0.0357 | 6 | 7 | 20 | 0.0782 | 11 | 12 | 20 | 0.0739 |
| 12 | 18 | 19 | 0.0406 | 1 | 8 | 20 | 0.0374 | 1 | 13 | 20 | 0.0224 |
| 13 | 18 | 19 | 0.0470 | 2 | 8 | 20 | 0.0450 | 2 | 13 | 20 | 0.0269 |
| 14 | 18 | 19 | 0.0557 | 3 | 8 | 20 | 0.0511 | 3 | 13 | 20 | 0.0307 |
| 15 | 18 | 19 | 0.0684 | 4 | 8 | 20 | 0.0570 | 4 | 13 | 20 | 0.0342 |
| 16 | 18 | 19 | 0.0890 | 5 | 8 | 20 | 0.0629 | 5 | 13 | 20 | 0.0378 |
| 17 | 18 | 19 | 0.1292 | 6 | 8 | 20 | 0.0690 | 6 | 13 | 20 | 0.0416 |
| 1 | 19 | 19 | 0.0015 | 7 | 8 | 20 | 0.0756 | 7 | 13 | 20 | 0.0456 |
| 2 | 19 | 19 | 0.0018 | 1 | 9 | 20 | 0.0333 | 8 | 13 | 20 | 0.0500 |
| 3 | 19 | 19 | 0.0021 | 2 | 9 | 20 | 0.0401 | 9 | 13 | 20 | 0.0550 |

TABLE 3

OUTLIER MEAN 3.0 OUTLIER STANDARD DEVIATION 1.0

| R | S | N | COVAR | R | S | N | COVAR | R | S | N | COVAR |
|---|---|---|---|---|---|---|---|---|---|---|---|
| 10 | 13 | 20 | 0.0607 | 10 | 16 | 20 | 0.0455 | 1 | 19 | 20 | 0.0103 |
| 11 | 13 | 20 | 0.0674 | 11 | 16 | 20 | 0.0505 | 2 | 19 | 20 | 0.0124 |
| 12 | 13 | 20 | 0.0754 | 12 | 16 | 20 | 0.0566 | 3 | 19 | 20 | 0.0141 |
| 1 | 14 | 20 | 0.0204 | 13 | 16 | 20 | 0.0642 | 4 | 19 | 20 | 0.0157 |
| 2 | 14 | 20 | 0.0245 | 14 | 16 | 20 | 0.0739 | 5 | 19 | 20 | 0.0174 |
| 3 | 14 | 20 | 0.0279 | 15 | 16 | 20 | 0.0871 | 6 | 19 | 20 | 0.0192 |
| 4 | 14 | 20 | 0.0312 | 1 | 17 | 20 | 0.0149 | 7 | 19 | 20 | 0.0210 |
| 5 | 14 | 20 | 0.0344 | 2 | 17 | 20 | 0.0179 | 8 | 19 | 20 | 0.0231 |
| 6 | 14 | 20 | 0.0378 | 3 | 17 | 20 | 0.0204 | 9 | 19 | 20 | 0.0255 |
| 7 | 14 | 20 | 0.0415 | 4 | 17 | 20 | 0.0228 | 10 | 19 | 20 | 0.0282 |
| 8 | 14 | 20 | 0.0456 | 5 | 17 | 20 | 0.0252 | 11 | 19 | 20 | 0.0314 |
| 9 | 14 | 20 | 0.0501 | 6 | 17 | 20 | 0.0277 | 12 | 19 | 20 | 0.0352 |
| 10 | 14 | 20 | 0.0553 | 7 | 17 | 20 | 0.0304 | 13 | 19 | 20 | 0.0400 |
| 11 | 14 | 20 | 0.0615 | 8 | 17 | 20 | 0.0334 | 14 | 19 | 20 | 0.0462 |
| 12 | 14 | 20 | 0.0688 | 9 | 17 | 20 | 0.0367 | 15 | 19 | 20 | 0.0547 |
| 13 | 14 | 20 | 0.0779 | 10 | 17 | 20 | 0.0406 | 16 | 19 | 20 | 0.0671 |
| 1 | 15 | 20 | 0.0185 | 11 | 17 | 20 | 0.0451 | 17 | 19 | 20 | 0.0874 |
| 2 | 15 | 20 | 0.0223 | 12 | 17 | 20 | 0.0506 | 18 | 19 | 20 | 0.1268 |
| 3 | 15 | 20 | 0.0254 | 13 | 17 | 20 | 0.0574 | 1 | 20 | 20 | 0.0015 |
| 4 | 15 | 20 | 0.0283 | 14 | 17 | 20 | 0.0661 | 2 | 20 | 20 | 0.0018 |
| 5 | 15 | 20 | 0.0313 | 15 | 17 | 20 | 0.0780 | 3 | 20 | 20 | 0.0020 |
| 6 | 15 | 20 | 0.0344 | 16 | 17 | 20 | 0.0951 | 4 | 20 | 20 | 0.0023 |
| 7 | 15 | 20 | 0.0377 | 1 | 18 | 20 | 0.0129 | 5 | 20 | 20 | 0.0025 |
| 8 | 15 | 20 | 0.0414 | 2 | 18 | 20 | 0.0155 | 6 | 20 | 20 | 0.0028 |
| 9 | 15 | 20 | 0.0456 | 3 | 18 | 20 | 0.0177 | 7 | 20 | 20 | 0.0031 |
| 10 | 15 | 20 | 0.0503 | 4 | 18 | 20 | 0.0198 | 8 | 20 | 20 | 0.0034 |
| 11 | 15 | 20 | 0.0559 | 5 | 18 | 20 | 0.0218 | 9 | 20 | 20 | 0.0038 |
| 12 | 15 | 20 | 0.0626 | 6 | 18 | 20 | 0.0240 | 10 | 20 | 20 | 0.0043 |
| 13 | 15 | 20 | 0.0709 | 7 | 18 | 20 | 0.0264 | 11 | 20 | 20 | 0.0049 |
| 14 | 15 | 20 | 0.0816 | 8 | 18 | 20 | 0.0290 | 12 | 20 | 20 | 0.0056 |
| 1 | 16 | 20 | 0.0167 | 9 | 18 | 20 | 0.0319 | 13 | 20 | 20 | 0.0066 |
| 2 | 16 | 20 | 0.0201 | 10 | 18 | 20 | 0.0353 | 14 | 20 | 20 | 0.0080 |
| 3 | 16 | 20 | 0.0229 | 11 | 18 | 20 | 0.0392 | 15 | 20 | 20 | 0.0102 |
| 4 | 16 | 20 | 0.0255 | 12 | 18 | 20 | 0.0440 | 16 | 20 | 20 | 0.0136 |
| 5 | 16 | 20 | 0.0282 | 13 | 18 | 20 | 0.0499 | 17 | 20 | 20 | 0.0200 |
| 6 | 16 | 20 | 0.0310 | 14 | 18 | 20 | 0.0576 | 18 | 20 | 20 | 0.0348 |
| 7 | 16 | 20 | 0.0341 | 15 | 18 | 20 | 0.0680 | 19 | 20 | 20 | 0.0862 |
| 8 | 16 | 20 | 0.0374 | 16 | 18 | 20 | 0.0832 | | | | |
| 9 | 16 | 20 | 0.0412 | 17 | 18 | 20 | 0.1076 | | | | |

TABLE 3

OUTLIER MEAN 4.0 OUTLIER STANDARD DEVIATION 1.0

| R | S | N | COVAR | R | S | N | COVAR | R | S | N | COVAR |
|---|---|---|---|---|---|---|---|---|---|---|---|
| 1 | 2 | 2 | 0.0039 | 1 | 6 | 7 | 0.0560 | 1 | 5 | 9 | 0.0748 |
| 1 | 2 | 3 | 0.3174 | 2 | 6 | 7 | 0.0768 | 2 | 5 | 9 | 0.0976 |
| 1 | 3 | 3 | 0.0011 | 3 | 6 | 7 | 0.1017 | 3 | 5 | 9 | 0.1209 |
| 2 | 3 | 3 | 0.0064 | 4 | 6 | 7 | 0.1384 | 4 | 5 | 9 | 0.1492 |
| 1 | 2 | 4 | 0.2756 | 5 | 6 | 7 | 0.2067 | 1 | 6 | 9 | 0.0602 |
| 1 | 3 | 4 | 0.1642 | 1 | 7 | 7 | 0.0005 | 2 | 6 | 9 | 0.0787 |
| 2 | 3 | 4 | 0.2744 | 2 | 7 | 7 | 0.0007 | 3 | 6 | 9 | 0.0978 |
| 1 | 4 | 4 | 0.0007 | 3 | 7 | 7 | 0.0009 | 4 | 6 | 9 | 0.1209 |
| 2 | 4 | 4 | 0.0016 | 4 | 7 | 7 | 0.0015 | 5 | 6 | 9 | 0.1523 |
| 3 | 4 | 4 | 0.0085 | 5 | 7 | 7 | 0.0030 | 1 | 7 | 9 | 0.0482 |
| 1 | 2 | 5 | 0.2456 | 6 | 7 | 7 | 0.0132 | 2 | 7 | 9 | 0.0632 |
| 1 | 3 | 5 | 0.1579 | 1 | 2 | 8 | 0.1962 | 3 | 7 | 9 | 0.0786 |
| 2 | 3 | 5 | 0.2358 | 1 | 3 | 8 | 0.1321 | 4 | 7 | 9 | 0.0974 |
| 1 | 4 | 5 | 0.1042 | 2 | 3 | 8 | 0.1745 | 5 | 7 | 9 | 0.1231 |
| 2 | 4 | 5 | 0.1572 | 1 | 4 | 8 | 0.0985 | 6 | 7 | 9 | 0.1629 |
| 3 | 4 | 5 | 0.2441 | 2 | 4 | 8 | 0.1307 | 1 | 8 | 9 | 0.0365 |
| 1 | 5 | 5 | 0.0006 | 3 | 4 | 8 | 0.1655 | 2 | 8 | 9 | 0.0479 |
| 2 | 5 | 5 | 0.0010 | 1 | 5 | 8 | 0.0765 | 3 | 8 | 9 | 0.0597 |
| 3 | 5 | 5 | 0.0021 | 2 | 5 | 8 | 0.1019 | 4 | 8 | 9 | 0.0741 |
| 4 | 5 | 5 | 0.0103 | 3 | 5 | 8 | 0.1296 | 5 | 8 | 9 | 0.0939 |
| 1 | 2 | 6 | 0.2243 | 4 | 5 | 8 | 0.1655 | 6 | 8 | 9 | 0.1248 |
| 1 | 3 | 6 | 0.1481 | 1 | 6 | 8 | 0.0599 | 7 | 8 | 9 | 0.1842 |
| 2 | 3 | 6 | 0.2084 | 2 | 6 | 8 | 0.0799 | 1 | 9 | 9 | 0.0004 |
| 1 | 4 | 6 | 0.1057 | 3 | 6 | 8 | 0.1018 | 2 | 9 | 9 | 0.0005 |
| 2 | 4 | 6 | 0.1498 | 4 | 6 | 8 | 0.1306 | 3 | 9 | 9 | 0.0007 |
| 3 | 4 | 6 | 0.2083 | 5 | 6 | 8 | 0.1742 | 4 | 9 | 9 | 0.0009 |
| 1 | 5 | 6 | 0.0738 | 1 | 7 | 8 | 0.0445 | 5 | 9 | 9 | 0.0012 |
| 2 | 5 | 6 | 0.1051 | 2 | 7 | 8 | 0.0595 | 6 | 9 | 9 | 0.0019 |
| 3 | 5 | 6 | 0.1472 | 3 | 7 | 8 | 0.0760 | 7 | 9 | 9 | 0.0038 |
| 4 | 5 | 6 | 0.2227 | 4 | 7 | 8 | 0.0977 | 8 | 9 | 9 | 0.0157 |
| 1 | 6 | 6 | 0.0005 | 5 | 7 | 8 | 0.1310 | 1 | 2 | 10 | 0.1781 |
| 2 | 6 | 6 | 0.0008 | 6 | 7 | 8 | 0.1942 | 1 | 3 | 10 | 0.1207 |
| 3 | 6 | 6 | 0.0013 | 1 | 8 | 8 | 0.0004 | 2 | 3 | 10 | 0.1541 |
| 4 | 6 | 6 | 0.0026 | 2 | 8 | 8 | 0.0006 | 1 | 4 | 10 | 0.0913 |
| 5 | 6 | 6 | 0.0118 | 3 | 8 | 8 | 0.0008 | 2 | 4 | 10 | 0.1170 |
| 1 | 2 | 7 | 0.2085 | 4 | 8 | 8 | 0.0011 | 3 | 4 | 10 | 0.1421 |
| 1 | 3 | 7 | 0.1394 | 5 | 8 | 8 | 0.0017 | 1 | 5 | 10 | 0.0727 |
| 2 | 3 | 7 | 0.1890 | 6 | 8 | 8 | 0.0034 | 2 | 5 | 10 | 0.0934 |
| 1 | 4 | 7 | 0.1024 | 7 | 8 | 8 | 0.0145 | 3 | 5 | 10 | 0.1138 |
| 2 | 4 | 7 | 0.1396 | 1 | 2 | 9 | 0.1863 | 4 | 5 | 10 | 0.1370 |
| 3 | 4 | 7 | 0.1832 | 1 | 3 | 9 | 0.1260 | 1 | 6 | 10 | 0.0595 |
| 1 | 5 | 7 | 0.0773 | 2 | 3 | 9 | 0.1632 | 2 | 6 | 10 | 0.0765 |
| 2 | 5 | 7 | 0.1058 | 1 | 4 | 9 | 0.0947 | 3 | 6 | 10 | 0.0933 |
| 3 | 5 | 7 | 0.1395 | 2 | 4 | 9 | 0.1233 | 4 | 6 | 10 | 0.1126 |
| 4 | 5 | 7 | 0.1888 | 3 | 4 | 9 | 0.1524 | 5 | 6 | 10 | 0.1370 |

TABLE 3

OUTLIER MEAN 4.0 OUTLIER STANDARD DEVIATION 1.0

| R | S | N | COVAR | R | S | N | COVAR | R | S | N | COVAR |
|---|---|---|---|---|---|---|---|---|---|---|---|
| 1 | 7 | 10 | 0.0491 | 1 | 7 | 11 | 0.0489 | 3 | 4 | 12 | 0.1270 |
| 2 | 7 | 10 | 0.0632 | 2 | 7 | 11 | 0.0622 | 1 | 5 | 12 | 0.0688 |
| 3 | 7 | 10 | 0.0772 | 3 | 7 | 11 | 0.0749 | 2 | 5 | 12 | 0.0864 |
| 4 | 7 | 10 | 0.0933 | 4 | 7 | 11 | 0.0889 | 3 | 5 | 12 | 0.1026 |
| 5 | 7 | 10 | 0.1137 | 5 | 7 | 11 | 0.1058 | 4 | 5 | 12 | 0.1199 |
| 6 | 7 | 10 | 0.1420 | 6 | 7 | 11 | 0.1275 | 1 | 6 | 12 | 0.0572 |
| 1 | 8 | 10 | 0.0400 | 1 | 8 | 11 | 0.0411 | 2 | 6 | 12 | 0.0719 |
| 2 | 8 | 10 | 0.0516 | 2 | 8 | 11 | 0.0523 | 3 | 6 | 12 | 0.0855 |
| 3 | 8 | 10 | 0.0631 | 3 | 8 | 11 | 0.0630 | 4 | 6 | 12 | 0.1000 |
| 4 | 8 | 10 | 0.0764 | 4 | 8 | 11 | 0.0749 | 5 | 6 | 12 | 0.1167 |
| 5 | 8 | 10 | 0.0933 | 5 | 8 | 11 | 0.0892 | 1 | 7 | 12 | 0.0484 |
| 6 | 8 | 10 | 0.1168 | 6 | 8 | 11 | 0.1077 | 2 | 7 | 12 | 0.0609 |
| 7 | 8 | 10 | 0.1538 | 7 | 8 | 11 | 0.1337 | 3 | 7 | 12 | 0.0725 |
| 1 | 9 | 10 | 0.0308 | 1 | 9 | 11 | 0.0340 | 4 | 7 | 12 | 0.0849 |
| 2 | 9 | 10 | 0.0397 | 2 | 9 | 11 | 0.0433 | 5 | 7 | 12 | 0.0992 |
| 3 | 9 | 10 | 0.0486 | 3 | 9 | 11 | 0.0522 | 6 | 7 | 12 | 0.1167 |
| 4 | 9 | 10 | 0.0589 | 4 | 9 | 11 | 0.0621 | 1 | 8 | 12 | 0.0412 |
| 5 | 9 | 10 | 0.0720 | 5 | 9 | 11 | 0.0741 | 2 | 8 | 12 | 0.0519 |
| 6 | 9 | 10 | 0.0904 | 6 | 9 | 11 | 0.0896 | 3 | 8 | 12 | 0.0619 |
| 7 | 9 | 10 | 0.1195 | 7 | 9 | 11 | 0.1115 | 4 | 8 | 12 | 0.0725 |
| 8 | 9 | 10 | 0.1760 | 8 | 9 | 11 | 0.1463 | 5 | 8 | 12 | 0.0849 |
| 1 | 10 | 10 | 0.0004 | 1 | 10 | 11 | 0.0264 | 6 | 8 | 12 | 0.1000 |
| 2 | 10 | 10 | 0.0005 | 2 | 10 | 11 | 0.0337 | 7 | 8 | 12 | 0.1198 |
| 3 | 10 | 10 | 0.0006 | 3 | 10 | 11 | 0.0407 | 1 | 9 | 12 | 0.0351 |
| 4 | 10 | 10 | 0.0008 | 4 | 10 | 11 | 0.0484 | 2 | 9 | 12 | 0.0442 |
| 5 | 10 | 10 | 0.0010 | 5 | 10 | 11 | 0.0578 | 3 | 9 | 12 | 0.0527 |
| 6 | 10 | 10 | 0.0013 | 6 | 10 | 11 | 0.0700 | 4 | 9 | 12 | 0.0619 |
| 7 | 10 | 10 | 0.0020 | 7 | 10 | 11 | 0.0873 | 5 | 9 | 12 | 0.0724 |
| 8 | 10 | 10 | 0.0041 | 8 | 10 | 11 | 0.1149 | 6 | 9 | 12 | 0.0855 |
| 9 | 10 | 10 | 0.0168 | 9 | 10 | 11 | 0.1690 | 7 | 9 | 12 | 0.1026 |
| 1 | 2 | 11 | 0.1713 | 1 | 11 | 11 | 0.0004 | 8 | 9 | 12 | 0.1269 |
| 1 | 3 | 11 | 0.1163 | 2 | 11 | 11 | 0.0005 | 1 | 10 | 12 | 0.0294 |
| 2 | 3 | 11 | 0.1466 | 3 | 11 | 11 | 0.0006 | 2 | 10 | 12 | 0.0370 |
| 1 | 4 | 11 | 0.0882 | 4 | 11 | 11 | 0.0007 | 3 | 10 | 12 | 0.0442 |
| 2 | 4 | 11 | 0.1117 | 5 | 11 | 11 | 0.0008 | 4 | 10 | 12 | 0.0518 |
| 3 | 4 | 11 | 0.1338 | 6 | 11 | 11 | 0.0011 | 5 | 10 | 12 | 0.0608 |
| 1 | 5 | 11 | 0.0707 | 7 | 11 | 11 | 0.0014 | 6 | 10 | 12 | 0.0718 |
| 2 | 5 | 11 | 0.0897 | 8 | 11 | 11 | 0.0022 | 7 | 10 | 12 | 0.0863 |
| 3 | 5 | 11 | 0.1077 | 9 | 11 | 11 | 0.0045 | 8 | 10 | 12 | 0.1069 |
| 4 | 5 | 11 | 0.1275 | 10 | 11 | 11 | 0.0178 | 9 | 10 | 12 | 0.1399 |
| 1 | 6 | 11 | 0.0584 | 1 | 2 | 12 | 0.1654 | 1 | 11 | 12 | 0.0231 |
| 2 | 6 | 11 | 0.0742 | 1 | 3 | 12 | 0.1124 | 2 | 11 | 12 | 0.0291 |
| 3 | 6 | 11 | 0.0892 | 2 | 3 | 12 | 0.1403 | 3 | 11 | 12 | 0.0347 |
| 4 | 6 | 11 | 0.1058 | 1 | 4 | 12 | 0.0855 | 4 | 11 | 12 | 0.0408 |
| 5 | 6 | 11 | 0.1256 | 2 | 4 | 12 | 0.1071 | 5 | 11 | 12 | 0.0478 |

TABLE 3

OUTLIER MEAN 4.0 OUTLIER STANDARD DEVIATION 1.0

| R | S | N | COVAR | R | S | N | COVAR | R | S | N | COVAR |
|---|---|---|---|---|---|---|---|---|---|---|---|
| 6 | 11 | 12 | 0.0566 | 2 | 9 | 13 | 0.0443 | 9 | 13 | 13 | 0.0017 |
| 7 | 11 | 12 | 0.0681 | 3 | 9 | 13 | 0.0523 | 10 | 13 | 13 | 0.0025 |
| 8 | 11 | 12 | 0.0845 | 4 | 9 | 13 | 0.0606 | 11 | 13 | 13 | 0.0051 |
| 9 | 11 | 12 | 0.1110 | 5 | 9 | 13 | 0.0701 | 12 | 13 | 13 | 0.0197 |
| 10 | 11 | 12 | 0.1630 | 6 | 9 | 13 | 0.0812 | 1 | 2 | 14 | 0.1557 |
| 1 | 12 | 12 | 0.0003 | 7 | 9 | 13 | 0.0951 | 1 | 3 | 14 | 0.1059 |
| 2 | 12 | 12 | 0.0004 | 8 | 9 | 13 | 0.1135 | 2 | 3 | 14 | 0.1302 |
| 3 | 12 | 12 | 0.0005 | 1 | 10 | 13 | 0.0305 | 1 | 4 | 14 | 0.0809 |
| 4 | 12 | 12 | 0.0006 | 2 | 10 | 13 | 0.0381 | 2 | 4 | 14 | 0.0997 |
| 5 | 12 | 12 | 0.0008 | 3 | 10 | 13 | 0.0450 | 3 | 4 | 14 | 0.1163 |
| 6 | 12 | 12 | 0.0009 | 4 | 10 | 13 | 0.0522 | 1 | 5 | 14 | 0.0655 |
| 7 | 12 | 12 | 0.0012 | 5 | 10 | 13 | 0.0604 | 2 | 5 | 14 | 0.0809 |
| 8 | 12 | 12 | 0.0016 | 6 | 10 | 13 | 0.0701 | 3 | 5 | 14 | 0.0945 |
| 9 | 12 | 12 | 0.0024 | 7 | 10 | 13 | 0.0822 | 4 | 5 | 14 | 0.1083 |
| 10 | 12 | 12 | 0.0048 | 8 | 10 | 13 | 0.0982 | 1 | 6 | 14 | 0.0548 |
| 11 | 12 | 12 | 0.0188 | 9 | 10 | 13 | 0.1211 | 2 | 6 | 14 | 0.0678 |
| 1 | 2 | 13 | 0.1602 | 1 | 11 | 13 | 0.0257 | 3 | 6 | 14 | 0.0793 |
| 1 | 3 | 13 | 0.1089 | 2 | 11 | 13 | 0.0322 | 4 | 6 | 14 | 0.0910 |
| 2 | 3 | 13 | 0.1349 | 3 | 11 | 13 | 0.0380 | 5 | 6 | 14 | 0.1037 |
| 1 | 4 | 13 | 0.0831 | 4 | 11 | 13 | 0.0442 | 1 | 7 | 14 | 0.0469 |
| 2 | 4 | 13 | 0.1032 | 5 | 11 | 13 | 0.0511 | 2 | 7 | 14 | 0.0580 |
| 3 | 4 | 13 | 0.1212 | 6 | 11 | 13 | 0.0593 | 3 | 7 | 14 | 0.0679 |
| 1 | 5 | 13 | 0.0671 | 7 | 11 | 13 | 0.0696 | 4 | 7 | 14 | 0.0780 |
| 2 | 5 | 13 | 0.0835 | 8 | 11 | 13 | 0.0833 | 5 | 7 | 14 | 0.0890 |
| 3 | 5 | 13 | 0.0983 | 9 | 11 | 13 | 0.1029 | 6 | 7 | 14 | 0.1017 |
| 4 | 5 | 13 | 0.1136 | 10 | 11 | 13 | 0.1345 | 1 | 8 | 14 | 0.0406 |
| 1 | 6 | 13 | 0.0560 | 1 | 12 | 13 | 0.0204 | 2 | 8 | 14 | 0.0503 |
| 2 | 6 | 13 | 0.0698 | 2 | 12 | 13 | 0.0255 | 3 | 8 | 14 | 0.0589 |
| 3 | 6 | 13 | 0.0822 | 3 | 12 | 13 | 0.0301 | 4 | 8 | 14 | 0.0677 |
| 4 | 6 | 13 | 0.0952 | 4 | 12 | 13 | 0.0350 | 5 | 8 | 14 | 0.0773 |
| 5 | 6 | 13 | 0.1096 | 5 | 12 | 13 | 0.0405 | 6 | 8 | 14 | 0.0884 |
| 1 | 7 | 13 | 0.0477 | 6 | 12 | 13 | 0.0471 | 7 | 8 | 14 | 0.1017 |
| 2 | 7 | 13 | 0.0595 | 7 | 12 | 13 | 0.0553 | 1 | 9 | 14 | 0.0354 |
| 3 | 7 | 13 | 0.0701 | 8 | 12 | 13 | 0.0663 | 2 | 9 | 14 | 0.0439 |
| 4 | 7 | 13 | 0.0812 | 9 | 12 | 13 | 0.0820 | 3 | 9 | 14 | 0.0514 |
| 5 | 7 | 13 | 0.0937 | 10 | 12 | 13 | 0.1075 | 4 | 9 | 14 | 0.0592 |
| 6 | 7 | 13 | 0.1084 | 11 | 12 | 13 | 0.1578 | 5 | 9 | 14 | 0.0676 |
| 1 | 8 | 13 | 0.0410 | 1 | 13 | 13 | 0.0003 | 6 | 9 | 14 | 0.0773 |
| 2 | 8 | 13 | 0.0512 | 2 | 13 | 13 | 0.0004 | 7 | 9 | 14 | 0.0890 |
| 3 | 8 | 13 | 0.0604 | 3 | 13 | 13 | 0.0005 | 8 | 9 | 14 | 0.1037 |
| 4 | 8 | 13 | 0.0701 | 4 | 13 | 13 | 0.0006 | 1 | 10 | 14 | 0.0309 |
| 5 | 8 | 13 | 0.0809 | 5 | 13 | 13 | 0.0007 | 2 | 10 | 14 | 0.0383 |
| 6 | 8 | 13 | 0.0937 | 6 | 13 | 13 | 0.0008 | 3 | 10 | 14 | 0.0450 |
| 7 | 8 | 13 | 0.1096 | 7 | 13 | 13 | 0.0010 | 4 | 10 | 14 | 0.0517 |
| 1 | 9 | 13 | 0.0354 | 8 | 13 | 13 | 0.0012 | 5 | 10 | 14 | 0.0591 |

TABLE 3

OUTLIER MEAN 4.0 OUTLIER STANDARD DEVIATION 1.0

| R | S | N | COVAR | R | S | N | COVAR | R | S | N | COVAR |
|---|---|---|---|---|---|---|---|---|---|---|---|
| 6 | 10 | 14 | 0.0677 | 9 | 14 | 14 | 0.0013 | 5 | 10 | 15 | 0.0576 |
| 7 | 10 | 14 | 0.0780 | 10 | 14 | 14 | 0.0018 | 6 | 10 | 15 | 0.0653 |
| 8 | 10 | 14 | 0.0910 | 11 | 14 | 14 | 0.0027 | 7 | 10 | 15 | 0.0742 |
| 9 | 10 | 14 | 0.1082 | 12 | 14 | 14 | 0.0054 | 8 | 10 | 15 | 0.0850 |
| 1 | 11 | 14 | 0.0268 | 13 | 14 | 14 | 0.0205 | 9 | 10 | 15 | 0.0988 |
| 2 | 11 | 14 | 0.0333 | 1 | 2 | 15 | 0.1517 | 1 | 11 | 15 | 0.0273 |
| 3 | 11 | 14 | 0.0390 | 1 | 3 | 15 | 0.1032 | 2 | 11 | 15 | 0.0337 |
| 4 | 11 | 14 | 0.0449 | 2 | 3 | 15 | 0.1261 | 3 | 11 | 15 | 0.0392 |
| 5 | 11 | 14 | 0.0514 | 1 | 4 | 15 | 0.0789 | 4 | 11 | 15 | 0.0448 |
| 6 | 11 | 14 | 0.0589 | 2 | 4 | 15 | 0.0967 | 5 | 11 | 15 | 0.0508 |
| 7 | 11 | 14 | 0.0679 | 3 | 4 | 15 | 0.1120 | 6 | 11 | 15 | 0.0576 |
| 8 | 11 | 14 | 0.0792 | 1 | 5 | 15 | 0.0640 | 7 | 11 | 15 | 0.0655 |
| 9 | 11 | 14 | 0.0944 | 2 | 5 | 15 | 0.0785 | 8 | 11 | 15 | 0.0751 |
| 10 | 11 | 14 | 0.1161 | 3 | 5 | 15 | 0.0911 | 9 | 11 | 15 | 0.0873 |
| 1 | 12 | 14 | 0.0228 | 4 | 5 | 15 | 0.1037 | 10 | 11 | 15 | 0.1036 |
| 2 | 12 | 14 | 0.0283 | 1 | 6 | 15 | 0.0537 | 1 | 12 | 15 | 0.0239 |
| 3 | 12 | 14 | 0.0332 | 2 | 6 | 15 | 0.0660 | 2 | 12 | 15 | 0.0294 |
| 4 | 12 | 14 | 0.0383 | 3 | 6 | 15 | 0.0767 | 3 | 12 | 15 | 0.0343 |
| 5 | 12 | 14 | 0.0438 | 4 | 6 | 15 | 0.0874 | 4 | 12 | 15 | 0.0392 |
| 6 | 12 | 14 | 0.0502 | 5 | 6 | 15 | 0.0988 | 5 | 12 | 15 | 0.0445 |
| 7 | 12 | 14 | 0.0579 | 1 | 7 | 15 | 0.0461 | 6 | 12 | 15 | 0.0504 |
| 8 | 12 | 14 | 0.0676 | 2 | 7 | 15 | 0.0567 | 7 | 12 | 15 | 0.0574 |
| 9 | 12 | 14 | 0.0807 | 3 | 7 | 15 | 0.0659 | 8 | 12 | 15 | 0.0659 |
| 10 | 12 | 14 | 0.0995 | 4 | 7 | 15 | 0.0752 | 9 | 12 | 15 | 0.0766 |
| 11 | 12 | 14 | 0.1298 | 5 | 7 | 15 | 0.0851 | 10 | 12 | 15 | 0.0910 |
| 1 | 13 | 14 | 0.0182 | 6 | 7 | 15 | 0.0961 | 11 | 12 | 15 | 0.1118 |
| 2 | 13 | 14 | 0.0226 | 1 | 8 | 15 | 0.0401 | 1 | 13 | 15 | 0.0205 |
| 3 | 13 | 14 | 0.0265 | 2 | 8 | 15 | 0.0494 | 2 | 13 | 15 | 0.0252 |
| 4 | 13 | 14 | 0.0305 | 3 | 8 | 15 | 0.0574 | 3 | 13 | 15 | 0.0294 |
| 5 | 13 | 14 | 0.0350 | 4 | 8 | 15 | 0.0655 | 4 | 13 | 15 | 0.0336 |
| 6 | 13 | 14 | 0.0401 | 5 | 8 | 15 | 0.0742 | 5 | 13 | 15 | 0.0381 |
| 7 | 13 | 14 | 0.0463 | 6 | 8 | 15 | 0.0840 | 6 | 13 | 15 | 0.0432 |
| 8 | 13 | 14 | 0.0541 | 7 | 8 | 15 | 0.0953 | 7 | 13 | 15 | 0.0492 |
| 9 | 13 | 14 | 0.0646 | 1 | 9 | 15 | 0.0352 | 8 | 13 | 15 | 0.0565 |
| 10 | 13 | 14 | 0.0798 | 2 | 9 | 15 | 0.0434 | 9 | 13 | 15 | 0.0658 |
| 11 | 13 | 14 | 0.1044 | 3 | 9 | 15 | 0.0505 | 10 | 13 | 15 | 0.0783 |
| 12 | 13 | 14 | 0.1532 | 4 | 9 | 15 | 0.0576 | 11 | 13 | 15 | 0.0964 |
| 1 | 14 | 14 | 0.0003 | 5 | 9 | 15 | 0.0653 | 12 | 13 | 15 | 0.1256 |
| 2 | 14 | 14 | 0.0004 | 6 | 9 | 15 | 0.0739 | 1 | 14 | 15 | 0.0164 |
| 3 | 14 | 14 | 0.0005 | 7 | 9 | 15 | 0.0840 | 2 | 14 | 15 | 0.0202 |
| 4 | 14 | 14 | 0.0005 | 8 | 9 | 15 | 0.0961 | 3 | 14 | 15 | 0.0236 |
| 5 | 14 | 14 | 0.0006 | 1 | 10 | 15 | 0.0310 | 4 | 14 | 15 | 0.0270 |
| 6 | 14 | 14 | 0.0007 | 2 | 10 | 15 | 0.0382 | 5 | 14 | 15 | 0.0306 |
| 7 | 14 | 14 | 0.0009 | 3 | 10 | 15 | 0.0445 | 6 | 14 | 15 | 0.0348 |
| 8 | 14 | 14 | 0.0010 | 4 | 10 | 15 | 0.0508 | 7 | 14 | 15 | 0.0396 |

TABLE 3

OUTLIER MEAN 4.0 OUTLIER STANDARD DEVIATION 1.0

| R | S | N | COVAR | R | S | N | COVAR | R | S | N | COVAR |
|---|---|---|---|---|---|---|---|---|---|---|---|
| 8 | 14 | 15 | 0.0455 | 5 | 8 | 16 | 0.0714 | 5 | 13 | 16 | 0.0390 |
| 9 | 14 | 15 | 0.0530 | 6 | 8 | 16 | 0.0801 | 6 | 13 | 16 | 0.0439 |
| 10 | 14 | 15 | 0.0631 | 7 | 8 | 16 | 0.0900 | 7 | 13 | 16 | 0.0494 |
| 11 | 14 | 15 | 0.0778 | 1 | 9 | 16 | 0.0349 | 8 | 13 | 16 | 0.0560 |
| 12 | 14 | 15 | 0.1017 | 2 | 9 | 16 | 0.0427 | 9 | 13 | 16 | 0.0640 |
| 13 | 14 | 15 | 0.1492 | 3 | 9 | 16 | 0.0494 | 10 | 13 | 16 | 0.0742 |
| 1 | 15 | 15 | 0.0003 | 4 | 9 | 16 | 0.0561 | 11 | 13 | 16 | 0.0881 |
| 2 | 15 | 15 | 0.0004 | 5 | 9 | 16 | 0.0631 | 12 | 13 | 16 | 0.1081 |
| 3 | 15 | 15 | 0.0004 | 6 | 9 | 16 | 0.0709 | 1 | 14 | 16 | 0.0185 |
| 4 | 15 | 15 | 0.0005 | 7 | 9 | 16 | 0.0797 | 2 | 14 | 16 | 0.0227 |
| 5 | 15 | 15 | 0.0006 | 8 | 9 | 16 | 0.0900 | 3 | 14 | 16 | 0.0263 |
| 6 | 15 | 15 | 0.0007 | 1 | 10 | 16 | 0.0310 | 4 | 14 | 16 | 0.0298 |
| 7 | 15 | 15 | 0.0008 | 2 | 10 | 16 | 0.0379 | 5 | 14 | 16 | 0.0336 |
| 8 | 15 | 15 | 0.0009 | 3 | 10 | 16 | 0.0439 | 6 | 14 | 16 | 0.0378 |
| 9 | 15 | 15 | 0.0011 | 4 | 10 | 16 | 0.0498 | 7 | 14 | 16 | 0.0426 |
| 10 | 15 | 15 | 0.0014 | 5 | 10 | 16 | 0.0561 | 8 | 14 | 16 | 0.0483 |
| 11 | 15 | 15 | 0.0018 | 6 | 10 | 16 | 0.0630 | 9 | 14 | 16 | 0.0553 |
| 12 | 15 | 15 | 0.0028 | 7 | 10 | 16 | 0.0709 | 10 | 14 | 16 | 0.0642 |
| 13 | 15 | 15 | 0.0057 | 8 | 10 | 16 | 0.0801 | 11 | 14 | 16 | 0.0762 |
| 14 | 15 | 15 | 0.0213 | 9 | 10 | 16 | 0.0915 | 12 | 14 | 16 | 0.0936 |
| 1 | 2 | 16 | 0.1481 | 1 | 11 | 16 | 0.0275 | 13 | 14 | 16 | 0.1219 |
| 1 | 3 | 16 | 0.1007 | 2 | 11 | 16 | 0.0337 | 1 | 15 | 16 | 0.0149 |
| 2 | 3 | 16 | 0.1224 | 3 | 11 | 16 | 0.0390 | 2 | 15 | 16 | 0.0183 |
| 1 | 4 | 16 | 0.0771 | 4 | 11 | 16 | 0.0443 | 3 | 15 | 16 | 0.0212 |
| 2 | 4 | 16 | 0.0939 | 5 | 11 | 16 | 0.0499 | 4 | 15 | 16 | 0.0241 |
| 3 | 4 | 16 | 0.1082 | 6 | 11 | 16 | 0.0561 | 5 | 15 | 16 | 0.0271 |
| 1 | 5 | 16 | 0.0626 | 7 | 11 | 16 | 0.0631 | 6 | 15 | 16 | 0.0305 |
| 2 | 5 | 16 | 0.0764 | 8 | 11 | 16 | 0.0714 | 7 | 15 | 16 | 0.0344 |
| 3 | 5 | 16 | 0.0882 | 9 | 11 | 16 | 0.0816 | 8 | 15 | 16 | 0.0390 |
| 4 | 5 | 16 | 0.0997 | 10 | 11 | 16 | 0.0945 | 9 | 15 | 16 | 0.0447 |
| 1 | 6 | 16 | 0.0527 | 1 | 12 | 16 | 0.0244 | 10 | 15 | 16 | 0.0519 |
| 2 | 6 | 16 | 0.0643 | 2 | 12 | 16 | 0.0299 | 11 | 15 | 16 | 0.0617 |
| 3 | 6 | 16 | 0.0743 | 3 | 12 | 16 | 0.0346 | 12 | 15 | 16 | 0.0760 |
| 4 | 6 | 16 | 0.0842 | 4 | 12 | 16 | 0.0393 | 13 | 15 | 16 | 0.0992 |
| 5 | 6 | 16 | 0.0945 | 5 | 12 | 16 | 0.0443 | 14 | 15 | 16 | 0.1455 |
| 1 | 7 | 16 | 0.0453 | 6 | 12 | 16 | 0.0498 | 1 | 16 | 16 | 0.0003 |
| 2 | 7 | 16 | 0.0554 | 7 | 12 | 16 | 0.0561 | 2 | 16 | 16 | 0.0004 |
| 3 | 7 | 16 | 0.0641 | 8 | 12 | 16 | 0.0635 | 3 | 16 | 16 | 0.0004 |
| 4 | 7 | 16 | 0.0726 | 9 | 12 | 16 | 0.0726 | 4 | 16 | 16 | 0.0005 |
| 5 | 7 | 16 | 0.0816 | 10 | 12 | 16 | 0.0841 | 5 | 16 | 16 | 0.0005 |
| 6 | 7 | 16 | 0.0915 | 11 | 12 | 16 | 0.0997 | 6 | 16 | 16 | 0.0006 |
| 1 | 8 | 16 | 0.0396 | 1 | 13 | 16 | 0.0215 | 7 | 16 | 16 | 0.0007 |
| 2 | 8 | 16 | 0.0484 | 2 | 13 | 16 | 0.0263 | 8 | 16 | 16 | 0.0008 |
| 3 | 8 | 16 | 0.0560 | 3 | 13 | 16 | 0.0305 | 9 | 16 | 16 | 0.0010 |
| 4 | 8 | 16 | 0.0635 | 4 | 13 | 16 | 0.0346 | 10 | 16 | 16 | 0.0011 |

TABLE 3

OUTLIER MEAN 4.0 OUTLIER STANDARD DEVIATION 1.0

| R | S | N | COVAR | R | S | N | COVAR | R | S | N | COVAR |
|---|---|---|---|---|---|---|---|---|---|---|---|
| 11 | 16 | 16 | 0.0014 | 5 | 10 | 17 | 0.0546 | 8 | 14 | 17 | 0.0484 |
| 12 | 16 | 16 | 0.0019 | 6 | 10 | 17 | 0.0608 | 9 | 14 | 17 | 0.0546 |
| 13 | 16 | 16 | 0.0030 | 7 | 10 | 17 | 0.0679 | 10 | 14 | 17 | 0.0623 |
| 14 | 16 | 16 | 0.0060 | 8 | 10 | 17 | 0.0760 | 11 | 14 | 17 | 0.0721 |
| 15 | 16 | 16 | 0.0221 | 9 | 10 | 17 | 0.0856 | 12 | 14 | 17 | 0.0854 |
| 1 | 2 | 17 | 0.1449 | 1 | 11 | 17 | 0.0275 | 13 | 14 | 17 | 0.1047 |
| 1 | 3 | 17 | 0.0985 | 2 | 11 | 17 | 0.0336 | 1 | 15 | 17 | 0.0168 |
| 2 | 3 | 17 | 0.1191 | 3 | 11 | 17 | 0.0387 | 2 | 15 | 17 | 0.0205 |
| 1 | 4 | 17 | 0.0754 | 4 | 11 | 17 | 0.0437 | 3 | 15 | 17 | 0.0237 |
| 2 | 4 | 17 | 0.0914 | 5 | 11 | 17 | 0.0489 | 4 | 15 | 17 | 0.0267 |
| 3 | 4 | 17 | 0.1049 | 6 | 11 | 17 | 0.0545 | 5 | 15 | 17 | 0.0300 |
| 1 | 5 | 17 | 0.0613 | 7 | 11 | 17 | 0.0608 | 6 | 15 | 17 | 0.0335 |
| 2 | 5 | 17 | 0.0745 | 8 | 11 | 17 | 0.0681 | 7 | 15 | 17 | 0.0374 |
| 3 | 5 | 17 | 0.0855 | 9 | 11 | 17 | 0.0768 | 8 | 15 | 17 | 0.0419 |
| 4 | 5 | 17 | 0.0963 | 10 | 11 | 17 | 0.0874 | 9 | 15 | 17 | 0.0474 |
| 1 | 6 | 17 | 0.0517 | 1 | 12 | 17 | 0.0246 | 10 | 15 | 17 | 0.0540 |
| 2 | 6 | 17 | 0.0628 | 2 | 12 | 17 | 0.0300 | 11 | 15 | 17 | 0.0626 |
| 3 | 6 | 17 | 0.0722 | 3 | 12 | 17 | 0.0346 | 12 | 15 | 17 | 0.0742 |
| 4 | 6 | 17 | 0.0813 | 4 | 12 | 17 | 0.0391 | 13 | 15 | 17 | 0.0911 |
| 5 | 6 | 17 | 0.0908 | 5 | 12 | 17 | 0.0438 | 14 | 15 | 17 | 0.1186 |
| 1 | 7 | 17 | 0.0446 | 6 | 12 | 17 | 0.0489 | 1 | 16 | 17 | 0.0136 |
| 2 | 7 | 17 | 0.0542 | 7 | 12 | 17 | 0.0546 | 2 | 16 | 17 | 0.0166 |
| 3 | 7 | 17 | 0.0624 | 8 | 12 | 17 | 0.0611 | 3 | 16 | 17 | 0.0192 |
| 4 | 7 | 17 | 0.0703 | 9 | 12 | 17 | 0.0689 | 4 | 16 | 17 | 0.0217 |
| 5 | 7 | 17 | 0.0785 | 10 | 12 | 17 | 0.0785 | 5 | 16 | 17 | 0.0243 |
| 6 | 7 | 17 | 0.0875 | 11 | 12 | 17 | 0.0908 | 6 | 16 | 17 | 0.0271 |
| 1 | 8 | 17 | 0.0390 | 1 | 13 | 17 | 0.0220 | 7 | 16 | 17 | 0.0303 |
| 2 | 8 | 17 | 0.0475 | 2 | 13 | 17 | 0.0268 | 8 | 16 | 17 | 0.0340 |
| 3 | 8 | 17 | 0.0547 | 3 | 13 | 17 | 0.0309 | 9 | 16 | 17 | 0.0385 |
| 4 | 8 | 17 | 0.0617 | 4 | 13 | 17 | 0.0349 | 10 | 16 | 17 | 0.0439 |
| 5 | 8 | 17 | 0.0689 | 5 | 13 | 17 | 0.0391 | 11 | 16 | 17 | 0.0509 |
| 6 | 8 | 17 | 0.0768 | 6 | 13 | 17 | 0.0436 | 12 | 16 | 17 | 0.0604 |
| 7 | 8 | 17 | 0.0856 | 7 | 13 | 17 | 0.0487 | 13 | 16 | 17 | 0.0743 |
| 1 | 9 | 17 | 0.0345 | 8 | 13 | 17 | 0.0546 | 14 | 16 | 17 | 0.0970 |
| 2 | 9 | 17 | 0.0421 | 9 | 13 | 17 | 0.0616 | 15 | 16 | 17 | 0.1422 |
| 3 | 9 | 17 | 0.0484 | 10 | 13 | 17 | 0.0703 | 1 | 17 | 17 | 0.0003 |
| 4 | 9 | 17 | 0.0547 | 11 | 13 | 17 | 0.0813 | 2 | 17 | 17 | 0.0003 |
| 5 | 9 | 17 | 0.0611 | 12 | 13 | 17 | 0.0962 | 3 | 17 | 17 | 0.0004 |
| 6 | 9 | 17 | 0.0682 | 1 | 14 | 17 | 0.0194 | 4 | 17 | 17 | 0.0005 |
| 7 | 9 | 17 | 0.0760 | 2 | 14 | 17 | 0.0237 | 5 | 17 | 17 | 0.0005 |
| 8 | 9 | 17 | 0.0850 | 3 | 14 | 17 | 0.0273 | 6 | 17 | 17 | 0.0006 |
| 1 | 10 | 17 | 0.0308 | 4 | 14 | 17 | 0.0309 | 7 | 17 | 17 | 0.0007 |
| 2 | 10 | 17 | 0.0375 | 5 | 14 | 17 | 0.0346 | 8 | 17 | 17 | 0.0007 |
| 3 | 10 | 17 | 0.0432 | 6 | 14 | 17 | 0.0386 | 9 | 17 | 17 | 0.0009 |
| 4 | 10 | 17 | 0.0488 | 7 | 14 | 17 | 0.0432 | 10 | 17 | 17 | 0.0010 |

TABLE 3

OUTLIER MEAN 4.0 OUTLIER STANDARD DEVIATION 1.0

| R | S | N | COVAR | R | S | N | COVAR | R | S | N | COVAR |
|---|---|---|---|---|---|---|---|---|---|---|---|
| 11 | 17 | 17 | 0.0012 | 4 | 10 | 18 | 0.0477 | 7 | 14 | 18 | 0.0429 |
| 12 | 17 | 17 | 0.0015 | 5 | 10 | 18 | 0.0531 | 8 | 14 | 18 | 0.0477 |
| 13 | 17 | 17 | 0.0020 | 6 | 10 | 18 | 0.0589 | 9 | 14 | 18 | 0.0533 |
| 14 | 17 | 17 | 0.0031 | 7 | 10 | 18 | 0.0653 | 10 | 14 | 18 | 0.0600 |
| 15 | 17 | 17 | 0.0062 | 8 | 10 | 18 | 0.0725 | 11 | 14 | 18 | 0.0682 |
| 16 | 17 | 17 | 0.0228 | 9 | 10 | 18 | 0.0808 | 12 | 14 | 18 | 0.0788 |
| 1 | 2 | 18 | 0.1419 | 1 | 11 | 18 | 0.0274 | 13 | 14 | 18 | 0.0931 |
| 1 | 3 | 18 | 0.0965 | 2 | 11 | 18 | 0.0333 | 1 | 15 | 18 | 0.0177 |
| 2 | 3 | 18 | 0.1162 | 3 | 11 | 18 | 0.0382 | 2 | 15 | 18 | 0.0215 |
| 1 | 4 | 18 | 0.0739 | 4 | 11 | 18 | 0.0429 | 3 | 15 | 18 | 0.0247 |
| 2 | 4 | 18 | 0.0892 | 5 | 11 | 18 | 0.0478 | 4 | 15 | 18 | 0.0278 |
| 3 | 4 | 18 | 0.1019 | 6 | 11 | 18 | 0.0530 | 5 | 15 | 18 | 0.0310 |
| 1 | 5 | 18 | 0.0601 | 7 | 11 | 18 | 0.0588 | 6 | 15 | 18 | 0.0344 |
| 2 | 5 | 18 | 0.0727 | 8 | 11 | 18 | 0.0653 | 7 | 15 | 18 | 0.0382 |
| 3 | 5 | 18 | 0.0831 | 9 | 11 | 18 | 0.0728 | 8 | 15 | 18 | 0.0424 |
| 4 | 5 | 18 | 0.0932 | 10 | 11 | 18 | 0.0818 | 9 | 15 | 18 | 0.0474 |
| 1 | 6 | 18 | 0.0507 | 1 | 12 | 18 | 0.0247 | 10 | 15 | 18 | 0.0534 |
| 2 | 6 | 18 | 0.0614 | 2 | 12 | 18 | 0.0300 | 11 | 15 | 18 | 0.0607 |
| 3 | 6 | 18 | 0.0703 | 3 | 12 | 18 | 0.0344 | 12 | 15 | 18 | 0.0702 |
| 4 | 6 | 18 | 0.0788 | 4 | 12 | 18 | 0.0387 | 13 | 15 | 18 | 0.0830 |
| 5 | 6 | 18 | 0.0876 | 5 | 12 | 18 | 0.0431 | 14 | 15 | 18 | 0.1017 |
| 1 | 7 | 18 | 0.0438 | 6 | 12 | 18 | 0.0478 | 1 | 16 | 18 | 0.0154 |
| 2 | 7 | 18 | 0.0531 | 7 | 12 | 18 | 0.0530 | 2 | 16 | 18 | 0.0187 |
| 3 | 7 | 18 | 0.0608 | 8 | 12 | 18 | 0.0589 | 3 | 16 | 18 | 0.0215 |
| 4 | 7 | 18 | 0.0682 | 9 | 12 | 18 | 0.0657 | 4 | 16 | 18 | 0.0242 |
| 5 | 7 | 18 | 0.0759 | 10 | 12 | 18 | 0.0739 | 5 | 16 | 18 | 0.0269 |
| 6 | 7 | 18 | 0.0840 | 11 | 12 | 18 | 0.0840 | 6 | 16 | 18 | 0.0299 |
| 1 | 8 | 18 | 0.0385 | 1 | 13 | 18 | 0.0223 | 7 | 16 | 18 | 0.0332 |
| 2 | 8 | 18 | 0.0466 | 2 | 13 | 18 | 0.0270 | 8 | 16 | 18 | 0.0369 |
| 3 | 8 | 18 | 0.0534 | 3 | 13 | 18 | 0.0310 | 9 | 16 | 18 | 0.0413 |
| 4 | 8 | 18 | 0.0600 | 4 | 13 | 18 | 0.0349 | 10 | 16 | 18 | 0.0465 |
| 5 | 8 | 18 | 0.0667 | 5 | 13 | 18 | 0.0388 | 11 | 16 | 18 | 0.0529 |
| 6 | 8 | 18 | 0.0739 | 6 | 13 | 18 | 0.0431 | 12 | 16 | 18 | 0.0612 |
| 7 | 8 | 18 | 0.0818 | 7 | 13 | 18 | 0.0478 | 13 | 16 | 18 | 0.0724 |
| 1 | 9 | 18 | 0.0341 | 8 | 13 | 18 | 0.0531 | 14 | 16 | 18 | 0.0889 |
| 2 | 9 | 18 | 0.0414 | 9 | 13 | 18 | 0.0593 | 15 | 16 | 18 | 0.1157 |
| 3 | 9 | 18 | 0.0475 | 10 | 13 | 18 | 0.0667 | 1 | 17 | 18 | 0.0125 |
| 4 | 9 | 18 | 0.0533 | 11 | 13 | 18 | 0.0758 | 2 | 17 | 18 | 0.0152 |
| 5 | 9 | 18 | 0.0593 | 12 | 13 | 18 | 0.0875 | 3 | 17 | 18 | 0.0175 |
| 6 | 9 | 18 | 0.0657 | 1 | 14 | 18 | 0.0200 | 4 | 17 | 18 | 0.0197 |
| 7 | 9 | 18 | 0.0728 | 2 | 14 | 18 | 0.0242 | 5 | 17 | 18 | 0.0219 |
| 8 | 9 | 18 | 0.0808 | 3 | 14 | 18 | 0.0278 | 6 | 17 | 18 | 0.0244 |
| 1 | 10 | 18 | 0.0305 | 4 | 14 | 18 | 0.0313 | 7 | 17 | 18 | 0.0270 |
| 2 | 10 | 18 | 0.0370 | 5 | 14 | 18 | 0.0348 | 8 | 17 | 18 | 0.0301 |
| 3 | 10 | 18 | 0.0425 | 6 | 14 | 18 | 0.0387 | 9 | 17 | 18 | 0.0336 |

TABLE 3

OUTLIER MEAN 4.0 OUTLIER STANDARD DEVIATION 1.0

| R | S | N | COVAR | R | S | N | COVAR | R | S | N | COVAR |
|---|---|---|---|---|---|---|---|---|---|---|---|
| 10 | 17 | 18 | 0.0379 | 1 | 8 | 19 | 0.0379 | 1 | 13 | 19 | 0.0224 |
| 11 | 17 | 18 | 0.0432 | 2 | 8 | 19 | 0.0458 | 2 | 13 | 19 | 0.0270 |
| 12 | 17 | 18 | 0.0500 | 3 | 8 | 19 | 0.0522 | 3 | 13 | 19 | 0.0309 |
| 13 | 17 | 18 | 0.0592 | 4 | 8 | 19 | 0.0584 | 4 | 13 | 19 | 0.0346 |
| 14 | 17 | 18 | 0.0727 | 5 | 8 | 19 | 0.0647 | 5 | 13 | 19 | 0.0384 |
| 15 | 17 | 18 | 0.0949 | 6 | 8 | 19 | 0.0713 | 6 | 13 | 19 | 0.0424 |
| 16 | 17 | 18 | 0.1392 | 7 | 8 | 19 | 0.0785 | 7 | 13 | 19 | 0.0467 |
| 1 | 18 | 18 | 0.0003 | 1 | 9 | 19 | 0.0337 | 8 | 13 | 19 | 0.0516 |
| 2 | 18 | 18 | 0.0003 | 2 | 9 | 19 | 0.0407 | 9 | 13 | 19 | 0.0571 |
| 3 | 18 | 18 | 0.0004 | 3 | 9 | 19 | 0.0465 | 10 | 13 | 19 | 0.0636 |
| 4 | 18 | 18 | 0.0004 | 4 | 9 | 19 | 0.0520 | 11 | 13 | 19 | 0.0713 |
| 5 | 18 | 18 | 0.0005 | 5 | 9 | 19 | 0.0577 | 12 | 13 | 19 | 0.0809 |
| 6 | 18 | 18 | 0.0005 | 6 | 9 | 19 | 0.0636 | 1 | 14 | 19 | 0.0202 |
| 7 | 18 | 18 | 0.0006 | 7 | 9 | 19 | 0.0700 | 2 | 14 | 19 | 0.0245 |
| 8 | 18 | 18 | 0.0007 | 8 | 9 | 19 | 0.0772 | 3 | 14 | 19 | 0.0280 |
| 9 | 18 | 18 | 0.0008 | 1 | 10 | 19 | 0.0303 | 4 | 14 | 19 | 0.0313 |
| 10 | 18 | 18 | 0.0009 | 2 | 10 | 19 | 0.0365 | 5 | 14 | 19 | 0.0348 |
| 11 | 18 | 18 | 0.0010 | 3 | 10 | 19 | 0.0417 | 6 | 14 | 19 | 0.0384 |
| 12 | 18 | 18 | 0.0012 | 4 | 10 | 19 | 0.0467 | 7 | 14 | 19 | 0.0423 |
| 13 | 18 | 18 | 0.0016 | 5 | 10 | 19 | 0.0518 | 8 | 14 | 19 | 0.0467 |
| 14 | 18 | 18 | 0.0021 | 6 | 10 | 19 | 0.0571 | 9 | 14 | 19 | 0.0518 |
| 15 | 18 | 18 | 0.0033 | 7 | 10 | 19 | 0.0629 | 10 | 14 | 19 | 0.0576 |
| 16 | 18 | 18 | 0.0065 | 8 | 10 | 19 | 0.0694 | 11 | 14 | 19 | 0.0647 |
| 17 | 18 | 18 | 0.0235 | 9 | 10 | 19 | 0.0767 | 12 | 14 | 19 | 0.0734 |
| 1 | 2 | 19 | 0.1393 | 1 | 11 | 19 | 0.0273 | 13 | 14 | 19 | 0.0847 |
| 1 | 3 | 19 | 0.0946 | 2 | 11 | 19 | 0.0330 | 1 | 15 | 19 | 0.0182 |
| 2 | 3 | 19 | 0.1135 | 3 | 11 | 19 | 0.0377 | 2 | 15 | 19 | 0.0220 |
| 1 | 4 | 19 | 0.0725 | 4 | 11 | 19 | 0.0422 | 3 | 15 | 19 | 0.0252 |
| 2 | 4 | 19 | 0.0872 | 5 | 11 | 19 | 0.0467 | 4 | 15 | 19 | 0.0282 |
| 3 | 4 | 19 | 0.0992 | 6 | 11 | 19 | 0.0516 | 5 | 15 | 19 | 0.0313 |
| 1 | 5 | 19 | 0.0590 | 7 | 11 | 19 | 0.0568 | 6 | 15 | 19 | 0.0346 |
| 2 | 5 | 19 | 0.0711 | 8 | 11 | 19 | 0.0627 | 7 | 15 | 19 | 0.0382 |
| 3 | 5 | 19 | 0.0810 | 9 | 11 | 19 | 0.0694 | 8 | 15 | 19 | 0.0421 |
| 4 | 5 | 19 | 0.0904 | 10 | 11 | 19 | 0.0772 | 9 | 15 | 19 | 0.0467 |
| 1 | 6 | 19 | 0.0499 | 1 | 12 | 19 | 0.0247 | 10 | 15 | 19 | 0.0520 |
| 2 | 6 | 19 | 0.0601 | 2 | 12 | 19 | 0.0298 | 11 | 15 | 19 | 0.0584 |
| 3 | 6 | 19 | 0.0685 | 3 | 12 | 19 | 0.0341 | 12 | 15 | 19 | 0.0663 |
| 4 | 6 | 19 | 0.0766 | 4 | 12 | 19 | 0.0382 | 13 | 15 | 19 | 0.0765 |
| 5 | 6 | 19 | 0.0847 | 5 | 12 | 19 | 0.0423 | 14 | 15 | 19 | 0.0903 |
| 1 | 7 | 19 | 0.0431 | 6 | 12 | 19 | 0.0467 | 1 | 16 | 19 | 0.0163 |
| 2 | 7 | 19 | 0.0520 | 7 | 12 | 19 | 0.0515 | 2 | 16 | 19 | 0.0197 |
| 3 | 7 | 19 | 0.0594 | 8 | 12 | 19 | 0.0568 | 3 | 16 | 19 | 0.0225 |
| 4 | 7 | 19 | 0.0664 | 9 | 12 | 19 | 0.0629 | 4 | 16 | 19 | 0.0252 |
| 5 | 7 | 19 | 0.0734 | 10 | 12 | 19 | 0.0700 | 5 | 16 | 19 | 0.0280 |
| 6 | 7 | 19 | 0.0809 | 11 | 12 | 19 | 0.0785 | 6 | 16 | 19 | 0.0309 |

TABLE 3

OUTLIER MEAN 4.0 OUTLIER STANDARD DEVIATION 1.0

| R | S | N | COVAR | R | S | N | COVAR | R | S | N | COVAR |
|---|---|---|---|---|---|---|---|---|---|---|---|
| 7 | 16 | 19 | 0.0341 | 4 | 19 | 19 | 0.0004 | 3 | 9 | 20 | 0.0456 |
| 8 | 16 | 19 | 0.0376 | 5 | 19 | 19 | 0.0005 | 4 | 9 | 20 | 0.0509 |
| 9 | 16 | 19 | 0.0417 | 6 | 19 | 19 | 0.0005 | 5 | 9 | 20 | 0.0561 |
| 10 | 16 | 19 | 0.0465 | 7 | 19 | 19 | 0.0006 | 6 | 9 | 20 | 0.0616 |
| 11 | 16 | 19 | 0.0522 | 8 | 19 | 19 | 0.0006 | 7 | 9 | 20 | 0.0675 |
| 12 | 16 | 19 | 0.0593 | 9 | 19 | 19 | 0.0007 | 8 | 9 | 20 | 0.0740 |
| 13 | 16 | 19 | 0.0684 | 10 | 19 | 19 | 0.0008 | 1 | 10 | 20 | 0.0300 |
| 14 | 16 | 19 | 0.0809 | 11 | 19 | 19 | 0.0009 | 2 | 10 | 20 | 0.0360 |
| 15 | 16 | 19 | 0.0990 | 12 | 19 | 19 | 0.0011 | 3 | 10 | 20 | 0.0410 |
| 1 | 17 | 19 | 0.0142 | 13 | 19 | 19 | 0.0013 | 4 | 10 | 20 | 0.0458 |
| 2 | 17 | 19 | 0.0172 | 14 | 19 | 19 | 0.0016 | 5 | 10 | 20 | 0.0505 |
| 3 | 17 | 19 | 0.0196 | 15 | 19 | 19 | 0.0022 | 6 | 10 | 20 | 0.0555 |
| 4 | 17 | 19 | 0.0220 | 16 | 19 | 19 | 0.0034 | 7 | 10 | 20 | 0.0608 |
| 5 | 17 | 19 | 0.0244 | 17 | 19 | 19 | 0.0067 | 8 | 10 | 20 | 0.0667 |
| 6 | 17 | 19 | 0.0270 | 18 | 19 | 19 | 0.0242 | 9 | 10 | 20 | 0.0733 |
| 7 | 17 | 19 | 0.0297 | 1 | 2 | 20 | 0.1368 | 1 | 11 | 20 | 0.0271 |
| 8 | 17 | 19 | 0.0329 | 1 | 3 | 20 | 0.0929 | 2 | 11 | 20 | 0.0326 |
| 9 | 17 | 19 | 0.0364 | 2 | 3 | 20 | 0.1111 | 3 | 11 | 20 | 0.0371 |
| 10 | 17 | 19 | 0.0406 | 1 | 4 | 20 | 0.0712 | 4 | 11 | 20 | 0.0414 |
| 11 | 17 | 19 | 0.0456 | 2 | 4 | 20 | 0.0853 | 5 | 11 | 20 | 0.0457 |
| 12 | 17 | 19 | 0.0518 | 3 | 4 | 20 | 0.0967 | 6 | 11 | 20 | 0.0502 |
| 13 | 17 | 19 | 0.0599 | 1 | 5 | 20 | 0.0580 | 7 | 11 | 20 | 0.0551 |
| 14 | 17 | 19 | 0.0708 | 2 | 5 | 20 | 0.0696 | 8 | 11 | 20 | 0.0604 |
| 15 | 17 | 19 | 0.0868 | 3 | 5 | 20 | 0.0790 | 9 | 11 | 20 | 0.0664 |
| 16 | 17 | 19 | 0.1130 | 4 | 5 | 20 | 0.0879 | 10 | 11 | 20 | 0.0733 |
| 1 | 18 | 19 | 0.0116 | 1 | 6 | 20 | 0.0490 | 1 | 12 | 20 | 0.0246 |
| 2 | 18 | 19 | 0.0140 | 2 | 6 | 20 | 0.0589 | 2 | 12 | 20 | 0.0296 |
| 3 | 18 | 19 | 0.0160 | 3 | 6 | 20 | 0.0669 | 3 | 12 | 20 | 0.0337 |
| 4 | 18 | 19 | 0.0180 | 4 | 6 | 20 | 0.0745 | 4 | 12 | 20 | 0.0376 |
| 5 | 18 | 19 | 0.0199 | 5 | 6 | 20 | 0.0821 | 5 | 12 | 20 | 0.0416 |
| 6 | 18 | 19 | 0.0220 | 1 | 7 | 20 | 0.0425 | 6 | 12 | 20 | 0.0457 |
| 7 | 18 | 19 | 0.0243 | 2 | 7 | 20 | 0.0510 | 7 | 12 | 20 | 0.0501 |
| 8 | 18 | 19 | 0.0269 | 3 | 7 | 20 | 0.0580 | 8 | 12 | 20 | 0.0550 |
| 9 | 18 | 19 | 0.0298 | 4 | 7 | 20 | 0.0646 | 9 | 12 | 20 | 0.0604 |
| 10 | 18 | 19 | 0.0332 | 5 | 7 | 20 | 0.0713 | 10 | 12 | 20 | 0.0667 |
| 11 | 18 | 19 | 0.0373 | 6 | 7 | 20 | 0.0782 | 11 | 12 | 20 | 0.0740 |
| 12 | 18 | 19 | 0.0425 | 1 | 8 | 20 | 0.0374 | 1 | 13 | 20 | 0.0224 |
| 13 | 18 | 19 | 0.0491 | 2 | 8 | 20 | 0.0450 | 2 | 13 | 20 | 0.0270 |
| 14 | 18 | 19 | 0.0581 | 3 | 8 | 20 | 0.0512 | 3 | 13 | 20 | 0.0307 |
| 15 | 18 | 19 | 0.0713 | 4 | 8 | 20 | 0.0570 | 4 | 13 | 20 | 0.0343 |
| 16 | 18 | 19 | 0.0930 | 5 | 8 | 20 | 0.0629 | 5 | 13 | 20 | 0.0379 |
| 17 | 18 | 19 | 0.1365 | 6 | 8 | 20 | 0.0690 | 6 | 13 | 20 | 0.0416 |
| 1 | 19 | 19 | 0.0003 | 7 | 8 | 20 | 0.0756 | 7 | 13 | 20 | 0.0457 |
| 2 | 19 | 19 | 0.0003 | 1 | 9 | 20 | 0.0333 | 8 | 13 | 20 | 0.0501 |
| 3 | 19 | 19 | 0.0004 | 2 | 9 | 20 | 0.0401 | 9 | 13 | 20 | 0.0551 |

TABLE 3

OUTLIER MEAN 4.0 OUTLIER STANDARD DEVIATION 1.0

| R | S | N | COVAR | R | S | N | COVAR | R | S | N | COVAR |
|---|---|---|---|---|---|---|---|---|---|---|---|
| 10 | 13 | 20 | 0.0608 | 10 | 16 | 20 | 0.0457 | 1 | 19 | 20 | 0.0108 |
| 11 | 13 | 20 | 0.0675 | 11 | 16 | 20 | 0.0508 | 2 | 19 | 20 | 0.0130 |
| 12 | 13 | 20 | 0.0756 | 12 | 16 | 20 | 0.0570 | 3 | 19 | 20 | 0.0148 |
| 1 | 14 | 20 | 0.0204 | 13 | 16 | 20 | 0.0646 | 4 | 19 | 20 | 0.0165 |
| 2 | 14 | 20 | 0.0246 | 14 | 16 | 20 | 0.0744 | 5 | 19 | 20 | 0.0182 |
| 3 | 14 | 20 | 0.0280 | 15 | 16 | 20 | 0.0878 | 6 | 19 | 20 | 0.0201 |
| 4 | 14 | 20 | 0.0312 | 1 | 17 | 20 | 0.0150 | 7 | 19 | 20 | 0.0220 |
| 5 | 14 | 20 | 0.0345 | 2 | 17 | 20 | 0.0181 | 8 | 19 | 20 | 0.0242 |
| 6 | 14 | 20 | 0.0379 | 3 | 17 | 20 | 0.0206 | 9 | 19 | 20 | 0.0267 |
| 7 | 14 | 20 | 0.0416 | 4 | 17 | 20 | 0.0230 | 10 | 19 | 20 | 0.0295 |
| 8 | 14 | 20 | 0.0457 | 5 | 17 | 20 | 0.0254 | 11 | 19 | 20 | 0.0328 |
| 9 | 14 | 20 | 0.0502 | 6 | 17 | 20 | 0.0279 | 12 | 19 | 20 | 0.0368 |
| 10 | 14 | 20 | 0.0555 | 7 | 17 | 20 | 0.0307 | 13 | 19 | 20 | 0.0418 |
| 11 | 14 | 20 | 0.0616 | 8 | 17 | 20 | 0.0337 | 14 | 19 | 20 | 0.0482 |
| 12 | 14 | 20 | 0.0690 | 9 | 17 | 20 | 0.0371 | 15 | 19 | 20 | 0.0571 |
| 13 | 14 | 20 | 0.0782 | 10 | 17 | 20 | 0.0410 | 16 | 19 | 20 | 0.0700 |
| 1 | 15 | 20 | 0.0185 | 11 | 17 | 20 | 0.0456 | 17 | 19 | 20 | 0.0913 |
| 2 | 15 | 20 | 0.0223 | 12 | 17 | 20 | 0.0511 | 18 | 19 | 20 | 0.1340 |
| 3 | 15 | 20 | 0.0254 | 13 | 17 | 20 | 0.0580 | 1 | 20 | 20 | 0.0003 |
| 4 | 15 | 20 | 0.0284 | 14 | 17 | 20 | 0.0668 | 2 | 20 | 20 | 0.0003 |
| 5 | 15 | 20 | 0.0314 | 15 | 17 | 20 | 0.0789 | 3 | 20 | 20 | 0.0004 |
| 6 | 15 | 20 | 0.0345 | 16 | 17 | 20 | 0.0965 | 4 | 20 | 20 | 0.0004 |
| 7 | 15 | 20 | 0.0379 | 1 | 18 | 20 | 0.0131 | 5 | 20 | 20 | 0.0004 |
| 8 | 15 | 20 | 0.0416 | 2 | 18 | 20 | 0.0158 | 6 | 20 | 20 | 0.0005 |
| 9 | 15 | 20 | 0.0457 | 3 | 18 | 20 | 0.0180 | 7 | 20 | 20 | 0.0005 |
| 10 | 15 | 20 | 0.0505 | 4 | 18 | 20 | 0.0201 | 8 | 20 | 20 | 0.0006 |
| 11 | 15 | 20 | 0.0561 | 5 | 18 | 20 | 0.0223 | 9 | 20 | 20 | 0.0007 |
| 12 | 15 | 20 | 0.0629 | 6 | 18 | 20 | 0.0245 | 10 | 20 | 20 | 0.0007 |
| 13 | 15 | 20 | 0.0713 | 7 | 18 | 20 | 0.0269 | 11 | 20 | 20 | 0.0008 |
| 14 | 15 | 20 | 0.0821 | 8 | 18 | 20 | 0.0295 | 12 | 20 | 20 | 0.0010 |
| 1 | 16 | 20 | 0.0168 | 9 | 18 | 20 | 0.0325 | 13 | 20 | 20 | 0.0011 |
| 2 | 16 | 20 | 0.0202 | 10 | 18 | 20 | 0.0359 | 14 | 20 | 20 | 0.0013 |
| 3 | 16 | 20 | 0.0230 | 11 | 18 | 20 | 0.0399 | 15 | 20 | 20 | 0.0017 |
| 4 | 16 | 20 | 0.0257 | 12 | 18 | 20 | 0.0448 | 16 | 20 | 20 | 0.0023 |
| 5 | 16 | 20 | 0.0284 | 13 | 18 | 20 | 0.0509 | 17 | 20 | 20 | 0.0035 |
| 6 | 16 | 20 | 0.0312 | 14 | 18 | 20 | 0.0587 | 18 | 20 | 20 | 0.0070 |
| 7 | 16 | 20 | 0.0343 | 15 | 18 | 20 | 0.0693 | 19 | 20 | 20 | 0.0248 |
| 8 | 16 | 20 | 0.0376 | 16 | 18 | 20 | 0.0849 | | | | |
| 9 | 16 | 20 | 0.0414 | 17 | 18 | 20 | 0.1105 | | | | |

TABLE 4

OUTLIER MEAN 0.0 OUTLIER STANDARD DEVIATION 0.5

| R | S | N | COVAR | R | S | N | COVAR | R | S | N | COVAR |
|---|---|---|---|---|---|---|---|---|---|---|---|
| 1 | 2 | 2 | 0.1989 | 3 | 8 | 8 | 0.0580 | 5 | 10 | 10 | 0.0546 |
| 1 | 3 | 3 | 0.1403 | 4 | 8 | 8 | 0.0686 | 6 | 10 | 10 | 0.0653 |
| 2 | 3 | 3 | 0.1949 | 5 | 8 | 8 | 0.0855 | 7 | 10 | 10 | 0.0821 |
| 2 | 3 | 4 | 0.1700 | 6 | 8 | 8 | 0.1157 | 8 | 10 | 10 | 0.1108 |
| 1 | 4 | 4 | 0.1017 | 7 | 8 | 8 | 0.1784 | 9 | 10 | 10 | 0.1683 |
| 2 | 4 | 4 | 0.1296 | 4 | 6 | 9 | 0.0957 | 5 | 7 | 11 | 0.0860 |
| 3 | 4 | 4 | 0.1951 | 5 | 6 | 9 | 0.1149 | 6 | 7 | 11 | 0.1005 |
| 2 | 4 | 5 | 0.1216 | 3 | 7 | 9 | 0.0702 | 4 | 8 | 11 | 0.0655 |
| 3 | 4 | 5 | 0.1599 | 4 | 7 | 9 | 0.0818 | 5 | 8 | 11 | 0.0750 |
| 1 | 5 | 5 | 0.0768 | 5 | 7 | 9 | 0.0979 | 6 | 8 | 11 | 0.0875 |
| 2 | 5 | 5 | 0.0959 | 6 | 7 | 9 | 0.1231 | 7 | 8 | 11 | 0.1052 |
| 3 | 5 | 5 | 0.1240 | 2 | 8 | 9 | 0.0517 | 3 | 9 | 11 | 0.0507 |
| 4 | 5 | 5 | 0.1929 | 3 | 8 | 9 | 0.0601 | 4 | 9 | 11 | 0.0576 |
| 3 | 4 | 6 | 0.1432 | 4 | 8 | 9 | 0.0697 | 5 | 9 | 11 | 0.0657 |
| 2 | 5 | 6 | 0.0935 | 5 | 8 | 9 | 0.0832 | 6 | 9 | 11 | 0.0765 |
| 3 | 5 | 6 | 0.1151 | 6 | 8 | 9 | 0.1045 | 7 | 9 | 11 | 0.0920 |
| 4 | 5 | 6 | 0.1541 | 7 | 8 | 9 | 0.1411 | 8 | 9 | 11 | 0.1155 |
| 1 | 6 | 6 | 0.0601 | 1 | 9 | 9 | 0.0339 | 2 | 10 | 11 | 0.0382 |
| 2 | 6 | 6 | 0.0746 | 2 | 9 | 9 | 0.0418 | 3 | 10 | 11 | 0.0440 |
| 3 | 6 | 6 | 0.0909 | 3 | 9 | 9 | 0.0485 | 4 | 10 | 11 | 0.0498 |
| 4 | 6 | 6 | 0.1208 | 4 | 9 | 9 | 0.0561 | 5 | 10 | 11 | 0.0567 |
| 5 | 6 | 6 | 0.1888 | 5 | 9 | 9 | 0.0668 | 6 | 10 | 11 | 0.0659 |
| 3 | 5 | 7 | 0.1074 | 6 | 9 | 9 | 0.0837 | 7 | 10 | 11 | 0.0792 |
| 4 | 5 | 7 | 0.1340 | 7 | 9 | 9 | 0.1132 | 8 | 10 | 11 | 0.0995 |
| 2 | 6 | 7 | 0.0748 | 8 | 9 | 9 | 0.1732 | 9 | 10 | 11 | 0.1334 |
| 3 | 6 | 7 | 0.0893 | 5 | 6 | 10 | 0.1065 | 1 | 11 | 11 | 0.0254 |
| 4 | 6 | 7 | 0.1107 | 4 | 7 | 10 | 0.0781 | 2 | 11 | 11 | 0.0311 |
| 5 | 6 | 7 | 0.1495 | 5 | 7 | 10 | 0.0911 | 3 | 11 | 11 | 0.0358 |
| 1 | 7 | 7 | 0.0485 | 6 | 7 | 10 | 0.1095 | 4 | 11 | 11 | 0.0405 |
| 2 | 7 | 7 | 0.0601 | 3 | 8 | 10 | 0.0591 | 5 | 11 | 11 | 0.0460 |
| 3 | 7 | 7 | 0.0712 | 4 | 8 | 10 | 0.0678 | 6 | 11 | 11 | 0.0534 |
| 4 | 7 | 7 | 0.0877 | 5 | 8 | 10 | 0.0789 | 7 | 11 | 11 | 0.0641 |
| 5 | 7 | 7 | 0.1182 | 6 | 8 | 10 | 0.0947 | 8 | 11 | 11 | 0.0806 |
| 6 | 7 | 7 | 0.1837 | 7 | 8 | 10 | 0.1191 | 9 | 11 | 11 | 0.1083 |
| 4 | 5 | 8 | 0.1224 | 2 | 9 | 10 | 0.0441 | 10 | 11 | 11 | 0.1637 |
| 3 | 6 | 8 | 0.0853 | 3 | 9 | 10 | 0.0510 | 6 | 7 | 12 | 0.0941 |
| 4 | 6 | 8 | 0.1020 | 4 | 9 | 10 | 0.0583 | 5 | 8 | 12 | 0.0718 |
| 5 | 6 | 8 | 0.1278 | 5 | 9 | 10 | 0.0677 | 6 | 8 | 12 | 0.0821 |
| 2 | 7 | 8 | 0.0615 | 6 | 9 | 10 | 0.0811 | 7 | 8 | 12 | 0.0959 |
| 3 | 7 | 8 | 0.0723 | 7 | 9 | 10 | 0.1019 | 4 | 9 | 12 | 0.0561 |
| 4 | 7 | 8 | 0.0859 | 8 | 9 | 10 | 0.1371 | 5 | 9 | 12 | 0.0634 |
| 5 | 7 | 8 | 0.1073 | 1 | 10 | 10 | 0.0292 | 6 | 9 | 12 | 0.0725 |
| 6 | 7 | 8 | 0.1452 | 2 | 10 | 10 | 0.0358 | 7 | 9 | 12 | 0.0846 |
| 1 | 8 | 8 | 0.0402 | 3 | 10 | 10 | 0.0413 | 8 | 9 | 12 | 0.1017 |
| 2 | 8 | 8 | 0.0496 | 4 | 10 | 10 | 0.0472 | 3 | 10 | 12 | 0.0441 |

TABLE 4

UUTLIER MEAN 0.0 OUTLIER STANDARD DEVIATION 0.5

| R | S | N | COVAR | R | S | N | COVAR | R | S | N | COVAR |
|---|---|---|---|---|---|---|---|---|---|---|---|
| 4 | 10 | 12 | 0.0497 | 10 | 11 | 13 | 0.1092 | 9 | 12 | 14 | 0.0711 |
| 5 | 10 | 12 | 0.0560 | 2 | 12 | 13 | 0.0297 | 10 | 12 | 14 | 0.0853 |
| 6 | 10 | 12 | 0.0639 | 3 | 12 | 13 | 0.0339 | 11 | 12 | 14 | 0.1064 |
| 7 | 10 | 12 | 0.0745 | 4 | 12 | 13 | 0.0380 | 2 | 13 | 14 | 0.0265 |
| 8 | 10 | 12 | 0.0896 | 5 | 12 | 13 | 0.0423 | 3 | 13 | 14 | 0.0302 |
| 9 | 10 | 12 | 0.1122 | 6 | 12 | 13 | 0.0475 | 4 | 13 | 14 | 0.0337 |
| 2 | 11 | 12 | 0.0335 | 7 | 12 | 13 | 0.0542 | 5 | 13 | 14 | 0.0373 |
| 3 | 11 | 12 | 0.0384 | 8 | 12 | 13 | 0.0632 | 6 | 13 | 14 | 0.0415 |
| 4 | 11 | 12 | 0.0432 | 9 | 12 | 13 | 0.0759 | 7 | 13 | 14 | 0.0466 |
| 5 | 11 | 12 | 0.0486 | 10 | 12 | 13 | 0.0950 | 8 | 13 | 14 | 0.0531 |
| 6 | 11 | 12 | 0.0553 | 11 | 12 | 13 | 0.1265 | 9 | 13 | 14 | 0.0620 |
| 7 | 11 | 12 | 0.0645 | 1 | 13 | 13 | 0.0200 | 10 | 13 | 14 | 0.0744 |
| 8 | 11 | 12 | 0.0775 | 2 | 13 | 13 | 0.0243 | 11 | 13 | 14 | 0.0929 |
| 9 | 11 | 12 | 0.0972 | 3 | 13 | 13 | 0.0278 | 12 | 13 | 14 | 0.1234 |
| 10 | 11 | 12 | 0.1298 | 4 | 13 | 13 | 0.0311 | 1 | 14 | 14 | 0.0180 |
| 1 | 12 | 12 | 0.0224 | 5 | 13 | 13 | 0.0346 | 2 | 14 | 14 | 0.0218 |
| 2 | 12 | 12 | 0.0274 | 6 | 13 | 13 | 0.0389 | 3 | 14 | 14 | 0.0249 |
| 3 | 12 | 12 | 0.0314 | 7 | 13 | 13 | 0.0443 | 4 | 14 | 14 | 0.0277 |
| 4 | 12 | 12 | 0.0352 | 8 | 13 | 13 | 0.0516 | 5 | 14 | 14 | 0.0307 |
| 5 | 12 | 12 | 0.0396 | 9 | 13 | 13 | 0.0620 | 6 | 14 | 14 | 0.0341 |
| 6 | 12 | 12 | 0.0451 | 10 | 13 | 13 | 0.0778 | 7 | 14 | 14 | 0.0382 |
| 7 | 12 | 12 | 0.0524 | 11 | 13 | 13 | 0.1038 | 8 | 14 | 14 | 0.0436 |
| 8 | 12 | 12 | 0.0630 | 12 | 13 | 13 | 0.1556 | 9 | 14 | 14 | 0.0508 |
| 9 | 12 | 12 | 0.0792 | 7 | 8 | 14 | 0.0842 | 10 | 14 | 14 | 0.0611 |
| 10 | 12 | 12 | 0.1060 | 6 | 9 | 14 | 0.0662 | 11 | 14 | 14 | 0.0764 |
| 11 | 12 | 12 | 0.1595 | 7 | 9 | 14 | 0.0747 | 12 | 14 | 14 | 0.1017 |
| 6 | 8 | 13 | 0.0780 | 8 | 9 | 14 | 0.0854 | 13 | 14 | 14 | 0.1520 |
| 7 | 8 | 13 | 0.0892 | 5 | 10 | 14 | 0.0532 | 7 | 9 | 15 | 0.0712 |
| 5 | 9 | 13 | 0.0612 | 6 | 10 | 14 | 0.0593 | 8 | 9 | 15 | 0.0802 |
| 6 | 9 | 13 | 0.0691 | 7 | 10 | 14 | 0.0668 | 6 | 10 | 15 | 0.0573 |
| 7 | 9 | 13 | 0.0790 | 8 | 10 | 14 | 0.0763 | 7 | 10 | 15 | 0.0639 |
| 8 | 9 | 13 | 0.0922 | 9 | 10 | 14 | 0.0891 | 8 | 10 | 15 | 0.0719 |
| 4 | 10 | 13 | 0.0489 | 4 | 11 | 14 | 0.0430 | 9 | 10 | 15 | 0.0822 |
| 5 | 10 | 13 | 0.0546 | 5 | 11 | 14 | 0.0478 | 5 | 11 | 15 | 0.0468 |
| 6 | 10 | 13 | 0.0615 | 6 | 11 | 14 | 0.0532 | 6 | 11 | 15 | 0.0518 |
| 7 | 10 | 13 | 0.0703 | 7 | 11 | 14 | 0.0599 | 7 | 11 | 15 | 0.0576 |
| 8 | 10 | 13 | 0.0821 | 8 | 11 | 14 | 0.0684 | 8 | 11 | 15 | 0.0648 |
| 9 | 10 | 13 | 0.0986 | 9 | 11 | 14 | 0.0798 | 9 | 11 | 15 | 0.0741 |
| 3 | 11 | 13 | 0.0388 | 10 | 11 | 14 | 0.0958 | 10 | 11 | 15 | 0.0864 |
| 4 | 11 | 13 | 0.0435 | 3 | 12 | 14 | 0.0345 | 4 | 12 | 15 | 0.0383 |
| 5 | 11 | 13 | 0.0486 | 4 | 12 | 14 | 0.0385 | 5 | 12 | 15 | 0.0423 |
| 6 | 11 | 13 | 0.0546 | 5 | 12 | 14 | 0.0427 | 6 | 12 | 15 | 0.0467 |
| 7 | 11 | 13 | 0.0623 | 6 | 12 | 14 | 0.0475 | 7 | 12 | 15 | 0.0520 |
| 8 | 11 | 13 | 0.0727 | 7 | 12 | 14 | 0.0534 | 8 | 12 | 15 | 0.0584 |
| 9 | 11 | 13 | 0.0873 | 8 | 12 | 14 | 0.0609 | 9 | 12 | 15 | 0.0668 |

TABLE 4

OUTLIER MEAN 0.0 OUTLIER STANDARD DEVIATION 0.5

| R | S | N | COVAR | R | S | N | COVAR | R | S | N | COVAR |
|---|---|---|---|---|---|---|---|---|---|---|---|
| 10 | 12 | 15 | 0.0778 | 9 | 11 | 16 | 0.0696 | 4 | 16 | 16 | 0.0225 |
| 11 | 12 | 15 | 0.0933 | 10 | 11 | 16 | 0.0795 | 5 | 16 | 16 | 0.0247 |
| 3 | 13 | 15 | 0.0309 | 5 | 12 | 16 | 0.0416 | 6 | 16 | 16 | 0.0271 |
| 4 | 13 | 15 | 0.0344 | 6 | 12 | 16 | 0.0457 | 7 | 16 | 16 | 0.0298 |
| 5 | 13 | 15 | 0.0379 | 7 | 12 | 16 | 0.0505 | 8 | 16 | 16 | 0.0331 |
| 6 | 13 | 15 | 0.0419 | 8 | 12 | 16 | 0.0561 | 9 | 16 | 16 | 0.0372 |
| 7 | 13 | 15 | 0.0465 | 9 | 12 | 16 | 0.0631 | 10 | 16 | 16 | 0.0424 |
| 8 | 13 | 15 | 0.0523 | 10 | 12 | 16 | 0.0721 | 11 | 16 | 16 | 0.0495 |
| 9 | 13 | 15 | 0.0597 | 11 | 12 | 16 | 0.0840 | 12 | 16 | 16 | 0.0593 |
| 10 | 13 | 15 | 0.0696 | 4 | 13 | 16 | 0.0344 | 13 | 16 | 16 | 0.0739 |
| 11 | 13 | 15 | 0.0834 | 5 | 13 | 16 | 0.0378 | 14 | 16 | 16 | 0.0978 |
| 12 | 13 | 15 | 0.1038 | 6 | 13 | 16 | 0.0415 | 15 | 16 | 16 | 0.1456 |
| 2 | 14 | 15 | 0.0239 | 7 | 13 | 16 | 0.0457 | 8 | 10 | 17 | 0.0655 |
| 3 | 14 | 15 | 0.0271 | 8 | 13 | 16 | 0.0508 | 9 | 10 | 17 | 0.0728 |
| 4 | 14 | 15 | 0.0302 | 9 | 13 | 16 | 0.0571 | 7 | 11 | 17 | 0.0538 |
| 5 | 14 | 15 | 0.0333 | 10 | 13 | 16 | 0.0652 | 8 | 11 | 17 | 0.0593 |
| 6 | 14 | 15 | 0.0367 | 11 | 13 | 16 | 0.0760 | 9 | 11 | 17 | 0.0660 |
| 7 | 14 | 15 | 0.0407 | 12 | 13 | 16 | 0.0909 | 10 | 11 | 17 | 0.0742 |
| 8 | 14 | 15 | 0.0457 | 3 | 14 | 16 | 0.0279 | 6 | 12 | 17 | 0.0447 |
| 9 | 14 | 15 | 0.0522 | 4 | 14 | 16 | 0.0309 | 7 | 12 | 17 | 0.0490 |
| 10 | 14 | 15 | 0.0609 | 5 | 14 | 16 | 0.0340 | 8 | 12 | 17 | 0.0540 |
| 11 | 14 | 15 | 0.0730 | 6 | 14 | 16 | 0.0373 | 9 | 12 | 17 | 0.0601 |
| 12 | 14 | 15 | 0.0910 | 7 | 14 | 16 | 0.0411 | 10 | 12 | 17 | 0.0676 |
| 13 | 14 | 15 | 0.1205 | 8 | 14 | 16 | 0.0456 | 11 | 12 | 17 | 0.0771 |
| 1 | 15 | 15 | 0.0163 | 9 | 14 | 16 | 0.0513 | 5 | 13 | 17 | 0.0373 |
| 2 | 15 | 15 | 0.0197 | 10 | 14 | 16 | 0.0585 | 6 | 13 | 17 | 0.0408 |
| 3 | 15 | 15 | 0.0224 | 11 | 14 | 16 | 0.0682 | 7 | 13 | 17 | 0.0447 |
| 4 | 15 | 15 | 0.0249 | 12 | 14 | 16 | 0.0816 | 8 | 13 | 17 | 0.0493 |
| 5 | 15 | 15 | 0.0274 | 13 | 14 | 16 | 0.1014 | 9 | 13 | 17 | 0.0548 |
| 6 | 15 | 15 | 0.0302 | 2 | 15 | 16 | 0.0216 | 10 | 13 | 17 | 0.0616 |
| 7 | 15 | 15 | 0.0335 | 3 | 15 | 16 | 0.0246 | 11 | 13 | 17 | 0.0703 |
| 8 | 15 | 15 | 0.0377 | 4 | 15 | 16 | 0.0272 | 12 | 13 | 17 | 0.0818 |
| 9 | 15 | 15 | 0.0430 | 5 | 15 | 16 | 0.0299 | 4 | 14 | 17 | 0.0310 |
| 10 | 15 | 15 | 0.0501 | 6 | 15 | 16 | 0.0328 | 5 | 14 | 17 | 0.0340 |
| 11 | 15 | 15 | 0.0602 | 7 | 15 | 16 | 0.0361 | 6 | 14 | 17 | 0.0372 |
| 12 | 15 | 15 | 0.0751 | 8 | 15 | 16 | 0.0400 | 7 | 14 | 17 | 0.0407 |
| 13 | 15 | 15 | 0.0997 | 9 | 15 | 16 | 0.0450 | 8 | 14 | 17 | 0.0448 |
| 14 | 15 | 15 | 0.1487 | 10 | 15 | 16 | 0.0513 | 9 | 14 | 17 | 0.0498 |
| 8 | 9 | 16 | 0.0762 | 11 | 15 | 16 | 0.0598 | 10 | 14 | 17 | 0.0559 |
| 7 | 10 | 16 | 0.0614 | 12 | 15 | 16 | 0.0717 | 11 | 14 | 17 | 0.0639 |
| 8 | 10 | 16 | 0.0684 | 13 | 15 | 16 | 0.0892 | 12 | 14 | 17 | 0.0743 |
| 9 | 10 | 16 | 0.0770 | 14 | 15 | 16 | 0.1178 | 13 | 14 | 17 | 0.0888 |
| 6 | 11 | 16 | 0.0503 | 1 | 16 | 16 | 0.0148 | 3 | 15 | 17 | 0.0253 |
| 7 | 11 | 16 | 0.0556 | 2 | 16 | 16 | 0.0179 | 4 | 15 | 17 | 0.0280 |
| 8 | 11 | 16 | 0.0618 | 3 | 16 | 16 | 0.0203 | 5 | 15 | 17 | 0.0307 |

TABLE 4

OUTLIER MEAN 0.0 OUTLIER STANDARD DEVIATION 0.5

| R | S | N | COVAR | R | S | N | COVAR | R | S | N | COVAR |
|---|---|---|---|---|---|---|---|---|---|---|---|
| 6 | 15 | 17 | 0.0335 | 9 | 12 | 18 | 0.0575 | 4 | 17 | 18 | 0.0226 |
| 7 | 15 | 17 | 0.0367 | 10 | 12 | 18 | 0.0639 | 5 | 17 | 18 | 0.0247 |
| 8 | 15 | 17 | 0.0404 | 11 | 12 | 18 | 0.0719 | 6 | 17 | 18 | 0.0268 |
| 9 | 15 | 17 | 0.0448 | 6 | 13 | 18 | 0.0401 | 7 | 17 | 18 | 0.0292 |
| 10 | 15 | 17 | 0.0503 | 7 | 13 | 18 | 0.0437 | 8 | 17 | 18 | 0.0319 |
| 11 | 15 | 17 | 0.0574 | 8 | 13 | 18 | 0.0478 | 9 | 17 | 18 | 0.0350 |
| 12 | 15 | 17 | 0.0669 | 9 | 13 | 18 | 0.0527 | 10 | 17 | 18 | 0.0388 |
| 13 | 15 | 17 | 0.0799 | 10 | 13 | 18 | 0.0585 | 11 | 17 | 18 | 0.0436 |
| 14 | 15 | 17 | 0.0992 | 11 | 13 | 18 | 0.0658 | 12 | 17 | 18 | 0.0498 |
| 2 | 16 | 17 | 0.0197 | 12 | 13 | 18 | 0.0750 | 13 | 17 | 18 | 0.0579 |
| 3 | 16 | 17 | 0.0224 | 5 | 14 | 18 | 0.0338 | 14 | 17 | 18 | 0.0692 |
| 4 | 16 | 17 | 0.0247 | 6 | 14 | 18 | 0.0368 | 15 | 17 | 18 | 0.0858 |
| 5 | 16 | 17 | 0.0271 | 7 | 14 | 18 | 0.0401 | 16 | 17 | 18 | 0.1129 |
| 6 | 16 | 17 | 0.0296 | 8 | 14 | 18 | 0.0438 | 1 | 18 | 18 | 0.0126 |
| 7 | 16 | 17 | 0.0323 | 9 | 14 | 18 | 0.0482 | 2 | 18 | 18 | 0.0151 |
| 8 | 16 | 17 | 0.0355 | 10 | 14 | 18 | 0.0536 | 3 | 18 | 18 | 0.0170 |
| 9 | 16 | 17 | 0.0394 | 11 | 14 | 18 | 0.0602 | 4 | 18 | 18 | 0.0188 |
| 10 | 16 | 17 | 0.0443 | 12 | 14 | 18 | 0.0687 | 5 | 18 | 18 | 0.0205 |
| 11 | 16 | 17 | 0.0505 | 13 | 14 | 18 | 0.0798 | 6 | 18 | 18 | 0.0223 |
| 12 | 16 | 17 | 0.0589 | 4 | 15 | 18 | 0.0282 | 7 | 18 | 18 | 0.0243 |
| 13 | 16 | 17 | 0.0704 | 5 | 15 | 18 | 0.0309 | 8 | 18 | 18 | 0.0265 |
| 14 | 16 | 17 | 0.0874 | 6 | 15 | 18 | 0.0336 | 9 | 18 | 18 | 0.0291 |
| 15 | 16 | 17 | 0.1153 | 7 | 15 | 18 | 0.0366 | 10 | 18 | 18 | 0.0323 |
| 1 | 17 | 17 | 0.0136 | 8 | 15 | 18 | 0.0400 | 11 | 18 | 18 | 0.0363 |
| 2 | 17 | 17 | 0.0164 | 9 | 15 | 18 | 0.0440 | 12 | 18 | 18 | 0.0414 |
| 3 | 17 | 17 | 0.0186 | 10 | 15 | 18 | 0.0488 | 13 | 18 | 18 | 0.0482 |
| 4 | 17 | 17 | 0.0205 | 11 | 15 | 18 | 0.0549 | 14 | 18 | 18 | 0.0576 |
| 5 | 17 | 17 | 0.0225 | 12 | 15 | 18 | 0.0626 | 15 | 18 | 18 | 0.0716 |
| 6 | 17 | 17 | 0.0245 | 13 | 15 | 18 | 0.0728 | 16 | 18 | 18 | 0.0944 |
| 7 | 17 | 17 | 0.0268 | 14 | 15 | 18 | 0.0868 | 17 | 18 | 18 | 0.1402 |
| 8 | 17 | 17 | 0.0295 | 3 | 16 | 18 | 0.0231 | 9 | 11 | 19 | 0.0606 |
| 9 | 17 | 17 | 0.0327 | 4 | 16 | 18 | 0.0256 | 10 | 11 | 19 | 0.0667 |
| 10 | 17 | 17 | 0.0367 | 5 | 16 | 18 | 0.0279 | 8 | 12 | 19 | 0.0506 |
| 11 | 17 | 17 | 0.0419 | 6 | 16 | 18 | 0.0304 | 9 | 12 | 19 | 0.0554 |
| 12 | 17 | 17 | 0.0488 | 7 | 16 | 18 | 0.0330 | 10 | 12 | 19 | 0.0609 |
| 13 | 17 | 17 | 0.0585 | 8 | 16 | 18 | 0.0361 | 11 | 12 | 19 | 0.0677 |
| 14 | 17 | 17 | 0.0727 | 9 | 16 | 18 | 0.0397 | 7 | 13 | 19 | 0.0427 |
| 15 | 17 | 17 | 0.0961 | 10 | 16 | 18 | 0.0440 | 8 | 13 | 19 | 0.0465 |
| 16 | 17 | 17 | 0.1428 | 11 | 16 | 18 | 0.0495 | 9 | 13 | 19 | 0.0508 |
| 9 | 10 | 18 | 0.0695 | 12 | 16 | 18 | 0.0564 | 10 | 13 | 19 | 0.0559 |
| 8 | 11 | 18 | 0.0572 | 13 | 16 | 18 | 0.0657 | 11 | 13 | 19 | 0.0621 |
| 9 | 11 | 18 | 0.0630 | 14 | 16 | 18 | 0.0784 | 12 | 13 | 19 | 0.0698 |
| 10 | 11 | 18 | 0.0701 | 15 | 16 | 18 | 0.0970 | 6 | 14 | 19 | 0.0362 |
| 7 | 12 | 18 | 0.0477 | 2 | 17 | 18 | 0.0181 | 7 | 14 | 19 | 0.0393 |
| 8 | 12 | 18 | 0.0522 | 3 | 17 | 18 | 0.0205 | 8 | 14 | 19 | 0.0428 |

TABLE 4

OUTLIER MEAN 0.0 OUTLIER STANDARD DEVIATION 0.5

| R | S | N | COVAR | R | S | N | COVAR | R | S | N | COVAR |
|---|---|---|---|---|---|---|---|---|---|---|---|
| 9 | 14 | 19 | 0.0467 | 6 | 18 | 19 | 0.0245 | 13 | 14 | 20 | 0.0679 |
| 10 | 14 | 19 | 0.0514 | 7 | 18 | 19 | 0.0265 | 6 | 15 | 20 | 0.0330 |
| 11 | 14 | 19 | 0.0571 | 8 | 18 | 19 | 0.0288 | 7 | 15 | 20 | 0.0357 |
| 12 | 14 | 19 | 0.0642 | 9 | 18 | 19 | 0.0314 | 8 | 15 | 20 | 0.0386 |
| 13 | 14 | 19 | 0.0731 | 10 | 18 | 19 | 0.0345 | 9 | 15 | 20 | 0.0419 |
| 5 | 15 | 19 | 0.0307 | 11 | 18 | 19 | 0.0383 | 10 | 15 | 20 | 0.0458 |
| 6 | 15 | 19 | 0.0334 | 12 | 18 | 19 | 0.0430 | 11 | 15 | 20 | 0.0503 |
| 7 | 15 | 19 | 0.0362 | 13 | 18 | 19 | 0.0491 | 12 | 15 | 20 | 0.0558 |
| 8 | 15 | 19 | 0.0393 | 14 | 18 | 19 | 0.0571 | 13 | 15 | 20 | 0.0627 |
| 9 | 15 | 19 | 0.0430 | 15 | 18 | 19 | 0.0680 | 14 | 15 | 20 | 0.0714 |
| 10 | 15 | 19 | 0.0472 | 16 | 18 | 19 | 0.0842 | 5 | 16 | 20 | 0.0281 |
| 11 | 15 | 19 | 0.0525 | 17 | 18 | 19 | 0.1108 | 6 | 16 | 20 | 0.0304 |
| 12 | 15 | 19 | 0.0589 | 1 | 19 | 19 | 0.0116 | 7 | 16 | 20 | 0.0329 |
| 13 | 15 | 19 | 0.0672 | 2 | 19 | 19 | 0.0139 | 8 | 16 | 20 | 0.0356 |
| 14 | 15 | 19 | 0.0780 | 3 | 19 | 19 | 0.0157 | 9 | 16 | 20 | 0.0386 |
| 4 | 16 | 19 | 0.0258 | 4 | 19 | 19 | 0.0173 | 10 | 16 | 20 | 0.0422 |
| 5 | 16 | 19 | 0.0282 | 5 | 19 | 19 | 0.0189 | 11 | 16 | 20 | 0.0464 |
| 6 | 16 | 19 | 0.0306 | 6 | 19 | 19 | 0.0205 | 12 | 16 | 20 | 0.0514 |
| 7 | 16 | 19 | 0.0331 | 7 | 19 | 19 | 0.0221 | 13 | 16 | 20 | 0.0577 |
| 8 | 16 | 19 | 0.0360 | 8 | 19 | 19 | 0.0240 | 14 | 16 | 20 | 0.0658 |
| 9 | 16 | 19 | 0.0393 | 9 | 19 | 19 | 0.0262 | 15 | 16 | 20 | 0.0763 |
| 10 | 16 | 19 | 0.0432 | 10 | 19 | 19 | 0.0288 | 4 | 17 | 20 | 0.0237 |
| 11 | 16 | 19 | 0.0479 | 11 | 19 | 19 | 0.0320 | 5 | 17 | 20 | 0.0258 |
| 12 | 16 | 19 | 0.0538 | 12 | 19 | 19 | 0.0359 | 6 | 17 | 20 | 0.0280 |
| 13 | 16 | 19 | 0.0614 | 13 | 19 | 19 | 0.0410 | 7 | 17 | 20 | 0.0302 |
| 14 | 16 | 19 | 0.0713 | 14 | 19 | 19 | 0.0476 | 8 | 17 | 20 | 0.0326 |
| 15 | 16 | 19 | 0.0849 | 15 | 19 | 19 | 0.0569 | 9 | 17 | 20 | 0.0354 |
| 3 | 17 | 19 | 0.0212 | 16 | 19 | 19 | 0.0705 | 10 | 17 | 20 | 0.0386 |
| 4 | 17 | 19 | 0.0234 | 17 | 19 | 19 | 0.0929 | 11 | 17 | 20 | 0.0425 |
| 5 | 17 | 19 | 0.0255 | 18 | 19 | 19 | 0.1378 | 12 | 17 | 20 | 0.0471 |
| 6 | 17 | 19 | 0.0277 | 10 | 11 | 20 | 0.0639 | 13 | 17 | 20 | 0.0529 |
| 7 | 17 | 19 | 0.0300 | 9 | 12 | 20 | 0.0535 | 14 | 17 | 20 | 0.0603 |
| 8 | 17 | 19 | 0.0326 | 10 | 12 | 20 | 0.0585 | 15 | 17 | 20 | 0.0699 |
| 9 | 17 | 19 | 0.0355 | 11 | 12 | 20 | 0.0643 | 16 | 17 | 20 | 0.0832 |
| 10 | 17 | 19 | 0.0391 | 8 | 13 | 20 | 0.0452 | 3 | 18 | 20 | 0.0196 |
| 11 | 17 | 19 | 0.0433 | 9 | 13 | 20 | 0.0492 | 4 | 18 | 20 | 0.0216 |
| 12 | 17 | 19 | 0.0487 | 10 | 13 | 20 | 0.0537 | 5 | 18 | 20 | 0.0235 |
| 13 | 17 | 19 | 0.0555 | 11 | 13 | 20 | 0.0591 | 6 | 18 | 20 | 0.0254 |
| 14 | 17 | 19 | 0.0645 | 12 | 13 | 20 | 0.0656 | 7 | 18 | 20 | 0.0274 |
| 15 | 17 | 19 | 0.0769 | 7 | 14 | 20 | 0.0386 | 8 | 18 | 20 | 0.0296 |
| 16 | 17 | 19 | 0.0951 | 8 | 14 | 20 | 0.0418 | 9 | 18 | 20 | 0.0321 |
| 2 | 18 | 19 | 0.0167 | 9 | 14 | 20 | 0.0454 | 10 | 18 | 20 | 0.0350 |
| 3 | 18 | 19 | 0.0188 | 10 | 14 | 20 | 0.0496 | 11 | 18 | 20 | 0.0385 |
| 4 | 18 | 19 | 0.0208 | 11 | 14 | 20 | 0.0545 | 12 | 18 | 20 | 0.0427 |
| 5 | 18 | 19 | 0.0226 | 12 | 14 | 20 | 0.0605 | 13 | 18 | 20 | 0.0479 |

TABLE 4

OUTLIER MEAN 0.0 OUTLIER STANDARD DEVIATION 0.5

| R | S | N | COVAR | R | S | N | COVAR | R | S | N | COVAR |
|---|---|---|---|---|---|---|---|---|---|---|---|
| 14 | 18 | 20 | 0.0546 | 12 | 19 | 20 | 0.0378 | 8 | 20 | 20 | 0.0220 |
| 15 | 18 | 20 | 0.0634 | 13 | 19 | 20 | 0.0425 | 9 | 20 | 20 | 0.0238 |
| 16 | 18 | 20 | 0.0755 | 14 | 19 | 20 | 0.0484 | 10 | 20 | 20 | 0.0260 |
| 17 | 18 | 20 | 0.0932 | 15 | 19 | 20 | 0.0562 | 11 | 20 | 20 | 0.0285 |
| 2 | 19 | 20 | 0.0155 | 16 | 19 | 20 | 0.0670 | 12 | 20 | 20 | 0.0316 |
| 3 | 19 | 20 | 0.0174 | 17 | 19 | 20 | 0.0828 | 13 | 20 | 20 | 0.0355 |
| 4 | 19 | 20 | 0.0192 | 18 | 19 | 20 | 0.1087 | 14 | 20 | 20 | 0.0405 |
| 5 | 19 | 20 | 0.0208 | 1 | 20 | 20 | 0.0108 | 15 | 20 | 20 | 0.0471 |
| 6 | 19 | 20 | 0.0225 | 2 | 20 | 20 | 0.0129 | 16 | 20 | 20 | 0.0561 |
| 7 | 19 | 20 | 0.0243 | 3 | 20 | 20 | 0.0146 | 17 | 20 | 20 | 0.0695 |
| 8 | 19 | 20 | 0.0263 | 4 | 20 | 20 | 0.0160 | 18 | 20 | 20 | 0.0914 |
| 9 | 19 | 20 | 0.0285 | 5 | 20 | 20 | 0.0174 | 19 | 20 | 20 | 0.1355 |
| 10 | 19 | 20 | 0.0310 | 6 | 20 | 20 | 0.0188 | | | | |
| 11 | 19 | 20 | 0.0341 | 7 | 20 | 20 | 0.0203 | | | | |

TABLE 4

OUTLIER MEAN 0.0 OUTLIER STANDARD DEVIATION 2.0

| R | S | N | COVAR | R | S | N | COVAR | R | S | N | COVAR |
|---|---|---|---|---|---|---|---|---|---|---|---|
| 1 | 2 | 2 | 0.7958 | 3 | 8 | 8 | 0.0948 | 5 | 10 | 10 | 0.0823 |
| 1 | 3 | 3 | 0.4237 | 4 | 8 | 8 | 0.1088 | 6 | 10 | 10 | 0.0965 |
| 2 | 3 | 3 | 0.4775 | 5 | 8 | 8 | 0.1323 | 7 | 10 | 10 | 0.1185 |
| 2 | 3 | 4 | 0.3187 | 6 | 8 | 8 | 0.1741 | 8 | 10 | 10 | 0.1566 |
| 1 | 4 | 4 | 0.2871 | 7 | 8 | 8 | 0.2670 | 9 | 10 | 10 | 0.2412 |
| 2 | 4 | 4 | 0.2672 | 4 | 6 | 9 | 0.1277 | 5 | 7 | 11 | 0.1096 |
| 3 | 4 | 4 | 0.3886 | 5 | 6 | 9 | 0.1545 | 6 | 7 | 11 | 0.1286 |
| 2 | 4 | 5 | 0.1976 | 3 | 7 | 9 | 0.0905 | 4 | 8 | 11 | 0.0814 |
| 3 | 4 | 5 | 0.2646 | 4 | 7 | 9 | 0.1073 | 5 | 8 | 11 | 0.0944 |
| 1 | 5 | 5 | 0.2173 | 5 | 7 | 9 | 0.1296 | 6 | 8 | 11 | 0.1109 |
| 2 | 5 | 5 | 0.1816 | 6 | 7 | 9 | 0.1620 | 7 | 8 | 11 | 0.1329 |
| 3 | 5 | 5 | 0.2261 | 2 | 8 | 9 | 0.0689 | 3 | 9 | 11 | 0.0618 |
| 4 | 5 | 5 | 0.3399 | 3 | 8 | 9 | 0.0780 | 4 | 9 | 11 | 0.0707 |
| 3 | 4 | 6 | 0.2207 | 4 | 8 | 9 | 0.0913 | 5 | 9 | 11 | 0.0817 |
| 2 | 5 | 6 | 0.1388 | 5 | 8 | 9 | 0.1097 | 6 | 9 | 11 | 0.0957 |
| 3 | 5 | 6 | 0.1736 | 6 | 8 | 9 | 0.1368 | 7 | 9 | 11 | 0.1147 |
| 4 | 5 | 6 | 0.2326 | 7 | 8 | 9 | 0.1820 | 8 | 9 | 11 | 0.1425 |
| 1 | 6 | 6 | 0.1753 | 1 | 9 | 9 | 0.1120 | 2 | 10 | 11 | 0.0504 |
| 2 | 6 | 6 | 0.1362 | 2 | 9 | 9 | 0.0765 | 3 | 10 | 11 | 0.0551 |
| 3 | 6 | 6 | 0.1567 | 3 | 9 | 9 | 0.0787 | 4 | 10 | 11 | 0.0621 |
| 4 | 6 | 6 | 0.2025 | 4 | 9 | 9 | 0.0875 | 5 | 10 | 11 | 0.0712 |
| 5 | 6 | 6 | 0.3079 | 5 | 9 | 9 | 0.1019 | 6 | 10 | 11 | 0.0831 |
| 3 | 5 | 7 | 0.1536 | 6 | 9 | 9 | 0.1247 | 7 | 10 | 11 | 0.0993 |
| 4 | 5 | 7 | 0.1941 | 7 | 9 | 9 | 0.1645 | 8 | 10 | 11 | 0.1233 |
| 2 | 6 | 7 | 0.1051 | 8 | 9 | 9 | 0.2529 | 9 | 10 | 11 | 0.1635 |
| 3 | 6 | 7 | 0.1257 | 5 | 6 | 10 | 0.1396 | 1 | 11 | 11 | 0.0909 |
| 4 | 6 | 7 | 0.1577 | 4 | 7 | 10 | 0.1002 | 2 | 11 | 11 | 0.0589 |
| 5 | 6 | 7 | 0.2107 | 5 | 7 | 10 | 0.1183 | 3 | 11 | 11 | 0.0583 |
| 1 | 7 | 7 | 0.1472 | 6 | 7 | 10 | 0.1424 | 4 | 11 | 11 | 0.0622 |
| 2 | 7 | 7 | 0.1084 | 3 | 8 | 10 | 0.0738 | 5 | 11 | 11 | 0.0688 |
| 3 | 7 | 7 | 0.1186 | 4 | 8 | 10 | 0.0857 | 6 | 11 | 11 | 0.0784 |
| 4 | 7 | 7 | 0.1424 | 5 | 8 | 10 | 0.1010 | 7 | 11 | 11 | 0.0922 |
| 5 | 7 | 7 | 0.1863 | 6 | 8 | 10 | 0.1214 | 8 | 11 | 11 | 0.1134 |
| 6 | 7 | 7 | 0.2848 | 7 | 8 | 10 | 0.1512 | 9 | 11 | 11 | 0.1501 |
| 4 | 5 | 8 | 0.1707 | 2 | 9 | 10 | 0.0584 | 10 | 11 | 11 | 0.2314 |
| 3 | 6 | 8 | 0.1151 | 3 | 9 | 10 | 0.0648 | 6 | 7 | 12 | 0.1181 |
| 4 | 6 | 8 | 0.1400 | 4 | 9 | 10 | 0.0743 | 5 | 8 | 12 | 0.0891 |
| 5 | 6 | 8 | 0.1757 | 5 | 9 | 10 | 0.0868 | 6 | 8 | 12 | 0.1028 |
| 2 | 7 | 8 | 0.0836 | 6 | 9 | 10 | 0.1040 | 7 | 8 | 12 | 0.1201 |
| 3 | 7 | 8 | 0.0969 | 7 | 9 | 10 | 0.1294 | 4 | 9 | 12 | 0.0678 |
| 4 | 7 | 8 | 0.1167 | 8 | 9 | 10 | 0.1719 | 5 | 9 | 12 | 0.0776 |
| 5 | 7 | 8 | 0.1460 | 1 | 10 | 10 | 0.1003 | 6 | 9 | 12 | 0.0895 |
| 6 | 7 | 8 | 0.1945 | 2 | 10 | 10 | 0.0666 | 7 | 9 | 12 | 0.1047 |
| 1 | 8 | 8 | 0.1271 | 3 | 10 | 10 | 0.0670 | 8 | 9 | 12 | 0.1252 |
| 2 | 8 | 8 | 0.0898 | 4 | 10 | 10 | 0.0728 | 3 | 10 | 12 | 0.0529 |

TABLE 4

OUTLIER MEAN 0.0 OUTLIER STANDARD DEVIATION 2.0

| R | S | N | COVAR | R | S | N | COVAR | R | S | N | COVAR |
|---|---|---|---|---|---|---|---|---|---|---|---|
| 4 | 10 | 12 | 0.0597 | 10 | 11 | 13 | 0.1290 | 9 | 12 | 14 | 0.0840 |
| 5 | 10 | 12 | 0.0679 | 2 | 12 | 13 | 0.0394 | 10 | 12 | 14 | 0.1000 |
| 6 | 10 | 12 | 0.0781 | 3 | 12 | 13 | 0.0420 | 11 | 12 | 14 | 0.1238 |
| 7 | 10 | 12 | 0.0912 | 4 | 12 | 13 | 0.0462 | 2 | 13 | 14 | 0.0354 |
| 8 | 10 | 12 | 0.1090 | 5 | 12 | 13 | 0.0515 | 3 | 13 | 14 | 0.0374 |
| 9 | 10 | 12 | 0.1352 | 6 | 12 | 13 | 0.0581 | 4 | 13 | 14 | 0.0408 |
| 2 | 11 | 12 | 0.0443 | 7 | 12 | 13 | 0.0663 | 5 | 13 | 14 | 0.0450 |
| 3 | 11 | 12 | 0.0478 | 8 | 12 | 13 | 0.0770 | 6 | 13 | 14 | 0.0501 |
| 4 | 11 | 12 | 0.0531 | 9 | 12 | 13 | 0.0918 | 7 | 13 | 14 | 0.0564 |
| 5 | 11 | 12 | 0.0600 | 10 | 12 | 13 | 0.1136 | 8 | 13 | 14 | 0.0643 |
| 6 | 11 | 12 | 0.0686 | 11 | 12 | 13 | 0.1505 | 9 | 13 | 14 | 0.0746 |
| 7 | 11 | 12 | 0.0798 | 1 | 13 | 13 | 0.0768 | 10 | 13 | 14 | 0.0887 |
| 8 | 11 | 12 | 0.0952 | 2 | 13 | 13 | 0.0478 | 11 | 13 | 14 | 0.1098 |
| 9 | 11 | 12 | 0.1181 | 3 | 13 | 13 | 0.0460 | 12 | 13 | 14 | 0.1454 |
| 10 | 11 | 12 | 0.1565 | 4 | 13 | 13 | 0.0479 | 1 | 14 | 14 | 0.0714 |
| 1 | 12 | 12 | 0.0832 | 5 | 13 | 13 | 0.0514 | 2 | 14 | 14 | 0.0436 |
| 2 | 12 | 12 | 0.0527 | 6 | 13 | 13 | 0.0565 | 3 | 14 | 14 | 0.0416 |
| 3 | 12 | 12 | 0.0515 | 7 | 13 | 13 | 0.0632 | 4 | 14 | 14 | 0.0428 |
| 4 | 12 | 12 | 0.0542 | 8 | 13 | 13 | 0.0725 | 5 | 14 | 14 | 0.0455 |
| 5 | 12 | 12 | 0.0589 | 9 | 13 | 13 | 0.0855 | 6 | 14 | 14 | 0.0494 |
| 6 | 12 | 12 | 0.0657 | 10 | 13 | 13 | 0.1053 | 7 | 14 | 14 | 0.0544 |
| 7 | 12 | 12 | 0.0752 | 11 | 13 | 13 | 0.1396 | 8 | 14 | 14 | 0.0611 |
| 8 | 12 | 12 | 0.0886 | 12 | 13 | 13 | 0.2157 | 9 | 14 | 14 | 0.0701 |
| 9 | 12 | 12 | 0.1091 | 7 | 8 | 14 | 0.1023 | 10 | 14 | 14 | 0.0828 |
| 10 | 12 | 12 | 0.1445 | 6 | 9 | 14 | 0.0802 | 11 | 14 | 14 | 0.1020 |
| 11 | 12 | 12 | 0.2230 | 7 | 9 | 14 | 0.0908 | 12 | 14 | 14 | 0.1353 |
| 6 | 8 | 13 | 0.0960 | 8 | 9 | 14 | 0.1039 | 13 | 14 | 14 | 0.2093 |
| 7 | 8 | 13 | 0.1102 | 5 | 10 | 14 | 0.0631 | 7 | 9 | 15 | 0.0854 |
| 5 | 9 | 13 | 0.0743 | 6 | 10 | 14 | 0.0712 | 8 | 9 | 15 | 0.0963 |
| 6 | 9 | 13 | 0.0846 | 7 | 10 | 14 | 0.0808 | 6 | 10 | 15 | 0.0684 |
| 7 | 9 | 13 | 0.0972 | 8 | 10 | 14 | 0.0925 | 7 | 10 | 15 | 0.0767 |
| 8 | 9 | 13 | 0.1133 | 9 | 10 | 14 | 0.1076 | 8 | 10 | 15 | 0.0866 |
| 4 | 10 | 13 | 0.0576 | 4 | 11 | 14 | 0.0498 | 9 | 10 | 15 | 0.0988 |
| 5 | 10 | 13 | 0.0653 | 5 | 11 | 14 | 0.0558 | 5 | 11 | 15 | 0.0544 |
| 6 | 10 | 13 | 0.0743 | 6 | 11 | 14 | 0.0629 | 6 | 11 | 15 | 0.0609 |
| 7 | 10 | 13 | 0.0853 | 7 | 11 | 14 | 0.0713 | 7 | 11 | 15 | 0.0684 |
| 8 | 10 | 13 | 0.0995 | 8 | 11 | 14 | 0.0816 | 8 | 11 | 15 | 0.0774 |
| 9 | 10 | 13 | 0.1187 | 9 | 11 | 14 | 0.0949 | 9 | 11 | 15 | 0.0885 |
| 3 | 11 | 13 | 0.0461 | 10 | 11 | 14 | 0.1130 | 10 | 11 | 15 | 0.1027 |
| 4 | 11 | 13 | 0.0514 | 3 | 12 | 14 | 0.0408 | 4 | 12 | 15 | 0.0436 |
| 5 | 11 | 13 | 0.0578 | 4 | 12 | 14 | 0.0451 | 5 | 12 | 15 | 0.0485 |
| 6 | 11 | 13 | 0.0655 | 5 | 12 | 14 | 0.0501 | 6 | 12 | 15 | 0.0542 |
| 7 | 11 | 13 | 0.0751 | 6 | 12 | 14 | 0.0561 | 7 | 12 | 15 | 0.0607 |
| 8 | 11 | 13 | 0.0874 | 7 | 12 | 14 | 0.0633 | 8 | 12 | 15 | 0.0686 |
| 9 | 11 | 13 | 0.1042 | 8 | 12 | 14 | 0.0723 | 9 | 12 | 15 | 0.0783 |

TABLE 4

OUTLIER MEAN 0.0 OUTLIER STANDARD DEVIATION 2.0

| R | S | N | COVAR | R | S | N | CCVAR | R | S | N | COVAR |
|---|---|---|---|---|---|---|---|---|---|---|---|
| 10 | 12 | 15 | 0.0909 | 9 | 11 | 16 | 0.0829 | 4 | 16 | 16 | 0.0353 |
| 11 | 12 | 15 | 0.1081 | 10 | 11 | 16 | 0.0945 | 5 | 16 | 16 | 0.0369 |
| 3 | 13 | 15 | 0.0366 | 5 | 12 | 16 | 0.0473 | 6 | 16 | 16 | 0.0392 |
| 4 | 13 | 15 | 0.0401 | 6 | 12 | 16 | 0.0527 | 7 | 16 | 16 | 0.0423 |
| 5 | 13 | 15 | 0.0441 | 7 | 12 | 16 | 0.0588 | 8 | 16 | 16 | 0.0462 |
| 6 | 13 | 15 | 0.0489 | 8 | 12 | 16 | 0.0659 | 9 | 16 | 16 | 0.0512 |
| 7 | 13 | 15 | 0.0545 | 9 | 12 | 16 | 0.0744 | 10 | 16 | 16 | 0.0576 |
| 8 | 13 | 15 | 0.0614 | 10 | 12 | 16 | 0.0848 | 11 | 16 | 16 | 0.0662 |
| 9 | 13 | 15 | 0.0699 | 11 | 12 | 16 | 0.0983 | 12 | 16 | 16 | 0.0782 |
| 10 | 13 | 15 | 0.0811 | 4 | 13 | 16 | 0.0388 | 13 | 16 | 16 | 0.0965 |
| 11 | 13 | 15 | 0.0964 | 5 | 13 | 16 | 0.0428 | 14 | 16 | 16 | 0.1281 |
| 12 | 13 | 15 | 0.1192 | 6 | 13 | 16 | 0.0474 | 15 | 16 | 16 | 0.1984 |
| 2 | 14 | 15 | 0.0320 | 7 | 13 | 16 | 0.0526 | 8 | 10 | 17 | 0.0768 |
| 3 | 14 | 15 | 0.0335 | 8 | 13 | 16 | 0.0587 | 9 | 10 | 17 | 0.0854 |
| 4 | 14 | 15 | 0.0363 | 9 | 13 | 16 | 0.0661 | 7 | 11 | 17 | 0.0632 |
| 5 | 14 | 15 | 0.0398 | 10 | 13 | 16 | 0.0753 | 8 | 11 | 17 | 0.0699 |
| 6 | 14 | 15 | 0.0439 | 11 | 13 | 16 | 0.0872 | 9 | 11 | 17 | 0.0778 |
| 7 | 14 | 15 | 0.0489 | 12 | 13 | 16 | 0.1036 | 10 | 11 | 17 | 0.0873 |
| 8 | 14 | 15 | 0.0549 | 3 | 14 | 16 | 0.0332 | 6 | 12 | 17 | 0.0518 |
| 9 | 14 | 15 | 0.0625 | 4 | 14 | 16 | 0.0360 | 7 | 12 | 17 | 0.0573 |
| 10 | 14 | 15 | 0.0724 | 5 | 14 | 16 | 0.0394 | 8 | 12 | 17 | 0.0636 |
| 11 | 14 | 15 | 0.0861 | 6 | 14 | 16 | 0.0432 | 9 | 12 | 17 | 0.0710 |
| 12 | 14 | 15 | 0.1064 | 7 | 14 | 16 | 0.0478 | 10 | 12 | 17 | 0.0799 |
| 13 | 14 | 15 | 0.1409 | 8 | 14 | 16 | 0.0531 | 11 | 12 | 17 | 0.0909 |
| 1 | 15 | 15 | 0.0667 | 9 | 14 | 16 | 0.0596 | 5 | 13 | 17 | 0.0416 |
| 2 | 15 | 15 | 0.0402 | 10 | 14 | 16 | 0.0678 | 6 | 13 | 17 | 0.0461 |
| 3 | 15 | 15 | 0.0379 | 11 | 14 | 16 | 0.0785 | 7 | 13 | 17 | 0.0511 |
| 4 | 15 | 15 | 0.0387 | 12 | 14 | 16 | 0.0933 | 8 | 13 | 17 | 0.0568 |
| 5 | 15 | 15 | 0.0408 | 13 | 14 | 16 | 0.1152 | 9 | 13 | 17 | 0.0635 |
| 6 | 15 | 15 | 0.0438 | 2 | 15 | 16 | 0.0291 | 10 | 13 | 17 | 0.0716 |
| 7 | 15 | 15 | 0.0477 | 3 | 15 | 16 | 0.0303 | 11 | 13 | 17 | 0.0815 |
| 8 | 15 | 15 | 0.0527 | 4 | 15 | 16 | 0.0326 | 12 | 13 | 17 | 0.0944 |
| 9 | 15 | 15 | 0.0593 | 5 | 15 | 16 | 0.0355 | 4 | 14 | 17 | 0.0350 |
| 10 | 15 | 15 | 0.0680 | 6 | 15 | 16 | 0.0389 | 5 | 14 | 17 | 0.0383 |
| 11 | 15 | 15 | 0.0804 | 7 | 15 | 16 | 0.0429 | 6 | 14 | 17 | 0.0420 |
| 12 | 15 | 15 | 0.0991 | 8 | 15 | 16 | 0.0477 | 7 | 14 | 17 | 0.0462 |
| 13 | 15 | 15 | 0.1315 | 9 | 15 | 16 | 0.0536 | 8 | 14 | 17 | 0.0511 |
| 14 | 15 | 15 | 0.2036 | 10 | 15 | 16 | 0.0609 | 9 | 14 | 17 | 0.0568 |
| 8 | 9 | 16 | 0.0902 | 11 | 15 | 16 | 0.0705 | 10 | 14 | 17 | 0.0638 |
| 7 | 10 | 16 | 0.0728 | 12 | 15 | 16 | 0.0837 | 11 | 14 | 17 | 0.0725 |
| 8 | 10 | 16 | 0.0812 | 13 | 15 | 16 | 0.1034 | 12 | 14 | 17 | 0.0839 |
| 9 | 10 | 16 | 0.0914 | 14 | 15 | 16 | 0.1369 | 13 | 14 | 17 | 0.0996 |
| 6 | 11 | 16 | 0.0592 | 1 | 16 | 16 | 0.0627 | 3 | 15 | 17 | 0.0304 |
| 7 | 11 | 16 | 0.0658 | 2 | 16 | 16 | 0.0372 | 4 | 15 | 17 | 0.0327 |
| 8 | 11 | 16 | 0.0736 | 3 | 16 | 16 | 0.0349 | 5 | 15 | 17 | 0.0355 |

TABLE 4

OUTLIER MEAN 0.0 OUTLIER STANDARD DEVIATION 2.0

| R | S | N | COVAR | R | S | N | COVAR | R | S | N | COVAR |
|---|---|---|---|---|---|---|---|---|---|---|---|
| 6 | 15 | 17 | 0.0387 | 9 | 12 | 18 | 0.0674 | 4 | 17 | 18 | 0.0268 |
| 7 | 15 | 17 | 0.0424 | 10 | 12 | 18 | 0.0749 | 5 | 17 | 18 | 0.0289 |
| 8 | 15 | 17 | 0.0467 | 11 | 12 | 18 | 0.0840 | 6 | 17 | 18 | 0.0314 |
| 9 | 15 | 17 | 0.0518 | 6 | 13 | 18 | 0.0456 | 7 | 17 | 18 | 0.0342 |
| 10 | 15 | 17 | 0.0581 | 7 | 13 | 18 | 0.0503 | 8 | 17 | 18 | 0.0374 |
| 11 | 15 | 17 | 0.0660 | 8 | 13 | 18 | 0.0556 | 9 | 17 | 18 | 0.0412 |
| 12 | 15 | 17 | 0.0763 | 9 | 13 | 18 | 0.0616 | 10 | 17 | 18 | 0.0457 |
| 13 | 15 | 17 | 0.0906 | 10 | 13 | 18 | 0.0687 | 11 | 17 | 18 | 0.0512 |
| 14 | 15 | 17 | 0.1118 | 11 | 13 | 18 | 0.0772 | 12 | 17 | 18 | 0.0581 |
| 2 | 16 | 17 | 0.0266 | 12 | 13 | 18 | 0.0878 | 13 | 17 | 18 | 0.0672 |
| 3 | 16 | 17 | 0.0275 | 5 | 14 | 18 | 0.0368 | 14 | 17 | 18 | 0.0797 |
| 4 | 16 | 17 | 0.0295 | 6 | 14 | 18 | 0.0406 | 15 | 17 | 18 | 0.0983 |
| 5 | 16 | 17 | 0.0320 | 7 | 14 | 18 | 0.0449 | 16 | 17 | 18 | 0.1301 |
| 6 | 16 | 17 | 0.0348 | 8 | 14 | 18 | 0.0496 | 1 | 18 | 18 | 0.0560 |
| 7 | 16 | 17 | 0.0381 | 9 | 14 | 18 | 0.0550 | 2 | 18 | 18 | 0.0325 |
| 8 | 16 | 17 | 0.0420 | 10 | 14 | 18 | 0.0613 | 3 | 18 | 18 | 0.0300 |
| 9 | 16 | 17 | 0.0467 | 11 | 14 | 18 | 0.0689 | 4 | 18 | 18 | 0.0300 |
| 10 | 16 | 17 | 0.0523 | 12 | 14 | 18 | 0.0784 | 5 | 18 | 18 | 0.0309 |
| 11 | 16 | 17 | 0.0594 | 13 | 14 | 18 | 0.0907 | 6 | 18 | 18 | 0.0324 |
| 12 | 16 | 17 | 0.0687 | 4 | 15 | 18 | 0.0321 | 7 | 18 | 18 | 0.0345 |
| 13 | 16 | 17 | 0.0816 | 5 | 15 | 18 | 0.0348 | 8 | 18 | 18 | 0.0370 |
| 14 | 16 | 17 | 0.1008 | 6 | 15 | 18 | 0.0378 | 9 | 18 | 18 | 0.0401 |
| 15 | 16 | 17 | 0.1333 | 7 | 15 | 18 | 0.0412 | 10 | 18 | 18 | 0.0439 |
| 1 | 17 | 17 | 0.0591 | 8 | 15 | 18 | 0.0451 | 11 | 18 | 18 | 0.0487 |
| 2 | 17 | 17 | 0.0347 | 9 | 15 | 18 | 0.0497 | 12 | 18 | 18 | 0.0548 |
| 3 | 17 | 17 | 0.0322 | 10 | 15 | 18 | 0.0551 | 13 | 18 | 18 | 0.0630 |
| 4 | 17 | 17 | 0.0324 | 11 | 15 | 18 | 0.0617 | 14 | 18 | 18 | 0.0745 |
| 5 | 17 | 17 | 0.0336 | 12 | 15 | 18 | 0.0701 | 15 | 18 | 18 | 0.0920 |
| 6 | 17 | 17 | 0.0355 | 13 | 15 | 18 | 0.0810 | 16 | 18 | 18 | 0.1222 |
| 7 | 17 | 17 | 0.0380 | 14 | 15 | 18 | 0.0960 | 17 | 18 | 18 | 0.1895 |
| 8 | 17 | 17 | 0.0411 | 3 | 16 | 18 | 0.0279 | 9 | 11 | 19 | 0.0697 |
| 9 | 17 | 17 | 0.0450 | 4 | 16 | 18 | 0.0299 | 10 | 11 | 19 | 0.0768 |
| 10 | 17 | 17 | 0.0499 | 5 | 16 | 18 | 0.0323 | 8 | 12 | 19 | 0.0584 |
| 11 | 17 | 17 | 0.0562 | 6 | 16 | 18 | 0.0350 | 9 | 12 | 19 | 0.0639 |
| 12 | 17 | 17 | 0.0645 | 7 | 16 | 18 | 0.0380 | 10 | 12 | 19 | 0.0703 |
| 13 | 17 | 17 | 0.0763 | 8 | 16 | 18 | 0.0416 | 11 | 12 | 19 | 0.0779 |
| 14 | 17 | 17 | 0.0942 | 9 | 16 | 18 | 0.0457 | 7 | 13 | 19 | 0.0494 |
| 15 | 17 | 17 | 0.1250 | 10 | 16 | 18 | 0.0507 | 8 | 13 | 19 | 0.0541 |
| 16 | 17 | 17 | 0.1938 | 11 | 16 | 18 | 0.0568 | 9 | 13 | 19 | 0.0593 |
| 9 | 10 | 18 | 0.0808 | 12 | 16 | 18 | 0.0644 | 10 | 13 | 19 | 0.0654 |
| 8 | 11 | 18 | 0.0665 | 13 | 16 | 18 | 0.0744 | 11 | 13 | 19 | 0.0726 |
| 9 | 11 | 18 | 0.0733 | 14 | 16 | 18 | 0.0882 | 12 | 13 | 19 | 0.0813 |
| 10 | 11 | 18 | 0.0813 | 15 | 16 | 18 | 0.1089 | 6 | 14 | 19 | 0.0402 |
| 7 | 12 | 18 | 0.0556 | 2 | 17 | 18 | 0.0244 | 7 | 14 | 19 | 0.0443 |
| 8 | 12 | 18 | 0.0611 | 3 | 17 | 18 | 0.0251 | 8 | 14 | 19 | 0.0489 |

TABLE 4

OUTLIER MEAN 0.0 OUTLIER STANDARD DEVIATION 2.0

| R | S | N | COVAR | R | S | N | COVAR | R | S | N | COVAR |
|---|---|---|---|---|---|---|---|---|---|---|---|
| 9 | 14 | 19 | 0.0539 | 6 | 18 | 19 | 0.0285 | 13 | 14 | 20 | 0.0791 |
| 10 | 14 | 19 | 0.0598 | 7 | 18 | 19 | 0.0309 | 6 | 15 | 20 | 0.0356 |
| 11 | 14 | 19 | 0.0666 | 8 | 18 | 19 | 0.0336 | 7 | 15 | 20 | 0.0392 |
| 12 | 14 | 19 | 0.0748 | 9 | 18 | 19 | 0.0367 | 8 | 15 | 20 | 0.0431 |
| 13 | 14 | 19 | 0.0849 | 10 | 18 | 19 | 0.0404 | 9 | 15 | 20 | 0.0475 |
| 5 | 15 | 19 | 0.0330 | 11 | 18 | 19 | 0.0448 | 10 | 15 | 20 | 0.0524 |
| 6 | 15 | 19 | 0.0362 | 12 | 18 | 19 | 0.0501 | 11 | 15 | 20 | 0.0580 |
| 7 | 15 | 19 | 0.0398 | 13 | 18 | 19 | 0.0569 | 12 | 15 | 20 | 0.0646 |
| 8 | 15 | 19 | 0.0437 | 14 | 18 | 19 | 0.0657 | 13 | 15 | 20 | 0.0725 |
| 9 | 15 | 19 | 0.0481 | 15 | 18 | 19 | 0.0779 | 14 | 15 | 20 | 0.0822 |
| 10 | 15 | 19 | 0.0532 | 16 | 18 | 19 | 0.0961 | 5 | 16 | 20 | 0.0299 |
| 11 | 15 | 19 | 0.0592 | 17 | 18 | 19 | 0.1271 | 6 | 16 | 20 | 0.0327 |
| 12 | 15 | 19 | 0.0664 | 1 | 19 | 19 | 0.0533 | 7 | 16 | 20 | 0.0356 |
| 13 | 15 | 19 | 0.0754 | 2 | 19 | 19 | 0.0305 | 8 | 16 | 20 | 0.0389 |
| 14 | 15 | 19 | 0.0871 | 3 | 19 | 19 | 0.0280 | 9 | 16 | 20 | 0.0425 |
| 4 | 16 | 19 | 0.0298 | 4 | 19 | 19 | 0.0278 | 10 | 16 | 20 | 0.0466 |
| 5 | 16 | 19 | 0.0320 | 5 | 19 | 19 | 0.0286 | 11 | 16 | 20 | 0.0514 |
| 6 | 16 | 19 | 0.0345 | 6 | 19 | 19 | 0.0298 | 12 | 16 | 20 | 0.0571 |
| 7 | 16 | 19 | 0.0373 | 7 | 19 | 19 | 0.0315 | 13 | 16 | 20 | 0.0639 |
| 8 | 16 | 19 | 0.0404 | 8 | 19 | 19 | 0.0336 | 14 | 16 | 20 | 0.0725 |
| 9 | 16 | 19 | 0.0441 | 9 | 19 | 19 | 0.0361 | 15 | 16 | 20 | 0.0837 |
| 10 | 16 | 19 | 0.0484 | 10 | 19 | 19 | 0.0391 | 4 | 17 | 20 | 0.0280 |
| 11 | 16 | 19 | 0.0535 | 11 | 19 | 19 | 0.0429 | 5 | 17 | 20 | 0.0298 |
| 12 | 16 | 19 | 0.0599 | 12 | 19 | 19 | 0.0476 | 6 | 17 | 20 | 0.0318 |
| 13 | 16 | 19 | 0.0678 | 13 | 19 | 19 | 0.0536 | 7 | 17 | 20 | 0.0341 |
| 14 | 16 | 19 | 0.0783 | 14 | 19 | 19 | 0.0616 | 8 | 17 | 20 | 0.0367 |
| 15 | 16 | 19 | 0.0928 | 15 | 19 | 19 | 0.0729 | 9 | 17 | 20 | 0.0397 |
| 3 | 17 | 19 | 0.0257 | 16 | 19 | 19 | 0.0900 | 10 | 17 | 20 | 0.0432 |
| 4 | 17 | 19 | 0.0274 | 17 | 19 | 19 | 0.1196 | 11 | 17 | 20 | 0.0472 |
| 5 | 17 | 19 | 0.0295 | 18 | 19 | 19 | 0.1856 | 12 | 17 | 20 | 0.0521 |
| 6 | 17 | 19 | 0.0318 | 10 | 11 | 20 | 0.0733 | 13 | 17 | 20 | 0.0582 |
| 7 | 17 | 19 | 0.0344 | 9 | 12 | 20 | 0.0610 | 14 | 17 | 20 | 0.0659 |
| 8 | 17 | 19 | 0.0374 | 10 | 12 | 20 | 0.0666 | 15 | 17 | 20 | 0.0760 |
| 9 | 17 | 19 | 0.0408 | 11 | 12 | 20 | 0.0732 | 16 | 17 | 20 | 0.0900 |
| 10 | 17 | 19 | 0.0449 | 8 | 13 | 20 | 0.0521 | 3 | 18 | 20 | 0.0237 |
| 11 | 17 | 19 | 0.0497 | 9 | 13 | 20 | 0.0566 | 4 | 18 | 20 | 0.0252 |
| 12 | 17 | 19 | 0.0556 | 10 | 13 | 20 | 0.0618 | 5 | 18 | 20 | 0.0270 |
| 13 | 17 | 19 | 0.0630 | 11 | 13 | 20 | 0.0678 | 6 | 18 | 20 | 0.0291 |
| 14 | 17 | 19 | 0.0727 | 12 | 13 | 20 | 0.0750 | 7 | 18 | 20 | 0.0313 |
| 15 | 17 | 19 | 0.0862 | 7 | 14 | 20 | 0.0442 | 8 | 18 | 20 | 0.0339 |
| 16 | 17 | 19 | 0.1063 | 8 | 14 | 20 | 0.0482 | 9 | 18 | 20 | 0.0368 |
| 2 | 18 | 19 | 0.0224 | 9 | 14 | 20 | 0.0527 | 10 | 18 | 20 | 0.0402 |
| 3 | 18 | 19 | 0.0230 | 10 | 14 | 20 | 0.0578 | 11 | 18 | 20 | 0.0441 |
| 4 | 18 | 19 | 0.0245 | 11 | 14 | 20 | 0.0637 | 12 | 18 | 20 | 0.0488 |
| 5 | 18 | 19 | 0.0264 | 12 | 14 | 20 | 0.0707 | 13 | 18 | 20 | 0.0545 |

TABLE 4

OUTLIER MEAN 0.0 OUTLIER STANDARD DEVIATION 2.0

| R | S | N | COVAR | R | S | N | CCVAR | R | S | N | COVAR |
|---|---|---|---|---|---|---|---|---|---|---|---|
| 14 | 18 | 20 | 0.0618 | 12 | 19 | 20 | 0.0440 | 8 | 20 | 20 | 0.0307 |
| 15 | 18 | 20 | 0.0713 | 13 | 19 | 20 | 0.0492 | 9 | 20 | 20 | 0.0328 |
| 16 | 18 | 20 | 0.0845 | 14 | 19 | 20 | 0.0557 | 10 | 20 | 20 | 0.0353 |
| 17 | 18 | 20 | 0.1041 | 15 | 19 | 20 | 0.0643 | 11 | 20 | 20 | 0.0383 |
| 2 | 19 | 20 | 0.0207 | 16 | 19 | 20 | 0.0762 | 12 | 20 | 20 | 0.0420 |
| 3 | 19 | 20 | 0.0212 | 17 | 19 | 20 | 0.0941 | 13 | 20 | 20 | 0.0466 |
| 4 | 19 | 20 | 0.0225 | 18 | 19 | 20 | 0.1244 | 14 | 20 | 20 | 0.0525 |
| 5 | 19 | 20 | 0.0241 | 1 | 20 | 20 | 0.0508 | 15 | 20 | 20 | 0.0604 |
| 6 | 19 | 20 | 0.0260 | 2 | 20 | 20 | 0.0289 | 16 | 20 | 20 | 0.0714 |
| 7 | 19 | 20 | 0.0280 | 3 | 20 | 20 | 0.0263 | 17 | 20 | 20 | 0.0882 |
| 8 | 19 | 20 | 0.0304 | 4 | 20 | 20 | 0.0260 | 18 | 20 | 20 | 0.1172 |
| 9 | 19 | 20 | 0.0330 | 5 | 20 | 20 | 0.0266 | 19 | 20 | 20 | 0.1820 |
| 10 | 19 | 20 | 0.0361 | 6 | 20 | 20 | 0.0276 | | | | |
| 11 | 19 | 20 | 0.0397 | 7 | 20 | 20 | 0.0290 | | | | |

TABLE 4

OUTLIER MEAN 0.0 OUTLIER STANDARD DEVIATION 3.0

| R | S | N | COVAR | R | S | N | COVAR | R | S | N | COVAR |
|---|---|---|---|---|---|---|---|---|---|---|---|
| 1 | 2 | 2 | 1.5915 | 3 | 8 | 8 | 0.1529 | 5 | 10 | 10 | 0.1220 |
| 1 | 3 | 3 | 0.9954 | 4 | 8 | 8 | 0.1615 | 6 | 10 | 10 | 0.1373 |
| 2 | 3 | 3 | 0.6937 | 5 | 8 | 8 | 0.1871 | 7 | 10 | 10 | 0.1642 |
| 2 | 3 | 4 | 0.3587 | 6 | 8 | 8 | 0.2403 | 8 | 10 | 10 | 0.2150 |
| 1 | 4 | 4 | 0.7627 | 7 | 8 | 8 | 0.3748 | 9 | 10 | 10 | 0.3403 |
| 2 | 4 | 4 | 0.4087 | 4 | 6 | 9 | 0.1336 | 5 | 7 | 11 | 0.1134 |
| 3 | 4 | 4 | 0.5482 | 5 | 6 | 9 | 0.1611 | 6 | 7 | 11 | 0.1329 |
| 2 | 4 | 5 | 0.2237 | 3 | 7 | 9 | 0.0975 | 4 | 8 | 11 | 0.0854 |
| 3 | 4 | 5 | 0.2904 | 4 | 7 | 9 | 0.1136 | 5 | 8 | 11 | 0.0982 |
| 1 | 5 | 5 | 0.6356 | 5 | 7 | 9 | 0.1364 | 6 | 8 | 11 | 0.1150 |
| 2 | 5 | 5 | 0.2946 | 6 | 7 | 9 | 0.1703 | 7 | 8 | 11 | 0.1378 |
| 3 | 5 | 5 | 0.3257 | 2 | 8 | 9 | 0.0831 | 3 | 9 | 11 | 0.0674 |
| 4 | 5 | 5 | 0.4761 | 3 | 8 | 9 | 0.0878 | 4 | 9 | 11 | 0.0754 |
| 3 | 4 | 6 | 0.2360 | 4 | 8 | 9 | 0.0999 | 5 | 9 | 11 | 0.0861 |
| 2 | 5 | 6 | 0.1593 | 5 | 8 | 9 | 0.1184 | 6 | 9 | 11 | 0.1004 |
| 3 | 5 | 6 | 0.1905 | 6 | 8 | 9 | 0.1470 | 7 | 9 | 11 | 0.1201 |
| 4 | 5 | 6 | 0.2530 | 7 | 8 | 9 | 0.1965 | 8 | 9 | 11 | 0.1496 |
| 1 | 6 | 6 | 0.5542 | 1 | 9 | 9 | 0.4206 | 2 | 10 | 11 | 0.0627 |
| 2 | 6 | 6 | 0.2333 | 2 | 9 | 9 | 0.1497 | 3 | 10 | 11 | 0.0633 |
| 3 | 6 | 6 | 0.2343 | 3 | 9 | 9 | 0.1313 | 4 | 10 | 11 | 0.0689 |
| 4 | 6 | 6 | 0.2848 | 4 | 9 | 9 | 0.1337 | 5 | 10 | 11 | 0.0775 |
| 5 | 6 | 6 | 0.4308 | 5 | 9 | 9 | 0.1474 | 6 | 10 | 11 | 0.0894 |
| 3 | 5 | 7 | 0.1641 | 6 | 9 | 9 | 0.1742 | 7 | 10 | 11 | 0.1063 |
| 4 | 5 | 7 | 0.2056 | 7 | 9 | 9 | 0.2262 | 8 | 10 | 11 | 0.1319 |
| 2 | 6 | 7 | 0.1226 | 8 | 9 | 9 | 0.3558 | 9 | 10 | 11 | 0.1764 |
| 3 | 6 | 7 | 0.1389 | 5 | 6 | 10 | 0.1449 | 1 | 11 | 11 | 0.3711 |
| 4 | 6 | 7 | 0.1710 | 4 | 7 | 10 | 0.1049 | 2 | 11 | 11 | 0.1236 |
| 5 | 6 | 7 | 0.2282 | 5 | 7 | 10 | 0.1231 | 3 | 11 | 11 | 0.1036 |
| 1 | 7 | 7 | 0.4969 | 6 | 7 | 10 | 0.1480 | 4 | 11 | 11 | 0.1005 |
| 2 | 7 | 7 | 0.1950 | 3 | 8 | 10 | 0.0799 | 5 | 11 | 11 | 0.1044 |
| 3 | 7 | 7 | 0.1843 | 4 | 8 | 10 | 0.0910 | 6 | 11 | 11 | 0.1137 |
| 4 | 7 | 7 | 0.2054 | 5 | 8 | 10 | 0.1063 | 7 | 11 | 11 | 0.1297 |
| 5 | 7 | 7 | 0.2588 | 6 | 8 | 10 | 0.1274 | 8 | 11 | 11 | 0.1563 |
| 6 | 7 | 7 | 0.3989 | 7 | 8 | 10 | 0.1588 | 9 | 11 | 11 | 0.2057 |
| 4 | 5 | 8 | 0.1790 | 2 | 9 | 10 | 0.0715 | 10 | 11 | 11 | 0.3273 |
| 3 | 6 | 8 | 0.1233 | 3 | 9 | 10 | 0.0737 | 6 | 7 | 12 | 0.1219 |
| 4 | 6 | 8 | 0.1480 | 4 | 9 | 10 | 0.0818 | 5 | 8 | 12 | 0.0921 |
| 5 | 6 | 8 | 0.1853 | 5 | 9 | 10 | 0.0940 | 6 | 8 | 12 | 0.1059 |
| 2 | 7 | 8 | 0.0992 | 6 | 9 | 10 | 0.1117 | 7 | 8 | 12 | 0.1237 |
| 3 | 7 | 8 | 0.1080 | 7 | 9 | 10 | 0.1387 | 4 | 9 | 12 | 0.0714 |
| 4 | 7 | 8 | 0.1270 | 8 | 9 | 10 | 0.1855 | 5 | 9 | 12 | 0.0809 |
| 5 | 7 | 8 | 0.1574 | 1 | 10 | 10 | 0.3935 | 6 | 9 | 12 | 0.0929 |
| 6 | 7 | 8 | 0.2103 | 2 | 10 | 10 | 0.1352 | 7 | 9 | 12 | 0.1084 |
| 1 | 8 | 8 | 0.4541 | 3 | 10 | 10 | 0.1156 | 8 | 9 | 12 | 0.1296 |
| 2 | 8 | 8 | 0.1688 | 4 | 10 | 10 | 0.1145 | 3 | 10 | 12 | 0.0580 |

TABLE 4

OUTLIER MEAN 0.0 OUTLIER STANDARD DEVIATION 3.0

| R | S | N | COVAR | R | S | N | CCVAR | R | S | N | COVAR |
|---|---|---|---|---|---|---|---|---|---|---|---|
| 4 | 10 | 12 | 0.0639 | 10 | 11 | 13 | 0.1356 | 9 | 12 | 14 | 0.0879 |
| 5 | 10 | 12 | 0.0719 | 2 | 12 | 13 | 0.0503 | 10 | 12 | 14 | 0.1047 |
| 6 | 10 | 12 | 0.0821 | 3 | 12 | 13 | 0.0491 | 11 | 12 | 14 | 0.1301 |
| 7 | 10 | 12 | 0.0956 | 4 | 12 | 13 | 0.0520 | 2 | 13 | 14 | 0.0459 |
| 8 | 10 | 12 | 0.1141 | 5 | 12 | 13 | 0.0566 | 3 | 13 | 14 | 0.0441 |
| 9 | 10 | 12 | 0.1420 | 6 | 12 | 13 | 0.0629 | 4 | 13 | 14 | 0.0461 |
| 2 | 11 | 12 | 0.0558 | 7 | 12 | 13 | 0.0712 | 5 | 13 | 14 | 0.0497 |
| 3 | 11 | 12 | 0.0553 | 8 | 12 | 13 | 0.0823 | 6 | 13 | 14 | 0.0545 |
| 4 | 11 | 12 | 0.0593 | 9 | 12 | 13 | 0.0978 | 7 | 13 | 14 | 0.0607 |
| 5 | 11 | 12 | 0.0656 | 10 | 12 | 13 | 0.1213 | 8 | 13 | 14 | 0.0687 |
| 6 | 11 | 12 | 0.0741 | 11 | 12 | 13 | 0.1623 | 9 | 13 | 14 | 0.0794 |
| 7 | 11 | 12 | 0.0855 | 1 | 13 | 13 | 0.3359 | 10 | 13 | 14 | 0.0943 |
| 8 | 11 | 12 | 0.1017 | 2 | 13 | 13 | 0.1066 | 11 | 13 | 14 | 0.1170 |
| 9 | 11 | 12 | 0.1262 | 3 | 13 | 13 | 0.0864 | 12 | 13 | 14 | 0.1567 |
| 10 | 11 | 12 | 0.1688 | 4 | 13 | 13 | 0.0813 | 1 | 14 | 14 | 0.3218 |
| 1 | 12 | 12 | 0.3522 | 5 | 13 | 13 | 0.0816 | 2 | 14 | 14 | 0.1001 |
| 2 | 12 | 12 | 0.1143 | 6 | 13 | 13 | 0.0852 | 3 | 14 | 14 | 0.0801 |
| 3 | 12 | 12 | 0.0941 | 7 | 13 | 13 | 0.0920 | 4 | 14 | 14 | 0.0745 |
| 4 | 12 | 12 | 0.0898 | 8 | 13 | 13 | 0.1025 | 5 | 14 | 14 | 0.0738 |
| 5 | 12 | 12 | 0.0915 | 9 | 13 | 13 | 0.1184 | 6 | 14 | 14 | 0.0759 |
| 6 | 12 | 12 | 0.0973 | 10 | 13 | 13 | 0.1442 | 7 | 14 | 14 | 0.0805 |
| 7 | 12 | 12 | 0.1075 | 11 | 13 | 13 | 0.1913 | 8 | 14 | 14 | 0.0878 |
| 8 | 12 | 12 | 0.1235 | 12 | 13 | 13 | 0.3066 | 9 | 14 | 14 | 0.0983 |
| 9 | 12 | 12 | 0.1497 | 7 | 8 | 14 | 0.1053 | 10 | 14 | 14 | 0.1141 |
| 10 | 12 | 12 | 0.1980 | 6 | 9 | 14 | 0.0822 | 11 | 14 | 14 | 0.1394 |
| 11 | 12 | 12 | 0.3163 | 7 | 9 | 14 | 0.0930 | 12 | 14 | 14 | 0.1855 |
| 6 | 8 | 13 | 0.0988 | 8 | 9 | 14 | 0.1064 | 13 | 14 | 14 | 0.2982 |
| 7 | 8 | 13 | 0.1133 | 5 | 10 | 14 | 0.0653 | 7 | 9 | 15 | 0.0876 |
| 5 | 9 | 13 | 0.0767 | 6 | 10 | 14 | 0.0732 | 8 | 9 | 15 | 0.0989 |
| 6 | 9 | 13 | 0.0870 | 7 | 10 | 14 | 0.0828 | 6 | 10 | 15 | 0.0699 |
| 7 | 9 | 13 | 0.0998 | 8 | 10 | 14 | 0.0947 | 7 | 10 | 15 | 0.0782 |
| 8 | 9 | 13 | 0.1162 | 9 | 10 | 14 | 0.1101 | 8 | 10 | 15 | 0.0882 |
| 4 | 10 | 13 | 0.0610 | 4 | 11 | 14 | 0.0530 | 9 | 10 | 15 | 0.1007 |
| 5 | 10 | 13 | 0.0683 | 5 | 11 | 14 | 0.0587 | 5 | 11 | 15 | 0.0566 |
| 6 | 10 | 13 | 0.0772 | 6 | 11 | 14 | 0.0656 | 6 | 11 | 15 | 0.0628 |
| 7 | 10 | 13 | 0.0884 | 7 | 11 | 14 | 0.0740 | 7 | 11 | 15 | 0.0702 |
| 8 | 10 | 13 | 0.1029 | 8 | 11 | 14 | 0.0845 | 8 | 11 | 15 | 0.0792 |
| 9 | 10 | 13 | 0.1229 | 9 | 11 | 14 | 0.0983 | 9 | 11 | 15 | 0.0903 |
| 3 | 11 | 13 | 0.0507 | 10 | 11 | 14 | 0.1172 | 10 | 11 | 15 | 0.1049 |
| 4 | 11 | 13 | 0.0552 | 3 | 12 | 14 | 0.0450 | 4 | 12 | 15 | 0.0467 |
| 5 | 11 | 13 | 0.0613 | 4 | 12 | 14 | 0.0485 | 5 | 12 | 15 | 0.0513 |
| 6 | 11 | 13 | 0.0690 | 5 | 12 | 14 | 0.0532 | 6 | 12 | 15 | 0.0567 |
| 7 | 11 | 13 | 0.0787 | 6 | 12 | 14 | 0.0591 | 7 | 12 | 15 | 0.0633 |
| 8 | 11 | 13 | 0.0914 | 7 | 12 | 14 | 0.0665 | 8 | 12 | 15 | 0.0712 |
| 9 | 11 | 13 | 0.1091 | 8 | 12 | 14 | 0.0757 | 9 | 12 | 15 | 0.0812 |

TABLE 4

OUTLIER MEAN 0.0 OUTLIER STANDARD DEVIATION 3.0

| R | S | N | COVAR | R | S | N | COVAR | R | S | N | COVAR |
|---|---|---|---|---|---|---|---|---|---|---|---|
| 10 | 12 | 15 | 0.0943 | 9 | 11 | 16 | 0.0842 | 4 | 16 | 16 | 0.0641 |
| 11 | 12 | 15 | 0.1123 | 10 | 11 | 16 | 0.0959 | 5 | 16 | 16 | 0.0623 |
| 3 | 13 | 15 | 0.0403 | 5 | 12 | 16 | 0.0497 | 6 | 16 | 16 | 0.0626 |
| 4 | 13 | 15 | 0.0431 | 6 | 12 | 16 | 0.0547 | 7 | 16 | 16 | 0.0648 |
| 5 | 13 | 15 | 0.0469 | 7 | 12 | 16 | 0.0606 | 8 | 16 | 16 | 0.0685 |
| 6 | 13 | 15 | 0.0516 | 8 | 12 | 16 | 0.0676 | 9 | 16 | 16 | 0.0738 |
| 7 | 13 | 15 | 0.0572 | 9 | 12 | 16 | 0.0760 | 10 | 16 | 16 | 0.0813 |
| 8 | 13 | 15 | 0.0643 | 10 | 12 | 16 | 0.0866 | 11 | 16 | 16 | 0.0918 |
| 9 | 13 | 15 | 0.0731 | 11 | 12 | 16 | 0.1004 | 12 | 16 | 16 | 0.1071 |
| 10 | 13 | 15 | 0.0848 | 4 | 13 | 16 | 0.0416 | 13 | 16 | 16 | 0.1316 |
| 11 | 13 | 15 | 0.1009 | 5 | 13 | 16 | 0.0454 | 14 | 16 | 16 | 0.1758 |
| 12 | 13 | 15 | 0.1253 | 6 | 13 | 16 | 0.0498 | 15 | 16 | 16 | 0.2840 |
| 2 | 14 | 15 | 0.0421 | 7 | 13 | 16 | 0.0550 | 8 | 10 | 17 | 0.0787 |
| 3 | 14 | 15 | 0.0401 | 8 | 13 | 16 | 0.0612 | 9 | 10 | 17 | 0.0877 |
| 4 | 14 | 15 | 0.0414 | 9 | 13 | 16 | 0.0688 | 7 | 11 | 17 | 0.0642 |
| 5 | 14 | 15 | 0.0442 | 10 | 13 | 16 | 0.0783 | 8 | 11 | 17 | 0.0711 |
| 6 | 14 | 15 | 0.0480 | 11 | 13 | 16 | 0.0908 | 9 | 11 | 17 | 0.0792 |
| 7 | 14 | 15 | 0.0528 | 12 | 13 | 16 | 0.1080 | 10 | 11 | 17 | 0.0890 |
| 8 | 14 | 15 | 0.0588 | 3 | 14 | 16 | 0.0365 | 6 | 12 | 17 | 0.0529 |
| 9 | 14 | 15 | 0.0666 | 4 | 14 | 16 | 0.0387 | 7 | 12 | 17 | 0.0583 |
| 10 | 14 | 15 | 0.0769 | 5 | 14 | 16 | 0.0418 | 8 | 12 | 17 | 0.0644 |
| 11 | 14 | 15 | 0.0913 | 6 | 14 | 16 | 0.0455 | 9 | 12 | 17 | 0.0718 |
| 12 | 14 | 15 | 0.1133 | 7 | 14 | 16 | 0.0501 | 10 | 12 | 17 | 0.0806 |
| 13 | 14 | 15 | 0.1517 | 8 | 14 | 16 | 0.0556 | 11 | 12 | 17 | 0.0918 |
| 1 | 15 | 15 | 0.3093 | 9 | 14 | 16 | 0.0623 | 5 | 13 | 17 | 0.0441 |
| 2 | 15 | 15 | 0.0945 | 10 | 14 | 16 | 0.0708 | 6 | 13 | 17 | 0.0483 |
| 3 | 15 | 15 | 0.0748 | 11 | 14 | 16 | 0.0820 | 7 | 13 | 17 | 0.0531 |
| 4 | 15 | 15 | 0.0688 | 12 | 14 | 16 | 0.0975 | 8 | 13 | 17 | 0.0586 |
| 5 | 15 | 15 | 0.0675 | 13 | 14 | 16 | 0.1211 | 9 | 13 | 17 | 0.0652 |
| 6 | 15 | 15 | 0.0686 | 2 | 15 | 16 | 0.0390 | 10 | 13 | 17 | 0.0733 |
| 7 | 15 | 15 | 0.0718 | 3 | 15 | 16 | 0.0366 | 11 | 13 | 17 | 0.0834 |
| 8 | 15 | 15 | 0.0769 | 4 | 15 | 16 | 0.0376 | 12 | 13 | 17 | 0.0965 |
| 9 | 15 | 15 | 0.0843 | 5 | 15 | 16 | 0.0398 | 4 | 14 | 17 | 0.0375 |
| 10 | 15 | 15 | 0.0948 | 6 | 15 | 16 | 0.0428 | 5 | 14 | 17 | 0.0405 |
| 11 | 15 | 15 | 0.1104 | 7 | 15 | 16 | 0.0466 | 6 | 14 | 17 | 0.0442 |
| 12 | 15 | 15 | 0.1352 | 8 | 15 | 16 | 0.0513 | 7 | 14 | 17 | 0.0484 |
| 13 | 15 | 15 | 0.1804 | 9 | 15 | 16 | 0.0571 | 8 | 14 | 17 | 0.0534 |
| 14 | 15 | 15 | 0.2907 | 10 | 15 | 16 | 0.0647 | 9 | 14 | 17 | 0.0593 |
| 8 | 9 | 16 | 0.0927 | 11 | 15 | 16 | 0.0747 | 10 | 14 | 17 | 0.0666 |
| 7 | 10 | 16 | 0.0744 | 12 | 15 | 16 | 0.0886 | 11 | 14 | 17 | 0.0757 |
| 8 | 10 | 16 | 0.0831 | 13 | 15 | 16 | 0.1100 | 12 | 14 | 17 | 0.0876 |
| 9 | 10 | 16 | 0.0935 | 14 | 15 | 16 | 0.1473 | 13 | 14 | 17 | 0.1043 |
| 6 | 11 | 16 | 0.0604 | 1 | 16 | 16 | 0.2983 | 3 | 15 | 17 | 0.0333 |
| 7 | 11 | 16 | 0.0670 | 2 | 16 | 16 | 0.0897 | 4 | 15 | 17 | 0.0350 |
| 8 | 11 | 16 | 0.0748 | 3 | 16 | 16 | 0.0702 | 5 | 15 | 17 | 0.0376 |

TABLE 4

OUTLIER MEAN 0.0 OUTLIER STANDARD DEVIATION 3.0

| R | S | N | COVAR | R | S | N | COVAR | R | S | N | COVAR |
|---|---|---|---|---|---|---|---|---|---|---|---|
| 6 | 15 | 17 | 0.0407 | 9 | 12 | 18 | 0.0682 | 4 | 17 | 18 | 0.0316 |
| 7 | 15 | 17 | 0.0444 | 10 | 12 | 18 | 0.0758 | 5 | 17 | 18 | 0.0330 |
| 8 | 15 | 17 | 0.0488 | 11 | 12 | 18 | 0.0851 | 6 | 17 | 18 | 0.0350 |
| 9 | 15 | 17 | 0.0540 | 6 | 13 | 18 | 0.0469 | 7 | 17 | 18 | 0.0375 |
| 10 | 15 | 17 | 0.0605 | 7 | 13 | 18 | 0.0513 | 8 | 17 | 18 | 0.0405 |
| 11 | 15 | 17 | 0.0688 | 8 | 13 | 18 | 0.0563 | 9 | 17 | 18 | 0.0442 |
| 12 | 15 | 17 | 0.0795 | 9 | 13 | 18 | 0.0622 | 10 | 17 | 18 | 0.0487 |
| 13 | 15 | 17 | 0.0946 | 10 | 13 | 18 | 0.0691 | 11 | 17 | 18 | 0.0543 |
| 14 | 15 | 17 | 0.1174 | 11 | 13 | 18 | 0.0776 | 12 | 17 | 18 | 0.0615 |
| 2 | 16 | 17 | 0.0363 | 12 | 13 | 18 | 0.0882 | 13 | 17 | 18 | 0.0708 |
| 3 | 16 | 17 | 0.0338 | 5 | 14 | 18 | 0.0396 | 14 | 17 | 18 | 0.0841 |
| 4 | 16 | 17 | 0.0344 | 6 | 14 | 18 | 0.0431 | 15 | 17 | 18 | 0.1044 |
| 5 | 16 | 17 | 0.0361 | 7 | 14 | 18 | 0.0470 | 16 | 17 | 18 | 0.1398 |
| 6 | 16 | 17 | 0.0385 | 8 | 14 | 18 | 0.0516 | 1 | 18 | 18 | 0.2794 |
| 7 | 16 | 17 | 0.0416 | 9 | 14 | 18 | 0.0568 | 2 | 18 | 18 | 0.0817 |
| 8 | 16 | 17 | 0.0453 | 10 | 14 | 18 | 0.0631 | 3 | 18 | 18 | 0.0629 |
| 9 | 16 | 17 | 0.0499 | 11 | 14 | 18 | 0.0709 | 4 | 18 | 18 | 0.0565 |
| 10 | 16 | 17 | 0.0556 | 12 | 14 | 18 | 0.0805 | 5 | 18 | 18 | 0.0542 |
| 11 | 16 | 17 | 0.0630 | 13 | 14 | 18 | 0.0931 | 6 | 18 | 18 | 0.0536 |
| 12 | 16 | 17 | 0.0726 | 4 | 15 | 18 | 0.0340 | 7 | 18 | 18 | 0.0544 |
| 13 | 16 | 17 | 0.0862 | 5 | 15 | 18 | 0.0366 | 8 | 18 | 18 | 0.0564 |
| 14 | 16 | 17 | 0.1071 | 6 | 15 | 18 | 0.0396 | 9 | 18 | 18 | 0.0594 |
| 15 | 16 | 17 | 0.1434 | 7 | 15 | 18 | 0.0431 | 10 | 18 | 18 | 0.0635 |
| 1 | 17 | 17 | 0.2884 | 8 | 15 | 18 | 0.0472 | 11 | 18 | 18 | 0.0691 |
| 2 | 17 | 17 | 0.0854 | 9 | 15 | 18 | 0.0520 | 12 | 18 | 18 | 0.0766 |
| 3 | 17 | 17 | 0.0663 | 10 | 15 | 18 | 0.0577 | 13 | 18 | 18 | 0.0867 |
| 4 | 17 | 17 | 0.0600 | 11 | 15 | 18 | 0.0647 | 14 | 18 | 18 | 0.1017 |
| 5 | 17 | 17 | 0.0579 | 12 | 15 | 18 | 0.0734 | 15 | 18 | 18 | 0.1254 |
| 6 | 17 | 17 | 0.0577 | 13 | 15 | 18 | 0.0849 | 16 | 18 | 18 | 0.1680 |
| 7 | 17 | 17 | 0.0591 | 14 | 15 | 18 | 0.1010 | 17 | 18 | 18 | 0.2724 |
| 8 | 17 | 17 | 0.0618 | 3 | 16 | 18 | 0.0306 | 9 | 11 | 19 | 0.0715 |
| 9 | 17 | 17 | 0.0658 | 4 | 16 | 18 | 0.0320 | 10 | 11 | 19 | 0.0788 |
| 10 | 17 | 17 | 0.0713 | 5 | 16 | 18 | 0.0341 | 8 | 12 | 19 | 0.0594 |
| 11 | 17 | 17 | 0.0788 | 6 | 16 | 18 | 0.0367 | 9 | 12 | 19 | 0.0652 |
| 12 | 17 | 17 | 0.0891 | 7 | 16 | 18 | 0.0397 | 10 | 12 | 19 | 0.0719 |
| 13 | 17 | 17 | 0.1042 | 8 | 16 | 18 | 0.0433 | 11 | 12 | 19 | 0.0798 |
| 14 | 17 | 17 | 0.1283 | 9 | 16 | 18 | 0.0476 | 7 | 13 | 19 | 0.0497 |
| 15 | 17 | 17 | 0.1717 | 10 | 16 | 18 | 0.0527 | 8 | 13 | 19 | 0.0543 |
| 16 | 17 | 17 | 0.2779 | 11 | 16 | 18 | 0.0590 | 9 | 13 | 19 | 0.0596 |
| 9 | 10 | 18 | 0.0829 | 12 | 16 | 18 | 0.0669 | 10 | 13 | 19 | 0.0657 |
| 8 | 11 | 18 | 0.0679 | 13 | 16 | 18 | 0.0773 | 11 | 13 | 19 | 0.0729 |
| 9 | 11 | 18 | 0.0751 | 14 | 16 | 18 | 0.0919 | 12 | 13 | 19 | 0.0818 |
| 10 | 11 | 18 | 0.0834 | 15 | 16 | 18 | 0.1141 | 6 | 14 | 19 | 0.0420 |
| 7 | 12 | 18 | 0.0562 | 2 | 17 | 18 | 0.0340 | 7 | 14 | 19 | 0.0457 |
| 8 | 12 | 18 | 0.0618 | 3 | 17 | 18 | 0.0313 | 8 | 14 | 19 | 0.0499 |

TABLE 4

OUTLIER MEAN 0.0 OUTLIER STANDARD DEVIATION 3.0

| R | S | N | COVAR | R | S | N | COVAR | R | S | N | COVAR |
|---|---|---|---|---|---|---|---|---|---|---|---|
| 9 | 14 | 19 | 0.0546 | 6 | 18 | 19 | 0.0320 | 13 | 14 | 20 | 0.0788 |
| 10 | 14 | 19 | 0.0602 | 7 | 18 | 19 | 0.0341 | 6 | 15 | 20 | 0.0380 |
| 11 | 14 | 19 | 0.0668 | 8 | 18 | 19 | 0.0366 | 7 | 15 | 20 | 0.0411 |
| 12 | 14 | 19 | 0.0749 | 9 | 18 | 19 | 0.0396 | 8 | 15 | 20 | 0.0446 |
| 13 | 14 | 19 | 0.0850 | 10 | 18 | 19 | 0.0432 | 9 | 15 | 20 | 0.0485 |
| 5 | 15 | 19 | 0.0358 | 11 | 18 | 19 | 0.0476 | 10 | 15 | 20 | 0.0531 |
| 6 | 15 | 19 | 0.0388 | 12 | 18 | 19 | 0.0532 | 11 | 15 | 20 | 0.0584 |
| 7 | 15 | 19 | 0.0421 | 13 | 18 | 19 | 0.0600 | 12 | 15 | 20 | 0.0648 |
| 8 | 15 | 19 | 0.0458 | 14 | 18 | 19 | 0.0691 | 13 | 15 | 20 | 0.0725 |
| 9 | 15 | 19 | 0.0502 | 15 | 18 | 19 | 0.0821 | 14 | 15 | 20 | 0.0821 |
| 10 | 15 | 19 | 0.0552 | 16 | 18 | 19 | 0.1020 | 5 | 16 | 20 | 0.0327 |
| 11 | 15 | 19 | 0.0613 | 17 | 18 | 19 | 0.1366 | 6 | 16 | 20 | 0.0352 |
| 12 | 15 | 19 | 0.0687 | 1 | 19 | 19 | 0.2713 | 7 | 16 | 20 | 0.0380 |
| 13 | 15 | 19 | 0.0779 | 2 | 19 | 19 | 0.0784 | 8 | 16 | 20 | 0.0412 |
| 14 | 15 | 19 | 0.0900 | 3 | 19 | 19 | 0.0599 | 9 | 16 | 20 | 0.0448 |
| 4 | 16 | 19 | 0.0310 | 4 | 19 | 19 | 0.0534 | 10 | 16 | 20 | 0.0489 |
| 5 | 16 | 19 | 0.0332 | 5 | 19 | 19 | 0.0510 | 11 | 16 | 20 | 0.0537 |
| 6 | 16 | 19 | 0.0358 | 6 | 19 | 19 | 0.0502 | 12 | 16 | 20 | 0.0596 |
| 7 | 16 | 19 | 0.0388 | 7 | 19 | 19 | 0.0504 | 13 | 16 | 20 | 0.0667 |
| 8 | 16 | 19 | 0.0421 | 8 | 19 | 19 | 0.0518 | 14 | 16 | 20 | 0.0755 |
| 9 | 16 | 19 | 0.0461 | 9 | 19 | 19 | 0.0542 | 15 | 16 | 20 | 0.0873 |
| 10 | 16 | 19 | 0.0507 | 10 | 19 | 19 | 0.0573 | 4 | 17 | 20 | 0.0285 |
| 11 | 16 | 19 | 0.0561 | 11 | 19 | 19 | 0.0615 | 5 | 17 | 20 | 0.0304 |
| 12 | 16 | 19 | 0.0629 | 12 | 19 | 19 | 0.0672 | 6 | 17 | 20 | 0.0326 |
| 13 | 16 | 19 | 0.0713 | 13 | 19 | 19 | 0.0745 | 7 | 17 | 20 | 0.0351 |
| 14 | 16 | 19 | 0.0824 | 14 | 19 | 19 | 0.0845 | 8 | 17 | 20 | 0.0380 |
| 15 | 16 | 19 | 0.0980 | 15 | 19 | 19 | 0.0994 | 9 | 17 | 20 | 0.0412 |
| 3 | 17 | 19 | 0.0283 | 16 | 19 | 19 | 0.1227 | 10 | 17 | 20 | 0.0450 |
| 4 | 17 | 19 | 0.0294 | 17 | 19 | 19 | 0.1647 | 11 | 17 | 20 | 0.0494 |
| 5 | 17 | 19 | 0.0311 | 18 | 19 | 19 | 0.2674 | 12 | 17 | 20 | 0.0548 |
| 6 | 17 | 19 | 0.0333 | 10 | 11 | 20 | 0.0749 | 13 | 17 | 20 | 0.0613 |
| 7 | 17 | 19 | 0.0359 | 9 | 12 | 20 | 0.0626 | 14 | 17 | 20 | 0.0694 |
| 8 | 17 | 19 | 0.0389 | 10 | 12 | 20 | 0.0685 | 15 | 17 | 20 | 0.0801 |
| 9 | 17 | 19 | 0.0423 | 11 | 12 | 20 | 0.0753 | 16 | 17 | 20 | 0.0954 |
| 10 | 17 | 19 | 0.0465 | 8 | 13 | 20 | 0.0526 | 3 | 18 | 20 | 0.0263 |
| 11 | 17 | 19 | 0.0514 | 9 | 13 | 20 | 0.0573 | 4 | 18 | 20 | 0.0272 |
| 12 | 17 | 19 | 0.0576 | 10 | 13 | 20 | 0.0628 | 5 | 18 | 20 | 0.0286 |
| 13 | 17 | 19 | 0.0652 | 11 | 13 | 20 | 0.0690 | 6 | 18 | 20 | 0.0305 |
| 14 | 17 | 19 | 0.0752 | 12 | 13 | 20 | 0.0766 | 7 | 18 | 20 | 0.0327 |
| 15 | 17 | 19 | 0.0895 | 7 | 14 | 20 | 0.0445 | 8 | 18 | 20 | 0.0352 |
| 16 | 17 | 19 | 0.1111 | 8 | 14 | 20 | 0.0483 | 9 | 18 | 20 | 0.0381 |
| 2 | 18 | 19 | 0.0320 | 9 | 14 | 20 | 0.0526 | 10 | 18 | 20 | 0.0415 |
| 3 | 18 | 19 | 0.0292 | 10 | 14 | 20 | 0.0576 | 11 | 18 | 20 | 0.0454 |
| 4 | 18 | 19 | 0.0293 | 11 | 14 | 20 | 0.0634 | 12 | 18 | 20 | 0.0503 |
| 5 | 18 | 19 | 0.0304 | 12 | 14 | 20 | 0.0704 | 13 | 18 | 20 | 0.0563 |

TABLE 4

OUTLIER MEAN 0.0 OUTLIER STANDARD DEVIATION 3.0

| R | S | N | COVAR | R | S | N | COVAR | R | S | N | COVAR |
|---|---|---|---|---|---|---|---|---|---|---|---|
| 14 | 18 | 20 | 0.0636 | 12 | 19 | 20 | 0.0466 | 8 | 20 | 20 | 0.0480 |
| 15 | 18 | 20 | 0.0734 | 13 | 19 | 20 | 0.0521 | 9 | 20 | 20 | 0.0499 |
| 16 | 18 | 20 | 0.0874 | 14 | 19 | 20 | 0.0587 | 10 | 20 | 20 | 0.0524 |
| 17 | 18 | 20 | 0.1084 | 15 | 19 | 20 | 0.0675 | 11 | 20 | 20 | 0.0555 |
| 2 | 19 | 20 | 0.0302 | 16 | 19 | 20 | 0.0803 | 12 | 20 | 20 | 0.0599 |
| 3 | 19 | 20 | 0.0274 | 17 | 19 | 20 | 0.0998 | 13 | 20 | 20 | 0.0656 |
| 4 | 19 | 20 | 0.0273 | 18 | 19 | 20 | 0.1336 | 14 | 20 | 20 | 0.0727 |
| 5 | 19 | 20 | 0.0281 | 1 | 20 | 20 | 0.2639 | 15 | 20 | 20 | 0.0825 |
| 6 | 19 | 20 | 0.0295 | 2 | 20 | 20 | 0.0754 | 16 | 20 | 20 | 0.0973 |
| 7 | 19 | 20 | 0.0312 | 3 | 20 | 20 | 0.0572 | 17 | 20 | 20 | 0.1203 |
| 8 | 19 | 20 | 0.0333 | 4 | 20 | 20 | 0.0507 | 18 | 20 | 20 | 0.1615 |
| 9 | 19 | 20 | 0.0358 | 5 | 20 | 20 | 0.0482 | 19 | 20 | 20 | 0.2627 |
| 10 | 19 | 20 | 0.0388 | 6 | 20 | 20 | 0.0472 | | | | |
| 11 | 19 | 20 | 0.0423 | 7 | 20 | 20 | 0.0471 | | | | |

TABLE 4

OUTLIER MEAN 0.0 OUTLIER STANDARD DEVIATION 4.0

| R | S | N | COVAR | R | S | N | COVAR | R | S | N | COVAR |
|---|---|---|---|---|---|---|---|---|---|---|---|
| 1 | 2 | 2 | 2.7056 | 3 | 8 | 8 | 0.2656 | 5 | 10 | 10 | 0.1683 |
| 1 | 3 | 3 | 2.0279 | 4 | 8 | 8 | 0.2221 | 6 | 10 | 10 | 0.1495 |
| 2 | 3 | 3 | 0.8426 | 5 | 8 | 8 | 0.2042 | 7 | 10 | 10 | 0.1437 |
| 2 | 3 | 4 | 0.3810 | 6 | 8 | 8 | 0.2173 | 8 | 10 | 10 | 0.1616 |
| 1 | 4 | 4 | 1.7633 | 7 | 8 | 8 | 0.3175 | 9 | 10 | 10 | 0.2579 |
| 2 | 4 | 4 | 0.5624 | 4 | 6 | 9 | 0.1368 | 5 | 7 | 11 | 0.1155 |
| 3 | 4 | 4 | 0.6050 | 5 | 6 | 9 | 0.1646 | 6 | 7 | 11 | 0.1352 |
| 2 | 4 | 5 | 0.2389 | 3 | 7 | 9 | 0.1014 | 4 | 8 | 11 | 0.0875 |
| 3 | 4 | 5 | 0.3045 | 4 | 7 | 9 | 0.1170 | 5 | 8 | 11 | 0.1003 |
| 1 | 5 | 5 | 1.6184 | 5 | 7 | 9 | 0.1399 | 6 | 8 | 11 | 0.1172 |
| 2 | 5 | 5 | 0.4599 | 6 | 7 | 9 | 0.1747 | 7 | 8 | 11 | 0.1404 |
| 3 | 5 | 5 | 0.3960 | 2 | 8 | 9 | 0.0925 | 3 | 9 | 11 | 0.0706 |
| 4 | 5 | 5 | 0.4869 | 3 | 8 | 9 | 0.0938 | 4 | 9 | 11 | 0.0780 |
| 3 | 4 | 6 | 0.2442 | 4 | 8 | 9 | 0.1050 | 5 | 9 | 11 | 0.0885 |
| 2 | 5 | 6 | 0.1718 | 5 | 8 | 9 | 0.1234 | 6 | 9 | 11 | 0.1029 |
| 3 | 5 | 6 | 0.2000 | 6 | 8 | 9 | 0.1526 | 7 | 9 | 11 | 0.1229 |
| 4 | 5 | 6 | 0.2641 | 7 | 8 | 9 | 0.2044 | 8 | 9 | 11 | 0.1533 |
| 1 | 6 | 6 | 1.5255 | 1 | 9 | 9 | 1.3732 | 2 | 10 | 11 | 0.0711 |
| 2 | 6 | 6 | 0.4090 | 2 | 9 | 9 | 0.3482 | 3 | 10 | 11 | 0.0685 |
| 3 | 6 | 6 | 0.3213 | 3 | 9 | 9 | 0.2539 | 4 | 10 | 11 | 0.0731 |
| 4 | 6 | 6 | 0.3118 | 4 | 9 | 9 | 0.2071 | 5 | 10 | 11 | 0.0812 |
| 5 | 6 | 6 | 0.4119 | 5 | 9 | 9 | 0.1804 | 6 | 10 | 11 | 0.0931 |
| 3 | 5 | 7 | 0.1698 | 6 | 9 | 9 | 0.1702 | 7 | 10 | 11 | 0.1102 |
| 4 | 5 | 7 | 0.2117 | 7 | 9 | 9 | 0.1865 | 8 | 10 | 11 | 0.1367 |
| 2 | 6 | 7 | 0.1336 | 8 | 9 | 9 | 0.2849 | 9 | 10 | 11 | 0.1835 |
| 3 | 6 | 7 | 0.1465 | 5 | 6 | 10 | 0.1476 | 1 | 11 | 11 | 1.3173 |
| 4 | 6 | 7 | 0.1784 | 4 | 7 | 10 | 0.1074 | 2 | 11 | 11 | 0.3329 |
| 5 | 6 | 7 | 0.2378 | 5 | 7 | 10 | 0.1257 | 3 | 11 | 11 | 0.2420 |
| 1 | 7 | 7 | 1.4601 | 6 | 7 | 10 | 0.1509 | 4 | 11 | 11 | 0.1941 |
| 2 | 7 | 7 | 0.3795 | 3 | 8 | 10 | 0.0834 | 5 | 11 | 11 | 0.1621 |
| 3 | 7 | 7 | 0.2854 | 4 | 8 | 10 | 0.0940 | 6 | 11 | 11 | 0.1396 |
| 4 | 7 | 7 | 0.2506 | 5 | 8 | 10 | 0.1091 | 7 | 11 | 11 | 0.1253 |
| 5 | 7 | 7 | 0.2570 | 6 | 8 | 10 | 0.1305 | 8 | 11 | 11 | 0.1220 |
| 6 | 7 | 7 | 0.3584 | 7 | 8 | 10 | 0.1628 | 9 | 11 | 11 | 0.1408 |
| 4 | 5 | 8 | 0.1833 | 2 | 9 | 10 | 0.0803 | 10 | 11 | 11 | 0.2351 |
| 3 | 6 | 8 | 0.1278 | 3 | 9 | 10 | 0.0792 | 6 | 7 | 12 | 0.1238 |
| 4 | 6 | 8 | 0.1523 | 4 | 9 | 10 | 0.0863 | 5 | 8 | 12 | 0.0937 |
| 5 | 6 | 8 | 0.1903 | 5 | 9 | 10 | 0.0982 | 6 | 8 | 12 | 0.1076 |
| 2 | 7 | 8 | 0.1092 | 6 | 9 | 10 | 0.1161 | 7 | 8 | 12 | 0.1257 |
| 3 | 7 | 8 | 0.1147 | 7 | 9 | 10 | 0.1438 | 4 | 9 | 12 | 0.0733 |
| 4 | 7 | 8 | 0.1329 | 8 | 9 | 10 | 0.1929 | 5 | 9 | 12 | 0.0827 |
| 5 | 7 | 8 | 0.1637 | 1 | 10 | 10 | 1.3426 | 6 | 9 | 12 | 0.0946 |
| 6 | 7 | 8 | 0.2188 | 2 | 10 | 10 | 0.3393 | 7 | 9 | 12 | 0.1103 |
| 1 | 8 | 8 | 1.4113 | 3 | 10 | 10 | 0.2466 | 8 | 9 | 12 | 0.1320 |
| 2 | 8 | 8 | 0.3608 | 4 | 10 | 10 | 0.1988 | 3 | 10 | 12 | 0.0610 |

TABLE 4

OUTLIER MEAN 0.0 OUTLIER STANDARD DEVIATION 4.0

| R | S | N | COVAR | R | S | N | COVAR | R | S | N | COVAR |
|---|---|---|---|---|---|---|---|---|---|---|---|
| 4 | 10 | 12 | 0.0663 | 10 | 11 | 13 | 0.1388 | 9 | 12 | 14 | 0.0898 |
| 5 | 10 | 12 | 0.0740 | 2 | 12 | 13 | 0.0582 | 10 | 12 | 14 | 0.1070 |
| 6 | 10 | 12 | 0.0842 | 3 | 12 | 13 | 0.0539 | 11 | 12 | 14 | 0.1332 |
| 7 | 10 | 12 | 0.0978 | 4 | 12 | 13 | 0.0557 | 2 | 13 | 14 | 0.0535 |
| 8 | 10 | 12 | 0.1167 | 5 | 12 | 13 | 0.0598 | 3 | 13 | 14 | 0.0487 |
| 9 | 10 | 12 | 0.1454 | 6 | 12 | 13 | 0.0659 | 4 | 13 | 14 | 0.0497 |
| 2 | 11 | 12 | 0.0640 | 7 | 12 | 13 | 0.0741 | 5 | 13 | 14 | 0.0527 |
| 3 | 11 | 12 | 0.0603 | 8 | 12 | 13 | 0.0853 | 6 | 13 | 14 | 0.0573 |
| 4 | 11 | 12 | 0.0632 | 9 | 12 | 13 | 0.1012 | 7 | 13 | 14 | 0.0634 |
| 5 | 11 | 12 | 0.0690 | 10 | 12 | 13 | 0.1256 | 8 | 13 | 14 | 0.0714 |
| 6 | 11 | 12 | 0.0773 | 11 | 12 | 13 | 0.1689 | 9 | 13 | 14 | 0.0823 |
| 7 | 11 | 12 | 0.0889 | 1 | 13 | 13 | 1.2779 | 10 | 13 | 14 | 0.0976 |
| 8 | 11 | 12 | 0.1053 | 2 | 13 | 13 | 0.3245 | 11 | 13 | 14 | 0.1212 |
| 9 | 11 | 12 | 0.1307 | 3 | 13 | 13 | 0.2374 | 12 | 13 | 14 | 0.1631 |
| 10 | 11 | 12 | 0.1756 | 4 | 13 | 13 | 0.1903 | 1 | 14 | 14 | 1.2622 |
| 1 | 12 | 12 | 1.2961 | 5 | 13 | 13 | 0.1577 | 2 | 14 | 14 | 0.3218 |
| 2 | 12 | 12 | 0.3281 | 6 | 13 | 13 | 0.1327 | 3 | 14 | 14 | 0.2364 |
| 3 | 12 | 12 | 0.2392 | 7 | 13 | 13 | 0.1131 | 4 | 14 | 14 | 0.1900 |
| 4 | 12 | 12 | 0.1915 | 8 | 13 | 13 | 0.0981 | 5 | 14 | 14 | 0.1575 |
| 5 | 12 | 12 | 0.1590 | 9 | 13 | 13 | 0.0887 | 6 | 14 | 14 | 0.1323 |
| 6 | 12 | 12 | 0.1348 | 10 | 13 | 13 | 0.0882 | 7 | 14 | 14 | 0.1119 |
| 7 | 12 | 12 | 0.1168 | 11 | 13 | 13 | 0.1077 | 8 | 14 | 14 | 0.0952 |
| 8 | 12 | 12 | 0.1054 | 12 | 13 | 13 | 0.1983 | 9 | 14 | 14 | 0.0823 |
| 9 | 12 | 12 | 0.1038 | 7 | 8 | 14 | 0.1067 | 10 | 14 | 14 | 0.0743 |
| 10 | 12 | 12 | 0.1231 | 6 | 9 | 14 | 0.0834 | 11 | 14 | 14 | 0.0746 |
| 11 | 12 | 12 | 0.2155 | 7 | 9 | 14 | 0.0943 | 12 | 14 | 14 | 0.0942 |
| 6 | 8 | 13 | 0.1002 | 8 | 9 | 14 | 0.1079 | 13 | 14 | 14 | 0.1831 |
| 7 | 8 | 13 | 0.1149 | 5 | 10 | 14 | 0.0666 | 7 | 9 | 15 | 0.0887 |
| 5 | 9 | 13 | 0.0782 | 6 | 10 | 14 | 0.0745 | 8 | 9 | 15 | 0.1000 |
| 6 | 9 | 13 | 0.0884 | 7 | 10 | 14 | 0.0840 | 6 | 10 | 15 | 0.0709 |
| 7 | 9 | 13 | 0.1013 | 8 | 10 | 14 | 0.0960 | 7 | 10 | 15 | 0.0793 |
| 8 | 9 | 13 | 0.1180 | 9 | 10 | 14 | 0.1117 | 8 | 10 | 15 | 0.0894 |
| 4 | 10 | 13 | 0.0628 | 4 | 11 | 14 | 0.0546 | 9 | 10 | 15 | 0.1020 |
| 5 | 10 | 13 | 0.0698 | 5 | 11 | 14 | 0.0601 | 5 | 11 | 15 | 0.0577 |
| 6 | 10 | 13 | 0.0787 | 6 | 11 | 14 | 0.0669 | 6 | 11 | 15 | 0.0639 |
| 7 | 10 | 13 | 0.0899 | 7 | 11 | 14 | 0.0753 | 7 | 11 | 15 | 0.0713 |
| 8 | 10 | 13 | 0.1047 | 8 | 11 | 14 | 0.0860 | 8 | 11 | 15 | 0.0803 |
| 9 | 10 | 13 | 0.1251 | 9 | 11 | 14 | 0.0999 | 9 | 11 | 15 | 0.0916 |
| 3 | 11 | 13 | 0.0536 | 10 | 11 | 14 | 0.1192 | 10 | 11 | 15 | 0.1064 |
| 4 | 11 | 13 | 0.0575 | 3 | 12 | 14 | 0.0477 | 4 | 12 | 15 | 0.0482 |
| 5 | 11 | 13 | 0.0633 | 4 | 12 | 14 | 0.0506 | 5 | 12 | 15 | 0.0526 |
| 6 | 11 | 13 | 0.0708 | 5 | 12 | 14 | 0.0550 | 6 | 12 | 15 | 0.0579 |
| 7 | 11 | 13 | 0.0805 | 6 | 12 | 14 | 0.0608 | 7 | 12 | 15 | 0.0644 |
| 8 | 11 | 13 | 0.0935 | 7 | 12 | 14 | 0.0681 | 8 | 12 | 15 | 0.0724 |
| 9 | 11 | 13 | 0.1115 | 8 | 12 | 14 | 0.0774 | 9 | 12 | 15 | 0.0825 |

TABLE 4

OUTLIER MEAN 0.0 OUTLIER STANDARD DEVIATION 4.0

| R | S | N | COVAR | R | S | N | COVAR | R | S | N | COVAR |
|---|---|---|---|---|---|---|---|---|---|---|---|
| 10 | 12 | 15 | 0.0958 | 9 | 11 | 16 | 0.0852 | 4 | 16 | 16 | 0.1908 |
| 11 | 12 | 15 | 0.1142 | 10 | 11 | 16 | 0.0971 | 5 | 16 | 16 | 0.1590 |
| 3 | 13 | 15 | 0.0430 | 5 | 12 | 16 | 0.0507 | 6 | 16 | 16 | 0.1340 |
| 4 | 13 | 15 | 0.0451 | 6 | 12 | 16 | 0.0557 | 7 | 16 | 16 | 0.1132 |
| 5 | 13 | 15 | 0.0486 | 7 | 12 | 16 | 0.0616 | 8 | 16 | 16 | 0.0954 |
| 6 | 13 | 15 | 0.0531 | 8 | 12 | 16 | 0.0685 | 9 | 16 | 16 | 0.0800 |
| 7 | 13 | 15 | 0.0587 | 9 | 12 | 16 | 0.0771 | 10 | 16 | 16 | 0.0669 |
| 8 | 13 | 15 | 0.0658 | 10 | 12 | 16 | 0.0878 | 11 | 16 | 16 | 0.0566 |
| 9 | 13 | 15 | 0.0747 | 11 | 12 | 16 | 0.1018 | 12 | 16 | 16 | 0.0504 |
| 10 | 13 | 15 | 0.0865 | 4 | 13 | 16 | 0.0431 | 13 | 16 | 16 | 0.0518 |
| 11 | 13 | 15 | 0.1031 | 5 | 13 | 16 | 0.0466 | 14 | 16 | 16 | 0.0714 |
| 12 | 13 | 15 | 0.1283 | 6 | 13 | 16 | 0.0509 | 15 | 16 | 16 | 0.1571 |
| 2 | 14 | 15 | 0.0495 | 7 | 13 | 16 | 0.0560 | 8 | 10 | 17 | 0.0796 |
| 3 | 14 | 15 | 0.0445 | 8 | 13 | 16 | 0.0623 | 9 | 10 | 17 | 0.0886 |
| 4 | 14 | 15 | 0.0448 | 9 | 13 | 16 | 0.0699 | 7 | 11 | 17 | 0.0650 |
| 5 | 14 | 15 | 0.0471 | 10 | 13 | 16 | 0.0795 | 8 | 11 | 17 | 0.0719 |
| 6 | 14 | 15 | 0.0506 | 11 | 13 | 16 | 0.0922 | 9 | 11 | 17 | 0.0801 |
| 7 | 14 | 15 | 0.0552 | 12 | 13 | 16 | 0.1099 | 10 | 11 | 17 | 0.0900 |
| 8 | 14 | 15 | 0.0613 | 3 | 14 | 16 | 0.0391 | 6 | 12 | 17 | 0.0538 |
| 9 | 14 | 15 | 0.0691 | 4 | 14 | 16 | 0.0406 | 7 | 12 | 17 | 0.0591 |
| 10 | 14 | 15 | 0.0796 | 5 | 14 | 16 | 0.0434 | 8 | 12 | 17 | 0.0653 |
| 11 | 14 | 15 | 0.0945 | 6 | 14 | 16 | 0.0470 | 9 | 12 | 17 | 0.0726 |
| 12 | 14 | 15 | 0.1173 | 7 | 14 | 16 | 0.0515 | 10 | 12 | 17 | 0.0816 |
| 13 | 14 | 15 | 0.1579 | 8 | 14 | 16 | 0.0569 | 11 | 12 | 17 | 0.0929 |
| 1 | 15 | 15 | 1.2485 | 9 | 14 | 16 | 0.0637 | 5 | 13 | 17 | 0.0451 |
| 2 | 15 | 15 | 0.3197 | 10 | 14 | 16 | 0.0723 | 6 | 13 | 17 | 0.0492 |
| 3 | 15 | 15 | 0.2360 | 11 | 14 | 16 | 0.0837 | 7 | 13 | 17 | 0.0539 |
| 4 | 15 | 15 | 0.1902 | 12 | 14 | 16 | 0.0997 | 8 | 13 | 17 | 0.0595 |
| 5 | 15 | 15 | 0.1580 | 13 | 14 | 16 | 0.1240 | 9 | 13 | 17 | 0.0661 |
| 6 | 15 | 15 | 0.1329 | 2 | 15 | 16 | 0.0462 | 10 | 13 | 17 | 0.0742 |
| 7 | 15 | 15 | 0.1121 | 3 | 15 | 16 | 0.0410 | 11 | 13 | 17 | 0.0845 |
| 8 | 15 | 15 | 0.0947 | 4 | 15 | 16 | 0.0409 | 12 | 13 | 17 | 0.0979 |
| 9 | 15 | 15 | 0.0801 | 5 | 15 | 16 | 0.0425 | 4 | 14 | 17 | 0.0388 |
| 10 | 15 | 15 | 0.0686 | 6 | 15 | 16 | 0.0452 | 5 | 14 | 17 | 0.0417 |
| 11 | 15 | 15 | 0.0616 | 7 | 15 | 16 | 0.0489 | 6 | 14 | 17 | 0.0452 |
| 12 | 15 | 15 | 0.0626 | 8 | 15 | 16 | 0.0535 | 7 | 14 | 17 | 0.0494 |
| 13 | 15 | 15 | 0.0822 | 9 | 15 | 16 | 0.0594 | 8 | 14 | 17 | 0.0544 |
| 14 | 15 | 15 | 0.1694 | 10 | 15 | 16 | 0.0670 | 9 | 14 | 17 | 0.0603 |
| 8 | 9 | 16 | 0.0938 | 11 | 15 | 16 | 0.0772 | 10 | 14 | 17 | 0.0676 |
| 7 | 10 | 16 | 0.0753 | 12 | 15 | 16 | 0.0917 | 11 | 14 | 17 | 0.0769 |
| 8 | 10 | 16 | 0.0840 | 13 | 15 | 16 | 0.1139 | 12 | 14 | 17 | 0.0891 |
| 9 | 10 | 16 | 0.0946 | 14 | 15 | 16 | 0.1534 | 13 | 14 | 17 | 0.1061 |
| 6 | 11 | 16 | 0.0613 | 1 | 16 | 16 | 1.2363 | 3 | 15 | 17 | 0.0358 |
| 7 | 11 | 16 | 0.0679 | 2 | 16 | 16 | 0.3181 | 4 | 15 | 17 | 0.0369 |
| 8 | 11 | 16 | 0.0757 | 3 | 16 | 16 | 0.2359 | 5 | 15 | 17 | 0.0391 |

TABLE 4

OUTLIER MEAN 0.0 OUTLIER STANDARD DEVIATION 4.0

| R | S | N | COVAR | R | S | N | COVAR | R | S | N | COVAR |
|---|---|---|---|---|---|---|---|---|---|---|---|
| 6 | 15 | 17 | 0.0421 | 9 | 12 | 18 | 0.0690 | 4 | 17 | 18 | 0.0347 |
| 7 | 15 | 17 | 0.0457 | 10 | 12 | 18 | 0.0766 | 5 | 17 | 18 | 0.0356 |
| 8 | 15 | 17 | 0.0500 | 11 | 12 | 18 | 0.0860 | 6 | 17 | 18 | 0.0372 |
| 9 | 15 | 17 | 0.0553 | 6 | 13 | 18 | 0.0477 | 7 | 17 | 18 | 0.0396 |
| 10 | 15 | 17 | 0.0618 | 7 | 13 | 18 | 0.0521 | 8 | 17 | 18 | 0.0425 |
| 11 | 15 | 17 | 0.0702 | 8 | 13 | 18 | 0.0571 | 9 | 17 | 18 | 0.0461 |
| 12 | 15 | 17 | 0.0812 | 9 | 13 | 18 | 0.0630 | 10 | 17 | 18 | 0.0506 |
| 13 | 15 | 17 | 0.0966 | 10 | 13 | 18 | 0.0700 | 11 | 17 | 18 | 0.0563 |
| 14 | 15 | 17 | 0.1202 | 11 | 13 | 18 | 0.0785 | 12 | 17 | 18 | 0.0635 |
| 2 | 16 | 17 | 0.0434 | 12 | 13 | 18 | 0.0892 | 13 | 17 | 18 | 0.0732 |
| 3 | 16 | 17 | 0.0380 | 5 | 14 | 18 | 0.0405 | 14 | 17 | 18 | 0.0869 |
| 4 | 16 | 17 | 0.0375 | 6 | 14 | 18 | 0.0439 | 15 | 17 | 18 | 0.1080 |
| 5 | 16 | 17 | 0.0387 | 7 | 14 | 18 | 0.0478 | 16 | 17 | 18 | 0.1457 |
| 6 | 16 | 17 | 0.0409 | 8 | 14 | 18 | 0.0523 | 1 | 18 | 18 | 1.2157 |
| 7 | 16 | 17 | 0.0438 | 9 | 14 | 18 | 0.0576 | 2 | 18 | 18 | 0.3160 |
| 8 | 16 | 17 | 0.0474 | 10 | 14 | 18 | 0.0640 | 3 | 18 | 18 | 0.2365 |
| 9 | 16 | 17 | 0.0520 | 11 | 14 | 18 | 0.0717 | 4 | 18 | 18 | 0.1928 |
| 10 | 16 | 17 | 0.0577 | 12 | 14 | 18 | 0.0815 | 5 | 18 | 18 | 0.1618 |
| 11 | 16 | 17 | 0.0652 | 13 | 14 | 18 | 0.0944 | 6 | 18 | 18 | 0.1373 |
| 12 | 16 | 17 | 0.0751 | 4 | 15 | 18 | 0.0353 | 7 | 18 | 18 | 0.1167 |
| 13 | 16 | 17 | 0.0892 | 5 | 15 | 18 | 0.0377 | 8 | 18 | 18 | 0.0988 |
| 14 | 16 | 17 | 0.1108 | 6 | 15 | 18 | 0.0406 | 9 | 18 | 18 | 0.0829 |
| 15 | 16 | 17 | 0.1493 | 7 | 15 | 18 | 0.0440 | 10 | 18 | 18 | 0.0687 |
| 1 | 17 | 17 | 1.2255 | 8 | 15 | 18 | 0.0481 | 11 | 18 | 18 | 0.0560 |
| 2 | 17 | 17 | 0.3169 | 9 | 15 | 18 | 0.0528 | 12 | 18 | 18 | 0.0451 |
| 3 | 17 | 17 | 0.2361 | 10 | 15 | 18 | 0.0586 | 13 | 18 | 18 | 0.0364 |
| 4 | 17 | 17 | 0.1917 | 11 | 15 | 18 | 0.0656 | 14 | 18 | 18 | 0.0313 |
| 5 | 17 | 17 | 0.1603 | 12 | 15 | 18 | 0.0745 | 15 | 18 | 18 | 0.0333 |
| 6 | 17 | 17 | 0.1355 | 13 | 15 | 18 | 0.0863 | 16 | 18 | 18 | 0.0526 |
| 7 | 17 | 17 | 0.1148 | 14 | 15 | 18 | 0.1027 | 17 | 18 | 18 | 0.1357 |
| 8 | 17 | 17 | 0.0969 | 3 | 16 | 18 | 0.0331 | 9 | 11 | 19 | 0.0722 |
| 9 | 17 | 17 | 0.0811 | 4 | 16 | 18 | 0.0338 | 10 | 11 | 19 | 0.0795 |
| 10 | 17 | 17 | 0.0673 | 5 | 16 | 18 | 0.0356 | 8 | 12 | 19 | 0.0600 |
| 11 | 17 | 17 | 0.0554 | 6 | 16 | 18 | 0.0380 | 9 | 12 | 19 | 0.0658 |
| 12 | 17 | 17 | 0.0460 | 7 | 16 | 18 | 0.0409 | 10 | 12 | 19 | 0.0725 |
| 13 | 17 | 17 | 0.0404 | 8 | 16 | 18 | 0.0445 | 11 | 12 | 19 | 0.0805 |
| 14 | 17 | 17 | 0.0421 | 9 | 16 | 18 | 0.0487 | 7 | 13 | 19 | 0.0504 |
| 15 | 17 | 17 | 0.0615 | 10 | 16 | 18 | 0.0538 | 8 | 13 | 19 | 0.0550 |
| 16 | 17 | 17 | 0.1459 | 11 | 16 | 18 | 0.0602 | 9 | 13 | 19 | 0.0603 |
| 9 | 10 | 18 | 0.0837 | 12 | 16 | 18 | 0.0682 | 10 | 13 | 19 | 0.0664 |
| 8 | 11 | 18 | 0.0686 | 13 | 16 | 18 | 0.0789 | 11 | 13 | 19 | 0.0737 |
| 9 | 11 | 18 | 0.0758 | 14 | 16 | 18 | 0.0939 | 12 | 13 | 19 | 0.0826 |
| 10 | 11 | 18 | 0.0842 | 15 | 16 | 18 | 0.1168 | 6 | 14 | 19 | 0.0427 |
| 7 | 12 | 18 | 0.0569 | 2 | 17 | 18 | 0.0409 | 7 | 14 | 19 | 0.0464 |
| 8 | 12 | 18 | 0.0625 | 3 | 17 | 18 | 0.0354 | 8 | 14 | 19 | 0.0505 |

TABLE 4

OUTLIER MEAN 0.0 OUTLIER STANDARD DEVIATION 4.0

| R | S | N | COVAR | R | S | N | CCVAR | R | S | N | COVAR |
|---|---|---|---|---|---|---|---|---|---|---|---|
| 9 | 14 | 19 | 0.0553 | 6 | 18 | 19 | 0.0342 | 13 | 14 | 20 | 0.0795 |
| 10 | 14 | 19 | 0.0609 | 7 | 18 | 19 | 0.0361 | 6 | 15 | 20 | 0.0386 |
| 11 | 14 | 19 | 0.0676 | 8 | 18 | 19 | 0.0385 | 7 | 15 | 20 | 0.0417 |
| 12 | 14 | 19 | 0.0757 | 9 | 18 | 19 | 0.0414 | 8 | 15 | 20 | 0.0452 |
| 13 | 14 | 19 | 0.0860 | 10 | 18 | 19 | 0.0450 | 9 | 15 | 20 | 0.0491 |
| 5 | 15 | 19 | 0.0367 | 11 | 18 | 19 | 0.0494 | 10 | 15 | 20 | 0.0537 |
| 6 | 15 | 19 | 0.0395 | 12 | 18 | 19 | 0.0549 | 11 | 15 | 20 | 0.0590 |
| 7 | 15 | 19 | 0.0428 | 13 | 18 | 19 | 0.0620 | 12 | 15 | 20 | 0.0654 |
| 8 | 15 | 19 | 0.0465 | 14 | 18 | 19 | 0.0715 | 13 | 15 | 20 | 0.0732 |
| 9 | 15 | 19 | 0.0509 | 15 | 18 | 19 | 0.0849 | 14 | 15 | 20 | 0.0831 |
| 10 | 15 | 19 | 0.0559 | 16 | 18 | 19 | 0.1055 | 5 | 16 | 20 | 0.0335 |
| 11 | 15 | 19 | 0.0620 | 17 | 18 | 19 | 0.1424 | 6 | 16 | 20 | 0.0359 |
| 12 | 15 | 19 | 0.0695 | 1 | 19 | 19 | 1.2070 | 7 | 16 | 20 | 0.0386 |
| 13 | 15 | 19 | 0.0789 | 2 | 19 | 19 | 0.3153 | 8 | 16 | 20 | 0.0418 |
| 14 | 15 | 19 | 0.0913 | 3 | 19 | 19 | 0.2370 | 9 | 16 | 20 | 0.0454 |
| 4 | 16 | 19 | 0.0323 | 4 | 19 | 19 | 0.1939 | 10 | 16 | 20 | 0.0495 |
| 5 | 16 | 19 | 0.0343 | 5 | 19 | 19 | 0.1634 | 11 | 16 | 20 | 0.0544 |
| 6 | 16 | 19 | 0.0367 | 6 | 19 | 19 | 0.1392 | 12 | 16 | 20 | 0.0603 |
| 7 | 16 | 19 | 0.0396 | 7 | 19 | 19 | 0.1188 | 13 | 16 | 20 | 0.0675 |
| 8 | 16 | 19 | 0.0430 | 8 | 19 | 19 | 0.1010 | 14 | 16 | 20 | 0.0766 |
| 9 | 16 | 19 | 0.0469 | 9 | 19 | 19 | 0.0852 | 15 | 16 | 20 | 0.0886 |
| 10 | 16 | 19 | 0.0515 | 10 | 19 | 19 | 0.0708 | 4 | 17 | 20 | 0.0298 |
| 11 | 16 | 19 | 0.0570 | 11 | 19 | 19 | 0.0578 | 5 | 17 | 20 | 0.0314 |
| 12 | 16 | 19 | 0.0638 | 12 | 19 | 19 | 0.0460 | 6 | 17 | 20 | 0.0335 |
| 13 | 16 | 19 | 0.0724 | 13 | 19 | 19 | 0.0358 | 7 | 17 | 20 | 0.0359 |
| 14 | 16 | 19 | 0.0838 | 14 | 19 | 19 | 0.0277 | 8 | 17 | 20 | 0.0388 |
| 15 | 16 | 19 | 0.0997 | 15 | 19 | 19 | 0.0230 | 9 | 17 | 20 | 0.0420 |
| 3 | 17 | 19 | 0.0307 | 16 | 19 | 19 | 0.0252 | 10 | 17 | 20 | 0.0458 |
| 4 | 17 | 19 | 0.0311 | 17 | 19 | 19 | 0.0443 | 11 | 17 | 20 | 0.0502 |
| 5 | 17 | 19 | 0.0326 | 18 | 19 | 19 | 0.1263 | 12 | 17 | 20 | 0.0555 |
| 6 | 17 | 19 | 0.0346 | 10 | 11 | 20 | 0.0756 | 13 | 17 | 20 | 0.0621 |
| 7 | 17 | 19 | 0.0371 | 9 | 12 | 20 | 0.0631 | 14 | 17 | 20 | 0.0705 |
| 8 | 17 | 19 | 0.0400 | 10 | 12 | 20 | 0.0690 | 15 | 17 | 20 | 0.0815 |
| 9 | 17 | 19 | 0.0434 | 11 | 12 | 20 | 0.0760 | 16 | 17 | 20 | 0.0970 |
| 10 | 17 | 19 | 0.0475 | 8 | 13 | 20 | 0.0531 | 3 | 18 | 20 | 0.0286 |
| 11 | 17 | 19 | 0.0525 | 9 | 13 | 20 | 0.0579 | 4 | 18 | 20 | 0.0289 |
| 12 | 17 | 19 | 0.0586 | 10 | 13 | 20 | 0.0633 | 5 | 18 | 20 | 0.0300 |
| 13 | 17 | 19 | 0.0665 | 11 | 13 | 20 | 0.0697 | 6 | 18 | 20 | 0.0317 |
| 14 | 17 | 19 | 0.0769 | 12 | 13 | 20 | 0.0773 | 7 | 18 | 20 | 0.0338 |
| 15 | 17 | 19 | 0.0914 | 7 | 14 | 20 | 0.0451 | 8 | 18 | 20 | 0.0362 |
| 16 | 17 | 19 | 0.1138 | 8 | 14 | 20 | 0.0489 | 9 | 18 | 20 | 0.0391 |
| 2 | 18 | 19 | 0.0387 | 9 | 14 | 20 | 0.0533 | 10 | 18 | 20 | 0.0425 |
| 3 | 18 | 19 | 0.0332 | 10 | 14 | 20 | 0.0583 | 11 | 18 | 20 | 0.0465 |
| 4 | 18 | 19 | 0.0323 | 11 | 14 | 20 | 0.0641 | 12 | 18 | 20 | 0.0513 |
| 5 | 18 | 19 | 0.0329 | 12 | 14 | 20 | 0.0710 | 13 | 18 | 20 | 0.0573 |

TABLE 4

OUTLIER MEAN 0.0 OUTLIER STANDARD DEVIATION 4.0

| R | S | N | COVAR | R | S | N | COVAR | R | S | N | COVAR |
|---|---|---|---|---|---|---|---|---|---|---|---|
| 14 | 18 | 20 | 0.0649 | 12 | 19 | 20 | 0.0483 | 8 | 20 | 20 | 0.1034 |
| 15 | 18 | 20 | 0.0750 | 13 | 19 | 20 | 0.0537 | 9 | 20 | 20 | 0.0876 |
| 16 | 18 | 20 | 0.0892 | 14 | 19 | 20 | 0.0607 | 10 | 20 | 20 | 0.0732 |
| 17 | 18 | 20 | 0.1110 | 15 | 19 | 20 | 0.0699 | 11 | 20 | 20 | 0.0601 |
| 2 | 19 | 20 | 0.0368 | 16 | 19 | 20 | 0.0830 | 12 | 20 | 20 | 0.0480 |
| 3 | 19 | 20 | 0.0313 | 17 | 19 | 20 | 0.1032 | 13 | 20 | 20 | 0.0370 |
| 4 | 19 | 20 | 0.0302 | 18 | 19 | 20 | 0.1393 | 14 | 20 | 20 | 0.0274 |
| 5 | 19 | 20 | 0.0305 | 1 | 20 | 20 | 1.1990 | 15 | 20 | 20 | 0.0197 |
| 6 | 19 | 20 | 0.0316 | 2 | 20 | 20 | 0.3147 | 16 | 20 | 20 | 0.0154 |
| 7 | 19 | 20 | 0.0332 | 3 | 20 | 20 | 0.2377 | 17 | 20 | 20 | 0.0178 |
| 8 | 19 | 20 | 0.0351 | 4 | 20 | 20 | 0.1952 | 18 | 20 | 20 | 0.0367 |
| 9 | 19 | 20 | 0.0375 | 5 | 20 | 20 | 0.1650 | 19 | 20 | 20 | 0.1175 |
| 10 | 19 | 20 | 0.0405 | 6 | 20 | 20 | 0.1411 | | | | |
| 11 | 19 | 20 | 0.0440 | 7 | 20 | 20 | 0.1210 | | | | |

Selected Tables in Mathematical Statistics
Volume V, 1977

# TABLES FOR OBTAINING OPTIMAL CONFIDENCE INTERVALS INVOLVING THE CHI-SQUARE DISTRIBUTION

G. Randall Murdock[1]
and
William O. Williford[2]
University of Georgia

ABSTRACT

Tables of critical values necessary for obtaining optimal confidence intervals in situations involving the chi-square distribution are presented. Included in such situations are optimal confidence intervals, from both the Bayesian and classical sampling theory approaches to estimation, for such parameters as the variance or precision (the reciprocal of the variance) of a normal distribution and the mean of a Poisson distribution. Tables of critical values are provided for $1-\alpha$ = .60, .70, .80, .90, .95, .975, .99, .995, .999 where $1-\alpha$ is the size of the confidence chi-square region and for $\nu = 1(1)100$ (except for one table I-2 where $\nu = 3(1)100$ is appropriate) where $\nu$ is the degrees of freedom of the chi-square distribution.

## 1. INTRODUCTION

The application of confidence interval estimation is too often limited to the so-called equal-tails intervals. If the sampling distribution of an estimator of the unknown parameter $\Theta$, or the posterior distribution of $\Theta$, is symmetric and has a unique mode, such as for example the normal

Received by the editors September 1975, and in revised form July 1976, January 1977 and February 1977.
AMS (MOS) Subject Classifications (1970): Primary 62Q05: Secondary 62F25.
[1]Present address: Texas Instruments, Dallas, Texas 75222
[2]Present address: Cooperative Studies Program Coordinating Center, VA Hospital, Perry Point, MD 21902

distribution, then optimal confidence intervals for $\Theta$, in the sense of shortness, are identical to the equal-tails interval. However, when the distribution is non-symmetric and has a unique mode, such as the chi-square distribution with which we will be concerned in this paper, then the equal-tails and optimal confidence intervals do not coincide. In the particular case of the chi-square distribution conditions for obtaining optimal confidence intervals are easily derived, however the construction of these intervals is hampered presently by the lack of appropriate comprehensive tables. In order to correct this deficiency extensive tables of critical values are provided in this paper for obtaining optimal confidence intervals in situations involving the chi-square distribution. Confidence intervals involving the variance or precision (the reciprocal of the variance) of a normal distribution and the mean of a Poisson distribution will be discussed in section 2. The best tables previously available were published between 1958 and 1961 and are each somewhat limited in usefulness due to the lack of range in either degrees of freedom, probability content or both. These tables were produced by Ramachandran (1958), Tate and Klett (1959), Lindley, East and Hamilton (1960), and Pachares (1961).

## 2. OPTIMAL CONFIDENCE INTERVALS

### 2.1 Bayesian Approach

We will initially consider the Bayesian approach. Box and Tiao (1965) defined a highest posterior density (H.P.D.) region to be a region with the property that the posterior probability of any point inside the region is greater than for any point outside of it. It then follows that for a given probability content the region has minimum volume. We would then say, speaking in terms of an interval, that for a given probability content the H.P.D. interval is that interval with minimum length. It further follows that if the posterior distribution is non-uniform and has a unique mode then the H.P.D. interval is unique. The posterior distribution to be considered in this paper is the chi-square (or gamma) distribution.

We will employ the following distributions and notation in the remainder of this paper.

Normal Distribution:

$$N(t|\mu,\sigma^2) = (2\pi\sigma^2)^{-1/2}\exp[-(t-\mu)^2/2\sigma^2], \ -\infty<t<\infty, \ \sigma^2>0.$$

Chi-Square Distribution:

$$CS(t|\nu) = \{2^{\nu/2}\Gamma(\nu/2)\}^{-1} \ t^{(\nu/2)-1} \ \exp(-t/2), \ t\geq 0, \ \nu>0.$$

Gamma Distribution:

$$Ga(t|\alpha,\beta) = \beta^{\alpha}t^{\alpha-1} \ \exp(-\beta t)/\Gamma(\alpha), \ \alpha>0, \ \beta>0, \ t\geq 0.$$

Note: if $t$ is distributed as $Ga(t|\alpha,\beta)$ then $2\beta t$ is distributed as $CS(2\beta t|2\alpha)$.

Poisson Distribution:

$$Po(t|\lambda) = e^{-\lambda}\lambda^t/t!, \ t=0,1,2,\cdots, \ \lambda>0.$$

We will consider the following two cases: one (case A) involving the variance ($\sigma^2$) or precision ($\tau=1/\sigma^2$) of a normal distribution, and the other (case B) the mean ($\lambda$) of a Poisson distribution. Each of these cases results in a gamma (chi-square) posterior distribution.

A. p.d.f. for $X_i$: $N(X_i|\mu,\sigma^2=1/\tau)$, $\mu$ and $\tau$ are unknown

(i) Priors: $Ga(\tau|\alpha,\beta)$ and $N(\mu|\mu^*,\sigma^{*2}=1/\tau\tau^*)$ CONJUGATE PRIORS

Posterior:
$$Ga(\tau|\alpha+n/2,\beta+\sum_{i=1}^{n}(X_i-\overline{X})^2/2 + [\tau^*n(\overline{X}-\mu^*)^2]/2(\tau^*+n))$$
$$= CS([2\beta+\sum_{i=1}^{n}(X_i-\overline{X})^2 + \{\tau^*n(\overline{X}-\mu^*)^2\}/(\tau^*+n)]\tau|2\alpha+n)$$

(ii) Priors: $p(\sigma)\propto 1/\sigma$ $[p(\tau)\propto k_1], p(\mu)\propto k_2$ DIFFUSE PRIORS

Posterior:
$$Ga(\tau|(n-1)/2, \sum_{i=1}^{n}(X_i-\overline{X})^2/2)$$
$$= CS([\sum_{i=1}^{n}(X_i-\overline{X})^2]\tau|n-1)$$

B. p.d.f. for $X_i$: $Po(X_i|\lambda)$, $\lambda$ is unknown

(i) Prior: $Ga(\lambda|\alpha,\beta)$ CONJUGATE PRIOR

$$\text{Posterior:} \quad Ga(\lambda|\alpha + \sum_{i=1}^{n} X_i, \beta + n)$$

$$= CS(2[\beta+n]\lambda | 2[\alpha + \sum_{i=1}^{n} X_i])$$

(ii) Prior: $p(\lambda) \propto 1/\lambda^{1/2}$ DIFFUSE PRIOR

$$\text{Posterior:} \quad Ga(\lambda | \sum_{i=1}^{n} X_i + 1/2, n)$$

$$= CS(2n\lambda | 2 \sum_{i=1}^{n} X_i + 1)$$

($k_i$ is a constant.)

Two types of prior distributions have been used in each of the above cases: conjugate and diffuse. A prior distribution is said to be a conjugate prior if both the prior and resulting posterior distributions of a parameter belong to the same family of distributions. The usefulness of the conjugate gamma distribution used here is twofold. First because of its general mathematical tractibility, and second and more importantly because of its "richness" or versatility in its capability of expressing our prior information through a proper choice of the parameters $\alpha$ and $\beta$. However, when our prior knowledge concerning the parameter is scant or vague, we employ what is called a vague or diffuse prior distribution. For a more complete discussion of the problem of choosing a prior distribution see Lindley (1965, section 5.2), Zellner (1971, section 2.3) and Box and Tiao (1973, section 1.3).

First consider the precision $\tau$ and either of the posterior distributions, which we will call $p_n(t)$, from case A. We want to find an H.P.D. interval with probability content $1-\alpha$, that is we want to find values for a and b such that $\int_a^b p_n(t)dt = 1-\alpha$, $a < 2\beta'\tau < b$ (where $\beta'$ is the second parameter in the posterior distribution), and hence where (b-a) is minimized. Taking the partial derivative of $(b-a) + \lambda[\int_a^b p_n(t) - (1-\alpha)]$ ($\lambda$ is a Lagrange multiplier), with respect to a and b and setting the derivatives equal to zero leads to the condition $p_n(a) = p_n(b)$. (It is easy to see that H.P.D. intervals for $\lambda$ (in case B) can be found in the same manner as the H.P.D. intervals found for $\tau$.)

Now consider case A(ii) and the variance $\sigma^2$, or more precisely $\ln\sigma^2$. Since, in this case, it was assumed that $P(\sigma)\propto 1/\sigma$, that is $p(\ln\sigma^2)\propto$ constant, we consider the posterior distribution of $\ln\sigma^2$. Then in order to find the H.P.D. interval in $\ln\sigma^2$ with probability content $1-\alpha$ it turns out that a and b must be determined such that $\int_a^b p_n(t)dt = 1-\alpha$ and $p_{n+2}(a) = p_{n+2}(b)$ (see Lindley, et al. (1960) and Lindley (1965)).

2.2 Sampling Theory Approach

As opposed to obtaining optimum confidence intervals using the Bayesian approach we will now consider the standard sampling theory approach. Sampling theory provides several alternative types of optimal confidence intervals. Besides requiring the confidence interval of size $1-\alpha$ to be shortest or have minimum length a useful and natural extension from point estimation is the requirement that the confidence interval of size $1-\alpha$ be unbiased. A confidence interval of size $1-\alpha$ is said to be unbiased if the probability that the random interval covers the parameter of interest, is at least as great as the probability that the interval covers any other constant value.

Since the sample statistic $(n-1)S^2/\sigma^2$ (with $(n-1)S^2 = \sum_{i=1}^{n}(X_i-\bar{X})^2$) has a chi-square distribution, say $p_n(t)$, with $\nu=n-1$ degrees of freedom the problem of obtaining a shortest unbiased confidence interval of size $\alpha$ reduces to finding values for a and b such the $\int_a^b p_n(t)dt = 1-\alpha$ and minimizing $\left(\frac{(n-1)S^2}{a} - \frac{(n-1)S^2}{b}\right)$ or $\left(\frac{1}{a} - \frac{1}{b}\right)$. Following the procedure outlined for $\tau$ when using the Bayesian approach we find these latter two conditions reduce to $\int_a^b p_n(t)dt = 1-\alpha$ and $p_{n+2}(a) = p_{n+2}(b)$. Thus computationally the classical sampling theory problem of finding a shortest unbiased confidence interval for $\sigma^2$ of size $\alpha$ is equivalent to finding an H.P.D. interval for $\ln\sigma^2$ with probability content $1-\alpha$ (see Lindley, et al. (1960) and Guenther (1971)).

2.3 Summary

Summarizing, consider the following two conditions:

$$\int_a^b p_n(t)dt = 1-\alpha \text{ and } p_{n+k}(a) = p_{n+k}(b), \tag{2.1}$$

where k is a non-negative integer. H.P.D. intervals of probability content

$1-\alpha$ are found for $\tau$ by finding values of a and b satisfying the conditions in (2.1) when k=0. H.P.D. intervals of probability content $1-\alpha$ are found for $\ln\sigma^2$, as well as shortest unbiased intervals for $\sigma^2$, by obtaining a and b satisfying the conditions in (2.1) when k=2. It is also true that values of a and b which satisfy (2.1) when (1) k=3 provides an interval which satisfies the likelihood ratio principle, when applied to a two-sided test of $\sigma^2 = \sigma_1^2$; and when (2) k=4 provides a minimum length (in the sense of sampling theory) confidence interval for $\sigma^2$ (see Tate and Klett (1959)).

## 3. DESCRIPTION OF TABLES

The three tables of critical values presented in the Appendix of this paper are denoted I-1, I-2 and II. Tables I-1, I-2 and II correspond respectively to the situations described in the last paragraph of section 2 when k=2, k=0 and k=4. The following chart presents a summary of the information contained in those tables previously published (which are portions of the present tables I-1 and II so similarly denoted here) in the four papers mentioned at the end of section 1. Following the chart is a brief description of the purpose of each of the three tables I-1, I-2 and II. Finally a second chart is provided, which may be compared with the first chart, which summarizes the information contained in the three tables presented in the Appendix of this paper.

| | Table | $1-\alpha$: size of confidence chi-square region | $\nu$-degrees of freedom | Significant figures |
|---|---|---|---|---|
| Ramachandran (1958) | I-1 | .95 | 2(1)8(2)24, 30,40,60 | 5 |
| Tate and Klett (1959) | I-1 | .90,.95,.99 .995,.999 | 2(1)29 | 4-6 |
| | II | .90,.95,.99 .995,.999 | 2(1)29 | 4-6 |
| Lindley, et al. (1960) | I-1 | .95,.99,.999 | 1(1)100 | 5 |
| Pachares (1961) | I-1 | .90,.95,.99 | 1(1)20,24 30,40,60,120 | 5 |

Table I-1(k=2): These tables provide critical values for finding the shortest interval in $\ln\sigma^2$, for a confidence region of size $1-\alpha$, of the posterior distribution of $\sigma^2$ the variance of a normal distribution. These tables can also be used to obtain shortest unbiased confidence intervals for the variance of a normal distribution (see Guenther (1971)). Pachares (1961) in his paper, also presents values of $\alpha_1$ and $\alpha_2$, the tail areas associated with the critical values. He also provides, for each of his $\alpha$'s, a set of four coefficients which may be used to approximate certain upper and lower critical values.

Table I-2(k=0): By reducing the degrees of freedom ($\nu$) by 2 the same tables, as those described above in I-1, may be used to find the shortest interval in $\tau$ for a confidence region of size $1-\alpha$ of a posterior distribution of $\tau$ the precision of a normal distribution. (In a similar manner, the shortest interval in $\lambda$ may be found for a critical region of size $\alpha$ of the posterior distribution of $\lambda$ the mean of a Poisson distribution.) The values of the tail areas $\alpha_1$ and $\alpha_2$, where $\alpha_1 + \alpha_2 = \alpha$ the critical region associated with the critical values, however are not the same for the two types of optimum confidence intervals and must be calculated separately if needed.

Table II(k=4): This table is not useful for any of the posterior distributions with which this paper is concerned. It does provide critical values for finding the shortest confidence interval in $\sigma^2$ for the usual classical sampling theory situation (see Tate and Klett (1959) and Guenther (1969)).

The tables presented in this paper are summarized as follows:

| Table | $1-\alpha$: size of confidence chi-square region | $\nu$-degrees of freedom | Significant figures |
|---|---|---|---|
| I-1 | .60,.70,.80 | 1(1)100 | 5-7 |
| I-2 | .90,.95,.975 | 3(1)100 | 5-7 |
| II | .99,.995,.999 | 1(1)100 | 5-7 |

The reason for presenting separately the two tables of critical values, described in I-1 and I-2, is in order to also give the values of the tail areas, $\alpha_1$ and $\alpha_2$, corresponding to the appropriate degrees of freedom.

## 4. COMPUTATIONAL ASPECTS

Recall, from section 2, that there are two conditions which must be satisfied in order to obtain the minimum length confidence intervals. They are

$$p_{n+k}(\chi_1^2) = p_{n+k}(\chi_2^2) \tag{4.1}$$

and

$$\int_{\chi_1^2}^{\chi_2^2} p_n(t)dt = 1-\alpha \tag{4.2}$$

Therefore we are interested in finding the unique pair, say $(\chi_1^{2\prime}, \chi_2^{2\prime})$, which satisfy (4.1) and (4.2) simultaneously.

In order to find the unique pair $(\chi_1^{2\prime}, \chi_2^{2\prime})$, for the cases where k=2 and k=4, $L=\chi_2^{2\prime}-\chi_1^{2\prime}$ must be maximized. For a given significance level $\alpha$, and since the chi-square distribution has positive skewness, $\chi_1^{2\prime}$ can be found near $\chi_\alpha^2$- for low degrees of freedom and within the range $(\chi_{\alpha/2}^2-\chi_\alpha^2)$ for large degrees of freedom. This latter approximation results from the fact that the chi-square distribution approaches the normal distribution as the degrees of freedom tend to infinity. With this basis for a first approximation of $\chi_1^{2\prime}$ it follows that the corresponding $\chi_2^{2\prime}$ will be rather large for small degrees of freedom and will approach $\chi_{1-\alpha/2}^2$ for large degrees of freedom.

In order to find the unique pair $(\chi_1^{2\prime}, \chi_2^{2\prime})$, for the case where k=0, it is simply a matter of minimizing $L=\chi_2^{2\prime} - \chi_1^{2\prime}$. Again, due to the positive skewness of the chi-square distribution, $\chi_1^{2\prime}$ is near $0^+$ for low degrees of freedom and approaches $\chi_{\alpha/2}^2$ for larger degrees of freedom. It follows that the corresponding $\chi_2^{2\prime}$ will lie in the range $(\chi_{1-\alpha}^2 - \chi_{1-\alpha/2}^2)$. We make note here of a particular relationship which exists between the cases when k=0 and k=2. Namely, the $(\chi_1^{2\prime}, \chi_2^{2\prime})$ values for the k=2 case at $\nu$ degrees of freedom are equal to the $(\chi_1^{2\prime}, \chi_2^{2\prime})$ values for the k=0 case at $(\nu+2)$ degrees of freedom. That this is true can easily be seen on integrating by parts. For the k=2 case, the mode is at the origin for one and two degrees of freedom hence the shortest interval is one-sided. (See Lindley, et al. (1960), pg. 434.)

The computer program was written to find the unique pair $(\chi_1^{2'}, \chi_2^{2'})$ for all three values of k(0,2,4). For each n within each confidence region $\varepsilon=1-\alpha$, 11 triples $(\chi_1^{2'}, \chi_2^{2'}, \varepsilon')$ were generated by choosing $\chi_1^{2'}$ from the appropriate range $(\chi_{\alpha/2}^2, \chi_\alpha^2)$, approximating $\chi_2^{2'}$ by Newton's method, and determining $\varepsilon'$ by using chi-square functions, from the S.D.P. (Statistical Distribution Package) Bargmann (1968). Then the 11 pairs $(\chi_1^{2'}, \varepsilon')$ were interpolated by using divided differences, yielding a final $\chi_1^{2'}$. This value was used to determine the final $\chi_2^{2'}$ by Newton's method. The final pair $(\chi_1^{2'}, \chi_2^{2'})$ was checked for accuracy by use of the S.D.P. chi-square functions. Using this approach, and double precision computations on the IBM SYSTEM/370, all results for confidence regions greater than or equal to 90% had precision to 10 decimal places (approximately 17 decimal digits used in computations), while the results for confidence regions less than 90% were slightly less precise. All the values given in Tables I-1, I-2, II are significant with an error of at most one in the last place printed. For confidence regions less than 90% a separate technique (Muller's, see Conte and deBoor (1965)) was used to approximate the $\chi_2^{2'}$. Since the range for these approximations were so much greater, finding an initial estimate for Newton's method became extremely difficult at low degrees of freedom. The approximations found by Muller's technique were then refined by Newton's method.

## 5. EXAMPLES

Example 1. (Box and Tiao (1973), pg. 90) The following information concerns the observed breaking strength of 20 samples of yarn obtained randomly. $SS_1 = \sum_{i=1}^{20} (X_i-\overline{X})^2=348$, $\overline{X}_1=50$, $S_1^2=348/19=18.3$. Assume the data is distributed normally with unknown variance $\sigma^2$ (or precision $\tau$). For the Bayesian approach we assume that our prior information concerning $\sigma^2$ is vague, therefore case A(ii) of section 2 is the appropriate one to apply here. The resulting posterior distribution is given by $Ga(\tau|19/2, 348/2)$ or $CS(348\tau|19)$, that is, $348\tau$ is chi-square distributed with 19 degrees of freedom. For $1-\alpha= .999$, the equal tails and certain optimum confidence intervals for $\sigma^2$ were calculated. For example, the equal tails result is given by (348/46=7.5652, 348/4.91=70.8757) and the shortest unbiased by (348/47.36987=7.3464, 348/5.12810=67.8613). The results are summarized as follows:

| Type of Interval | Table, Critical Values, $(\alpha_1,\alpha_2)$ | Interval $\begin{pmatrix}1-\alpha=.999\\ \alpha=\alpha_1+\alpha_2\end{pmatrix}$ | Length |
|---|---|---|---|
| Equal Tails $(\sigma^2)$ | * (4.91, 46) (.0005,.0005) | (7.5652,70.8757) | 63.3105 |
| H.P.D. $(\ln\sigma^2)$ or Shortest Unbiased $(\sigma^2)$ | I-1 (5.12810,47.36987) (.0006839,.0003161) | (7.3464,67.8613) | 60.5149 |
| Minimum Length $(\sigma^2)$ | II (5.37853,53.69916) (.0009633,.0000367) | (6.4805,64.7016) | 58.2211 |
| $(\tau)$; Reciprocal of Equal Tails $(\sigma^2)$ | * (4.91,46) (.0005,.0005) | (0.0141,0.1321) | 0.1180 |
| H.P.D. $(\tau)$ | I-2 (4.17865,44.32527) (.0001488,.0008512) | (0.0120,0.1273) | 0.1153 |

*See Ostle and Mensing (1975)

Suppose a second sample of size 20 is drawn such that $SS_2=316$, $\overline{X}_2=51$ and $S_2^2=16.6$. Then we may use the posterior distribution from the initial sample as the prior distribution of the second sample; that is, as Lindley (1971, p.2) says "today's posterior is tomorrow's prior." Hence, for the informative case A(i) section 2, if $\mu^*=50$, $1/\tau^*=18.3$, $\alpha=19/2$, $\beta=348/2$

then the new posterior distribution of $\tau$ is $Ga(\tau|39/2,664.0544/2)$ or $CS\ (664.0544\tau|39)$. Hence we have the following results:

| Type of Interval | Table, Critical Values, $(\alpha_1,\alpha_2)$ | Interval $\begin{pmatrix}1-\alpha=.999\\ \alpha=\alpha_1+\alpha_2\end{pmatrix}$ | Length |
|---|---|---|---|
| H.P.D.$(\ln\sigma^2)$ | I-1 (16.59496,75.87395) (.0006313,.0003687) | (8.7520,40.0154) | 31.2634 |
| H.P.D.$(\tau)$ | I-2 (15.33500,73.13553) (.0002427,.0007573) | (0.0230,0.1101) | 0.0871 |

Example 2. (Parzen (1960), pg. 256) From 1837 to 1932 vacancies occurred in the United States Supreme Court, either by death or by resignation of members as follows:

| j = number of vacancies during the year | $N_j$ = number of years with j vacancies |
|---|---|
| 0 | 59 |
| 1 | 27 |
| 2 | 9 |
| 3 | 1 |
| over 3 | 0 |

$N = \sum_{j=0}^{\infty} N_j = 96$, $\sum_{j=0}^{\infty} jN_j = 48$, $\hat{\lambda} = 48/96 = 0.50$. Assume the above data is in agreement with the Poisson distribution. For the Bayesian approach we assume that our prior information concerning $\lambda$ is vague, therefore case B(ii) of section 2 is the appropriate one to apply here. The posterior distribution is then given by $Ga(\lambda|97/2,192/2)$. For $1-\alpha=.999$ the results are summarized as follows:

| Type of Interval | Table, Critical Values, $(\alpha_1,\alpha_2)$ | Interval $\begin{pmatrix}1-\alpha=.999\\ \alpha=\alpha_1+\alpha_2\end{pmatrix}$ | Length |
|---|---|---|---|
| Equal Tails $(\lambda)$ | * (57.6,149.5) (.0005,.0005) | (.3000,.7786) | .4786 |
| H.P.D. $(\lambda)$ | I-2 (56.5009,147.95038) (.0003326,.0006674) | (.2942,.7705) | .4763 |

*See Ostle and Mensing (1975)

## REFERENCES

Abromowitz, M. and Stegun, I.A. (1964), *Handbook of Mathematical Functions*, National Bureau of Standards, Applied Mathematics, Series 55.

Bargmann, R.E. (1968), *A Statistics Distribution Computer Package*, Department of Statistics and Computer Science, University of Georgia.

Box, G.E.P. and Tiao, G.C. (1965), Multiparameter Problems from a Bayesian Point of View, *Annals of Mathematical Statistics, 36*, 1468-1482.

Box. G.E.P. and Tiao, G.C. (1973), *Bayesian Inference in Statistical Analysis*, Addison-Wesley.

Conte, S.D. and deBoor, C. (1965), *Elementary Numerical Analysis: An Algorithmic Approach*, Second Ed., McGraw-Hill.

Guenther, William C. (1969), Shortest Confidence Intervals, *The American Statistician*, Vol. 23, No. 1, 22-25.

Guenther, William C. (1971), Unbiased Confidence Intervals, *The American Statistician*, Vol. 25, No. 1, 51-53.

Lindley, D.V., East, D.A. and Hamilton, P.A. (1960), Tables for Making Inferences about the Variance of a Normal Distribution, *Biometrika*, 47, 433-437.

Lindley, D.V. (1965), *Introduction to Probability and Statistics from a Bayesian Viewpoint, Part 2, Inference*, Cambridge University Press.

Lindley, D.V. (1971), *Bayesian Statistics, a Review*, Philadelphia, Soc. Ind. Appl. Math.

Ostle, B. and Mensing, R.W. (1975), *Statistics in Research*, Third Ed., Iowa State University Press.

Pachares, James (1961), Tables for Unbiased Tests on the Variance of a Normal Population, *Annals of Mathematical Statistics*, 32, 84-87.

Parzen, E. (1960), *Modern Probability Theory and Its Applications*, John Wiley.

Ramachandran, K.V. (1958), A test of variances, *Journal of the American Statistical Association*, 53, 741-747.

Tate, R.F. and Klett, G.W. (1959), Optimal Confidence Intervals for the Variance of a Normal Distribution, *Journal of the American Statistical Association*, 54, 674-682.

Zellner, A. (1971), *An Introduction to Bayesian Inference in Econometrics*, John Wiley & Sons, Inc.

## APPENDIX

## Tables of Critical Values (and tail areas) For The Chi-Square Distribution

Table I-1 (K=2)

$1-\alpha$ = .999, .995, .990, .975, .950, .900, .800, .700, .600

$\nu$ = 1(1)100

Table I-2 (K=0)

$1-\alpha$ = .999, .995, .990, .975, .950, .900, .800, .700, .600

$\nu$ = 3(1)100

Table II (K=4)

$1-\alpha$ = .999, .995, .990, .975, .950, .900, .800, .700, .600

$\nu$ = 1(1)100

TABLE I-1 (K=2)

| | .999 | .995 | .990 |
|---|---|---|---|
| 1 | $0.0^{5}14027$(.0009450) | $0.0^{4}34117$(.0046604) | $0.0^{3}13422$(.0092435) |
| | 16.26624 (.0000550) | 12.83818 (.0003396) | 11.34496 (.0007565) |
| 2 | $0.0^{2}18055$(.0009023) | $0.0^{2}88359$(.0044082) | 0.017469(.0086965) |
| | 18.46773 (.0000977) | 14.86468 (.0005918) | 13.28545 (.0013035) |
| 3 | 0.022097(.0008678) | 0.063893(.0042139) | 0.10105 (.0082887) |
| | 20.52385 (.0001322) | 16.77525 (.0007861) | 15.12695 (.0017113) |
| 4 | 0.083097(.0008396) | 0.18593 (.0040627) | 0.26396 (.0079798) |
| | 22.48557 (.0001604) | 18.61024 (.0009373) | 16.90131 (.0020202) |
| 5 | 0.19336 (.0008163) | 0.37234 (.0039427) | 0.49623 (.0077394) |
| | 24.37772 (.0001837) | 20.38641 (.0010573) | 18.62131 (.0022605) |
| 6 | 0.35203 (.0007969) | 0.61445 (.0038453) | 0.78565 (.0075470) |
| | 26.21411 (.0002031) | 22.11382 (.0011547) | 20.29555 (.0024530) |
| 7 | 0.55491 (.0007805) | 0.90376 (.0037646) | 1.12211 (.0073889) |
| | 28.00391 (.0002195) | 23.79988 (.0012354) | 21.93086 (.0026111) |
| 8 | 0.79722 (.0007664) | 1.23315 (.0036965) | 1.49785 (.0072565) |
| | 29.75386 (.0002336) | 25.45042 (.0013035) | 23.53276 (.0027435) |
| 9 | 1.07446 (.0007541) | 1.59692 (.0036381) | 1.90684 (.0071435) |
| | 31.46918 (.0002459) | 27.07011 (.0013619) | 25.10564 (.0028565) |
| 10 | 1.38268 (.0007434) | 1.99047 (.0035873) | 2.34441 (.0070457) |
| | 33.15408 (.0002566) | 28.66272 (.0014127) | 26.65312 (.0029543) |
| 11 | 1.71852 (.0007339) | 2.41012 (.0035427) | 2.80685 (.0069600) |
| | 34.81200 (.0002661) | 30.23135 (.0014573) | 28.17816 (.0030400) |
| 12 | 2.07910 (.0007254) | 2.85283 (.0035031) | 3.29117 (.0068842) |
| | 36.44579 (.0002746) | 31.77863 (.0014969) | 29.68321 (.0031158) |
| 13 | 2.46197 (.0007177) | 3.31610 (.0034677) | 3.79493 (.0068165) |
| | 38.05783 (.0002823) | 33.30667 (.0015323) | 31.17032 (.0031835) |
| 14 | 2.86505 (.0007108) | 3.79786 (.0034357) | 4.31610 (.0067555) |
| | 39.65018 (.0002892) | 34.81734 (.0015643) | 32.64124 (.0032445) |
| 15 | 3.28653 (.0007045) | 4.29633 (.0034067) | 4.85297 (.0067003) |
| | 41.22456 (.0002955) | 36.31221 (.0015933) | 34.09741 (.0032997) |
| 16 | 3.72485 (.0006987) | 4.81002 (.0033802) | 5.40412 (.0066499) |
| | 42.78249 (.0003013) | 37.79262 (.0016198) | 35.54015 (.0033501) |
| 17 | 4.17865 (.0006934) | 5.33762 (.0033559) | 5.96828 (.0066038) |
| | 44.32527 (.0003066) | 39.25974 (.0016441) | 36.97052 (.0033962) |
| 18 | 4.64675 (.0006884) | 5.87803 (.0033334) | 6.54441 (.0065613) |
| | 45.85406 (.0003116) | 40.71460 (.0016666) | 38.38951 (.0034387) |
| 19 | 5.12810 (.0006839) | 6.43024 (.0033127) | 7.13157 (.0065220) |
| | 47.36987 (.0003161) | 42.15811 (.0016873) | 39.79797 (.0034780) |
| 20 | 5.62176 (.0006796) | 6.99342 (.0032934) | 7.72894 (.0064855) |
| | 48.87360 (.0003204) | 43.59108 (.0017066) | 41.19662 (.0035145) |

TABLE I-1 (K=2)

| | .999 | .995 | .990 |
|---|---|---|---|
| 21 | 6.12692 (.0006756)<br>50.36603 (.0003244) | 7.56678 (.0032754)<br>45.01419 (.0017246) | 8.33580 (.0064515)<br>42.58615 (.0035485) |
| 22 | 6.64283 (.0006719)<br>51.84790 (.0003281) | 8.14966 (.0032586)<br>46.42809 (.0017414) | 8.95152 (.0064197)<br>43.96713 (.0035803) |
| 23 | 7.16883 (.0006684)<br>53.31985 (.0003316) | 8.74144 (.0032429)<br>47.83334 (.0017571) | 9.57551 (.0063899)<br>45.34012 (.0036101) |
| 24 | 7.70432 (.0006651)<br>54.78246 (.0003349) | 9.34158 (.0032280)<br>49.23047 (.0017720) | 10.20728 (.0063620)<br>46.70558 (.0036380) |
| 25 | 8.24875 (.0006620)<br>56.23625 (.0003380) | 9.94959 (.0032141)<br>50.61993 (.0017859) | 10.84634 (.0063356)<br>48.06395 (.0036644) |
| 26 | 8.80163 (.0006591)<br>57.68172 (.0003409) | 10.56502 (.0032009)<br>52.00217 (.0017991) | 11.49230 (.0063107)<br>49.41562 (.0036893) |
| 27 | 9.36251 (.0006563)<br>59.11928 (.0003437) | 11.18747 (.0031884)<br>53.37755 (.0018116) | 12.14476 (.0062872)<br>50.76094 (.0037128) |
| 28 | 9.93098 (.0006536)<br>60.54936 (.0003463) | 11.81657 (.0031765)<br>54.74643 (.0018235) | 12.80339 (.0062648)<br>52.10027 (.0037352) |
| 29 | 10.50667 (.0006511)<br>61.97232 (.0003489) | 12.45197 (.0031653)<br>56.10913 (.0018347) | 13.46786 (.0062436)<br>53.43388 (.0037564) |
| 30 | 11.08921 (.0006487)<br>63.38849 (.0003513) | 13.09337 (.0031545)<br>57.46597 (.0018455) | 14.13789 (.0062235)<br>54.76207 (.0037765) |
| 31 | 11.67830 (.0006465)<br>64.79820 (.0003535) | 13.74049 (.0031443)<br>58.81720 (.0018557) | 14.81321 (.0062042)<br>56.08507 (.0037958) |
| 32 | 12.27364 (.0006443)<br>66.20172 (.0003557) | 14.39305 (.0031346)<br>60.16307 (.0018654) | 15.49356 (.0061859)<br>57.40314 (.0038141) |
| 33 | 12.87494 (.0006422)<br>67.59935 (.0003578) | 15.05081 (.0031252)<br>61.50385 (.0018748) | 16.17873 (.0061683)<br>58.71648 (.0038317) |
| 34 | 13.48196 (.0006402)<br>68.99132 (.0003598) | 15.71355 (.0031163)<br>62.83972 (.0018837) | 16.86850 (.0061516)<br>60.02530 (.0038484) |
| 35 | 14.09445 (.0006388)<br>70.37785 (.0003617) | 16.38106 (.0031078)<br>64.17090 (.0018922) | 17.56268 (.0061355)<br>61.32980 (.0038645) |
| 36 | 14.71220 (.0006364)<br>71.75920 (.0003636) | 17.05313 (.0030995)<br>65.49759 (.0019005) | 18.26109 (.0061201)<br>62.63014 (.0038799) |
| 37 | 15.33500 (.0006346)<br>73.13553 (.0003654) | 17.72958 (.0030917)<br>66.81996 (.0019083) | 18.96355 (.0061053)<br>63.92650 (.0038947) |
| 38 | 15.96265 (.0006329)<br>74.50705 (.0003671) | 18.41026 (.0030841)<br>68.13817 (.0019159) | 19.66991 (.0060910)<br>65.21902 (.0039090) |
| 39 | 16.59496 (.0006313)<br>75.87395 (.0003687) | 19.09499 (.0030768)<br>69.45238 (.0019232) | 20.38000 (.0060773)<br>66.50786 (.0039227) |
| 40 | 17.23178 (.0006297)<br>77.23637 (.0003703) | 19.78362 (.0030697)<br>70.76274 (.0019303) | 21.09372 (.0060641)<br>67.79314 (.0039359) |

TABLE I-1 (K=2)

| | .999 | .995 | .990 |
|---|---|---|---|
| 41 | 17.87294 (.0006282) | 20.47601 (.0030630) | 21.81090 (.0060514) |
| | 78.59450 (.0003718) | 72.06938 (.0019370) | 69.07500 (.0039486) |
| 42 | 18.51828 (.0006267) | 21.17204 (.0030564) | 22.53143 (.0060391) |
| | 79.94847 (.0003733) | 73.37244 (.0019436) | 70.35355 (.0039609) |
| 43 | 19.16766 (.0006253) | 21.87157 (.0030501) | 23.25519 (.0060272) |
| | 81.29843 (.0003747) | 74.67204 (.0019499) | 71.62889 (.0039728) |
| 44 | 19.82097 (.0006239) | 22.57448 (.0030440) | 23.98209 (.0060158) |
| | 82.64450 (.0003761) | 75.96829 (.0019560) | 72.90117 (.0039842) |
| 45 | 20.47806 (.0006226) | 23.28067 (.0030381) | 24.71201 (.0060047) |
| | 83.98682 (.0003774) | 77.26131 (.0019619) | 74.17046 (.0039953) |
| 46 | 21.13879 (.0006213) | 23.99002 (.0030324) | 25.44484 (.0059940) |
| | 85.32550 (.0003787) | 78.55119 (.0019676) | 75.43684 (.0040060) |
| 47 | 21.80309 (.0006200) | 24.70244 (.0030268) | 26.18050 (.0059836) |
| | 86.66066 (.0003800) | 79.83804 (.0019732) | 76.70045 (.0040164) |
| 48 | 22.47081 (.0006188) | 25.41783 (.0030214) | 26.91890 (.0059735) |
| | 87.99240 (.0003812) | 81.12193 (.0019786) | 77.96133 (.0040265) |
| 49 | 23.14186 (.0006177) | 26.13611 (.0030162) | 27.65997 (.0059637) |
| | 89.32083 (.0003823) | 82.40298 (.0019838) | 79.21959 (.0040363) |
| 50 | 23.81616 (.0006165) | 26.85718 (.0030112) | 28.40361 (.0059542) |
| | 90.64604 (.0003835) | 83.68127 (.0019888) | 80.47530 (.0040458) |
| 51 | 24.49361 (.0006154) | 27.58096 (.0030062) | 29.14977 (.0059450) |
| | 91.96812 (.0003846) | 84.95686 (.0019938) | 81.72852 (.0040550) |
| 52 | 25.17409 (.0006143) | 28.30739 (.0030015) | 29.89835 (.0059361) |
| | 93.28719 (.0003857) | 86.22983 (.0019985) | 82.97934 (.0040639) |
| 53 | 25.85756 (.0006133) | 29.03638 (.0029968) | 30.64929 (.0059274) |
| | 94.60329 (.0003867) | 87.50027 (.0020032) | 84.22781 (.0040726) |
| 54 | 26.54390 (.0006123) | 29.76785 (.0029923) | 31.40253 (.0059189) |
| | 95.91650 (.0003877) | 88.76823 (.0020077) | 85.47401 (.0040811) |
| 55 | 27.23306 (.0006113) | 30.50175 (.0029879) | 32.15800 (.0059107) |
| | 97.22694 (.0003887) | 90.03880 (.0020121) | 86.71800 (.0040893) |
| 56 | 27.92494 (.0006103) | 31.23802 (.0029836) | 32.91565 (.0059026) |
| | 98.53465 (.0003897) | 91.29701 (.0020164) | 87.95984 (.0040974) |
| 57 | 28.61951 (.0006094) | 31.97658 (.0029795) | 33.67542 (.0058948) |
| | 99.83972 (.0003906) | 92.55795 (.0020205) | 89.19957 (.0041052) |
| 58 | 29.31667 (.0006084) | 32.71738 (.0029754) | 34.43726 (.0058872) |
| | 101.14220 (.0003916) | 93.81667 (.0020246) | 90.43726 (.0041128) |
| 59 | 30.01634 (.0006075) | 33.46036 (.0029714) | 35.20110 (.0058798) |
| | 102.44214 (.0003925) | 95.07321 (.0020286) | 91.67294 (.0041202) |
| 60 | 30.71851 (.0006067) | 34.20547 (.0029676) | 35.96690 (.0058726) |
| | 103.73964 (.0003933) | 96.32765 (.0020324) | 92.90668 (.0041274) |

TABLE I-1 (K=2)

| | .999 | .995 | .990 |
|---|---|---|---|
| 61 | 31.42307 (.0006058) | 34.95265 (.0029638) | 36.73462 (.0058655) |
| | 105.03474 (.0003942) | 97.58003 (.0020362) | 94.13853 (.0041345) |
| 62 | 32.12999 (.0006050) | 35.70187 (.0029601) | 37.50420 (.0058586) |
| | 106.32750 (.0003950) | 98.83038 (.0020399) | 95.36852 (.0041414) |
| 63 | 32.83920 (.0006042) | 36.45306 (.0029565) | 38.27560 (.0058519) |
| | 107.61797 (.0003958) | 100.07878 (.0020435) | 96.59669 (.0041481) |
| 64 | 33.55066 (.0006034) | 37.20619 (.0029530) | 39.04880 (.0058453) |
| | 108.90620 (.0003966) | 101.32526 (.0020470) | 97.82312 (.0041547) |
| 65 | 34.26430 (.0006026) | 37.96120 (.0029496) | 39.82372 (.0058389) |
| | 110.19226 (.0003974) | 102.56987 (.0020504) | 99.04781 (.0041611) |
| 66 | 34.98010 (.0006018) | 38.71806 (.0029462) | 40.60034 (.0058326) |
| | 111.47618 (.0003982) | 103.81264 (.0020538) | 100.27080 (.0041674) |
| 67 | 35.69798 (.0006011) | 39.47672 (.0029429) | 41.37863 (.0058265) |
| | 112.75801 (.0003989) | 105.05360 (.0020571) | 101.49216 (.0041735) |
| 68 | 36.41792 (.0006004) | 40.23715 (.0029397) | 42.15855 (.0058205) |
| | 114.03780 (.0003996) | 106.29282 (.0020603) | 102.71188 (.0041795) |
| 69 | 37.13988 (.0005996) | 40.99930 (.0029366) | 42.94005 (.0058146) |
| | 115.31558 (.0004004) | 107.53032 (.0020634) | 103.93004 (.0041854) |
| 70 | 37.86378 (.0005989) | 41.76315 (.0029335) | 43.72311 (.0058088) |
| | 116.59142 (.0004011) | 108.76614 (.0020665) | 105.14665 (.0041912) |
| 71 | 38.58961 (.0005983) | 42.52864 (.0029305) | 44.50769 (.0058032) |
| | 117.86534 (.0004017) | 110.00032 (.0020695) | 106.36174 (.0041968) |
| 72 | 39.31732 (.0005976) | 43.29576 (.0029276) | 45.29376 (.0057977) |
| | 119.13737 (.0004024) | 111.23289 (.0020724) | 107.57535 (.0042023) |
| 73 | 40.04689 (.0005969) | 44.06447 (.0029247) | 46.08130 (.0057923) |
| | 120.40758 (.0004031) | 112.46388 (.0020753) | 108.78751 (.0042077) |
| 74 | 40.77826 (.0005963) | 44.83473 (.0029218) | 46.87027 (.0057870) |
| | 121.67598 (.0004037) | 113.69331 (.0020782) | 109.99825 (.0042130) |
| 75 | 41.51140 (.0005957) | 45.60652 (.0029191) | 47.66064 (.0057818) |
| | 122.94261 (.0004043) | 114.92125 (.0020809) | 111.20758 (.0042182) |
| 76 | 42.24628 (.0005951) | 46.37979 (.0029163) | 48.45239 (.0057767) |
| | 124.20752 (.0004049) | 116.14769 (.0020837) | 112.41556 (.0042233) |
| 77 | 42.98286 (.0005944) | 47.15454 (.0029137) | 49.24548 (.0057717) |
| | 125.47072 (.0004056) | 117.37267 (.0020863) | 113.62219 (.0042283) |
| 78 | 43.72112 (.0005939) | 47.93071 (.0029110) | 50.03990 (.0057668) |
| | 126.73225 (.0004061) | 118.59621 (.0020890) | 114.82750 (.0042332) |
| 79 | 44.46100 (.0005933) | 48.70830 (.0029085) | 50.83560 (.0057620) |
| | 127.99216 (.0004067) | 119.81836 (.0020915) | 116.03152 (.0042380) |
| 80 | 45.20251 (.0005927) | 49.48727 (.0029060) | 51.63260 (.0057573) |
| | 129.25046 (.0004073) | 121.03914 (.0020940) | 117.23428 (.0042427) |

TABLE I-1 (K=2)

| | .999 | | .995 | | .990 | |
|---|---|---|---|---|---|---|
| 81 | 45.94559 | (.0005921) | 50.26761 | (.0029035) | 52.43083 | (.0057526) |
| | 130.50717 | (.0004079) | 122.25854 | (.0020965) | 118.43581 | (.0042474) |
| 82 | 46.69023 | (.0005916) | 51.04927 | (.0029010) | 53.23030 | (.0057481) |
| | 131.76234 | (.0004084) | 123.47664 | (.0020990) | 119.63609 | (.0042519) |
| 83 | 47.43639 | (.0005910) | 51.83224 | (.0028987) | 54.03096 | (.0057436) |
| | 133.01601 | (.0004090) | 124.69344 | (.0021013) | 120.83519 | (.0042564) |
| 84 | 48.18405 | (.0005905) | 52.61652 | (.0028963) | 54.83282 | (.0057392) |
| | 134.26817 | (.0004095) | 125.90894 | (.0021037) | 122.03310 | (.0042607) |
| 85 | 48.93317 | (.0005900) | 53.40204 | (.0028940) | 55.63583 | (.0057349) |
| | 135.51886 | (.0004100) | 127.12318 | (.0021060) | 123.22986 | (.0042651) |
| 86 | 49.68375 | (.0005895) | 54.18880 | (.0028917) | 56.43999 | (.0057307) |
| | 136.76811 | (.0004105) | 128.33620 | (.0021083) | 124.42548 | (.0042693) |
| 87 | 50.43575 | (.0005890) | 54.97679 | (.0028895) | 57.24527 | (.0057265) |
| | 138.01596 | (.0004110) | 129.54800 | (.0021105) | 125.61998 | (.0042735) |
| 88 | 51.18913 | (.0005885) | 55.76598 | (.0028873) | 58.05165 | (.0057225) |
| | 139.26241 | (.0004115) | 130.75861 | (.0021127) | 126.81339 | (.0042775) |
| 89 | 51.94389 | (.0005880) | 56.55634 | (.0028852) | 58.85912 | (.0057184) |
| | 140.50748 | (.0004120) | 131.96803 | (.0021148) | 128.00572 | (.0042816) |
| 90 | 52.70001 | (.0005875) | 57.34785 | (.0028831) | 59.66766 | (.0057145) |
| | 141.75121 | (.0004125) | 133.17630 | (.0021169) | 129.19698 | (.0042855) |
| 91 | 53.45746 | (.0005870) | 58.14052 | (.0028810) | 60.47725 | (.0057106) |
| | 142.99361 | (.0004130) | 134.38344 | (.0021190) | 130.38719 | (.0042894) |
| 92 | 54.21620 | (.0005866) | 58.93431 | (.0028789) | 61.28786 | (.0057067) |
| | 144.23471 | (.0004134) | 135.58946 | (.0021211) | 131.57639 | (.0042932) |
| 93 | 54.97624 | (.0005861) | 59.72920 | (.0028769) | 62.09949 | (.0057030) |
| | 145.47452 | (.0004139) | 136.79437 | (.0021231) | 132.76456 | (.0042970) |
| 94 | 55.73753 | (.0005856) | 60.52518 | (.0028749) | 62.91212 | (.0056993) |
| | 146.71307 | (.0004144) | 137.99821 | (.0021251) | 133.95175 | (.0043007) |
| 95 | 56.50009 | (.0005852) | 61.32222 | (.0028730) | 63.72574 | (.0056956) |
| | 147.95038 | (.0004148) | 139.20099 | (.0021270) | 135.13794 | (.0043044) |
| 96 | 57.26387 | (.0005848) | 62.12033 | (.0028711) | 64.54033 | (.0056920) |
| | 149.18646 | (.0004152) | 140.40271 | (.0021289) | 136.32318 | (.0043080) |
| 97 | 58.02885 | (.0005843) | 62.91946 | (.0028692) | 65.35587 | (.0056885) |
| | 150.42134 | (.0004157) | 141.60339 | (.0021308) | 137.50748 | (.0043115) |
| 98 | 58.79501 | (.0005839) | 63.71962 | (.0028673) | 66.17235 | (.0056850) |
| | 151.65504 | (.0004161) | 142.80307 | (.0021327) | 138.69083 | (.0043150) |
| 99 | 59.56236 | (.0005835) | 64.52078 | (.0028655) | 66.98975 | (.0056816) |
| | 152.88756 | (.0004165) | 144.00174 | (.0021345) | 139.87325 | (.0043184) |
| 100 | 60.33087 | (.0005831) | 65.32294 | (.0028636) | 67.80806 | (.0056782) |
| | 154.11894 | (.0004169) | 145.19942 | (.0021364) | 141.05478 | (.0043218) |

TABLE I-1 (K=2)

| | .975 | .950 | .900 |
|---|---|---|---|
| 1 | $0.0^{3}81456$(.0227689)<br>9.34895 (.0022311) | $0.0^{2}31593$(.0448238)<br>7.81683 (.0051762) | 0.012116(.0876468)<br>6.25947 (.0123531) |
| 2 | 0.042930(.0212365)<br>11.16481 (.0037635) | 0.084727(.0414785)<br>9.53034 (.0085215) | 0.16763 (.0803984)<br>7.86429 (.0196016) |
| 3 | 0.18585 (.0201596)<br>12.90776 (.0048404) | 0.29624 (.0392656)<br>11.19146 (.0107344) | 0.47639 (.0759541)<br>9.43382 (.0240459) |
| 4 | 0.42217 (.0193782)<br>14.59399 (.0056218) | 0.60700 (.0377175)<br>12.80244 (.0122825) | 0.88265 (.0729638)<br>10.95835 (.0270361) |
| 5 | 0.73118 (.0187864)<br>16.23137 (.0062136) | 0.98923 (.0365701)<br>14.36861 (.0134299) | 1.35469 (.0707955)<br>12.44235 (.0292044) |
| 6 | 1.09671 (.0183210)<br>17.82713 (.0066790) | 1.42500 (.0356801)<br>15.89659 (.0143199) | 1.87459 (.0691365)<br>13.89223 (.0308635) |
| 7 | 1.50729 (.0179437)<br>19.38750 (.0070563) | 1.90259 (.0349654)<br>17.39226 (.0150346) | 2.43126 (.0678164)<br>15.31356 (.0321836) |
| 8 | 1.95469 (.0176303)<br>20.91756 (.0073697) | 2.41392 (.0343759)<br>18.86043 (.0156241) | 3.01733 (.0667345)<br>16.71078 (.0332655) |
| 9 | 2.43285 (.0173649)<br>22.42142 (.0076351) | 2.95321 (.0338790)<br>20.30495 (.0161210) | 3.62759 (.0658274)<br>18.08742 (.0341726) |
| 10 | 2.93720 (.0171364)<br>23.90244 (.0078636) | 3.51616 (.0334531)<br>21.72890 (.0165469) | 4.25822 (.0650527)<br>19.44624 (.0349473) |
| 11 | 3.46416 (.0169371)<br>25.36333 (.0080629) | 4.09944 (.0330828)<br>23.13478 (.0169172) | 4.90631 (.0643812)<br>20.78947 (.0356188) |
| 12 | 4.01091 (.0167614)<br>26.80634 (.0082386) | 4.70046 (.0327571)<br>24.52469 (.0172429) | 5.56959 (.0637919)<br>22.11896 (.0362080) |
| 13 | 4.57517 (.0166048)<br>28.23334 (.0083952) | 5.31713 (.0324676)<br>25.90030 (.0175324) | 6.24623 (.0632695)<br>23.43616 (.0367305) |
| 14 | 5.15505 (.0164643)<br>29.64592 (.0085357) | 5.94773 (.0322082)<br>27.26305 (.0177918) | 6.93475 (.0628021)<br>24.74234 (.0371979) |
| 15 | 5.74898 (.0163372)<br>31.04541 (.0086628) | 6.59084 (.0319740)<br>28.61414 (.0180260) | 7.63393 (.0623808)<br>26.03856 (.0376192) |
| 16 | 6.35564 (.0162216)<br>32.43295 (.0087784) | 7.24527 (.0317612)<br>29.95464 (.0182388) | 8.34274 (.0619984)<br>27.32573 (.0380016) |
| 17 | 6.97391 (.0161158)<br>33.80954 (.0088842) | 7.91000 (.0315667)<br>31.28540 (.0184333) | 9.06031 (.0616493)<br>28.60460 (.0383507) |
| 18 | 7.06282 (.0160185)<br>35.17606 (.0089815) | 8.58416 (.0313880)<br>32.60722 (.0186120) | 9.78588 (.0613290)<br>29.87585 (.0386710) |
| 19 | 8.24151 (.0159286)<br>36.53325 (.0090714) | 9.26700 (.0312231)<br>33.92079 (.0187769) | 10.51882 (.0610337)<br>31.14008 (.0389663) |
| 20 | 8.88926 (.0158453)<br>37.88177 (.0091547) | 9.95786 (.0310704)<br>35.22670 (.0189296) | 11.25856 (.0607603)<br>32.39778 (.0392397) |

TABLE I-1 (K=2)

| | .975 | .950 | .900 |
|---|---|---|---|
| 21 | 9.54542 (.0157678) | 10.65616 (.0309284) | 12.00459 (.0605063) |
| | 39.22224 (.0092322) | 36.52547 (.0190716) | 33.64943 (.0394937) |
| 22 | 10.20941 (.0156954) | 11.36137 (.0307958) | 12.75649 (.0602694) |
| | 40.55519 (.0093046) | 37.81757 (.0192041) | 34.89543 (.0397306) |
| 23 | 10.88072 (.0156276) | 12.07306 (.0306718) | 13.51387 (.0600478) |
| | 41.88107 (.0093724) | 39.10344 (.0193282) | 36.13612 (.0399522) |
| 24 | 11.55889 (.0155639) | 12.79080 (.0305554) | 14.27637 (.0598400) |
| | 43.20032 (.0094361) | 40.38345 (.0194446) | 37.37186 (.0401600) |
| 25 | 12.24351 (.0155040) | 13.51423 (.0304459) | 15.04369 (.0596446) |
| | 44.51332 (.0094960) | 41.65796 (.0195541) | 38.60292 (.0403554) |
| 26 | 12.93420 (.0154474) | 14.24302 (.0303426) | 15.81554 (.0594603) |
| | 45.82042 (.0095526) | 42.92725 (.0196574) | 39.82957 (.0405397) |
| 27 | 13.63063 (.0153939) | 14.97687 (.0302450) | 16.59166 (.0592862) |
| | 47.12196 (.0096061) | 44.19164 (.0197550) | 41.05206 (.0407137) |
| 28 | 14.33249 (.0153433) | 15.71551 (.0301525) | 17.37183 (.0591214) |
| | 48.41820 (.0096567) | 45.45134 (.0198475) | 42.27060 (.0408786) |
| 29 | 15.03949 (.0152952) | 16.45868 (.0300647) | 18.15582 (.0589651) |
| | 49.70943 (.0097048) | 46.70663 (.0199353) | 43.48540 (.0410349) |
| 30 | 15.75138 (.0152494) | 17.20616 (.0299813) | 18.94347 (.0588165) |
| | 50.99588 (.0097506) | 47.95773 (.0200187) | 44.69662 (.0411835) |
| 31 | 16.46791 (.0152058) | 17.95773 (.0299019) | 19.73457 (.0586751) |
| | 52.27779 (.0097942) | 49.20480 (.0200981) | 45.90448 (.0413249) |
| 32 | 17.18889 (.0151643) | 18.71321 (.0298261) | 20.52896 (.0585402) |
| | 53.55536 (.0098357) | 50.44807 (.0201739) | 47.10910 (.0414598) |
| 33 | 17.91411 (.0151246) | 19.47243 (.0297538) | 21.32651 (.0584115) |
| | 54.82878 (.0098754) | 51.68768 (.0202462) | 48.31062 (.0415885) |
| 34 | 18.64336 (.0150866) | 20.23520 (.0296846) | 22.12706 (.0582884) |
| | 56.09825 (.0099134) | 52.92380 (.0203154) | 49.50920 (.0417116) |
| 35 | 19.37648 (.0150502) | 21.00136 (.0296183) | 22.93048 (.0581705) |
| | 57.36391 (.0099498) | 54.15657 (.0203817) | 50.70497 (.0418295) |
| 36 | 20.11331 (.0150153) | 21.77080 (.0295548) | 23.73665 (.0580576) |
| | 58.62593 (.0099847) | 55.38614 (.0204452) | 51.89801 (.0419424) |
| 37 | 20.85371 (.0149818) | 22.54337 (.0294938) | 24.54546 (.0579492) |
| | 59.88445 (.0100182) | 56.61264 (.0205062) | 53.08847 (.0420508) |
| 38 | 21.59752 (.0149496) | 23.31894 (.0294352) | 25.35681 (.0578450) |
| | 61.13960 (.0100504) | 57.83618 (.0205648) | 54.27641 (.0421550) |
| 39 | 22.34462 (.0149186) | 24.09741 (.0293789) | 26.17059 (.0577449) |
| | 62.39153 (.0100814) | 59.05687 (.0206211) | 55.46198 (.0422551) |
| 40 | 23.09488 (.0148888) | 24.87865 (.0293246) | 26.98672 (.0576485) |
| | 63.64034 (.0101112) | 60.27483 (.0206754) | 56.64522 (.0423515) |

TABLE I-1 (K=2)

| | .975 | .950 | .900 |
|---|---|---|---|
| 41 | 23.84818 (.0148601) | 25.66257 (.0292723) | 27.80511 (.0575556) |
| | 64.88615 (.0101399) | 61.49014 (.0207277) | 57.82623 (.0424444) |
| 42 | 24.60443 (.0148323) | 26.44908 (.0292219) | 28.62569 (.0574660) |
| | 66.12906 (.0101677) | 62.70291 (.0207781) | 59.00510 (.0425340) |
| 43 | 25.36351 (.0148055) | 27.23807 (.0291732) | 29.44835 (.0573795) |
| | 67.36917 (.0101945) | 63.91321 (.0208268) | 60.18188 (.0426205) |
| 44 | 26.12532 (.0147796) | 28.02948 (.0291261) | 30.27306 (.0572959) |
| | 68.60658 (.0102203) | 65.12115 (.0208739) | 61.35666 (.0427041) |
| 45 | 26.88979 (.0147546) | 28.82320 (.0290806) | 31.09970 (.0572152) |
| | 69.84138 (.0102454) | 66.32678 (.0209194) | 62.52951 (.0427848) |
| 46 | 27.65681 (.0147304) | 29.61919 (.0290366) | 31.92825 (.0571370) |
| | 71.07365 (.0102696) | 67.53020 (.0209634) | 63.70049 (.0428630) |
| 47 | 28.42633 (.0147069) | 30.41734 (.0289940) | 32.75864 (.0570613) |
| | 72.30348 (.0102931) | 68.73146 (.0210060) | 64.86964 (.0429387) |
| 48 | 29.19823 (.0146842) | 31.21761 (.0289527) | 33.59081 (.0569880) |
| | 73.53091 (.0103158) | 69.93063 (.0210473) | 66.03705 (.0430120) |
| 49 | 29.97247 (.0146622) | 32.01991 (.0289126) | 34.42470 (.0569170) |
| | 74.75607 (.0103378) | 71.12779 (.0210874) | 67.20274 (.0430830) |
| 50 | 30.74896 (.0146408) | 32.82420 (.0288737) | 35.26025 (.0568480) |
| | 75.97897 (.0103592) | 72.32300 (.0211262) | 68.36679 (.0431520) |
| 51 | 31.52765 (.0146200) | 33.63043 (.0288360) | 36.09743 (.0567811) |
| | 77.19972 (.0103800) | 73.51628 (.0211640) | 69.52924 (.0432189) |
| 52 | 32.30846 (.0145998) | 34.43851 (.0287994) | 36.93617 (.0567161) |
| | 78.41835 (.0104002) | 74.70772 (.0212006) | 70.69011 (.0432839) |
| 53 | 33.09134 (.0145802) | 35.24841 (.0287638) | 37.77646 (.0566529) |
| | 79.63495 (.0104198) | 75.89737 (.0212362) | 71.84947 (.0433471) |
| 54 | 33.87624 (.0145611) | 36.06007 (.0287291) | 38.61823 (.0565915) |
| | 80.84955 (.0104389) | 77.08525 (.0212709) | 73.00737 (.0434085) |
| 55 | 34.66309 (.0145426) | 36.87346 (.0286954) | 39.46144 (.0565317) |
| | 82.06219 (.0104574) | 78.27145 (.0213046) | 74.16382 (.0434683) |
| 56 | 35.45183 (.0145245) | 37.68852 (.0286627) | 40.30606 (.0564736) |
| | 83.27298 (.0104755) | 79.45599 (.0213373) | 75.31888 (.0435264) |
| 57 | 36.24243 (.0145069) | 38.50520 (.0286307) | 41.15204 (.0564170) |
| | 84.48192 (.0104931) | 80.63892 (.0213693) | 76.47260 (.0435830) |
| 58 | 37.03485 (.0144898) | 39.32346 (.0285996) | 41.99936 (.0563618) |
| | 85.68906 (.0105102) | 81.82027 (.0214004) | 77.62497 (.0436382) |
| 59 | 37.82903 (.0144731) | 40.14326 (.0285693) | 42.84799 (.0563080) |
| | 86.89447 (.0105269) | 83.00009 (.0214307) | 78.77606 (.0436920) |
| 60 | 38.62491 (.0144568) | 40.96458 (.0285397) | 43.69788 (.0562556) |
| | 88.09818 (.0105432) | 84.17842 (.0214603) | 79.92589 (.0437444) |

TABLE I-1 (K=2)

| | .975 | .950 | .900 |
|---|---|---|---|
| 61 | 39.42245 (.0144408) | 41.78737 (.0285108) | 44.54900 (.0562045) |
| | 89.30022 (.0105591) | 85.35530 (.0214892) | 81.07451 (.0437955) |
| 62 | 40.22165 (.0144253) | 42.61157 (.0284827) | 45.40134 (.0561545) |
| | 90.50064 (.0105747) | 86.53075 (.0215173) | 82.22191 (.0438454) |
| 63 | 41.02243 (.0144102) | 43.43719 (.0284552) | 46.25484 (.0561058) |
| | 91.69948 (.0105898) | 87.70482 (.0215448) | 83.36815 (.0438942) |
| 64 | 41.82477 (.0143954) | 44.26418 (.0284283) | 47.10950 (.0560582) |
| | 92.89677 (.0106046) | 88.87753 (.0215717) | 84.51326 (.0439418) |
| 65 | 42.62863 (.0143809) | 45.09248 (.0284021) | 47.96527 (.0560118) |
| | 94.09256 (.0106191) | 90.04890 (.0215979) | 85.65724 (.0439882) |
| 66 | 43.43398 (.0143668) | 45.92209 (.0283765) | 48.82214 (.0559663) |
| | 95.28688 (.0106332) | 91.21901 (.0216235) | 86.80014 (.0440337) |
| 67 | 44.24077 (.0143530) | 46.75298 (.0283514) | 49.68010 (.0559219) |
| | 96.47974 (.0106470) | 92.38783 (.0216486) | 87.94197 (.0440781) |
| 68 | 45.04900 (.0143394) | 47.58511 (.0283269) | 50.53909 (.0558785) |
| | 97.67120 (.0106606) | 93.55542 (.0216731) | 89.08275 (.0441215) |
| 69 | 45.85860 (.0143262) | 48.41847 (.0283029) | 51.39911 (.0558360) |
| | 98.86128 (.0106738) | 94.72180 (.0216971) | 90.22252 (.0441640) |
| 70 | 46.66957 (.0143133) | 49.25302 (.0282794) | 52.26013 (.0557944) |
| | 100.05000 (.0106867) | 95.88699 (.0217206) | 91.36130 (.0442056) |
| 71 | 47.48187 (.0143006) | 50.08873 (.0282564) | 53.12215 (.0557536) |
| | 101.23741 (.0106994) | 97.05103 (.0217436) | 92.49908 (.0442463) |
| 72 | 48.29547 (.0142882) | 50.92558 (.0282339) | 53.98511 (.0557138) |
| | 102.42352 (.0107118) | 98.21393 (.0217661) | 93.63593 (.0442862) |
| 73 | 49.11035 (.0142760) | 51.76357 (.0282119) | 54.84903 (.0556747) |
| | 103.60837 (.0107240) | 99.37570 (.0217881) | 94.77184 (.0443253) |
| 74 | 49.92648 (.0142641) | 52.60263 (.0281903) | 55.71385 (.0556365) |
| | 104.79196 (.0107359) | 100.53641 (.0218097) | 95.90681 (.0443635) |
| 75 | 50.74384 (.0142524) | 53.44278 (.0281691) | 56.57959 (.0555990) |
| | 105.97435 (.0107476) | 101.69603 (.0218309) | 97.04089 (.0444010) |
| 76 | 51.56239 (.0142410) | 54.28398 (.0281483) | 57.44621 (.0555622) |
| | 107.15553 (.0107590) | 102.85460 (.0218517) | 98.17410 (.0444378) |
| 77 | 52.38213 (.0142297) | 55.12621 (.0281280) | 58.31371 (.0555262) |
| | 108.33554 (.0107703) | 104.01216 (.0218720) | 99.30644 (.0444738) |
| 78 | 53.20302 (.0142187) | 55.96945 (.0281080) | 59.18205 (.0554908) |
| | 109.51442 (.0107813) | 105.16870 (.0218920) | 100.43794 (.0445092) |
| 79 | 54.02504 (.0142079) | 56.81369 (.0280884) | 60.05124 (.0554561) |
| | 110.69215 (.0107921) | 106.32426 (.0219116) | 101.56859 (.0445439) |
| 80 | 54.84818 (.0141973) | 57.65891 (.0280692) | 60.92123 (.0554221) |
| | 111.86879 (.0108027) | 107.47884 (.0219308) | 102.69844 (.0445779) |

TABLE I-1 (K=2)

| | .975 | .950 | .900 |
|---|---|---|---|
| 81 | 55.67241 (.0141869)<br>113.04433 (.0108131) | 58.50508 (.0280503)<br>108.63248 (.0219497) | 61.79204 (.0553887)<br>103.82748 (.0446113) |
| 82 | 56.49771 (.0141767)<br>114.21881 (.0108233) | 59.35219 (.0280318)<br>109.78517 (.0219682) | 62.66364 (.0553559)<br>104.95573 (.0446441) |
| 83 | 57.32405 (.0141666)<br>115.39224 (.0108333) | 60.20023 (.0280136)<br>110.93695 (.0219864) | 63.53601 (.0553237)<br>106.08322 (.0446763) |
| 84 | 58.15144 (.0141568)<br>116.56465 (.0108432) | 61.04916 (.0279957)<br>112.08784 (.0220043) | 64.40915 (.0552921)<br>107.20995 (.0447079) |
| 85 | 58.97984 (.0141471)<br>117.73604 (.0108529) | 61.89899 (.0279782)<br>113.23784 (.0220218) | 65.28302 (.0552610)<br>108.33592 (.0447390) |
| 86 | 59.80923 (.0141376)<br>118.90645 (.0108624) | 62.74969 (.0279609)<br>114.38696 (.0220390) | 66.15764 (.0552305)<br>109.46118 (.0447695) |
| 87 | 60.63962 (.0141282)<br>120.07587 (.0108718) | 63.60126 (.0279440)<br>115.53523 (.0220560) | 67.03299 (.0552004)<br>110.58571 (.0447996) |
| 88 | 61.47095 (.0141190)<br>121.24434 (.0108810) | 64.45366 (.0279273)<br>116.68265 (.0220727) | 67.90903 (.0551709)<br>111.70953 (.0448290) |
| 89 | 62.30324 (.0141100)<br>122.41187 (.0108900) | 65.30690 (.0279109)<br>117.82925 (.0220890) | 68.78578 (.0551420)<br>112.83266 (.0448580) |
| 90 | 63.13646 (.0141011)<br>123.57846 (.0108989) | 66.16095 (.0278948)<br>118.97505 (.0221052) | 69.66322 (.0551134)<br>113.95509 (.0448866) |
| 91 | 63.97060 (.0140923)<br>124.74414 (.0109076) | 67.01581 (.0278790)<br>120.12004 (.0221210) | 70.54132 (.0550854)<br>115.07687 (.0449146) |
| 92 | 64.80563 (.0140837)<br>125.90894 (.0109163) | 67.87146 (.0278634)<br>121.26424 (.0221366) | 71.42010 (.0550578)<br>116.19798 (.0449422) |
| 93 | 65.64156 (.0140753)<br>127.07283 (.0109247) | 68.72789 (.0278481)<br>122.40767 (.0221519) | 72.29953 (.0550306)<br>117.31844 (.0449694) |
| 94 | 66.47836 (.0140669)<br>128.23587 (.0109331) | 69.58507 (.0278330)<br>123.55034 (.0221670) | 73.17960 (.0550039)<br>118.43825 (.0449961) |
| 95 | 67.31601 (.0140587)<br>129.39806 (.0109413) | 70.44301 (.0278181)<br>124.69225 (.0221819) | 74.06030 (.0549776)<br>119.55743 (.0450224) |
| 96 | 68.15451 (.0140507)<br>130.55939 (.0109493) | 71.30170 (.0278035)<br>125.83344 (.0221965) | 74.94162 (.0549518)<br>120.67601 (.0450482) |
| 97 | 68.99385 (.0140427)<br>131.71989 (.0109573) | 72.16110 (.0277891)<br>126.97389 (.0222109) | 75.82356 (.0549263)<br>121.79396 (.0450737) |
| 98 | 69.83400 (.0140349)<br>132.87959 (.0109651) | 73.02122 (.0277749)<br>128.11362 (.0222251) | 76.70612 (.0549012)<br>122.91132 (.0450988) |
| 99 | 70.67496 (.0140272)<br>134.03848 (.0109728) | 73.88206 (.0277609)<br>129.25266 (.0222391) | 77.58925 (.0548765)<br>124.02808 (.0451235) |
| 100 | 71.51671 (.0140196)<br>135.19658 (.0109804) | 74.74361 (.0277472)<br>130.39099 (.0222528) | 78.47298 (.0548521)<br>125.14426 (.0451479) |

TABLE I-1 (K=2)

| | .800 | .700 | .600 |
|---|---|---|---|
| 1 | 0.045737(.1693456)<br>4.67221 (.0306544) | 0.098691(.2465941)<br>3.73117 (.0534059) | 0.16992 (.3198150)<br>3.06115 (.0801849) |
| 2 | 0.33460 (.1540541)<br>6.16058 (.0459459) | 0.50616 (.2235941)<br>5.14339 (.0764059) | 0.68442 (.2898026)<br>4.41096 (.1101973) |
| 3 | 0.77886 (.1454843)<br>7.62157 (.0545157) | 1.05233 (.2114073)<br>6.52728 (.0885926) | 1.31535 (.2745096)<br>5.73038 (.1254903) |
| 4 | 1.30776 (.1399425)<br>9.04203 (.0600575) | 1.66922 (.2036998)<br>7.87411 (.0963001) | 2.00400 (.2649778)<br>7.01644 (.1350222) |
| 5 | 1.89115 (.1360081)<br>10.42741 (.0639918) | 2.33096 (.1982902)<br>9.19031 (.1017097) | 2.72911 (.2583374)<br>8.27608 (.1416625) |
| 6 | 2.51341 (.1330363)<br>11.78404 (.0669636) | 3.02449 (.1942326)<br>10.48199 (.1057673) | 3.48006 (.2533789)<br>9.51495 (.1466211) |
| 7 | 3.16523 (.1306921)<br>13.11703 (.0693078) | 3.74213 (.1910466)<br>11.75375 (.1089534) | 4.25060 (.2494970)<br>10.73714 (.1505029) |
| 8 | 3.84053 (.1287828)<br>14.43028 (.0712171) | 4.47891 (.1884602)<br>13.00902 (.1115397) | 5.03669 (.2463524)<br>11.94560 (.1536475) |
| 9 | 4.53507 (.1271893)<br>15.72679 (.0728107) | 5.23134 (.1863069)<br>14.25039 (.1136931) | 5.83551 (.2437385)<br>13.14255 (.1562614) |
| 10 | 5.24578 (.1258333)<br>17.00891 (.0741667) | 5.99692 (.1844780)<br>15.47985 (.1155218) | 6.64501 (.2415213)<br>14.32966 (.1584787) |
| 11 | 5.97030 (.1246613)<br>18.27852 (.0753386) | 6.77373 (.1828999)<br>16.69897 (.1171001) | 7.46365 (.2396098)<br>15.50826 (.1603902) |
| 12 | 6.70682 (.1236353)<br>19.53711 (.0763646) | 7.56029 (.1815199)<br>17.90904 (.1184800) | 8.29021 (.2379397)<br>16.67940 (.1620603) |
| 13 | 7.45388 (.1227273)<br>20.78589 (.0772727) | 8.35541 (.1803000)<br>19.11107 (.1197000) | 9.12373 (.2364643)<br>17.84392 (.1635357) |
| 14 | 8.21032 (.1219164)<br>22.02589 (.0780836) | 9.15814 (.1792114)<br>20.30588 (.1207886) | 9.96344 (.2351484)<br>19.00255 (.1648515) |
| 15 | 8.97514 (.1211863)<br>23.25800 (.0788136) | 9.96768 (.1782321)<br>21.49422 (.1217678) | 10.80868 (.2339653)<br>20.15585 (.1660347) |
| 16 | 9.74754 (.1205246)<br>24.48288 (.0794753) | 10.78337 (.1773451)<br>22.67667 (.1226549) | 11.65893 (.2328940)<br>21.30435 (.1671059) |
| 17 | 10.52683 (.1199212)<br>25.70122 (.0800787) | 11.60465 (.1765367)<br>23.85374 (.1234633) | 12.51372 (.2319181)<br>22.44844 (.1680819) |
| 18 | 11.31241 (.1193680)<br>26.91351 (.0806319) | 12.43104 (.1757959)<br>25.02591 (.1242040) | 13.37266 (.2310241)<br>23.58852 (.1689758) |
| 19 | 12.10378 (.1188585)<br>28.12024 (.0811415) | 13.26212 (.1751139)<br>26.19353 (.1248860) | 14.23542 (.2302012)<br>24.72487 (.1697987) |
| 20 | 12.90048 (.1183870)<br>29.32184 (.0816129) | 14.09753 (.1744832)<br>27.35695 (.1255168) | 15.10169 (.2294405)<br>25.85780 (.1705595) |

TABLE I-1 (K=2)

| | .800 | .700 | .600 |
|---|---|---|---|
| 21 | 13.70213 (.1179493)<br>30.51863 (.0820506) | 14.93695 (.1738977)<br>28.51646 (.1261022) | 15.97123 (.2287344)<br>26.98755 (.1712655) |
| 22 | 14.50838 (.1175414)<br>31.71098 (.0824585) | 15.78010 (.1733523)<br>29.67235 (.1266476) | 16.84380 (.2280769)<br>28.11432 (.1719231) |
| 23 | 15.31893 (.1171601)<br>32.89915 (.0828398) | 16.62672 (.1728426)<br>30.82484 (.1271573) | 17.71919 (.2274625)<br>29.23833 (.1725375) |
| 24 | 16.13348 (.1168026)<br>34.08342 (.0831974) | 17.47662 (.1723649)<br>31.97414 (.1276351) | 18.59724 (.2268867)<br>30.35973 (.1731133) |
| 25 | 16.95183 (.1164665)<br>35.26399 (.0835334) | 18.32957 (.1719159)<br>33.12044 (.1280840) | 19.47778 (.2263456)<br>31.47868 (.1736543) |
| 26 | 17.77371 (.1161498)<br>36.44110 (.0838501) | 19.18539 (.1714929)<br>34.26393 (.1285070) | 20.36066 (.2258360)<br>32.59534 (.1741639) |
| 27 | 18.59895 (.1158507)<br>37.61493 (.0841492) | 20.04395 (.1710935)<br>35.40474 (.1289064) | 21.24574 (.2253548)<br>33.70982 (.1746451) |
| 28 | 19.42737 (.1155677)<br>38.78566 (.0844322) | 20.90506 (.1707156)<br>36.54305 (.1292844) | 22.13292 (.2248995)<br>34.82225 (.1751004) |
| 29 | 20.25877 (.1152992)<br>39.95345 (.0847007) | 21.76862 (.1703572)<br>37.67894 (.1296427) | 23.02206 (.2244679)<br>35.93274 (.1755320) |
| 30 | 21.09305 (.1150442)<br>41.11844 (.0849557) | 22.63448 (.1700169)<br>38.81258 (.1299831) | 23.91309 (.2240580)<br>37.04135 (.1759419) |
| 31 | 21.93001 (.1148015)<br>42.28078 (.0851984) | 23.50252 (.1696930)<br>39.94405 (.1303069) | 24.80591 (.2236681)<br>38.14822 (.1763319) |
| 32 | 22.76956 (.1145703)<br>43.44057 (.0854297) | 24.37267 (.1693844)<br>41.07346 (.1306155) | 25.70042 (.2232965)<br>39.25340 (.1767035) |
| 33 | 23.61157 (.1143494)<br>44.59795 (.0856505) | 25.24481 (.1690898)<br>42.20088 (.1309101) | 26.59656 (.2229418)<br>40.35698 (.1770581) |
| 34 | 24.45593 (.1141384)<br>45.75299 (.0858616) | 26.11885 (.1688083)<br>43.32645 (.1311917) | 27.49423 (.2226029)<br>41.45903 (.1773971) |
| 35 | 25.30255 (.1139364)<br>46.90584 (.0860636) | 26.99472 (.1685389)<br>44.45021 (.1314611) | 28.39340 (.2222785)<br>42.55962 (.1777214) |
| 36 | 26.15132 (.1137428)<br>48.05655 (.0862572) | 27.87233 (.1682807)<br>45.57224 (.1317193) | 29.29398 (.2219678)<br>43.65880 (.1780322) |
| 37 | 27.00215 (.1135570)<br>49.20523 (.0864429) | 28.75162 (.1680330)<br>46.69263 (.1319669) | 30.19592 (.2216697)<br>44.75662 (.1783302) |
| 38 | 27.85497 (.1133786)<br>50.35196 (.0866213) | 29.63251 (.1677951)<br>47.81142 (.1322048) | 31.09917 (.2213834)<br>45.85316 (.1786166) |
| 39 | 28.70970 (.1132071)<br>51.49680 (.0867929) | 30.51495 (.1675664)<br>48.92868 (.1324335) | 32.00368 (.2211081)<br>46.94846 (.1788918) |
| 40 | 29.56625 (.1130419)<br>52.63982 (.0869580) | 31.39888 (.1673463)<br>50.04446 (.1326537) | 32.90938 (.2208433)<br>48.04256 (.1791567) |

TABLE I-1 (K=2)

| | .800 | .700 | .600 |
|---|---|---|---|
| 41 | 30.42458 (.1128829)<br>53.78110 (.0871171) | 32.28423 (.1671342)<br>51.15884 (.1328657) | 33.81625 (.2205881)<br>49.13550 (.1794119) |
| 42 | 31.28461 (.1127295)<br>54.92070 (.0872704) | 33.17096 (.1669298)<br>52.27185 (.1330702) | 34.72424 (.2203420)<br>50.22733 (.1796579) |
| 43 | 32.14627 (.1125814)<br>56.05867 (.0874185) | 34.05902 (.1667324)<br>53.38354 (.1332675) | 35.63333 (.2201047)<br>51.31808 (.1798953) |
| 44 | 33.00952 (.1124384)<br>57.19507 (.0875615) | 34.94836 (.1665419)<br>54.49396 (.1334581) | 36.54344 (.2198754)<br>52.40781 (.1801246) |
| 45 | 33.87431 (.1123002)<br>58.32996 (.0876997) | 35.83896 (.1663577)<br>55.60315 (.1336423) | 37.45457 (.2196538)<br>53.49654 (.1803462) |
| 46 | 34.74057 (.1121665)<br>59.46338 (.0878335) | 36.73074 (.1661795)<br>56.71114 (.1338204) | 38.36667 (.2194394)<br>54.58427 (.1805605) |
| 47 | 35.60828 (.1120371)<br>60.59537 (.0879629) | 37.62370 (.1660070)<br>57.81799 (.1339929) | 39.27972 (.2192319)<br>55.67110 (.1807680) |
| 48 | 36.47737 (.1119117)<br>61.72598 (.0880883) | 38.51778 (.1658399)<br>58.92371 (.1341600) | 40.19368 (.2190309)<br>56.75700 (.1809690) |
| 49 | 37.34781 (.1117901)<br>62.85527 (.0882099) | 39.41295 (.1656780)<br>60.02835 (.1343220) | 41.10852 (.2188361)<br>57.84203 (.1811638) |
| 50 | 38.21956 (.1116722)<br>63.98326 (.0883278) | 40.30916 (.1655209)<br>61.13196 (.1344790) | 42.02423 (.2186472)<br>58.92619 (.1813528) |
| 51 | 39.09256 (.1115577)<br>65.11000 (.0884423) | 41.20642 (.1653684)<br>62.23453 (.1346315) | 42.94078 (.2184638)<br>60.00954 (.1815361) |
| 52 | 39.96680 (.1114466)<br>66.23552 (.0885534) | 42.10466 (.1652204)<br>63.33612 (.1347796) | 43.85812 (.2182857)<br>61.09207 (.1817142) |
| 53 | 40.84224 (.1113386)<br>67.35985 (.0886614) | 43.00388 (.1650765)<br>64.43674 (.1349235) | 44.77626 (.2181126)<br>62.17383 (.1818873) |
| 54 | 41.71883 (.1112335)<br>68.48302 (.0887664) | 43.90402 (.1649367)<br>65.53644 (.1350633) | 45.69516 (.2179445)<br>63.25482 (.1820555) |
| 55 | 42.59656 (.1111314)<br>69.60507 (.0888686) | 44.80510 (.1648006)<br>66.63522 (.1351993) | 46.61481 (.2177809)<br>64.33508 (.1822191) |
| 56 | 43.47539 (.1110320)<br>70.72604 (.0889680) | 45.70705 (.1646682)<br>67.73312 (.1353317) | 47.53517 (.2176217)<br>65.41461 (.1823783) |
| 57 | 44.35527 (.1109352)<br>71.84595 (.0890647) | 46.60988 (.1645393)<br>68.83015 (.1354606) | 48.45624 (.2174667)<br>66.49345 (.1825333) |
| 58 | 45.23621 (.1108409)<br>72.96481 (.0891590) | 47.51353 (.1644138)<br>69.92635 (.1355861) | 49.37799 (.2173157)<br>67.57161 (.1826842) |
| 59 | 46.11816 (.1107490)<br>74.08267 (.0892509) | 48.41803 (.1642914)<br>71.02173 (.1357085) | 50.30042 (.2171685)<br>68.64909 (.1828314) |
| 60 | 47.00110 (.1106594)<br>75.19954 (.0893405) | 49.32332 (.1641721)<br>72.11632 (.1358278) | 51.22350 (.2170251)<br>69.72594 (.1829749) |

TABLE I-1 (K=2)

| | .800 | .700 | .600 |
|---|---|---|---|
| 61 | 47.88501 (.1105720)<br>76.31546 (.0894279) | 50.22940 (.1640558)<br>73.21013 (.1359442) | 52.14720 (.2168851)<br>70.80214 (.1831148) |
| 62 | 48.76987 (.1104867)<br>77.43044 (.0895132) | 51.13623 (.1639422)<br>74.30318 (.1360577) | 53.07153 (.2167487)<br>71.87773 (.1832513) |
| 63 | 49.65562 (.1104035)<br>78.54451 (.0895964) | 52.04381 (.1638314)<br>75.39548 (.1361685) | 53.99646 (.2166154)<br>72.95273 (.1833845) |
| 64 | 50.54230 (.1103222)<br>79.65768 (.0896777) | 52.95212 (.1637232)<br>76.48706 (.1362768) | 54.92198 (.2164853)<br>74.02713 (.1835147) |
| 65 | 51.42984 (.1102428)<br>80.76999 (.0897571) | 53.86113 (.1636174)<br>77.57794 (.1363825) | 55.84808 (.2163581)<br>75.10095 (.1836418) |
| 66 | 52.31824 (.1101652)<br>81.88144 (.0898347) | 54.77084 (.1635141)<br>78.66812 (.1364858) | 56.77475 (.2162339)<br>76.17422 (.1837660) |
| 67 | 53.20747 (.1100894)<br>82.99205 (.0899106) | 55.68123 (.1634131)<br>79.75764 (.1365868) | 57.70197 (.2161125)<br>77.24693 (.1838875) |
| 68 | 54.09752 (.1100152)<br>84.10185 (.0899848) | 56.59227 (.1633143)<br>80.84650 (.1366856) | 58.62971 (.2159938)<br>78.31912 (.1840062) |
| 69 | 54.98837 (.1099426)<br>85.21086 (.0900574) | 57.50397 (.1632177)<br>81.93469 (.1367822) | 59.55800 (.2158776)<br>79.39078 (.1841224) |
| 70 | 55.88002 (.1098715)<br>86.31909 (.0901284) | 58.41631 (.1631232)<br>83.02228 (.1368768) | 60.48680 (.2157639)<br>80.46191 (.1842360) |
| 71 | 56.77243 (.1098020)<br>87.42654 (.0901979) | 59.32925 (.1630306)<br>84.10924 (.1369693) | 61.41609 (.2156526)<br>81.53255 (.1843473) |
| 72 | 57.66557 (.1097339)<br>88.53326 (.0902660) | 60.24281 (.1629400)<br>85.19559 (.1370599) | 62.34589 (.2155437)<br>82.60271 (.1844562) |
| 73 | 58.55948 (.1096672)<br>89.63925 (.0903327) | 61.15698 (.1628513)<br>86.28134 (.1371487) | 63.27617 (.2154370)<br>83.67236 (.1845629) |
| 74 | 59.45409 (.1096019)<br>90.74452 (.0903980) | 62.07172 (.1627643)<br>87.36652 (.1372356) | 64.20692 (.2153324)<br>84.74156 (.1846675) |
| 75 | 60.34940 (.1095379)<br>91.84909 (.0904621) | 62.98703 (.1626791)<br>88.45113 (.1373209) | 65.13814 (.2152300)<br>85.81029 (.1847699) |
| 76 | 61.24541 (.1094751)<br>92.95296 (.0905248) | 63.90289 (.1625956)<br>89.53519 (.1374044) | 66.06981 (.2151296)<br>86.87857 (.1848703) |
| 77 | 62.14211 (.1094136)<br>94.05615 (.0905864) | 64.81932 (.1625137)<br>90.61868 (.1374863) | 67.00192 (.2150312)<br>87.94640 (.1849688) |
| 78 | 63.03946 (.1093532)<br>95.15869 (.0906467) | 65.73628 (.1624333)<br>91.70164 (.1375666) | 67.93448 (.2149346)<br>89.01379 (.1850653) |
| 79 | 63.93745 (.1092940)<br>96.26059 (.0907059) | 66.65376 (.1623545)<br>92.78409 (.1376454) | 68.86746 (.2148399)<br>90.08076 (.1851600) |
| 80 | 64.83611 (.1092359)<br>97.36183 (.0907640) | 67.57178 (.1622772)<br>93.86601 (.1377227) | 69.80087 (.2147470)<br>91.14731 (.1852530) |

TABLE I-1 (K=2)

| | .800 | .700 | .600 |
|---|---|---|---|
| 81 | 65.73538 (.1091789) | 68.49028 (.1622013) | 70.73468 (.2146558) |
| | 98.46246 (.0908210) | 94.94743 (.1377986) | 92.21344 (.1853442) |
| 82 | 66.63527 (.1091229) | 69.40930 (.1621268) | 71.66891 (.2145662) |
| | 99.56247 (.0908770) | 96.02835 (.1378731) | 93.27917 (.1854337) |
| 83 | 67.53577 (.1090680) | 70.32880 (.1620537) | 72.60353 (.2144783) |
| | 100.66188 (.0909320) | 97.10878 (.1379462) | 94.34451 (.1855216) |
| 84 | 68.43686 (.1090140) | 71.24878 (.1619819) | 73.53854 (.2143919) |
| | 101.76070 (.0909860) | 98.18874 (.1380181) | 95.40945 (.1856080) |
| 85 | 69.33853 (.1089609) | 72.16925 (.1619113) | 74.47394 (.2143071) |
| | 102.85893 (.0910390) | 99.26822 (.1380886) | 96.47401 (.1856928) |
| 86 | 70.24078 (.1089088) | 73.09016 (.1618420) | 75.40971 (.2142238) |
| | 103.95659 (.0910911) | 100.34723 (.1381580) | 97.53819 (.1857761) |
| 87 | 71.14360 (.1088576) | 74.01155 (.1617739) | 76.34586 (.2141419) |
| | 105.05370 (.0911423) | 101.42580 (.1382261) | 98.60201 (.1858580) |
| 88 | 72.04697 (.1088073) | 74.93336 (.1617069) | 77.28236 (.2140614) |
| | 106.15024 (.0911927) | 102.50391 (.1382931) | 99.66547 (.1859385) |
| 89 | 72.95088 (.1087578) | 75.85564 (.1616410) | 78.21924 (.2139823) |
| | 107.24625 (.0912421) | 103.58159 (.1383589) | 100.72856 (.1860176) |
| 90 | 73.85533 (.1087092) | 76.77834 (.1615763) | 79.15646 (.2139044) |
| | 108.34171 (.0912908) | 104.65883 (.1384237) | 101.79129 (.1860955) |
| 91 | 74.76031 (.1086613) | 77.70148 (.1615126) | 80.09402 (.2138279) |
| | 109.43666 (.0913386) | 105.73564 (.1384873) | 102.85370 (.1861720) |
| 92 | 75.66580 (.1086142) | 78.62502 (.1614500) | 81.03194 (.2137526) |
| | 110.53108 (.0913857) | 106.81204 (.1385500) | 103.91576 (.1862473) |
| 93 | 76.57181 (.1085679) | 79.54898 (.1613883) | 81.97018 (.2136786) |
| | 111.62500 (.0914320) | 107.88803 (.1386116) | 104.97748 (.1863214) |
| 94 | 77.47832 (.1085223) | 80.47334 (.1613277) | 82.90875 (.2136057) |
| | 112.71841 (.0914776) | 108.96361 (.1386722) | 106.03886 (.1863942) |
| 95 | 78.38531 (.1084775) | 81.39812 (.1612680) | 83.84766 (.2135340) |
| | 113.81134 (.0915225) | 110.03880 (.1387319) | 107.09993 (.1864660) |
| 96 | 79.29280 (.1084333) | 82.32327 (.1612093) | 84.78687 (.2134634) |
| | 114.90379 (.0915666) | 111.11359 (.1387907) | 108.16068 (.1865366) |
| 97 | 80.20076 (.1083898) | 83.24883 (.1611514) | 85.72639 (.2133939) |
| | 115.99576 (.0916101) | 112.18799 (.1388485) | 109.22112 (.1866061) |
| 98 | 81.10919 (.1083470) | 84.17476 (.1610945) | 86.66624 (.2133254) |
| | 117.08725 (.0916529) | 113.26202 (.1389055) | 110.28125 (.1866745) |
| 99 | 82.01810 (.1083049) | 85.10106 (.1610384) | 87.60638 (.2132580) |
| | 118.17828 (.0916951) | 114.33566 (.1389616) | 111.34106 (.1867419) |
| 100 | 82.92746 (.1082633) | 86.02773 (.1609831) | 88.54683 (.2131916) |
| | 119.26886 (.0917366) | 115.40895 (.1390168) | 112.40059 (.1868083) |

TABLE I-2 (K=0)

| | .999 | .995 | .990 |
|---|---|---|---|
| 1 | -------<br>------- | -------<br>------- | -------<br>------- |
| 2 | -------<br>------- | -------<br>------- | -------<br>------- |
| 3 | $0.0^{5}14027$(.0000000)<br>16.26624 (.0010000) | $0.0^{4}34117$(.0000001)<br>12.83818 (.0049999) | $0.0^{3}13422$(.0000004)<br>11.34496 (.0099996) |
| 4 | $0.0^{2}18055$(.0000004)<br>18.46773 (.0009996) | $0.0^{2}88359$(.0000097)<br>14.86468 (.0049903) | 0.017469(.0000379)<br>13.28545 (.0099621) |
| 5 | 0.022097(.0000038)<br>20.52385 (.0009962) | 0.063893(.0000537)<br>16.77525 (.0049463) | 0.10105 (.0001665)<br>15.12695 (.0098335) |
| 6 | 0.083097(.0000116)<br>22.48557 (.0009884) | 0.18593 (.0001249)<br>18.61024 (.0048751) | 0.26396 (.0003472)<br>16.90131 (.0096528) |
| 7 | 0.19336 (.0000224)<br>24.37772 (.0009776) | 0.37234 (.0002072)<br>20.38641 (.0047928) | 0.49623 (.0005398)<br>18.62131 (.0094602) |
| 8 | 0.35203 (.0000348)<br>26.21411 (.0009652) | 0.61445 (.0002907)<br>22.11382 (.0047093) | 0.78565 (.0007261)<br>20.29555 (.0092739) |
| 9 | 0.55491 (.0000476)<br>28.00391 (.0009524) | 0.90376 (.0003709)<br>23.79988 (.0046291) | 1.12211 (.0008994)<br>21.93086 (.0091006) |
| 10 | 0.79722 (.0000603)<br>29.75386 (.0009397) | 1.23315 (.0004460)<br>25.45042 (.0045540) | 1.49785 (.0010580)<br>23.53276 (.0089420) |
| 11 | 1.07446 (.0000725)<br>31.46918 (.0009275) | 1.59692 (.0005155)<br>27.07011 (.0044845) | 1.90684 (.0012025)<br>25.10564 (.0087975) |
| 12 | 1.38268 (.0000842)<br>33.15408 (.0009158) | 1.99047 (.0005797)<br>28.66272 (.0044203) | 2.34441 (.0013341)<br>26.65312 (.0086659) |
| 13 | 1.71852 (.0000952)<br>34.81200 (.0009048) | 2.41012 (.0006389)<br>30.23135 (.0043611) | 2.80685 (.0014542)<br>28.17816 (.0085458) |
| 14 | 2.07910 (.0001055)<br>36.44579 (.0008945) | 2.85283 (.0006934)<br>31.77863 (.0043065) | 3.29117 (.0015641)<br>29.68321 (.0084359) |
| 15 | 2.46197 (.0001153)<br>38.05783 (.0008847) | 3.31610 (.0007439)<br>33.30667 (.0042561) | 3.79493 (.0016649)<br>31.17032 (.0083351) |
| 16 | 2.86505 (.0001245)<br>39.65018 (.0008755) | 3.79786 (.0007906)<br>34.81734 (.0042094) | 4.31610 (.0017578)<br>32.64124 (.0082422) |
| 17 | 3.28653 (.0001331)<br>41.22456 (.0008669) | 4.29633 (.0008340)<br>36.31221 (.0041660) | 4.85297 (.0018437)<br>34.09741 (.0081563) |
| 18 | 3.72485 (.0001412)<br>42.78249 (.0008588) | 4.81002 (.0008745)<br>37.79262 (.0041255) | 5.40412 (.0019233)<br>35.54015 (.0080767) |
| 19 | 4.17865 (.0001488)<br>44.32527 (.0008512) | 5.33762 (.0009122)<br>39.25974 (.0040878) | 5.96828 (.0019974)<br>36.97052 (.0080026) |
| 20 | 4.64675 (.0001560)<br>45.85406 (.0008440) | 5.87803 (.0009475)<br>40.71460 (.0040525) | 6.54441 (.0020665)<br>38.38951 (.0079335) |

TABLE I-2 (K=0)

| | .999 | .995 | .990 |
|---|---|---|---|
| 21 | 5.12810 (.0001628) | 6.43024 (.0009807) | 7.13157 (.0021312) |
| | 47.36987 (.0008372) | 42.15811 (.0040193) | 39.79797 (.0078688) |
| 22 | 5.62176 (.0001692) | 6.99342 (.0010119) | 7.72894 (.0021919) |
| | 48.87360 (.0008308) | 43.59108 (.0039881) | 41.19662 (.0078081) |
| 23 | 6.12692 (.0001753) | 7.56678 (.0010412) | 8.33580 (.0022489) |
| | 50.36603 (.0008247) | 45.01419 (.0039588) | 42.58615 (.0077511) |
| 24 | 6.64283 (.0001811) | 8.14966 (.0010690) | 8.95152 (.0023027) |
| | 51.84790 (.0008189) | 46.42809 (.0039310) | 43.96713 (.0076973) |
| 25 | 7.16883 (.0001866) | 8.74144 (.0010952) | 9.57551 (.0023535) |
| | 53.31985 (.0008134) | 47.83334 (.0039048) | 45.34012 (.0076465) |
| 26 | 7.70432 (.0001918) | 9.34158 (.0011201) | 10.20728 (.0024015) |
| | 54.78246 (.0008082) | 49.23047 (.0038799) | 46.70558 (.0075984) |
| 27 | 8.24875 (.0001968) | 9.94959 (.0011438) | 10.84634 (.0024471) |
| | 56.23625 (.0008032) | 50.61993 (.0038562) | 48.06395 (.0075529) |
| 28 | 8.80163 (.0002016) | 10.56502 (.0011663) | 11.49230 (.0024904) |
| | 57.68172 (.0007984) | 52.00217 (.0038337) | 49.41562 (.0075096) |
| 29 | 9.36251 (.0002061) | 11.18747 (.0011877) | 12.14476 (.0025316) |
| | 59.11928 (.0007939) | 53.37755 (.0038123) | 50.76094 (.0074684) |
| 30 | 9.93098 (.0002105) | 11.81657 (.0012081) | 12.80339 (.0025708) |
| | 60.54936 (.0007895) | 54.74643 (.0037919) | 52.10027 (.0074292) |
| 31 | 10.50667 (.0002146) | 12.45197 (.0012276) | 13.46786 (.0026082) |
| | 61.97232 (.0007854) | 56.10913 (.0037724) | 53.43388 (.0073918) |
| 32 | 11.08921 (.0002186) | 13.09337 (.0012463) | 14.13789 (.0026440) |
| | 63.38849 (.0007814) | 57.46597 (.0037537) | 54.76207 (.0073560) |
| 33 | 11.67830 (.0002225) | 13.74049 (.0012642) | 14.81321 (.0026782) |
| | 64.79820 (.0007775) | 58.81720 (.0037358) | 56.08507 (.0073218) |
| 34 | 12.27364 (.0002262) | 14.39305 (.0012813) | 15.49356 (.0027110) |
| | 66.20172 (.0007738) | 60.16307 (.0037187) | 57.40314 (.0072890) |
| 35 | 12.87494 (.0002297) | 15.05081 (.0012978) | 16.17873 (.0027425) |
| | 67.59935 (.0007703) | 61.50385 (.0037022) | 58.71648 (.0072575) |
| 36 | 13.48196 (.0002331) | 15.71355 (.0013136) | 16.86850 (.0027727) |
| | 68.99132 (.0007669) | 62.83972 (.0036864) | 60.02530 (.0072273) |
| 37 | 14.09445 (.0002364) | 16.38106 (.0013288) | 17.56268 (.0028017) |
| | 70.37785 (.0007636) | 64.17090 (.0036712) | 61.32980 (.0071983) |
| 38 | 14.71220 (.0002396) | 17.05313 (.0013434) | 18.26109 (.0028296) |
| | 71.75920 (.0007604) | 65.49759 (.0036566) | 62.63014 (.0071704) |
| 39 | 15.33500 (.0002427) | 17.72958 (.0013575) | 18.96355 (.0028565) |
| | 73.13553 (.0007573) | 66.81996 (.0036425) | 63.92650 (.0071435) |
| 40 | 15.96265 (.0002456) | 18.41026 (.0013712) | 19.66991 (.0028824) |
| | 74.50705 (.0007544) | 68.13817 (.0036288) | 65.21902 (.0071176) |

TABLE I-2 (K=0)

| | .999 | .995 | .990 |
|---|---|---|---|
| 41 | 16.59496 (.0002485)<br>75.87395 (.0007515) | 19.09499 (.0013843)<br>69.45238 (.0036157) | 20.38000 (.0029074)<br>66.50786 (.0070926) |
| 42 | 17.23178 (.0002512)<br>77.23637 (.0007488) | 19.78362 (.0013970)<br>70.76274 (.0036030) | 21.09372 (.0029316)<br>67.79314 (.0070684) |
| 43 | 17.87294 (.0002539)<br>78.59450 (.0007461) | 20.47601 (.0014093)<br>72.06938 (.0035907) | 21.81090 (.0029549)<br>69.07500 (.0070451) |
| 44 | 18.51828 (.0002565)<br>79.94847 (.0007435) | 21.17204 (.0014211)<br>73.37244 (.0035789) | 22.53143 (.0029775)<br>70.35355 (.0070225) |
| 45 | 19.16766 (.0002590)<br>81.29843 (.0007410) | 21.87157 (.0014326)<br>74.67204 (.0035674) | 23.25519 (.0029993)<br>71.62889 (.0070007) |
| 46 | 19.82097 (.0002615)<br>82.64450 (.0007385) | 22.57448 (.0014438)<br>75.96829 (.0035562) | 23.98209 (.0030205)<br>72.90117 (.0069795) |
| 47 | 20.47806 (.0002638)<br>83.98682 (.0007362) | 23.28067 (.0014546)<br>77.26131 (.0035454) | 24.71201 (.0030409)<br>74.17046 (.0069591) |
| 48 | 21.13879 (.0002661)<br>85.32550 (.0007339) | 23.99002 (.0014650)<br>78.55119 (.0035350) | 25.44484 (.0030608)<br>75.43684 (.0069392) |
| 49 | 21.80309 (.0002684)<br>86.66066 (.0007316) | 24.70244 (.0014752)<br>79.83804 (.0035248) | 26.18050 (.0030801)<br>76.70045 (.0069199) |
| 50 | 22.47081 (.0002705)<br>87.99240 (.0007295) | 25.41783 (.0014851)<br>81.12193 (.0035149) | 26.91890 (.0030988)<br>77.96133 (.0069012) |
| 51 | 23.14186 (.0002726)<br>89.32083 (.0007274) | 26.13611 (.0014947)<br>82.40298 (.0035053) | 27.65997 (.0031170)<br>79.21959 (.0068830) |
| 52 | 23.81616 (.0002747)<br>90.64604 (.0007253) | 26.85718 (.0015040)<br>83.68127 (.0034960) | 28.40361 (.0031347)<br>80.47530 (.0068653) |
| 53 | 24.49361 (.0002767)<br>91.96812 (.0007233) | 27.58096 (.0015131)<br>84.95686 (.0034869) | 29.14977 (.0031519)<br>81.72852 (.0068481) |
| 54 | 25.17409 (.0002787)<br>93.28719 (.0007213) | 28.30739 (.0015219)<br>86.22983 (.0034781) | 29.89835 (.0031686)<br>82.97934 (.0068314) |
| 55 | 25.85756 (.0002806)<br>94.60329 (.0007194) | 29.03638 (.0015305)<br>87.50027 (.0034695) | 30.64929 (.0031849)<br>84.22781 (.0068151) |
| 56 | 26.54390 (.0002824)<br>95.91650 (.0007176) | 29.76785 (.0015389)<br>88.76823 (.0034611) | 31.40253 (.0032007)<br>85.47401 (.0067993) |
| 57 | 27.23306 (.0002842)<br>97.22694 (.0007158) | 30.50175 (.0015471)<br>90.03380 (.0034529) | 32.15800 (.0032161)<br>86.71800 (.0067839) |
| 58 | 27.92494 (.0002860)<br>98.53465 (.0007140) | 31.23802 (.0015550)<br>91.29701 (.0034450) | 32.91565 (.0032312)<br>87.95984 (.0067688) |
| 59 | 28.61951 (.0002877)<br>99.83972 (.0007123) | 31.97658 (.0015628)<br>92.55795 (.0034372) | 33.67542 (.0032459)<br>89.19957 (.0067541) |
| 60 | 29.31667 (.0002894)<br>101.14220 (.0007106) | 32.71738 (.0015704)<br>93.81667 (.0034296) | 34.43726 (.0032602)<br>90.43726 (.0067398) |

TABLE I-2 (K=0)

| | .999 | .995 | .990 |
|---|---|---|---|
| 61 | 30.01634 (.0002910) | 33.46036 (.0015778) | 35.20110 (.0032742) |
| | 102.44214 (.0007090) | 95.07321 (.0034222) | 91.67294 (.0067258) |
| 62 | 30.71851 (.0002926) | 34.20547 (.0015850) | 35.96690 (.0032878) |
| | 103.73964 (.0007074) | 96.32765 (.0034150) | 92.90668 (.0067122) |
| 63 | 31.42307 (.0002942) | 34.95265 (.0015921) | 36.73462 (.0033011) |
| | 105.03474 (.0007058) | 97.58003 (.0034079) | 94.13853 (.0066989) |
| 64 | 32.12999 (.0002957) | 35.70187 (.0015990) | 37.50420 (.0033142) |
| | 106.32750 (.0007043) | 98.83038 (.0034010) | 95.36852 (.0066858) |
| 65 | 32.83920 (.0002972) | 36.45306 (.0016057) | 38.27560 (.0033269) |
| | 107.61797 (.0007028) | 100.07878 (.0033943) | 96.59669 (.0066731) |
| 66 | 33.55066 (.0002987) | 37.20619 (.0016123) | 39.04880 (.0033393) |
| | 108.90620 (.0007013) | 101.32526 (.0033877) | 97.82312 (.0066607) |
| 67 | 34.26430 (.0003001) | 37.96120 (.0016188) | 39.82372 (.0033515) |
| | 110.19226 (.0006999) | 102.56987 (.0033812) | 99.04781 (.0066485) |
| 68 | 34.98010 (.0003015) | 38.71806 (.0016251) | 40.60034 (.0033634) |
| | 111.47618 (.0006985) | 103.81264 (.0033749) | 100.27080 (.0066366) |
| 69 | 35.69798 (.0003029) | 39.47672 (.0016312) | 41.37863 (.0033751) |
| | 112.75801 (.0006971) | 105.05360 (.0033688) | 101.49216 (.0066249) |
| 70 | 36.41792 (.0003042) | 40.23715 (.0016373) | 42.15855 (.0033865) |
| | 114.03780 (.0006958) | 106.29282 (.0033627) | 102.71188 (.0066135) |
| 71 | 37.13988 (.0003055) | 40.99930 (.0016432) | 42.94005 (.0033977) |
| | 115.31558 (.0006945) | 107.53032 (.0033568) | 103.93004 (.0066023) |
| 72 | 37.86378 (.0003068) | 41.76315 (.0016490) | 43.72311 (.0034086) |
| | 116.59142 (.0006932) | 108.76614 (.0033510) | 105.14665 (.0065914) |
| 73 | 38.58961 (.0003081) | 42.52864 (.0016547) | 44.50769 (.0034193) |
| | 117.86534 (.0006919) | 110.00032 (.0033453) | 106.36174 (.0065807) |
| 74 | 39.31732 (.0003094) | 43.29576 (.0016603) | 45.29376 (.0034298) |
| | 119.13737 (.0006906) | 111.23289 (.0033397) | 107.57535 (.0065702) |
| 75 | 40.04689 (.0003106) | 44.06447 (.0016658) | 46.08130 (.0034401) |
| | 120.40758 (.0006894) | 112.46388 (.0033342) | 108.78751 (.0065599) |
| 76 | 40.77826 (.0003118) | 44.83473 (.0016711) | 46.87027 (.0034502) |
| | 121.67598 (.0006882) | 113.69331 (.0033289) | 109.99825 (.0065498) |
| 77 | 41.51140 (.0003129) | 45.60652 (.0016764) | 47.66064 (.0034601) |
| | 122.94261 (.0006871) | 114.92125 (.0033236) | 111.20758 (.0065399) |
| 78 | 42.24628 (.0003141) | 46.37979 (.0016816) | 48.45239 (.0034699) |
| | 124.20752 (.0006859) | 116.14769 (.0033184) | 112.41556 (.0065301) |
| 79 | 42.98286 (.0003152) | 47.15454 (.0016866) | 49.24548 (.0034794) |
| | 125.47072 (.0006848) | 117.37267 (.0033134) | 113.62219 (.0065206) |
| 80 | 43.72112 (.0003163) | 47.93071 (.0016916) | 50.03990 (.0034888) |
| | 126.73225 (.0006837) | 118.59621 (.0033084) | 114.82750 (.0065112) |

TABLE I-2 (K=0)

| | .999 | .995 | .990 |
|---|---|---|---|
| 81 | 44.46100 (.0003174) | 48.70830 (.0016965) | 50.83560 (.0034980) |
| | 127.99216 (.0006826) | 119.81836 (.0033035) | 116.03152 (.0065020) |
| 82 | 45.20251 (.0003185) | 49.48727 (.0017013) | 51.63260 (.0035070) |
| | 129.25046 (.0006815) | 121.03914 (.0032987) | 117.23428 (.0064930) |
| 83 | 45.94559 (.0003196) | 50.26761 (.0017060) | 52.43083 (.0035159) |
| | 130.50717 (.0006804) | 122.25854 (.0032940) | 118.43581 (.0064841) |
| 84 | 46.69023 (.0003206) | 51.04927 (.0017106) | 53.23030 (.0035246) |
| | 131.76234 (.0006794) | 123.47664 (.0032894) | 119.63609 (.0064754) |
| 85 | 47.43639 (.0003216) | 51.83224 (.0017152) | 54.03096 (.0035331) |
| | 133.01601 (.0006784) | 124.69344 (.0032848) | 120.83519 (.0064669) |
| 86 | 48.18405 (.0003226) | 52.61652 (.0017196) | 54.83282 (.0035415) |
| | 134.26817 (.0006774) | 125.90894 (.0032804) | 122.03310 (.0064585) |
| 87 | 48.93317 (.0003236) | 53.40204 (.0017240) | 55.63583 (.0035498) |
| | 135.51886 (.0006764) | 127.12318 (.0032760) | 123.22986 (.0064502) |
| 88 | 49.68375 (.0003246) | 54.18880 (.0017284) | 56.43999 (.0035579) |
| | 136.76811 (.0006754) | 128.33620 (.0032716) | 124.42548 (.0064421) |
| 89 | 50.43575 (.0003255) | 54.97679 (.0017326) | 57.24527 (.0035659) |
| | 138.01596 (.0006745) | 129.54800 (.0032674) | 125.61998 (.0064341) |
| 90 | 51.18913 (.0003265) | 55.76598 (.0017368) | 58.05165 (.0035738) |
| | 139.26241 (.0006735) | 130.75861 (.0032632) | 126.81339 (.0064262) |
| 91 | 51.94389 (.0003274) | 56.55634 (.0017409) | 58.85912 (.0035815) |
| | 140.50748 (.0006726) | 131.96803 (.0032591) | 128.00572 (.0064185) |
| 92 | 52.70001 (.0003283) | 57.34785 (.0017450) | 59.66766 (.0035891) |
| | 141.75121 (.0006717) | 133.17630 (.0032550) | 129.19698 (.0064109) |
| 93 | 53.45746 (.0003292) | 58.14052 (.0017489) | 60.47725 (.0035966) |
| | 142.99361 (.0006708) | 134.38344 (.0032511) | 130.38719 (.0064034) |
| 94 | 54.21620 (.0003301) | 58.93431 (.0017529) | 61.28786 (.0036040) |
| | 144.23471 (.0006699) | 135.58946 (.0032471) | 131.57639 (.0063960) |
| 95 | 54.97624 (.0003309) | 59.72920 (.0017567) | 62.09949 (.0036113) |
| | 145.47452 (.0006691) | 136.79437 (.0032433) | 132.76456 (.0063887) |
| 96 | 55.73753 (.0003318) | 60.52518 (.0017605) | 62.91212 (.0036184) |
| | 146.71307 (.0006682) | 137.99821 (.0032395) | 133.95175 (.0063816) |
| 97 | 56.50009 (.0003326) | 61.32222 (.0017643) | 63.72574 (.0036254) |
| | 147.95038 (.0006674) | 139.20099 (.0032357) | 135.13794 (.0063746) |
| 98 | 57.26387 (.0003334) | 62.12033 (.0017680) | 64.54033 (.0036324) |
| | 149.18646 (.0006666) | 140.40271 (.0032320) | 136.32318 (.0063676) |
| 99 | 58.02885 (.0003343) | 62.91946 (.0017716) | 65.35587 (.0036392) |
| | 150.42134 (.0006657) | 141.60339 (.0032284) | 137.50748 (.0063608) |
| 100 | 58.79501 (.0003351) | 63.71962 (.0017752) | 66.17235 (.0036459) |
| | 151.65504 (.0006649) | 142.80307 (.0032248) | 138.69083 (.0063541) |

TABLE I-2 (K=0)

| | .975 | .950 | .900 |
|---|---|---|---|
| 1 | -------<br>------- | -------<br>------- | -------<br>------- |
| 2 | -------<br>------- | -------<br>------- | -------<br>------- |
| 3 | $0.\overset{3}{0}81456$(.0000062)<br>9.34895 (.0249938) | $0.\overset{2}{0}31593$(.0000472)<br>7.81683 (.0499528) | 0.012116(.0003534)<br>6.25947 (.0996466) |
| 4 | 0.042930(.0002271)<br>11.16481 (.0247729) | 0.084727(.0008724)<br>9.53034 (.0491276) | 0.16763 (.0033222)<br>7.86429 (.0966777) |
| 5 | 0.18585 (.0007413)<br>12.90776 (.0242587) | 0.29624 (.0022868)<br>11.19146 (.0477132) | 0.47639 (.0070379)<br>9.43382 (.0929621) |
| 6 | 0.42217 (.0013392)<br>14.59399 (.0236608) | 0.60700 (.0037174)<br>12.80244 (.0462826) | 0.88265 (.0103277)<br>10.95835 (.0896722) |
| 7 | 0.73118 (.0019155)<br>16.23137 (.0230845) | 0.98923 (.0049996)<br>14.36861 (.0450004) | 1.35469 (.0130813)<br>12.44235 (.0869187) |
| 8 | 1.09671 (.0024396)<br>17.82713 (.0225604) | 1.42500 (.0061157)<br>15.89659 (.0438843) | 1.87459 (.0153819)<br>13.89223 (.0846180) |
| 9 | 1.50729 (.0029077)<br>19.38750 (.0220923) | 1.90259 (.0070840)<br>17.39226 (.0429160) | 2.43126 (.0173247)<br>15.31356 (.0826753) |
| 10 | 1.95469 (.0033242)<br>20.91756 (.0216758) | 2.41392 (.0079284)<br>18.86043 (.0420716) | 3.01733 (.0189866)<br>16.71078 (.0810133) |
| 11 | 2.43285 (.0036957)<br>22.42142 (.0213043) | 2.95321 (.0086702)<br>20.30495 (.0413298) | 3.62759 (.0204261)<br>18.08742 (.0795739) |
| 12 | 2.93720 (.0040286)<br>23.90244 (.0209714) | 3.51616 (.0093270)<br>21.72890 (.0406730) | 4.25822 (.0216868)<br>19.44624 (.0783132) |
| 13 | 3.46416 (.0043284)<br>25.36333 (.0206716) | 4.09944 (.0099132)<br>23.13478 (.0400868) | 4.90631 (.0228020)<br>20.78947 (.0771980) |
| 14 | 4.01091 (.0045998)<br>26.80634 (.0204002) | 4.70046 (.0104399)<br>24.52469 (.0395601) | 5.56959 (.0237971)<br>22.11896 (.0762029) |
| 15 | 4.57517 (.0048468)<br>28.23334 (.0201532) | 5.31713 (.0109163)<br>25.90030 (.0390837) | 6.24623 (.0246918)<br>23.43616 (.0753082) |
| 16 | 5.15505 (.0050727)<br>29.64592 (.0199273) | 5.94773 (.0113496)<br>27.26305 (.0386504) | 6.93475 (.0255017)<br>24.74234 (.0744983) |
| 17 | 5.74898 (.0052803)<br>31.04541 (.0197197) | 6.59084 (.0117460)<br>28.61414 (.0382540) | 7.63393 (.0262393)<br>26.03856 (.0737607) |
| 18 | 6.35564 (.0054717)<br>32.43295 (.0195283) | 7.24527 (.0121101)<br>29.95464 (.0378899) | 8.34274 (.0269145)<br>27.32573 (.0730854) |
| 19 | 6.97391 (.0056489)<br>33.80954 (.0193511) | 7.91000 (.0124462)<br>31.28540 (.0375538) | 9.06031 (.0275358)<br>28.60460 (.0724642) |
| 20 | 7.60282 (.0058136)<br>35.17606 (.0191864) | 8.58416 (.0127576)<br>32.60722 (.0372424) | 9.78588 (.0281097)<br>29.87585 (.0718902) |

TABLE I-2 (K=0)

| | .975 | .950 | .900 |
|---|---|---|---|
| 21 | 8.24151 (.0059672) | 9.26700 (.0130471) | 10.51882 (.0286420) |
| | 36.53325 (.0190328) | 33.92079 (.0369528) | 31.14008 (.0713579) |
| 22 | 8.88926 (.0061108) | 9.95786 (.0133172) | 11.25856 (.0291375) |
| | 37.88177 (.0188892) | 35.22670 (.0366828) | 32.39778 (.0708624) |
| 23 | 9.54542 (.0062454) | 10.65616 (.0135700) | 12.00459 (.0296002) |
| | 39.22224 (.0187546) | 36.52547 (.0364300) | 33.64943 (.0703997) |
| 24 | 10.20941 (.0063719) | 11.36137 (.0138071) | 12.75649 (.0300336) |
| | 40.55519 (.0186281) | 37.81757 (.0361929) | 34.89543 (.0699664) |
| 25 | 10.88072 (.0064911) | 12.07306 (.0140301) | 13.51387 (.0304406) |
| | 41.88107 (.0185089) | 39.10344 (.0359699) | 36.13612 (.0695594) |
| 26 | 11.55889 (.0066037) | 12.79080 (.0142404) | 14.27637 (.0308237) |
| | 43.20032 (.0183963) | 40.38345 (.0357596) | 37.37186 (.0691763) |
| 27 | 12.24351 (.0067103) | 13.51423 (.0144391) | 15.04369 (.0311852) |
| | 44.51332 (.0182897) | 41.65796 (.0355609) | 38.60292 (.0688148) |
| 28 | 12.93420 (.0068113) | 14.24302 (.0146272) | 15.81554 (.0315271) |
| | 45.82042 (.0181887) | 42.92725 (.0353728) | 39.82957 (.0684729) |
| 29 | 13.63063 (.0069072) | 14.97687 (.0148057) | 16.59166 (.0318511) |
| | 47.12196 (.0180928) | 44.19164 (.0351943) | 41.05206 (.0681489) |
| 30 | 14.33249 (.0069985) | 15.71551 (.0149753) | 17.37183 (.0321586) |
| | 48.41820 (.0180015) | 45.45134 (.0350247) | 42.27060 (.0678414) |
| 31 | 15.03949 (.0070855) | 16.45868 (.0151368) | 18.15582 (.0324511) |
| | 49.70943 (.0179145) | 46.70663 (.0348632) | 43.48540 (.0675489) |
| 32 | 15.75138 (.0071685) | 17.20616 (.0152907) | 18.94347 (.0327296) |
| | 50.99588 (.0178315) | 47.95773 (.0347093) | 44.69662 (.0672703) |
| 33 | 16.46791 (.0072478) | 17.95773 (.0154376) | 19.73457 (.0329954) |
| | 52.27779 (.0177522) | 49.20480 (.0345624) | 45.90448 (.0670046) |
| 34 | 17.18889 (.0073237) | 18.71321 (.0155781) | 20.52896 (.0332492) |
| | 53.55536 (.0176763) | 50.44807 (.0344219) | 47.10910 (.0667508) |
| 35 | 17.91411 (.0073964) | 19.47243 (.0157126) | 21.32651 (.0334920) |
| | 54.82878 (.0176036) | 51.68768 (.0342874) | 48.31062 (.0665079) |
| 36 | 18.64336 (.0074662) | 20.23520 (.0158416) | 22.12706 (.0337246) |
| | 56.09825 (.0175338) | 52.92380 (.0341584) | 49.50920 (.0662754) |
| 37 | 19.37648 (.0075331) | 21.00136 (.0159653) | 22.93048 (.0339477) |
| | 57.36391 (.0174669) | 54.15657 (.0340347) | 50.70497 (.0660523) |
| 38 | 20.11331 (.0075975) | 21.77080 (.0160841) | 23.73665 (.0341618) |
| | 58.62593 (.0174025) | 55.38614 (.0339159) | 51.89801 (.0658382) |
| 39 | 20.85371 (.0076595) | 22.54337 (.0161984) | 24.54546 (.0343676) |
| | 59.88445 (.0173405) | 56.61264 (.0338016) | 53.08847 (.0656324) |
| 40 | 21.59752 (.0077191) | 23.31894 (.0163084) | 25.35681 (.0345656) |
| | 61.13960 (.0172809) | 57.83618 (.0336916) | 54.27641 (.0654344) |

TABLE I-2 (K=0)

| | .975 | .950 | .900 |
|---|---|---|---|
| 41 | 22.34462 (.0077766)<br>62.39153 (.0172234) | 24.09741 (.0164144)<br>59.05687 (.0335856) | 25.17059 (.0347562)<br>55.46198 (.0652437) |
| 42 | 23.09488 (.0078321)<br>63.64034 (.0171679) | 24.87865 (.0165166)<br>60.27483 (.0334834) | 26.98672 (.0349400)<br>56.64522 (.0650600) |
| 43 | 23.84818 (.0078857)<br>64.88615 (.0171143) | 25.66257 (.0166152)<br>61.49014 (.0333848) | 27.80511 (.0351173)<br>57.82623 (.0648826) |
| 44 | 24.60443 (.0079375)<br>66.12906 (.0170625) | 26.44908 (.0167105)<br>62.70291 (.0332895) | 28.62569 (.0352885)<br>59.00510 (.0647115) |
| 45 | 25.36351 (.0079876)<br>67.36917 (.0170124) | 27.23807 (.0168027)<br>63.91321 (.0331973) | 29.44835 (.0354539)<br>60.18188 (.0645460) |
| 46 | 26.12532 (.0080361)<br>68.60658 (.0169639) | 28.02948 (.0168918)<br>65.12115 (.0331082) | 30.27306 (.0356139)<br>61.35666 (.0643861) |
| 47 | 26.88979 (.0080830)<br>69.84138 (.0169170) | 28.82320 (.0169781)<br>66.32678 (.0330219) | 31.09970 (.0357687)<br>62.52951 (.0642313) |
| 48 | 27.65681 (.0081285)<br>71.07365 (.0168715) | 29.61919 (.0170617)<br>67.53020 (.0329383) | 31.92825 (.0359186)<br>63.70049 (.0640814) |
| 49 | 28.42633 (.0081727)<br>72.30348 (.0168273) | 30.41734 (.0171427)<br>68.73146 (.0328573) | 32.75864 (.0360638)<br>64.86964 (.0639361) |
| 50 | 29.19823 (.0082155)<br>73.53091 (.0167845) | 31.21761 (.0172213)<br>69.93063 (.0327787) | 33.59081 (.0362047)<br>66.03705 (.0637953) |
| 51 | 29.97247 (.0082571)<br>74.75607 (.0167429) | 32.01991 (.0172976)<br>71.12779 (.0327024) | 34.42470 (.0363414)<br>67.20274 (.0636585) |
| 52 | 30.74896 (.0082975)<br>75.97897 (.0167025) | 32.82420 (.0173717)<br>72.32300 (.0326283) | 35.26025 (.0364741)<br>68.36679 (.0635259) |
| 53 | 31.52765 (.0083367)<br>77.19972 (.0166633) | 33.63043 (.0174437)<br>73.51628 (.0325563) | 36.09743 (.0366030)<br>69.52924 (.0633969) |
| 54 | 32.30846 (.0083749)<br>78.41835 (.0166251) | 34.43851 (.0175137)<br>74.70772 (.0324863) | 36.93617 (.0367284)<br>70.69011 (.0632716) |
| 55 | 33.09134 (.0084121)<br>79.63495 (.0165879) | 35.24841 (.0175818)<br>75.89737 (.0324182) | 37.77646 (.0368502)<br>71.84947 (.0631498) |
| 56 | 33.87624 (.0084482)<br>80.84955 (.0165518) | 36.06007 (.0176481)<br>77.08525 (.0323519) | 38.61823 (.0369688)<br>73.00737 (.0630312) |
| 57 | 34.66309 (.0084835)<br>82.06219 (.0165165) | 36.87346 (.0177126)<br>78.27145 (.0322874) | 39.46144 (.0370842)<br>74.16382 (.0629158) |
| 58 | 35.45183 (.0085178)<br>83.27298 (.0164822) | 37.68852 (.0177755)<br>79.45599 (.0322245) | 40.30606 (.0371966)<br>75.31888 (.0628034) |
| 59 | 36.24243 (.0085513)<br>84.48192 (.0164487) | 38.50520 (.0178367)<br>80.63892 (.0321633) | 41.15204 (.0373061)<br>76.47260 (.0626939) |
| 60 | 37.03485 (.0085839)<br>85.68906 (.0164161) | 39.32346 (.0178965)<br>81.82027 (.0321035) | 41.99936 (.0374129)<br>77.62497 (.0625871) |

TABLE I-2 (K=0)

| | .975 | .950 | .900 |
|---|---|---|---|
| 61 | 37.82901 (.0086157) | 40.14326 (.0179547) | 42.84799 (.0375170) |
| | 86.89447 (.0163843) | 83.00009 (.0320453) | 78.77606 (.0624830) |
| 62 | 38.62491 (.0086468) | 40.96458 (.0180116) | 43.69788 (.0376185) |
| | 88.09818 (.0163532) | 84.17842 (.0319884) | 79.92589 (.0623815) |
| 63 | 39.42245 (.0086772) | 41.78737 (.0180671) | 44.54900 (.0377177) |
| | 89.30022 (.0163228) | 85.35530 (.0319329) | 81.07451 (.0622823) |
| 64 | 40.22165 (.0087068) | 42.61157 (.0181213) | 45.40134 (.0378145) |
| | 90.50064 (.0162932) | 86.53075 (.0318787) | 82.22191 (.0621855) |
| 65 | 41.02243 (.0087357) | 43.43719 (.0181742) | 46.25484 (.0379090) |
| | 91.69948 (.0162642) | 87.70482 (.0318258) | 83.36815 (.0620910) |
| 66 | 41.82477 (.0087641) | 44.26418 (.0182260) | 47.10950 (.0380014) |
| | 92.89677 (.0162359) | 88.87753 (.0317740) | 84.51326 (.0619986) |
| 67 | 42.62863 (.0087917) | 45.09248 (.0182766) | 47.96527 (.0380917) |
| | 94.09256 (.0162083) | 90.04890 (.0317234) | 85.65724 (.0619083) |
| 68 | 43.43398 (.0088188) | 45.92209 (.0183260) | 48.82214 (.0381799) |
| | 95.28688 (.0161812) | 91.21901 (.0316740) | 86.80014 (.0618201) |
| 69 | 44.24077 (.0088453) | 46.75298 (.0183744) | 49.68010 (.0382663) |
| | 96.47974 (.0161547) | 92.38783 (.0316256) | 87.94197 (.0617337) |
| 70 | 45.04900 (.0088712) | 47.58511 (.0184218) | 50.53909 (.0383508) |
| | 97.67120 (.0161288) | 93.55542 (.0315782) | 89.08275 (.0616492) |
| 71 | 45.85860 (.0088966) | 48.41847 (.0184681) | 51.39911 (.0384335) |
| | 98.86128 (.0161034) | 94.72180 (.0315319) | 90.22252 (.0615665) |
| 72 | 46.66957 (.0089215) | 49.25302 (.0185135) | 52.26013 (.0385144) |
| | 100.05000 (.0160785) | 95.88699 (.0314865) | 91.36130 (.0614856) |
| 73 | 47.48187 (.0089458) | 50.08873 (.0185580) | 53.12215 (.0385937) |
| | 101.23741 (.0160542) | 97.05103 (.0314420) | 92.49908 (.0614063) |
| 74 | 48.29547 (.0089697) | 50.92558 (.0186016) | 53.98511 (.0386713) |
| | 102.42352 (.0160303) | 98.21393 (.0313984) | 93.63593 (.0613287) |
| 75 | 49.11035 (.0089931) | 51.76357 (.0186442) | 54.84903 (.0387474) |
| | 103.60837 (.0160069) | 99.37570 (.0313558) | 94.77184 (.0612526) |
| 76 | 49.92648 (.0090160) | 52.60263 (.0186861) | 55.71385 (.0388220) |
| | 104.79196 (.0159840) | 100.53641 (.0313139) | 95.90681 (.0611780) |
| 77 | 50.74384 (.0090385) | 53.44278 (.0187271) | 56.57959 (.0388951) |
| | 105.97435 (.0159615) | 101.69603 (.0312729) | 97.04089 (.0611049) |
| 78 | 51.56239 (.0090605) | 54.28398 (.0187674) | 57.44621 (.0389668) |
| | 107.15553 (.0159395) | 102.85460 (.0312326) | 98.17410 (.0610332) |
| 79 | 52.38213 (.0090822) | 55.12621 (.0188068) | 58.31371 (.0390371) |
| | 108.33554 (.0159178) | 104.01216 (.0311932) | 99.30644 (.0609628) |
| 80 | 53.20302 (.0091034) | 55.96945 (.0188456) | 59.18205 (.0391061) |
| | 109.51442 (.0158966) | 105.16870 (.0311544) | 100.43794 (.0608938) |

TABLE I-2 (K=0)

| | .975 | .950 | .900 |
|---|---|---|---|
| 81 | 54.02504 (.0091243) | 56.81369 (.0188836) | 60.05124 (.0391739) |
| | 110.69215 (.0158757) | 106.32426 (.0311164) | 101.56859 (.0608261) |
| 82 | 54.84818 (.0091447) | 57.65891 (.0189209) | 60.92123 (.0392403) |
| | 111.86879 (.0158553) | 107.47884 (.0310791) | 102.69844 (.0607597) |
| 83 | 55.67241 (.0091648) | 58.50508 (.0189576) | 61.79204 (.0393056) |
| | 113.04433 (.0158352) | 108.63248 (.0310424) | 103.82748 (.0606944) |
| 84 | 56.49771 (.0091846) | 59.35219 (.0189935) | 62.66364 (.0393697) |
| | 114.21881 (.0158154) | 109.78517 (.0310064) | 104.95573 (.0606303) |
| 85 | 57.32405 (.0092040) | 60.20023 (.0190289) | 63.53601 (.0394326) |
| | 115.39224 (.0157960) | 1[illegible].93695 (.0309711) | 106.08322 (.0605674) |
| 86 | 58.15144 (.0092230) | 61.04916 (.0190636) | 64.40915 (.0394945) |
| | 116.56465 (.0157770) | 112.08784 (.0309363) | 107.20995 (.0605055) |
| 87 | 58.97984 (.0092418) | 61.89899 (.0190978) | 65.28302 (.0395552) |
| | 117.73604 (.0157582) | 113.23784 (.0309022) | 108.33592 (.0604448) |
| 88 | 59.80923 (.0092602) | 62.74969 (.0191314) | 66.15764 (.0396150) |
| | 118.90645 (.0157398) | 114.38696 (.0308686) | 109.46118 (.0603850) |
| 89 | 60.63962 (.0092783) | 63.60126 (.0191643) | 67.03299 (.0396737) |
| | 120.07587 (.0157217) | 115.53523 (.0308357) | 110.58571 (.0603263) |
| 90 | 61.47095 (.0092961) | 64.45366 (.0191968) | 67.90903 (.0397314) |
| | 121.24434 (.0157039) | 116.68265 (.0308032) | 111.70953 (.0602686) |
| 91 | 62.30324 (.0093136) | 65.30690 (.0192287) | 68.78578 (.0397882) |
| | 122.41187 (.0156864) | 117.82925 (.0307713) | 112.83266 (.0602118) |
| 92 | 63.13646 (.0093308) | 66.16095 (.0192601) | 69.66322 (.0398440) |
| | 123.57846 (.0156692) | 118.97505 (.0307399) | 113.95509 (.0601560) |
| 93 | 63.97060 (.0093478) | 67.01581 (.0192910) | 70.54132 (.0398990) |
| | 124.74414 (.0156522) | 120.12004 (.0307090) | 115.07687 (.0601010) |
| 94 | 64.80563 (.0093645) | 67.87146 (.0193214) | 71.42010 (.0399530) |
| | 125.90894 (.0156355) | 121.26424 (.0306786) | 116.19798 (.0600470) |
| 95 | 65.64156 (.0093809) | 68.72789 (.0193513) | 72.29953 (.0400062) |
| | 127.07283 (.0156191) | 122.40767 (.0306487) | 117.31844 (.0599938) |
| 96 | 66.47836 (.0093971) | 69.58507 (.0193807) | 73.17960 (.0400586) |
| | 128.23587 (.0156029) | 123.55034 (.0306193) | 118.43825 (.0599414) |
| 97 | 67.31601 (.0094130) | 70.44301 (.0194097) | 74.06030 (.0401101) |
| | 129.39806 (.0155870) | 124.69225 (.0305903) | 119.55743 (.0598899) |
| 98 | 68.15451 (.0094287) | 71.30170 (.0194382) | 74.94162 (.0401609) |
| | 130.55939 (.0155713) | 125.83344 (.0305618) | 120.67601 (.0598391) |
| 99 | 68.99385 (.0094441) | 72.16110 (.0194663) | 75.82356 (.0402109) |
| | 131.71989 (.0155559) | 126.97389 (.0305336) | 121.79396 (.0597891) |
| 100 | 69.83400 (.0094593) | 73.02122 (.0194940) | 76.70612 (.0402601) |
| | 132.87959 (.0155407) | 128.11362 (.0305060) | 122.91132 (.0597399) |

TABLE I-2 (K=C)

| | .800 | .700 | .600 |
|---|---|---|---|
| 1 | ------- | ------- | ------- |
| | ------- | ------- | ------- |
| 2 | ------- | ------- | ------- |
| | ------- | ------- | ------- |
| 3 | 0.045737(.0025661) | 0.098691(.0080060) | 0.16992 (.0177070) |
| | 4.67221 (.1974339) | 3.73117 (.2919940) | 3.06115 (.3822929) |
| 4 | 0.33460 (.0125275) | 0.50616 (.0271014) | 0.68442 (.0467643) |
| | 6.16058 (.1874725) | 5.14339 (.2728986) | 4.41096 (.3532357) |
| 5 | 0.77886 (.0216381) | 1.05233 (.0417644) | 1.31535 (.0666561) |
| | 7.62157 (.1783618) | 6.52728 (.2582356) | 5.73038 (.3333438) |
| 6 | 1.30776 (.0287720) | 1.66922 (.0525276) | 2.00400 (.0806701) |
| | 9.04203 (.1712280) | 7.87411 (.2474723) | 7.01644 (.3193299) |
| 7 | 1.89115 (.0343817) | 2.33096 (.0607204) | 2.72911 (.0911195) |
| | 10.42741 (.1656183) | 9.19031 (.2392796) | 8.27608 (.3088804) |
| 8 | 2.51341 (.0388974) | 3.02449 (.0671884) | 3.48006 (.0992681) |
| | 11.78404 (.1611025) | 10.48199 (.2328115) | 9.51495 (.3007319) |
| 9 | 3.16523 (.0426180) | 3.74213 (.0724495) | 4.25060 (.1058421) |
| | 13.11703 (.1573820) | 11.75375 (.2275504) | 10.73714 (.2941579) |
| 10 | 3.84053 (.0457457) | 4.47891 (.0768319) | 5.03669 (.1112863) |
| | 14.43028 (.1542543) | 13.00902 (.2231681) | 11.94560 (.2887136) |
| 11 | 4.53507 (.0484196) | 5.23134 (.0805531) | 5.83551 (.1158893) |
| | 15.72679 (.1515804) | 14.25039 (.2194469) | 13.14255 (.2841107) |
| 12 | 5.24578 (.0507381) | 5.99692 (.0837628) | 6.64501 (.1198461) |
| | 17.00891 (.1492618) | 15.47985 (.2162372) | 14.32966 (.2801538) |
| 13 | 5.97030 (.0527728) | 6.77373 (.0865675) | 7.46365 (.1232946) |
| | 18.27852 (.1472272) | 16.69897 (.2134324) | 15.50826 (.2767054) |
| 14 | 6.70682 (.0545764) | 7.56029 (.0890455) | 8.29021 (.1263345) |
| | 19.53711 (.1454235) | 17.90904 (.2109545) | 16.67940 (.2736654) |
| 15 | 7.45388 (.0561895) | 8.35541 (.0912552) | 9.12373 (.1290405) |
| | 20.78589 (.1438105) | 19.11107 (.2087448) | 17.84392 (.2709594) |
| 16 | 8.21032 (.0576430) | 9.15814 (.0932416) | 9.96344 (.1314693) |
| | 22.02589 (.1423570) | 20.30588 (.2067583) | 19.00255 (.2685306) |
| 17 | 8.97514 (.0589616) | 9.96768 (.0950399) | 10.80868 (.1336652) |
| | 23.25800 (.1410384) | 21.49422 (.2049600) | 20.15585 (.2663348) |
| 18 | 9.74754 (.0601647) | 10.78337 (.0966779) | 11.65893 (.1356630) |
| | 24.48288 (.1398352) | 22.67667 (.2033221) | 21.30435 (.2643369) |
| 19 | 10.52683 (.0612683) | 11.60465 (.0981780) | 12.51372 (.1374909) |
| | 25.70122 (.1387317) | 23.85374 (.2018219) | 22.44844 (.2625090) |
| 20 | 11.31241 (.0622853) | 12.43104 (.0995585) | 13.37266 (.1391715) |
| | 26.91351 (.1377147) | 25.02591 (.2004415) | 23.58852 (.2608284) |

TABLE I-2 (K=0)

| | .800 | .700 | .600 |
|---|---|---|---|
| 21 | 12.10378 (.0632263)<br>28.12024 (.1367736) | 13.26212 (.1008344)<br>26.19353 (.1991655) | 14.23542 (.1407237)<br>24.72487 (.2592762) |
| 22 | 12.90048 (.0641005)<br>29.32184 (.1358994) | 14.09753 (.1020184)<br>27.35695 (.1979816) | 15.10169 (.1421630)<br>25.85780 (.2578370) |
| 23 | 13.70213 (.0649153)<br>30.51863 (.1350846) | 14.93695 (.1031208)<br>28.51646 (.1968791) | 15.97123 (.1435023)<br>26.98755 (.2564976) |
| 24 | 14.50838 (.0656772)<br>31.71098 (.1343228) | 15.78010 (.1041507)<br>29.67235 (.1958492) | 16.84380 (.1447528)<br>28.11432 (.2552471) |
| 25 | 15.31893 (.0663916)<br>32.89915 (.1336084) | 16.62672 (.1051157)<br>30.82484 (.1948843) | 17.71919 (.1459239)<br>29.23833 (.2540761) |
| 26 | 16.13348 (.0670632)<br>34.08342 (.1329367) | 17.47662 (.1060222)<br>31.97414 (.1939777) | 18.59724 (.1470236)<br>30.35973 (.2529764) |
| 27 | 16.95183 (.0676962)<br>35.26399 (.1323037) | 18.32957 (.1068760)<br>33.12044 (.1931239) | 19.47778 (.1480588)<br>31.47868 (.2519411) |
| 28 | 17.77371 (.0682941)<br>36.44110 (.1317058) | 19.18539 (.1076820)<br>34.26393 (.1923179) | 20.36066 (.1490357)<br>32.59534 (.2509642) |
| 29 | 18.59895 (.0688601)<br>37.61493 (.1311399) | 20.04395 (.1084445)<br>35.40474 (.1915554) | 21.24574 (.1499595)<br>33.70982 (.2500405) |
| 30 | 19.42737 (.0693967)<br>38.78566 (.1306032) | 20.90506 (.1091672)<br>36.54305 (.1908327) | 22.13292 (.1508349)<br>34.82225 (.2491651) |
| 31 | 20.25877 (.0699067)<br>39.95345 (.1300933) | 21.76862 (.1098536)<br>37.67894 (.1901464) | 23.02206 (.1516658)<br>35.93274 (.2483341) |
| 32 | 21.09305 (.0703920)<br>41.11844 (.1296080) | 22.63448 (.1105064)<br>38.81258 (.1894935) | 23.91309 (.1524560)<br>37.04135 (.2475439) |
| 33 | 21.93001 (.0708545)<br>42.28078 (.1291454) | 23.50252 (.1111284)<br>39.94405 (.1888715) | 24.80591 (.1532088)<br>38.14822 (.2467911) |
| 34 | 22.76956 (.0712961)<br>43.44057 (.1287038) | 24.37267 (.1117220)<br>41.07346 (.1882780) | 25.70042 (.1539269)<br>39.25340 (.2460731) |
| 35 | 23.61157 (.0717182)<br>44.59795 (.1282818) | 25.24481 (.1122892)<br>42.20088 (.1877108) | 26.59656 (.1546129)<br>40.35698 (.2453870) |
| 36 | 24.45593 (.0721222)<br>45.75299 (.1278777) | 26.11885 (.1128319)<br>43.32645 (.1871681) | 27.49423 (.1552692)<br>41.45903 (.2447308) |
| 37 | 25.30255 (.0725094)<br>46.90584 (.1274905) | 26.99472 (.1133518)<br>44.45021 (.1866481) | 28.39340 (.1558978)<br>42.55962 (.2441021) |
| 38 | 26.15132 (.0728810)<br>48.05655 (.1271190) | 27.87233 (.1138506)<br>45.57224 (.1861494) | 29.29398 (.1565006)<br>43.65880 (.2434993) |
| 39 | 27.00215 (.0732378)<br>49.20523 (.1267621) | 28.75162 (.1143295)<br>46.69263 (.1856704) | 30.19592 (.1570795)<br>44.75662 (.2429205) |
| 40 | 27.85497 (.0735810)<br>50.35196 (.1264189) | 29.63251 (.1147898)<br>47.81142 (.1852100) | 31.09917 (.1576357)<br>45.85316 (.2423642) |

TABLE I-2 (K=0)

| | .800 | .700 | .600 |
|---|---|---|---|
| 41 | 28.70970 (.0739112) | 30.51495 (.1152329) | 32.00368 (.1581710) |
| | 51.49680 (.1260887) | 48.92868 (.1847671) | 46.94846 (.2418290) |
| 42 | 29.56625 (.0742295) | 31.39888 (.1156596) | 32.90938 (.1586864) |
| | 52.63982 (.1257704) | 50.04446 (.1843404) | 48.04256 (.2413135) |
| 43 | 30.42458 (.0745363) | 32.28423 (.1160710) | 33.81625 (.1591833) |
| | 53.78110 (.1254636) | 51.15884 (.1839290) | 49.13550 (.2408167) |
| 44 | 31.28461 (.0748325) | 33.17096 (.1164680) | 34.72424 (.1596626) |
| | 54.92070 (.1251674) | 52.27185 (.1835320) | 50.22733 (.2403373) |
| 45 | 32.14627 (.0751185) | 34.05902 (.1168513) | 35.63333 (.1601254) |
| | 56.05867 (.1248814) | 53.38354 (.1831487) | 51.31808 (.2398745) |
| 46 | 33.00952 (.0753950) | 34.94836 (.1172218) | 36.54344 (.1605727) |
| | 57.19507 (.1246049) | 54.49396 (.1827782) | 52.40781 (.2394272) |
| 47 | 33.87431 (.0756626) | 35.83896 (.1175801) | 37.45457 (.1610053) |
| | 58.32996 (.1243374) | 55.60315 (.1824198) | 53.49654 (.2389947) |
| 48 | 34.74057 (.0759215) | 36.73074 (.1179270) | 38.36667 (.1614239) |
| | 59.46338 (.1240784) | 56.71114 (.1820730) | 54.58427 (.2385760) |
| 49 | 35.60828 (.0761725) | 37.62370 (.1182629) | 39.27972 (.1618294) |
| | 60.59537 (.1238275) | 57.81799 (.1817371) | 55.67110 (.2381706) |
| 50 | 36.47737 (.0764157) | 38.51778 (.1185885) | 40.19368 (.1622223) |
| | 61.72598 (.1235843) | 58.92371 (.1814114) | 56.75700 (.2377777) |
| 51 | 37.34781 (.0766516) | 39.41295 (.1189043) | 41.10852 (.1626033) |
| | 62.85527 (.1233484) | 60.02835 (.1810957) | 57.84203 (.2373967) |
| 52 | 38.21956 (.0768805) | 40.30916 (.1192108) | 42.02423 (.1629730) |
| | 63.98326 (.1231194) | 61.13196 (.1807892) | 58.92619 (.2370269) |
| 53 | 39.09256 (.0771029) | 41.20642 (.1195083) | 42.94078 (.1633320) |
| | 65.11000 (.1228970) | 62.23453 (.1804916) | 60.00954 (.2366679) |
| 54 | 39.96680 (.0773190) | 42.10466 (.1197975) | 43.85812 (.1636807) |
| | 66.23552 (.1226810) | 63.33612 (.1802025) | 61.09207 (.2363192) |
| 55 | 40.84224 (.0775291) | 43.00388 (.1200785) | 44.77626 (.1640197) |
| | 67.35985 (.1224709) | 64.43674 (.1799214) | 62.17383 (.2359802) |
| 56 | 41.71883 (.0777334) | 43.90402 (.1203519) | 45.69516 (.1643494) |
| | 68.48302 (.1222665) | 65.53644 (.1796481) | 63.25482 (.2356505) |
| 57 | 42.59656 (.0779323) | 44.80510 (.1206178) | 46.61481 (.1646701) |
| | 69.60507 (.1220677) | 66.63522 (.1793821) | 64.33508 (.2353298) |
| 58 | 43.47539 (.0781259) | 45.70705 (.1208768) | 47.53517 (.1649824) |
| | 70.72604 (.1218740) | 67.73312 (.1791232) | 65.41461 (.2350175) |
| 59 | 44.35527 (.0783145) | 46.60988 (.1211290) | 48.45624 (.1652865) |
| | 71.84595 (.1216854) | 68.83015 (.1788709) | 66.49345 (.2347134) |
| 60 | 45.23621 (.0784984) | 47.51353 (.1213748) | 49.37799 (.1655828) |
| | 72.96481 (.1215016) | 69.92635 (.1786252) | 67.57161 (.2344171) |

TABLE I-2 (K=0)

| | .800 | .700 | .600 |
|---|---|---|---|
| 61 | 46.11816 (.0786776)<br>74.08267 (.1213223) | 48.41803 (.1216144)<br>71.02173 (.1783855) | 50.30042 (.1658717)<br>68.64909 (.2341282) |
| 62 | 47.00110 (.0788525)<br>75.19954 (.1211475) | 49.32332 (.1218481)<br>72.11632 (.1781518) | 51.22350 (.1661534)<br>69.72594 (.2338465) |
| 63 | 47.88501 (.0790231)<br>76.31546 (.1209769) | 50.22940 (.1220762)<br>73.21013 (.1779238) | 52.14720 (.1664283)<br>70.80214 (.2335716) |
| 64 | 48.76987 (.0791896)<br>77.43044 (.1208103) | 51.13623 (.1222987)<br>74.30318 (.1777012) | 53.07153 (.1666966)<br>71.87773 (.2333034) |
| 65 | 49.65562 (.0793523)<br>78.54451 (.1206477) | 52.04381 (.1225160)<br>75.39548 (.1774839) | 53.99646 (.1669585)<br>72.95273 (.2330414) |
| 66 | 50.54230 (.0795111)<br>79.65768 (.1204888) | 52.95212 (.1227283)<br>76.48706 (.1772716) | 54.92198 (.1672144)<br>74.02713 (.2327856) |
| 67 | 51.42984 (.0796664)<br>80.76999 (.1203336) | 53.86113 (.1229358)<br>77.57794 (.1770641) | 55.84808 (.1674644)<br>75.10095 (.2325355) |
| 68 | 52.31824 (.0798182)<br>81.88144 (.1201817) | 54.77084 (.1231386)<br>78.66812 (.1768613) | 56.77475 (.1677088)<br>76.17422 (.2322912) |
| 69 | 53.20747 (.0799666)<br>82.99205 (.1200333) | 55.68123 (.1233369)<br>79.75764 (.1766630) | 57.70197 (.1679477)<br>77.24693 (.2320523) |
| 70 | 54.09752 (.0801119)<br>84.10185 (.1198881) | 56.59227 (.1235309)<br>80.84650 (.1764691) | 58.62971 (.1681814)<br>78.31912 (.2318185) |
| 71 | 54.98837 (.0802540)<br>85.21086 (.1197460) | 57.50397 (.1237206)<br>81.93469 (.1762792) | 59.55800 (.1684101)<br>79.39078 (.2315899) |
| 72 | 55.88002 (.0803931)<br>86.31909 (.1196069) | 58.41631 (.1239064)<br>83.02228 (.1760935) | 60.48680 (.1686339)<br>80.46191 (.2313660) |
| 73 | 56.77243 (.0805293)<br>87.42654 (.1194707) | 59.32925 (.1240883)<br>84.10924 (.1759116) | 61.41609 (.1688530)<br>81.53255 (.2311469) |
| 74 | 57.66557 (.0806627)<br>88.53326 (.1193373) | 60.24281 (.1242665)<br>85.19559 (.1757334) | 62.34589 (.1690677)<br>82.60271 (.2309323) |
| 75 | 58.55948 (.0807934)<br>89.63925 (.1192065) | 61.15698 (.1244410)<br>86.28134 (.1755589) | 63.27617 (.1692779)<br>83.67236 (.2307221) |
| 76 | 59.45409 (.0809215)<br>90.74452 (.1190785) | 62.07172 (.1246120)<br>87.36652 (.1753879) | 64.20692 (.1694839)<br>84.74156 (.2305161) |
| 77 | 60.34940 (.0810470)<br>91.84909 (.1189529) | 62.98703 (.1247796)<br>88.45113 (.1752203) | 65.13814 (.1696858)<br>85.81029 (.2303141) |
| 78 | 61.24541 (.0811701)<br>92.95296 (.1188298) | 63.90239 (.1249440)<br>89.53519 (.1750559) | 66.06981 (.1698837)<br>86.87857 (.2301162) |
| 79 | 62.14211 (.0812908)<br>94.05615 (.1187091) | 64.81932 (.1251051)<br>90.61868 (.1748948) | 67.00192 (.1700779)<br>87.94640 (.2299221) |
| 80 | 63.03946 (.0814093)<br>95.15869 (.1185906) | 65.73628 (.1252633)<br>91.70164 (.1747367) | 67.93448 (.1702683)<br>89.01379 (.2297316) |

TABLE I-2 (K=0)

| | .800 | .700 | .600 |
|---|---|---|---|
| 81 | 63.93745 (.0815256)<br>96.26059 (.1184744) | 66.65376 (.1254184)<br>92.78409 (.1745815) | 68.86746 (.1704552)<br>90.08076 (.2295448) |
| 82 | 64.83611 (.0816396)<br>97.36183 (.1183603) | 67.57178 (.1255707)<br>93.86601 (.1744293) | 69.80087 (.1706385)<br>91.14731 (.2293615) |
| 83 | 65.73538 (.0817516)<br>98.46246 (.1182483) | 68.49028 (.1257201)<br>94.94743 (.1742798) | 70.73468 (.1708184)<br>92.21344 (.2291815) |
| 84 | 66.63527 (.0818616)<br>99.56247 (.1181384) | 69.40930 (.1258669)<br>96.02835 (.1741331) | 71.66891 (.1709952)<br>93.27917 (.2290047) |
| 85 | 67.53577 (.0819696)<br>100.66188 (.1180304) | 70.32880 (.1260110)<br>97.10878 (.1739889) | 72.60353 (.1711687)<br>94.34451 (.2288312) |
| 86 | 68.43686 (.0820757)<br>101.76070 (.1179243) | 71.24878 (.1261526)<br>98.18874 (.1738474) | 73.53854 (.1713392)<br>95.40945 (.2286608) |
| 87 | 69.33853 (.0821799)<br>102.85893 (.1178200) | 72.16925 (.1262916)<br>99.26822 (.1737083) | 74.47394 (.1715066)<br>96.47401 (.2284933) |
| 88 | 70.24078 (.0822824)<br>103.95659 (.1177176) | 73.09016 (.1264284)<br>100.34723 (.1735716) | 75.40971 (.1716712)<br>97.53819 (.2283287) |
| 89 | 71.14360 (.0823831)<br>105.05370 (.1176169) | 74.01155 (.1265627)<br>101.42580 (.1734372) | 76.34586 (.1718330)<br>98.60201 (.2281669) |
| 90 | 72.04697 (.0824821)<br>106.15024 (.1175179) | 74.93336 (.1266947)<br>102.50391 (.1733052) | 77.28236 (.1719920)<br>99.66547 (.2280080) |
| 91 | 72.95088 (.0825794)<br>107.24625 (.1174205) | 75.85564 (.1268246)<br>103.58159 (.1731753) | 78.21924 (.1721483)<br>100.72856 (.2278516) |
| 92 | 73.85533 (.0826752)<br>108.34171 (.1173248) | 76.77834 (.1269524)<br>104.65883 (.1730476) | 79.15646 (.1723021)<br>101.79129 (.2276978) |
| 93 | 74.76031 (.0827693)<br>109.43666 (.1172306) | 77.70148 (.1270780)<br>105.73564 (.1729220) | 80.09402 (.1724533)<br>102.85370 (.2275466) |
| 94 | 75.66580 (.0828620)<br>110.53108 (.1171379) | 78.62502 (.1272016)<br>106.81204 (.1727983) | 81.03194 (.1726021)<br>103.91576 (.2273978) |
| 95 | 76.57181 (.0829532)<br>111.62500 (.1170467) | 79.54898 (.1273232)<br>107.88803 (.1726767) | 81.97018 (.1727486)<br>104.97748 (.2272514) |
| 96 | 77.47832 (.0830430)<br>112.71841 (.1169569) | 80.47334 (.1274429)<br>108.96361 (.1725570) | 82.90875 (.1728926)<br>106.03886 (.2271073) |
| 97 | 78.38531 (.0831313)<br>113.81134 (.1168686) | 81.39812 (.1275607)<br>110.03880 (.1724392) | 83.84766 (.1730345)<br>107.09993 (.2269655) |
| 98 | 79.29280 (.0832183)<br>114.90379 (.1167817) | 82.32327 (.1276767)<br>111.11359 (.1723232) | 84.78687 (.1731741)<br>108.16068 (.2268258) |
| 99 | 80.20076 (.0833039)<br>115.99576 (.1166960) | 83.24883 (.1277909)<br>112.18799 (.1722090) | 85.72639 (.1733115)<br>109.22112 (.2266884) |
| 100 | 81.10919 (.0833883)<br>117.08725 (.1166117) | 84.17476 (.1279034)<br>113.26202 (.1720965) | 86.66624 (.1734470)<br>110.28123 (.2265530) |

TABLE II (K=4)

| | .999 | .995 | .990 |
|---|---|---|---|
| 1 | $0.\overset{5}{0}15708$(.0010000) | $0.\overset{4}{0}39270$(.0050000) | $0.\overset{3}{0}15709$(.0100000) |
| | 51.94217 (.0000000) | 41.62096 (.0000000) | 37.11862 (.0000000) |
| 2 | $0.\overset{2}{0}20010$(.0010000) | 0.010025(.0049999) | 0.020100(.0099995) |
| | 39.57080 (.0000000) | 32.32399 (.0000001) | 29.13622 (.0000005) |
| 3 | 0.024297(.0009999) | 0.071710(.0049988) | 0.11480 (.0099954) |
| | 36.61345 (.0000001) | 30.30365 (.0000012) | 27.51067 (.0000046) |
| 4 | 0.090791(.0009997) | 0.20689 (.0049953) | 0.29686 (.0099840) |
| | 35.98451 (.0000003) | 30.08438 (.0000047) | 27.46030 (.0000160) |
| 5 | 0.21014 (.0009992) | 0.41134 (.0049886) | 0.55343 (.0099640) |
| | 36.26622 (.0000008) | 30.56969 (.0000114) | 28.02689 (.0000360) |
| 6 | 0.38083 (.0009982) | 0.67468 (.0049787) | 0.87001 (.0099363) |
| | 36.98898 (.0000018) | 31.39635 (.0000213) | 28.89270 (.0000637) |
| 7 | 0.59793 (.0009969) | 0.98709 (.0049659) | 1.23502 (.0099019) |
| | 37.95383 (.0000031) | 32.41031 (.0000341) | 29.92284 (.0000981) |
| 8 | 0.85598 (.0009952) | 1.34059 (.0049507) | 1.63972 (.0098624) |
| | 39.06271 (.0000048) | 33.53558 (.0000493) | 31.05063 (.0001376) |
| 9 | 1.15000 (.0009932) | 1.72885 (.0049335) | 2.07755 (.0098190) |
| | 40.26213 (.0000068) | 34.73068 (.0000665) | 32.23962 (.0001810) |
| 10 | 1.47566 (.0009909) | 2.14691 (.0049149) | 2.54346 (.0097728) |
| | 41.52066 (.0000091) | 35.97118 (.0000851) | 33.46835 (.0002272) |
| 11 | 1.82931 (.0009883) | 2.59080 (.0048950) | 3.03354 (.0097248) |
| | 42.81877 (.0000117) | 37.24188 (.0001050) | 34.72345 (.0002752) |
| 12 | 2.20787 (.0009856) | 3.05732 (.0048744) | 3.54467 (.0096756) |
| | 44.14378 (.0000144) | 38.53294 (.0001256) | 35.99622 (.0003244) |
| 13 | 2.60874 (.0009826) | 3.54389 (.0048532) | 4.07436 (.0096257) |
| | 45.48720 (.0000174) | 39.83772 (.0001468) | 37.28082 (.0003743) |
| 14 | 3.02971 (.0009796) | 4.04834 (.0048316) | 4.62056 (.0095756) |
| | 46.84312 (.0000204) | 41.15167 (.0001684) | 38.57324 (.0004244) |
| 15 | 3.46889 (.0009765) | 4.56887 (.0048099) | 5.18156 (.0095257) |
| | 48.20740 (.0000235) | 42.47153 (.0001901) | 39.87062 (.0004743) |
| 16 | 3.92468 (.0009732) | 5.10396 (.0047881) | 5.75594 (.0094761) |
| | 49.57706 (.0000268) | 43.79500 (.0002119) | 41.17093 (.0005239) |
| 17 | 4.39567 (.0009700) | 5.65233 (.0047663) | 6.34249 (.0094271) |
| | 50.94991 (.0000300) | 45.12041 (.0002337) | 42.47269 (.0005729) |
| 18 | 4.88064 (.0009666) | 6.21284 (.0047447) | 6.94017 (.0093788) |
| | 52.32434 (.0000334) | 46.44649 (.0002553) | 43.77480 (.0006212) |
| 19 | 5.37853 (.0009633) | 6.78454 (.0047233) | 7.54809 (.0093313) |
| | 53.69916 (.0000367) | 47.77234 (.0002767) | 45.07648 (.0006687) |
| 20 | 5.88839 (.0009600) | 7.36658 (.0047022) | 8.16546 (.0092847) |
| | 55.07347 (.0000400) | 49.09726 (.0002978) | 46.37709 (.0007153) |

TABLE II (K=4)

| | .999 | .995 | .990 |
|---|---|---|---|
| 21 | 6.40939 (.0009566) | 7.95822 (.0046813) | 8.79160 (.0092390) |
| | 56.44658 (.0000434) | 50.42075 (.0003187) | 47.67624 (.0007610) |
| 22 | 6.94080 (.0009533) | 8.55880 (.0046607) | 9.42590 (.0091943) |
| | 57.81799 (.0000467) | 51.74243 (.0003392) | 48.97354 (.0008057) |
| 23 | 7.48194 (.0009500) | 9.16773 (.0046405) | 10.06784 (.0091506) |
| | 59.18732 (.0000500) | 53.06200 (.0003595) | 50.26880 (.0008494) |
| 24 | 8.03222 (.0009467) | 9.78450 (.0046207) | 10.71692 (.0091079) |
| | 60.55426 (.0000533) | 54.37927 (.0003793) | 51.56181 (.0008921) |
| 25 | 8.59111 (.0009434) | 10.40862 (.0046012) | 11.37271 (.0090662) |
| | 61.91862 (.0000566) | 55.69408 (.0003988) | 52.85248 (.0009338) |
| 26 | 9.15811 (.0009402) | 11.03968 (.0045821) | 12.03484 (.0090254) |
| | 63.28024 (.0000598) | 57.00632 (.0004179) | 54.14067 (.0009746) |
| 27 | 9.73278 (.0009370) | 11.67729 (.0045634) | 12.70295 (.0089857) |
| | 64.63901 (.0000630) | 58.31593 (.0004366) | 55.42636 (.0010143) |
| 28 | 10.31472 (.0009338) | 12.32111 (.0045450) | 13.37671 (.0089468) |
| | 65.99486 (.0000662) | 59.62285 (.0004550) | 56.70952 (.0010532) |
| 29 | 10.90356 (.0009307) | 12.97080 (.0045271) | 14.05584 (.0089089) |
| | 67.34773 (.0000693) | 60.92708 (.0004729) | 57.99010 (.0010911) |
| 30 | 11.49896 (.0009276) | 13.62609 (.0045095) | 14.74007 (.0088719) |
| | 68.69760 (.0000724) | 62.22859 (.0004905) | 59.26814 (.0011281) |
| 31 | 12.10060 (.0009246) | 14.28670 (.0044922) | 15.42915 (.0088358) |
| | 70.04448 (.0000754) | 63.52739 (.0005078) | 60.54362 (.0011642) |
| 32 | 12.70821 (.0009216) | 14.95238 (.0044753) | 16.12286 (.0088006) |
| | 71.38835 (.0000784) | 64.82350 (.0005247) | 61.81659 (.0011994) |
| 33 | 13.32151 (.0009186) | 15.62290 (.0044588) | 16.82100 (.0087662) |
| | 72.72923 (.0000814) | 66.11694 (.0005412) | 63.08705 (.0012338) |
| 34 | 13.94025 (.0009157) | 16.29807 (.0044426) | 17.52336 (.0087326) |
| | 74.06715 (.0000843) | 67.40775 (.0005574) | 64.35504 (.0012674) |
| 35 | 14.56421 (.0009129) | 16.97766 (.0044268) | 18.22978 (.0086998) |
| | 75.40215 (.0000871) | 68.69595 (.0005732) | 65.62059 (.0013002) |
| 36 | 15.19317 (.0009101) | 17.66151 (.0044113) | 18.94008 (.0086678) |
| | 76.73425 (.0000899) | 69.98160 (.0005887) | 66.88376 (.0013322) |
| 37 | 15.82693 (.0009073) | 18.34944 (.0043961) | 19.65410 (.0086365) |
| | 78.06351 (.0000927) | 71.26471 (.0006039) | 68.14456 (.0013635) |
| 38 | 16.46530 (.0009046) | 19.04131 (.0043812) | 20.37170 (.0086060) |
| | 79.38994 (.0000954) | 72.54535 (.0006188) | 69.40305 (.0013940) |
| 39 | 17.10812 (.0009019) | 19.73694 (.0043666) | 21.09277 (.0085761) |
| | 80.71361 (.0000981) | 73.82353 (.0006334) | 70.65924 (.0014239) |
| 40 | 17.75523 (.0008992) | 20.43622 (.0043524) | 21.81714 (.0085469) |
| | 82.03456 (.0001008) | 75.09933 (.0006476) | 71.91322 (.0014531) |

TABLE II (K=4)

| | .999 | .995 | .990 |
|---|---|---|---|
| 41 | 18.40645 (.0008966)<br>83.35284 (.0001034) | 21.13899 (.0043384)<br>76.37277 (.0006616) | 22.54472 (.0085184)<br>73.16501 (.0014816) |
| 42 | 19.06166 (.0008941)<br>84.66849 (.0001059) | 21.84517 (.0043247)<br>77.64392 (.0006753) | 23.27539 (.0084904)<br>74.41466 (.0015096) |
| 43 | 19.72070 (.0008916)<br>85.98157 (.0001084) | 22.55460 (.0043113)<br>78.91280 (.0006887) | 24.00905 (.0084631)<br>75.66219 (.0015369) |
| 44 | 20.38347 (.0008891)<br>87.29210 (.0001109) | 23.26718 (.0042981)<br>80.17944 (.0007019) | 24.74559 (.0084364)<br>76.90765 (.0015636) |
| 45 | 21.04982 (.0008866)<br>88.60016 (.0001134) | 23.98282 (.0042852)<br>81.44392 (.0007148) | 25.48491 (.0084103)<br>78.15111 (.0015897) |
| 46 | 21.71967 (.0008842)<br>89.90578 (.0001158) | 24.70142 (.0042726)<br>82.70628 (.0007274) | 26.22694 (.0083847)<br>79.39258 (.0016153) |
| 47 | 22.39287 (.0008819)<br>91.20900 (.0001181) | 25.42288 (.0042602)<br>83.96654 (.0007398) | 26.97159 (.0083597)<br>80.63211 (.0016403) |
| 48 | 23.06934 (.0008796)<br>92.50989 (.0001204) | 26.14713 (.0042480)<br>85.22476 (.0007520) | 27.71877 (.0083351)<br>81.86975 (.0016648) |
| 49 | 23.74898 (.0008773)<br>93.80847 (.0001227) | 26.87405 (.0042361)<br>86.48096 (.0007639) | 28.46841 (.0083111)<br>83.10551 (.0016889) |
| 50 | 24.43169 (.0008750)<br>95.10480 (.0001250) | 27.60359 (.0042244)<br>87.73520 (.0007756) | 29.22044 (.0082876)<br>84.33946 (.0017124) |
| 51 | 25.11739 (.0008728)<br>96.39893 (.0001272) | 28.33568 (.0042129)<br>88.98750 (.0007871) | 29.97479 (.0082645)<br>85.57161 (.0017355) |
| 52 | 25.80598 (.0008706)<br>97.69087 (.0001294) | 29.07022 (.0042017)<br>90.23793 (.0007983) | 30.73140 (.0082419)<br>86.80200 (.0017581) |
| 53 | 26.49741 (.0008685)<br>98.98068 (.0001315) | 29.80717 (.0041906)<br>91.48648 (.0008094) | 31.49020 (.0082198)<br>88.03069 (.0017802) |
| 54 | 27.19157 (.0008664)<br>100.26840 (.0001336) | 30.54645 (.0041798)<br>92.73323 (.0008202) | 32.25113 (.0081980)<br>89.25769 (.0018020) |
| 55 | 27.88840 (.0008643)<br>101.55408 (.0001357) | 31.28799 (.0041691)<br>93.97820 (.0008309) | 33.01414 (.0081767)<br>90.48303 (.0018233) |
| 56 | 28.58784 (.0008623)<br>102.83775 (.0001377) | 32.03177 (.0041586)<br>95.22141 (.0008414) | 33.77917 (.0081558)<br>91.70676 (.0018442) |
| 57 | 29.28981 (.0008603)<br>104.11945 (.0001397) | 32.77768 (.0041483)<br>96.46291 (.0008517) | 34.54617 (.0081353)<br>92.92889 (.0018647) |
| 58 | 29.99425 (.0008583)<br>105.39920 (.0001417) | 33.52570 (.0041382)<br>97.70273 (.0008618) | 35.31509 (.0081152)<br>94.14948 (.0018848) |
| 59 | 30.70110 (.0008563)<br>106.67706 (.0001437) | 34.27576 (.0041283)<br>98.94089 (.0008717) | 36.08589 (.0080954)<br>95.36853 (.0019046) |
| 60 | 31.41029 (.0008544)<br>107.95305 (.0001456) | 35.02782 (.0041186)<br>100.17744 (.0008814) | 36.85852 (.0080760)<br>96.58609 (.0019240) |

TABLE II (K=4)

| | .999 | .995 | .990 |
|---|---|---|---|
| 61 | 32.12178 (.0008525) | 35.78183 (.0041090) | 37.63292 (.0080570) |
| | 109.22720 (.0001475) | 101.41240 (.0008910) | 97.80217 (.0019430) |
| 62 | 32.83551 (.0008506) | 36.53773 (.0040995) | 38.40907 (.0080383) |
| | 110.49956 (.0001494) | 102.64580 (.0009005) | 99.01680 (.0019617) |
| 63 | 33.55142 (.0008488) | 37.29552 (.0040903) | 39.18692 (.0080199) |
| | 111.77016 (.0001512) | 103.87766 (.0009097) | 100.23003 (.0019801) |
| 64 | 34.26947 (.0008470) | 38.05510 (.0040811) | 39.96643 (.0080018) |
| | 113.03902 (.0001530) | 105.10802 (.0009188) | 101.44186 (.0019982) |
| 65 | 34.98961 (.0008452) | 38.81647 (.0040722) | 40.74756 (.0079841) |
| | 114.30617 (.0001548) | 106.33690 (.0009278) | 102.65233 (.0020159) |
| 66 | 35.71179 (.0008434) | 39.57956 (.0040634) | 41.53029 (.0079666) |
| | 115.57164 (.0001566) | 107.56435 (.0009366) | 103.86145 (.0020334) |
| 67 | 36.43596 (.0008417) | 40.34436 (.0040547) | 42.31456 (.0079495) |
| | 116.83548 (.0001583) | 108.79034 (.0009453) | 105.06926 (.0020505) |
| 68 | 37.16209 (.0008400) | 41.11082 (.0040461) | 43.10036 (.0079326) |
| | 118.09769 (.0001600) | 110.01495 (.0009539) | 106.27577 (.0020674) |
| 69 | 37.89014 (.0008383) | 41.87891 (.0040377) | 43.88765 (.0079160) |
| | 119.35832 (.0001617) | 111.23819 (.0009623) | 107.48102 (.0020840) |
| 70 | 38.62004 (.0008366) | 42.64859 (.0040295) | 44.67639 (.0078997) |
| | 120.61737 (.0001634) | 112.46007 (.0009705) | 108.68500 (.0021003) |
| 71 | 39.35179 (.0008350) | 43.41983 (.0040213) | 45.46657 (.0078837) |
| | 121.87491 (.0001650) | 113.68062 (.0009787) | 109.88777 (.0021163) |
| 72 | 40.08533 (.0008334) | 44.19260 (.0040133) | 46.25813 (.0078679) |
| | 123.13092 (.0001666) | 114.89986 (.0009867) | 111.08932 (.0021321) |
| 73 | 40.82062 (.0008318) | 44.96686 (.0040054) | 47.05107 (.0078523) |
| | 124.38545 (.0001682) | 116.11781 (.0009946) | 112.28969 (.0021477) |
| 74 | 41.55765 (.0008302) | 45.74260 (.0039976) | 47.84537 (.0078370) |
| | 125.63852 (.0001698) | 117.33452 (.0010024) | 113.48889 (.0021630) |
| 75 | 42.29636 (.0008287) | 46.51976 (.0039899) | 48.64096 (.0078220) |
| | 126.89014 (.0001713) | 118.54997 (.0010100) | 114.68695 (.0021780) |
| 76 | 43.03673 (.0008271) | 47.29834 (.0039824) | 49.43787 (.0078072) |
| | 128.14035 (.0001729) | 119.76419 (.0010176) | 115.88388 (.0021928) |
| 77 | 43.77873 (.0008256) | 48.07831 (.0039750) | 50.23602 (.0077926) |
| | 129.38918 (.0001744) | 120.97722 (.0010250) | 117.07968 (.0022074) |
| 78 | 44.52232 (.0008241) | 48.85963 (.0039676) | 51.03543 (.0077782) |
| | 130.63663 (.0001759) | 122.18906 (.0010324) | 118.27441 (.0022218) |
| 79 | 45.26749 (.0008226) | 49.64229 (.0039604) | 51.83606 (.0077640) |
| | 131.88272 (.0001774) | 123.39973 (.0010396) | 119.46805 (.0022360) |
| 80 | 46.01419 (.0008212) | 50.42625 (.0039533) | 52.63788 (.0077501) |
| | 133.12750 (.0001788) | 124.60927 (.0010467) | 120.66064 (.0022499) |

TABLE II (K=4)

| | .999 | | .995 | | .990 | |
|---|---|---|---|---|---|---|
| 81 | 46.76239 | (.0008198) | 51.21152 | (.0039463) | 53.44089 | (.0077364) |
| | 134.37097 | (.0001802) | 125.81767 | (.0010537) | 121.85219 | (.0022636) |
| 82 | 47.51208 | (.0008183) | 51.99803 | (.0039393) | 54.24504 | (.0077228) |
| | 135.61314 | (.0001817) | 127.02496 | (.0010607) | 123.04271 | (.0022772) |
| 83 | 48.26323 | (.0008169) | 52.78578 | (.0039325) | 55.05034 | (.0077095) |
| | 136.85405 | (.0001831) | 128.23116 | (.0010675) | 124.23221 | (.0022905) |
| 84 | 49.01581 | (.0008156) | 53.57475 | (.0039258) | 55.85675 | (.0076963) |
| | 138.09372 | (.0001844) | 129.43626 | (.0010742) | 125.42073 | (.0023037) |
| 85 | 49.76979 | (.0008142) | 54.36491 | (.0039191) | 56.66426 | (.0076833) |
| | 139.33214 | (.0001858) | 130.64032 | (.0010809) | 126.60826 | (.0023167) |
| 86 | 50.52516 | (.0008129) | 55.15627 | (.0039126) | 57.47284 | (.0076706) |
| | 140.56935 | (.0001871) | 131.84332 | (.0010874) | 127.79483 | (.0023294) |
| 87 | 51.28188 | (.0008115) | 55.94876 | (.0039061) | 58.28249 | (.0076580) |
| | 141.80537 | (.0001885) | 133.04530 | (.0010939) | 128.98045 | (.0023420) |
| 88 | 52.03995 | (.0008102) | 56.74240 | (.0038998) | 59.09317 | (.0076455) |
| | 143.04022 | (.0001898) | 134.24628 | (.0011002) | 130.16513 | (.0023545) |
| 89 | 52.79932 | (.0008089) | 57.53717 | (.0038935) | 59.90488 | (.0076333) |
| | 144.27391 | (.0001911) | 135.44623 | (.0011065) | 131.34889 | (.0023667) |
| 90 | 53.55998 | (.0008076) | 58.33302 | (.0038873) | 60.71761 | (.0076212) |
| | 145.50644 | (.0001924) | 136.64520 | (.0011127) | 132.53174 | (.0023788) |
| 91 | 54.32193 | (.0008064) | 59.12997 | (.0038811) | 61.53131 | (.0076092) |
| | 146.73785 | (.0001936) | 137.84320 | (.0011189) | 133.71368 | (.0023908) |
| 92 | 55.08511 | (.0008051) | 59.92798 | (.0038751) | 62.34601 | (.0075975) |
| | 147.96815 | (.0001949) | 139.04025 | (.0011249) | 134.89476 | (.0024025) |
| 93 | 55.84955 | (.0008039) | 60.72704 | (.0038691) | 63.16165 | (.0075859) |
| | 149.19736 | (.0001961) | 140.23634 | (.0011309) | 136.07495 | (.0024141) |
| 94 | 56.61517 | (.0008027) | 61.52713 | (.0038632) | 63.97826 | (.0075744) |
| | 150.42548 | (.0001973) | 141.43152 | (.0011368) | 137.25429 | (.0024256) |
| 95 | 57.38200 | (.0008014) | 62.32823 | (.0038574) | 64.79578 | (.0075631) |
| | 151.65253 | (.0001986) | 142.62576 | (.0011426) | 138.43277 | (.0024369) |
| 96 | 58.15001 | (.0008003) | 63.13034 | (.0038517) | 65.61423 | (.0075519) |
| | 152.87852 | (.0001997) | 143.81909 | (.0011483) | 139.61043 | (.0024481) |
| 97 | 58.91917 | (.0007991) | 63.93344 | (.0038460) | 66.43358 | (.0075409) |
| | 154.10349 | (.0002009) | 145.01154 | (.0011540) | 140.78725 | (.0024591) |
| 98 | 59.68948 | (.0007979) | 64.73750 | (.0038404) | 67.25381 | (.0075300) |
| | 155.32741 | (.0002021) | 146.20311 | (.0011596) | 141.96326 | (.0024700) |
| 99 | 60.46091 | (.0007968) | 65.54253 | (.0038348) | 68.07494 | (.0075193) |
| | 156.55034 | (.0002032) | 147.39380 | (.0011652) | 143.13847 | (.0024807) |
| 100 | 61.23344 | (.0007956) | 66.34850 | (.0038294) | 68.89691 | (.0075086) |
| | 157.77226 | (.0002044) | 148.58362 | (.0011706) | 144.31290 | (.0024914) |

TABLE II (K=4)

| | .975 | .950 | .900 |
|---|---|---|---|
| 1 | $0.0^{3}98207$(.0250000)<br>31.08910 (.0000000) | $0.0^{2}39321$(.0499997)<br>26.44485 (.0000003) | 0.015790(.0999967)<br>21.69176 (.0000032) |
| 2 | 0.050627(.0249959)<br>24.83224 (.0000041) | 0.10254 (.0499783)<br>21.48122 (.0000216) | 0.21045 (.0998771)<br>18.00771 (.0001229) |
| 3 | 0.21562 (.0249714)<br>23.71788 (.0000286) | 0.35125 (.0498810)<br>20.74382 (.0001190) | 0.58208 (.0994777)<br>17.63812 (.0005223) |
| 4 | 0.48353 (.0249157)<br>23.88216 (.0000843) | 0.70825 (.0496923)<br>21.06314 (.0003077) | 1.05608 (.0988235)<br>18.10623 (.0011765) |
| 5 | 0.82865 (.0248298)<br>24.54909 (.0001702) | 1.13925 (.0494285)<br>21.80006 (.0005715) | 1.59379 (.0980006)<br>18.90813 (.0019994) |
| 6 | 1.23188 (.0247195)<br>25.46056 (.0002805) | 1.62329 (.0491120)<br>22.74097 (.0008880) | 2.17505 (.0970838)<br>19.87390 (.0029162) |
| 7 | 1.68019 (.0245912)<br>26.50658 (.0004088) | 2.14732 (.0487619)<br>23.79436 (.0012381) | 2.78826 (.0961252)<br>20.93025 (.0038747) |
| 8 | 2.16448 (.0244506)<br>27.63263 (.0005494) | 2.70272 (.0483929)<br>24.91470 (.0016071) | 3.42620 (.0951586)<br>22.04050 (.0048414) |
| 9 | 2.67829 (.0243021)<br>28.80881 (.0006979) | 3.28355 (.0480154)<br>26.07692 (.0019846) | 4.08403 (.0942047)<br>23.18436 (.0057953) |
| 10 | 3.21682 (.0241490)<br>30.01743 (.0008510) | 3.88553 (.0476367)<br>27.26619 (.0023633) | 4.75836 (.0932758)<br>24.34979 (.0067241) |
| 11 | 3.77644 (.0239939)<br>31.24747 (.0010061) | 4.50546 (.0472617)<br>28.47322 (.0027383) | 5.44670 (.0923790)<br>25.52931 (.0076210) |
| 12 | 4.35432 (.0238386)<br>32.49178 (.0011614) | 5.14090 (.0468936)<br>29.69199 (.0031064) | 6.14716 (.0915176)<br>26.71803 (.0084824) |
| 13 | 4.94822 (.0236846)<br>33.74551 (.0013154) | 5.78992 (.0465346)<br>30.91843 (.0034654) | 6.85826 (.0906929)<br>27.91263 (.0093071) |
| 14 | 5.55631 (.0235327)<br>35.00537 (.0014673) | 6.45097 (.0461858)<br>32.14973 (.0038142) | 7.57882 (.0899048)<br>29.11089 (.0100952) |
| 15 | 6.17712 (.0233836)<br>36.26901 (.0016164) | 7.12280 (.0458479)<br>33.38397 (.0041521) | 8.30787 (.0891525)<br>30.31122 (.0108474) |
| 16 | 6.80939 (.0232377)<br>37.53473 (.0017623) | 7.80434 (.0455214)<br>34.61971 (.0044786) | 9.04460 (.0884347)<br>31.51247 (.0115653) |
| 17 | 7.45208 (.0230954)<br>38.80135 (.0019046) | 8.49473 (.0452061)<br>35.85593 (.0047939) | 9.78835 (.0877497)<br>32.71385 (.0122503) |
| 18 | 8.10430 (.0229567)<br>40.06793 (.0020433) | 9.19319 (.0449021)<br>37.09189 (.0050979) | 10.53853 (.0870959)<br>33.91478 (.0129041) |
| 19 | 8.76526 (.0228219)<br>41.33385 (.0021781) | 9.89909 (.0446089)<br>38.32706 (.0053911) | 11.29465 (.0864714)<br>35.11482 (.0135286) |
| 20 | 9.43431 (.0226909)<br>42.59862 (.0023091) | 10.61187 (.0443263)<br>39.56104 (.0056737) | 12.05627 (.0858746)<br>36.31367 (.0141254) |

TABLE II (K=4)

| | .975 | .950 | .900 |
|---|---|---|---|
| 21 | 10.11086 (.0225637) | 11.33103 (.0440539) | 12.82302 (.0853038) |
| | 43.86186 (.0024363) | 40.79352 (.0059461) | 37.51109 (.0146962) |
| 22 | 10.79439 (.0224402) | 12.05615 (.0437913) | 13.59456 (.0847574) |
| | 45.12335 (.0025598) | 42.02431 (.0062087) | 38.70694 (.0152426) |
| 23 | 11.48444 (.0223203) | 12.78682 (.0435381) | 14.37058 (.0842340) |
| | 46.38286 (.0026797) | 43.25325 (.0064619) | 39.90109 (.0157660) |
| 24 | 12.18061 (.0222040) | 13.52272 (.0432938) | 15.15082 (.0837322) |
| | 47.64027 (.0027960) | 44.48022 (.0067062) | 41.09348 (.0162678) |
| 25 | 12.88252 (.0220912) | 14.26353 (.0430580) | 15.93505 (.0832506) |
| | 48.89548 (.0029088) | 45.70517 (.0069420) | 42.28404 (.0167494) |
| 26 | 13.58986 (.0219817) | 15.00898 (.0428304) | 16.72302 (.0827881) |
| | 50.14844 (.0030183) | 46.92804 (.0071696) | 43.47278 (.0172119) |
| 27 | 14.30231 (.0218754) | 15.75881 (.0426105) | 17.51457 (.0823434) |
| | 51.39909 (.0031246) | 48.14882 (.0073895) | 44.65964 (.0176566) |
| 28 | 15.01961 (.0217723) | 16.51279 (.0423979) | 18.30949 (.0819156) |
| | 52.64743 (.0032277) | 49.36748 (.0076021) | 45.84465 (.0180844) |
| 29 | 15.74151 (.0216721) | 17.27072 (.0421924) | 19.10764 (.0815037) |
| | 53.89345 (.0033279) | 50.58403 (.0078076) | 47.02782 (.0184963) |
| 30 | 16.46777 (.0215748) | 18.03241 (.0419936) | 19.90884 (.0811067) |
| | 55.13715 (.0034252) | 51.79849 (.0080064) | 48.20915 (.0188932) |
| 31 | 17.19821 (.0214803) | 18.79767 (.0418011) | 20.71298 (.0807239) |
| | 56.37856 (.0035197) | 53.01088 (.0081989) | 49.38867 (.0192760) |
| 32 | 17.93262 (.0213884) | 19.56635 (.0416146) | 21.51991 (.0803545) |
| | 57.61769 (.0036116) | 54.22122 (.0083853) | 50.56641 (.0196455) |
| 33 | 18.67082 (.0212991) | 20.33829 (.0414340) | 22.32951 (.0799977) |
| | 58.85457 (.0037009) | 55.42953 (.0085660) | 51.74240 (.0200023) |
| 34 | 19.41266 (.0212123) | 21.11336 (.0412588) | 23.14168 (.0796528) |
| | 60.08923 (.0037877) | 56.63586 (.0087412) | 52.91667 (.0203471) |
| 35 | 20.15799 (.0211278) | 21.89142 (.0410889) | 23.95630 (.0793193) |
| | 61.32172 (.0038722) | 57.84024 (.0089111) | 54.08925 (.0206807) |
| 36 | 20.90663 (.0210456) | 22.67236 (.0409240) | 24.77328 (.0789965) |
| | 62.55206 (.0039544) | 59.04269 (.0090760) | 55.26016 (.0210035) |
| 37 | 21.65849 (.0209655) | 23.45607 (.0407639) | 25.59254 (.0786839) |
| | 63.78030 (.0040345) | 60.24327 (.0092361) | 56.42946 (.0213161) |
| 38 | 22.41344 (.0208875) | 24.24243 (.0406083) | 26.41400 (.0783809) |
| | 65.00647 (.0041125) | 61.44199 (.0093917) | 57.59715 (.0216191) |
| 39 | 23.17136 (.0208116) | 25.03134 (.0404571) | 27.23755 (.0780872) |
| | 66.23062 (.0041884) | 62.63890 (.0095429) | 58.76331 (.0219128) |
| 40 | 23.93211 (.0207375) | 25.82274 (.0403101) | 28.06314 (.0778022) |
| | 67.45277 (.0042625) | 63.83405 (.0096899) | 59.92793 (.0221978) |

TABLE II (K=4)

| | .975 | .950 | .900 |
|---|---|---|---|
| 41 | 24.69565 (.0206653)<br>68.67299 (.0043347) | 26.61650 (.0401671)<br>65.02745 (.0098329) | 28.89070 (.0775256)<br>61.09106 (.0224744) |
| 42 | 25.46182 (.0205949)<br>69.89128 (.0044051) | 27.41258 (.0400278)<br>66.21915 (.0099722) | 29.72015 (.0772569)<br>62.25275 (.0227431) |
| 43 | 26.23058 (.0205263)<br>71.10773 (.0044737) | 28.21088 (.0398923)<br>67.40919 (.0101077) | 30.55145 (.0769958)<br>63.41299 (.0230041) |
| 44 | 27.00182 (.0204592)<br>72.32233 (.0045408) | 29.01131 (.0397602)<br>68.59761 (.0102398) | 31.38452 (.0767420)<br>64.57185 (.0232580) |
| 45 | 27.77547 (.0203938)<br>73.53514 (.0046062) | 29.81384 (.0396315)<br>69.78444 (.0103685) | 32.21930 (.0764951)<br>65.72934 (.0235049) |
| 46 | 28.55144 (.0203299)<br>74.74620 (.0046701) | 30.61838 (.0395061)<br>70.96970 (.0104939) | 33.05576 (.0762548)<br>66.88550 (.0237452) |
| 47 | 29.32967 (.0202674)<br>75.95554 (.0047326) | 31.42487 (.0393837)<br>72.15343 (.0106163) | 33.89383 (.0760208)<br>68.04036 (.0239792) |
| 48 | 30.11009 (.0202064)<br>77.16319 (.0047936) | 32.23325 (.0392644)<br>73.33568 (.0107356) | 34.73347 (.0757929)<br>69.19394 (.0242071) |
| 49 | 30.89264 (.0201468)<br>78.36920 (.0048532) | 33.04347 (.0391478)<br>74.51645 (.0108522) | 35.57463 (.0755708)<br>70.34627 (.0244292) |
| 50 | 31.67725 (.0200885)<br>79.57359 (.0049115) | 33.85548 (.0390341)<br>75.69580 (.0109659) | 36.41728 (.0753543)<br>71.49738 (.0246457) |
| 51 | 32.46385 (.0200314)<br>80.77640 (.0049686) | 34.66924 (.0389230)<br>76.87375 (.0110770) | 37.26137 (.0751431)<br>72.64729 (.0248569) |
| 52 | 33.25240 (.0199756)<br>81.97766 (.0050244) | 35.48468 (.0388144)<br>78.05032 (.0111856) | 38.10686 (.0749370)<br>73.79604 (.0250629) |
| 53 | 34.04285 (.0199210)<br>83.17741 (.0050790) | 36.30176 (.0387083)<br>79.22556 (.0112917) | 38.95370 (.0747359)<br>74.94363 (.0252640) |
| 54 | 34.83514 (.0198675)<br>84.37567 (.0051325) | 37.12044 (.0386046)<br>80.39946 (.0113954) | 39.80188 (.0745395)<br>76.09010 (.0254604) |
| 55 | 35.62923 (.0198152)<br>85.57246 (.0051848) | 37.94066 (.0385031)<br>81.57208 (.0114969) | 40.65134 (.0743477)<br>77.23549 (.0256522) |
| 56 | 36.42505 (.0197639)<br>86.76784 (.0052361) | 38.76241 (.0384039)<br>82.74344 (.0115961) | 41.50206 (.0741603)<br>78.37979 (.0258397) |
| 57 | 37.22260 (.0197136)<br>87.96181 (.0052864) | 39.58563 (.0383067)<br>83.91356 (.0116933) | 42.35402 (.0739771)<br>79.52304 (.0260229) |
| 58 | 38.02179 (.0196644)<br>89.15440 (.0053356) | 40.41031 (.0382117)<br>85.08247 (.0117883) | 43.20717 (.0737979)<br>80.66525 (.0262020) |
| 59 | 38.82260 (.0196161)<br>90.34566 (.0053839) | 41.23639 (.0381186)<br>86.25018 (.0118814) | 44.06149 (.0736228)<br>81.80646 (.0263772) |
| 60 | 39.62498 (.0195688)<br>91.53560 (.0054312) | 42.06383 (.0380274)<br>87.41673 (.0119726) | 44.91695 (.0734513)<br>82.94666 (.0265487) |

TABLE II (K=4)

| | .975 | .950 | .900 |
|---|---|---|---|
| 61 | 40.42892 (.0195224) | 42.89261 (.0379380) | 45.77353 (.0732835) |
| | 92.72423 (.0054776) | 88.58212 (.0120620) | 84.08589 (.0267164) |
| 62 | 41.23438 (.0194769) | 43.72272 (.0378505) | 46.63120 (.0731193) |
| | 93.91159 (.0055231) | 89.74640 (.0121495) | 85.22418 (.0268807) |
| 63 | 42.04129 (.0194322) | 44.55409 (.0377647) | 47.48993 (.0729584) |
| | 95.09770 (.0055678) | 90.90958 (.0122353) | 86.36153 (.0270416) |
| 64 | 42.84964 (.0193884) | 45.38670 (.0376806) | 48.34969 (.0728008) |
| | 96.28259 (.0056116) | 92.07167 (.0123194) | 87.49796 (.0271991) |
| 65 | 43.65939 (.0193454) | 46.22055 (.0375980) | 49.21048 (.0726464) |
| | 97.46629 (.0056546) | 93.23270 (.0124019) | 88.63348 (.0273535) |
| 66 | 44.47054 (.0193031) | 47.05560 (.0375171) | 50.07227 (.0724951) |
| | 98.64880 (.0056969) | 94.39270 (.0124829) | 89.76813 (.0275049) |
| 67 | 45.28300 (.0192616) | 47.89182 (.0374377) | 50.93501 (.0723467) |
| | 99.83015 (.0057384) | 95.55167 (.0125623) | 90.90190 (.0276532) |
| 68 | 46.09680 (.0192209) | 48.72916 (.0373598) | 51.79872 (.0722013) |
| | 101.01036 (.0057791) | 96.70960 (.0126402) | 92.03482 (.0277987) |
| 69 | 46.91188 (.0191809) | 49.56764 (.0372833) | 52.66338 (.0720586) |
| | 102.18945 (.0058191) | 97.86661 (.0127167) | 93.16692 (.0279413) |
| 70 | 47.72824 (.0191415) | 50.40721 (.0372081) | 53.52893 (.0719187) |
| | 103.36745 (.0058585) | 99.02261 (.0127919) | 94.29817 (.0280813) |
| 71 | 48.54582 (.0191029) | 51.24786 (.0371344) | 54.39539 (.0717813) |
| | 104.54437 (.0058971) | 100.17766 (.0128656) | 95.42863 (.0282187) |
| 72 | 49.36461 (.0190649) | 52.08957 (.0370619) | 55.26273 (.0716465) |
| | 105.72023 (.0059351) | 101.33179 (.0129381) | 96.55830 (.0283535) |
| 73 | 50.18459 (.0190275) | 52.93231 (.0369907) | 56.13092 (.0715141) |
| | 106.89503 (.0059725) | 102.48499 (.0130093) | 97.68718 (.0284859) |
| 74 | 51.00574 (.0189908) | 53.77606 (.0369208) | 56.99995 (.0713841) |
| | 108.06882 (.0060092) | 103.63727 (.0130792) | 98.81531 (.0286159) |
| 75 | 51.82802 (.0189547) | 54.62080 (.0368520) | 57.86983 (.0712565) |
| | 109.24161 (.0060453) | 104.78868 (.0131480) | 99.94267 (.0287435) |
| 76 | 52.65143 (.0189191) | 55.46652 (.0367844) | 58.74051 (.0711310) |
| | 110.41339 (.0060809) | 105.93921 (.0132156) | 101.06929 (.0288690) |
| 77 | 53.47594 (.0188842) | 56.31319 (.0367180) | 59.61198 (.0710078) |
| | 111.58420 (.0061158) | 107.08888 (.0132820) | 102.19519 (.0289922) |
| 78 | 54.30151 (.0188498) | 57.16081 (.0366526) | 60.48424 (.0708867) |
| | 112.75406 (.0061502) | 108.23772 (.0133474) | 103.32037 (.0291133) |
| 79 | 55.12816 (.0188159) | 58.00934 (.0365884) | 61.35727 (.0707676) |
| | 113.92296 (.0061841) | 109.38571 (.0134116) | 104.44485 (.0292324) |
| 80 | 55.95584 (.0187826) | 58.85878 (.0365251) | 62.23106 (.0706506) |
| | 115.09094 (.0062174) | 110.53288 (.0134749) | 105.56865 (.0293494) |

TABLE II (K=4)

| | .975 | .950 | .900 |
|---|---|---|---|
| 81 | 56.78453 (.0187497) | 59.70911 (.0364629) | 63.10559 (.0705355) |
| | 116.25800 (.0062503) | 111.67926 (.0135371) | 106.69174 (.0294645) |
| 82 | 57.61423 (.0187174) | 60.56030 (.0364017) | 63.98083 (.0704222) |
| | 117.42416 (.0062826) | 112.82483 (.0135983) | 107.81418 (.0295777) |
| 83 | 58.44492 (.0186856) | 61.41235 (.0363414) | 64.85681 (.0703109) |
| | 118.58945 (.0063144) | 113.96964 (.0136586) | 108.93596 (.0296891) |
| 84 | 59.27658 (.0186543) | 62.26526 (.0362821) | 65.73349 (.0702013) |
| | 119.75385 (.0063457) | 115.11366 (.0137179) | 110.05708 (.0297986) |
| 85 | 60.10918 (.0186234) | 63.11899 (.0362237) | 66.61086 (.0700935) |
| | 120.91739 (.0063766) | 116.25693 (.0137763) | 111.17757 (.0299065) |
| 86 | 60.94272 (.0185930) | 63.97353 (.0361662) | 67.48889 (.0699874) |
| | 122.08009 (.0064070) | 117.39946 (.0138338) | 112.29741 (.0300126) |
| 87 | 61.77718 (.0185630) | 64.82887 (.0361096) | 68.36761 (.0698829) |
| | 123.24196 (.0064370) | 118.54126 (.0138904) | 113.41664 (.0301170) |
| 88 | 62.61253 (.0185335) | 65.68501 (.0360538) | 69.24698 (.0697801) |
| | 124.40300 (.0064665) | 119.68233 (.0139462) | 114.53526 (.0302199) |
| 89 | 63.44879 (.0185044) | 66.54192 (.0359988) | 70.12701 (.0696788) |
| | 125.56323 (.0064956) | 120.82268 (.0140012) | 115.65329 (.0303212) |
| 90 | 64.28592 (.0184756) | 67.39958 (.0359446) | 71.00766 (.0695791) |
| | 126.72267 (.0065243) | 121.96234 (.0140554) | 116.77071 (.0304209) |
| 91 | 65.12390 (.0184474) | 68.25800 (.0358913) | 71.88895 (.0694808) |
| | 127.88132 (.0065526) | 123.10130 (.0141087) | 117.88754 (.0305191) |
| 92 | 65.96274 (.0184194) | 69.11716 (.0358387) | 72.77084 (.0693840) |
| | 129.03920 (.0065805) | 124.23958 (.0141613) | 119.00381 (.0306159) |
| 93 | 66.80240 (.0183919) | 69.97704 (.0357868) | 73.65335 (.0692887) |
| | 130.19630 (.0066081) | 125.37720 (.0142132) | 120.11951 (.0307113) |
| 94 | 67.64290 (.0183648) | 70.83763 (.0357357) | 74.53647 (.0691947) |
| | 131.35266 (.0066352) | 126.51414 (.0142643) | 121.23465 (.0308053) |
| 95 | 68.48419 (.0183380) | 71.69894 (.0356853) | 75.42017 (.0691021) |
| | 132.50829 (.0066620) | 127.65044 (.0143147) | 122.34923 (.0308979) |
| 96 | 69.32628 (.0183116) | 72.56094 (.0356356) | 76.30444 (.0690108) |
| | 133.66316 (.0066884) | 128.78609 (.0143644) | 123.46327 (.0309892) |
| 97 | 70.16916 (.0182856) | 73.42361 (.0355866) | 77.18929 (.0689208) |
| | 134.81732 (.0067144) | 129.92110 (.0144134) | 124.57677 (.0310792) |
| 98 | 71.01282 (.0182598) | 74.28697 (.0355382) | 78.07469 (.0688321) |
| | 135.97076 (.0067402) | 131.05548 (.0144618) | 125.68974 (.0311679) |
| 99 | 71.85722 (.0182345) | 75.15100 (.0354905) | 78.96066 (.0687446) |
| | 137.12350 (.0067655) | 132.18925 (.0145095) | 126.80220 (.0312554) |
| 100 | 72.70238 (.0182094) | 76.01567 (.0354434) | 79.84717 (.0686583) |
| | 138.27556 (.0067906) | 133.32242 (.0145566) | 127.91414 (.0313417) |

TABLE II (K=4)

| | .800 | .700 | .600 |
|---|---|---|---|
| 1 | 0.064157(.1999575) | 0.14825 (.2997894) | 0.27394 (.3992966) |
| | 16.76050 (.0000424) | 13.73449 (.0002106) | 11.48045 (.0007033) |
| 2 | 0.44437 (.1992312) | 0.70647 (.2975872) | 1.00273 (.3942964) |
| | 14.34139 (.0007688) | 12.05394 (.0024128) | 10.33332 (.0057036) |
| 3 | 0.99491 (.1975152) | 1.39571 (.2934585) | 1.80678 (.3865393) |
| | 14.33341 (.0024847) | 12.26070 (.0065414) | 10.70050 (.0134607) |
| 4 | 1.62211 (.1951869) | 2.13215 (.2885323) | 2.62846 (.3782104) |
| | 14.94668 (.0048131) | 12.96105 (.0114677) | 11.46681 (.0217896) |
| 5 | 2.29223 (.1925923) | 2.89273 (.2834814) | 3.45785 (.3702246) |
| | 15.81010 (.0074077) | 13.86074 (.0165186) | 12.39357 (.0297754) |
| 6 | 2.99033 (.1899400) | 3.66884 (.2786122) | 4.29274 (.3628728) |
| | 16.79674 (.0100600) | 14.85832 (.0213878) | 13.39846 (.0371271) |
| 7 | 3.70863 (.1873412) | 4.45644 (.2740445) | 5.13250 (.3562027) |
| | 17.85139 (.0126587) | 15.90957 (.0259555) | 14.44579 (.0437973) |
| 8 | 4.44258 (.1848506) | 5.25333 (.2698110) | 5.97684 (.3501742) |
| | 18.94630 (.0151494) | 16.99232 (.0301890) | 15.51778 (.0498257) |
| 9 | 5.18925 (.1824914) | 6.05806 (.2659061) | 6.82553 (.3447217) |
| | 20.06613 (.0175086) | 18.09441 (.0340939) | 16.60471 (.0552783) |
| 10 | 5.94660 (.1802704) | 6.86964 (.2623081) | 7.67833 (.3397759) |
| | 21.20178 (.0197296) | 19.20859 (.0376918) | 17.70087 (.0602241) |
| 11 | 6.71315 (.1781856) | 7.68731 (.2589902) | 8.53501 (.3352729) |
| | 22.34760 (.0218144) | 20.33043 (.0410098) | 18.80276 (.0647270) |
| 12 | 7.48779 (.1762305) | 8.51048 (.2559248) | 9.39534 (.3311567) |
| | 23.49991 (.0237694) | 21.45705 (.0440751) | 19.90816 (.0688432) |
| 13 | 8.26961 (.1743969) | 9.33864 (.2530862) | 10.25911 (.3273789) |
| | 24.65627 (.0256030) | 22.58655 (.0469138) | 21.01556 (.0726210) |
| 14 | 9.05789 (.1726760) | 10.17138 (.2504507) | 11.12611 (.3238981) |
| | 25.81499 (.0273240) | 23.71762 (.0495493) | 22.12401 (.0761018) |
| 15 | 9.85206 (.1710587) | 11.00834 (.2479973) | 11.99615 (.3206793) |
| | 26.97491 (.0289413) | 24.84941 (.0520026) | 23.23279 (.0793206) |
| 16 | 10.65159 (.1695366) | 11.84922 (.2457076) | 12.86905 (.3176924) |
| | 28.13522 (.0304634) | 25.98126 (.0542924) | 24.34145 (.0823076) |
| 17 | 11.45606 (.1681020) | 12.69374 (.2435651) | 13.74464 (.3149117) |
| | 29.29533 (.0318980) | 27.11275 (.0564348) | 25.44963 (.0850883) |
| 18 | 12.26511 (.1667476) | 13.54165 (.2415556) | 14.62278 (.3123153) |
| | 30.45482 (.0332524) | 28.24355 (.0584443) | 26.55714 (.0876847) |
| 19 | 13.07841 (.1654668) | 14.39275 (.2396665) | 15.50332 (.3098841) |
| | 31.61339 (.0345331) | 29.37344 (.0603335) | 27.66379 (.0901158) |
| 20 | 13.89567 (.1642538) | 15.24684 (.2378865) | 16.38614 (.3076017) |
| | 32.77081 (.0357462) | 30.50227 (.0621134) | 28.76949 (.0923983) |

TABLE II (K=4)

| | .800 | .700 | .600 |
|---|---|---|---|
| 21 | 14.71664 (.1631030) | 16.10374 (.2362061) | 17.27110 (.3054538) |
| | 33.92694 (.0368970) | 31.62996 (.0637938) | 29.87415 (.0945461) |
| 22 | 15.54110 (.1620097) | 16.96330 (.2346165) | 18.15813 (.3034279) |
| | 35.08168 (.0379903) | 32.75641 (.0653835) | 30.97772 (.0965720) |
| 23 | 16.36884 (.1609694) | 17.82539 (.2331099) | 19.04710 (.3015131) |
| | 36.23495 (.0390305) | 33.88156 (.0668901) | 32.08020 (.0984868) |
| 24 | 17.19968 (.1599783) | 18.68987 (.2316796) | 19.93793 (.2996997) |
| | 37.38673 (.0400217) | 35.00543 (.0683203) | 33.18155 (.1003003) |
| 25 | 18.03345 (.1590326) | 19.55659 (.2303196) | 20.83051 (.2979791) |
| | 38.53696 (.0409673) | 36.12796 (.0696803) | 34.28177 (.1020208) |
| 26 | 18.87000 (.1581292) | 20.42549 (.2290243) | 21.72479 (.2963439) |
| | 39.68565 (.0418707) | 37.24919 (.0709757) | 35.38087 (.1036561) |
| 27 | 19.70920 (.1572651) | 21.29643 (.2277888) | 22.62067 (.2947872) |
| | 40.83279 (.0427349) | 38.36911 (.0722111) | 36.47887 (.1052128) |
| 28 | 20.55092 (.1564376) | 22.16934 (.2266088) | 23.51810 (.2933030) |
| | 41.97839 (.0435624) | 39.48772 (.0733911) | 37.57578 (.1066970) |
| 29 | 21.39503 (.1556442) | 23.04413 (.2254803) | 24.41699 (.2918859) |
| | 43.12248 (.0443558) | 40.60506 (.0745196) | 38.67160 (.1081141) |
| 30 | 22.24144 (.1548828) | 23.92070 (.2243997) | 25.31731 (.2905310) |
| | 44.26508 (.0451172) | 41.72113 (.0756002) | 39.76637 (.1094689) |
| 31 | 23.09004 (.1541513) | 24.79900 (.2233638) | 26.21898 (.2892340) |
| | 45.40617 (.0458487) | 42.83597 (.0766362) | 40.86011 (.1107660) |
| 32 | 23.94075 (.1534477) | 25.67894 (.2223695) | 27.12196 (.2879908) |
| | 46.54584 (.0465523) | 43.94958 (.0776305) | 41.95284 (.1120091) |
| 33 | 24.79346 (.1527705) | 26.56046 (.2214142) | 28.02620 (.2867979) |
| | 47.68407 (.0472295) | 45.06201 (.0785857) | 43.04456 (.1132020) |
| 34 | 25.64810 (.1521180) | 27.44351 (.2204955) | 28.93164 (.2856520) |
| | 48.82089 (.0478819) | 46.17329 (.0795044) | 44.13533 (.1143479) |
| 35 | 26.50459 (.1514889) | 28.32802 (.2196110) | 29.83824 (.2845501) |
| | 49.95636 (.0485111) | 47.28343 (.0803889) | 45.22514 (.1154498) |
| 36 | 27.36288 (.1508817) | 29.21394 (.2187589) | 30.74597 (.2834895) |
| | 51.09050 (.0491183) | 48.39246 (.0812411) | 46.31404 (.1165105) |
| 37 | 28.22290 (.1502953) | 30.10124 (.2179370) | 31.65480 (.2824675) |
| | 52.22331 (.0497047) | 49.50040 (.0820630) | 47.40205 (.1175324) |
| 38 | 29.08456 (.1497284) | 30.98984 (.2171437) | 32.56465 (.2814822) |
| | 53.35486 (.0502716) | 50.60727 (.0828562) | 48.48917 (.1185178) |
| 39 | 29.94783 (.1491801) | 31.87970 (.2163774) | 33.47552 (.2805311) |
| | 54.48514 (.0508199) | 51.71312 (.0836226) | 49.57544 (.1194688) |
| 40 | 30.81265 (.1486493) | 32.77080 (.2156366) | 34.38739 (.2796125) |
| | 55.61421 (.0513507) | 52.81796 (.0843634) | 50.66087 (.1203874) |

TABLE II (K=4)

| | .800 | .700 | .600 |
|---|---|---|---|
| 41 | 31.67896 (.1481352)<br>56.74208 (.0518648) | 33.66309 (.2149199)<br>53.92181 (.0850801) | 35.30019 (.2787245)<br>51.74550 (.1212754) |
| 42 | 32.54671 (.1476369)<br>57.86879 (.0523631) | 34.55650 (.2142261)<br>55.02470 (.0857739) | 36.21391 (.2778655)<br>52.82933 (.1221345) |
| 43 | 33.41586 (.1471537)<br>58.99435 (.0528463) | 35.45105 (.2135538)<br>56.12665 (.0864460) | 37.12852 (.2770339)<br>53.91238 (.1229661) |
| 44 | 34.28636 (.1466847)<br>60.11882 (.0533153) | 36.34666 (.2129022)<br>57.22769 (.0870977) | 38.04399 (.2762282)<br>54.99467 (.1237717) |
| 45 | 35.15819 (.1462293)<br>61.24217 (.0537706) | 37.24333 (.2122702)<br>58.32784 (.0877298) | 38.96031 (.2754474)<br>56.07623 (.1245526) |
| 46 | 36.03128 (.1457869)<br>62.36449 (.0542130) | 38.14101 (.2116567)<br>59.42709 (.0883432) | 39.87743 (.2746899)<br>57.15707 (.1253101) |
| 47 | 36.90561 (.1453570)<br>63.48575 (.0546430) | 39.03967 (.2110610)<br>60.52551 (.0889390) | 40.79535 (.2739547)<br>58.23721 (.1260452) |
| 48 | 37.78114 (.1449388)<br>64.60599 (.0550612) | 39.93930 (.2104821)<br>61.62309 (.0895178) | 41.71404 (.2732408)<br>59.31667 (.1267591) |
| 49 | 38.65784 (.1445318)<br>65.72525 (.0554681) | 40.83984 (.2099193)<br>62.71985 (.0900806) | 42.63347 (.2725471)<br>60.39545 (.1274529) |
| 50 | 39.53568 (.1441357)<br>66.84352 (.0558642) | 41.74130 (.2093719)<br>63.81581 (.0906281) | 43.55362 (.2718728)<br>61.47357 (.1281272) |
| 51 | 40.41460 (.1437499)<br>67.96086 (.0562501) | 42.64365 (.2088391)<br>64.91100 (.0911608) | 44.47449 (.2712168)<br>62.55107 (.1287832) |
| 52 | 41.29462 (.1433740)<br>69.07727 (.0566260) | 43.54684 (.2083204)<br>66.00542 (.0916795) | 45.39604 (.2705784)<br>63.62793 (.1294215) |
| 53 | 42.17566 (.1430076)<br>70.19276 (.0569924) | 44.45088 (.2078151)<br>67.09911 (.0921848) | 46.31827 (.2699569)<br>64.70418 (.1300431) |
| 54 | 43.05774 (.1426502)<br>71.30736 (.0573498) | 45.35573 (.2073227)<br>68.19205 (.0926772) | 47.24115 (.2693514)<br>65.77983 (.1306485) |
| 55 | 43.94080 (.1423016)<br>72.42110 (.0576984) | 46.26137 (.2068427)<br>69.28429 (.0931573) | 48.16467 (.2687614)<br>66.85490 (.1312385) |
| 56 | 44.82483 (.1419613)<br>73.53398 (.0580387) | 47.16779 (.2063744)<br>70.37582 (.0936255) | 49.08882 (.2681862)<br>67.92940 (.1318138) |
| 57 | 45.70981 (.1416290)<br>74.64603 (.0583709) | 48.07495 (.2059175)<br>71.46669 (.0940824) | 50.01358 (.2676251)<br>69.00333 (.1323748) |
| 58 | 46.59570 (.1413046)<br>75.75725 (.0586954) | 48.98286 (.2054715)<br>72.55687 (.0945284) | 50.93892 (.2670777)<br>70.07672 (.1329223) |
| 59 | 47.48250 (.1409875)<br>76.86768 (.0590125) | 49.89148 (.2050360)<br>73.64641 (.0949640) | 51.86485 (.2665433)<br>71.14957 (.1334567) |
| 60 | 48.37016 (.1406776)<br>77.97731 (.0593224) | 50.80081 (.2046105)<br>74.73531 (.0953894) | 52.79134 (.2660214)<br>72.22189 (.1339785) |

TABLE II (K=4)

|  | .800 | .700 | .600 |
|---|---|---|---|
| 61 | 49.25870 (.1403745) | 51.71083 (.2041948) | 53.71838 (.2655116) |
|  | 79.08620 (.0596254) | 75.82356 (.0958052) | 73.29370 (.1344883) |
| 62 | 50.14807 (.1400782) | 52.62154 (.2037884) | 54.64598 (.2650135) |
|  | 80.19431 (.0599218) | 76.91121 (.0962116) | 74.36501 (.1349865) |
| 63 | 51.03825 (.1397883) | 53.53288 (.2033910) | 55.57410 (.2645265) |
|  | 81.30168 (.0602117) | 77.99825 (.0966090) | 75.43582 (.1354734) |
| 64 | 51.92924 (.1395045) | 54.44487 (.2030022) | 56.50273 (.2640502) |
|  | 82.40834 (.0604955) | 79.08470 (.0969978) | 76.50615 (.1359497) |
| 65 | 52.82101 (.1392267) | 55.35750 (.2026218) | 57.43188 (.2635844) |
|  | 83.51428 (.0607733) | 80.17056 (.0973782) | 77.57600 (.1364155) |
| 66 | 53.71356 (.1389546) | 56.27072 (.2022493) | 58.36153 (.2631286) |
|  | 84.61952 (.0610453) | 81.25587 (.0977505) | 78.64539 (.1368713) |
| 67 | 54.60686 (.1386881) | 57.18457 (.2018848) | 59.29166 (.2626824) |
|  | 85.72408 (.0613118) | 82.34061 (.0981151) | 79.71431 (.1373175) |
| 68 | 55.50089 (.1384270) | 58.09900 (.2015277) | 60.22226 (.2622455) |
|  | 86.82797 (.0615729) | 83.42480 (.0984722) | 80.78279 (.1377544) |
| 69 | 56.39565 (.1381711) | 59.01401 (.2011779) | 61.15334 (.2618176) |
|  | 87.93120 (.0618289) | 84.50845 (.0988221) | 81.85083 (.1381823) |
| 70 | 57.29109 (.1379203) | 59.92958 (.2008351) | 62.08485 (.2613985) |
|  | 89.03377 (.0620797) | 85.59158 (.0991648) | 82.91844 (.1386015) |
| 71 | 58.18724 (.1376742) | 60.84572 (.2004991) | 63.01685 (.2609877) |
|  | 90.13571 (.0623257) | 86.67418 (.0995008) | 83.98563 (.1390122) |
| 72 | 59.08406 (.1374329) | 61.76239 (.2001696) | 63.94926 (.2605851) |
|  | 91.23703 (.0625670) | 87.75629 (.0998303) | 85.05240 (.1394149) |
| 73 | 59.98155 (.1371962) | 62.67960 (.1998466) | 64.88211 (.2601902) |
|  | 92.33774 (.0628037) | 88.83788 (.1001534) | 86.11876 (.1398097) |
| 74 | 60.87970 (.1369638) | 63.59734 (.1995296) | 65.81538 (.2598031) |
|  | 93.43784 (.0630361) | 89.91898 (.1004704) | 87.18471 (.1401969) |
| 75 | 61.77849 (.1367359) | 64.51558 (.1992186) | 66.74907 (.2594233) |
|  | 94.53734 (.0632641) | 90.99960 (.1007813) | 88.25027 (.1405767) |
| 76 | 62.67790 (.1365120) | 65.43434 (.1989134) | 67.68317 (.2590506) |
|  | 95.63626 (.0634879) | 92.07974 (.1010866) | 89.31544 (.1409493) |
| 77 | 63.57793 (.1362922) | 66.35359 (.1986138) | 68.61766 (.2586849) |
|  | 96.73462 (.0637078) | 93.15942 (.1013861) | 90.38023 (.1413150) |
| 78 | 64.47856 (.1360762) | 67.27333 (.1983196) | 69.55255 (.2583258) |
|  | 97.83240 (.0639237) | 94.23865 (.1016803) | 91.44468 (.1416741) |
| 79 | 65.37979 (.1358641) | 68.19354 (.1980307) | 70.48784 (.2579733) |
|  | 98.92963 (.0641358) | 95.31741 (.1019692) | 92.50870 (.1420267) |
| 80 | 66.28160 (.1356556) | 69.11423 (.1977469) | 71.42349 (.2576271) |
|  | 100.02632 (.0643443) | 96.39574 (.1022530) | 93.57239 (.1423728) |

TABLE II (K=4)

| | .800 | .700 | .600 |
|---|---|---|---|
| 81 | 67.18399 (.1354508) | 70.03537 (.1974681) | 72.35951 (.2572870) |
| | 101.12247 (.0645491) | 97.47363 (.1025319) | 94.63571 (.1427130) |
| 82 | 68.08694 (.1352494) | 70.95697 (.1971940) | 73.29591 (.2569528) |
| | 102.21809 (.0647505) | 98.55109 (.1028059) | 95.69870 (.1430472) |
| 83 | 68.99045 (.1350515) | 71.87901 (.1969247) | 74.23267 (.2566244) |
| | 103.31320 (.0649484) | 99.62813 (.1030752) | 96.76132 (.1433755) |
| 84 | 69.89449 (.1348568) | 72.80150 (.1966600) | 75.16978 (.2563016) |
| | 104.40779 (.0651432) | 100.70474 (.1033400) | 97.82362 (.1436983) |
| 85 | 70.79907 (.1346653) | 73.72441 (.1963996) | 76.10724 (.2559843) |
| | 105.50188 (.0653346) | 101.78096 (.1036003) | 98.88559 (.1440156) |
| 86 | 71.70418 (.1344769) | 74.64775 (.1961436) | 77.04504 (.2556723) |
| | 106.59547 (.0655230) | 102.85677 (.1038563) | 99.94722 (.1443276) |
| 87 | 72.60980 (.1342916) | 75.57150 (.1958917) | 77.98318 (.2553654) |
| | 107.68858 (.0657083) | 103.93217 (.1041082) | 101.00851 (.1446345) |
| 88 | 73.51595 (.1341092) | 76.49567 (.1956440) | 78.92165 (.2550636) |
| | 108.78120 (.0658907) | 105.00719 (.1043560) | 102.06950 (.1449363) |
| 89 | 74.42258 (.1339297) | 77.42023 (.1954001) | 79.86044 (.2547666) |
| | 109.87335 (.0660703) | 106.08182 (.1045998) | 103.13019 (.1452333) |
| 90 | 75.32971 (.1337529) | 78.34520 (.1951602) | 80.79956 (.2544744) |
| | 110.96504 (.0662470) | 107.15608 (.1048397) | 104.19055 (.1455255) |
| 91 | 76.23734 (.1335790) | 79.27055 (.1949240) | 81.73900 (.2541868) |
| | 112.05626 (.0664210) | 108.22997 (.1050760) | 105.25061 (.1458131) |
| 92 | 77.14543 (.1334077) | 80.19629 (.1946915) | 82.67874 (.2539037) |
| | 113.14703 (.0665923) | 109.30347 (.1053085) | 106.31036 (.1460962) |
| 93 | 78.05399 (.1332389) | 81.12241 (.1944625) | 83.61879 (.2536250) |
| | 114.23737 (.0667610) | 110.37662 (.1055374) | 107.36983 (.1463749) |
| 94 | 78.96301 (.1330727) | 82.04890 (.1942371) | 84.55914 (.2533506) |
| | 115.32726 (.0669273) | 111.44942 (.1057629) | 108.42900 (.1466493) |
| 95 | 79.87250 (.1329089) | 82.97577 (.1940150) | 85.49980 (.2530804) |
| | 116.41670 (.0670910) | 112.52185 (.1059850) | 109.48788 (.1469195) |
| 96 | 80.78244 (.1327476) | 83.90298 (.1937962) | 86.44073 (.2528142) |
| | 117.50574 (.0672523) | 113.59395 (.1062038) | 110.54649 (.1471857) |
| 97 | 81.69281 (.1325886) | 84.83057 (.1935806) | 87.38197 (.2525520) |
| | 118.59435 (.0674113) | 114.66570 (.1064193) | 111.60481 (.1474480) |
| 98 | 82.60362 (.1324319) | 85.75850 (.1933682) | 88.32349 (.2522936) |
| | 119.68253 (.0675680) | 115.73711 (.1066317) | 112.66286 (.1477063) |
| 99 | 83.51488 (.1322774) | 86.68677 (.1931589) | 89.26529 (.2520390) |
| | 120.77031 (.0677225) | 116.80820 (.1068410) | 113.72064 (.1479610) |
| 100 | 84.42654 (.1321251) | 87.61539 (.1929526) | 90.20737 (.2517881) |
| | 121.85768 (.0678748) | 117.87894 (.1070474) | 114.77814 (.1482118) |